Plant Physiology

A TREATISE

EDITED BY

F. C. STEWARD

Professor of Botany
Cornell University, Ithaca, New York

Volume IVA: Metabolism: Organic Nutrition and
Nitrogen Metabolism

1965

ACADEMIC PRESS, *New York and London*

ACADEMIC PRESS INC.
111 Fifth Avenue
New York, New York 10003

United Kingdom Edition
Published by
ACADEMIC PRESS INC. (London) Ltd.
Berkeley Square House, London W. 1

Library of Congress Catalog Card Number: 59-7689

PRINTED IN THE UNITED STATES OF AMERICA

CONTRIBUTORS TO VOLUME IVA

D. J. Durzan, *Petawawa Forest Experiment Station, Department of Forestry, Chalk River, Ontario, Canada*

Dorothy F. Forward, *Professor of Botany, University of Toronto, Toronto, Canada*

Edgar Lemon, *U.S. Department of Agriculture, Agricultural Research Service and Cornell University, Ithaca, New York*

F. C. Steward, *Professor of Botany, Cornell University, Ithaca, New York*

Moyer D. Thomas, *Air Pollution Research Center, University of California. Riverside, California*

E. W. Yemm, *Professor of Botany, University of Bristol, Bristol, England*

PREFACE TO VOLUME IV

In the original plan of this treatise, Volume IV had as its objective a full treatment of the important subject of metabolism. The intention was to provide opportunity for the discussion of certain topics as these are to be seen and understood through the behavior of whole organs or organisms. This was planned even though the same problem had previously been considered in earlier volumes at the cellular and sub-cellular level, or from the standpoint of reaction mechanisms within cells. Thus a chapter on photosynthesis, as a property not only of green cells and leaves but also of plants and populations in the field, was contemplated, and another on respiration to describe this important function through the behavior of organs and organisms. Moreover, intermediary metabolism of carbon compounds (i.e., the metabolism of carbohydrates, organic acids, and lipids) as well as the nature of the nitrogenous compounds in plants and the part they play in plant metabolism were all to be discussed in this volume. After the chemical working of plants was described, there was also to be a place in this volume where it is appropriate to discuss the contrasted behavior of plants in health and disease.

Problems, however, arose as the original design unfolded. One of these the editor could anticipate, though not readily solve, for it concerned the now familiar difficulty of securing these sufficiently comprehensive chapters simultaneously for their inclusion in the volume. To those authors who met deadlines the editor is especially grateful; to those readers who have awaited a volume which seemed too long overdue it should be explained that it seemed better to have the topics covered in the requisite detail than to curtail unduly the work of the authors who undertook this difficult task. As a result, therefore, the volume outgrew its planned size, even as it fell behind its projected timetable. Thus Volume IV now appears in two parts, IVA and IVB, each with its separate pagination and complete with its own indexes to subjects, authors, and plant names. Insofar as a belated division of one projected volume into two, Parts A and B, can be logical, this has been attempted under the titles:

Volume IVA—Metabolism: Organic Nutrition and Nitrogen Metabolism.

Volume IVB—Metabolism: Intermediary Metabolism and Pathology.

Somewhat liberal cross-referencing of chapters to others, both in this and other volumes of the treatise, may overcome to some extent the inevitable separation of topics which, under other arrangements, might appropriately have been more nearly consecutive. Although a comprehensive understanding of plant metabolism requires the study of the entire volume, and the materials there cited, the hope is that the Parts A and B each have a sufficient degree of unity to render them separately useful.

To appreciate the task which authors have undertaken in the volume and the range of material to be summarized so that a synthesis of knowledge of plant metabolism might be achieved, one may note that even the *selected works* here referred to by title number about 3000, despite the expressed disinclination of editor and authors alike to extend the bibliographic treatment beyond that which reasonably documents the salient facts which are presented. Thus the price being paid for the increasing body of recorded information about plants is the ever increasing difficulty of distilling out of it an understanding of plants. In no branch of plant physiology is this more true than in the subject matter of Volume IV. Thus, authors who have attempted the task of summary and of synthesis, and on whose work the merit of the volume rests, have the grateful thanks of the editor; but the editor alone must necessarily accept responsibility for both the delays in the execution of the original plans and for the changes in them which, with time, have been necessary.

The general style and organization of the treatise remains as consistent with earlier volumes as possible. References cited in the text are given in full (though this is not always so when references are cited in tables). The approved abbreviations for journals remain those of the American Chemical Society in its list of 1956, even though a later approved list adopts less fastidious conventions in the abbreviations of German periodicals. The increasingly telegraphic style of modern biochemical and physiological writing, with its liberal use of symbols like ATP, IAA, DNP, etc., presents problems. The hope is that abbreviations will usually be self-evident, but where possible they are to be cross-referenced in the Subject Index which also serves as a glossary of abbreviations, viz. ATP (see adenosine triphosphate). Revisions in the recommended nomenclature of enzymes and coenzymes, that resulted from a Commission of the International Union of Biochemistry (cf. *Science*, 1962, **137**, 405–408), were merely mentioned in Volume III but had to be faced in Volume IV. Each author first used the nomenclature that he preferred and, significantly, the older terminology predominated. Therefore, instead of rigorously and exclusively adopting one or the other terminology, the old or the new, both will be found in the text; and in the Subject Index (or by occasional parenthetical explanations) the older and the new designations, viz. DPN or NAD, TPN or NADP, will be related. It seems that more time must elapse before one usage or the other will predominate in the literature of plant physiology and plant biochemistry.

In fact, as biochemical systems multiply, the science faces problems of nomenclature long familiar in biology, but with which chemists have hitherto been somewhat impatient. Indeed, in a work like this, it is now a task for a specialist on the nomenclature of enzymes to see that all such names are correctly used! The editor can only refer readers, in cases of doubt, to such sources as Report of the Commission on Enzymes of the International Union of Biochemistry published by The Macmillan Company (New York) and Pergamon Press (London) (1961).

But this is a proper place to record, gratefully, that Dr. W. J. Dress has again checked the use of, and has prepared the index to, the plant names which are cited in the text. Subject Indexes have been prepared by Dr. R. D. Holsten, and for the discharge of this arduous but necessary task the editor's thanks are due. Last, but by no means least, the staff of Academic Press have discharged their task with sympathetic understanding of the problems with which the editor was so frequently faced.

Consistent with the plan in previous volumes, authors have been encouraged to include in their respective chapters material in the form of tables, diagrams, and figures which illustrates the salient points which are discussed. This permits the chapters to be read without the otherwise inevitable interruptions as cited sources are consulted. Thus, each author has sought and obtained permission for the use of the materials he has obtained from other published works. While collective thanks are rendered here to the large number who have readily given this permission, the form of citation in the text and bibliography carries with it the thanks of each author for each usage.

F. C. STEWARD

Ithaca, New York
June, 1965

PLANT PHYSIOLOGY

The Plan of the Treatise

The treatise is planned in three main sections, as follows:

Section on *Cell Physiology and Problems Relating to Water and Solutes*

The purpose of this section is to present the properties of cells, their energy relations (Volume I) and behavior toward water and solutes with the closely related problems of the movement of solutes within the plant body and the economy of water in plants (Volume II).

The underlying theme of Volumes I and II is the basis of plant physiology in cell physiology.

Section on *Nutrition and Metabolism*

In this section the detailed facts and knowledge of nutrition and metabolism are presented, first with reference to the need for, and utilization of, inorganic nutrients (Volume III), and second with respect to the processes of organic nutrition (Volume IV). The treatment of organic nutrition leads to a reconsideration of photosynthesis and respiration at the level of organs and organisms. Volume IV describes the intermediary metabolism of carbon and nitrogenous compounds and presents a brief comparison of plants in health and in disease.

The main theme of Volumes III and IV is the nutrition, organic and inorganic, of plants and the biochemical steps by which these processes are achieved.

Section on *Growth and Development*

The purpose of the last section is to present the problems of plant physiology as seen through the analysis of growth and development, mainly with reference to flowering plants. This entails (Volume V) a reappraisal of the main events of growth and development from the standpoint of morphology and leads to a consideration of growth of cells and of organs. Tropisms and the role of hormones and the effects of synthetic growth regulators are discussed. In Volume VI the attention is focused upon the quantitative analysis of growth and development, the physiology of reproduction, the development of fruits and seeds, the problems of dormancy and perennation. The role of environmental factors in the control of growth and development merits separate treatment. Finally the problems of growth and development are examined from the standpoint of genetic control and from the interpretation of abnormal growth as seen in the formation of tumors. Throughout this treatment the controlling mechanisms of growth are evaluated.

Thus the last section of the work provides a synthesis of knowledge about plants since all their physiological processes converge upon growth and development.

The fulfillment of these objectives is possible only through the co-operation of many authors. The scope and treatment of individual chapters reflects the special interests of the contributors. While each volume is a complete unit, with its own table of contents and indexes, it is also an integral part of the whole plan.

Outline of the Plan

Section on *Cell Physiology and Problems Relating to Water and Solutes*

Volume IA.   Cellular Organization and Respiration
Volume IB.   Photosynthesis and Chemosynthesis
Volume II.   Plants in Relation to Water and Solutes

Section on *Nutrition and Metabolism*

Volume III.   Inorganic Nutrition of Plants
Volume IVA. Metabolism: Organic Nutrition and Nitrogen Metabolism
Volume IVB. Metabolism: Intermediary Metabolism and Pathology

Section on *Growth and Development*

Volume V.   Analysis of Growth
Volume VI.   The Physiology of Development

NOTE ON THE USE OF PLANT NAMES

The policy has been to identify by its scientific name, whenever possible, any plant mentioned by a vernacular name by the contributors to this work. In general, this has been done on the first occasion in each chapter when a vernacular name has been used. Particular care was taken to ensure the correct designation of plants mentioned in tables and figures which record actual observations. Sometimes, when reference has been made by an author to work done by others, it has not been possible to ascertain the exact identity of the plant material originally used, because the original workers did not identify their material except by generic or common name.

It should be unnecessary to state that the precise identification of plant material used in experimental work is as important for the enduring value of the work as the precise definition of any other variables in the work. "Warm" or "cold" would not usually be considered an acceptable substitute for a precisely stated temperature, nor could a general designation of "sugar" take the place of the precise molecular configuration of the substance used; "sunflower" and "*Helianthus*" are no more acceptable as plant names, considering how many diverse species are covered by either designation. Plant physiologists are becoming increasingly aware that different species of one genus (even different varieties or cultivars of one species) may differ in their physiological responses as well as in their external morphology, and that experimental plants should therefore be identified as precisely as possible if the observations made are to be verified by others.

On the assumption that such common names as lettuce and bean are well understood, it may appear pedantic to append the scientific names to them—but such an assumption cannot safely be made. Workers in the United States who use the unmodified word "bean" almost invariably are referring to some form of *Phaseolus vulgaris;* whereas in Britain *Vicia faba*, a plant of another genus entirely, might be implied. "Artichoke" is another such name that comes to mind, sometimes used for *Helianthus tuberosus* (properly, the Jerusalem artichoke), though the true artichoke is *Cynara scolymus*.

By the frequent interpolation of scientific names, consideration has also been given to the difficulties that any vernacular English name alone may present to a reader whose native tongue is not English. Even some American and most British botanists would be led into a misinterpretation of the identity of "yellow poplar," for instance, if this vernacular American name were not supplemented by its scientific equivalent *Liriodendron tulipifera*, for this is not a species of *Populus* as might be expected, but a member of the quite unrelated magnolia family.

When reference has been made to the work of another investigator who, in his published papers, has used a plant name not now accepted by the nomenclatural authorities followed in the present work, that name ordinarily has been included in parentheses, as a synonym, immediately after the accepted name. In a few instances, when it seemed expedient to employ a plant name as it was used by an original author, even though that name is not now recognized as the valid one, the valid name, preceded by the sign $=$, has been supplied in parentheses: e.g., *Betula verrucosa* ($= B. pendula$). Synonyms have occasionally been added elsewhere also, as in the case of a plant known and frequently reported upon in the literature under more than one name: e.g., *Pseudotsuga menziesii* (*P. taxifolia*); species of *Elodea* (*Anacharis*).

Having adopted these conventions, their implementation rested first with each contributor to this work; but all outstanding problems of nomenclature have been referred to Dr. W. J. Dress of the Bailey Hortorium, Cornell University. The authorities for the nomenclature employed in this work have been Bailey's *Hortus Second* and Bailey's *Manual of Cultivated Plants* for cultivated plants. For bacteria Bergey's *Manual of Determinative Bacteriology*, for fungi Ainsworth and Bisbee's *Dictionary of the Fungi* have been used as reference sources; other names have been checked where necessary against Engler's *Syllabus der Pflanzenfamilien*. Recent taxonomic monographs and floras have been consulted where necessary. Dr. Dress' work in ensuring consistency and accuracy in the use of plant names is deeply appreciated.

THE EDITOR

CONTENTS

CONTRIBUTORS TO VOLUME IVA v
PREFACE vii
THE PLAN OF THE TREATISE xi
NOTE ON THE USE OF PLANT NAMES xiii
CONTENTS OF VOLUMES IA, IB, II, AND III xix

INTRODUCTION

The Concept of Metabolism 1

Preamble to Chapters 1 and 2 7

CHAPTER ONE

Photosynthesis (Carbon Assimilation): Environmental and Metabolic Relationships by MOYER D. THOMAS 9

    I. Introduction 10
    II. The Carbon Dioxide Supply and Exchange 27
    III. Effect of Environmental Factors on Photosynthesis 63
    IV. Nutritional Factors in Photosynthesis 102
    V. Photosynthesis under Field Conditions 135
    VI. Algal Culture 168
    References 183

CHAPTER TWO

Micrometeorology and the Physiology of Plants in Their Natural Environment by EDGAR LEMON 203

    I. Introduction. 203
    II. Radiation Exchange at the Earth's Surface 204
    III. Exchange of Mass and Momentum at the Earth's Surface . . 206
    IV. Airflow and Turbulent Exchange above the Vegetation Canopy 209
    V. Airflow and Turbulent Exchange within the Vegetation Canopy 214
    VI. The Diurnal Cycle of Microclimate 216
    VII. The Energy Balance 220
    References 225

Preamble to Chapter 3 229

CHAPTER THREE

Part 1. The Respiration of Plants and Their Organs by E. W. YEMM 231

    I. Introduction 231
    II. The Respiratory Processes and Their Regulation 232

III. Respiration of Germinating Seeds 246
IV. The Respiration of Leaves 269
V. The Respiration of Roots 291
VI. Conclusion 303
    References 304

Part 2. The Respiration of Bulky Organs *by* DOROTHY F. FORWARD   311

I. Introduction 311
II. The Ontogenetic Pattern of Respiration in Fleshy Fruits . . . 312
III. The Respiration of Fleshy Roots and Tubers 324
IV. Respiration and the Internal Atmosphere of Bulky Organs . . 328
V. Oxygen Dependence 343
VI. Respiration and Fermentation 350
VII. The Relation of Respiration to Other Metabolism; the Role of
     Oxygen 360
VIII. Regulation of Respiration Imposed by Organization 368
     References 369

Preamble to Chapter 4 377

CHAPTER FOUR

Metabolism of Nitrogenous Compounds *by* F. C. STEWARD AND
D. B. DURZAN 379

Part 1. Nitrogen in Nature and in Plants: Historical Perspective   381

I. Introduction 381
II. The Determination of the Forms of Nitrogen That Occur in
    Plants 389
III. Trends in the Understanding of Nitrogen Metabolism . . . 401

Part 2. Free Nitrogen Compounds: Their Nature, Reactions, and
        Determination 451

IV. The Range of Soluble Nitrogen Compounds Now Found in
    Plants 451
V. Factors That Affect the Complement of Soluble Nitrogen
   Compounds in Plants 480

Part 3. Proteins and Protein Metabolism: Regulatory Control   . 533

VI. The Complement of Protein Nitrogen in Plants 533
VII. Ribonucleic Acids, Nucleoproteins, and Virus Protein . . . 548
VIII. The Mechanism of Protein Synthesis in Plants 555
IX. The Regulation of Nitrogen Metabolism and of Protein Synthe-
    sis *in Vivo* 579
X. Epilogue 602
   References 604

Appendix I. Amino Acids and Amides Found in the Protein of Plants
and Their Free Occurrence 637
Appendix II. Amino, Imino Acids and Their Derivatives That Occur
Free in Plants 644
Appendix III. The Keto Acids That May Occur in Plants, Their
Amino Acid Analogs, and Their Metabolic Roles . . 680

Author Index. 687

Index to Plant Names 708

Subject Index 717

CONTENTS OF VOLUMES IA, IB, II, AND III

Historical Introduction—F. C. STEWARD

Volume IA. CELLULAR ORGANIZATION AND RESPIRATION

The Plant Cell and Its Inclusions—R. BROWN

Proteins, Enzymes, and the Mechanism of Enzyme Action—BRIGIT VENNESLAND

Cellular Respiration—DAVID R. GODDARD AND WALTER D. BONNER

Volume IB. PHOTOSYNTHESIS AND CHEMOSYNTHESIS

Energy Storage: Photosynthesis—HANS GAFFRON

Chemosynthesis: The Energy Relations of Chemoautotrophic Organisms —MARTIN GIBBS AND JEROME A. SCHIFF

Volume II. PLANTS IN RELATION TO WATER AND SOLUTES

Cell Membranes: Their Resistance to Penetration and Their Capacity for Transport—RUNAR COLLANDER

Water Relations of Cells—T. A. BENNET-CLARK

The Water Relations to Stomatal Cells and the Mechanisms of Stomatal Movement—O. V. S. HEATH

Plants in Relation to Inorganic Salts—F. C. STEWARD AND J. F. SUTCLIFFE

Translocation of Organic Solutes—C. A. SWANSON

Translocation of Inorganic Solutes—O. BIDDULPH

Transpiration and the Water Economy of Plants—PAUL J. KRAMER

Volume III. INORGANIC NUTRITION OF PLANTS

Mineral Nutrition of Plants in Soils and in Culture Media—C. BOULD AND E. J. HEWITT

The Essential Nutrient Elements: Requirements and Interactions in Plants—E. J. HEWITT

Inorganic Nutrient Nutrition of Microorganisms—D. J. D. NICHOLAS

Modes of Action of the Essential Mineral Elements—ALVIN NASON AND WILLIAM D. McELROY

Biological Nitrogen Fixation—ARTTURI I. VIRTANEN AND JORMA K. MIETTINEN

Microbial Activities of Soil as They Affect Plant Nutrition—J. H. QUASTEL

The Concept of Metabolism

F. C. Steward

The word "metabolism" (fr. Greek *metabolē*, meaning change) has become so familiar in modern biology that its origin and first use are rarely commented upon. The concept is at least as old as the cell theory, which was formulated by, if not originated by, Schleiden and Schwann in about 1839. Schwann first discussed an "organismal" view of growth and vitality, the cause of which resides in the totality of the organism such that it may take up molecules from the environment and appropriate them to the formation of new parts or to the growth of those that already exist. In this phrase, quoted from Schwann (1839)* and cited from a translation by Henry Smith (see Schwann, 1847, p. 191), Schwann obviously introduced the idea of nutrition, but he then went on to outline also what was in effect a "cellular view" of nutrition and metabolism. Schwann pointed out that the "power of growth" is not diffusely present in the organism as a whole, for each of its separate or elementary units possesses, as it were, an independent life, each such unit consisting of molecules so arranged that they "set free a power of attracting new molecules," and so they increase by the "reciprocal action" (or interaction) of one "elementary part" upon another. One may pardonably read into this passage from Schwann a preview of the concept of the specialization of function and "division of labor" between what we would now describe as the cellular organelles. And again Schwann states that the "manifestation of the power which resides in the cell depends upon conditions to which *it is subject only when in connection with the whole organism*" (cf. 1847 translation, p. 192). Schwann postulated two "powers" which exist in "cells." These were (a) those powers which relate to the combination of the molecules to form a cell and which may be denominated by what he called the "plastic phenomena of the cell"; (b) those which result from the *chemical changes* either in the component particles of cells or in the surroundings, and which he called *metabolic* phenomena (τὸ μετϙβολικόν), which implied

---

* In this historical introduction references are cited by author and date, instead of by number as in other chapters, in order to emphasize the historical perspective.

"that which is liable to occasion or to suffer change." Again Schwann recognized that cells must have the faculty of producing chemical change in their constituent particles, which became altered during the course of vegetation (i.e., growth), and it was the unknown cause of these *metabolic* phenomena which Schwann referred to as the "metabolic power" of cells (cf. 1847 translation of Schwann). In essence, therefore, Schwann visualized under (a) the purely structural features required for the growth of cells and under (b) the dynamic, or what we now recognize as the metabolic, features of their growth.

Thus the general concept of metabolism and its designation as such was implicit in the cell theory as it was conceived in the first part of the nineteenth century. However, the use of the noun *metabolism*, as distinct from the adjective *metabolic*, as in "metabolic power" or "metabolic action," seems to have appeared at a somewhat later date and to have been used first by M. Foster (1876, cf. p. 2). The concepts of "anabolism" and "catabolism" came even later. This required that a distinction should first be drawn between those features of the chemical changes in organisms which are concerned with the synthetic metabolism, or the building-up of more complex from simpler molecules, and the chemical processes of breakdown in which the more complex molecules give rise to simpler ones. These concepts seem to derive from W. H. Gaskell (1886, cf. p. 46). The relevant passage is: "When a tissue such as a muscle is said to be at rest, we know that in reality a continual interchange of material or metabolism is all the time taking place, the condition of equilibrium which we denote by the term rest being brought about by the counterbalancing of the two opposite processes of destructive and constructive metabolism, or, as Hering has called them of assimilation and dissimilation. In other words, metabolism includes the two opposite processes of destruction and construction, *or as they may be called of katabolism and anabolism.*"

By the end of the nineteenth century these ideas were well known in plant physiology. Reference may here be made to such works as Jost's *Lectures on Plant Physiology* (1907, cf. p. 147 *et seq.*) and Pfeffer's *Physiology of Plants* (1900, Vol. I, cf. pp. 290 and 442–443). However, it may be said that in these early references anabolism and catabolism were usually regarded as somewhat distinct and separated features of the organism which might indeed be separated in both time and place. Anabolism predominates in, and is the main feature of, the cells and organs that are still in active growth and development; whereas catabolism predominates in, and is emphasized by, those cells and organs the development of which has been completed and which approach senescence. This is in fact reflected in the dictionary meaning of the words, quoted as follows from "Webster's New International Dictionary" (2nd ed., unabridged): "Metabolism may be regarded as including two aspects: constructive (termed *anabolism* or *assimilation*) or destructive (*catabolism* or *dissimilation*) . . . . In anabolism these [steps] are in the main synthetic, resulting in the building up of the nutritive substances into the more complex living protoplasm. In catabolism they are mainly destructive, . . . "

The idea of cells as intensely dynamic systems, with their metabolism consisting of innumerable interlocking and reversible cycles, with an intimate and balanced interplay of some processes which are rightly regarded as anabolic with others that are properly regarded as catabolic, required more knowledge of modern biochemistry and enzymology than was generally available at the turn of the century. This, however, was to be ushered in by the rapidly expanding knowledge of the specificity of enzymes as proteins, which followed after Sumner had prepared the first crystalline enzyme-protein in 1926; furthermore understanding of their coenzymes and their cofactors emerged in the 1930's. The book entitled "Dynamic Aspects of Biochemistry," by Baldwin, first published in 1947 (cf. Baldwin, 1963) conveniently summarizes this point of view.

Ideas upon metabolism have encountered other trends and have passed through other phases of development. Though geneticists were at first preoccupied with demonstrating the *existence* of genetic determinants, they did not remain content to treat these as abstractions. Indeed, to render the means of gene action intelligible, the gene-enzyme idea became prominent in the mid-1940's. This led to the fruitful partnership of biochemistry and genetics, now known as biochemical genetics; this was to lead in turn to the more current concept of "molecular biology."

Although the phrase "molecular biology" is probably older, and originated in a somewhat different manner than its present use might indicate, it nevertheless now marks a trend in our knowledge of, and research upon, metabolism. The term "molecular biology" owes much to W. T. Astbury, who, in the 1930's and early 1940's, was seized with the need to interpret such complex macromolecules as cellulose, starch, and the structural proteins like keratin in terms of their molecular architecture. Astbury wished to denote the study of these relationships as "molecular biology" [cf. Bernal on Astbury (1963), pp. 1–35]. The elucidation of the molecular structure of nucleic acids; the determination of the amino acid composition and the amino acid sequence of proteins; the recognition of the role of the DNA of the nucleus in self reproduction and of the relationship to its own molecular configuration; the ideas of templates and of the DNA–RNA–protein sequence, and of the genetic determination of the structure of particular proteins; all comprise the doctrine known as molecular biology.

Thus "molecular biology" has come to mean the working out of the implications for the behavior of the organism of the structure of the particular DNA which it inherits in its cell nuclei. Of necessity, these ideas, which have rapidly acquired the validity of a dogma, pervade all areas of metabolism that consist essentially of enzyme-catalyzed reactions. Though this vigorous branch of biochemical biology has come to stay, all its tenets need not yet be regarded as sacrosanct. In fact, molecular biology as such does not as yet have final answers for some of the most apparent problems of metabolism. It is one thing to endow cells and organisms with the ability to synthesize a given enzyme protein and to display its activity in an isolated preparation; this obviously tells us what is biochemically feasible. It is still another problem, however, to determine the features of the cell or the factors in its immediate environment which call forth this enzymatic activity and make it effective. This latter problem involves yet another trend in the development of our ideas of metabolism.

This trend first became evident in a borderline subject that first came to be known as "biochemical cytology," which attempted to locate (often by histochemical methods) the site in cells of given reactions or substances (Brachet, 1957; Brachet and Mirsky, 1959–1964). But with the rapid advances that have been made in the understanding of the fine structure of cells and of their organelles, the biology of metabolism is no longer to be studied only at the molecular level, for it involves organization of a much higher order. The recognition of isolated, active and inactive pools of metabolites in cells is now an increasingly conspicuous feature of the current scene. In fact it is now a major dilemma and a serious problem to determine how the flow of metabolites in the intact cell through such isolated soluble pools may be traced, for in this respect the behavior of the cells may be studied only while the entire organization is intact and functioning. These considerations are obviously relevant to a full understanding of the intermediary metabolism of photosynthesis, respiration, nitrogen metabolism, etc. (cf. Bidwell et al., 1964). Techniques of high resolution autoradiography, coupled to electron microscopy, may be valuable here.

In fact, one may anticipate that the next great advances to be made will re-emphasize the *bio* part of biochemistry, as it is found increasingly necessary to interpret the reactions of metabolism not only in terms of the genes and the hereditary mechanisms that dictate their feasibility, but in terms of the factors which act upon the organization in cells to modulate the activity of the genes and, during development, determine what shall be allowed to occur. Thus, whether in this volume or in the future, one can look for the metabolism of cells, organs, and organisms to be increasingly studied in terms of their organization; and in this activity the plant physiologist as well as the biochemist must needs come into his own and the biochemist will also be concerned increasingly with form, as shown in the work on mitochondria by Lehninger (1964).

Concepts of comparative biochemistry [which followed an initial stimulus given by the work and writings of Kluyver and of Van Niel in the early 1930's (cf. Kluyver and Van Niel, 1956)] have rightly stressed that the basic biochemical events may often be similar in kind even in organisms that are very different. Nevertheless, the fact remains that different organisms do make use of these basic attributes in very different ways. We have thus arrived at the point where these differences, which may often relate to the organization of the cells, organs, and organisms in which the reactions occur, also require to be interpreted, for they are part of the essential biological problem of distinguishing one organism from another.

In short, metabolism comprises the whole pattern of the synthetic (anabolic) reactions and of the destructive (catabolic) reactions which together comprise the chemical working of the organism visualized as a molecular machine. The cellular organelles now represent the major working parts of this machine, but the innumerable molecules and their reactions represent the individual bits of the mechanism which must "mesh" with each other in order that the whole system can operate smoothly and in a coordinated way.

The elegant methods of specific labeling by C^{14} and by other isotopes, which is a feature of developments since World War II, were needed to trace elements and radicals through this maze of metabolic reaction. Nevertheless, when all this has been done and said, there will still remain the problem of visualizing what "winds up" the cellular machinery in the first place and releases the propensities for metabolism, for growth and for development which distinguish a living organism. The treatment of metabolism is, therefore, the prelude to, and essentially sets the stage for, the treatment of growth and development which is to follow.

References

Baldwin, Ernest. "Dynamic Aspects of Biochemistry," 4th ed. Cambridge Univ. Press, London and New York, 1963.

Bernal, J. D., on William Thomas Astbury, in *Biograph. Mem. Fellows Roy. Soc.* 9, 1–35 (1963).

Bidwell, R. G. S., R. A. Barr, and F. C. Steward. Protein synthesis and turn-over in cultured plant tissue: sources of carbon for synthesis and the fate of the protein breakdown products. *Nature* **203**, 367–373 (1964).

Brachet, Jean. "Biochemical Cytology." Academic Press, New York, 1957.

Brachet, Jean, and Alfred E. Mirsky, eds. "The Cell," Vols. 1–6. Academic Press, New York, 1959–1964.

Foster, M. "A Textbook of Physiology." Macmillan, New York, 1876.

Gaskell, W. H. On the structure, distribution and function of the nerves which innervate the visceral and vascular systems. *J. Physiol.* **7**, 1–80 (1886).

Jost, Ludwig. "Lectures on Plant Physiology" (1903), translated by R. J. Harvey Gibson. Oxford Univ. Press (Clarendon), London and New York, 1907.

Kluyver, A. J., and C. B. Van Niel. "The Microbes Contribution to Biology." Harvard Univ. Press, Cambridge, Massachusetts, 1956.

Lehninger, Albert L. "The Mitochondrion." W. A. Benjamin, New York, 1964.

Pfeffer, Wilhelm. "The Physiology of Plants," 2nd ed. (1897), translated and edited by A. J. Ewart. Oxford Univ. Press (Clarendon), London and New York, 1900.

Schleiden, M. J. Beiträge zur Phytogenesis. *Arch. Anat. u. Physiol.* Part II, pp. 137–176 (1838).

Schleiden, M. J. "Contributions to Phytogenesis," translated from the German by Henry Smith and bound with the Sydenham Soc. translation of T. Schwann's "Microscopical Researches," pp. 231–268 (1847).

Schwann, T. "Microscopische Untersuchungen über die Uebereinstimmung in der Structur und dem Wachsthume der Thiere und Pflanzen." Berlin, 1839.

Schwann, T. "Microscopical Researches into the Accordance in the Structure and Growth of Animal and Plants," translated from the German by Henry Smith for the Sydenham Soc., London, 1847.

PREAMBLE TO CHAPTERS 1 AND 2

In Chapter 4 of Volume IB photosynthesis is presented as a problem of cell physiology, of photochemistry, and of biochemistry; carbohydrate metabolism, per se, is to be discussed in Chapter 5 of Volume IVB. But many topics still remain to be considered according as the photosynthetic process occurs in free cells, in organs, in plants, in natural communities, or in stands of agricultural plants. At any of these various levels of organization, the over-all process may be restricted or conditioned by external variables. Chapters 1 and 2 of this volume deal with these topics.

The study of photosynthesis in relation to environmental factors is the general area of plant physiological interpretation which gave rise in 1905 to the concept of limiting factors. Since then the principle of limiting factors has provoked various discussions of the extent to which the photosynthetic process is normally restricted, or limited, by either the carbon dioxide content of the air or by the intensity of natural sunlight. This area of interpretation also looms large as man now considers how to push the photosynthetic process and the biological use of solar energy to the upper limit of its effectiveness; this gives special significance to the discussion in Chapter 1 of photosynthesis of plants as they grow in the field.

However, the precise definition and interpretation of the environment of photosynthesing plants as they grow in the field is by no means easy to achieve—for the plants themselves alter and affect that environment in the immediate vicinity of leaves. Within a plant community the atmospheric conditions may be very different from those which may be physically characterized by measuring the various properties of the atmosphere in ways that are familiar in meteorology. Therefore, to specify more precisely the conditions at, or adjacent to, the surface of green leaves as they exist within a population, or a stand of plants, it is now necessary to resort to the techniques and thinking of micrometeorology, i.e., the rapidly growing study of microclimatic conditions in the immediate vicinity of plants. Thus Chapter 2, which is a sequel to Chapter 1, summarizes the way these problems are being approached, the kind of observations that are to be made in this field, and the sort of analysis to which the data need to be subjected.

CHAPTER ONE

Photosynthesis (Carbon Assimilation): Environmental and Metabolic Relationships

Moyer D. Thomas

I. Introduction 10
  A. History of Photosynthesis 10
  B. Photosynthesis on the Earth 23
II. The Carbon Dioxide Supply and Exchange 27
  A. Photosynthetic and Respiratory Quotients 27
  B. Concentration of Carbon Dioxide in the Atmosphere 32
  C. Mechanism of Absorption of Carbon Dioxide by Leaves . . . 40
  D. Stomatal Movements and Photosynthesis 49
  E. Respiration during Photosynthesis 55
III. Effect of Environmental Factors on Photosynthesis 63
  A. Light Intensity and Carbon Dioxide Concentration. 63
  B. Light Tolerance 74
  C. Light Quality. 82
  D. Effect of Temperature 88
  E. Effect of Hydrogen Ion Concentration 95
  F. Age 97
  G. Drought 99
  H. Oxygen and Isotopes 101
IV. Nutritional Factors in Photosynthesis 102
  Mineral Nutrition 102
V. Photosynthesis under Field Conditions 135
  A. General Considerations 135
  B. Growth Analysis 135
  C. Carbon Dioxide Exchange Measurements in the Field 148
VI. Algal Culture 168
  A. Objectives 168
  B. Growth of Algae 169
  C. Theoretical Yields 175
  D. Culture Techniques 176
  E. Nitrogen-Fixing Algae 179
  F. Algal Culture in Sewage 180
  G. Pilot Plant Studies 180
  References 183

I. Introduction

A. HISTORY OF PHOTOSYNTHESIS

1. General Survey

Solar energy is converted into the potential energy of organic compounds by photochemical oxidation-reduction reactions between carbon dioxide and water in the presence of the chlorophyll in green plants. This process is fundamental to the existence of life on the earth. The fossil fuels, peat, coal, oil and oil shale, and natural gas, are the products of photosynthetic reactions that occurred ages ago. Current reactions produce the crops, forage, forests, and water plants on which all animals now living depend for food. The development of nuclear energy and practicable methods of converting solar energy may reduce our dependence on photosynthesis for power in the near future, but it is likely that our dependence on photosynthesis for food will continue for a long time. Though the process of photosynthesis in nature is relatively inefficient in that only a small percentage of the available light energy is effectively utilized for the production of organic matter, it still exceeds by several orders of magnitude any other chemical process on the earth.

The composition of the primitive atmosphere has been the subject of much speculation. Great ingenuity has been displayed in inventing hypotheses to explain the origin of the earth and its atmosphere. Clarke (81), Mason (234), and Vernadskii (366) have given critical discussions of the various geochemical hypotheses dealing with this subject. For example, one theory postulates the formation of carbides, nitrides, and silicides in the molten globe, a process that would imply that the primitive atmosphere might have been rich in oxygen. Another theory suggests that practically all the oxygen was combined with carbon, iron, and other metals. In this case, the atmosphere would consist principally of nitrogen, carbon dioxide, water, and possibly some hydrogen, but little or no oxygen. The latter composition appears more probable, because mixtures of this kind are evolved from volcanoes and also from the heating of many kinds of rocks.

With the development of plant life on earth, the process of photosynthesis was evolved by which the energy of sunlight was used to promote chlorophyll-activated oxidation-reduction reactions between carbon dioxide and water that resulted in the formation of organic matter and the liberation of oxygen. The evolution of photosynthesis has been discussed by Van Niel (360). With an initial high concentration of carbon dioxide in the atmosphere, this process increased in intensity with the increasing development of plant life to a maximum in the carboniferous period. At that time the intensity of photosynthesis began to fall off owing to excessive utilization of carbon dioxide. The decline has continued to the present time. Now, with only 300 ppm carbon dioxide in the atmosphere, photosynthetic reactions generally become light saturated at about 10–50% of full sunlight intensity, depending on the type of vegetation. Only a limited number of plants can utilize full sunlight.

2. Photosynthesis Studies before 1800

Historically, a correct understanding of the basic over-all reactions of photosynthesis dates from the last half of the eighteenth century, when the chemistry of oxygen and carbon dioxide was clarified and the phlogiston theory was superseded by the modern theory of combustion. A comprehensive account of the early history of photosynthesis is given by Sachs (295). Harvey-Gibson (166) has reexamined some of the judgments and conclusions of Sachs regarding the pioneer investigators of photosynthesis. The latter are also reviewed in monographs by Aykroydt (17), Holt (178), Reed (281), and Wiesner (391).

The Aristotelian dictum that the substance of plants is derived from the soil was first denied in 1660 by van Helmont, whose classic study of the willow tree remains a model of quantitative experimentation. This investigator planted a small tree in a large pot of soil that had been carefully oven-dried and weighed. He watered the tree with rain water and kept the container covered to avoid gain or loss of dust. After five years, the tree was completely removed. It had gained 164 pounds in weight. The soil had lost 2 ounces. Lacking any knowledge of carbon dioxide in the atmosphere, van Helmont was forced to conclude that the substance of the tree was derived from the water. It certainly did not come from the soil. He made no attempt to explain the small loss from the soil. This was done shortly afterward by Mariotte, who thus called attention to mineral nutrition in plants.

The Italian physiologist Malpighi (227), using the newly developed lenses and microscopes that were beginning to be available, devoted the decade from 1662 to 1672 to a study of plants and published his "Anatomia Plantarum Idea" in 1674. He observed the xylem, phloem, and stomata, and he deduced with remarkable insight the uptake of crude sap through the roots and distribution of "food" from the leaves. He suggested that "the active leaves seem to have been contrived by nature for the digestion of food which is their chief function, for that part of the nutrient sap which enters the roots . . . at length slowly reaches the leaves by way of the woody veins. This is necessary so that the sap may linger in the adjacent vesicles . . . and be fermented. In this the warmth of the surrounding atmosphere is of no little assistance for it helps . . . to evaporate that which is of no service. For this purpose, nature has provided the leaf with numerous special glands or bellows for the sweating forth and gradual elimination of moisture so that the sap being thereby condensed may be the more readily digested in the leaves." He recognized that plants must have air to breathe and associated the stomata with this process, but he did not suspect intake of carbon dioxide from the atmosphere. Grew (1671–1682) (156) also observed the stomata.

The publication a half century later in 1727 of "Vegetable Staticks" by Stephen Hales (160) brought forth evidence, based on experimental absorption and evolution of gas by plants, that "plants very probably draw through their leaves some part of their nourishment from the air." Hales also suggested that light, which he regarded as a substance, contributed to the "ennobling" and "refining" of the "principles of vegetables." Direct observation of the evolution of "vital air" (oxygen) by submerged leaves illuminated by sunlight and cessation of this evolution in the dark was recorded by Bonnet (48) in 1754, who also noted that this process did not occur if the water was previously boiled.

Nevertheless, clarification of gas exchange in photosynthesis and respiration had still to await the discovery and elucidation of the reactions of oxygen, water vapor, and carbon dioxide in the atmosphere. This was accomplished in the thirty-year period before 1785. Black discovered "fixed air" or carbon dioxide in 1754, noting its presence in the atmosphere, but he did not determine its concentration. Priestley (1771–1773) (17, 178, 276) discovered "vital air" or oxygen. Scheele (1773–1777) separated air into two gases, one of which supported combustion, "fire air"; the other, "vitiated air," did not. Cavendish (1776) (17) discovered hydrogen and its combination with oxygen to form water. Lavoisier (1772–1785) (17) showed that combustion and rusting involved a combination of various elements with oxygen and that respiration produced carbon dioxide, which he proved to be a compound of carbon and oxygen. The phlogiston theory was thus discredited and the foundation was laid for an understanding of photosynthesis. In this environment, the observations of Priestley, Ingen-Housz, Senebier, and de Saussure could have meaning and a rational interpretation of the over-all reactions of photosynthesis could be established.

In 1771 Priestley (276) observed that a "sprig of mint in a glass jar standing inverted in a vessel of water" continued to grow for some months. Nor would the air in the jar "extinguish a candle or inconvenience a mouse even after the plant had grown in it for a long time." He also found that if he put the mint in air in which a candle had burned itself out, after 10 days another candle burned perfectly well in that air. These effects were exactly opposite to the known effects of animals in air, and in 1773 Priestley was awarded the Copley Medal of the Royal Society for his discovery. On that occasion John Pringle, the President of the Royal Society, delivered an address in which he emphasized the mutual advantages of the coexistence of plants and animals.

When Priestley resumed work on plants in 1777, however, he was unable to get the same results uniformly from his experiments, probably because of insufficient light in some cases, and he became somewhat confused. Even an experiment in which he observed the evolution of oxygen in sunlight from green algae on the walls of glass vessels raised some doubts because his microscopic examination of the algal deposit did not reveal definite organic form, and he thought it might be mineral rather than vegetable material. However, the evolution of oxygen occurred only in the light, and he soon became convinced that he was dealing with a plant response.

Priestley discussed this phase of his work in Volume IX of his "Experiments and Observations on Different Kinds of Air" (1790) (276). He relates that Mr. Walker (of English dictionary fame) told him about a horse trough at an inn in Harwich which the owner refused to have cleaned out because the water remained sweet when the sides and bottom were "covered with a green substance known to be of vegetable nature." He himself observed "the spontaneous emission of dephlogisticated air from water containing vegetative green matter, . . . only in circumstances in which the water had been exposed to light." He first supposed the green substance to be a plant "but was unable to discover the form of one. Several of my friends better skilled in botany than myself never entertained any doubt of its being a plant and I had afterward the fullest conviction that it must be one. Mr. Bewley has lately observed the regular form of it by a microscope. My own eyes having always been weak, I have as much as possible avoided the use of the microscope."

In 1779, Ingen-Housz (184, 281, 391), a Dutch physician impressed by the address of Pringle, came to England and carried out a remarkable series of experiments, published the same year, in which he confirmed Priestley's early work and answered the questions that arose later. He pointed out that plants can change the "air" they undoubtedly absorb from the common atmosphere into "fine dephlogisticated air" (oxygen) which renders the atmosphere more fit for animal life. He emphasized strongly that this "purification" process begins after sunrise and ends at sunset and is in proportion to the light intensity. Dense shade prevents the plants from "performing this office." It is performed only by the leaves and green stalks and only in adequate light. The flowers and roots vitiate the air at all times, while the leaves vitiate the air in the dark. The production of vital "dephlogisticated" air, or oxygen, in photosynthesis and its disappearance in respiration was thus clearly established.

Neither Priestley nor Ingen-Housz seemed to understand at first the role of carbon dioxide in photosynthesis. They spoke of the poisonous and "mephitic" properties of the air after respiration, greatly exaggerating the toxicity. They were vague in describing what was absorbed by the leaves in the light reaction and what was the "poisonous" product of dark respiration. It was Senebier (1782–1783) (303) who, having confirmed the results of Ingen-Housz by independent experiments, insisted that the absorption of "fixed air" or carbon dioxide from the atmosphere was the source of the "vital air." However, Senebier did not believe that the fixed air was absorbed by the leaves directly; he held that it entered through the roots in solution and was then transported to the leaves. Even in "Physiologie Végétale," Volume III, 1800 (305) he stated, "Carbonic acid present in the air favors vegetation . . . . The acid enters into the plant dissolved in water." He also questioned the production of carbon dioxide by respiration. The polemics between Ingen-Housz and Senebier (304) during the 1780's were due partly to these misunderstandings and partly to the inherent complexity of the subject together with the ambiguity of the phlogiston theory.

By 1784, Lavoisier had completely discredited phlogiston and had established his theory of combustion. The latter was not immediately understood and accepted, but in 1796 Ingen-Housz (185) published his "Food of Plants and Renovation of the Soil," in which respiration and photosynthesis were described correctly in terms of the new chemistry. The two separate processes were clearly distinguished. Absorption of carbon dioxide through the leaves was advocated, without perhaps an appreciation of how small the concentration in the air actually was. Black had observed a very small amount, but Lavoisier had found none at all; von Humbolt (in the preface to the German edition of the Ingen-Housz treatise) reported about 1%. In any case Ingen-Housz was definitely opposed to Senebier's suggestion that the carbon dioxide was dissolved in water and was taken up through the roots.

As a final major step in this epoch of the history of photosynthesis, de Saussure (1804) (296) published his "Recherches Chimiques sur la Végétation," which is an account of quantitative experimentation on the subject. He showed that (a) plants cannot live in the absence of oxygen; (b) they not only cannot grow but are soon injured when they are deprived of carbon dioxide by being confined in a closed space containing quick lime or alkali; (c) excess of carbon dioxide above the normal atmospheric content up to 8% increases the growth rate of peas but 12% or more is harmful and 75–100% kills the plant in a few days; (d) carbon monoxide cannot be utilized; (e) plants growing in the open atmosphere with their roots in pure water obtain their carbon from the small amount of carbonic acid gas present in the air.

The combined photosynthesis and respiration quotients of five species of plants were determined. A brief description of these experiments follows.

The plants were enclosed under large glass jars 8–16 cm in diameter, standing in mercury which was covered with a small amount of water to avoid the toxic action of mercury vapor. The plant roots were placed in separate glass vessels containing distilled water. The plants were grown in carbon dioxide-enriched air for 6–18 days in "moderated" sunlight for 6 hours each day. The gases were analyzed by reliable methods for oxygen and carbon dioxide before and after the exposure, and the total gas volumes were read with good precision. There was practically no change in the total gas volumes in nearly all the experiments. However, the volume of oxygen liberated was only 68–81% of the volume of carbon dioxide absorbed. The difference between carbon dioxide absorbed and the oxygen evolved was assumed to be nitrogen liberated from the plants; the assumption appears to be an inadequate explanation because even the total nitrogen in the plants was insufficient to make up the difference. Evidently there were some rather large unidentified errors in this work. Later studies indicate that the photosynthetic quotient should be about unity. However, the results are of interest because they are the first measurements of this ratio (see Section II, A).

De Saussure made careful determinations of the increase in dry matter in two of the foregoing species. The carbon, estimated from the amount of carbon dioxide absorbed by the plants, was found to be 41 and 49% of the dry weight increase in the two species, respectively. As oxygen evolution had accounted for a large part of the oxygen in the carbon dioxide, it was postulated that the elements of water had participated to an equivalent extent in the assimilation and respiration reactions. The author stated: "In no case do plants directly decompose water in assimilating hydrogen or eliminating oxygen as a gas. They never exhale oxygen save as the consequence of the direct decomposition of carbonic acid." Photosynthesis was thus placed on a quantitative foundation though only the over-all reactions were understood. De Saussure could not know that in all probability the exhaled oxygen actually came from water instead of from carbon dioxide.

The importance of nitrogen and the minerals in the nutrition of plants was also established by de Saussure in quantitative experiments. It was shown that these materials entered through the roots, though the proof that atmospheric nitrogen is not absorbed in the leaves was not as clear as might be desired. This proof was supplied later by Boussingault (1854), who developed a water culture technique.

Summarizing the work of Priestley, Ingen-Housz, Senebier, and de Saussure, the over-all reactions of photosynthesis and respiration in plants can be written in modern notation as follows:

$$CO_2 + H_2O + \text{light} + \text{green leaves or stems} \rightarrow \text{organic matter } (CH_2O) + O_2$$
$$\text{Organic matter} + O_2 \text{ (absence of light or green tissue)} \rightarrow CO_2 + H_2O + \text{heat}$$

These four authors came to understand quite well the over-all chemistry involved in these reactions, but they made no claim to an understanding of the energy relations; these relations were worked out later for photosynthesis, starting with the recognition by J. R. Mayer (237) in 1845 of the conversion of light energy into the potential energy of organic compounds and continuing for more than a century to the present time.

3. Photosynthesis Studies 1800–1860

During nearly forty years following the studies of de Saussure, little was done to advance the knowledge of photosynthesis and many serious errors and misconceptions were perpetuated, particularly among the botanists. This was due to a rather loose reliance on a theory of "vital force" to explain phenomena difficult to explain by chemistry and physics. For example, Treveranus (1835) conceived as the true food of plants, a "vital" matter like albumin or gelatin which was supposed to be present in all organic bodies and was even extractable from the humus of soil. A "humus" theory was advanced for the nutrition and growth of plants. No importance was attached to the uptake of carbon dioxide by the leaves. Similarly, Meyer (1838) rejected the uptake of carbon dioxide by the leaves, though de Candolle (1832) and Dutrochet (1837) (105, 106) made it the central theme of their theories of plant nutrition. The latter also showed that reduction of carbon dioxide and absorption of oxygen can take place simultaneously in the light. Further, Dutrochet performed a useful service in emphasizing that carbon dioxide is reduced only in cells containing chlorophyll. On this matter de Saussure had been confused because he observed photosynthesis in the red garden plant orache (*Atriplex hortensis*), not realizing that the chlorophyll was masked by the red color.

Experiments carried out by Senebier (1800), de Saussure (1822), and Vrolik and de Vriss (1836–1837) showed that in flowers there was production of heat due to respiration. These experiments were considered to be inconclusive because it was thought that flowers as organs of reproduction possessed the "vital force" and therefore could generate heat spontaneously. Dutrochet (1837–1840) showed that growing shoots generate heat due to respiration. He discussed at length the analogies between respiration of plants and animals and concluded that heat production in plants, as in animals, is the result of their respiration. J. R. Mayer (237) stated, "Plants are able only to convert energy not to create it."

In 1840 Liebig (213) first published his "Organic Chemistry in Its Application to Agriculture and Physiology," which with its subsequent editions, finally brought about general acceptance of the principles underlying the photosynthetic process as enunciated by Ingen-Housz and de Saussure and answered the questions raised by both the "vital force" theory and the "humus" theory. The syntheses of organic compounds in the laboratory, which previously were thought to require the "vital force" to produce them, did much to discredit the "vital force" theory. Liebig showed that the amount of "humus" in the soil increased rather than decreased with crop production. It therefore could not be the "true" food of plants.

Boussingault (1855) observed that plants grew well in soil that had been calcined to destroy all organic matter or in water cultures containing nitrates or ammonium salts and other minerals but no carbon or organic nitrogen compounds. The carbon dioxide content of the air was thus the only source of carbon to consider, and Liebig's calculations based on gas analysis indicated that this source was adequate to account for all plant growth. The participation of the elements of water in the assimilation process as shown by de Saussure was now more clearly understood in the light of the new organic chemistry. Finally, the source of nitrogen as nitrates and ammonium salts in the soil rather than as nitrogen in the atmosphere was well established.

4. Photosynthesis Studies 1860–1930

During the latter half of the nineteenth century, serious consideration was given to the chemistry, mechanisms, and environments of photosynthesis. Many new lines of investigation were initiated or reactivated, most of which did not come to fruition until later. These included chemical and physical studies of chlorophyll and other leaf pigments; the primary reaction of photosynthesis; energy relations and spectral responses in assimilation; the role of the stomata; and external factors such as light, carbon dioxide concentration, and temperature.

One of the most important investigations initiated in this period was the study of the organic chemistry of chlorophyll and the other leaf pigments. Berzelius (1837) (29) was one of the first to attempt to analyze chlorophyll, but the complexity and lability of the substance made the task extremely difficult. The similarity of the molecule to that of hematin was shown by Verdeil (1851) (362) and Hoppe-Seyler (1879–1881) (181), who converted chlorophyll to a red porphyrin similar to one given by hematin.

The principal features of the structural formulas of chlorophyll a and chlorophyll b, together with the reactions that produce phyllins, phytins, phorbins, chlorines, and porphines were worked out in twenty papers by Willstätter and associates (396) between 1905 and 1913. Willstätter and Stoll summarized these and later studies in their books "Chlorophyll" (1913) (399) and "Carbon Assimilation" (1918) (398) (English edition, 1928) (397). Additional studies on chlorophyll structure were done later by Conant (14 papers between 1929 and 1934) and by Hans Fischer and co-workers (128). The latter published 100 papers on the chlorophylls between 1928 and 1940, and a comparable series of papers on the porphyrins.

This voluminous literature established the chemistry of the chlorophylls and their derivatives, though complete synthesis was not accomplished until 1960 (400). Essentially, the chlorophyll molecule (Fig. 1) consists of four pyrrole rings joined together by four reduced methylene groups between the carbons adjacent to the nitrogen in the various pyrrole rings, thus giving an 18-member conjugated ring structure with a magnesium atom in the center attached to the four nitrogens by primary and secondary valence bonds. The substituents attached to this configuration have been very difficult to locate, but it appears most probable that methyl groups are attached to carbons 1, 3, 5, and 8, vinyl to 2, ethyl to 4, and phytyl propionate to 7. The cyclopentanone ring V with oxygen at C-9 and methylester at C-10 completes the structure of chlorophyll a. Chlorophyll b differs from chlorophyll a by having an aldehyde group in place of the methyl on C-3.

Three different patterns of an 18-membered closed conjugated ring, indicated by the heavy line in Fig. 1, can be drawn through this structure depending on the location of the primary Mg-N bonds and of the isolated double bond shown in ring II, which can be shifted to ring

I or ring III. The three patterns are presumably "mesomeric," involving only electron shifts. If they are truly isomeric, the arrangement shown in Fig. 1 seems most probable because of its symmetry. This structure is quite similar to that suggested by Fischer for protochlorophyll from purple bacteria, which is reduced at C-3 and C-4. Chlorophyll is optically active due to carbon atoms 7 and 8. Atom 10 does not have practical asymmetry because of the ease of enolization of the carbonyl in position 9.

The energy levels, fluorescence, and configurations of this remarkable substance and its unique properties for energy transformation and storage have been discussed in detail by Gaffron in Volume 1B.

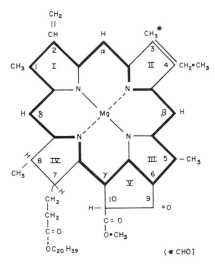

Fig. 1. Structure of chlorophyll a. Chlorophyll b is formed by replacing CH₃ in position 3 with CHO. From Willstätter and Stoll (397).

Following is a brief account of the efforts made during the 1850–1930 period to work out the chemical reactions of carbon assimilation. J. R. Green (151), Spoehr (321), Stiles (336), and Rabinowitch (279) have discussed these efforts. Contemporaneous historians include Hansen (162, 163), Meyer (239), Sachs (295), Pfeffer (270), and Vines (368).

Liebig (213) suggested that the conversion of carbon dioxide to sugar takes place stepwise starting with the reduction of carbonic acid to oxalic acid, then to tartaric acid, malic acid, etc., thus gradually reducing the amount of oxygen and increasing the hydrogen in the molecule until carbohydrates are formed. In support of this theory, he cited the lowering of the acidity of fruit during ripening and the high acid content of succulent leaves—reactions that are actually unrelated to the photosynthetic process itself.

The dominating hypothesis during this period of over half a century was that of von Baeyer (1870) (18), who suggested that carbon dioxide and water could react in the presence of chlorophyll, under the influence of sunlight, to liberate oxygen and form formaldehyde, which was then condensed to hexoses and other carbohydrates. Carbon monoxide was an intermediate product. This hypothesis involves four assumptions: (a) carbon dioxide is dissociated in the light to carbon monoxide and oxygen; (b) carbon monoxide is fixed by chlorophyll; (c) carbon monoxide is reduced to formaldehyde by hydrogen, which is produced by the action of light on water; (d) formaldehyde is condensed to sugars under the conditions that obtain in the plant. Since no significant photosynthesis occurs when carbon monoxide is substituted for carbon dioxide, a variation of the theory assumes that carbonic acid is reduced by hydrogen directly to formaldehyde. The appealing simplicity of this hypothesis and the many tantalizing indications, particularly from *in vitro* experiments, that such reactions could actually occur, led to the performance of a large amount of work which must be considered as essentially futile so far as the elucidation of photosynthesis in the plant is concerned.

Attempts to identify formaldehyde as an intermediate in the process yielded contradictory results. No more than trace amounts have been found, and evidence for its formation is questionable. Oxygen evolution was not observed. Feeding experiments using formaldehyde or less toxic compounds that are readily converted to formaldehyde, such as methylal, have also yielded contradictory results. Some workers obtained increased dry weight in the dark; others had negative results even under the same apparent conditions. Positive results from feeding formaldehyde would not necessarily substantiate the Baeyer hypothesis anyway, since a number of compounds of low molecular weight, such as glycol and acetate, can be similarly utilized by plants.

A great many *in vitro* studies of photosynthesis have been made. Chemical reduction of carbon dioxide requires powerful reducing agents like sodium amalgam or sodium azide. Even ultraviolet light does not reduce the gas unless organic impurities are present (273). Much effort has been expended in the attempt to simulate photosynthesis by irradiating suspensions of colored solids in carbon dioxide-saturated water with visible light. Baly (20) used specially prepared green nickel oxides containing thorium oxide and obtained small, variable amounts of unidentified organic compounds that reduced Fehling's solution after acid hydroylsis. On this evidence he considered that he had achieved photosynthesis *in vitro*, even though oxygen evolution was not established.

Dhar and Ram (101, 102) likewise obtained small yields of formaldehyde by irradiating solutions of the salts of many metals, e.g., iron, nickel, cobalt, copper, manganese, chromium, and uranium, containing carbon dioxide or sodium bicarbonate. Formaldehyde concentrations from 8 to 40 ppm were reported. Some reducing sugars were also obtained.

Baur and co-workers (25–27) carried out many experiments, for the most part with negative results, over the thirty-year period before 1943, in the attempt to produce photosynthesis *in vitro*. They finally achieved some measure of carbon dioxide reduction by the use of a two-phase system having a sensitizer (e.g., chlorophyll, eosin, rhodamine) plus an oxidant (carbon dioxide bound in a lipophilic carbonic acid ester) in the nonaqueous phase, and the reductant plus an "auxiliary oxidation-reduction system" (e.g., methylene blue, ascorbic acid) in the aqueous phase. Some formaldehyde was formed by strong illumination (25,000 lux), but definite evidence of oxygen evolution was not obtained. In later experiments (27), using a nonaqueous two-phase emulsion, both formaldehyde and oxygen formation were claimed, but the claim could not be substantiated by Linstead *et al.* (215). The latter noted that the substrate was photolyzed, giving an aldehyde but no formaldehyde.

According to Rabinowitch (279), the small yields of organic compounds obtained by Baly, Dhar and Ram, and Baur could have been due to chance side reactions involving carbon dioxide reduction by light. "So long as quantum yields remain extremely small (of the order of 10^{-5} or 10^{-6}), everything is possible in photochemistry." In any case these researches and most of the others in the 1850–1930 period represent essentially futile attempts to advance our knowledge of the detailed photosynthetic reactions in plants. Though some of the experiments, particularly those employing extreme ultraviolet light, may have a bearing on the origin of organic matter on the earth, it is evident that almost insuperable difficulties attended these early inquiries because of the complexity of the initial reactions and their association with the subsequent reaction of metabolism. Their solution had to await the development of new chemical techniques such as chromatography, polarography, and especially the advent of the tracer elements, radioactive and inert, by which our knowledge of the subject has been materially advanced.

5. *Photosynthesis Studies since 1930*

This subject will be treated here only in outline. Detailed treatments are given by Gaffron (Volume IB), and by Rabinowitch (279), Arnon

(7) has presented a penetrating discussion of the present status of photosynthetic theory with its historical background. Many reviews of various phases of the subject are also available (13, 386).

(a) Chlorophyll studies have been carried out based on the known structure of the molecules and their distribution as monolayers on the laminated proteins and lipoproteins in the chloroplasts, as shown by the electron microscope.

(b) Van Niel (360), using the comparative biochemical approach of Kluiver (202), emphasized that photosynthesis is a chlorophyll-catalyzed oxidation-reduction process involving the oxidation of water to furnish hydrogen for the reduction of carbon dioxide, with evolution of molecular oxygen. With the advent of O^{18} Ruben and co-workers (292) submitted strong evidence that this oxygen came from water, not from carbon dioxide. However, they recognized a possibility that isotopic exchange in the plant might have vitiated their conclusion (192).

There is close similarity between photosynthesis in green plants, and photosynthesis in green and purple bacteria which utilize hydrogen sulfide instead of water as the hydrogen donor. The process in bacteria may be further generalized. The reducing agent can be a reduced selenium compound, molecular hydrogen, or any one of a number of organic compounds such as isopropanol (131) and some organic acids. Conversely, other oxidants besides carbon dioxide can participate in the photosynthetic reactions as shown by the Hill reactions using isolated chloroplasts and various inorganic and organic oxidants which serve for the study of the photochemical aspects of photosynthesis.

(c) Studies based on the formaldehyde theory were gradually abandoned and the Liebig notion of stepwise reactions following uptake of carbon dioxide in a carboxylation was exploited. Radioactive carbon was employed along with chromatographic methods of separation and analysis. A primary enzymatically mediated reaction product was identified by Calvin and co-workers (75) and Gaffron and co-workers (Volume IB) as carboxylated ribulose diphosphate, which splits rapidly to form two moles of phosphoglyceric acid. Then follows a complex series of enzyme-controlled oxidation-reduction reactions, isomerizations, dismutations, phosphorylations, etc., resulting in the formation of sugars and starch on the one hand and the cyclic regeneration of ribulose diphosphate on the other, together with evolution of oxygen. Other pathways are also suggested.

(d) Fluorescence and phosphorescence studies (279, 375) demonstrated the singlet and triplet energy states in chlorophyll and their relationships to energy transfer in photosynthesis. The quantum yield in photosynthesis has been exhaustively studied. The Warburg school

(370) still supports the minimum theoretical value of 3 to 4 quanta per molecule of carbon dioxide reduced while the majority of investigators favor a value 2 or 3 times larger (13, 121, 229, 326).

B. Photosynthesis on the Earth

1. Insolation

Considerable information is available as to the amount of light received on the earth's surface and the utilization of that light for photo-

TABLE I

WORLDWIDE INSOLATION, PHOTOSYNTHESIS, RESPIRATION, CARBON RESERVE, AND UTILIZATION OF ORGANIC MATTER

A. Insolation		
1. To outer atmosphere (solar constant)	1300×10^{21}	Cal/year
2. To earth's surface	650×10^{21}	Cal/year
3. Less infrared (60%)		
On land	80×10^{21}	Cal/year
In oceans	90×10^{21}	Cal/year
4. Probable absorption by green plants		
On land	25×10^{21}	Cal/year
In oceans	90×10^{21}	Cal/year
5. Photosynthesis (2% efficiency)		
Carbon equivalent, land	50×10^{9}	Tons/year
Carbon equivalent, oceans	180×10^{9}	Tons/year
B. Carbon fixation		
Land (1.5×10^{8} km²)	$16^a–24^b \times 10^{9}$	Tons/year
Oceans (3.6×10^{8} km²)	$20^c–155^d \times 10^{9}$	Tons/year
C. Respiration		
Land plants	$4–6 \times 10^{9}$	Tons/year
Ocean plants	$4–30 \times 10^{9}$	Tons/year
Animals (man $= 0.2 \times 10^{9}$ tons/year)	0.4×10^{9}	Tons/year
Microorganisms	$10–15 \times 10^{9}$	Tons/year
D. Carbon reserves		
Lithosphere to depth of 16 km	200×10^{18}	Tons
Hydrosphere	500×10^{11}	Tons
Troposphere to height of 11 km	600×10^{9}	Tons
E. Energy utilization in 1955 coal equivalent[e]		
Coal, lignite, etc.	1.7×10^{9}	Tons/year
Crude petroleum	1.0×10^{9}	Tons/year
Natural gas	0.4×10^{9}	Tons/year
Hydroelectric	0.3×10^{9}	Tons/year
Total	3.4×10^{9}	Tons/year

[a] Estimate of Schroeder (301).
[b] Estimate of Ebermayer (110).
[c] Estimate of Steemann-Nielsen (334).
[d] Estimate of Riley (290) and Rabinowitch (279).
[e] From United Nations summaries for 1955.

synthesis. The solar constant is defined as the energy flux at normal incidence received from the sun outside the earth's atmosphere. It has been measured by pyrheliometers, indirectly from the ground and directly from rockets at 100 km elevation. Its value approaches 2.0 cal/cm^2/min, equivalent to about 1300×10^{21} cal/year on a flat circular surface 4000 miles in diameter. This energy is partly depleted in passing through the atmosphere to the earth's surface, where the undepleted

TABLE II

Average Daily Insolation by Months (1950) at Three Locations in the United States[a]

Month	Blue Hills, Massachusetts	El Paso, Texas	Phoenix, Arizona
	cal/cm²/day		
January	103	350	313
February	157	427	391
March	331	586	525
April	332	675	666
May	453	715	751
June	545	739	774
July	490	650	657
August	398	688	662
September	305	560	572
October	232	502	491
November	143	411	370
December	102	348	311
Average insolation (cal/cm²/day)	299	554	539
Kcal/m²/day × 40%	1200	2216	2156
Theoretical yield (9 quanta basis) gm/m²/day	63	118	115

[a] Data by U.S. Weather Bureau.

portion is absorbed on a surface of double the incident area. However, owing to the earth's rotation, the total area of absorption is four times the incident area. Table I summarizes the data for insolation, photosynthesis, and respiration. It also includes data on the carbon reserves and their current utilization for energy.

According to Byers (73) the depletion factor for light passing through the earth's atmosphere is 31% with clear skies and 65% with cloudy skies. It is cloudy during about half the daytime periods, so that the total radiation reaching the earth's surface is about half the incident radiation. This amounts to about 650×10^{21} cal/year, equivalent to a daytime average of 0.5 cal/cm²/min. Confirmation of this value is obtained from continuous measurement of insolation. For example, in Washington, D.C., in 1925 such measurement indicated an average of 490 cal/cm²/day in clear weather and 320 cal/cm²/day under average sky conditions. These figures represent about 0.7 and 0.45 cal/cm²/min. There is a threefold range of insolation between winter and summer due to angle of incidence. Table II gives insolation data near Boston, El Paso, and Phoenix. Boston receives a daily average of 300 cal/cm² as compared to 550 cal/cm² in the Southwest.

2. Fixation of Carbon Dioxide on Land and in the Oceans

Of the radiation received on the earth's surface, about 60% is far red and infrared (above 7000 A) and does not participate in photosynthesis, leaving about 260×10^{21} cal/year that might be utilized. This is divided into 180×10^{21} cal/year for the oceans and 80×10^{21} cal/year for the land areas. More than half the land (estimate of Schroeder, 1919) (301), is desert or steppe on which very little vegetation is produced. This leaves about 40×10^{21} cal/year for wooded $(44 \times 10^{6}$ km²$)$ and farming areas $(27 \times 10^{6}$ km²$)$. Vegetation on the latter areas is dormant about one-third of the year on the average, reducing the effective insolation to about 25×10^{21} cal/year. If the efficiency of photosynthesis is 2%, this represents about 0.5×10^{21} cal/year, equivalent to 50×10^{9} tons of organic carbon. This may be compared with Schroeder's estimate of 16×19^{9} tons fixed on land. Liebig's estimate (213) was 30×10^{9} tons, and Ebermayer's (110) 24×10^{9} tons. Evidently the two independent methods of estimating the photosynthetic yield agree reasonably well for land plants, considering the approximations involved.

Regarding the 180×10^{21} cal/year received by the oceans, if about half this radiation is absorbed by the plankton, again with photosynthetic efficiency of 2%, 1.8×10^{21} cal/year or 180×10^{9} tons of organic carbon would be involved, compared with an estimate of Rabinowitch (279) of 155×10^{9} tons or of Steemann-Nielsen (332–334) of 20×10^{9} tons/year. The former estimate is based on experiments by Riley (290, 291), who measured photosynthesis in phytoplankton at a few locations in the ocean by a number of indirect methods, such as oxygen liberation and consumption of phosphate, which yielded an average assimilation value of about 1-gram calorie (gcal)/m²/day. Steemann-Nielsen showed that these methods tend to give high results compared with a more direct method using carbon-14 (330). From traverses of the South Atlantic, Indian, Pacific, and North Atlantic Oceans by the Galathea Expedition, he concluded that the average rate of assimilation in the bathosphere is about 0.15 gcal/m²/day, equivalent to about 20×10^9 tons of carbon per year.

Ryther (294) believes that Steemann-Nielsen's estimate is low because it is largely based on single observations at a time of year when the rates were below the annual average. Ryther measured photosynthesis in the infertile Sargasso Sea throughout the year and found an average assimilation of 0.4 gcal/m²/day. This is compared with Steemann-Nielsen's value in June, with which Ryther agrees, of 0.07 gcal/m²/day. The latter value is a seasonal minimum. Much higher yields are found in April. In the North Sea in April and May, Cushing (87) found algal production of 4–16 mg-cal/m³/day by the carbon-14 method representing about 20–80% of the standing crop. Depth of the photosynthetically active water layer is not stated, but if 20 meters is taken, production would be about 0.20 cal/m²/day. If the Sargasso Sea is representative of the less productive areas of the oceans, Ryther considers that the seas are at least twice as productive as the land, a total yield of 40×10^9 tons of carbon being fixed.

The assumption that the phytoplankton absorb half the visible light that enters the more dilute suspensions of these organisms may be optimistic because, owing to the transparency of the water, a significant portion of the photosynthesis must occur from near the surface to depths of 100 meters. These dilute suspensions represent large areas of the oceans wherever there is only slight local circulation of the water. Greater than average upwelling from the depths of the ocean is required to bring to the surface sufficient nutrients to support appreciable plant growth.

The carbon content of the earth's crust is very large. However, it is largely fixed in the rocks as carbonate and is for the most part unavailable for photosynthesis. The carbon content of the atmosphere and the hydrosphere is nearly all in available form as carbon dioxide or soluble carbonates. The oceans have about 100 times as much carbon dioxide as the atmosphere, and the large supply in the ocean serves to regulate the atmospheric content within narrow limits near 300 ppm. The quantity of carbon dioxide in the atmosphere is sufficient to support the photosynthesis on land for about thirty years at the current rate of consumption.

The 20×10^9 tons of carbon removed annually from the atmosphere by land plants is replenished by the respiration of these plants to the extent of about 20% or 4×10^9 tons. Respiration, like photosynthesis, in the oceans is uncertain by a factor of 7. As most of the organic matter currently produced is eventually decomposed to carbon dioxide by microorganisms, it is reasonable to assume that 10 to 15×10^9 tons of carbon are oxidized in this way on land. However, there are no independent data on which to base such an estimate. Finally, current industrial activity oxidizes another 3×10^9 tons in the fossil fuels.

Power for industry is increasing from year to year. Data for 1955 from United Nations summaries are listed in Table I. According to Sievers (311), total solid fuels used in 1920 were equivalent to about 1.0×10^9 metric tons of carbon. Other fuels were relatively insignificant, so that the 1955 total is about 3 times the 1920 value. Hydroelectric power in 1955 represented 35% of the electrical energy, but only 9% of the total energy. Total energy employed in industry was less than 20% of the photosynthesis on land in 1955.

II. The Carbon Dioxide Supply and Exchange

A. Photosynthetic and Respiratory Quotients

In photosynthesis, carbon dioxide is absorbed and oxygen is evolved. The reverse process occurs in respiration. The ratios of the gas volumes involved in the two processes may be expressed as quotients:

$$Q_P = \frac{O_2}{CO_2}; \; Q_R = \frac{CO_2}{O_2}$$

These quotients are good indexes of the over-all chemical processes involved in the plant. A value of unity for Q_P suggests carbohydrate production; values below unity suggest the production of simple organic acids, such as citric or malic; values above unity, protein and fat production. Similarly Q_R values of unity suggest combustion of carbohydrates whereas high and low Q_R values indicate combustion of simple organic acids and proteins or fats, respectively. A few of the historically important determinations of these quotients are listed in Table III and discussed below.

The net exchange ratio of oxygen to carbon dioxide in the leaf was found by de Saussure (296) to range from about 0.7 to 0.8 for the average of the photosynthesis and respiration covering 6–18 successive days with about 6 hours of bright light each day. Subsequent studies have usually yielded ratios nearer unity. An appreciable amount of inert gas (nitrogen?) appeared to be liberated along with the oxygen.

Boussingault (51) evaluated the experiments of de Saussure and concluded that the residual nitrogen reported by the latter could not have been derived from the plants because the amounts were greater than even the total combined nitrogen in the plant samples. He observed that an appreciable amount of nitrogen was entrained in the leaves and stems of similar plants; also that the residual gas contained some carbon monoxide. Possibly the low oxygen:carbon dioxide ratio was due to side reactions during the 6–18 days of the experiment.

Boussingault devised apparatus to determine the photosynthetic quotient under rigid control. Each unit of the system consisted of a 550–700 cc flask filled with water to which carbon dioxide and a leaf sample were added after all dissolved gases had been boiled out. The

TABLE III
STEADY STATE PHOTOSYNTHETIC AND RESPIRATORY QUOTIENTS

Plant type	Photosynthetic quotients $\Delta O_2/\Delta CO_2$		Respiratory quotients $\Delta CO_2/\Delta O_2$		Reference
	Range	Average	Range	Average	
Broad leaf	0.68–0.81	0.75	—	—	de Saussure (296)
Broad leaf	0.89–1.06	0.99 ± 0.04	—	—	Boussingault (51)
Broad leaf	1.10–1.30	1.18	0.73–0.96	0.85	Bonnier and Mangin (49)
Broad leaf	0.97–1.12	1.04 ± 0.04	0.92–1.12	1.05 ± 0.05	Maquenne and Demoussy (230, 231)
Broad leaf	0.95–1.02	1.00	—	—	Willstätter and Stoll (398)
Broad leaf	—	1.13	—	—	Kostytschew (207)
Diatoms	—	1.05 ± 0.03	—	0.79 ± 0.03[a] 0.93 ± 0.04[a]	Barker (21)
Algae	—	1.3 to 1.4	—	—	Myers and Johnston (254)
Algae	0.3–1.5	0.98	—	—	Manning et al. (229)
Algae	0.92–1.07	1.00	—	—	Van der Paauw (358a)

[a] Different species.

flask was closed and the leaves were subjected to light or dark treatments. The flask was then placed in communication, through a smaller flask containing a small amount of water, with an eudiometer tube filled with mercury. With the connection between the two flasks closed and the long delivery tube dipping into mercury, the water in the second flask was boiled to remove all fixed gas. Then with the connection open and the delivery tube connected to the eudiometer, the large flask was boiled under vacuum to drive all the gas contained in that flask into the eudiometer, where it was measured, then treated with phosphorus and alkali for the determination of oxygen, carbon dioxide, and residual nitrogen.

Three units of the apparatus, as nearly identical as possible in dimensions and manipulation, were used in each experiment. One was a control blank run. The other two had the same weight (about 10 gm) of fresh leaves in the large flask. Each leaf was sponged lightly with boiled water to remove adhering gas before being placed quickly in the large flasks, which were then filled with water containing dissolved carbon dioxide and closed. The second unit remained in the dark; the third was exposed for 1–10 hours to sunlight. Gas exchange of the third unit was calculated with allowance for the gases found to be present in the system without illumination in the first two units. In 41 experiments with 25 species or varieties of plants, the average photosynthetic quotient was 0.99 with a standard deviation of 0.04 and a range from 0.89 to 1.06.

Another series of experiments was carried out by illuminating the leaves in carbon dioxide-enriched air in eudiometer tubes and carefully determining the gas exchange. With about 30% carbon dioxide, the average Q_P value in 12 experiments was 1.00 ± 0.04. Measurements made in pure carbon dioxide gave the same average Q_P value, but the range was much larger ($\pm 16\%$) owing to the reaction rates being only one-fifth as rapid and the gas volumes correspondingly smaller.

In the foregoing study no attempt was made to determine the ratios for assimilation and respiration separately. This was done by Bonnier and Mangin (49), who "corrected the assimilation values by adding night respiration rates to obtain true photosynthesis." This involves the questionable assumption that light respiration is equal to dark respiration. Three other indirect methods of evaluating true photosynthesis were also used (see discussion of respiration during photosynthesis, Section II, E). The photosynthetic quotients Q_P by the four methods ranged from 1.05 to 1.13, and respiratory quotients Q_R ranged from 0.73 to 0.96. Table III summarizes the steady state values of Q_P and Q_R obtained by the various investigators.

Maquenne and Demoussy (230–231) also corrected their assimilation values by adding night respiration. Their results for 27 species of broadleaf plants average a little more than unity with a small range: $Q_P = 1.03$, range 0.97–1.12; $Q_R = 1.05$, range 0.94–1.12.

Willstätter and Stoll (398) worked with high light intensity (45,000 lux). Assimilation exceeded respiration by a wide margin, and the respiration correction was small. Their results with such plants as *Sambucus*, *Cyclamen*, *Pelargonium*, *Ilex*, and *Aesculus* averaged $Q_P = 0.99$, and ranged from 0.95 to 1.02. A wide range of experimental conditions was used, including high carbon dioxide concentration, 5–7%; oxygen, 0.01 to 21%; temperatures from 10° to 35°; and light exposure times from 0.5 to 10 hours. There appeared to be a slight tendency for Q_P values to decrease with time in prolonged experiments.

TABLE IV

Effect of Duration and Intensity of Exposure to Light, After a Dark Period, on the Photosynthesis Quotient $Q_P = \Delta O_2/\Delta CO_2$

Plant	Light	\multicolumn{9}{c}{Exposure period (minutes)}
		3	5–6	10	15	20	30–40	60–120	180–240	300–360
Syringa[a]	Sun			0.73						
Syringa[a]	Diffuse						0.99			
Betula[a]	Sun	0.40	0.80	1.00	1.13	1.01				
Betula[a]	Shade						0.72[d]	0.67[e]		
Lamium[a]	Sun		0.76 then diffuse light ⟶ 0.99							
Lamium[a]	Shade							0.54		
Achillea[a]	Sun			0.78			0.95			
Epilobium[a]	Sun		0.70		0.97					
Potentilla[a]	Sun	0.58					1.00			
Salix[a]	Sun		0.75			0.98				
Algae[a,c]	Sun		0.22			0.79				
Algae[a,c]	Sun		0.31							
Algae[a,c]	Sun	0.56 then dark ⟶ 0.52								
Algae[a,c]	Diffuse						0.96			
"Phyllocactus"[b]										
25°C	45,000 lux						1.56	1.45	1.45	
35°C	45,000 lux						1.42		1.18	
Opuntia[b]										
25°C	45,000 lux						1.49		1.25	1.12
30°C	45,000 lux						2.28	1.96		1.18

[a] Data by Kostytschew (207).
[b] Data by Willstätter and Stoll (398).
[c] Spirogyra communis and Zygnema stellinum.
[d] 20.5°C.
[e] 17.0°C.

When various forms of cactus were measured, however, the initial photosynthetic quotient (Table IV) was much greater than unity after a period of darkness. In 0.5 hour, Q_P ranged from 1.42 to 2.28. After about 6 hours it had fallen to 1.12–1.18. There is a considerable accumulation of organic acids from respiration in the fleshy cactus leaves during the dark period. These acids react first after illumination, reducing the intake of carbon dioxide below the volume of oxygen evolved. Later, as the accumulated organic acids are used up, Q_P approaches unity.

Kostytschew (207) (Table IV) found that Q_P values in many leaves increased appreciably during the first few minutes of illumination by direct sunlight. *Syringa vulgaris* gave a quotient of 0.73 in the first 10 minutes of sunlight and 0.99 in 40 minutes. *Betula pendula* (*B. verrucosa*) gave 0.40 in 3 minutes; 0.79 in 6 minutes, 1.13 in the next 10 minutes, or an average of 1.00 for 16 minutes in sunlight. One hour in the shade gave 0.67. The other broad-leaf plants behaved similarly. Evidently these leaves require carbon dioxide in excess initially, whereas the cacti require an excess of oxygen initially. It is of great interest that when a steady state of assimilation is established, Q_P approximates unity. This indicates that carbohydrates are the primary products of photosynthesis in both fleshy- and thin-leaved plants.

Barker (21) studied the diatom *Nitzschia*, which stores oil instead of carbohydrates. The Q_P value approximated 1.05 under high and low light intensity and for a temperature range from 12° to 30°C, in both the marine variety, *N. closterium* and the freshwater variety, *N. palea*. Q_R differed in the two diatoms, averaging 0.93 in the former and 0.79 in the latter. The Q_P value of 1.05 suggests that only about 10% of the total photosynthate can appear as stored fat. Direct experiments indicated that only small amounts of carbohydrates were present in the photosynthate, which was rapidly converted into soluble cell constituents. The respiratory quotients of the two diatoms differed markedly. In *N. closterium*, Q_P and Q_R were not greatly different, a result indicating that the composition of the materials synthesized and respired was similar. The low Q_R value of 0.79 in *N. palea* suggests that the respired material was about 70% fat and 30% carbohydrate, assuming negligible protein respiration.

It seems reasonable to conclude that the primary photosynthetic process in all plants yields carbohydrates, whereas the production of organic acids, proteins, and fats occurs in intermediate steady state reactions or in subsequent dark reactions.

Myers and Johnston (254) noted that the value of the photosynthetic quotient for *Chlorella pyrenoidosa* usually found during a short time (1–2 hours) light-saturating exposure in the Warburg apparatus was about 1.1. If the light exposure was more prolonged, Q_P increased considerably. They stirred 1 liter of a *Chlorella* culture solution for 2–4 days in a closed 10-liter bottle filled with a 5% carbon dioxide gas mixture and illuminated at 50 or 500 foot-candles (fc). Photosynthesis was followed by Haldane gas analysis. Nitrate utilization was also measured. The results recorded in Table V indicate that nearly all the nitrate and carbon dioxide which disappeared from the liquid and gas phases could be accounted for in new cells. Excretion of ammonia or organic compounds was almost negligible. The lower Q_P value of 1.1 in short-time experiments may represent a temporary adjustment to an increased carbohydrate synthesis. The higher Q_P value of 1.3 in the steady state of photosynthesis is the result of nitrate

TABLE V

CARBON AND NITROGEN RECOVERY AND Q_P VALUES DURING PHOTOSYNTHESIS IN
Chlorella pyrenoidosa[a]

		Recovery in cells		$\Delta O_2/\Delta CO_2$ with light exposure	
		---	---	---	---
No. Exp.	Light (fc)	Nitrogen (%)	Carbon (%)	Short	Prolonged
3	50	98	95	1.45	1.41
3	500	95	97	1.10	1.32

[a] From Myers and Johnston (254).

reduction. Evidently the nitrate content of the nutrient solution is a factor to be considered in interpreting the photosynthetic quotient.

B. CONCENTRATION OF CARBON DIOXIDE IN THE ATMOSPHERE

1. Natural Levels

The carbon dioxide content of the atmosphere appears to fluctuate within rather narrow limits. Benedict (28) in 1913 reported a value of 310 ppm by volume as the average of 212 determinations in Boston, Massachusetts. Ten analyses on the North Atlantic Ocean and nine on Pike's Peak gave the same average. Carpenter (76) in 1937, using the same analytical method as Benedict, reported 310 ± 16 ppm as the mean of 1156 analyses in Baltimore, Boston, and Durham, New Hampshire. However, local conditions can produce wide divergences from this average, particularly near the ground, where industrial, biological, and meteorological factors cause major changes. It has been suggested (285) that the oceans, which contain about 100 times as much carbon dioxide as the atmosphere, serve as reservoirs that regulate, in large measure, this constituent of the air. Tables VI and VII summarize representative data in the literature.

Most of the sea (211) and high altitude data (146, 255) approximate the values of Benedict (28) except those in the Arctic and Antarctic Oceans (249), where the water temperature was near freezing. Presumably the equilibrium concentration of carbon dioxide in the air above cold ocean water is reduced to about half its usual amount.

TABLE VI

Carbon Dioxide Content of the Atmosphere

Location	Number	CO$_2$ Range (ppm)	CO$_2$ Avg. (ppm)	Reference	Year
Sea air					
North Atlantic Ocean	51	—	295	Thorpe (353a)	1866
North Atlantic Ocean	10	—	310	Benedict (28)	1913
North Atlantic Ocean	81	—	319	Buch (68a)	1932–1935
Dieppe, France	169	—	291	Reiset[b]	1872–1880
Coast of France	—	308–362	335 ± 27	Legendre (211)	1907
Petsamo, Finland	95	—	321	Buch (68a)	1934–1935
Spitzenberg	—	—	152	Buch (68a)	1934–1935
Antarctic, 70° latitude	—	145–170	158	Muntz (249)	1912
High altitude					
Pike's Peak, Colorado 4.3 km	9	—	310	Benedict (28)	1913
Balloon over England 4 to 10 km	—	240–300	250 ± 10	Glueckauf (146)	1944
Balloon over Minnesota 21.5 km	—	270–310	290	National Geographic Society (255)	1938
City air					
London streets (no wind)	18	425–900	650	Haldane (159)	1936
Paris streets	—	340–620	480	Florentin (130)	1927
London: Kew and Imperial College	—	310–370	340	Glueckauf (146)	1944
Durham, N.H.; Baltimore; Boston	1156	270–350[a]	310 ± 16	Carpenter (76)	1937
Boston, Carnegie Laboratory	212	—	310	Benedict (28)	1913

[a] One analysis 600 ppm.
[b] Quoted by Callendar (74).

The very high altitude data are insufficient to justify discussion of the small observed differences from low altitude concentrations. Evidently they are practically identical.

Callendar (74) made a critical survey of the literature in 1940

TABLE VII

CARBON DIOXIDE CONTENT OF RURAL AIR AS AFFECTED BY VEGETATION

Location	Vegetation	Season	Time of day	CO_2 Range (ppm)	CO_2 Avg. (ppm)	Reference	Year
England	Field crops	July–Oct.	Day	210——	324	Haldane (159)	1936
England	Field crops	July–Oct.	Night	320–440	386	Haldane (159)	1936
Iowa	Maize	Summer	Day	—	250[a]	Verduin and Loomis (366)	1953
Nebraska	Potato	Summer	Day	180–410	250[a]	Chapman and Loomis (79)	1944
Utah	Field crops	July–Oct.[b]	Day	272–321	297 ± 4	Thomas and Hill (349a)	1933–1935
Utah	Field crops	July–Oct.[b]	Night	302–489	358 ± 20	Thomas and Hill (349a)	1933–1935
Utah	None	Jan.–Feb.[c]	Day	300–356	321 ± 5	Thomas and Hill (349a)	1933–1935
Utah	None	Jan.–Feb.[c]	Night	—	328 ± 7	Thomas and Hill (349a)	1933–1935
Utah	Barren area	Sept.[d]	Day	—	308 ± 3	Thomas and Hill (349a)	1933–1935
Utah	Barren area	Sept.[d]	Night	303–367	310 ± 2	Thomas and Hill (349a)	1933–1935
New York	Field crops	Summer	Day	250–320	280	Moss (247)	1961
New York	Field crops	Summer	Night	300–390	320	Moss (247)	1961
Italy	Field crops	All year	8 A.M.	270–730	440	Tonzig (355)	1950
Italy	Field crops	All year	Noon	240–630	380	Tonzig (355)	1950
Italy	Field crops	All year	8 P.M.	250–620	430	Tonzig (355)	1950
Nigeria	Rain forest[e]	March–May	7–8 A.M.	410–580	510	Evans (122)	1939
Nigeria	Rain forest	March–May	P.M.	330–410	380	Evans (122)	1939
Java	Botanical gardens	Nov.–Feb.	8–9 A.M.	290–390	360	Stocker (338)	1935
Java	Botanical gardens	Nov.–Feb.	1–2 P.M.	270–310	290	Stocker (338)	1935

[a] Midday concentration = 25% less than early morning.
[b] Continuous record averaged by hours: 81 days.
[c] Continuous record averaged by hours: 32 days.
[d] Continuous record averaged by hours: 3 days.
[e] Light intensity less than 1% of sunlight, 0.3–1.2 meters above ground

to evaluate the carbon dioxide concentrations representing large air masses with a minimum of local contamination. Only the most reliable data were considered. The mean of 3050 determinations by 10 observers before 1900 was 292 ppm, the range for the averages found by the different observers being 287–296 ppm. Benedict (28) from 1909 to 1912, made 645 analyses which averaged 303 ppm. Buch in Finland, and Haldane (159) in Great Britain (1934–1935), found 321 ppm in 228 analyses. Callendar thought it significant that whereas the increase from about 292 to 321 ppm represents 2×10^{11} tons of carbon dioxide, 1.5×10^{11} tons had been added to the atmosphere in 35 years from combustion processes. There had evidently been no significant change in the concentration in 40 years before 1900. The increase suggested by Callendar is questionable. The reliable analyses at Carnegie Laboratory in Boston reported in 1913 and 1937 showed no change. Huber (182) states that when trees are dated by growth rings and by radiocarbon, much better correspondence is obtained if the atmospheric radiocarbon of 1840 is used as the basis of calculating age instead of present-day radiocarbon. The C^{14} content of the air has fallen considerably since 1840 owing to the combustion of fossil fuels that are free of C^{14}. At the same time the total carbon dioxide has increased by about 3% (10 ppm).

Marine and arctic air masses (211) reaching the north coast of France from southwest to north contained 291 ppm carbon dioxide; continental air from the northeast to southeast had 298 ppm; and subtropical air from the south had 301 ppm. At Kew Gardens (61, 62) west of London, westerly winds carried 286 ppm and easterly winds 313 ppm, a divergence possibly representing the effect of city smoke. The foregoing data were obtained about 1900.

2. Effect of Industry and Vegetation

On the downtown streets of large cities (130, 159) carbon dioxide concentration may be two to three times the normal concentration as a result of automobile traffic and other combustion processes, especially when there is little or no wind to ventilate these areas. High carbon dioxide values on city streets are accompanied by high carbon monoxide values. On the other hand, concentrations are about normal and fluctuations are small in the outlying sections of the cities (28, 76, 145, 146), thus emphasizing the very restricted effects of the high concentrations on the city streets (Table VI).

In rural and agricultural areas, fluctuations of carbon dioxide concentrations are caused by photosynthesis and respiration of the vegetation and by microbiological activity in the soil. Data by Haldane

(159) in England indicate higher concentrations in the fields at night than in the daytime. The observed night concentrations were sometimes greater than 380 ppm. In the daytime the concentrations averaged 324 ppm. Midday analyses (78, 79, 365) above maize (*Zea mays*) and potato (*Solanum tuberosum*) plants in fields in Iowa and Nebraska gave an average of 250 ppm during each of five summers. The concentration usually fell off about 25% between early morning and noon. It increased gradually in the late afternoon to about 300 ppm, a value that was seldom exceeded in the daytime. No measurements at night were reported. Continuous analyses (247) near field corn (*Zea mays*) in New York showed that a maximum concentration of about 330 ppm usually occurred at 5 A.M. and a minimum of about 265 ppm at 4 P.M. Extreme values were 390 and 250 ppm, respectively. Concentrations were higher than average during storms and in very cloudy weather. The low daytime concentrations suggest a significant uptake of carbon dioxide by plants due to photosynthesis.

Extensive continuous carbon dioxide measurements in Utah (350) have shown that definite diurnal patterns of atmospheric concentrations recur consistently as a result of photosynthesis and respiration, although these patterns may be modified by meteorological factors. Daytime levels near vegetation in the field were generally in the range from 280 to 320 ppm. At night the concentrations approached 400 ppm and occasionally exceeded 500 ppm, when there was no wind. Figure 2 shows two typical 2-day periods in an agricultural area in August and September. Detailed carbon dioxide concentrations and wind speeds are given.

At night there is generally a strong temperature inversion that prevents mixing of the air near the ground with the higher layers. This causes a buildup of carbon dioxide at low levels due to respiration of plants and microorganisms, the amount depending on the plant cover, the soil type, the temperature, and the wind velocity. Rich, open soils of low pH contribute more carbon dioxide than heavy calcareous soils. The temperature coefficient of respiration is large, and output of carbon dioxide from the plant and soil is favored by elevated temperatures. If mixing is minimized by lack of air movement, concentrations up to 600 ppm can be built up at night. However, the ground layer of high carbon dioxide concentration is shallow, and it can be readily dispersed by wind, as shown during the night of September 21–22 (Fig. 2).

Conditions are largely reversed in the daytime. The plant absorbs carbon dioxide in photosynthesis, but root and microbiological respiration may also proceed at increased rates owing to higher temperatures.

Moreover, the temperature inversion is seldom present from mid-morning to late afternoon, and the ground layer of air tends to rise as a result of solar heating. The carbon dioxide concentration remains at a low level because of this mixing. Accordingly, although the rate of uptake of carbon dioxide by the plant in the daytime is usually several times greater than the output at night, the concentration of carbon dioxide in the ambient air near the leaves falls much less in the light than it rises in the dark. The effect of the shorter day length and reduced biological activity in September as compared with

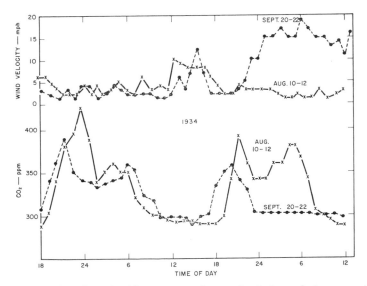

Fig. 2. Diurnal carbon dioxide concentrations and wind speeds in an agricultural area in Utah, August 10–12 and September 20–22, 1934. From Thomas and Hill (349a).

August is shown in Fig. 2. With comparable wind speeds, the carbon dioxide concentrations were higher at night and lower during the day in August than in September. On the Utah soils the contribution from soil microorganisms was probably small. The respiration levels were about the same as those observed with similar plants growing in sand cultures where microbiological activity was absent and root respiration was measured separately. By contrast, Katz et al. (196) carried out similar studies on a sandy loam soil in British Columbia and observed much higher soil respiration levels. Because of the large amount of carbon dioxide coming from the soil, net assimilation levels were unduly small unless corrected for soil respiration. Egle and Schenk (114) measured the daily march of carbon dioxide concentrations in a greenhouse. Figure 3 shows typical results on a nearly clear day for temperature, light, and carbon dioxide. The carbon dioxide levels were lowest at the top of the plants during the mid-day period of maximum light and temperature as a result of photosynthesis. They were highest at night in the greenhouse, a result of respiration. Intermediate carbon dioxide levels were observed outside the greenhouse.

Tonzig (355) in Italy measured the carbon dioxide concentration in fields of six different crops throughout one year. He sampled the air at the top of the plants three times each day, at 8:00 A.M., noon, and 8:00 P.M. The same fields were sampled at intervals of 3–4 days. The data are summarized in Table VIII. Monthly averages are given

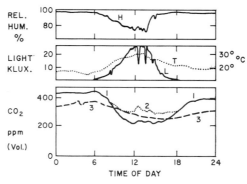

Fig. 3. Diurnal relative humidity (*H*), light intensity (*L*), temperature (*T*), and carbon dioxide concentrations in a greenhouse in Germany. Curve *1*, at mean height of plants (80 cm); curve *2*, 30 cm above floor; curve *3*, outside above greenhouse. From Egle and Schenk (114).

except when a significant change occurred during the month. Two or more crops are averaged together when the corresponding values were nearly identical.

During February and March the carbon dioxide concentrations were almost constant at 280 ppm. During late October, November, April, and the first half of May, concentrations were about 300 ppm, but the noon values were 10–50 ppm lower than those in the morning and evening, which were approximately equal. A sharp increase near all plants occurred at the middle of May, the noon level reaching 350–400 ppm and the average difference between morning or evening and noon being about 80 ppm. Evidently the background carbon dioxide was considerably higher than normal. It rose still higher, to 400–600 ppm, later in the summer and autumn except for the stormy week of July 1st. The differences between the morning and noon values suggest a progressively increasingly amount of photosynthesis for the crops to a maximum for rape (*Brassica napus*) seed early in May; for the grains late in May, for maize in July; and for clover (*Trifolium* sp.) late in May and in September. The data for the grains in July indicate that these crops were ripe at that time.

TABLE VIII

Carbon Dioxide Content of the Atmosphere in Cultivated Fields in Milan, Italy at the Level of the Top of the Plants[a]

| Month | Plants[b] | Carbon dioxide in air (ppm) | | | Difference (ppm) |
		8 A.M.	Noon	8 P.M.	8 A.M.–noon
October 30	G R	340	320	330	20
November	G	280	250	290	30
November	R	340	300	330	40
February–March	G R	280	280	280	0
April	C G R	320	290	320	30
May 1–12	C G R	350	300	350	50
May 13–31	C G	470	350	440	120
May 13–31	M	450	400	430	50
June	C G M	500	430	460	70
June 28–July 5[c]	C M	390	330	390	60
July	G	510	500	520	10
July	M	500	350	420	150
July	C	500	450	490	50
August	C M	630	530	600	100
September	C	620	460	570	160
October	C	600	530	590	70

[a] From Tonzig (355).

[b] Clover (C); grains (G), wheat, oats, and rye; maize (M); rape seed (R).

[c] A period of strong wind.

These summer daytime analyses are much higher than those of Haldane, Loomis, Moss, and Thomas though they show the same diurnal trends. Accepting them at face value it would be of great interest to know whether and how much the development and yield of the crops was stimulated by their growth in the carbon dioxide-enriched atmospheres.

In the rain forest of Nigeria, G. C. Evans (122) found concentrations of 450–600 ppm in the morning with a decline to about 360 ppm in the afternoon. The cover in this forest was dense, cutting off about 99% of the incident light. With the resultant reduced photosynthesis, the concentration of the carbon dioxide was always greater than normal. Stocker's (337, 338) data in the more open botanical gardens in Java showed lower concentrations but a similar decline during the day from about 360 ppm at 8:00 A.M. to 229 ppm at 1:30 P.M.

In an extreme case, McLean (226) reported two analyses of carbon dioxide in the hot, humid, dense rain forest of southeast Brazil. The values were 1400 ppm at 5:30 P.M. December 29, and 3400 ppm at 6:30 A.M. December 30 (2 hours after sunrise). The day was clear, hot, and calm. The night was warm without wind or rain. The samples were taken 1.5 meters above ground in a dense shrubby undergrowth having a double canopy of vines and trees. The analyses were done by the Pettenkoffer method. The values are very high but probably not unreasonable for the conditions described. Evidently there was rapid decay of organic matter on the forest floor in addition to a high rate of respiration at night, with resulting accumulation of carbon dioxide in the calm air. During the day, assimilation was limited because the plants were "immersed in gloom" and in a poor soil as well.

In the oceans, the carbon dioxide content of the water is at a minimum at 50 meters below the surface owing to photosynthesis of phytoplankton. The maximum, at 500 meters, is due to decay of organic matter. Off the west coast of Africa this maximum reached 1100 ppm. Owing to upwelling of this water, the overlying atmosphere may be enriched to 700 ppm.

C. MECHANISM OF ABSORPTION OF CARBON DIOXIDE BY LEAVES

1. Early Studies

The mechanism of absorption of carbon dioxide from the atmosphere into leaves of plants has been the subject of much study and some controversy. "Nourishment" of the plant was associated with stomatal apertures at an early date by Malpighi, Grew, and Stephen Hales, although their ideas were necessarily vague. Following recognition of the fact that carbon dioxide and oxygen are involved in an exchange of gases between the atmosphere and the extensive absorptive surfaces of the intercellular spaces within the plant leaves, Dutrochet (1832) (105) advocated the open stomata as the pathway for this exchange since they seemed to afford direct openings through the otherwise impervious leaf cuticle. Garreau (1850) (138) observed that the amount of carbon dioxide respired from leaves was roughly proportional to the number of stomata. Sachs (1865–1882) (295) advocated the stomatal theory, but he also thought there could be some exchange through the cuticle. The major difficulty was to explain how so much carbon dioxide (about 0.1 cc/cm²/hour) could be absorbed by the leaves from an atmosphere containing only 300 ppm of the gas, when the openings in the leaf surface represented only about 1% of the total surface.

Boussingault (1864–1869) (52, 53) attempted to answer the question experimentally. He covered the lower stomata-bearing surface of an oleander (*Nerium oleander*) leaf with lard, also the upper stomata-free surface of a similar leaf. The two leaves were confined in an atmosphere containing 30% carbon dioxide and illuminated equally for 8 hours with bright diffuse light. The second leaf with uncovered stomata absorbed 10 cc of the carbon dioxide; the other leaf, 17 cc. It was concluded that the gas diffused into the leaf through the epidermis, the upper surface being more pervious to carbon dioxide than the lower, despite the latter's uncovered stomata.

Barthélemy (1866) (22, 23) assumed that the cuticular layers of the epidermis were well fitted to transmit carbon dioxide because of the solubility of the latter in the cutin. He suggested that the stomata served to allow gases such as carbon dioxide, oxygen, and water vapor to leave the interior of the leaf, but not to enter. Merget (238) opposed this position by many diffusion experiments involving mercury vapor, hydrogen sulfide, sulfur dioxide, ammonia, and hydrogen, all of which required the open stomata for entrance into the leaf.

In the thirty-year controversy that followed, the cuticular theory was strongly favored at first, but later the weight of evidence seemed to favor the stomatal theory, although it was assumed even by Sachs (295) and Pfeffer (270) that some cuticular absorption could occur. Wiesner (1879) (389) found that the rate of passage of gases through stomata conformed to the theory of gaseous diffusion, and Wiesner and Molisch (1889) (390) showed that a stomata-free epidermis, detached from the leaf by the action of bacteria, effectively resists the passage of gas under pressure. Mangin (228) also worked with detached epidermis and concluded that in photosynthesis it would be necessary for most of the carbon dioxide to enter through open stomata. Boehm (1889) (39) and Stahl (1894) (324), using the formation of starch in the chloroplasts as an index of carbon dioxide absorption and photosynthesis, observed ready entrance of the gas through the stomata, but little or none through the epidermal cells, from ordinary air.

The unfortunate experiment of Boussingault was not properly evaluated until 1895 when F. F. Blackman (31) repeated it and showed that the results were due to the high concentrations of carbon dioxide employed in the gas mixture. With lower concentrations exactly opposite results were obtained. It was suggested that an excessive amount of gas entered through the stomata, poisoning the photosynthetic process, whereas the lesser amount which diffused through the upper epidermis could be more effectively utilized by the leaf. Godlewski (1873) (147) had observed interference with starch formation in the chloroplasts due to high carbon dioxide concentrations.

F. F. Blackman (31) constructed an apparatus in which the two sides of a leaf were mounted in separate chambers through which independent gas streams could be passed. Simultaneous measurements of the carbon dioxide absorbed or evolved on the two sides were roughly proportional to the stomatal densities. Table IX summarizes Blackman's

TABLE IX

ASSIMILATION OF CARBON DIOXIDE AS AFFECTED BY GAS CONCENTRATION, WITH AND WITHOUT A VASELINE (PETROLEUM JELLY) COATING ON THE LOWER STOMATOUS SURFACE OF *Nerium oleander* LEAVES[a]

Conc. CO_2 (%)	Carbon dioxide absorbed by[b]		Ratio $A:B$
	Normal leaf (A)	Coated leaf (B)	
6	0.07	0.01	1:0.14
6.3	0.055	0.01	1:0.18
7.5	0.046	0.017	1:0.37
14	0.18	0.04	1:0.22
55	0.049	0.067	1:1.4
50	0.043	0.069	1:1.6
97	0.033	0.060	1:1.8
26	0.020	0.019[c]	1:0.95

[a] From F. F. Blackman (31).

[b] Values stated as cubic centimeters CO_2 per square centimeter of leaf area per hour.

[c] Upper astomatous surface coated.

data giving the assimilation of carbon dioxide by oleander leaves with and without coating the stomata with Vaseline petroleum jelly. The last entry indicates that there was only slight absorption at moderate carbon dioxide levels by the stomata-free surface. Covering the stomata reduced assimilation to 14–37% of that of the normal leaves, when less than 6–14% carbon dioxide was present in the gas; with 50% carbon dioxide, assimilation was greater in the leaves with the coated stomata than in normal leaves. A six- to sevenfold increase in penetration through the astomatous upper surface of leaves having the lower surface covered with Vaseline petroleum jelly resulted from increasing the carbon dioxide concentration from 6 to 97%. Blackman concluded that, with normal atmospheric concentrations, the "sole pathway of carbon dioxide into or out of the leaf is by the stomata." This conclusion is reexamined later.

H. T. Brown and Escombe (59) repeated Blackman's experiments in similar but larger apparatus and arrived at esssentially the same conclusions, viz: In hypo- or hyperstomatous leaves "the respiratory and assimilatory exchanges of gases take place only through . . . the side to which the stomatic openings are confined . . . . In amphistomatous leaves . . . exchanges are carried on by both surfaces of the leaf" . . . and there is "a rough quantitative relation to the distribution of stomata on the two leaf surfaces . . . ." "In bright sunlight there is a greater intake of carbon dioxide into the upper leaf surface," whereas "in light of a less degree of intensity, there is a nearer approach to a correspondence of the assimilatory and stomatic ratios, but it is seldom a very close one."

H. T. Brown and Escombe (59) next made quantitative determinations of the diffusion of carbon dioxide through the stomata. For example, the hypostomatous leaf of *Catalpa bignonioides* assimilated 0.07 cc $CO_2/cm^2/hour$ from ordinary air under favorable conditions. The leaf carried 145 stomata/mm^2, each having an area when fully open of 6.18×10^{-5} mm^2. The maximum stomatal area was therefore 0.9% of total leaf area. If all the carbon dioxide entered through the stomata, the rate of flow through the openings was 7.77 cc/cm^2/hour. By comparison, a strong solution of caustic soda absorbed from moderately still ordinary air 0.12 cc $CO_2/cm^2/hour$ of 0.18 cc/cm^2/hour from turbulent air. Evidently the leaf absorbed about half as much gas as an equal area of strong sodium hydroxide, while the stomatal areas alone absorbed about 50 times as much.

The authors state, "A consideration of such facts as these leads us to demand more evidence before we can unreservedly accept the proposition that the gaseous exchanges of assimilation take place exclusively through the fine openings of the leaf stomata." The subsequent study was directed toward the application of the laws of diffusion to this special problem.

2. The Diameter Law

Stefan (335) in 1881 had already derived an equation, based on the diffusion studies of Graham (150), for evaporation from a circular or elliptical plane surface; this equation indicated that if the air was perfectly quiet, the steady-state rate of evaporation was proportional to the diameter, rather than the area, of the surface. Stefan suggested that the vapor tends to diffuse outward from the plane in "uniform pressure surfaces" or "shells" ellipsoidal in shape, with the stream lines of diffusion perpendicular to a family of ellipsoids, as illustrated in Fig. 4 and defined by a, the distance from center to focus, and r the radius of the corresponding circular plane. Total evaporation V is given by the equation:

$$V = K(a - \sqrt{a^2 - r^2}) \log \frac{P - p_0}{P - p_1}$$

where K is a proportionality constant, P is atmospheric pressure, and p_1 and p_0 are the vapor pressures at the liquid surface and at a great distance, respectively.

For linear proportionality, a and r must be equal. In Fig. 4 it is evident that the ellipsoids approach circles as they expand. Of course, such a condition could exist only in the absence of any turbulent disturbance of the flow.

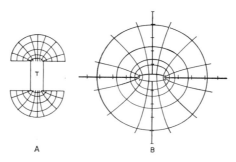

A   B

FIG. 4. Diagram of the diffusion flow pattern of gases through a stoma. (A) Transverse section through stomatal tube T, with flow lines from outside to intercellular space. (B) Stream lines of gas into elliptical stoma in center. From Spoehr (321).

Brown and Escombe recognized that their problem was exactly the opposite of Stefan's since they were concerned with static diffusion into the leaf rather than evaporation from the leaf. Considering the low concentration of carbon dioxide in the atmosphere, the diffusion gradients would necessarily be very small. Free diffusion of atmospheric carbon dioxide through 2-cm glass tubes, mounted horizontally and attached at one end by a right-angle bend to a flask containing normal sodium hydroxide solution, was first studied. Convection currents in the tubes were carefully excluded. The results conformed to the equation

$$Q = \frac{K(\rho - \rho_0)A \cdot t}{L}$$

that is, the quantity of carbon dioxide absorbed was proportional to the concentration gradient, the time t, the area of the tube A, and inversely proportional to its length L. If the efficiency of the absorbing surface is perfect, $\rho = 0$. The average value found for the diffusion coefficient was $K = 0.157$ cgs units. Other investigators, using much higher concentrations of carbon dioxide in air, found $K = 0.131$ to 0.142 cgs units.

If the mouth of the flask was obstructed by a septum having a circular aperture, the diffusive flow into the flask per unit area of aperture increased rapidly as the size of the hole decreased. The rate of diffusion became directly proportional to the linear dimensions of the apertures below a certain size. This is illustrated in Table X. Apertures of 6 mm and smaller conform to the diameter ratio; those of 12 mm fall halfway between the area and diameter ratios. This

TABLE X

DIFFUSION OF ATMOSPHERIC CARBON DIOXIDE INTO SODIUM HYDROXIDE SOLUTION THROUGH AN APERTURE IN A SEPTUM COVERING THE MOUTH OF A FLASK[a]

Diameter of aperture (mm)	Ratio		
	Area of apertures	Diameter of apertures	CO_2 diffused in unit time
22.7	1.00	1.00	1.00
12.06	0.28	0.53	0.39
6.03	0.07	0.26	0.26
3.23	0.02	0.14	0.16
2.00	0.007	0.09	0.10

[a] From H. T. Brown and Escombe (59).

experiment has been repeated in various ways and fully confirmed by later work (363).

The reality of the "shells" postulated by Stefan was demonstrated visually by preparing 0.05% methylene blue in 5% gelatin and placing it in juxtaposition with clear gelatin of the same strength, the two gels being separated by a thin septum perforated with a hole of known dimensions. The aperture was the only point of contact for the two gels. The limiting surface of the dyed and colorless gelatin assumed an ellipsoidal shape, clearly evident in the early stages, with the major axes of the ellipsoid in the plane of septum and their common foci on the edges of the hole. The surfaces of equal density were closer together near the diaphragm than further away. As the diffusion progressed, the spheroid of colored gelatin more nearly approached the dimensions of a hemisphere, which, however, it never quite reached. Using different-sized apertures, the volumes which the spheroids assumed in equal times were proportional to the diameters rather than to the areas of the openings.

3. Interference between Apertures of Multiperforate Septa

H. T. Brown and Escombe (59) made multiperforate septa by punching 0.38 mm holes in photographic film. In the different septa the holes were spaced from 1 to 6 mm or 2.6 to 15.7 diameters apart. The relative rates of diffusion of atmospheric carbon dioxide through these septa, mounted on vertical glass tubes of 34.6 mm diameter,

TABLE XI

DIFFUSION OF CARBON DIOXIDE THROUGH MULTIPERFORATE SEPTA WITH 0.38-MM HOLES, COVERING A 34.6-MM TUBE CONTAINING SODIUM HYDROXIDE[a]

Number of holes	Distance between holes' diameters	Ratio			
		Areas	$\sqrt{\text{Areas}}$	Absorption of CO_2 by NaOH at distance below septa:	
				40 mm	10 mm
Open	—	8.80[b]	2.97	0.96[c]	1.00[d]
935	2.6	1.00	1.00	1.00	0.56
230	5.3	0.24	0.49	0.41	0.52
105	7.8	0.113	0.34	0.36	0.41
60	10.5	0.064	0.26	0.30	0.32
37	13.1	0.040	0.20	—	0.21
26	15.6	0.027	0.156	0.19	0.14

[a] From H. T. Brown and Escombe (59).
[b] Area = 934 mm².
[c] 0.35 cc CO_2/hour.
[d] 0.76 cc CO_2/hour.

containing sodium hydroxide solution at 10 or 40 mm below the septa, are given in Table XI. These rates are compared with the rate through the open tube, and with the relative areas and linear dimensions (square root of the area) of the apertures. The relative diffusion rates agree fairly well with the relative diameters, showing that the amount of carbon dioxide which passed across the septum was more nearly a function of the circumference than of the area of the pores. With close spacing of the holes, the total diffusion may approach the rate for the open tube. Evidently this value cannot be significantly exceeded.

The authors concluded that when the apertures were 10 diameters or more apart there was no mutual interference with the diffusion through adjacent holes. This conclusion has been questioned because when size of the openings is extrapolated to stomatal dimensions, rates of diffusion up to 50 times greater than for an open surface can be calculated for the leaf. Verduin (363) derived an expression for the mutual interference of the apertures in multiperforate septa

$$\log Q = \log Q_1 - \frac{K}{D^2}$$

where Q is the diffusion rate per pore, Q_1 is the rate for an isolated pore, D is distance between pores, and K is a proportionality constant.

Diffusion experiments were made using septa with holes 0.8, 0.4, and 0.2 mm in diameter and spacings of various dimensions ranging from 10 to 40 pore diameters. When the logarithm of the diffusion was plotted against the inverse square of the distance between holes, straight lines were obtained for each pore size, confirming the equation. Maximum diffusion or minimum interference was attained at the different pore sizes with spacing of the holes as follows: 0.8 mm holes, 20 pore diameters; 0.4 mm holes, 30 pore diameters; and 0.2 mm holes, 40 pore diameters. The data permit a reliable extrapolation to stomatal dimensions, but they also suggest that interference between stomata is probably large.

4. Diffusion into Stomata

Eckerson (111) measured the length ($2a$) and width ($2b$) of wide-open stomata, also the number (n) per square millimeter on the leaves of 37 plants representing 31 species. Verduin (364) calculated from these data the average distances in microns between stomatal centers, $A = 1000/(\sqrt{n} - 1)$, also the effective diameters of the pores, $B = 2[(a^2 + b^2)/2]^{1/2}$. A/B is the stomatal spacing measured in effective pore diameters. Table XII gives the data for the lower surface stomata of representative plants from Eckerson's study, plus a *Quercus falcata* (*Q. triloba*) and a second sunflower (*Helinathus annuus*) sample. Stomatal density ranged from 1192 to 14 per square millimeter, and stomatal size ranged, more or less reciprocally to $\sqrt{n}$, from 5×1 to $40 \times 7 \ \mu$, or from 3.6 to 28.8 μ effective pore diameters. Upper surface stomata were present on only 16 of 38 plants. The stomata were approximately the same size on both surfaces, but those on the upper surface were much more sparse, ranging in most cases from 13 to 50% of the number on the lower surface. Oats (*Avena sativa*), tulips (*Tulipa*), and vetch (*Vicia*) had about the same number, and wheat (*Triticum sativum*) over twice as many, on the upper as on the lower surface.

The spacing ratio A/B had a narrow range from 5.3 to 15.7 pore diameters on the lower surface. The average value was 10.5 with a standard deviation of 1.6. Evidently there is a tendency for stomatal size and number to vary reciprocally on the lower surface. Sunflower is an ex-

TABLE XII

SIZE AND DENSITY OF FULLY OPEN STOMATA AND THE CALCULATED RELATIVE RATES OF DIFFUSION OF CARBON DIOXIDE INTO THEM FROM THE ATMOSPHERE[a]

Plant		Lower surface stomata				
	Number (n)/mm^2	Length $(2a)$ $\times$ width $(2b)$ (μ)	Spacing[b] A (μ)	Diameter[c] B (μ)	Ratio[d] A/B	Diffusion[e] rate of CO_2 (cc/cm^2/hr)
Quercus falcata[f]						
(*Q. triloba*)	1192	5 × 1	30	3.6	8.4	0.88
Ficus repens	282	5 × 3	64	4.1	15.7	0.30
Phaseolus vulgaris	281	7 × 3	64	5.4	11.8	0.49
Fagopyrum esculentum	152	12 × 6	89	9.5	9.4	0.83
Helianthus annuus[g]	330	—	59	10.7	5.5	2.40
Helianthus annuus	156	22 × 8	87	16.5	5.3	2.60
Pelargonium hortorum						
("*P. zonale*")	118	19 × 12	102	15.9	6.4	1.82
Zea mays	68	19 × 5	138	13.9	9.9	0.82
Vicia faba	48	19 × 8	169	14.1	12.0	0.63
Primula sinensis	31	30 × 9	218	22.2	9.9	1.06
Avena sativa	23	38 × 8	265	27.5	9.7	1.20
Triticum sativum	14	38 × 7	368	27.4	13.5	0.80
		Upper surface stomata				
Triticum sativum	33	40 × 7	211	28.8	7.4	1.90

[a] From data of S. F. Eckerson (111).
[b] A = Spacing, distance between centers of stomata: $(1000/\sqrt{n} - 1)$.
[c] B = Effective diameter: $2[(a^2 + b^2)/2]^{1/2}$.
[d] A/B = Distance between stomata in effective diameters.
[e] $Q = (K \cdot \rho \cdot a \cdot N \ 3600)/(L + X) = K_1 \cdot A \cdot N$ cc/cm^2/hour.
[f] Handbook Biological Data (W. S. Spector, ed.), Natl. Acad. Sci.–Natl. Res. Council, Washington, D.C., 1956.
[g] H. T. Brown and Escombe (59).

ception. Its stomata are about twice as large as those of other plants with the same number per square millimeter. Two samples of sunflower, one measured by Brown and Escombe, differed in stomatal density by a factor of 2.1, yet the distances between stomata measured in effective diameters (A/B) were nearly identical.

In the last column of Table XII are calculated the rates of diffusion of carbon dioxide from the atmosphere at 300 ppm ($\rho = 3 \times 10^{-4}$ cc/cc) into the wide-open stomata, assuming the stomatal channels have uniform cross-sectional areas (a in cm^2) and the same length ($L + X = 20\,\mu$) in all the different plants. The diffusion coefficient of CO_2 is taken as $K = 0.145$ cgs units, and the stomatal density in N/cm^2 = $100n$/mm^2. The quantity of the CO_2 diffused (Q) is given by the equation

$$Q = \frac{3600K \cdot \rho \cdot a \cdot N}{L + X} = 80a \cdot N \text{ cc/cm}^2/\text{hour}$$

The calculated rates of diffusion have, over-all, a 9-fold range from 0.3 to 2.6 cc/cm^2 hour compared with a 3-fold range in the case of the A/B values. However, most of the rates fall in a much narrower range than the maximum, as is the case with the A/B values. The rates would not be affected greatly by considerable variation in the length of the stomatal tube. It may be noted that the two samples of sunflower gave the same rates of diffusion. On the other hand, three varieties of pelargonium (*Pelargonium hortorum*) gave rates of 0.46, 1.2, and 1.8 cc/cm^2/hour.

Assimilation rates of 0.3 to 2.6 cc/cm^2/hour are considerably higher than the maximum rates that have been actually measured. They would be much greater still if they were calculated from the stomatal diameters instead of areas. The maximum measured rates range from about 0.1 to 0.4 cc/cm^2/hour. Presumably if the gas diffuses into the stomata at the calculated rates it is only partially absorbed onto the cell surfaces. This would reduce the gradient from the outside air with resultant lowering of the rate of diffusion. Further, interference with diffusion probably occurs because of the average close spacing of the holes at about 10 pore diameters. As the stomata close, this spacing will increase and interference between stomata will be reduced. Thus closure will have less effect in reducing absorption than the reduction in the size of the opening would suggest. This might explain why considerable photosynthesis occurs in some plants (e.g., corn) when the stomata appear to be closed.

It is therefore of interest that although the stomatal apparatus of different kinds of plants varies widely in the number and size of pores on the leaves, the leaves have, nevertheless, quantitatively very similar potentialities for admitting carbon dioxide.

D. Stomatal Movements and Photosynthesis

There is evidence that photosynthesis and stomatal movements are either mutually dependent on each other to a considerable extent or are controlled by the same external and internal factors. The latter include light intensity, concentration of carbon dioxide inside the leaf, temperature, relative humidity, available moisture in the leaves, and some chemical treatments. The water relations of stomatal cells and the mechanisms of stomatal movement are discussed by Heath in Volume II, Chapter 3. Many methods of measuring the stomatal apertures are described, and the various hypotheses advanced to explain stomatal movement in terms of light, carbon dioxide, and water supply are also summarized. Evidently these relationships have an important bearing on photosynthesis in the plant.

One of the earlier studies of the relationships between photosynthesis and stomatal movements is that of Maskell (233), who mounted a leaf

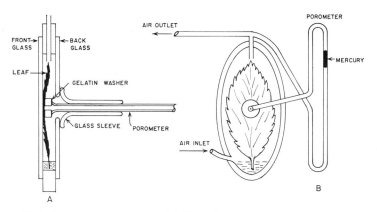

FIG. 5. Maskell's porometer. Rate of fall of the mercury slug is measured to indicate extent of stomatal opening. From Maskell (233).

of cherry laurel (*Prunus laurocerasus*) in a chamber consisting of an oval rim of iron, to which two glass plates were fastened with hard wax. Intake and outlet tubes were provided so that air could be drawn through the chamber (Fig. 5). A porometer was also provided, consisting of a glass tube having a washer of stiff gelatin on one end which could be pressed against the leaf surface inside the chamber to make a gas-tight contact. This tube, emerging through a gas-tight adaptor cemented on the lower glass plate, communicated with a simple mercury pump. The latter consisted of a 15 mm slug of mercury in a glass tube, which could be turned horizontally to position the slug near one end. When the tube was turned upright, the mercury fell, drawing air through the leaf. The rate of fall was timed with a stop watch. The readings were calibrated to indicate the volume of air drawn through the leaf into the porometer. Stomatal counts were made on all leaves studied, and the air volume per stoma was calculated. Calibration was also attempted by measuring wax impressions of the stomata with the mircroscope. The data were insufficient to establish an exact relationship between porometer rate and pore area, but they indicated definitely that the variations in porometer rates were due primarily to variations in the stomatal openings. More sophisticated porometers are described by Heath.

Many experiments were carried out over a two-year period in which simultaneous assimilation and porometer measurements were made repeatedly throughout 24-hour periods or longer. Assimilation was expressed as the ratio of the apparent assimilation (AA) to the average carbon dioxide concentration (AA:CO_2). "High" or "low" light intensity was used and the carbon dioxide concentration was usually low enough to be limiting (1000–4000 ppm). With higher carbon dioxide levels, light became limiting and assimilation remained practically constant throughout the experiment.

"Carbon-dioxide-limited" assimilation curves followed the porometer curves closely, suggesting that the stomata were controlling the rate of assimilation. The diurnal patterns of both processes were also reproducible. The stomata were open and assimilation was high during the morning hours. During mid and late afternoon the stomata closed and assimilation fell to a low value. This condition prevailed under illumination during the night, but the next morning the stomata opened as on the previous day and assimilation usually rose. This pattern was observed in the summer and winter months. In the spring and autumn the curves did not always rise in the morning. A long late-season drought in 1921 reduced all activity to a low level, but the expected pattern returned shortly after it rained in October.

The close correspondence between the assimilation and the porometer curves indicated strongly that the "diurnal rhythm of assimilation is due to a diurnal rhythm of stomatal opening." Further, "an approximate estimate of stomatal diffusive capacity from the data of air flow per stoma, suggests that the stomatal changes observed are of about the order required to produce the observed changes in assimilation rates."

Heath (167, 168, 170) found that the stomata under his porometer cup opened abnormally if the air flow through the cup was stopped while the leaf was being illuminated. He determined that this effect was caused by a depletion, due to photosynthesis, of carbon dioxide under the cup and in the intercellular spaces in the leaf to 100 ppm or less. He also observed that when the stomata were closed, external carbon dioxide had no effect on them. However, if the internal concentration fell, the pores opened and the apertures reached a maximum width at about 100 ppm carbon dioxide. Reducing the carbon dioxide below 100 ppm had no further effect.

Scarth and Shaw (297) studied *Pelargonium hortorum* ("*P. zonale*") which usually closes its stomata in the dark. They used infrared absorption to measure carbon dioxide exchange, and both the epidermal stripping and the porometer methods to measure the stomatal apertures. They found that when plants with closed stomata were illuminated, photosynthesis commenced concurrently with the opening of the stomata, whether the opening was prompt or delayed. Then the two processes followed each other faithfully as they increased or decreased in response to increasing or decreasing light intensity. Even anomalous irregularities in the course of the aperture curve were reflected in corresponding irregularities in the rate of photosynthesis.

Scarth and Shaw (297) confirmed the observations of Heath that the carbon dioxide concentration inside the leaf provides the controlling mechanism for the mutual dependence of photosynthesis and stomatal aperture. They also observed that reduction of this concentration caused the stomata to open, and vice versa. In the dark, carbon dioxide from respiration caused the stomata to close. With illumination, cellular carbon dioxide was depleted by photosynthesis in proportion to the light intensity. This caused the stomata to open in proportion to the depletion. The dependence of stomatal activity on cellular carbon dioxide was indicated also by exposing the plants to carbon dioxide-free air when the stomata opened even in the dark, as had been observed earlier by Darwin (88), Lloyd (217), Loftfield (218), and Linsbauer (214). Treatment of the leaves with low concentrations of chloroform vapor inhibited photosynthesis, and the stomata closed to 20% of their full aperture. There was gradual reversal of these effects when the chloroform was removed. Evidently the chloroform caused a buildup of carbon dioxide in the cells. Higher chloroform concentrations caused irreversible effects.

C. C. Wilson (394) found a similar close correlation between stomatal apertures and photosynthesis in the first-year leaves of privet (*Ligustrum*) and camellia (*Camellia*). The pairs of diurnal curves were nearly coincident. In second-year leaves, the correlation was poor because the photosynthesis fell to a low level at midday or in the early afternoon when the stomata were wide open. Evidently the age of leaves may be an important factor in the relationships of stomata to photosynthesis, causing divergence when the assimilation apparatus begins to fail while the stomatal apparatus remains unimpaired.

Control of stomatal openings by the internal carbon dioxide concentration would depend on photosynthesis and respiration in the mesophyll and in the guard cells. The ratio of assimilation to respiration in the mesophyll is large (about 10 to 1) which would be expected to cause the stomata to be either fully open or fully closed except at very low light intensities. For this reason it is thought that the mesophyll reactions play only a minor role in stomatal control.

Guard cell studies (306, 307) have shown that they are low in chlorophyll and that they carry on photosynthesis at a much lower rate than the mesophyll. Assimilation in the guard cells is too slow to account for the rapid osmotic pressure changes observed during stomatal movements, but it might reduce the carbon dioxide concentration in the stomata enough to act as a "trigger" to keep them open and favor continuous uptake of the gas in the light (cf. Heath, Volume II, Chapter 3). However, slow assimilation in the guard cells might be offset by respiration in adjacent chlorophyll-free epidermal cells, which would cause closure. Possibly as the light and pH vary there is a reversible interconversion of starch and soluble carbohydrates in the guard cells, due to phosphorylase, as suggested by Alvim (4). W. T. Williams (393) postulated that the stomata are normally closed due to an "active, non-osmotic, energy-requiring, transfer of water from guard-cells to neighboring cells, possibly mediated by contractile structures of some type," in the presence of carbon dioxide (and presumably of oxygen also). Opening occurs when the concentration of carbon dioxide in the cells is lowered by light-induced reactions or by supplying carbon dioxide-free air. Carbohydrates play a secondary stabilizing role. However Heath and Orchard (170a) found that lack of oxygen increased rather than decreased the closing tendency for wheat stomata. Neither have "contractile" vacuoles been observed in the guard cells.

Temperature, like light, affects the two processes similarly. At about 0°C the stomata close and there is little or no photosynthesis. As the temperature is raised, the stomatal openings and photosynthesis increase together. With cotton (*Gossypium*) (394) the stomata open nearly linearly from about 15% of full opening at 5°C to 85% at 20°, then further to 100% at 25° and 30°, and finally close to 35% at 35°. Assimilation rates follow a similar course. The diurnal curves of light intensity and of stomatal apertures in privet are nearly coincident above 20°C, but the latter lag appreciably behind the former below 20°C.

A case in which photosynthesis and stomatal apertures do not correspond is described by Mitchell (244). He found nearly identical rates of assimilation in primula, cineraria (*Senecio cruentus*), pelargonium, and tomato (*Lycopersicon esculentum*) with high, medium, and low relative humidity in the air. Microscopic observation indicated that with high and medium humidity the stomata were open, but with low humidity (10–30% relative humidity) they were closed. Porometer readings were not made. The low humidity leaves had as much moisture as the others, until they showed signs of wilting. Then the assimilation rate fell off 30–40%. Even if the stomata were not tightly closed in the dry environment, it is difficult to understand why the assimilation was practically equal to that in the more humid atmosphere when the pores were fully open. Air slightly enriched with carbon dioxide was used in this study.

Dugger (103) repeated the classical experiments of F. F. Blackman (31) and H. T. Brown and Escombe (59) on the permeability of stomatous and stomata-free leaf surfaces to high concentrations of carbon dioxide. Either the upper or the lower surface of a hydrangea or coleus leaf was exposed to air enriched to 1, 5, or 10% with carbon dioxide containing radiocarbon, C^{14}. Experiments were carried out in both light and darkness. Uptake of carbon dioxide was determined by C^{14} counting and by autographs. In the light, the upper stomata-free surface of hydrangea absorbed only 3, 6, and 34% of the carbon dioxide taken up by the lower stomatous surface from 1, 5, and 10% gas, respectively. The corresponding absorptions by coleus were 103, 29, and 51%. In the dark, absorption was low, but the lower surface usually absorbed more than the upper surface.

The data indicate that the surfaces of hydrangea and coleus differ markedly in their penetrability to carbon dioxide. The upper epidermis of coleus is much more permeable than that of hydrangea, particularly to the 1% gas. Even the same plant may show wide variation judging from the different responses of pelargonium reported by Scarth (297) and Mitchell (244). In one case photosynthesis and stomatal apertures followed each other closely. In the other, there was no apparent relation. Freeland (133) pointed out similar discrepancies. In six species with stomata-free upper epidermis, the ratios of assimilation from normal air on upper to lower surfaces were coleus, 0.3; avocado (*Persea* sp.), 0.7; poinsettia (*Euphorbia pulcherrima*), 0.15; rubber plant (*Ficus elastica*), 0; *Begonia*, 0.7; *Rhoeo spathacea* (*R. discolor*), 0. In plants that have stomata on both surfaces, assimilation on the two surfaces may or may not be proportional to the relative number of stomata.

Evidently the stomata-free epidermis of some plants is more or less permeable to carbon dioxide. Presumably the stomatous epidermis on the same leaves would be similarly permeable with the stomata closed. It follows that Blackman's dictum about absorption from air occurring "solely through the open stomata" is not universally true. This might explain observations, like those of Mitchell, in which photosynthesis took place with the stomata apparently closed. The discrepancies between the results of Mitchell and Scarth with pelargonium raises a question whether the conditions under which a plant is grown can modify the permeability of the epidermis to carbon dioxide.

E. RESPIRATION DURING PHOTOSYNTHESIS

1. Older Studies

Plants respire continuously, not only in the dark when carbon dioxide is evolved, but also in the light when carbon dioxide uptake predominates. "Respiration" in the dark and "apparent assimilation" in the light can be determined readily by measuring the gas exchange for either carbon dioxide or oxygen (or both) in darkness and in light. "Net assimilation" is the difference between these values. Similarly, "apparent assimilation" is the difference between "true photosynthesis" and "light respiration." Unfortunately, neither of the latter entities can be measured directly because the gas exchange of one is opposed to that of the other. It is possible that the rate of respiration in the light might be modified considerably by the concurrent assimilation processes. An evaluation of light respiration assumes great importance when "true" photosynthesis is sought rather than apparent assimilation, in studies such as the determination of quantum efficiency (326, 370). Unequivocal determination of light respiration presents major difficulties.

It has generally been assumed that the rate of respiration is about the same in the light as in the dark, making allowance for differences in temperature. This assumption is based on a number of different types of evidence as reviewed by Weintraub (383). Nonchlorophyllous plants and plant parts usually show respiration in the light within about ±25% of the dark respiration, though wider deviations have been observed in some cases. The same or increased respiration rates in the light have been noted in yeasts, molds, some flowers, fruits, seeds, seedlings, buds, roots, tubers, rhizomes, and albino leaves. Decreased respiration in the light has been noted in mushrooms and some flowers, seedlings, roots, and rhizomes. Increased light respiration in these materials has been observed more frequently than has decreased respiration. Albino leaves respire at about the same rate in the dark and light and also at about the same rate as comparable green leaves in the dark. Leaf treatments with chemicals that inhibit photosynthesis completely, but appear to have little or no effect on respiration, suggest no consistent differences between light and dark respiration. The same is true of genetical variants that are incapable of photosynthesis (383).

These plants, though green, have about the same gas exchange rates in the dark and the light, whereas wild types respire similarly in the dark but exhibit strong photosynthesis in the light.

Photosynthesis can be completely inhibited by heating such plants as *Elodea* and lichens at 40°–55°C. After treatment with the lower temperatures in this range, the rates of respiration in light and darkness are about the same. But after treatment with the higher temperatures, the light respiration may become severalfold greater than dark respiration. The respiratory quotient is also changed.

Photosynthesis can also be completely inhibited by high light intensity. This is usually characterized by increased light respiration, which is greater the higher the light intensity. Subsequent dark respiration is greater than before illumination. Prolonged exposure of algae (252) to intense light can cause the absorption of oxygen to fall to zero, accompanied by bleaching and irreversible injury. This stage is reached after a definite amount of oxygen has been absorbed, regardless of the rate of uptake. It is probable that the intense light injures the photosynthetic and respiratory mechanisms by photooxidation processes unrelated to normal respiration.

When photosynthesis is nearly completely inhibited by reducing the carbon dioxide supply to as near zero as possible, e.g., by employing potassium hydroxide in the Warburg apparatus, or by supplying carbon dioxide-free air, light respiration is increased proportionately to the light intensity, at least to the point of injury by photooxidation. Blue light is more effective than red light under these conditions. Subsequent dark respiration may be increased above preillumination respiration, so that in this case also there may be photooxidation unrelated to normal respiration.

None of the foregoing methods give unequivocal evidence regarding light respiration when photosynthesis is occurring simultaneously because there is no assurance that, when photosynthesis is stopped by any of the methods employed, the respiratory process in the light is not modified to some extent also. Such modification would seem probable when the leaves are injured by intense light or high temperatures, or when the system fails to return rapidly to its preillumination level of dark respiration after the light is extinguished.

It is evident that an attempt should be made to define the term "respiration." So far as dark respiration is concerned, a reasonable definition might restrict the function to steady state uptake of oxygen and evolution of carbon dioxide in more or less "normal" vegetation. Plants that have been subjected to intense light, high temperature, drought, prolonged darkness, or chemical treatments frequently have abnormally high or low dark respiration, owing presumably to modification of the respiratory apparatus and function. They should probably be excluded even though it is not known how the reactions may differ from those in "normal" vegetation.

Light respiration is more difficult to define. The abnormal conditions mentioned for dark respiration would apply to light respiration. "Photo-oxidation" has been considered by many investigators. It should be excluded. Likewise, the transient "outburst and insurge" effects described later that occur when the plants with excess carbon dioxide are suddenly illuminated or darkened probably do not represent the normal light respiration process.

2. The Kok Effect

Kok (204) observed that the rate of oxygen evolution in algae with increasing light intensity above the compensation point, was about half

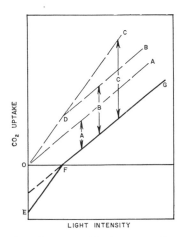

Fig. 6. Diagram illustrating three possible interpretations of the Kok effect regarding respiration in the light. *EFG* represents loss or gain of carbon dioxide, below and above the compensation point, *F*. True photosynthesis could be given by *OA*, *ODB*, or *OC* if light respiration were less than, equal to, or greater than dark respiration, *OE*. Adapted from Decker (95).

the rate of oxygen uptake below the compensation point. Van der Veen (359) made a similar observation on tobacco (*Nicotiana tabacum*) leaf disks. This was interpreted to mean that the rate of respiration in the light is about half that in the dark. Decker (95) has pointed out that this change of slope of the oxygen-light curves at compensation, referred to as the Kok effect, is ambiguous in defining respiration response to light. The curve *EFG* (Fig. 6) gives the experimental relationship between the oxygen evolution and light intensity. The displacement from *FG* to the hypothetical lines *OA*, *ODB*, and *OC* then represent light respiration that is less than, equal to, or greater than dark respiration indicated by the distance *OE*. Three alternative interpretations of the same data are thus possible. Steemann-Nielsen (328) pointed out that the Kok experiments were done with 5% carbon dioxide in the gas phase whereas those of Van der Veen employed 3.6%. With lower carbon dioxide concentrations, the change of slope of the photosynthesis curves at the compensation point became smaller and was negligible below 1% or in ordinary air. Presumably the Kok effect could not be used for a definitive decision between alternative light respiration processes.

3. *"Insurge" and "Outburst" of Carbon Dioxide*

The interesting anomalies in gas exchange that occur when the lights are turned on or off have attracted the attention of many investigators. Emerson and Lewis (121), using manometric methods, observed an "outburst" of carbon dioxide in *Chlorella* when the lights were turned on and an "insurge" when they were turned off. These results were confirmed by A. H. Brown and Whittingham (58) using the mass spectrograph and by Spruit and Kok (323) using the polarograph for oxygen and the pH recorder for carbon dioxide.

The "outburst" in algae seems to represent the emptying, during the induction period after illumination commences, of a carbon dioxide reservoir that was filled in the preceding dark period. Its magnitude depends on the carbon dioxide concentration in the gas phase before illumination. With 0.5% there is no "outburst"; rather, carbon dioxide absorption and oxygen production follow each other closely. With 2–5%, carbon dioxide production exceeds oxygen consumption by a wide margin. A dark period of at least 20 minutes is needed to maximize the "outburst." With shorter dark periods, it becomes progressively smaller. Addition of $C^{13}O_2$ to the gas immediately before illumination showed an appreciable amount of tracer in the "outburst" after a dark exposure of 2.5–4 minutes and full replacement in 20–40 minutes. Franck (132) suggested that the "outburst" is related to the induction phenomenon of photosynthesis and is due to a temporary deficiency of the oxygen-liberating catalyst in the presence of an abnormally large supply of carbon dioxide-saturated addition compounds. This process is not related to ordinary respiration, nor is the "insurge" when these plants are suddenly darkened.

These observation suggest the complex nature of gas exchange in the light. Spruit and Kok (323) conclude that in the transition periods of rapid change in illumination, the usual equality of ratio between carbon dioxide and oxygen is lost and the two gases appear to act independently of one another.

4. Use of Isotopic Oxygen

A. H. Brown and co-workers (56, 57, 189) have attacked the problem of light respiration using the mass spectrometer to measure the exchange of oxygen of mass 32 or mass 34 and carbon dioxide (mass 44). *Chlorella* was investigated in the first paper, the blue-green alga *Anabaena* in the second, and the photosynthetic bacterium *Rhodospirillum rubrum* in the third.

In preliminary work with *Chlorella* it was established that other gases, such as hydrogen, carbon monoxide, and nitrogen, were not involved in the reactions. When oxygen, enriched with O^{18} to give molecules of $O^{(16-18)}$ or mass 34, was supplied as the gas phase, it was assumed that respiration could be studied either in light or darkness by observing the decrease in mass 34. In most of the experiments with *Chlorella* and in an experiment with tobacco leaf disks, the mass 34 concentration in the gas phase decreased continuously and uniformly with time regardless of light or darkness, whereas the concentrations of O_2 mass 32 and CO_2 mass 44 rose or fell sharply in response to turning the lights on or off. On the other hand, *Anabaena* (57) showed a very significant photoinhibition of respiration at 0.5% oxygen, particularly when the light intensity was also low. In some experiments, respiration appeared to be completely inhibited. At higher light intensities (5–10 times compensation) and higher oxygen tensions (1.0–1.5% oxygen) there appeared to be stimulation of respiration up to 250%.

Rhodospirillum rubrum showed definite evidence of partial respiratory inhibition by light, up to 60–85%. It was confirmed that this bacterium does not evolve any oxygen in photosynthesis. Respiratory inhibition in this plant by light was suggested earlier by Van Niel (360).

The authors concluded that their experiments prove unequivocally that the respiration rate of *Chlorella* is the same in darkness and in light. In the case of *Anabaena*, while both photoinhibition and photostimulation were indicated under different treatments, it was suggested that the former might be spurious owing to reutilization of photosynthetically produced oxygen, but some other mechanism was needed to explain the photostimulation of respiration. The photoinhibition of respiration in *Rhodospirillum* could not be complicated by photosynthetic oxygen since none was produced.

The question of endogenous oxygen from photosynthesis is perhaps ing. This could be reutilized for respiration to a variable but unknown extent without actually appearing in the gas phase surrounding the leaf. Such an effect might vitiate measurements by the mass spectrograph using isotopic oxygen, and the rate of uptake of oxygen in the light from the gas phase could be either reduced, unchanged, or increased under different conditions. Thus the interpretation of the data for *Chlorella* as given by Brown might be incorrect, and there could be either inhibition or stimulation of photorespiration. However, in either of these cases at least a jog in the O_{34} curve might reasonably be expected. Likewise, the values for *Anabaena* would not necessarily have quantitative significance, but those for *Rhodospirillum* should be correct.

5. Studies with the Infrared Carbon Dioxide Analyzer

Decker (95, 97, 98) has sought additional evidence concerning the respiration of higher plants in the light. He used an infrared gas

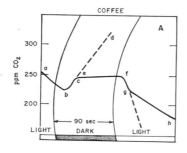
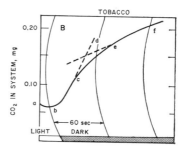

FIG. 7. Carbon dioxide recorder charts which show a marked change of slope shortly after the turning on or off of the light and suggest an elevated respiration rate in the light as compared with steady state dark respiration. From Decker (97) and Decker and Tio (98).

analyzer to monitor the carbon dioxide concentration of the air in a closed system containing a leaf attached to the plant. He observed that the rate of evolution of the gas was severalfold more rapid immediately after the light was turned off than a minute later; also that when the light was turned on there was a momentary surge of photosynthesis that subsided within about a minute. These effects are shown in Fig. 7 for coffee (*Coffea* sp.) and tobacco. Decker suggested that they represent the transition from a high level of photorespiration to a lower level of dark respiration on darkening and vice versa. A number of other species behaved similarly. These responses are the reverse of the "outburst" and "insurge" in *Chlorella*, described above.

Decker (96) employed the same method to study the effects of carbon dioxide concentration on assimilation of the gas at constant light and temperature. The rate of uptake of CO_2 was measured at 400, 300, 200, and 100 ppm, at the compensation point, and also at a concentration considerably below the compensation point. A nearly straight-line relationship was obtained between assimilation and carbon dioxide concentration which could be extrapolated to zero carbon dioxide. For example, in experiments with cotton (Fig. 8), three low light intensities were used, giving three straight-line carbon dioxide concentration-assimilation curves, the slopes of which increased with light intensity.

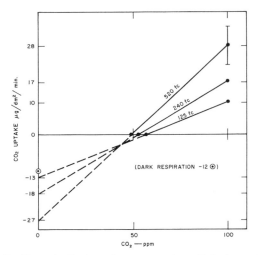

Fig. 8. Carbon dioxide assimilation of cotton at three light intensities and different carbon dioxide concentrations. Extrapolation to zero carbon dioxide suggests an appreciably greater rate of respiration in the light than in the dark. From Decker (97).

On extrapolation to zero carbon dioxide, the increases in light respiration above dark respiration at 30°C and 125, 240, and 525 fc were about 10, 50, and 125%, respectively. In an experiment with tobacco leaves at 2500 fc and 25.5°, assuming no direct reuse of endogenous carbon dioxide, minimum absorption (CO_2 = zero) was -57 μg $CO_2/dm^2/min$. Dark respiration was -16 μg $CO_2/dm^2/min$. Photorespiration was therefore at least -41 μg $CO_2/dm^2/min$, or 2.5 times greater than dark respiration. Total respiration in the light was at least 3.5 times dark respiration.

Photosynthesis-light experiments with *Mimulus* (Fig. 9), at 20°, 30°, and 40°C and 2000 fc gave three nearly parallel straight lines which extrapolated to indicate appreciable amounts of photorespiration. The fact that the curves were nearly parallel suggests that respiration was influenced by temperature much more than was photosynthesis. Respiration is independent of the carbon dioxide concentration, whereas photosynthesis is dependent. If photosynthesis were temperature dependent as well, the curves would diverge. If the displacement of the curves was due to the effect of temperature on respiration (assuming $Q_{10} = 2.0$ for respiration and $Q_{10} = 1.0$ for photosynthesis), the respiration rate in the light would have to be over three times that in the dark in order to account for the observed displacements. From Fig. 9, the excess of light respiration over dark respiration was —41, —47, and —47 μg CO_2/dm²/min at 20°, 30°, and 40°, respectively.

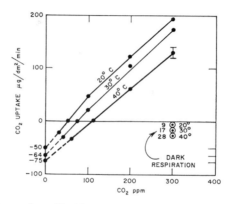

FIG. 9. Assimilation-carbon dioxide curves of *Mimulus* at three temperatures, and 2000 fc. Elevated light respiration is indicated by extrapolating the curves to zero carbon dioxide. From Decker (96).

Reutilization of endogenous carbon dioxide, if it occurs in these experiments, would mean that the values of the apparent assimilation are low. Presumably the correction for this effect would be greater, the higher the carbon dioxide concentration. If so, the slopes of the assimilation curves should be steeper, resulting in larger values of light respiration. The values as found are probably minimal.

It seems reasonable to conclude that the light respiration during normal photosynthesis of many species, particularly of the higher plants, may be up to severalfold greater than dark respiration, depending on the temperature and light intensity. This circumstance might have practical as well as theoretical significance. If light respiration could be reduced by any means, net assimilation would be correspondingly increased.

III. Effect of Environmental Factors on Photosynthesis

A. Light Intensity and Carbon Dioxide Concentration

1. Blackman-Type Curves

Light intensity and carbon dioxide concentration are the primary external parameters which control photosynthesis. Temperature is a secondary factor. F. F. Blackman (30) in 1905 proposed the "principle

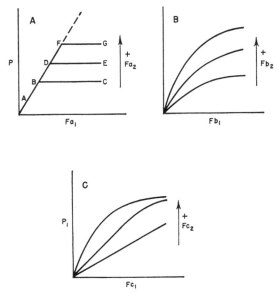

Fig. 10. Diagrammatic summary of the effects of various external and internal factors on photosynthesis. Chart A: Blackman-type curves (32) in which Fa_1 (light, CO_2, or temperature, etc.) is varied continuously while Fa_2 is varied stepwise. Chart B: Harder (or Bose) type curves (164) analogous to the Blackman type. Chart C: Types suggested by Rabinowitch (279) involving enzymes, etc.

of limiting factors" to describe the response of the plant to a number of variables acting simultaneously. He stated, "When a process is conditioned as to its rapidity by a number of separate factors, the rate of the process is limited by the pace of the 'slowest' factor." He drew his well-known diagram shown in Fig. 10A with different symbols, to illustrate this idea in which all factors except one, Fa_1, are held constant and this variable factor is allowed to increase from zero. Under these circumstances photosynthesis (P) increases linearly (AB) until another factor, Fa_2, becomes limiting. Then additional Fa_1 no longer increases photo synthesis and the curve suddenly becomes horizontal (BC). If Fa_2 is increased stepwise, the plateau is shifted to successively higher levels $(DE$ and $FG)$ indicating widely separated saturation values but coincident initial curves. F. F. Blackman and Smith (33) obtained their

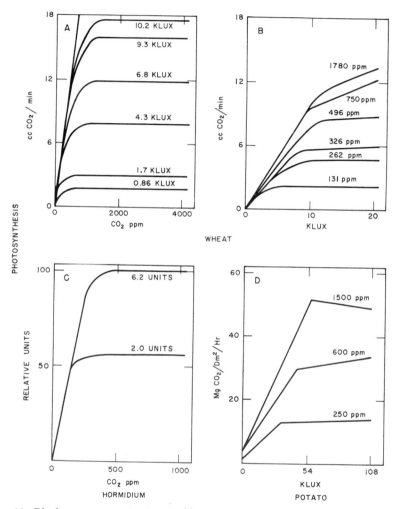

FIG. 11. Blackman-type assimilation-CO_2 curves for wheat (180) and *Hormidium* (358) and Harder-type assimilation-light curves for wheat and potato (79).

assimilation-carbon dioxide curves at constant light and temperature, having this sharp change of direction, with the water plants *Elodea* and *Fontinalis*.

In practice it has been found that the transition from the sloping linear portion of the curve to the horizontal saturation plateau is seldom sharp, as suggested by Blackman. Rather it is more or less gradual, and in the extreme case all curves of the family are completely separated commencing at the origin (Fig. 10B).

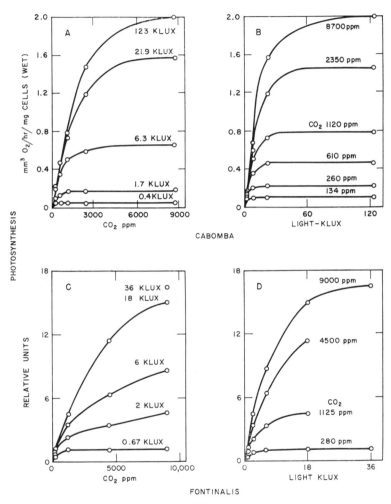

FIG. 12. Blackman-type curves for *Cabomba* and Harder-type curves for *Fontinalis* (164).

Figure 10C illustrates a third type of relationship suggested by Rabinowitch (279) involving enzymes as the Fc_1 and Fc_2 parameters. Fc_2 codetermines with Fc_1 for initial stages of photosynthesis but has no effect on the saturation plateau.

Figure 11A and C and Figure 12A show Blackman-type relationships between assimilation and carbon dioxide concentrations at different light intensities for wheat (180), *Hormidium flaccidum* (358), and *Cabomba caroliniana* (315, 316). The initial portion of the curves coincide at first, then turn more or less gradually to the horizontal. The assimilation light curves of *Cabomba* in Fig. 12B are quite similar to these. The *Hormidium* curves conform best to the Blackman model. It may be noted that Emerson and Green (118) have published an assimilation-carbon dioxide curve for *Chlorella* at low light intensity which shows a sharp change of direction with no transition range, in complete accord with the Blackman model.

2. Harder-Type Curves

In Fig. 11B and D, the initial slopes of the assimilation-light curves for wheat and potato (79) increase with increasing carbon dioxide concentration at least between 300 and 800 ppm, thus showing definite divergence from the Blackman model. However, the potato curves represent best-fit lines in scatter diagrams and are not very precise. It can be seen by inspection that the plateau heights in most of the charts are approximately proportional to the F_2 parameters except at the highest values of the latter. Table XIII brings out this relationship for the wheat and *Cabomba* data.

Harder's curves (164) for *Fontinalis antipyretica* in Fig. 12C and D do not conform well to the Blackman model. These curves diverge almost from the origin, suggesting that this plant cannot utilize the light and carbon dioxide as effectively as the other plants. F. F. Blackman and Smith (33) found that the initial slope of the assimilation-carbon dioxide curve is about twice as steep with *Elodea* as with *Fontinalis*. This suggested that the obstacles to the diffusion of carbon dioxide from the solution to the chloroplasts were greater in *Fontinalis* than in *Elodea*. This may be a class distinction between the bryophyte *Fontinalis* and the phanerogam *Elodea*. The latter has a well-developed system of air spaces containing an "internal atmosphere," which is lacking in the former.

Steemann-Nielsen (328) has questioned the Harder curves because the measurements were made in solutions that were shaken only once each 5 minutes. He found that *Myriophyllum spicatum* gave similar curves in stagnant water whereas in well-stirred solutions the initial portions of assimilation-light or carbon dioxide curves were coincident at all F_2 values, as found for *Cabomba*. In additional work with *Fontinalis*, he noted that the terminal buds of the plants grown with adequate free carbon dioxide consisted of many fully developed leaves forming a large compact bulk at the end of the shoot. Presumably the diffusion of carbon dioxide to these leaves was slow because of the length of the channels to the active surfaces. When the terminal buds were removed, the assimilation-carbon dioxide curves at constant light were much steeper and saturation was attained at half the carbon dioxide concentration required when these buds were present.

TABLE XIII

Stepwise Ratios of the Heights of the Assimilation Plateaux with Different Light Intensities Plus Excess Carbon Dioxide: Also with Different CO_2 Concentrations Plus Excess Light

Light (klux)	Max CO_2 (ppm)	Increment ratio		Max light (klux)	CO_2 (ppm)	Increment ratio	
		Light	Assimilation			CO_2	Assimilation
			Wheat[a]				
10	4000	—	—	20	131	—	—
23	4000	2.3	1.8	20	262	2.0	2.1
57	4000	2.5	2.6	20	326	1.2	1.2
91	4000	1.6	1.5	20	496	1.5	1.5
124	4000	1.4	1.4	20	750	1.5	1.4
136	4000	1.1	1.1	20	1270	1.7	1.0
			Cabomba[b]				
0.4	8700	—	—	130	134	—	—
1.7	8700	4.3	4.0	130	260	2.0	2.0
6.3	8700	3.6	3.1	130	610	2.4	2.1
22	8700	3.3	2.5	130	1120	1.8	1.6
123	—	5.6	1.3	130	2350	2.1	2.0
				130	8700	3.7	1.4

[a] From Hoover *et al.* (180).
[b] From F. F. Blackman and Smith (33).

Strict adherence to the Blackman form suggests that the factor F_2 controls the maximum rate of a partial process that does not depend on the variable parameter F_1. If F_1 is carbon dioxide supply and F_2 is light, it is evident that the former would vary from plant to plant depending on the ease of diffusion and the distance traveled by the carbon dioxide molecules to reach the chloroplasts. In unicellular algae like *Chlorella*, the distance of diffusion is about 1 μ; in some large water plants, about 40 μ. In higher plants, the carbon dioxide diffuses as a gas, which is a much more rapid process than diffusion in aqueous solution. In this case, the gas supply would be subject to control by the stomata. It may be noted in Fig. 12 that the carbon dioxide assimilation curves of *Fontinalis* and *Cabomba* involve carbon dioxide concentrations up to 9000 ppm, without complete saturation even at full sunlight intensities, whereas with wheat saturation is approached with 2000 ppm at 20 klux. Inadequate stirring of the solution may be partly responsible for this condition, but slow diffusion of carbon dioxide through the tissues is probably more important.

Not only the transport of the carbon dioxide to the chloroplast, but also its fixation after arrival, must be considered. This involves a series of reactions before and after combination with an acceptor "A" to form $A \cdot CO_2$. The $A \cdot CO_2$ is reduced in light reactions in the presence of photosynthetic catalysts, followed by "finishing" reactions that produce sugars, etc. The rates of these reactions will determine in part the characteristics of the carbon dioxide-assimilation curves. The carboxylation reaction could even be reversed if adequate amounts of catalysts are not present. It is suggested by Rabinowitch that "carbon dioxide curves that diverge from the origin can be expected if the carbon dioxide-acceptor complex, $A \cdot CO_2$ is not fully saturated with carbon dioxide at low $[CO_2]$ values, even in the equilibrium state . . . ," because of inadequate carbon dioxide supply and/or slow diffusion through the plant tissues.

Considering the light-assimilation curves where F_1 is light and F_2 is carbon dioxide supply, it must be expected that there will be a great range of light intensities reaching chloroplasts that are at different distances below the illuminated surface and even large variations within a single chloroplast. Moreover, factors like shading and orientation of the leaves will add to the variability. It would be difficult to average these inequalities so as to estimate the theoretical light effects on the system. The inequalities will certainly tend to cause more or less departure from Blackman-type light-assimilation curves. However, approximately the same relationships are seen in the light curves and in the carbon dioxide curves, including the Harder type with *Fontinalis*.

It is evident that the environmental and internal factors that affect photosynthesis are complex, and some departure from the simple Blackman relationship is generally to be expected. Blackman himself recognized this condition when he stated, "The way of those who set out to evaluate exactly the effects of changes in a single factor upon a multi-conditioned metabolic process is hard, and especially so when the process is being pushed toward the upper limits of its activity . . . at present our science entirely lacks data that will stand critical analysis from the point of view indicated in this article."

3. Field Measurements

Most of the foregoing experiments on the assimilation-light-carbon dioxide relationships were carried out under controlled conditions in the laboratory. Similar studies have been done in the field.

Loomis and co-workers (79, 365) measured photosynthesis in individual leaves of maize and potato in the field. Large differences were noted between carefully paired leaves on a single plant. With maize and potato the average differences between pairs of leaves were 25 and 28%, respectively. The maximum differences reached 100%. A large number of observations were therefore needed to obtain significant results. The scatter diagrams of assimilation vs. light intensity yield Blackman-type curves when the initial sloping portion and the plateau are calculated independently.

Figure 11D gives the best-fit regression lines for potato in the field in 1951. Ambient air at about 250 ppm carbon dioxide and air streams enriched to about 600 and 1500 ppm were employed. The sloping lines had significant correlation coefficients of 0.71, 0.75, and 0.89, respectively. Their slopes were 0.34, 0.58, and 0.90 mg CO_2/dm²/1000 lux. The plateaux were essentially horizontal, with elevations of 13.6 ± 1.0, 32.2 ± 2.6, and 50.0 ± 2.6 mg CO_2/dm²/hour. The second carbon dioxide concentration and assimilation levels were both 2.4 times the first, but the third assimilation level was only 3.7 instead of 6 times the first. The three curves appear to diverge from the origin, but this is probably due to the scattered data, a distribution that renders exact characterization of the curves impossible. Light saturation occurred at 32, 45, and 56 klux[1] with the three carbon dioxide levels. Other light saturation values of field crops are listed in Table XIV, including data for rice (*Oryza sativa*) by Shimizu and Tsuno (308) and for tree seedlings by Kramer and Decker (209).

Thomas and Hill (350) made assimilation measurements using carbon dioxide-enriched atmospheres on a number of crops in both sand culture and field plots. Experimental treatments were of short duration (2–3 hours). The data are shown in Fig. 13. Only the sloping portions of the curves were obtained. Evidently the carbon dioxide concentrations employed were insufficient to reach the plateaux. The slopes of the curves varied somewhat in the different plots. The sulfur-deficient sugar beets (*Beta vulgaris*) in the sand culture did not respond to additional carbon dioxide. Photosynthesis was doubled by multiplying the carbon dioxide concentration by the following factors:

[1] For the relations between different units in which light energy is measured see Section III, A.

TABLE XIV
LIGHT SATURATION VALUES OF FIELD CROPS

Crop	Season	CO_2 (ppm)	Light (klux)	Reference
Apple	Summer	300	47	Heinicke and Childers (172)
Apple	September	300	44	Heinicke and Childers (172)
Alfalfa	Summer	300	49	Thomas and Hill (350)
Alfalfa	October	300	35	Thomas and Hill (350)
Wheat	June–July	300	57	Thomas and Hill (350)
Sugar Beets	August–October	300	47	Thomas and Hill (350)
Maize[a]	Summer	250	27	Verduin and Loomis (365)
Potato[a]	Summer	250	32	Chapman and Loomis (79)
Potato[a]	Summer	600	45	Chapman and Loomis (79)
Potato[a]	Summer	1500	56	Chapman and Loomis (79)
Rice	March	300	43	Shimizu and Tsuno (308)
Rice	April	300	72	Shimizu and Tsuno (308)
Rice	May	300	110+	Shimizu and Tsuno (308)
Rice	June	300	51	Shimizu and Tsuno (308)
Pine seedlings	—	300	Up to 90	Kramer and Decker (209)
Hardwood seedlings	—	300	30	Kramer and Decker (209)

[a] Single leaf in direct sunlight.

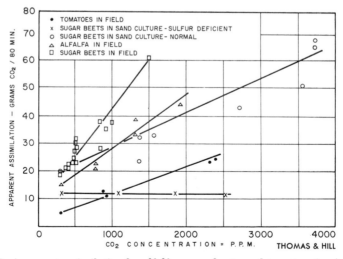

FIG. 13. Apparent assimilation by alfalfa, sugar beets, and tomatoes in air enriched by CO_2. From Thomas and Hill (349).

tomatoes 2.5–3.1; alfalfa (*Medicago sativa*) 2.7; sugar beets in the field 2.5–2.7; sulfur-deficient sugar beets ∞.

4. Carbon Dioxide Fertilization

It is evident that nearly all crops growing in the field or greenhouse in sunlight are carbon dioxide limited rather than light limited, because they can utilize only 10–50% of direct sunlight when supplied with the normal 300 ppm carbon dioxide. The problem of increasing crop production by increasing the carbon dioxide content of the air has been under investigation since 1898.

H. T. Brown and Escombe (60) carried out three series of experiments comparing carbon dioxide-enriched air with ordinary air. In the first series assimilation was measured on single sunflower and catalpa leaves mounted in glass containers, through which either normal air or air enriched four- to sevenfold with carbon dioxide was passed at about 1.3–1.5 liters per minute. In bright diffuse sky light, the rate of assimilation was directly proportional to the carbon dioxide concentration. In dull rainy weather the increase in rate of assimilation was about 60–70% of the increase in carbon dioxide concentration. These experiments lasted 4 hours. In 11-day experiments with potted *Vicia faba* or sunflower plants under bell jars, no significant difference in dry weight increment was observed in bright diffuse light due to two- and fourfold increments in carbon dioxide concentration.

Finally, two experiments were carried out in a pair of small (380 cubic feet) greenhouses. Potted plants of nine different species were treated with normal air and air enriched 3.5-fold, for 12 hours a day on 77 days by continuous addition of carbon dioxide. The greenhouses were whitewashed and kept closed during the day. At night they were opened for free ventilation. A similar experiment was carried out for 84 days, with air enriched 200-fold. The two experiments gave nearly the same results, in spite of the great difference in carbon dioxide concentration. The growth of the high-carbon-dioxide plants was unsatisfactory. Leaves were smaller and deeper green in color, axial growth was stimulated, internodes were shortened but more numerous, flowers failed to develop, no fruit was formed and dry weights were decreased. Compressed carbon dioxide in tanks was used in these experiments.

This work was followed shortly by that of Demoussy (99), who worked with 15 species of plants and obtained 22–162% heavier plants as a result of fivefold carbon dioxide enrichment applied during the daytime over a 2-month period. He used closed glass chambers, 1 cubic meter in volume, protected from direct sunlight by a cloth covering. Carbon dioxide, obtained by heating sodium bicarbonate, was introduced as needed. The chambers were opened at night and well aerated. Cerighelli (77) also obtained promising results.

Many studies made subsequently have indicated small to large increases in yield. Cummings and Jones in Vermont (86) found the closed chamber to be unsatisfactory for their experiments. They devised glass-sided hot-bed type boxes $26 \times 18 \times 26$ inches deep, or $52 \times 28 \times 12$ inches deep for small-stature plants like lettuce (*Lactuca sativa*), covering the soil or pots in which the plants were growing. The boxes were open at the top. Carbon dioxide was generated continuously by adding dilute sulfuric acid to a solution of sodium bicarbonate. The gas was led to the center of the box and allowed to diffuse about the plants. Concentrations were not measured, but regularly increasing yields were obtained by increasing the gas volumes up to 116 liters of carbon dioxide in 8 hours in the open boxes. However, much of this gas was lost by diffusion from the box.

Beans (*Phaseolus vulgaris*) and peas (*Pisum sativum*) gave yields of seeds up to 200–300% greater than the controls. Potatoes and lettuce showed large increases also. Strawberries (*Fragaria chiloensis var. ananassa*) developed 55% more plants and yielded 80% more fruit. The "gassed" plants generally had more carbohydrate and a little less protein than the controls.

O. Owen *et al.* (265) in England prepared their carbon dioxide in the greenhouse by treating a paste of potassium bicarbonate with concentrated sulfuric acid. A rapid decrease in the gas concentration was observed even when the greenhouse was closed, as soon as generation was stopped, so that a large excess of gas was required to maintain the desired concentration during treatment. Tomatoes, treated with 0.6 and 0.9% carbon dioxide for 2.7 hours per day from the seedling stage through harvest, gave about 25 and 16% greater yield of fruit than the control.

E. S. Johnston (188) obtained increased tillering of wheat, greater number and weight of heads and kernels, and greater weight of straw due to fertilization with 0.12% carbon dioxide. Heading out of the grain was delayed slightly, but the weight per kernel was not affected.

Lundegardh (222) burned denatured alcohol in lamps in the greenhouse and attained carbon dioxide concentrations 3–4 times normal. When the lamps were spontaneously extinguished, the residual dilute alcohol containing denaturant was discarded. Cucumber (*Cucumis sativus*) fruits showed 28–45% increase in weight. Strawberry plants and fruits were also larger and the number of fruits were 12% greater than on the controls. The extra fruit more than paid for the alcohol burned.

Bolas and Henderson (42) carried out a carefully controlled experiment with seedling cucumbers on a turntable illuminated by 1000 or 1500 watt Mazda lamps. They obtained a 40% increase in dry weight in 18 days using 0.3% carbon dioxide as compared with 0.04%.

Thomas and Hill (350) treated tomatoes in their sand culture plots with about 3000 ppm carbon dioxide, obtained from dry ice, for about 6–9 hours per day for 2 weeks. The leaves were injured and photosynthesis was reduced to about half its former level. The experiment was repeated after a month's recovery, with a similar result. The dry ice used was not analyzed but analysis of similar material did not reveal sulfur compounds or ethylene.

In Germany and Russia great interest in "carbon dioxide fertilization" has been manifested and claims for large increases in yield have been made. Hugo Fischer (127) studied the subject, starting in 1912, and became an enthusiastic advocate of the practice. Riedel (289) patented a water scrubbing tower to remove sulfur dioxide and other plant toxicants from the gases from coal and coke furnaces or even from metallurgical furnaces, so that carbon dioxide could be piped to nearby greenhouses or fields for large scale use. Riedel (288) and also Bömer and Rintelen (44) described field experiments using purified coke gas. Treatments were applied for 4–8 hours each fair day from May to October. The rate of application was 3.5–5.1 liters CO_2/m^2 hour. The season's fumigations represented 1400 hours on 210 days. Rye (*Secale cereale*), potatoes, turnips, vetch (*Vicia* sp.), and cabbage (*Brassica oleracea var. capitata*) were benefited; oats (*Avena sativa*), corn, and lupines were only slightly affected. Of the added carbon dioxide, 65–82% was accounted for as increased yield in rye, potatoes, and turnips equivalent to about 15–25 tons of dry matter per acre. Katunskii (194, 195) and Chesnokov (80) in Russia used carbon dioxide prepared by the catalytic combustion of various fuels and obtained substantial increases in yield. The former noted that the gas is most effective if applied when the rate of accumulation of organic substances is greatest, e.g., in oats during tillering and in peas and beans when the flower buds are beginning to form.

It is not apparent why some workers have obtained satisfactory results from carbon dioxide enrichment whereas others have failed. It may be noted that response should occur only when (a) the light intensity is greater than certain rather high values depending on the light saturation level of the plant; (b) the wind velocity is low in the field or ventilation is restricted in the greenhouse; (c) the photosynthetic apparatus is not injured by the high carbon dioxide concentration, and (d) toxic compounds are not present in the gas. The first two conditions are usually self-evident. For example, in the data of Cummings and Jones, the effect of treatment was small when the percentage of sunshine was less than about 20%. However, even with 40–60% of full sunlight the response was sometimes small. It has been suggested that the plants may be injured by excessive humidity and heat along with the high carbon dioxide concentration when treated in closed chambers. Presumably the photosynthetic apparatus may be overtaxed. A question also is raised, particularly in the experiments of Brown and Escombe and of Thomas and Hill, as to the possible presence of phytotoxic compounds in the carbon dioxide. Evidently the purity of the carbon dioxide should be examined critically.

More recently, the possibility of carbon dioxide fertilization through the roots as well as through the shoots, has been suggested. Grinfel'd (157) in Russia fertilized sugar beets with 30–50 kg CO_2 per hectare as ammonium carbonate. He attributed increased yields of 7–16% to the carbonate rather than to the ammonium ion. Kursanov (210a) reported increased yields of several crops up to 18% due to the use of soluble carbonates in addition to various fertilizers. It was estimated that the roots absorbed about 25% as much carbon dioxide as the shoots.

Carbon dioxide fertilization through the roots in the foregoing studies appears to be questionable. Stolwijk and Thimann (342) grew peas and oats in nutrient solutions aerated with concentrations of carbon dioxide up to 6%. Elongation of the pea roots was inhibited by more than 1% carbon dioxide in the gas. At 3%, inhibition was 90% and at 6% it was complete. There was a very slight stimulation at 0.5%. Beans and sunflower responded similarly to peas. Oats and barley (*Hordeum vulgare*) were unaffected by the carbon dioxide. Uptake of C^{14}-labeled carbon dioxide by the roots was found to be only a few per cent of the amount produced by respiration, and it ceased after about 8 hours. Peas took up about five times as much carbon dioxide as barley—an uptake that might account for the difference in response of the two species. It is likely that there was a specific toxic effect of the gas on the pea roots, but the barley did not take up enough to be affected. It was found that pea plants could not be grown successfully when the sole supply of carbon dioxide was through the roots.

B. LIGHT TOLERANCE

1. Photosynthesis by Sun and Shade Plants

The light exposure of a leaf during its formative period can affect markedly its subsequent light tolerance and photosynthetic capacity.

Burnside and Böhning (41, 72) give the light-assimilation curves of a number of typical sun plants grown in both light and shade. A summary of representative values from these curves is given in Table XV.

The light intensity required for maximum photosynthesis of sun plants that had been grown in the sun was about double that of sun plants grown in the shade (except with sunflower which was unaffected by the shade treatment) and about 3 times that of typical shade plants grown in the shade. The rates of photosynthesis of sun plants grown in the sun were about 1.6 times those of sun plants grown in

TABLE XV

Photosynthesis and Respiration of Typical Sun Plants Grown in Sunlight and Shade As Compared with Typical Shade Plants Grown in the Shade[a]

Plant	Growth conditions	Light saturation (fc)	Carbon dioxide absorbed (mg/dm²/hour)		
			Dark respiration	Photosynthesis at	
				2000–2500 fc	5000–6000 fc
7 Sun plants					
Tobacco and	Sun	2500–2800	−4	16	16–20
cotton	Shade	1000–1700	−2	7–13	7–13
Beans[b] and	Sun	2000–2500	−2	18–21	18–22
tomato	Shade	1000–1500	−2, −4	10–14	10–14
Sunflowers	Sun	2500	−2	20	20
	Shade	2500	−2	19	19
6 Shade plants[c]					
	Shade	400–1000	−0.4, −1.4	2.4–5.9	2.4–5.9

[a] From Burnside and Böhning (72) and Böhning and Burnside (41).
[b] Bean, castor bean, soybean.
[c] Dryopteris, Nephrolepis, Oxalis, Philodendron, Saintpaulia spp.

the shade and 4–5 times those of shade plants grown in the shade. Again, sunflower was practically unaffected by the shade treatment. The rates of respiration of sun plants grown in sun and shade were about the same, but they were 2–5 times the rates of typical shade plants grown in the shade. Evidently the conditioning of a sun plant to a shade environment does not reduce its photosynthetic activity to the level of a typical shade plant, at least not in one generation.

Nutman (261) found that photosynthesis of coffee leaves in the field was proportional to the light at low intensities but total assimilation for the day was greater in the shade than in the sun. A marked midday depression of photosynthesis occurred on clear days.

Rhykerd *et al.* (286) grew plots of legume seedlings under various controlled light intensities applied each day for 30 days. Then all plants were given a 15-minute treatment with $C^{14}O_2$ in full sunlight. Uptake of carbon dioxide was greatest in the plants that had received 2400–9600 fc for 12 hours per day and least in the plants with these intensities for 6 hours per day. Other intensities and durations of exposure resulted in intermediate assimilation.

The two sides of the same leaf may behave quite differently. Brun (68) made independent and simultaneous measurements of transpiration and photosynthesis or respiration on the two surfaces of a banana (*Musa acuminata* 'Gros Michel') leaf. The leaf was directly illuminated on the upper surface only. Intensities from 50 to 2900 fc were used. The light reaching the lower surface was about 7–10% of that incident on the upper surface. Air from a cylinder of compressed gas was humidified to about 40% by bubbling through dilute sulfuric acid, then was passed over the leaf surfaces enclosed in 100 cm² Lucite chambers attached to the two surfaces. Carbon dioxide was monitored with an infrared analyzer, and water was recorded with a dew point recorder.

Steady state transpiration and photosynthetic rates on each surface increased regularly with light intensity. At 2900 fc, transpiration rates were 820 mg $H_2O/dm^2/hr$ on the lower surface and 120 mg $H_2O/dm^2/hr$ on the upper surface. The corresponding values for photosynthesis were 8.2 and 0.6 mg $CO_2/dm^2/hr$. Transpiration rates were therefore 100 times greater on the lower surface and 200 times greater on the upper surface than assimilation rates. Further, both rates were about 7–14 times larger on the lower surface than on the upper surface, where the light intensity was over 10 times greater. It is probable that these effects were due to different stomatal densities and stomatal apertures on the two surfaces. Stomatal counts showed 345/cm² on the upper surface and 1428/cm² on the lower surface.

It was noted that, when the lights were turned on, transpiration and photosynthesis were very low for about 20–30 minutes. Then both increased somewhat abruptly to their steady state maxima in the brighter lights, owing probably to delayed opening of the stomata. When the lights were turned off photosynthesis was terminated abruptly and pronounced respiration became apparent. This subsequently decreased steadily, as did also transpiration, presumably as a result of closing of the stomata.

F. M. Eaton and Ergle (108) studied the effect of shading on the growth, fruiting, and carbohydrates of cotton. It was noted that self-shading in a cotton field in blossom reduced the light intensity at half the height of the crop to 30% and at ground level to 5% of the intensity above the plants. The half-height lighting was simulated on field plots by placing cloth shades above the plants, starting at the beginning of the blossoming period.

Table XVI lists some changes that occurred in these plants expressed as percentage of the unshaded control plants. The shading reduced the weight of leaves and stems by 13% and their carbohydrate content by 20–30%. The most important effect was a reduction in the number of bolls and weight of seed cotton to about half

TABLE XVI

The Effect on Well-Irrigated Cotton Plants in the Field of a Reduction in Light Intensity of 68% by Means of Cloth Shades after Start of Flowering[a]

Parameter	Per cent of full light control
Weight of leaves plus stems	87
Weight of tops (leaves, stems, and bolls)	66
Weight of seed cotton	53
Number of bolls per plant	49
Carbohydrates in leaves	81
Carbohydrates in main stems	69
Carbohydrates in 14-day bolls	99

[a] From F. M. Eaton and Ergle (108).

that of the controls, due probably to enzyme rather than to nutritional effects.

2. Effect of Intermittent Light

Plants are normally subjected to rather definite periods of illumination during each 24 hours depending on the time of year, latitude, and local weather and environmental conditions. Morphogenesis, flowering, etc. are controlled by these factors, which produce the "long-day" or the "short-day" plants. In these processes a relatively small expenditure of energy is required in comparison with photosynthesis, which is largely determined by the total quantity of white light available.

Garner and Allard (137) showed that the length of the light and dark periods, applied to give the same total illumination, had a profound effect on growth. They subjected cosmos (*Cosmos bipinnatus*), sugar beet, delphinium, buckwheat (*Fagopyrum esculentum*), and *Ipomoea* to successive light and dark periods of equal length ranging from 5 seconds to 12 hours for 5–6 weeks. Dry weight production was 2- to 100-fold greater with 12-hour periods than with periods of 1–15 minutes. Periods of 5 seconds approached in effectiveness the 12 hours' lighting.

Bonde (46) treated tomato and cocklebur (*Xanthium* sp.) plants similarly to Garner and Allard's plants for 2–3 weeks. Growth increased and decreased rhythmically as the light and dark periods were lengthened. Dry weight production was favored in tomato by periods of 6 seconds, 5 minutes, 4 hours, and 12 hours; in cocklebur by 6 seconds, 1 minute, 1 hour, and 12 hours. Intermediate periods gave reduced growth. Minimum growth occurred with periods of 15–30 minutes. Evidently the products of the light reaction enter a chain of dark reactions which have definite periodicity. The responses were not affected by spraying the plants with sucrose, auxin (indoleacetic acid), or antiauxin (2,3,5-triiodobenzoic acid).

3. Effect of Continuous Light

The plant exhibits limitations as to the length of its exposure to light of moderately high intensity just as it does to very high light intensity. Shade plants are particularly sensitive to prolonged exposures, and they may be bleached by relatively low intensities. The algae can utilize continuous light if thresholds are not exceeded. These plants form new cells by division. Arctic species have continuous light throughout the summer, subject to 12-hour diurnal variations in intensity. The assimilation–time-of-day curves of sun plants are usually quite symmetrical on cloudless days, indicating no appreciable change in the rate of photosynthesis except as the angle of incidence of the light changes.

Böhning (40) measured the photosynthesis of attached apple (*Malus sylvestris*) leaves by making carbon dioxide analyses on an air stream passing through a leaf mounted on a special cup. The carbon dioxide in the air supply was also determined. Tanks of compressed air were used in later experiments to avoid the fluctuations in carbon dioxide content of ambient air. The system was operated at constant temperature, 26°C. The samples were titrated at 4-hour intervals.

In one experiment (Fig. 14), the leaf was illuminated with 5800 foot-candles, day and night for 18 days. No effect on the rate of assimilation was noticed during the first 6 days. Later, marked fading of the green color occurred; and the rate finally fell about 39%, while the carbon dioxide content of the air supply fell 5%. At 3200 fc the rate inceased by about 30% during 17 days, owing possibly to a 14% increase in the carbon dioxide concentration of the atmosphere.

Only slight bleaching occurred in this 17-day exposure. When compressed air was used, an 8-day exposure to continuous light at 5200 fc showed no change in rate of photosynthesis.

When "shade" leaves which had been grown under low light conditions in the greenhouse in spring were substituted for "sun" leaves grown outside in summer, light tolerance was greatly reduced. At 3800 fc, there was an initial rate-lowering of about 33% in the first 2 days; then the rate remained nearly constant until the ninth day;

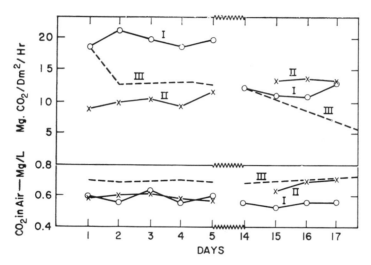

Fig. 14. Rate of photosynthesis of continuously illuminated apple leaves. Curve *I*, sun leaves at 5800 fc; curve *II*, sun leaves at 3200 fc; curve *III*, shade leaves at 3800 fc. From Böhning (40).

thereafter it fell steadily for 10 days to about one-sixth the initial rate. Marked bleaching and necrosis were produced.

4. High Light Intensity

Plants have an upper limit of tolerance for light. Exposures to higher intensities cause temporary or permanent injury to the photosynthetic mechanism. Intense and prolonged illumination can cause bleaching of chlorophyll due to photooxidation, although chlorophyll is known to be remarkably stable *in vivo*. Bleaching does not occur in the absence of oxygen. It is accelerated by all factors that inhibit photosynthesis, such as excess oxygen, carbon dioxide starvation, and poisons. Sun plants can tolerate sunlight for long periods, but shade plants are susceptible to lower intensities. *Chlorella* suspensions may be bleached by sunlight if they are so dilute that the individual cells are exposed continuously or if the more dense cultures are not adequately stirred.

Myers and Burr (252) made a critical study of the solarization effect of a wide range of light intensities from 1000 to 27,700 fc (1 to 27.7 kfc) on the photosynthesis of *Chlorella*, as shown in part in Fig. 15. The activity at 1 kfc is represented by 17.5 mm change of pressure due to oxygen evolution in the Warburg apparatus in 10 minutes. At 4 kfc the initial rate was somewhat greater than at 1 kfc (about 22 mm)

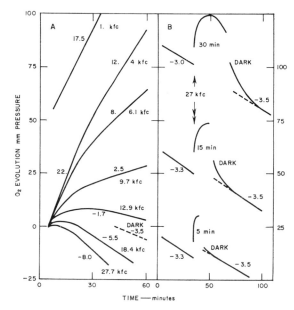

FIG. 15. Effect of high light intensities on rate of photosynthesis of *Chlorella*. Chart A: Inhibition due to 4–27.7 kfc. Chart B: Inhibition due to 5, 15, and 30 minutes at 27 kfc (27,000 fc). Numbers on curves give limiting pressure changes in 10 minutes. From Myers and Burr (252).

but after about 30 minutes the rate fell to 12 mm, where it remained constant for over half an hour. At higher light intensities the rates became progressively less, until at 12.9 kfc oxygen evolution was negative (−1.5 mm)—that is, photooxidation exceeded photosynthesis. At still higher intensities oxygen evolution became still more negative, reaching value of −8 at 27.7 kfc. In these treatments there was not only no net assimilation, but normal respiration (−3.5 mm) was increased by the addition of photooxidation.

In another series of experiments, the *Chlorella* was illuminated at 27 kfc for periods of 5, 15, and 30 minutes. Photosynthesis increased rapidly to a maximum, then was offset by photooxidation. During the last half of the 30-minute exposure, photooxidation predominated. When the light was turned off, the initial dark respiration rate was established almost immediately after the 5-minute exposure, but after the 30-minute exposure it was 20 minutes before photooxidation subsided. There is thus a limited reversibility of the solarization effect.

Steemann-Nielsen (327) observed similar effects with *Cladophora insignis*. Photosynthesis was temporarily doubled by increasing the light from 3 to 100 klux, then it fell steadily to about 55% in 1 hour (Fig. 16A). Photosynthesis at 3 klux fell below the compensation point following this 1-hour exposure to 100 klux, and recovery at 3 klux required many hours (Fig. 16B). Even 10 minutes at 100 klux

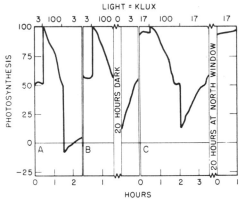

FIG. 16. Photosynthesis of *Cladophora insignis* at 3, 17, and 100 klux as percentage of maximum rate at 100 klux, at 18°C. From Steeman-Nielsen (327).

reduced subsequent assimilation at 3 klux to 40%, and complete recovery required more than an hour. Following exposure to 100 klux, smaller reductions and more rapid recoveries were observed at 17 and 35 klux than at 3 klux. An exposure of 80 minutes at 100 klux (Fig. 16C) reduced the level at 17 klux to 12%, but 90 minutes later the level at 17 klux was 62% of the initial 17 klux level. Similarly, 50 minutes at 100 klux reduced the level at 35 klux to about 48%; 40–80 minutes later the level at 35 klux was 75–80% of the initial level. After 1 hour at 100 klux and 20 hours in darkness, photosynthesis at 3 klux started at the 10% level, and the normal rate was established in about an hour. Moderate light seems to be needed to reverse the injurious effect of very bright light. It is supposed that chlorophyll-sensitized photooxidation in very bright light attacks one or more enzymes that participate in the chemical reactions and also damage part of the photochemical mechanism. The respiration system appears to be unaffected. Kok (204) also treated *Chlorella* with very high light intensity up to about 100–1000 times light saturation. Results similar to those described above were obtained; they involved injury to the photosynthetic apparatus and slow restoration in the dark.

C. LIGHT QUALITY

The effect of light quality on photosynthesis and growth of plants has been studied since the time of Senebier, who constructed double-walled bell jars or domes and placed colored liquids in the annular space. Plants grown under these jars received more or less limited portions of the spectrum. Most of the data obtained from these studies, and also from experiments using colored glass filters, have not been satisfactory because of the difficulty of obtaining sufficient light of narrow spectral range and measured quantum energy content for controlled comparisons of photosynthesis. The subject is further complicated by the fact that different percentages of the various incident wavelengths are reflected, transmitted, and absorbed by the leaves of different plants, depending on leaf morphology and pigment content. The relative effectiveness of the various colors in photosynthesis is also in part an indirect expression of the opening of the stomatal apertures. According to Paetz (269), this effectiveness is red > blue > green in causing the stomata to open whereas according to Sierp (310) and Harms (165) green = blue > red. The latter did not exclude the far red and infrared, which Paetz eliminated. Scarth and Shaw (297) found that the relative intensities of white, red, and green light required to cause equal photosynthesis at the compensation point of pelargonium leaves were 100, 88, and 173, respectively. The stomatal apertures were identical in the three light sources. Arthur and Stewart (15) found that equal intensities of illumination from mercury, Mazda, neon, and sodium lamps gave the following relative yields of buckwheat, respectively: 0.62, 1.00, 1.20, and 1.41.

Evidently the relationships between wavelength and photosynthesis are complex. For a fundamental study of the problem, assimilation per absorbed quantum is required, rather than assimilation per incident lumen or quantum. Environmental conditions also need to be controlled.

In perhaps the simplest case, Moore and Duggar (246) determined the quantum yields of photosynthesis of *Chlorella* in monochromatic red, green, blue, and violet light. Light intensities up to 800 lux were used. Oxygen production was measured using the dropping mercury electrode. The quantum yields were practically the same in the different-colored lights, when the correction for dark respiration was added.

The same was true if this correction was obviated by measuring oxygen production in light of one color as a reference then turning on additional light of another color. It is of great interest that a quantum of visible light absorbed in these green plants apparently causes the same amount of photosynthesis regardless of its wavelength.

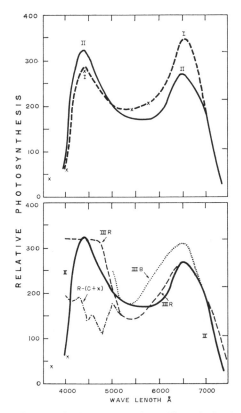

Fig. 17. Relative photosynthesis for equal incident light intensities for wheat. Curve *I* [Hoover (179)]. Curve *II*, quantized. Curve *III* [Burns (71)] incident light absorbed by white pine trees grown in red (*R*) and blue (*B*) light. Curve $R - (C + X)$ shows absorption without the carotin *C* and xanthophyll *X* [Burns (71)].

Hoover (179) measured the photosynthesis of a wheat plant in colored light on an equal incident intensity basis. A very sensitive infrared method was used to measure the photosynthesis by carbon dioxide exchange. Curve *I* in Fig. 17 shows the action spectrum obtained with two types of narrow-band Christiansen filters. Curve *II* gives the same data expressed as quanta of incident light. The data are corrected for stray light. The crosses represent light from isolated lines of the mercury arc. Evidently a high degree of experimental precision was attained in this work. The upper cut-off occurs at about 7400 A, but there is still some photosynthesis at 3650 A. The curve has maxima at 6600 A and 4400 A and a minimum in the yellow green, indicating that the incident red and blue lights are considerably more effective than the yellow or green in photosynthesis. The red light is more effective than the blue for photosynthesis, based on equal incident energy, but less effective based on the equivalent number of quanta.

When the relative amounts of absorption, reflection, and transmission by the leaves of the different wavelengths are considered, a more

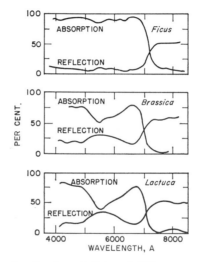

FIG. 18. Absorption and reflection of light of different wavelengths by leaves of *Ficus, Brassica,* and *Lactuca.* From Rabideau *et al.* (278).

complete picture emerges. Rabideaux, French, and Holt (278) have shown (Fig. 18) that, in fleshy leaves like fig (*Ficus* sp.) and pineapple (*Ananas comosus*), absorption of incident light below 7000 A is 80–90% and reflection is 10–20%. Transmission is slight. Above 7000 A the curves cross, and reflection and transmission exceed absorption. Thin leaves show a more or less pronounced minimum of absorption at about 5400 A and a corresponding maximum of reflection. Transmission is generally greater than in thick leaves also. The absorption curves of *Brassica* and *Lactuca* have general similarity to the action spectrum of Hoover for photosynthesis in wheat, based on incident light.

Burns (70, 71) measured the photosynthesis of 3–4-year-old white-pine (*Pinus strobus*) seedlings illuminated under controlled conditions with red, orange, yellow, and green light. He also measured the amount of light reflected from the needles at 45 degrees, and the transmission by an 80% acetone solution of the leaf pigments at the same pigment concentration as in the leaf. A "primary" absorption spectrum was estimated as the incident radiation minus the percentage of the light reflected by the needles and transmitted by the solution. Two groups of trees were used that had been grown in either red or blue light for a month before the experimental treatment began. There was a marked difference in the color of the two sets of needles. The percentages of the incident radiation absorbed by these two groups of trees, designated R and B and referred to as "primary absorption spectra" are superimposed on Hoover's curve in Fig. 17. The curve for the red-grown trees is continued into the blue, where total absorption minus absorption by carotene and xanthophyll is also given. The general shape of all the curves is similar above 5000 A.

It was found that the ratio of photosynthesis for any two spectral regions, except blue, agreed within a few per cent with the ratio of the quanta of energy absorbed in those regions on the basis of the "primary" absorption spectrum. It was concluded that the "primary" absorption curves approximate the actual absorption of the leaf and that the quantum efficiency of absorbed energy in photosynthesis is independent of the wavelength above 5000 A. The data suggest that not all the light absorbed by the yellow pigments in the blue and violet is effective in photosynthesis, so that the relationship between assimilation and energy absorbed does not hold closely in this region. It is known that only a small part of the blue light absorbed by most of the carotenoids, except fucoxanthol in the brown algae, is effective in photosynthesis.

Another quantized action spectrum of carbon dioxide assimilation by wheat leaves has been supplied by Stoy (343). This curve is shown in Fig. 19 together with Hoover's curve from Fig. 17, and the absorption spectrum of a living wheat leaf. The action spectrum of nitrate reduction is also given. This will be discussed later. The spectra of Hoover and Stoy differ somewhat in detail, particularly in the shape and location of the minimum, but they show general similarity. Presumably different varieties of a species can vary considerably in their pigment composition though the predominant chlorophyll spectrum is unmistakable.

Wassink and Stolwijk (373–375) have constructed narrow-band light chambers ($110 \times 30 \times 70$ cm) in which colored lights of fairly high intensity (about 3 klux) and satisfactory spectral purity were obtained from special fluorescent lamps of limited spectral output together with glass filters of large area. The different chambers included violet

4000 A; blue, 4600 A; green, 5400 A; yellow, 5900 A; red, 6600 A; also white and infrared plus red. All the chambers were adjusted to the same incident energy intensity of about 0.02 cal/cm²/min.

These chambers have been used principally to study the effect of narrow spectral regions on the exothermic reactions involved in growth and development of plants, including photostimulation effects, such as elongation of stems and leaves, initiation of germination and flowering, photoperiodic, and photomorphogenic reactions. Photosynthesis has been studied to a limited extent by growing the plants for prolonged periods in the different wavelengths and determining the

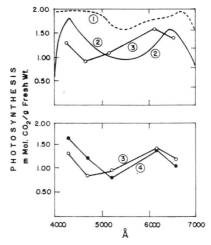

Fig. 19. Effect of light quality on photosynthesis and nitrate reduction in wheat. Curve *1* [Stoy (343)], absorption spectrum of living leaf. Curve *2* [Hoover (179)], quantized spectrum of photosynthesis. Curves *3* and *4* [Stoy (343)], quantized spectra of photosynthesis and nitrate reduction in the same leaf.

dry weight increments. Table XVII gives data for cosmos (*Cosmos bipinnatus*) (340, 373). The number of nodes was only slightly affected by color of the light, but their lengths varied considerably. Compared with their length in white light, the nodes were somewhat shorter in blue and longer in green, but in yellow and red they were twice as long. Yield was greatest in red. On a quantum basis, the yield in the green was 93% of the red, but in the blue and yellow only 40 and 58% of the red, respectively. In white light the yield was about 80% of the red. These results are not in close accord with those of Hoover and Burns. However, no data are available on light absorption in cosmos.

Stolwijk (341) observed the growth of tomato plants over a 31-day period. The intensity of the blue, green, yellow, red, or white light

TABLE XVII

Yield and Elongation of Cosmos Grown 33 Days in Light of Limited Spectral Band Width[a,b]

| | | | Internodes | | Yield | |
Color	Wavelength	Quanta: ratio to red	Number	Length (mm)	Weight (mg)	Quanta: ratio to red
Blue	3800–5000	0.65	7.25	30	259	0.40
Green	4900–5900	0.81	7.00	52	745	0.93
Yellow	5890	0.89	7.50	80	518	0.58
Red	6200–6900	1.00	7.75	77	1114	1.00
White	—	0.9	7.00	40	710	0.80

[a] From Wassink et al. (373).
[b] Conditions: Light: 16 hours/day at 14 kergs/cm²/sec. Temperature: 20°C, light; 15°C, dark.

was 35 kergs/sec on a sphere of 1 cm² cross section (about 2 klux on a flat surface) for 16 hours daily. The day and night temperatures were 20° and 17°C, respectively. Figure 20 gives the growth data. Elongation of stems and leaves and number of leaves per plant were increased

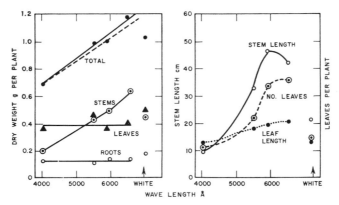

Fig. 20. Dry weight per plant of leaves, stems, and roots of tomato, also stem and leaf length as a function of the wavelength of the incident light. From Stolwijk (341).

two- to fourfold by the green, yellow, and red light as compared with the blue and white. Dry weights of leaves and roots were independent of the wavelength, but there was a linear increase in the weights of the stems such that the rate of increment of total plant substance was the same as the rate of wavelength increase. Evidently the quantum efficiencies of the incident light of the four colors for apparent assimilation were identical. However, data for absorbed light are lacking.

D. EFFECT OF TEMPERATURE

1. Temperature Coefficient of Photosynthesis

The effect of temperature on photosynthesis (32) varies with light intensity and the carbon dioxide supply. Both these factors control the photochemical reactions, which are generally independent of temperature. With low light intensity and adequate carbon dioxide or with

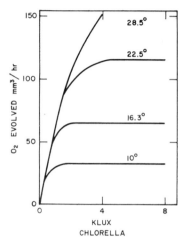

FIG. 21. Assimilation-light curves of *Chlorella* at different temperatures. From Wassink and Stolwijk (372).

low carbon dioxide and adequate light, the light reactions will be limiting and the temperature coefficients will approximate unity. Conversely, with high light and high carbon dioxide, the enzymatic reactions will predominate in the system and the large temperature coefficients characteristic of chemical reactions will be observed.

In Fig. 21, the assimilation-light curves for *Chlorella* at different temperatures show that the rate of assimilation is doubled by 6° rise of temperature ($Q_{10} = 2.7$). The CO_2 concentration was 7.8×10^{-5} moles/liter, equivalent to 2340 ppm, and the light intensity was saturating at all temperatures except the highest, 28.5°C. There are many other studies in the literature that indicate large temperature coefficient values for *Chlorella*: e.g., 2.07, Noddack and Kopp (260); 1.9–2.7, Emerson (116). An extreme value of 25 was also observed by Emerson in the temperature range 1° to 6°. For *Gigartina* (120), Q_{10} was 1.7 from 12° to 16° and 13.5 for 4° to 8°. In all these experiments a sufficient excess of carbon dioxide was employed to saturate the light reactions. The temperature-dependent enzymatic reactions were therefore in control. This was most striking when the system was so cold that some of the reactions were approaching their temperature threshold.

Experiments with higher plants growing in the normal atmosphere have indicated much smaller temperature coefficients for photosynthesis than the foregoing. Q_{10} values for McAlister's (224) wheat plant ranged from 1.0 to 1.9 with various light intensities, and carbon dioxide concentrations. Decker (94) found Q_{10} values of assimilation of 1.06 and 1.14 for red (*Pinus resinosa*) and loblolly (*P. taeda*) pines between 20° and 30°C in sunlight and normal carbon dioxide. Respiration doubled ($Q_{10} = 2.0$) in this temperature interval. Between 30° and 40°C, the coefficients were 0.72 and 0.66 for assimilation and about 1.5 for respiration. Evidently 40° is above the optimum for these trees. Moss *et al.* (247) found $Q_{10} = 1.2$ for field corn.

Table XVIII by Thomas and Hill (350) gives temperature coefficients for alfalfa growing in the field, obtained by spraying water over a small greenhouse to cool it while an unsprayed greenhouse served as control. Leaf temperatures as indicated by the thermocouples were reduced by the spray to about the same extent as the air temperature, but the light intensity was not measurably changed. Carbon dioxide exchange was recorded automatically. The Q_{10} values for photosynthesis of the alfalfa were about unity indicating a zero temperature effect under these conditions.

2. Responses at Very Low Carbon Dioxide Concentrations

Gabrielsen (136) treated leaves of *Sambucus nigra* in a glass chamber with a slow stream of carbon dioxide-free air in the light (10 klux) or in the dark. The carbon dioxide evolved was determined. Thick sun leaves with high respiratory rates (0.5–1.2 mg $CO_2/dm^2/hr$) reassimilated part of this carbon dioxide at 10 klux and reduced the concentration in the gas phase by 50–80%. Thin shade leaves evolving 0.2–0.3 mg $CO_2/dm^2/hr$ or less in the dark, showed no reassimilation in the light. Reassimilation in sun leaves was inhibited by reducing the respiration rate by lowering the temperature. Conversely reassimilation in shade leaves was increased by raising the temperature and respiration rate. Increasing the supply of CO_2-free air also stopped reassimilation by reducing the carbon dioxide concentration in the gas phase. Evidently a threshold concentration is required before reassimilation can occur. This appears to be about 60–90 ppm carbon dioxide. The author suggests that none of the intermediate products of respiration are reutilized for photosynthesis.

E. S. Miller and Burr (240) found that ten widely different types of plants produced the same limiting concentration of carbon dioxide when they were illuminated for many hours in a closed chamber. At 22 klux (200 fc) this concentration was about 100 ppm. In all plants this level was quickly reached and maintained at 4–5°. At 35–37° some plants behaved similarly; with others the level fell rapidly to 100 ppm, then rose to 300–600 ppm in about 24 hours. With *Zebrina pendula* the level

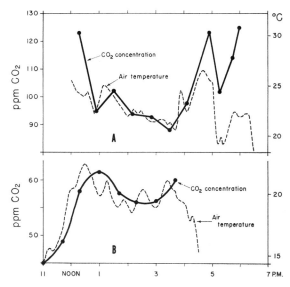

FIG. 22. Equilibrium concentrations of carbon dioxide due to photosynthesis and respiration at different temperatures in full sunlight in a closed chamber. Chart A: alfalfa, September 30. Chart B: sugar beet, November 4. From Thomas *et al.* (349).

fell initially to only 200 ppm. *Kalanchoe* (*Bryophyllum*) *pinnatum* and another of the Crassulaceae emitted a considerable amount of carbon dioxide from their tissues for several hours at the start of the experiment at 35°–37°. Subsequently the level fell and remained in the 300–600 ppm range for many hours. Evidently photosynthesis falls to a low rate in some plants at 35°–37°. Possibly the photosynthetic apparatus is injured by the high temperature. The maintainence of 100 ppm by other plants at both 5° and 35° suggests that the temperature coefficients of photosynthesis and respiration were equal in these plants.

This result is not supported by the work of Thomas *et al.* (349), who measured the apparent equilibrium between photosynthesis and respiration in an unrenewed atmosphere in sunlight at different temperatures. Carbon dioxide concentrations and temperatures were recorded continuously. The temperature was changed by water sprays on the outside of the cabinet. Similar experiments were carried out with alfalfa and sugar beets. Figure 22 indicates that with both plants the carbon dioxide concentrations and temperatures followed each other closely from about noon to 4 p.m. September 30 was a cloudless day. November 4 was partly cloudy but probably had adequate light. The pyrheliometer record fluctuated between 0.4 and 0.7 cal/cm²/min. Good agreement was found between the minimum carbon dioxide concentration (40 ppm) observed in the sugar beet experiment and that calculated from the equilibrium between the known rates of photosynthesis and dark respiration at 15° (30 ppm). The divergence in the alfalfa data was greater (45 ppm calculated, 88 ppm found at 21°),

TABLE XVIII

Temperature Coefficient of Photosynthesis of Alfalfa[a]

Month	Average temperature			Photosynthesis	
	Plot D_4 (°C)	Plot D_2 (°C)	Difference (°C)	Ratio $D_4:D_2$	Q_{10}
August	28.0	21.5	6.5	0.97 ± 0.023[b]	0.95
August	29.0	26.8	2.2	1.02 ± 0.045	1.10
September	23.0	16.4	6.6	0.92 ± 0.035	0.89
September	27.0	21.6	5.4	0.94 ± 0.033	0.90
September	23.2	20.0	3.2	1.00 ± 0.061	1.00

[a] From Thomas and Hill (350).
[b] Standard deviation.

but the equilibrium was somewhat questionable. The Q_{10} values ranged from about 1.4 to 1.6. It is reasonable to ascribe these temperature effects entirely to respiration, which usually has a Q_{10} of about 2.0. In that case the Q_{10} for assimilation would be unity or less, confirming the alfalfa data in Table XVIII.

3. Temperature as an Ecological Factor

The extensive "controlled environment" studies first carried out in Pasadena by Went (384, 385) have clarified the role of temperature as an ecological factor. Different plants have optimum day temperatures for assimilation and optimum night temperatures for translocation, stem elongation, and leaf expansion. The latter effects are the result of dark metabolic reactions which continue for as long as about 24 hours after photosynthesis. Growth and yield represent the total of all these reactions. Maximum yields are obtained under the conditions for which each plant is best adapted. According to the investigations of Hiesey and co-workers (174, 175), this is particularly true of the grasses, the different species of which have become adapted to a narrow range of temperature conditions due to altitude, latitude, and precipitation. It is also true of other widely distributed plants such as mimulus (*Mimulus cardinalis*) (243). In Table XIX two races of this plant, one from a coastal habitat, the other from a location in the Sierra Nevada mountain range are compared with respect to the light intensities required to saturate photosynthesis at different temperatures. Photosynthesis and dark respiration are also given. Between 10° and 40°C,

TABLE XIX

PHOTOSYNTHESIS, RESPIRATION, AND LIGHT INTENSITY REQUIRED TO SATURATE PHOTOSYNTHESIS OF TWO RACES OF *Mimulus* AT DIFFERENT TEMPERATURES[a]

Temperature (°C)	Light for saturation		Photosynthesis at light saturation[d]		Dark respiration[d]	
	Race L.T.[b] (fc)	Race Y[c] (fc)	Race L.T.[b]	Race Y[c]	Race L.T.[b]	Race Y[c]
10	1000	1550	14	15	0.9	1.1
20	2170	2750	22	25	2.3	3.3
30	2750	4730	27	32	4.0	5.1
35	3500	5800	—	—	—	—
40	3250	5800	22	24	5.3	6.3

[a] From Milner *et al.* (243).
[b] Race L.T. = Los Trancos coastal elevation.
[c] Race Y = Yosemite Mid-Sierran elevation.
[d] Values are stated as milligrams carbon dioxide per dm^2 per hour.

the coastal race could utilize only about two-thirds to one-half the light used by the mid-Sierran race. Furthermore, the mountain race was more responsive to increased temperatures than the coastal race. It also gave somewhat higher levels of photosynthesis and respiration, but not in proportion to the greater light saturation levels; see also (243a).

Went (385) found that the optimum day temperature for tomato was 26.5°C. If the night temperature was also 26.5, the plants grew well but fruited poorly. With 26.5° in the day and 17–20° at night, fruiting was most abundant and the stems were elongated 23–27 mm per day. If this extra growth represents extra dry weight, it cannot be entirely accounted for by reduced respiration at the lower night temperature. Some increase in photosynthesis probably occurred also.

Gist and Mott (144) have studied the growth of several legumes between 60° and 90°F. Figure 23 shows the dry weights of shoots and roots of 45-day-old plants grown in controlled climate chambers with constant temperature and light intensity. Alfalfa and red clover (*Trifolium pratense*) grew best at 60°F; the trefoil (*Lotus corniculatus*) at 70°F. Assimilation was materially reduced at 90°F. Light saturation

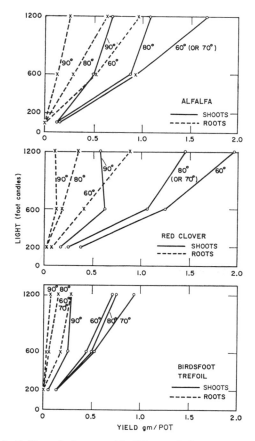

FIG. 23. Yield of alfalfa and clovers with different light intensities and temperatures. From Gist and Mott (144).

was reached at 600 fc at 90°F with the clovers, but at lower temperatures light saturation was greater than 1200 fc. Alfalfa was more light tolerant at 90°F than the clovers. Evidently, each plant had its own optimum conditions for photosynthesis.

4. Temperature and Carbohydrate Production

J. H. C. Smith (317, 318) carried out experiments that showed the effect of temperature on carbohydrate production in sunflower leaves.

TABLE XX

CARBOHYDRATES IN THE PHOTOSYNTHATE OF SUNFLOWER LEAVES AS AFFECTED BY TEMPERATURE DURING PHOTOSYNTHESIS[a]

Temp. (°C)	Duration of photosynthesis (min)	Duration of dark respiration (min)	Carbon absorbed (mg)	Percentage assimilated carbon found as carbohydrates						
				Glucose[b]	Sucrose	Other sugars	Starch	Soluble	Insoluble	Total
10	152	—	8.1	7	71	5	16	99	8	107 ± 5
20	58	8	7.8	10	52	4	26	92	6	98 ± 3
20	71	220	7.8	35	45	3	18	101	0	101 ± 3

[a] From J. H. C. Smith (317, 318).
[b] Monosaccharides.

The latter were first cut in two at the midrib, then the weight and area of each part was determined. One half was dried and analyzed as a control; the other half was allowed to perform a measured amount of photosynthesis and respiration at different temperatures before being dried and analyzed. The data are summarized in Table XX. The assimilated carbon was all accounted for in the various soluble and insoluble carbohydrate fractions. The relative amounts of the different compounds varied considerably with the temperature of assimilation and the length of the post-illumination dark period. The photoperiod at 10° had to be about 2.5 times as long as at 20° to produce the same amount of photosynthate. The 10° product had much more sucrose and correspondingly less starch and glucose (monosaccharides) than the 20° product. Similarly when the photoperiod at 20° was followed by 3.6 hours of dark respiration, the glucose in the latter treatment increased about 25% at the expense of the sucrose, starch, and insoluble carbohydrate.

E. Effect of Hydrogen Ion Concentration

The pH of the soil can vary over a wide range without affecting appreciably the growth of higher plants. Indirect effects may include an increase or a decrease in the supply of essential nutrients such as iron and phosphorus or bringing toxic elements like aluminum into solution. Water plants are affected directly through the carbon dioxide supply which is made available to them as dissolved carbon dioxide and carbonic acid molecules or as bicarbonate and carbonate ions. Some plants can utilize only the undissociated molecules, others can utilize the ions as well.

The concentrations of these molecular and ionic species in the nutrient solutions depend on several factors such as pH, cation concentrations in excess of those associated with neutral salts, and carbon dioxide in the gas phase. Equations can be written for all these relationships. Below pH 6.0 the bicarbonate concentration is negligible. The available carbon dioxide is determined primarily by the gas phase concentration and mechanical mixing. Above pH 6.0, equilibrium reactions involving CO_2, H_2CO_3, HCO_3^-, CO_3^{--}, and OH^- are established at each pH depending on the cation and gas phase CO_2 concentrations.

Emerson and Green (119), Whittingham (387), and others have shown that *Chlorella pyrenoidosa* can utilize only dissolved carbon dioxide for photosynthesis. Whittingham observed that when the alga was grown in a closed system with an infrared carbon dioxide analyzer to monitor the gas phase, the maximum rate of photosynthesis occurred at the same dissolved carbon dioxide concentration (about $5 \times 10^{-6} M$)

whether the medium was at pH 4.6, 6.4, or 8.2. The same result was obtained by Emerson and Green even though the [HCO_3^-] ranged from $5 \times 10^{-6} M$ at pH 4.6 to 92000×10^{-6} at pH 8.9. Evidently, the assimilation is independent of the bicarbonate ion concentration which varies with the pH.

Scenedesmus quadricauda behaves differently. Osterlind (264) aspirated the illuminated cells with air containing 10–1100 ppm carbon dioxide for 5 hours. Growth was slight at pH 5.5–6 or less. At higher pH values, assimilation began to increase. The richer carbon dioxide-air mixtures showed the greatest initial response. However, all the curves later approached the same maximum at a pH of about 9.0, a result suggesting that bicarbonate ion controls assimilation in this plant. Growth rate is directly proportional to the concentration of HCO_3^- below 10^{-5} moles per liter.

Steemann-Nielsen (329) observed that *Stigeoclonium* did not grow at pH 3.5–4.5. Assimilation started at pH 5.5 but was still insignificant at pH 6.5. There was vigorous growth at pH 7.5–8.0. In *Cladophora insignis* also, assimilation was vigorous at pH 7.4 but ceased after a short exposure at pH 5.5. Both these algae assimilate HCO_3^-.

Among the higher flowering water plants there is a similar division into those that can utilize bicarbonate and those that cannot. Examples of bicarbonate users include *Elodea (Anacharis) canadensis*, *Ceratophyllum demersum*, and *Myriophyllum spicatum*. Assimilation is unchanged from pH 4.5 to 8.2 so long as either free carbon dioxide or bicarbonate is in excess. These plants occur only in alkaline waters. Presumably, acid waters in general do not have enough carbon dioxide to support them. The water moss *Fontinalis dalecarlica* has an optimum range in nature from pH 4.2 to 6.6. On the acid side, assimilation is inhibited by exposure to pH 3 for more than a few hours, but the plant is not otherwise injured. An exposure of 22 hours at pH 2.1 caused bleaching of the leaves. On the alkaline side a 40-minute exposure to pH 10.5 caused a temporary reduction of assimilation at pH 4.5. Evidently, the high pH is harmful. This plant cannot utilize HCO_3^-.

Fontinalis antipyretica has a natural range from pH 5.5 to 8.2. It is found in alkaline waters whereas *F. dalecarlica* is restricted to acid waters. Steemann-Nielsen (328) carried out experiments with leaves of the water plant *Potamogeton lucens* clamped between the ground end of two glass cylinders. Lake water containing $2.7 \times 10^{-3} N$ HCO_3^- was placed on the morphologically lower side of the leaf and distilled water on the upper side. Glass plates closed the containers. When the cylinder assembly was rotated in a horizontal position to provide stirring and the leaf was illuminated on its upper surface with 24 klux, the following concentration changes in the distilled water occurred in 80 minutes (equivalents/liter $\times$ 10^{-5}): OH^- = 0.05 to 25; CO_3^{--} = 0 to 37; HCO_3^- = 3 to 10. The pH increased from 7.9 to 10.6, whereas the pH of the lake water on the lower surface decreased from 8.20 to 8.12. Absorption of HCO_3^- plus equivalent cations on the lower surface was 1.04 $\times$ 10^{-5} equivalent, from which the CO_2 from 0.65 $\times$ 10^{-5} equivalent was assimilated by the leaf, resulting in the excretion of the anions listed above, plus equivalent cations. The assimilation of CO_2 in this experiment was calculated to be 0.032 mg/cm^2/hour. A confirming value of 0.029 mg/cm^2/hour was found by measuring oxygen evolution.

High alkalinity in the surface water of a Danish lake has been described (329). A pH of 10 was reached in the daytime in summer, due to absorption of bicarbonate and release of hydroxyl by some plants. The algae in the surface layer were killed during the day, but a fresh supply was brought up from below at night by vertical mixing. The algal population of the lake is said to be definitely limited by this pH effect.

F. AGE

The photosynthetic potential of a leaf changes with time. Singh and Lal (312) found that the newly formed leaves of sugar cane (*Saccharum officinarum*), wheat, and flax (*Linum usitatissimum*) assimilated carbon dioxide very slowly but the rate increased to a maximum as the leaves expanded and matured, and declined later as senescence progressed. Reference has already been made to the observations of C. C. Wilson (394) on privet and camellia in which the curves for photosynthesis and stomatal apertures are quite similar with first year leaves, but entirely different with older leaves owing to the erratic photosynthetic response of the latter.

Evidently the chloroplasts themselves are responsible for these aging effects. If they are isolated and observed in the Hill reaction, as Clendenning and Gorham (82) have done with spinach (*Spinacia oleracea*) and wheat, they show the same rise and fall of activity as the plants from which they are isolated. Suspending chloroplasts from young leaves in the cell sap of old leaves and vice versa does not change the response of the chloroplasts. It is common knowledge that algae show a similar pattern of aging; for this reason, cells of a definite age are used in most experimental work.

Stålfelt, quoted by Freeland (135), determined the photosynthetic rates of spruce (*Picea*) and pine (*Pinus*) needles of different ages up to five years and more. Expressed on the fresh or dry weight of the needles, the activity declined somewhat from year to year. The rate, related however, that the needles increased in weight from year to year. Expressed on unit number of needles, the activity in spruce increased for five years and in pine for three years before declining. The increase was greatest with 30% of full sunlight, but in full sunlight the rate was about the same for needles of all ages.

Freeland (135) repeated these experiments at Northwestern University, near Chicago, using four species of pine and one each of spruce and fir (*Abies*). Photosynthesis measurements were carried with strict control of temperature (24°) and light (2200 fc). It was also considered best to express the results on unit number of needles. The data are summarized in Table XXI. All the plants declined in activity over

TABLE XXI

APPARENT PHOTOSYNTHESIS, NEAR CHICAGO, ILLINOIS, OF PINE, SPRUCE, AND FIR NEEDLES OF DIFFERENT AGES[a]

Plant	Number of trials	Photosynthesis in needles (mg CO_2/100 needles/hour)		
		1 Year old	2 Years old	3 Years old
Austrian pine	4	2.1	1.4	0 9
Western yellow pine	2	1.8	0.6	0.2
Scotch pine	4	1.4	1.1	1.0
White pine	5	1.1	0.8	0.7
White fir	4	0.6	0.4	0.2
Colorado spruce	2	0.4	0.3	0.2

[a] From Freeland (135).

a three-year period. Western yellow pine (*Pinus ponderosa*) decreased ninefold and white fir (*Abies concolor*) threefold. The others declined 50–100%.

The reason for the disagreement between Stälfelt and Freeland is not apparent. It was noted that Freeland's older needles were very dirty. However, their activity was changed only slightly by cleaning them. Possibly long exposure to the air pollution of a Chicago suburb reduced the activity of the chloroplasts, independent of accumulated dirt on the surface. In any event it is important to know that most of the three-year-old needles still retained a considerable part of their photosynthetic activity.

Nixon and Wedding (259) divided attached date (*Phoenix dactylifera*) leaves longitudinally into two halves at 6 A.M. One half was removed and dried immediately; the other half was allowed to photosynthesize until 4:30 P.M., when it also was detached and dried. Rate of photosynthesis was deduced both from the increase in weight and from the assimilation of leaf disks in the Warburg apparatus. One-year-old leaves had maximum activity. There was a steady decrease with age until at 4 years the leaves had about half the activity of one-year-old leaves.

G. DROUGHT

Lack of adequate water reduces the rate of photosynthesis. A very slight reduction may be noted in Fig. 34, on Sept. 24, where a small moisture stress on part of a plot was reflected in a lowering of the assimilation curve. P. C. Owen's (268) work at Rothamsted suggests that water stress somewhat above the wilting point did not interfere with the net assimilation rate (N.A.R.) in sugar beet, though the leaf area of the plant was curtailed by prolonged growth in the drier soils.

Many studies have been made of the relationships between soil moisture and rate of photosynthesis. In general it has been observed that if the soil is allowed to dry, there is very little effect on photosynthesis until the wilting percentage is approached. Schneider and Childers (300) noted that shortly before wilting became noticeable in apple leaves, the rate of photosynthesis was reduced 55%. When definite signs of wilting were seen, the reduction was 87%. Transpiration was also reduced, and respiration was increased. With the application of water, partial recovery was rapid at first, but complete recovery required 2–7 days.

Ashton (16) studied large potted sugar cane plants with shoots enclosed in plastic cabinets $72 \times 72 \times 4$ inches. He allowed the soil of one plant to dry out until the wilting point (25% water) was reached. Then the soil was raised to its saturation capacity (60%) and the drying process was repeated. Five cycles covering a period of about 65 days, or 13 days per cycle, were carried out. The control plant was well watered at all times. Photosynthesis was determined for 13 minutes of each half hour throughout the day, as described later. Figure 24 summarizes the results.

In each cycle photosynthesis remained above 75% of the normal level except during the 2 days before irrigation, when the wilting percentage was being closely approached, and on the day of the irrigation, when recovery was only partially attained. In this period the rate of photosynthesis fell from 90–100% of the control to about 25%, then rose to about 75%. It continued to rise uniformly to 100% during the following 5–10 days. The diurnal pattern of photosynthesis near the wilting point was characteristic, involving (a) a fairly high rate early in the morning, suggesting recovery during the preceding night; (b)

a minimum in midmorning, ranging from 9 A.M. in the first cycle to
11 A.M. in the last; (c) a maximum at about 1 P.M. followed by a
decline to nearly zero at 4 P.M. The midday maxima were due to
shading of the lower leaves by the upper leaves. Reduced light intensity
increased the relative photosynthesis near the wilting point. Presumably,
the reduced light intensity lowered the moisture stress in the leaves.

Allmendinger *et al.* (3) found similar drought effects with young
apple trees. The assimilation rate was reduced about 20% by allowing
the soil to dry just to the wilting point before the next irrigation. The
rate was unaffected if no more than 80% of the available soil moisture
was used. In this case however, growth of the leaves and of the trees
was somewhat retarded. Nearly identical relationships to these were

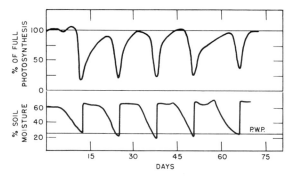

Fig. 24. Rate of photosynthesis of sugar cane as affected by soil moisture during
five drying cycles (P.W.P., permanent wilting percentage). From Ashton (16).

observed by Loustalot (219) with pecan (*Carya illinoinensis*) leaves
and by Upchurch *et al.* (357) with Ladino clover (*Trifolium repens*).
The latter did not find any reduction in photosynthesis until 90% of
the "available" moisture in the soil (between the "moisture equivalent"
and the permanent wilting percentage) was abstracted. When the
plants were definitely wilted, photosynthesis was reduced by about
75%. With irrigation, 50% recovery occurred in 6 hours. The drought
responses in these experiments were somewhat smaller than in the
Schneider and Childers study, though essentially similar.

Dastur and Desai (89, 90) measured photosynthesis and respiration of
single attached leaves of *Abutilon*, cineraria (*Senecio cruentus*),
Ricinus, Helianthus, etc. Normal or carbon dioxide-enriched air from
a large aspirator bottle was passed over the leaf, mounted in a thermostated glass envelope which in turn, along with the potted plant, was
enclosed in an independently thermostated chamber. The leaf was

illuminated by water-filtered light from a 1500-watt Osram lamp at a 32-inch distance. The air was humidified before reaching the leaf, then was dried well before absorption of the carbon dioxide in soda lime for gravimetric determination. One liter of air was passed over the leaf in 2 hours for each exchange measurement. Another analysis was made of the air supply. After several duplicate experiments in light and darkness, the leaf was removed; its area and fresh weight were quickly found, then it was dried for moisture determination, and also chlorophyll analysis. The data indicate that the moisture content of the leaf influenced markedly the rate of carbon assimilation. At constant temperature, photosynthesis of comparable leaves of each species was directly proportional to the moisture content of the tissue. Species with low normal moisture levels were more active than species with high normal levels. With variable temperature, an "assimilation number" (carbon dioxide absorbed/dm² ÷ water content/dm²) was found which increased steadily with increasing temperature to a maximum at 35–37°, then fell sharply. The "assimilation number" of Willstätter and Stoll, employing chlorophyll instead of water per unit area, indicated a much poorer correlation, if any, between assimilation and temperature. Evidently the water supply of some leaves was inadequate owing to excessive demand or insufficient function in the conducting elements. Stomatal openings were not considered, but it seems unlikely that carbon dioxide supply was a controlling factor since similar responses were obtained with air containing 300 ppm, or 5%, carbon dioxide.

It may be noted that only about 1% of the water entering the plant is used in the chemical reactions of photosynthesis. The remainder is needed as the medium in which most of the plant processes occur. Near the wilting point there is marked interference with these processes and photosnythesis is almost completely inhibited. Closure of the stomata is an important response of wilted leaves, but chemical changes in the leaf are very important also. For example, Gates and Bonner (140), have shown that accumulation of nitrogen and phosphorus is inhibited and that net increases of ribonucleic acid (RNA) are suppressed by water stress.

H. Oxygen and Isotopes

The role of oxygen in photosynthesis, if any, is obscure. It has been known for a long time that anaerobic conditions inhibit photosynthesis. This may be due to the formation of fermentation products in the leaves rather than to lack of oxygen per se. Wassink *et al.* (375) found no change in the photosynthesis of *Chlorella* in air as compared with

nitrogen, although Warburg had reported a 30% decrease in air. Mc-
Alister and Myers (224a) found a similar decrease with wheat. However, Wassink *et al.* found a 36–47% decrease with *Chlorella* in pure
oxygen as compared with air (Table XLVII).

Turner *et al.* (356) observed inhibition of photosynthesis by oxygen
in 25 species of plants. At high levels of light and carbon dioxide, inhibition was absent or slight with 0 to 20% oxygen except in wheat,
which showed a 23% reduction in air. With more than 20% oxygen,
inhibition became large and variable, reaching 20–50% in pure oxygen.
There was no change in the rate of the dark reactions.

Craig, Pratt, and Trelease (83, 274) found that the rate of photosynthesis of *Chlorella* in 98% heavy water (D_2O) was only 40% of the
rate in ordinary water owing to interference with the dark reactions.
Using flashing light of 4.5 milliseconds' duration, yield per flash was
maximum and identical in the two liquids when the minimum dark
period was 0.02 seconds in water and 0.07 seconds in D_2O. Evidently
heavy water has no effect on the light reaction but, like ordinary water,
enters into the dark reactions of photosynthesis, though at a reduced
rate.

When carbon-14 is used (331) for measurement of photosynthesis, a
discrimination correction of about 5% must be made to allow for the
fact that $C^{14}O_2$ is less readily assimilated than $C^{12}O_2$.

IV. Nutritional Factors in Photosynthesis

MINERAL NUTRITION

1. Significance

The level of photosynthesis depends on the general well-being of the
plant. Most of the factors involved in plant vigor are likely to be reflected in the rate of photosynthesis. However, the discussion of net
assimilation rates (N.A.R.) will show that assimilation per unit area
of leaf surface tends to be remarkably constant even on plants that
differ widely in size and vigor. Treatments, e.g., the supplying of
water or application of fertilizer, that give increased yields per acre
do not necessarily have much effect on the N.A.R. Rather, they may
change primarily the leaf area available for photosynthesis. Drought
effects have already been discussed. At or near the wilting point, photosynthesis is seriously inhibited, but the function is rapidly recovered
after the application of water, provided permanent injury to the leaves
has not occurred. Similarly many nutritional deficiencies reduce photosynthesis. Some of these are subject to rapid recovery when the deficient

nutrient is added; others cause changes in the leaves that can be corrected only slowly, if at all. Development of the new leaves may be required before complete recovery is possible.

Each of the essential nutrient elements is involved in some phase or many phases of the metabolic process and therefore influences photosynthesis either directly or indirectly. (The general problem of the role of essential nutrients is dealt with in Volume III.) Adverse effects are caused by deficiency, lack of balance, or excess of any of these elements. However, there is a fairly wide range of concentration within which each element can vary without changing the photosynthetic rate.

Nitrogen and magnesium, as constituent elements in chlorophyll, together with phosphorus, nitrogen, and sulfur as essential elements in the photosynthetic enzymes, are directly involved in the light reactions. The trace elements, iron and manganese, play a part in chlorophyll synthesis as well as in associated enzymatic light reactions. Potassium, magnesium, calcium, and the trace elements zinc, copper, manganese, boron, vanadium, molybdenum, chlorine, and perhaps cobalt, promote the dark reactions that dispose of the initial products of photosynthesis and maintain the supply of primary substrates for the light reactions.

2. Mineral Deficiencies and Excesses in Algae

Pirson (271) has reviewed the functional aspects of mineral nutrients in green plants with special reference to unicellular green algae. Table XXII lists the effects of slight-to-moderate deficiencies of the major and some of the trace elements on the carbon assimilation and respiration of algae. Photosynthesis is reduced by the deficiency of any one of the elements listed. If the deficiency is corrected, there is complete recovery within 1–2 hours in the case of "minus-manganese" and partial recovery in the same period from "minus-potassium" or "minus-phosphorus." The nitrogen response is less rapid. There is no short-term recovery of photosynthesis after correcting magnesium, iron, or sulfur deficiencies. The latter usually cause chlorosis which requires more time to overcome, owing presumably to the necessity of regenerating chlorophyll. Nitrate assimilation in algae is inhibited by deficiencies of potassium, phosphorus, manganese, molybdenum, and nitrogen, especially the latter. The effects are reversible.

Respiration may be either increased, decreased, or unchanged by these deficiencies. It is markedly increased during the deficiency period by lack of potassium, slightly increased by lack of phosphorus, magnesium, and iron; unchanged by lack of manganese and sulfur, and either increased or decreased slightly by lack of nitrogen. Normal respiration is rapidly restored when the missing elements are supplied, except phosphorus and nitrogen, which maintain an elevated respiration level during the recovery period.

TABLE XXII

Effect of Slight-to-Medium Mineral Deficiencies on the Metabolic Characteristics of Green Algae[a]

Deficient element	Photosynthesis at light saturation during		Dark respiration during		Nitrate reduction (assimilation)
	Deficiency	Early recovery[b] (1–2 hr)	Deficiency	Early recovery	
Mn	Inhibited	Completely restored	Normal	Unaltered	No specific effect
K	Inhibited	Partially restored	Markedly increased	Restored	Inhibited (slowly reversible)
P	Inhibited	Partially restored	Normal or slightly increased	Supernormal	Inhibited (slowly reversible)
N	Strongly inhibited	Sudden but slight increase after adding nitrate	Slightly increased or decreased	Far above normal for several hours	Very low, strong increase during recovery
Mg	Inhibited	Unaltered	Slightly increased	Unaltered	No specific effect
Fe	Inhibited	Unaltered	Normal or slightly increased	Unaltered	No specific effect
S	Inhibited	Unaltered	About normal	Unaltered	No specific effect

[a] From Pirson (271).
[b] Short-term recovery of photosynthesis is independent of chlorophyll formation.

Greenfield (153) studied the inhibitory effects of inorganic compounds on the photosynthesis of *Chlorella*. Standard cultures were employed containing 35×10^6 cells. Treatments consisted of suspending the cells for 20 minutes in solutions of many inorganic compounds, then rinsing centrifugally with water for 10 minutes, and measuring the photosynthesis in carbonate-bicarbonate buffers at different light intensities. Glucose, phenylurethane, and potassium cyanide were included in the study for comparison. Threshold concentrations were found below which there was no inhibition but above which inhibition increased rather rapidly. In general, inhibition at a given concentration was greater the higher the light intensity, a result suggesting interference with the dark reactions primarily. Table XXIII gives representative data at 0.7 and 22 klux and at concentrations that reduced

TABLE XXIII

RATES OF PHOTOSYNTHESIS OF *Chlorella* AT TWO LIGHT INTENSITIES WITH AND WITHOUT THE ADDITION OF SALTS, GLUCOSE, AND PHENYLURETHANE IN AMOUNT SUFFICIENT TO REDUCE PHOTOSYNTHESIS 25-50%[a]

	Photosynthesis (mm^3 O$_2$/35 × 10^6 cells/min) at					
	0.70 Klux			22 Klux		
Solute and conc, (M)	Control C	Treated T	T:C (%)	Control C	Treated T	T:C (%)
CuSO$_4$[b] 5 × 10^{-6}	0.43	0.34	79	1.26	0.68	54
HgCl$_2$ 1–2.5 × 10^{-5}	No effect to complete inhibition					
ZuSO$_4$[b] 0.2	0.40	0.30	75	1.34	0.64	48
CoSO$_4$[c] 0.2	0.50	0.26	52	1.65	0.88	54
KCl[b] 0.3	0.49	0.47	96	1.92	1.58	82
KI[b] 0.3	0.53	0.50	94	1.89	1.58	84
(NH$_4$)$_2$SO$_4$[c] 0.4	0.49	0.30	61	1.80	1.10	61
NiSO$_4$[b] 0.4	0.49	0.46	94	1.80	1.44	80
H$_3$BO$_3$[b] 0.5	0.47	0.37	78	1.25	0.66	53
MnSO$_4$ 1.0	—	—	—	1.32	0.91	69
MgSO$_4$ 1.0	—	—	—	1.80	1.45	81
KNO$_3$ 1.0	—	—	—	1.42	1.05	74
Glucose 0.7	0.34	0.28	83	1.44	0.80	56
Phenylurethane[d] 7 × 10^{-3}	0.56	0.37	66	1.90	1.50	80
KCN[b] 0.2	0.43	0.40	93	1.97	0.90	46

[a] From Greenfield (153).
[b] Inhibits the dark reactions primarily or solely.
[c] Inhibits the dark and light reactions about equally.
[d] Inhibits the light reactions.

photosynthesis about 25–50% with the stronger light. Great toxicity was observed with copper and mercury salts. Aluminum sulfate showed strong inhibition between 10^{-4} and 10^{-2} M, but the formation of precipitates rendered the results somewhat uncertain. Some of the salts, notably zinc sulfate, cobalt sulfate, and potassium cyanide caused more inhibition than could be explained by osmotic effects as indicated by equivalent glucose concentrations. Potassium chloride and iodide ammonium and nickel sulfates, and boric acid caused only limited additional inhibition whereas manganese and magnesium sulfates and potassium nitrate caused less inhibition than glucose. On the basis of the photosynthetic response of the treated cells to different light intensities as illustrated diagrammatically in Fig. 25, it may be concluded that zinc and nickel sulfates and potassium chloride retard only the dark reactions; copper sulfate, boric acid, and potassium iodide cause retardation of the dark reactions principally; and cobalt and ammonium sulfates inhibit the dark and light reactions about equally.

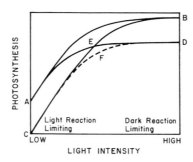

FIG. 25. Diagrammatic representation of inhibitory effects of inorganic compounds on photosynthesis in *Chlorella*. Curves: *AB*, normal curve; *AED*, dark reactions reduced; *CEB*, light reactions reduced; *CFD*, both light and dark reactions reduced. From Greenfield (153).

None of the compounds tested except phenylurethane retard the light reactions exclusively.

3. Nitrogen, Phosphorus, and Potassium in Higher Plants

These elements are required in larger amounts by plants. Nitrogen and phosphorus participate directly or indirectly in most of the reactions associated with photosynthesis. The former is a constituent of the various leaf pigments and also of all the active protein and peptide molecules that serve as enzymes in photosynthesis; the latter supplies the high energy bonds of adenosine triphosphate (ADP), diphosphopyridine nucleotide (DPN), phosphoglyceric acid (PGA), ribulose diphosphate (RDP), and the many other phosphorus compounds involved in the photosynthetic and Krebs cycles (cf. Volume IB, Chapter 4; and Volume IA, Chapter 3). The role of potassium is less apparent, being concerned primarily with the dark reactions of assimilation. The element obviously has a number of specific functions because it can be replaced to only a very limited extent by any of the other alkali metals, lithium, sodium, rubidium, or cesium. Lack of potassium (287) interferes seriously with nitrogen metabolism, causing an increase in soluble nitrogen compounds, particularly the more basic amides such as arginine, glutamine and putrescine due to reduced protein synthesis and/or proteolysis (cf. Volume III, Chapter 2). Lack of potassium also interferes seriously with assimilation and carbohydrate metabolism. S. V. Eaton (109) found that 42-day-old potassium-deficient sunflower plants had only 15% of the dry weight of the controls but the leaves contained 50% more total sugars while the stems contained less sugars. In the following discussion, several important researches involving deficiencies of the major elements are considered in detail with respect to the photosynthesis and respiration of specific plants.

a. *Barley* (*Hordeum vulgare*). Gregory and Richards (155), continuing earlier studies initiated by Briggs (55), measured the effects of moderate deficiencies of phosphorus, potassium, and nitrogen on assimilation and respiration rates of barley. The plants were grown in pots containing 30 pounds of sand. Disodium phosphate, sodium nitrate, potassium sulfate, calcium chloride, and magnesium sulfate were added at the rate of 2.52, 9.1, 1.85, 0.37, and 1.25 gm per pot, respectively, and trace elements were added to supply the complete nutrient. For deficient phosphorus and nitrogen, the indicated levels of those elements were reduced to one-fifth and for deficient potassium, to one-tenth. Assuming that half the pore space in the sand was filled with liquid, it is estimated that the full-strength solution contained about 300, 600, and 900 mg per liter (0.002, 0.006, and 0.06 M) of P_2O_5, K_2O, and N, respectively, and that the deficient solutions contained 60, 60, and 180 mg per liter.

The deficiency of potassium was somewhat more acute than those of phosphorus and nitrogen. All deficient plants exhibited typical specific symptoms of starvation that were very marked. The "minus-nitrogen" plants had fewer tillers and lighter green leaves than the controls; the "minus-phosphorus" plants also had fewer tillers, and the leaves showed characteristic reddening and rolling; the "minus-potassium" plants had thin, broad, yellow leaves, shortened stems, and sterile heads.

Starting in the third week of growth the most active leaf on a plant was removed; its respiration and assimilation rates were measured with the katharometer (369) (hot-wire carbon dioxide analyzer) at both high and low light intensities (5000 and 300 lux). The same measurements were made on all the fully manured and deficient plants each week until the leaves became too old to function normally.

Assimilation rates at 5000 lux, calculated for unit leaf area per hour are plotted for each week during the summer in Fig. 26. The rates fell off with age until about the sixth or seventh week then remained approximately constant until the leaves had passed maturity. The four curves lie fairly close together except when the potassium-deficient curve fell to a low level in the fourth and fifth weeks and the nitrogen-deficient curve fell similarly in the eighth week. The reason for these exceptions is not known. The phosphorus-deficient plants were a little more active (111%) than those in complete nutrient culture whereas the nitrogen- and potassium-deficient plants were a little less active (88 and 85%, respectively).

The rate of assimilation at 300 lux was about 10% of the rate at 5000 lux. There was no significant correlation with the age of leaves, and the

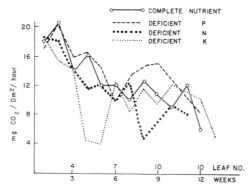

FIG. 26. Photosynthesis of barley grown in sand culture with complete nutrient solution and with solutions deficient in phosphorus, nitrogen, or potassium. Corresponding leaves were measured weekly. Light 5000 lux. From Gregory and Richards (155).

deficiency effects were not significant except in the "minus-potassium" leaves of the fourth and fifth weeks, which showed the same lowerings of the assimilation rates as at high light intensity.

Respiration rates per gram of dry weight of the leaves, recorded each week, are given in Fig. 27. Again the activity of the leaves fell off with age until the sixth or seventh week, then remained nearly constant or increased somewhat until maturity. The leaves from the complete nutrient and the phosphorus-deficient cultures had nearly identical respiration rates; those from the nitrate-deficient culture had about 83%, and those from potassium-deficient culture about 158%, of the respiration rates of the complete nutrient leaves. Results on a leaf area basis were about the same as on a leaf weight basis. The high respiration rate of the "minus-potassium" plants is striking. Evidently this was partly responsible for the lowered assimilation rate of these plants.

b. Bracken (Pteridium aquilinum). Schwabe (302) applied the methods of Gregory and Richards to determine the effects of phosphorus and potassium concentrations in the nutrient solution on the dry weight and on the net assimilation rate of bracken. Two levels of phosphorus and three levels of potassium were used. The highest levels in each case were considered adequate. The low phosphorus level was $\frac{1}{10}$ the high level, and the medium and low potassium levels were $\frac{1}{9}$ and $\frac{1}{81}$, respectively, of the highest level. These were mixed in three types of solution in which the cations were mainly sodium (*A*), calcium (*C*), or ammonium (*M*) to supply the other nutrients. The nitrate level was high in all solutions.

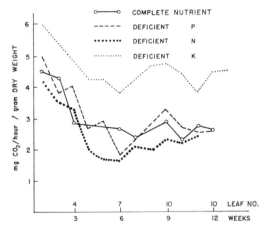

FIG. 27. Respiration rates of barley in complete nutrient solution and in solutions deficient in phosphorus, nitrogen or potassium. From Gregory and Richards (155).

The results are summarized in Table XXIV as percentage of the average dry weight and average net assimilation rate, N.A.R., for all the treatments. With both high phosphorus and high potassium, dry weights were about 2–5 times the average whereas the N.A.R. was increased by factors of only 1.2–1.7. With high phosphorus along with either medium or low potassium, or with low phosphorus at all potassium levels, N.A.R. values were less than average in nearly all cases, and the yields ranged from about one-half to one-ninth of the average dry weight. A notable exception was the high phosphorus, medium potassium treatment in the calcium-rich solution. Dry weight and N.A.R. were both about 1.5 times average in this case. A similar exception occurred with high phosphorus and medium potassium in the sodium-rich solution, in which N.A.R. was 1.9 times average.

The calcium solution gave higher yields and assimilation rates than the sodium or ammonium solutions in nine of the twelve comparisons. The minimum N.A.R. percentages in the calcium, ammonium, and sodium solutions were 79, 56, and 28, respectively. The corresponding minimum dry weight percentages were 34, 11, and 11, respectively.

The greatest range of net assimilation rates was observed in the sodium (A) solutions. With high phosphorus and medium potassium, the rate was 191% of the average rate; whereas with low phosphorus

TABLE XXIV

EFFECT OF THE COMPOSITION OF NUTRIENT SOLUTION ON THE RELATIVE DRY WEIGHT AND NET ASSIMILATION RATE (N.A.R.) OF BRACKEN[a]

| | | Percentage of average value per pot[b] | | | |
| | | High phosphorus (P) | | Low phosphorus (0.1 P) | |
Potassium	Cations	Dry weight	N.A.R.	Dry weight	N.A.R.
High (K₁)	Ca	460	171	46	85
High (K₁)	NH₄	493	142	36	93
High (K₁)	Na	234	119	45	89
Medium (K₂)[c]	Ca	154	143	34	79
Medium (K₂)[c]	NH₄	58	87	33	120
Medium (K₂)[c]	Na	56	191	29	88
Low (K₃)[d]	Ca	35	90	35	108
Low (K₃)[d]	NH₄	22	64	11	56
Low (K₃)[d]	Na	11	46	12	28

[a] From Schwabe (302).
[b] Average dry weight per pot, 4.10 gm; average N.A.R., 3.95 gm CO_2/100 cm^2/day.
[c] Level = $\frac{1}{9}$ of the high potassium level.
[d] Level = $\frac{1}{81}$ of the high potassium level.

and low potassium, the rate was only 28%. Similarly, the dry weight in the ammonium solution (M) ranged from 493% with high phosphorus and high potassium to 11% with low phosphorus and low potassium.

In general, the net assimilation rates show smaller deviations from the average rate than are shown by the dry weights with respect to the average dry weights. *This suggests that the plant growth is controlled primarily by its total leaf area rather than by the assimilation rate per unit leaf area.*

Schwabe (Table XXV) compared the dry weight data for barley, flax, and bracken under the same deficiency conditions with respect to nitrogen, phosphorus, and potassium. In most cases the three plants responded similarly to the various deficiencies.

c. *Wheat* (*Triticum aestivum*). The potassium nutrition of wheat was reinvestigated by Eckstein (113), employing the growth culture and photosynthetic techniques used earlier by Gassner and Goeze (139). The two studies gave consistent results. The plants were grown in aerated liter jars, mounted on uniformly illuminated and thermostated turntables. The jars contained glass sand plus 1.5% peat, which was kept at 40% of its water-holding capacity. Phosphorus (P_2O_5) and nitrogen (N), were added at either 20 or 100 mg each per jar. Potassium sulfate was added at several rates from 0 to 250 mg K_2O per jar. The

TABLE XXV

Effect of Similar Deficiencies of Nitrogen, Phosphorus, and Potassium in Reducing the Dry Weight of Barley, Bracken, and Flax[a]

	Dry weight as percentage of full nutrient controls		
Nutrient solution	Barley	Bracken	Flax
Nitrogen ($\frac{1}{9} N_1$)	83	67	—
Phosphorus ($\frac{1}{9} P_1$)	66	70	84
CK_3 ($\frac{1}{9} K_1 + Ca$) = C	57	65	37
MK_3 ($\frac{1}{9} K_1 + NH_4$) = M	68	81	55
AK_3 ($\frac{1}{9} K_1 + Na$) = A	50	77	50

[a] From Schwabe (302).

nutrient solutions also received calcium, magnesium, and trace elements in conventional amounts.

Photosynthesis measurements were made on groups of six basal leaves mounted in a glass boat containing water. The boat was placed in a narrow rectangular chamber with two glass sides, immersed in a thermostat, and illuminated from both sides with two 1000-watt lamps. Air of known carbon dioxide content was passed through the chamber, and several 5-minute outlet samples were taken to determine the amount of carbon dioxide absorbed. Leaf area was measured, and the results were expressed as milligrams CO_2 absorbed per 100 cm^2 per hour.

Representative data for the basal leaves on plants of different ages and nutritional levels are given in Table XXVI. The maximum assimilation rate with the lower N-P_2O_5 level was obtained with 25 mg K_2O. With 100 mg K_2O, assimilation fell off 30%. At the higher N-P_2O_5-

level, assimilation increased rapidly with increasing potassium to 25 mg K_2O, then continued to increase slowly to 250 mg K_2O. Evidently the plant can tolerate and utilize to some extent much more potassium at the higher N-P_2O_5 level.

The first leaf on the plant reached its maximum assimilation rate at 18 days in the high N-P_2O_5 solution. Thereafter, the rate fell off until at 30 days it was zero with the lowest potassium level and was reduced to 50% of the 18-day value at the three highest potassium levels. Singh and Lal (312) observed that potassium-deficient leaves cease to function much sooner than normal leaves.

TABLE XXVI

Effect of Potassium Levels in Solutions of Low and High Nitrogen and Phosphorus Concentrations on the Photosynthesis of the First Wheat Leaves of Different Ages[a]

Potassium, K_2O/jar (mg)	Photosynthesis (mg CO_2/100 cm²/hour)				
	N and P_2O_5, 20 mg each	N and P_2O_5, 100 mg each			
	25 Days[b]	13 Days[b]	18 Days[b]	25 Days[b]	30 Days[b]
0	4.7	3.4	2.7	1.2	—
5	6.5	4.3	6.7	3.0	0.0
10	8.2	6.5	11.1	6.6	3.1
25	9.1	8.9	12.9	9.2	6.3
100	6.3	(6.5)	13.9	11.4	6.8
250	—	8.6	14.9	12.5	7.6

[a] From Eckstein (113).
[b] Age of leaves.

In connection with the foregoing assimilation measurements, transpiration, chlorophyll, and protein in the leaves were also observed. In the 25-day-old basal leaves in the 100 mg N-P_2O_5 solution, there was an approximate doubling of all these values with increase of potassium from zero to the higher levels. The 25-day-old second leaves also showed a doubling of the transpiration and protein, but the chlorophyll was uniformly high in all the samples. Evidently a normal chlorophyll content does not imply normal assimilation function when potassium is low.

d. Tung (Aleurites fordii). Loustalot *et al.* (221) measured the effects of three levels of nitrogen and potassium in the leaves of tung seedlings on the apparent assimilation and carbohydrate composition of the plants. From many analyses of the leaves of tung plantings in the field that showed symptoms characteristic of various states of deficiency of these two elements, it was determined that adequate nutrition was attained if the nitrogen and potassium concentrations in the leaves were about 2.6 and 1.4%, respectively, whereas definite deficiency appeared when these levels fell to 1.4 and 0.5%, respectively. Experimentally these levels were attained by varying the composition of the nutrient

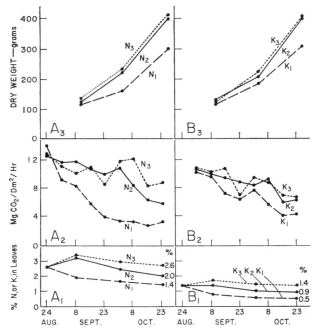

Fig. 28. Effect of three levels of nitrogen and three levels of potassium in the nutrient solution on the percentage of nitrogen (A_1) and potassium (B_1) in tung leaves; on the apparent assimilation $(A_2$ and $B_2)$ during the afternoons averaged each week; and on the dry weight per plant $(A_3$ and $B_3)$ at 3-week intervals. From Loustalot *et al.* (221).

solutions as dictated by leaf analyses. Photosynthesis was measured in the mornings and in the afternoons by determining the carbon dioxide exchange by Heinicke's method (171) in metered air streams passing through assimilation chambers attached to individual leaves.

The afternoon assimilation curves together with the yields and the nitrogen and potassium concentrations in the leaves are presented in Fig. 28. Similar assimilation curves were obtained in the morning. A summary of all the data is given in Table XXVII. Over the 2-month period there was a gradual decrease in the nitrogen and potassium levels in the leaves due to reducing the supply of these elements. At the end of the experiment, when minimum levels had been attained, assimilation with the lowest levels of nitrogen and potassium had fallen to 32 and 65% of the controls, respectively. The corresponding values for the intermediate nutrient levels were 80 and 91%. The nitrogen deficiency was accompanied by change in color of the leaves; later by a stoppage of linear growth and by leaf abscission. Starvation symptoms were not observed in the potassium-deficient plants.

It would be expected that the yields would reflect the average rather than the minimum assimilation rates, and there is fairly good agree-

TABLE XXVII

Assimilation and Yield of Tung Seedlings As Affected by Progressive Deficiency of Nitrogen and Potassium[a]

Element	Conc.	Conc. in leaves (dry basin)			Assimilation		Final dry weight (% of control)
		Max. (%)	Average (%)	Final (%)	Average (% of control)	Final (% of control)	
Nitrogen	High	3.4	2.9	2.6	100	100	100
	Medium	3.3	2.6	1.9	97	80	97
	Low	2.6	1.9	1.4	60	32	72
Potassium	High	1.7	1.5	1.4	100	100	100
	Medium	1.4	1.1	0.9	96	91	99
	Low	1.3	0.8	0.5	84	65	75

[a] From Loustalot et al. (221).

ment between the final dry weight value and the corresponding average assimilation value.

Potassium and nitrogen deficiencies modified the carbohydrate balance in the tung plants. The reduction in photosynthesis and growth due to lack of potassium was accompanied by an increase in reducing sugars ranging from 0.8 to 2.0% in the leaves; 0.3 to 2.9% in the stems; and 3.3 to 4.9% in the roots. There was a corresponding decrease in nonreducing sugars, suggesting that potassium assists in the condensation of hexoses to disaccharides. Lack of nitrogen caused similar effects in the leaves but did not react consistently in the stems and roots. The combinations of low nitrogen and high potassium or low potassium and high nitrogen (Table XXVIII) appear to check growth more than photosynthesis, thereby permitting starch to accumulate in the leaves.

TABLE XXVIII

Starch Content of Tung Leaves at Various Potassium
and Nitrogen Levels[a]

| | Percentage of starch in leaves | | |
| | Potassium level | | |
Nitrogen level	Low	Medium	High
Low	2.1	2.5	2.8
Medium	2.7	2.6	2.0
High	2.5	2.0	2.0

[a] From Loustalot et al. (221).

Maskell (232) observed that the starch concentration in potassium-deficient potato leaves was increased 33% by the addition of potassium sulfate to the soil but other potassium salts had little or no effect.

4. Nitrate Reduction and Photosynthesis

Nitrate and ammonium salts are the principal nitrogen compounds that serve as plant nutrients. Both are normally absorbed through the roots and furnish nitrogen to the leaves for the reactions leading to protein formation. Reduction of nitrate is a preliminary step in this process. (For a discussion of the mechanism of nitrate reduction see Volume III, Chapter 4.) It appears to be a photochemical reaction in which reduced phosphopyridine nucleotides (TPNH) are reducing agents and therefore, in the leaf, may be mediated by chlorophyll and other leaf pigments.

The quantized action spectra of carbonate and nitrate reduction measured simultaneously by Stoy (343) in the same wheat leaves are given in Fig. 19. The two curves are quite similar in shape, each having maxima at 4400 A and at 6100 A. The minimum for carbon assimilation is at 4900 A and for nitrate reduction at 5200 A. The absolute amount of nitrate reduction is only about 1% of the carbon dioxide assimilation. The curves show that the amount of nitrate assimilation in the blue-green as compared with the red is appreciably greater than for the comparable comparison of carbon assimilation. Stoy suggests that the yellow flavin and carotenoid pigments are responsible for these differences in response to the two processes. The chlorophyll reactions are evidently similar.

It has been observed that nitrate reduction in the light does not occur if carbon dioxide is excluded or even reduced appreciably. Possibly some of the immediate products of carbon assimilation are needed for the dark reactions involved in fixing the amino compounds resulting from nitrate reduction in the light. Nitrate reduction can occur at a very low rate in the dark, employing energy from respiration. In this case there appear to be sufficient respiration intermediates to fix the amino compounds and build proteins. The role of manganese and molybdenum in nitrate reduction is considered later. Eckerson (112) found that nitrate and carbohydrates accumulate in phosphorus-deficient tomato plants, even causing nitrogen starvation in the presence of abundant nitrate. Evidently phosphorus is needed to promote reductase activity.

5. Magnesium and Calcium

As a constituent of the chlorophyll molecule, magnesium is directly involved in photosynthesis. Any marked deficiency of the element

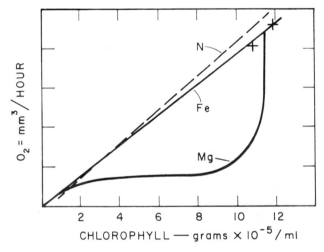

Fig. 29. Photosynthesis of *Chlorella* as affected by the chlorophyll content of the cells, which in turn was controlled by graded deficiencies of iron, nitrogen, or magnesium. Crosses represent full nutrient solutions. From Fleischer (129).

would be reflected in reduced chlorophyll and consequently in reduced photosynthesis. Magnesium is also indirectly involved in photosynthesis as a constituent of many of the important enzyme systems that function in carbohydrate metabolism. For example, the kinases and mutases generally require chelation with magnesium for the group transfer processes in which phosphorus participates. Magnesium plays a predominant role in promoting the formation of enzyme-substrate complexes and the resulting reaction intermediates. Pyrophosphates chelated with magnesium are present in many cofactors. The metal serves as a bridge between ATP, DPN, etc., and a protein molecule.

The dual role of magnesium in photosynthesis is well illustrated by experiments of Fleischer (129) in which the chlorophyll content of *Chlorella* was varied by controlled deficiencies of either iron, nitrogen, or magnesium. A linear relationship was found (Fig. 29) between rate of photosynthesis and the chlorophyll concentration if the latter was the result of graded deficiencies of iron or nitrogen. However, if magnesium was withheld from the cells until the chlorophyll content and the photosynthesis were both very low, increasing amounts of the metal caused chlorophyll formation but photosynthesis remained nearly

TABLE XXIX

Photosynthesis in Flashing Light of *Chlorella* Preparations Made Chlorotic by Deficiencies of Magnesium or Iron As Compared with Normal Cells[a]

Culture	Chlorophyll content (gm $\times 10^{-5}$/ml suspension)	Dark period (seconds)	Photosynthesis per flash per gram chlorophyll (mm^3 O$_2$)
Full nutrient[b]	10.7	0.02	13.8
Full nutrient	11.0	0.42	39.1
Magnesium, 1 mg/l	11.6	0.02	6.0
Magnesium, 1 mg/l	10.8	0.42	32.6
Iron, 0.75 mg/l	10.7	0.02	15.3
Iron, 0.75 mg/l	11.0	0.42	35.3

[a] From Kennedy (199).
[b] Solution diluted 2.5-fold to make chlorophyll content comparable with that of the other preparations.

constant at a low level until the pigment had almost reached its normal concentration; then photosynthesis increased rapidly while the chlorophyll increased slowly until both factors were at their normal levels. Evidently the chlorophyll cannot function fully in photosynthesis until enough magnesium is also present to mediate the dark reactions.

Further evidence of the participation of magnesium in the dark reactions is given by Kennedy (199) using flashing light. *Chlorella* was grown in full nutrient cultures to obtain normal cells and in cultures somewhat deficient in magnesium or iron to obtain chlorotic cells. The latter were compared in flashing light with the normal cells after diluting the full nutrient cultures until they had the same chlorophyll concentration as the deficient cultures. Table XXIX shows that the yield per flash increased approximately two- to threefold when the full-nutrient and iron-deficient cultures were given a dark period of 0.42 seconds instead of 0.02 seconds whereas the magnesium-deficient cultures increased about 5.5-fold, but the final yield was still a little below that of the other cultures. With the 0.02-second dark period, the reactions in the magnesium-deficient cultures were much slower than in the other cultures; the 0.42-second period was not quite long enough to complete the reactions with low magnesium.

Calcium, like magnesium (126), is required in the dark reactions of photosynthesis, but no direct relationship to the light reactions has been established even though many enzymes require calcium.

6. Sulfur

The sulfur requirements of crops are large, ranging from 0.1 to 0.6% of the dry weight of the leaves in different species, equivalent to about 10–50 pounds sulfur per acre per year. Sulfur and phosphorus are required in comparable amounts. Sulfur-deficient plants exhibit a chlorosis characterized by a light green or yellow color without noticeable differentiation of the veins. The younger leaves are involved first. In view of the fact that crop yields on sulfur-deficient soils in areas of low atmospheric sulfur (84) may be greatly improved by the addition of sulfur compounds to the soil, it can be concluded that the rate of photosynthesis in the chlorotic leaves is less than in normal leaves.

Thomas *et al.* (348a) grew alfalfa (*Medicago sativa*) in sand culture plots using nutrient solutions with several levels of sulfate (Tables XXX and XXXI). Continuous recording of assimilation and night respiration was maintained throughout the life of the crops. The lowest sulfur levels (0.3–0.6 ppm S) caused definite symptoms of sulfur deficiency, such as chlorotic leaves and restricted growth. These plants had reduced photosynthetic rates, as compared with plants supplied with 1.6 ppm or more sulfur in the nutrient solution or with plants fumigated with sublethal concentrations of sulfur dioxide (0.1 ppm, 6–7 hours per day, 6 days per week). This is shown by the yield data in Table XXX, which also indicates excellent agreement between the yield and assimilation values for each of the four plots of alfalfa.

The fumigation treatment increased assimilation and yield of the most deficient plants by 13–36%. A smaller increase is indicated for the medium-low sulfur plots (0.6 ppm S). However, the fumigation treatments were less effective for sulfur nutrition than the supplying of sulfate through the roots because the assimilation rates and yields from the latter method of supplying sulfur were not fully attained by using the sulfur dioxide, particularly at pH 7.0 or in the "high nutrient" solution (Table XXXI) (348a). The fumigations had little or no effect on the yields of the high sulfur plots, the differences ranging from −7 to +7%, except for one pair in which root disease was present. These differences are not statistically significant.

It may be noted that the first-year crops in 1939 and 1940 showed much larger percentage deficiency effects than were shown by the

TABLE XXX

Net Assimilation of Carbon Dioxide, Expressed as Equivalent Dry Matter, and Yield of Four Sulfur-Deficient Alfalfa Plots, As Affected by Repeated Fumigation with Sublethal Concentrations of Sulfur Dioxide[a]

| | High nutrient level | | | | | |
| | Net assimilation | | | Yield | | |
Parameter	Check (A) (kg)	Fum. (B) (kg)	$\dfrac{B-A}{A}$ (%)	Check (A) (kg)	Fum. (B) (kg)	$\dfrac{B-A}{A}$ (%)
1940 Crops	3.15	4.21	34	1.62	2.75	70
Winter respiration	−0.27	−0.27	0	—	—	—
1941 Crops	8.13	9.82	21	7.23	8.60	19
Roots	—	—	—	2.86	2.72	−5
Total	11.01	13.76	25	11.71	14.07	20

| | Low nutrient level | | | | | |
	(C)	(D)	$\dfrac{D-C}{C}$	(C)	(D)	$\dfrac{D-C}{C}$
1940 Crops	3.98	5.58	40	2.12	3.76	78
Winter respiration	−0.23	−0.23	0	—	—	—
1941 Crops	8.20	8.17	0	6.58	6.78	3
Roots	—	—	—	3.25	2.90	−11
Total	11.95	13.52	13	11.95	13.44	13

[a] From Thomas et al. (348a).

second-year 1941 crops (Table XXX). There is evidence that the large second-year crops had taken up an appreciable amount of sulfur from other sources besides the nutrient solution, possibly from the walls of the sand bed or from the atmosphere during the winter when it was too cold to operate the air cleaning equipment. A careful consideration of "unaccounted for" sulfur indicates that a concentration of at least 1.5 to 2.0 ppm S instead about 1.0 ppm would have been needed in the nutrient solution to avoid sulfur deficiency in these plants if the solution had been the only sulfur source.

Sulfur participates in many of the enzymatic reactions in the leaf through the labile sulfhydryl, SH⁻ group, also as a constituent of the essential amino acids cysteine, cystine, and methionine; the peptide glutathione; and the vitamins thiamine, biotin, etc. Calvin (75) has suggested that the transfer of electrons from light-excited chlorophyll (about 40/kcal/mole) to produce the reduced form of triphosphopyridine nucleotide (TPNH), which in turn is instrumental in operating the photosynthetic cycle, is accomplished by an intermediate reduction

TABLE XXXI

YIELD OF ALFALFA AS AFFECTED BY THE SULFUR LEVEL IN THE NUTRIENT
SOLUTION AND BY REPEATED FUMIGATION WITH SUBLETHAL
AMOUNTS OF SULFUR DIOXIDE[a]

| | | Plot yield | | | | | |
| | | pH 7.0 | | | pH 5.5 | | |
Year	Nutrient sulfate S (ppm)	Control (A) (kg)	Fum. (B) (kg)	$\dfrac{B-A}{A}$ (%)	Control (C) (kg)	Fum. (D) (kg)	$\dfrac{D-C}{C}$ (%)
1939	85	8.6	9.2	7	8.8	9.2	5
	4	8.1	8.0	−1	9.2	8.6	−7
	0.5	5.0	6.8	36	6.1	8.4	38
		High nutrient			Low nutrient		
1940–1941	9.1	19.1	18.0	−5	14.8	12.8[b]	−13
	1.6	18.6	18.2	−2	15.0	13.9	−7
	0.6	16.1	16.8	4	12.3	13.4	9
	0.3	11.7	14.1	20	11.9	13.4	13

[a] From Thomas et al. (348a).
[b] Plot affected by root disease.

of the disulfide form of 6,8-thioctic acid to the dithiol form. Further, the acetylating coenzyme A, which mediates a pyruvic acid oxidase reaction between the photosynthetic cycle and the Krebs cycle, contains sulfhydryl, disulfide, and possibly thiazoline groups. Sulfur thus appears to be directly involved in the light reactions of photosynthesis, though Arnon et al. (12) do not accept Calvin's thioctic acid hypothesis.

7. The Trace Elements

a. Nutrient and leaf concentrations. In addition to the foregoing six "major" elements it is now recognized that at least seven or eight other elements are essential for flowering plants in trace quantities (see Volume III, Chapter 2). These are all required for normal plant growth, and therefore may have implications directly or indirectly for photosynthesis, but the amounts needed can vary over a considerable range for each element. There is no "ideal" nutrient or soil solution suitable for all conditions. For example, the size and number of plants growing in a given container will affect the rate of depletion of the nutrients and necessitate different solute concentrations and/or frequencies of solution renewal.

Table XXXII gives the nutrient concentrations commonly used in solution cultures for optimum growth of plants. The higher values of the six major elements in most cases represent the recommendations of Arnon and Hoagland (8). The lower values were found to be adequate by Thomas et al. (349), in their large sand culture system. Scharrer and Jung (298) discussed micronutrient concentrations. The range for each element encompasses much of the published literature. Smaller amounts cause deficiency symptoms; larger amounts are either useless or harmful. Table XXXII also includes data by Reuther, Embleton, and Jones (284) giving analyses of orange (*Citrus sinensis*) leaves for all the nutrient elements, arranged to distinguish the trees with optimum growth from those with deficiency or excess symptoms. They pointed out that these leaf concentrations apply also to tung leaves in the marginal range between deficiency and adequacy. They suggested that "leaf analysis standards . . . may have rather general application in the range of soil and climatic conditions to which the crop is adapted." Many data in the literature support this generalization; some do not conform.

b. Iron. Deficiencies of iron have long been recognized as a common cause of chlorosis of plants growing on nonacid soils. Photosynthesis in such plants is usually greatly curtailed. The symptoms of chlorosis can often be relieved by spraying the leaves or injecting the stems with soluble iron salts. Iron is therefore an essential element involved in the synthesis of chlorophyll. However, its probable role in the process is still unknown, though many speculations have been made on the basis of the structural similarities of the chlorophyll and hematin molecules.

Analysis of green and chlorotic leaves for total iron did not reveal any obvious relationship between iron and chlorophyll until Oserkowsky (263) and later Jacobson (187) undertook to distinguish a fraction which they termed "active iron" and which they found was directly proportional to the chlorophyll content in the leaves of pear (*Pyrus communis*), peach (*Prunus persica*), apricot (*Prunus armeniaca*), tobacco, and possibly corn. In order to evaluate "active iron," leaf samples representing various stages of chlorosis were selected from a single chlorotic tree or group of chlorotic plants. Chlorophyll was determined in fresh leaf material. Then dry tissue was extracted with 1.0 N hydrochloric acid and "total acid soluble" iron determined in

TABLE XXXII

Optimum Concentrations of the Elements in Nutrient Solutions;[a] Also Tentative Leaf Analyses for Classification of the Nutrient Status of Valencia Orange Trees[b]

Element	Nutrient solutions[c] (ppm)	Orange leaves (ppm, dry basis)[d]		
		Deficient	Optimum	Excess
N	100–250	<19000	24000–27000	>33000
P	3–60	< 800	1200–1600	> 3000 ?
K	80–400	< 6000	12000–17000	>24000
Ca	50–200	<15000 ?	30000–55000	>70000 ?
Mg	20–50	< 1500	3000–6000	>12000 ?
S	2–60	< 1300	2000–3000	> 6000 ?
B	0.25–1.0	< 20	50–150	> 270 ?
Fe	0.5–1.0	< 35	60–120	> 250 ?
Mn	0.2–1.0	< 15	25–200 ?	> 1000 ?
Zn	0.05–0.5	< 15	25–100 ?	> 300 ?
Cu	0.02–0.1	< 3.5	5–16 ?	> 23 ?
Mo	0.01–1.0	< 0.05	0.1–0.3 ?	> — ?
Cl[g]	Trace	—	<3000	> 7000 ?
Na[e]	?	—	<1600	> 2500 ?
Li[e]	?	—	< 1 ?	> 10 ?
V	0.02–0.10			
Co[e]	<0.01	< 0.07[f]	0.07–0.30[f]	> 1.0[f]

[a] From various authors.
[b] From Reuther et al. (284) and Smith et al. (319).
[c] Concentrations commonly employed.
[d] Four- to seven-month-old bloom cycle leaves from nonfruiting terminals.
[e] Not known to be essential for plants.
[f] Limits for animal nutrition in pasture grass [Price et al. (275)].
[g] Arnon and Whatley (11).

the extract. These values were plotted against the corresponding chlorophyll concentrations. A linear relationship was usually observed originating with zero chlorophyll at a finite iron concentration which was termed "inactive acid soluble" iron. The latter was subtracted from the total "acid soluble" iron to give "active" iron. The correlation between "active" iron and chlorophyll content of the leaves is very striking until mid-season but breaks down during the latter part of the season, owing presumably to changes associated with leaf senescence.

Jacobson found that a threshold concentration of total iron must be exceeded before chlorophyll formation can occur. This threshold varies with the species and growth conditions. In tobacco, active iron is found solely in the chloroplasts but the other forms of iron are also present. The chlorophyll-protein complex prepared by ammonium sulfate precipitation of protein from tobacco and corn leaves, contains iron.

Hematins, cytochromes, and catalase are present in the leaves of plants, being concentrated particularly in the plastids and mitochondria. Cytochrome f has been found in the chloroplasts. Hill and Hartree (176) suggest that it may play a part in photosynthesis just as cytochrome c is known to play a role in animal respiration. It has been suggested also that catalase may be involved in the liberation of photosynthetic oxygen from hydrogen peroxide. Rabinowitch (279, Chapter 11) summarized the arguments for and against this hypothesis and concluded that it was untenable because certain strains of *Scenedesmus* can carry on photosynthesis when the catalase is completely inhibited by cyanide. Moreover, the oxygen evolution can be inhibited by the herbicide N-(3,4-dichlorophenyl) methacrylamide (DCMA) without interfering with catalase activity. Possibly the catalase protects the photosynthetic mechanism from low levels of hydrogen peroxide formed by certain side reactions. Peroxide destroys the carboxylation activity.

Oddo and Pollacci, in 1920 and earlier (262), described experiments that showed normal vigorous growth of corn and other species of plants in iron-free nutrient solutions containing per liter about 240 mg of the magnesium salt of α-pyrrolecarbonic acid. Without the pyrrole or iron the plants were chlorotic and were only about one-third their normal size. It was suggested that the iron mediated the formation of pyrrole derivatives that could be further synthesized to chlorophyll without the help of iron.

Deuber (100) repeated this work in 1925 with corn, cowpea (*Vigna sinensis*), soybean (*Glycine max*), and the water plant *Spirodela* but failed completely to confirm the reported results. The pyrrole not only did not promote the synthesis of chlorophyll, but it was definitely toxic to the plants even in concentrations far below those used in Italy. All attempts to reconcile the results failed, including the testing of a sample of pyrrole carbonate supplied by Oddo. In 1935, Pollacci (272) repeated the earlier claims based on studies with *Chlorella*, *Stichococcus*, and *Hormidium*, which after 10 days in iron-free nutrient solution containing 250 mg pyrrole carbonate per liter were fresh and green and growing rapidly. Concentrations of about 500 mg/liter were toxic. Another attempt by Aronoff and Mackinney (14) to confirm the results failed for the same reasons as those found by Deubar. The pyrrole aldehyde and mesoporphyrin were also synthesized, but they were found to be more toxic than the carbonate. Investigation of this strange disagreement appears to be at a standstill.

Somers and Shive (320), working with soybeans, found that deficiency and toxicity symptoms due to iron and manganese were dependent on the relative rather than the absolute amounts of the two elements in the plants. With high iron and low manganese or with low iron and high manganese the characteristic toxicity symptoms of either iron or manganese were manifested. Optimum growth of

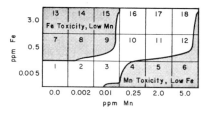

FIG. 30. Growth of soybeans in 18 solution cultures with different levels of iron and manganese. Shaded areas indicate iron or manganese toxicity (1.9 gm average dry weight per culture); unshaded area, normal plants (3.7 gm per culture). From Somers and Shive (320).

normal plants occurred when the levels of both elements were not greatly different, regardless of whether they were low or high. Figure 30 illustrates these relationships. In the unshaded portion of the graph where normal plants, all of approximately the same dry weight, were produced, the Fe:Mn ratio was within the range 0.1–10. The shaded areas had larger and smaller Fe:Mn ratios. The plants were injured and reduced in weight in both these areas.

There is an extensive literature dealing with iron deficiency chlorosis induced by excesses of many elements (cf. Volume III). Hewitt (173) and J. C. Brown and co-workers (63–67) have contributed a great deal to this subject in recent years. Bolle-Jones (43) showed that adequate potassium increased the uptake and mobility of iron in the potato (*Solanum tuberosum*) plant; also its efficiency in promoting chlorophyll formation. Iron deficiency chlorosis was accentuated by reducing the potassium level or increasing the lime or phosphate levels. Lime-induced chlorosis, which is an important agronomic ailment, may be caused by inability to absorb iron or to utilize it after absorption. Phosphate and carbonate ions may accentuate these effects, either outside or inside the plant. The subject is complex because both soil and plant factors are involved. The availability of the iron (242) may be reduced by excess of manganese, copper, nickel, zinc, cadmium, cobalt, molybdenum, or vanadium. Using radioiron, J. C. Brown and Holmes (64) showed that one variety of soybeans could absorb iron from a calcareous soil and another variety could not. Translocation of radioiron in the latter variety was slight. The leaves were chlorotic. Much work will be required to elucidate the complex relationships in iron nutrition, since both chlorophyll formation and the photosynthesis process are involved.

TABLE XXXIII

Decrease in the Apparent Assimilation of Oats Due to Manganese Deficiency[a]

Mn in plant (mg/kg dry wt.)		Assimilation		
		Mg CO_2/dm²/hour		Decrease (%)
+Mn	−Mn	+Mn	−Mn	
76.5	6.8	9.7	3.3[b]	66
63.4	2.1	10.1	4.0	60
47.6	1.4	13.9	3.5[c]	75
—	—	6.4	4.9	23
—	—	12.2	10.9	10
—	—	7.5	3.7	51

[a] From Gerretson (142).
[b] Leaves normal green color.
[c] Leaves chlorotic.

c. *Manganese.* Manganese and iron are probably the most important of the trace elements in relation to the light reactions of photosynthesis (cf. Volume IB, Chapter 4). Manganese is predominant in enzymatic decarboxylation and hydrolysis reactions although the mechanisms are not worked out (225). The element functions in many of the dark reactions in which magnesium predominates. Manganese is involved also in nitrogen metabolism as a constituent of the flavoprotein hydroxylamine reductase. Manganese-deficient plants are characterized by low levels of soluble carbohydrates (241). Photosynthesis is depressed in these plants, even before the leaves show any chlorosis, as illustrated in Table XXXIII.

Gerretson (142) studied the redox potentials of chloroplast suspensions from crushed, filtered, and centrifuged oat (*Avena sativa*) leaves. With free access of oxygen, these suspensions showed rapid redox potential changes in the light amounting to about 100 mv in a few minutes. Addition of about 0.1% manganese sulfate had no effect in the dark but raised the potentials to about 500 mv in the light, probably as a result of peroxide formation. Iron acted similarly, but less strongly. It is suggested the illuminated chloroplasts plus manganese split water into free hydrogen atoms and free hydroxyl radicals which are extremely reactive as reducing and oxidizing agents, respectively. Although they are short-lived themselves (0.01 to 0.1 sec) they can unite with organic or inorganic acceptors to produce reductants and oxidants of long life (up to several hours).

Symptoms of manganese deficiency in the cereals have been described by Eltinge (115). The chloroplasts are affected first, becoming yellow, then vacuolated, then granular and finally disintegrating into a mass of brown colloid. Low levels of starch and sugars are found in the leaves of oats and tomatoes. Cereals show decreased resistance to frost. The leaves are thinner, less rigid, and smaller than normal. They may have masses of calcium oxalate crystals. The stems and their xylem are smaller than normal, and the xylem may be plugged. In the potato, manganese deficiency appears as black spots on the growing leaves and brown spots on the older leaves. Slight chlorosis may develop on the margins of apical leaves, which show much lower manganese concentrations than the basal leaves. (For a fuller account of these symptoms see Volume III, Chapter 4.)

Ruck and Bolas (293) studied the effect of manganese on assimilation and respiration rates of rooted potato leaves and shoots growing in nutrient solutions having either 0.25 ppm or no added manganese (<0.001 ppm Mn). Some of the low manganese leaves were later transferred to the high manganese solution. Net assimilation rates and net respiration rates were determined over 2-week periods. Leaves with adequate manganese at rooting (47–77 ppm) showed an average increase of 170% in dry weight in 2 weeks as compared with 70% for the low manganese leaves (17–20 ppm). The net assimilation rates were 0.45 and 0.19 mg/cm²/day, respectively. The N.A.R. values for the high and low manganese rooted shoots were 0.42 and 0.30 mg/cm²/day. The low manganese leaves which were grown after rooting in the high manganese solution had N.A.R. of only 0.16 mg/cm²/day, in spite of a large uptake of manganese after rooting. Evidently manganese can promote the development of effective assimilation apparatus only during the early formative period of the development of the leaf. A deficiency at this time cannot subsequently be made good by the addition of more manganese. It should be noted that all the foregoing leaves, regardless of treatment, had the same chlorophyll content. Nor was there any statistically significant difference in rate of respiration due to the manganese level.

Symptoms of manganese deficiency in apple and apricot shoots were most severe on the tip leaves, and the basal leaves showed much less injury (388). Lateral shoots were light green with necrotic spots and soon ceased to grow. Evidently the supply of manganese was used up as the plant developed. Translocation from the older tissue did not occur. Older normal leaves remained green when younger leaves had become severely chlorotic. Table XXXIV shows that the growth of these plants was reduced to one-third by manganese deficiency. The manganese content of the leaves was drastically reduced whereas the soluble iron increased. The soluble Fe:Mn ratio in the

TABLE XXXIV

Effect of Manganese Deficiency on the Growth and Mineral Composition of Apple Seedlings Growing in Otherwise Complete Nutrient Solution[a,b]

Mn added to the nutrient solution (ppm):	0.25	0.001	0
Leaf area (cm²)	63	36	35
Dry weight, tops (gm)	107	44	31
Dry weight, roots (gm)	27	14	9
Mn in leaves (ppm)			
Soluble	8.8	1.1	0.3
Insoluble	24.6	1.9	1.1
Total	33.4	3.0	1.4
Fe in leaves (ppm)			
Soluble	2.8	3.2	4.9
Insoluble	51.8	47.5	44.0
Total	54.5	50.7	48.9

[a] From Wiederspahn (388).
[b] Iron, 0.5 ppm, as iron sequestrine, was added to the nutrient solution.

leaves increased from 0.32 to 16.0 in the three solutions. This might cause iron toxicity, as indicated by the data of Somers and Shive (320) discussed earlier.

Reuther and Burrows (283) made a field study of the photosynthesis of manganese-deficient tung foliage. The leaves described as frenched, showed lesions ranging from partial interveinal chlorosis to necrotic areas. Some of the nonfruiting terminals were dipped in 1% manganous sulfate solution plus 0.5% calcium carbonate. Others served as controls. Photosynthesis was measured on many leaves by Heinicke's cup method. The terminal leaves on the manganese-treated shoots showed a 20% increase in assimilation rate as compared with their controls, but basal leaves were unaffected. The former were much more chlorotic than the latter.

The role of manganese in relation to photosynthesis is complex and involves interaction with other metabolic processes. Jones *et al.* (190) have demonstrated the role of manganese in nitrogen metabolism by growing soybeans in a nitrate-rich, manganese-deficient, nutrient solution (Shive's R_5S_2) having its surface covered with mineral oil to exclude oxygen of the air. Nitrite was formed, dissolved oxygen in the solution was exhausted, the leaves became chlorotic, and the roots brown. Addition of manganese established normal growth conditions. Evidently the reduction of nitrate was blocked at the nitrite level by exclusion of oxygen from the roots or by lack of manganese. It is known (257) that the 8-electron reduction of nitrate to ammonia proceeds in four, 2-electron steps, mediated by flavoprotein enzymes requiring different metals for activation. Nitrate reductase employs molybdenum; nitrite reductase and hyponitrite reductase use copper and iron; and hydroxylamine reductase uses manganese (see Volume III). Molybdenum, iron, and copper were not specifically considered by Jones *et al.* but were probably adequate.

The optimum concentration of manganese in a balanced nutrient solution is about 0.5 ppm. Eyster *et al.* (125) found that photosynthesis, the Hill reaction, and growth of *Chlorella* were at a very low level with 5×10^{-5} ppm (0.05 ppb) or less manganese in the nutrient solution. With increase in manganese concentration, these responses increased steadily to a maximum between 50 and 500 ppb. The greatest rate of increase occurred between 0.5 and 5 ppb. At 5 ppb, which was about the minimum amount for maximum photosynthesis, it was estimated that the plants contained about 1.2×10^5 atoms of manganese and 70×10^6 chlorophyll molecules per chloroplast. The ratio was therefore 600 chlorophyll molecules per atom of manganese. Emerson and Arnold (117) found that about 2500 chlorophyll molecules were required for the evolution of one molecule of oxygen. At 4 quanta per molecule, each electron transfer would be associated with 625 molecules of chlorophyll and one atom of manganese. The suggestion was made that manganese can be the limiting factor in photochemical oxygen evolution.

Kessler (200) grew *Ankistrodesmus braunii* with hydrogen and 4% carbon dioxide and showed that manganese deficiency reduced photosynthesis by 75% but had no effect on photoreduction. He concluded that manganese is concerned mainly, if not exclusively, with oxygen evolution in photosynthesis, which is the only partial process not needed in photoreduction.

d. Copper, zinc, and cobalt. Trace concentrations of these elements are indispensable for plants because of the role they play in so many enzymatic reactions (see Volume III). For example, copper is a constituent of some oxidases such as ascorbic acid oxidase, polyphenol oxidase, and laccase. Arnon (6) found that polyphenol oxidase is localized in the chloroplasts of chard (*Beta vulgaris* var. *cicla*). The organic compounds (152) that inhibit the copper enzymes—thiourea, 8-hydroxyquinoline, sodium diethyl dithiocarbamate, etc.—also inhibit photosynthesis and respiration in *Chlorella*. Copper is present in nitrite and hyponitrite reductases, as already pointed out.

Zinc is a constituent of carbonic acid anhydrase (197), present in the chloroplasts, which catalyzes the decomposition of carbonic acid to carbon dioxide and water. Zinc is also found in aldolase (47, 277), which mediates the reversible transfer of phosphoglyceraldehyde to fructose 1,6-diphosphate. Reed (280) has discussed the role of zinc in phosphate and carbohydrate metabolism.

Cobalt is essential for animals, but it is not established as an essential element for higher plants. It is a constituent of B_{12} vitamin. Ahmed and Evans (1) report that cobalt salts and B_{12} can cause increases in the dry matter of soybeans of 30–88% due to 0.1–1.0 ppb cobalt in a highly purified nutrient solution. Nicholas and Thomas (258) found that higher levels of cobalt are toxic, causing symptoms of iron deficiency without changing the iron level, and depressing the uptake of phosphate. Molybdate accentuates these effects.

Copper deficiency symptoms in apple, pear, plum, citrus, and tung trees include cupping of the terminal leaves, which are also abnormally small and show interveinal chlorosis. Later, dieback of the branches occurs owing to cessation of terminal growth. A copper deficiency (143) that reduced the copper content of the leaves and fruit of tung trees by about 50% also reduced photosynthesis about 75%. Protein in the leaves was increased 25%, but starch and reducing sugars were correspondingly reduced. Yield of oil was reduced 34% because the fruit was small and the oil concentration was lower.

The root zones of a pair of abnormal tung trees showing copper deficiency were treated in July with 5 gm each of copper sulfate [Loustalot *et al.* (220)]. A month later all deficiency symptoms had disappeared. In October, assimilation measurement was made on representative leaves (a) on this pair of trees, (b) on a pair of normal trees and (c) on two pairs of untreated copper-deficient trees that had shown deficiency symptoms all summer. The following assimilation values were obtained: (a) 14.0; (b) 12.3; (c) 2.9 and 5.3 mg CO_2 per 100 cm² leaf area per hour. The leaves on the copper sulfate-treated trees contained 4.0 ppm copper as compared with 3.5 ppm in the normal, and 3.1 ppm in the copper-deficient, leaves.

In the foregoing tung plantings some of the trees showed symptoms of zinc rather than copper deficiency. Lack of zinc caused deformation, bronzing, crinkling, and one-sided development of the leaves. Some of the roots were stunted, giving unbalanced growth. The root zone of each copper- and zinc-deficient tree was treated with a complete nutrient solution to which was added 10 gm of copper and/or zinc sulfate. Table XXXV shows the photosynthesis of leaves on these trees, 3 weeks after their root zones were treated as indicated. Evidently, photosynthesis was drastically reduced by deficiencies of either zinc or copper even though the leaves appeared normal. However, the leaves

TABLE XXXV

PHOTOSYNTHESIS OF LATE-SEASON TUNG LEAVES FROM TREES SHOWING DEFICIENCY SYMPTOMS DUE TO LACK OF COPPER OR ZINC AS COMPARED WITH LEAVES FROM NORMAL TREES[a]

Tree	Leaf appearance	CO_2 absorbed[b] (mg/100 cm²/hour)
Normal[c]	Normal	11.6
Cu-deficient[d]	Normal	5.2
Cu-deficient[d]	With lesions	2.2
Zn-deficient[e]	Normal	8.2
Zn-deficient[e]	With lesions	4.5

[a] From Loustalot et al. (220).
[b] Each entry is the mean of 54 readings. $P = <0.01$.
[c] Root zone treated with complete nutrient solution plus 10 gm each of copper and zinc sulfates.
[d] Root zone treated with complete nutrient solution minus copper.
[e] Root zone treated with complete nutrient solution minus zinc.

showing deficiency symptoms were only half as active as the deficient but normal-appearing leaves. Lack of copper reduced assimilation more than lack of zinc.

Zinc deficiency has been frequently observed particularly in apples and the stone fruits, which develop rosettes of little leaves that are only 5–10% of normal size and contain 2–3 ppm zinc as compared with 10–17 ppm in normal leaves. On the newly irrigated land in central Washington state (38, 367), severe zinc deficiency has been observed on alfalfa, corn, and beans, particularly on areas where the top soil had been removed in leveling operations. The alfalfa was stunted so that the plants were only 4 inches high, as compared with healthy plants 14 inches high. Corn was also stunted; the leaves were chlorotic, and ears did not develop. Beans had chlorotic upper leaves and brown necrotic lower leaves. Pod setting was poor and maturity was delayed. In all cases, spraying with 0.5% zinc sulfate caused the symptoms to disappear and new vigorous growth to start.

e. Molybdenum and vanadium. Extensive recent work starting in 1939 (9) has served to develop considerable insight into the role of molybdenum in plant nutrition. Through its marked effects on yield, molybdenum may be regarded as implicated in photosynthesis, if only indirectly. The March 1956 issue of *Soil Science* (Vol. 81, pp. 159–258) is a symposium issue, comprising 9 papers with 392 references and 8 color plates, devoted to molybdenum. Vanadium (10) has recently been proved to be an essential plant nutrient. (Also see

TABLE XXXVI

ACCUMULATION OF NITRATE IN THE LEAVES OF CANTALOUPE GROWING IN A MOLYBDENUM-DEFICIENT SOIL[a]

Date, 1951	Mo in leaf sample	Nitrate N in dry leaves (ppm)
August 2	Normal	900
August 2	Deficient	5880
August 10	Normal	734
August 10	Deficient	3965
August 10	Deficient[b]	1040

[a] From H. J. Evans (123).
[b] Sprayed August 2 with 0.1% Na_2MoO_4.

Volume III for discussions of these elements in plant nutrition.) However, the scope of the vanadium studies to date precludes much discussion of the role of this element in photosynthesis.

Molybdenum (248, 257) plays a dual role in plant nutrition. First, it is a constituent element of the flavoprotein-nitrate enzyme reductase. As such it is essential for the reduction of nitrate to nitrite as an initial step in nitrogen metabolism. It is also concerned with nitrogen fixation by bacteria in legume nodules. Failure to reduce nitrate causes chlorosis, mottling, wilting, and necrotic symptoms due largely to excess of nitrate. Table XXXVI, from H. J. Evans (123), illustrates this type of nitrate accumulation. Failure to fix nitrogen in legumes can cause the yellowing, stunting, and unthriftiness characteristic of nitrogen starvation. In either case, photosynthesis is reduced.

A striking example of nitrogen starvation in a legume is given by Reisenauer (282), who observed molybdenum deficiency in alfalfa on the Couse soil in Washington state. The stand was patchy, having areas of healthy and deficient plants. The latter were stunted and pale green to yellow in color. Yellow interveinal spots appeared first and in extreme cases spread over the whole leaf, causing abscission. Chlorosis was most severe on the lower leaves, and on some plants only the upper leaves remained. The lesions were typical of nitrogen starvation. Fertilization of a number of deficient areas with 0.8 pound of molybdenum per acre increased the yield with respect to the unfertilized areas by about 40% when the unfertilized leaves had less than 0.1 ppm molybdenum. When 0.4 ppm molybdenum was present in the leaves, fertilization had no effect on yields. The increase in yield appears to have been due to an increase in the nitrogen content of the leaves accompanying the molybdenum increase. The phosphorus and manganese concentrations were unchanged.

In Australia (5) it is often necessary to correct phosphorus and sulfur deficiencies on legumes before molybdenum deficiency can be corrected. With the addition per acre of about 200 pounds of superphosphate containing about 1 pound of molybdenum salt per ton, nitrogen fixation was stimulated and yields of clover were increased twofold or more.

Molybdenum is also essential, independently of nitrate reduction or nitrogen fixation. Characteristic molybdenum deficiency symptoms, e.g., whiptail in the cabbage family, cupping of the younger leaves of broccoli and cauliflower (both forms of *Brassica oleracea* var. *botrytis*), and stunting, chlorosis, and necrosis of many crop plants, may be observed in the presence of adequate amounts of reduced nitrogen compounds, such as ammonium salts or urea, if the molybdenum level is sufficiently low ($<5 \times 10^{-6}$ to 10^{-3} ppm) in the nutrient solution; 0.02 ppm is an adequate level. More molybdenum is required if the nitrogen reduction reactions have to be mediated in addition to overcoming this basic deficiency.

Molybdenum deficiency effects, apart from nitrogen reduction and fixation, are poorly defined. Ascorbic acid formation is reduced and phosphatase activity is increased by molybdenum deficiency. The active phosphatases might be expected to reduce the level of phosphorylated metabolites with a resultant reduction in amounts of ascorbic acid and possibly of some of the intermediates of photosynthesis and respiration. Aldehyde oxidase is a flavoprotein that requires both iron and molybdenum as cofactors. H. J. Evans (123) suggested that a study of this system may lead to an understanding of the interaction of iron and molybdenum in plants.

Arnon and Wessel (10) found that vanadium is essential for

Scenedesmus obliquus. When the alga was grown in a purified basic nutrient solution having all known essential trace elements, including 1 ppm iron, growth rate was accelerated two- to threefold by the addition of 20 μg vanadium per liter. Increased initial growth rates were obtained up to 100 μg vanadium per liter. Much smaller quantities of molybdenum than of vanadium are required for growth. No more than 0.1 μg molybdenum per liter was effective in promoting growth. Also the chlorophyll content of the algae declined more rapidly with a deficiency of molybdenum than of vanadium. Vanadium can partially replace molybdenum in nitrogen fixation by some strains of *Azotobacter*. Warrington (371) found that vanadium (1–10 ppm) counteracted some of the symptoms of manganese toxicity in soybeans. Low vanadium (0.1 ppm) had no effect on growth or the iron nutrition with either high or low manganese content.

f. Boron. Boron appears to be an indirect contributor to the assimilation process. The extensive studies of F. M. Eaton (107), who grew 58 species of plants in outdoor solution cultures using a range of boron concentrations from deficiency to toxicity (0.03–25 ppm), have served to define plant responses to this element. A large group of plants were sensitive to boron, growing best with 1 ppm or less in the nutrient solution. They showed slight to severe deficiency symptoms at 0.03 ppm. A second more tolerant group grew best at 5 ppm, and a third very tolerant group preferred 10–15 ppm. Large accumulations of boron were found in the leaves with the higher concentrations. According to Skok and McIlrath (314), the boron in sunflower leaves, after maceration and differential centrifugation, was about equally divided between the liquid and solid fractions. In the former, the nondialyzable boron remained constant as the dialyzable boron decreased to zero with onset of deficiency symptoms. In the latter the plastids had much more boron than the mitochondria and microsomes.

Kelley *et al.* (198) grew carrots (*Daucus carota* var. *sativa*) in quartz sand nutrient culture, with and without the addition of 0.5 ppm boron. In a winter experiment from December 5 to April 24, the plants grew slowly until mid-March, when deficiency symptoms began to appear in the "minus-boron" plants. Anthocyanin formation and slight bronzing were first noted; then wilting of the leaves in bright light despite adequate moisture; deformation of leaves and failure of new leaves to form; and finally death of the older leaves. The "plus-boron" carrots had blunt ends; the "minus-boron" carrots were sharply pointed and reduced 25% in weight. In a summer experiment, boron deficiency produced small, sharply tapered roots that were flaccid but tough and woody when cut, with longitudinal cracking of the phloem and pericycle. These symptoms were corrected by the addition of 0.1 ppm boron to the nutrient solution. Table XXXVII gives the yields, carotene content, and boron accumulation of the carrots with boron additions up to 5 ppm. Maximum yield was obtained with 2 ppm. There was some inhibition at 5 ppm. The carotene content of the roots was raised 20–25% by 0.1 ppm, but additional boron had no significant effect. The boron content of the tops increased seventeenfold between 0 and 5 ppm, but the boron in the roots increased less than threefold.

It is definitely established that boron facilitates the translocation of sucrose and other photosynthates in the plant. Sisler *et al.* (313) used $C^{14}O_2$ for photosynthesis in tomato and found that the radioactive

TABLE XXXVII

Effect of the Boron Concentration in the Nutrient Solution on the Growth, Carotene Content, and Boron Accumulation in Carrots[a]

Boron added to nutrient (ppm)	Yield dry weight (gm/plant)		Carotene in roots (mg/100 gm)		Boron (ppm, dry basis)	
	Tops	Roots	Fresh	Dry	Tops	Roots
0	4.5	4.1	9.2	71	20	16
0.1	8.8	9.9	10.9	96	39	18
0.5	10.8	11.4	11.7	98	82	23
2.0	11.1	15.0	12.0	99	168	32
5.0	10.9	11.4	13.2	98	307	44

[a] From Kelley *et al.* (198).

products were translocated more readily in plants that had adequate boron than in boron-deficient plants. In parallel studies, a solution of labeled sucrose with and without 50 ppm boron was introduced into the plants through the stump of the petiole on the third node from the tip. In boron-deficient plants, showing no deficiency symptoms, the added boron definitely enhanced the translocation of the sugar. Plants adequately supplied with boron and plants injured by lack of boron showed no such enhancement. Many hypotheses have been advanced to account for these effects. A favored suggestion involves a complexing of borates with the hydroxyls of the sugar, a mechanism that is said to facilitate the passage of the sugar through protoplasmic membranes. Enzymatic reactions are also postulated. Dugger *et al.* (104) observed that conversion of glucose 1-phosphate to starch by starch phosphorylase in the leaves was inhibited by boron. This might result in a higher sugar concentration in the leaves and a consequent greater tendency for its translocation to other parts of the plant with adequate boron.

Baker *et al.* (19) found that transpiration is reduced in boron-deficient bean leaves owing to (a) high sugar and colloid content, (b) reduced water absorption, and (c) a high percentage of non-functional stomata as compared with normal leaves. The conditions that interfere with transpiration are often responsible for reduced assimilation also. This would be expected with the abnormal morphology of boron-deficient leaves involving distorted epidermal cells. Nonfunctional stomata often had only a single guard cell, and the irregular growth produced thickened spots within the generally thickened leaf.

V. Photosynthesis under Field Conditions

A. General Considerations

Photosynthesis in the field involves the interplay of all the interdependent environmental and internal factors that mutually affect the over-all process. Some field studies have been made (79, 247, 350) and the subject has been reviewed (347). Light and the concentrations of carbon dioxide or bicarbonate are the two external factors most directly concerned, but photosynthesis is influenced by all the environmental and nutritional factors that affect the well-being and functional activity of the plant. These include temperature, moisture supply, relative humidity of the air, fertility of the soil, concentration and pH of the nutrient solution, age of the leaf, its history, and toxic compounds in the soil or air. The morphology and turgidity of the leaf and the associated opening and closing of the stomata are important responses to environment that affect photosynthesis.

Two general methods have been employed to evaluate photosynthesis in the field, namely, growth analysis and measurement of gas exchange. These methods actually measure net assimilation, which is the excess of carbon dioxide uptake during the day over carbon dioxide output during the night. Respiration in the light must be estimated by indirect means if it is to be considered at all.

B. Growth Analysis

1. Net Assimilation Rate

V. H. Blackman (36) found that the growth of annual plants follows the compound interest law, at least in the early vegetative stages of development. He suggested a relative growth rate (R_w), which at any instant is

$$R_w = \frac{1}{W}\frac{dW}{dt} = \frac{d\ln W}{dt} \qquad (1)$$

Where W is the weight of dry matter present at any time per unit of ground area and dW/dt is the rate of increase in dry weight at that time. R_w serves as a criterion for indicating the relative efficiency of different plants in dry matter production. R_w was therefore called the "efficiency index."

Equation 1 integrates to:

$$W_2 = W_1 e^{Rt} \qquad (2)$$

where R is an assimilation rate:

$$R = \frac{\ln W_2 - \ln W_1}{t_2 - t_1} \qquad (3)$$

Since it is the leaves rather than the whole plant substance that produce the increase in weight, the "growing material" can be more accurately represented by the leaf area (A) and the efficiency index by

$$R_w = \frac{1}{A}\cdot\frac{dW}{dt} \qquad (4)$$

If A is proportional to W, Eq. 4 can be integrated to yield a net assimilation rate, designated N.A.R., for each time interval, $t_2 - t_1$:

$$\text{N.A.R.} = \frac{(W_2 - W_1)(\ln A_2 - \ln A_1)}{(A_2 - A_1)(t_2 - t_1)} \qquad (5)$$

In applying the method, representative areas or plants from uniform fields or field plots are harvested completely at intervals of one to several weeks. Roots as well as shoots are taken. The leaf area (A) and dry weight (W) of each sample is then determined. The N.A.R. is usually expressed as grams dry weight per 100 cm² leaf area per week. The growth increment per week is the product of N.A.R. and leaf area. The N.A.R. is also expressed in grams per square meter per day.

The method is laborious, and a number of replicate determinations must be made in order to achieve statistical significance. However, the method has the advantage of dealing with plants in the field under entirely natural conditions. Furthermore only the simplest equipment and manipulations are required.

As a measure of photosynthesis or apparent assimilation, the method has several limitations. Leaf area appears to be one of the best criteria for N.A.R. It is better than leaf weight but, according to R. F. Williams

(392), not as good as the protein content of the leaf. Photosynthesis occurs not only in the leaves, but in all the other green parts of the plant as well. Respiration, which is included in the N.A.R. value, occurs in all parts of the plant. The errors which might be ascribed to all these factors are thought to be minor compared with sampling errors and short-term climatic factors. The latter were analyzed by Gregory (154), who found that N.A.R. is correlated positively with day temperature and insolation and negatively with night temperature. About 80% of the variation of N.A.R. is due to external factors.

Heath and Gregory (169) concluded on the basis of data assembled in England, Germany, the Sudan (85), and South Africa (Table XXXVIII) that all the different species of temperate zone sun plants studied have about the same N.A.R. during the vegetative stage. If this conclusion is substantially correct, "the main determining factor leading to the great differences in dry weight accumulation in different types of plants is the rate of extension of leaf surface, in which is included both the size of individual leaves and the rate of development of new leaves." Later work (378) has shown that these generalizations are only approximate. Appreciable differences in N.A.R. can exist between species and even between varieties of a species under the same climatic conditions. For example, sugar beet in England had double the N.A.R. of wheat and barley during the period when their vegetative development was comparable. Smaller differences were noted between the different cereals and between sugar beet and potato. N.A.R. was also dependent on the weather. The lowest values were obtained with short days, low light intensities, and crowded plant environments. Cereals have maximum N.A.R. with high day temperature and low night temperature. This probably means high photosynthesis and low night respiration. The reverse is true of sugar beet, which seems to prefer a small daily temperature range indicating overcast skies. An explanation offered by Watson links N.A.R. with the moisture in the leaf. Sugar beet tends to wilt in bright sunshine, with resulting stomatal closure and reduction in photosynthesis. Comparison of a number of varieties of potatoes and sugar beets showed that the higher yields were associated with the higher mean leaf areas rather than with the higher N.A.R. values. Much work needs to be done under controlled conditions to explain adequately the variations, due to external and internal factors, that have been observed in N.A.R.

Heath and Gregory concluded, and Watson concurs, that "N.A.R. is the only really satisfactory measure of assimilation from an ecological and agricultural point of view since it integrates the gains due to assimilation and the losses due to respiration of the whole plant under

TABLE XXXVIII

NET ASSIMILATION RATES (N.A.R.) OF VEGETATION GROWN IN VARIOUS SEASONS AND LOCATIONS[a]

Date	Plant	Location	Average hours of daylight	N.A.R. (grams dry wt./dm²/week) Range	Ave	Reference
May–June 1885	*Brassica hirta (Sinapis alba)*	Bonn, Germany	14.7	0.31–0.67	0.51	Hornberger
1921–1924	Barley	Rothamsted	16.25	0.42–0.66	0.55	Gregory
1929	Ryegrass	Rothamsted	16.5		0.32[b]	Richards
1929	Cocksfoot grass	Rothamsted	16.5		0.24[b]	Richards
1929	Meadowgrass	Rothamsted	16.3		0.40[b]	Richards
June–Sept. 1934	Mangold	Rothamsted	14.5	0.51–0.69	0.60	Watson and Baptista
June–Sept. 1934	Sugar beet	Rothamsted	14.5	0.64–0.72	0.68	Watson and Baptista
Nov. 1916	Cucumber	Cheshunt	8.5		0.12[c]	Gregory
Feb. 1917	Cucumber	Cheshunt[e]	11.2		0.34	Gregory
May 1917	Cucumber	Cheshunt[e]	16.7		0.41	Gregory
August 1932	Tomato	Cheshunt[e]	14.		0.61	Bolas
Sept. 1932	Tomato	Cheshunt[e]	12.5		0.47	Bolas
Oct. 1932	Tomato	Cheshunt[e]	10.7		0.42	Bolas
April 1938	*Endymion nutans (Scilla nutans)*	London	14.		0.61	G. E. Blackman
Nov.–Feb. 1933–35	Cotton tops	South Africa	13.6	0.42[d]–0.57	0.52	Heath
Oct.–Nov. 1928	Cotton tops	Sudan	11.7		0.68	Crowther

[a] From Heath and Gregory (169).
[b] Very dense stand; CO₂ supply reduced.
[c] Low light intensity limited assimilation.
[d] Poor growth due to aphis.
[e] Under glass.

actual growing conditions and the result is obtained without interference with the normal functioning of the leaves." This sweeping conclusion may need to be reexamined in the light of a new method of
estimating assimilation in the field employing wind velocity and CO_2
concentration gradients (cf. Section V, C) (212).

2. Maximum Biological and Economic Yield

a. General considerations. The biological yield of a plant community
involves three principal variables: (a) the net assimilation rate, (b)
the total leaf area of the plants, and (c) the duration of the growth
period. The objective of maximum yield requires maximum values for
all these variables.

Nichiporovich and Strogonova (256) have made a strong plea for
more comprehensive studies of applied photosynthesis as the principal
single factor involved in plant production which is responsible for
accumulating not only 90–95% of the crop weight, but also the energy
needed for the functioning of all plant processes. Nutritional factors
such as nitrogen, phosphorus, potassium, and other nutrients, and water
supply can be effective only if they support and promote photosynthesis
which in turn supplies the organic substrates and the energy of the
vital system. It is suggested that "the course of growth and the state
and efficiency of the photosynthetic apparatus should underly our
approach to the problem of finding the optimal conditions of formation
of crops."

Biological yield and economic yield are different entities. Relatively,
they vary widely in different crops. For each crop and for each plant
there is a close correlation between biological and economic yields even
though there is a wide crop-to-crop and plant-to-plant variation under
different environmental and cultural conditions. Evidently good biological yields are a prerequisite to good economic yields.

b. Ratio of leaf area to ground area. The fundamental importance
of a large leaf area in producing a large crop is well established. The
ratio of the leaf area of the plants to the ground area they cover is
designated "leaf area ratio" (L.A.R.) or "leaf area index" (380). Its
maximum value can vary from about 1:1 to about 6:1 or more in
different plants. Plant communities therefore have maximum leaf areas
ranging from about 4000 to 25,000 m²/acre depending on soil, water,
and cultural factors. In some plants, e.g., wheat, the leaves develop
steadily up to the flowering stage, then they dry up over a short time
as the grain ripens. Early planting and stimulation of early growth
are important for high yields of cereals. In other plants, e.g., cotton,
the leaves remain functional and some new leaves are continually

produced over a long period of plant maturity. High yields require steady growth throughout as long a season as possible. The value to the plant of a particular leaf depends on its life span. In the conifers and other perennials this may amount to one to ten years or more. However, long-lived leaves may have appreciably lower net assimilation rates than short-lived leaves.

c. *Photosynthetic potential of a plant community.* The "photosynthetic potential of a plant community" is defined as the integral of

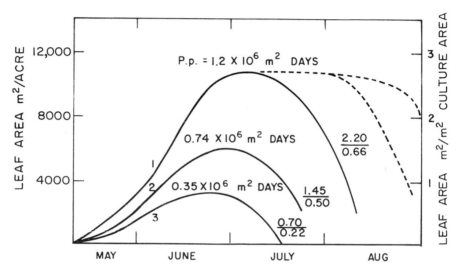

FIG. 31. Leaf area production in wheat. Curve *1*, irrigated twice and fertilized; curve 2, irrigated twice, not fertilized; curve 3, not irrigated or fertilized. N.A.R. of all crops, approximately 9.5 gm/m²/day. Photosynthetic potential (P.p.), in square meter days, and also biological yield (upper figure) and economic yield (lower figure) are given for each curve. Dashed curves are hypothetical (see text). From Nichiporovitch and Strogonova (256).

leaf area over the vegetative period, and may range from one half to four million "square meter days" per acre. The net assimilation rate is the biological yield divided by the square meter days. This can range from 2 to 10 gm/m²/day, but at certain periods in the growth cycle it may reach 15–18 gm/m²/day or even fall to negative values.

Figure 31 (256) illustrates the development and senescence of the leaf area of wheat under three cultural regimes together with the resultant biological and economic yields and the "m² days." A linear relationship is shown between both the maximum and integrated total areas and yield, but N.A.R. remains constant. There is a limit to which increase of leaf area can be profitably carried because too high a leaf density involves shading and perhaps other factors unfavorable to photosynthesis. According to Nichiporovich, it is desirable that the leaf area should reach as quickly as possible, but not exceed, about 12,000–24,000 m² per acre depending on the plant and the part that is utilized. In Watson's units (m²/m²) (380) this represents about 3–6 times as much leaf area as ground area covered by the plant. The lower limits would apply to the grains and other seed crops. Fodder crops would be benefited by the high limits. The optimum leaf area for sugar beet and potato appears to be about 20,000 and for kale about 12,000 m² per acre (Fig. 33).

In view of the limitations due to maximum leaf area, further increases in yield should be possible if the functional life of the leaves could be extended, without increasing their area, as illustrated by the dashed lines in Fig. 31. Attempts to modify the curves for grain in this way have not met with much success because ripening follows so soon after flowering, even with nitrogen fertilization. However, the prospects for larger square-meter-per-day values in the root and fodder crops are much better within the limitations imposed by the length of the growing season. In England, where the best light and temperature conditions for photosynthesis occur in June and July, it is desirable to adjust planting dates and fertilizing practices so that the maximum leaf areas are presented in these months.

Figure 32 by Thorne and Watson (353) gives data for leaf area of winter wheat in England in May and June, as affected by early and late nitrogenous fertilization. An application of solid fertilizer on April 25 doubled the leaf area by June 1, whereas the same amount of nitrogen in sprays applied during May and June caused only a slight increase in the leaf area above the control. However, final maturing of the early fertilized leaves was practically identical with the other two plots between about June 25 and July 15.

Still larger yields would result from an increase in net assimilation rate (N.A.R.) and also from an increase in the ratio of economic to biological yield. Opinion is divided as to the limitations of these factors. Watson (378, 379) considers that N.A.R. is dependent primarily on light and temperature, which cannot be regulated in the field. He therefore thinks that increasing the N.A.R. is an impracticable procedure. Nichiporovich is more optimistic. In contrast to Watson's observations, he finds that nitrogenous fertilizers, adequate water, and favorable genotype plants are capable of giving assimilation rates of 10 gm dry matter per square meter per day with a maximum leaf area of about 12,000 m² per acre. This is about double the usual rate and would give a biological yield of 13 tons dry weight per acre in 1.2×10^6 m² days. Work in this area is recommended. Increase in the ratio of economic to biological yields would involve timing of fertilizing and cultural treatments.

d. *Relative growth rate.* Watson (379) measured the net assimilation rate of kale and sugar beet as influenced by leaf area. Well-developed, rather densely planted plots of these crops were thinned by removing ¼, ½, or ¾ of the plants without unduly disturbing the others. Weight and leaf area measurements were made on the harvested plants. After 10–14

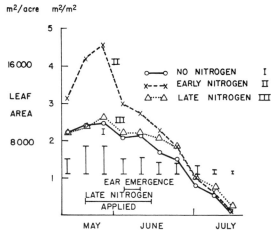

Fig. 32. Effect of nitrogenous fertilization on leaf area of winter wheat in England. Curve *I*, no nitrogen; curve *II*, solid fertilizer in soil April 25; curve *III*, same total nitrogen in 8 sprays to soil or leaves. Vertical lines indicate least significant differences ($P = 0.05$). From Thorne and Watson (353).

days, the plants remaining on the plots were similarly harvested. One experiment is illustrated in Fig. 33. Another experiment by Nichiporovich and Strogonova gives similar data for potato. The two experiments differ in that the latter represents the whole growth period of the developing plants, while the former represents only a 2-week growth period by well-developed plants.

The increase in growth rate with increasing leaf area was substantial in all the plants. However, owing to excessive shading by the upper leaves, the rate for kale became constant at 12,000 m² per acre whereas with larger leaf areas, rate per unit area declined. With sugar beet and potato the rates increased at least to 20,000 m² per acre. An interesting decrease in rate is shown in the potato curve just below 10,000 m² per acre, associated possibly with flowering of the plant. Subsequently the rate increased appreciably.

Net assimilation rate declined rapidly in kale and slowly in sugar beet with increasing leaf area. There was a large initial decline in potato followed by a small increase then a leveling off of the rate. These

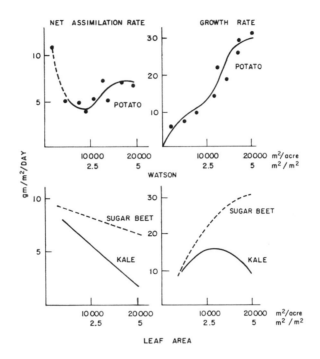

Fig. 33. Net assimilation and relative growth rates as dry weight (gm/m²/day) in relation to leaf area (m²/acre or m²/m² of ground surface), for potato, sugar beet, and kale. From Nichiporovitch and Strogonova (256) and Watson (380).

curves suggest that an inverse relationship between growth rate and net assimilation may often occur.

3. Assimilation Rate versus Log Light Intensity

During the past fifteen years, G. E. Blackman and associates (34, 35) have discussed many phases of photosynthesis in the field in their series of papers on "physiological and ecological studies in the analysis of plant environment."

In shading experiments they found a direct proportionality between the logarithm of the light intensity and both net assimilation rate, N.A.R. (gm/dm² leaf area per week) and relative growth rate, R.G.R.

(gm/gm weight of whole plant per week) in the range from full daylight to 12–20% of daylight intensity. In some plants this proportionality extended to 5% of daylight, but there was generally a departure from linearity in weak light. In the eightfold range from 12 to 100% of daylight the N.A.R. and R.G.R. increased three- to sixfold in different species. Similarly an inverse proportionality was found between loglight intensity and the leaf area ratio or ratio of leaf area to plant weight (cm²/gm). The eightfold range of light intensity reduced this ratio 10–40%.

The foregoing experiments, carried out in England, indicate that the light saturation was not attained with some of the plants in full sunlight, even though in a number of experiments the plants received 32,000 to 39,000 foot-candle-hours of light per day equivalent to an average daylight illumination of 1900–2200 fc. It was estimated that the attainment of optimal light levels for several plants would require multiplying the maximum light levels by the following factors: alfalfa (*Medicago sativa*) 2.51, white clover (*Trifolium repens*) 1.85, alsike clover (*T. hybridum*) 1.48, and sunflower (*Helianthus annuus*) 1.14. The other plants appeared to be approximately light saturated.

4. Effect of Water Stress on Leaf Area and N.A.R.

P. C. Owen at Rothamsted (268) measured the growth of sugar beets under different water regimes (Table XXXIX). Field plots protected from rain were used instead of pots because the amount of water available to the roots in a pot was too limited to permit a prolonged exposure to moisture stress near the wilting point without injury to the

TABLE XXXIX
GROWTH OF SUGAR BEETS UNDER THREE WATER REGIMES[a]

Irrigation treatment June 3 to August 20, total inches	Leaf area (m²/m² soil surface)			Dry weight (kg/m² leaf area)			N.A.R. (gm/m²/week)		Dry weight (kg/m² per inch water)
	July 9	Aug. 6	Aug. 20	July 9	Aug. 6	Aug. 20	July 9 to Aug. 6	Aug. 6 to Aug. 20	Aug. 20
A, 5.4[b]	5.4	7.8	7.4	0.47	1.25	1.55	33	20	0.29
B, 1.0[c]	2.5	5.2	5.0	0.35	0.92	1.17	36	19	1.17
C, 1.0[d]	3.1	5.1	6.0	0.42	0.98	1.26	38	25	1.26

[a] From P. C. Owen (268).
[b] Irrigated on 15 days: total water 5.4 inches.
[c] Irrigated once: July 10: total water 1 inch.
[d] Irrigated on 5 days: total water 1 inch.

plant. One set of plots was maintained at field moisture capacity, by applications of water totaling 5.4 inches (treatment A). A second set was allowed to dry after June 3 until the wilting point was approached on July 10. Then they were given a single large irrigation (treatment B). A third set was given a number of small irrigations throughout the season (treatment C). In B and C the same total amount of water, 1 inch, was applied. Leaf area and growth were reduced by about 20–50% in groups B and C as compared with A. However, the net assimilation rate, N.A.R., was not significantly different in the three treatment groups. The increased yield of the group subjected to treatment A was due to its greater leaf area. Evidently the leaves on the water-stressed plants had the same photosynthetic efficiency per unit area as those with adequate water. The water was used with greater relative effectiveness by the drier plots. They produced 80% of the maximum yield with only 20% of the water applied in treatment A. Treatment C was somewhat more effective than treatment B, presumably because of somewhat less water stress on the plants.

5. Photosynthesis under Optimal Natural Conditions

The N.A.R. values of Heath and Gregory in Table XXXVIII for a few herbaceous sun plants may be compared with data assembled from the literature by Verduin (364) (Table XL), giving the rates of photosynthesis under "optimal near natural" conditions of many different types of plants from many habitats. For sun plants except alpine plants, Verduin's rate is 2–4 μmoles CO_2/cm^2 leaf area per hour. For alpine herbaceous plants the rate is 7, for temperate shade plants 1, and for desert evergreen plants 0.1 μmoles CO_2/cm^2 per hour. The sun plant rate in N.A.R. units, assuming 70 hours per week and 44% carbon in the dry plant, would be 0.4–0.8 gm dry weight per 100 cm^2/per week; these values cover the sun plant range in Table XXXVIII. The latter values are equivalent to about 5–11 gm dry weight per square meter per day.

The high rate of assimilation by alpine plants quoted by Verduin occurred in spite of the fact that the total carbon dioxide pressure was only one-half to two-thirds the pressure at sea level. This is surprising when it is remembered that assimilation can be materially reduced by relatively slight depletion of the carbon dioxide in the air, as shown by Heinicke and Hoffman (171), by Decker (93), and by J. W. Wilson and Wadsworth (395). Evidently both the total pressure and the partial pressure affect the diffusion of carbon dioxide through the stomata.

G. F. Blackman and Black (Table XLI) (34) have summarized the maximal values recorded in the literature for net assimilation rate,

TABLE XL

AVERAGE RATES OF PHOTOSYNTHESIS UNDER OPTIMAL NEAR-NATURAL CONDITIONS[a]

		CO₂ absorbed (μmoles/hour)[b]			N.A.R., dry weight (gm/dm²/ week)
Location	Type	Per mg dry wt.	Per cm² leaf area	Per μmole/CO₂ respired	
Temperate	Herbaceous				
Zone	Sun plants	1	4	16	0.8
	Woody plants	—	3	9	0.6
	Shade plants	—	1	4	0.2
Tropic Zone	Trees and shrubs	—	2	3	0.4
Arctic Zone	Herbs	—	2	—	0.4
Alpine	Plants	—	7	—	1.4
Desert	Evergreen plants	—	0.1	2	0.02
	Mosses and lichens	0.05	—	5	—
	Submersed water plants	—	0.4	8	0.08
	Microscopic aquatics	2	0.1	11	0.02
	Chlorella	7–18	0.2	25	0.04

[a] From Verduin (364).
[b] Range routinely observed from ½ to 2 times the values listed.

TABLE XLI

MAXIMAL RECORDED NET ASSIMILATION AND GROWTH RATES OF SOME SPECIES UNDER DIVERSE ENVIRONMENTAL CONDITIONS[a]

Plant	Location	N.A.R. (gm/dm²/ week) A	Leaf area ratio (dm²/gm) B	Relative growth rate (gm/gm plant wt./week) A/B Observed	Calc.
Helianthus annuus	England	0.73–0.98	1.70–1.10	1.19–1.05	1.24–1.08
Helianthus annuus	Scotland	0.98–1.19	1.32–0.95	1.26–1.13	1.29–1.13
Helianthus annuus	U.S.A.	1.09	—	1.12	—
Helianthus annuus	Australia[b]	0.86–1.36	1.98–1.30	1.82–1.75	1.71–1.77
Trifolium subterraneum	Australia[b]	0.91	1.08	0.91	0.98
Zea mays	U.S.A.	0.86	—	1.68	—
Zea mays	Nigeria	1.52	1.69	2.31	2.57
"*Ipomoea caerulea*"	Nigeria	0.69–0.82	2.32–2.37	1.61–1.89	1.60–1.94
Solanum melongena var. *esculentum*	Nigeria	0.91	2.30	1.96	2.09
"*Vigna unguiculata*"	Nigeria	0.82	1.85	1.47	1.52

[a] From G. E. Blackman and Black (34).
[b] Adelaide and Brisbane.

leaf area ratio, and relative growth rate. The third item is the product of the first two. The values in tropical and subtropical locations appear to be considerably higher than in temperate climates, a result evidently of higher insolation and temperatures in the former. The leaf area ratio is increased to a greater extent than net assimilation. Table XLII gives some maximal dry matter production data in Europe and the United

TABLE XLII

Maximal Recorded Rates of Dry Matter Production and Estimates of Efficiency of Light Utilization under Diverse Environments[a]

| Plant | Location | Dry matter production | | Light utilization 4000–7000 A (%) |
		Gm/m²/ day	Tons/acre/ year	
Beta vulgaris (*B. maritima*), sugar beet	England	31	—	9.5
Beta vulgaris (*B. maritima*), sugar beet	U.S.A.	—	9.3[b]	—
Hordeum vulgare, barley	England	23	—	4.3
Brassica oleracea var. *acephala*, kale	England	21	—	4.9
Helianthus tuberosus, Jerusalem artichoke	England	—	9.0	—
Solanum tuberosum, potato	Ireland	—	8.0	—
Pinus sylvestris, pine	England	—	10.0	2.4
Fagus sylvatica, beech	Denmark	—	10.5	3.5
Zea mays, corn	U.S.A.	24–27	—	—
Gossypium sp., cotton	U.S.A.	27	—	—
Medicago sativa, alfalfa	U.S.A.	—	13.3[b]	—
Malus sp., apple	U.S.A.	—	15.5[c]	—
Saccharum officinarum, sugar cane	Hawaii	38	32–35	3.2

[a] From G. E. Blackman and Black (34).

[b] Under glass [Thomas and Hill (350)].

[c] Under glass [Heinicke and Childers (172)].

States including Hawaii; also some estimates of the efficiency of light utilization. All the yield data except those for sugar cane (*Saccharum officinarum*) in Hawaii, fall within a comparatively narrow range. The assimilation rate of the sugar cane was about 50% higher than the rates for the other crops, and its annual yield was 2–3 times greater. Evidently the long growing season in Hawaii along with more rapid assimilation can account for the large yield of sugar cane. These assimilation rates approach the maximum rates that have been attained experimentally in algal culture (see Table XLVIII). Utilization of light energy was generally higher than the 2% value usually quoted for the efficiency of photosynthesis.

The narrow range of N.A.R. values in widely separated localities may be explained in part by an observation of Shirley (309), who states that during the growing season the amount of light received per 24 hours is about the same in the tropics, the temperate zone, and the arctic regions.

C. CARBON DIOXIDE EXCHANGE MEASUREMENTS IN THE FIELD

1. Objective and Precautions

The gas exchange method of evaluating photosynthesis in the field is restricted to carbon dioxide studies because it is impracticable to

TABLE XLIII

NET ASSIMILATION OF CARBON DIOXIDE, EXPRESSED AS EQUIVALENT DRY MATTER FOR FOUR PLOTS OF SUGAR BEETS AS COMPARED WITH THE CORRESPONDING YIELD DATA (1944)[a]

Nutrient level	Yield			Net assimilation (kg)	Yield/assimilation (%)
	Roots (kg)	Tops (kg)	Total (kg)		
Low	4.94	1.94	6.88	7.02	98
Low	4.75	1.77	6.52	6.72	97
High	4.15	2.78	6.93	6.70	103
High	4.41	2.39	6.80	6.75	101
Total	—	—	27.13	27.19	100

[a] From Thomas et al. (349).

measure very small differences in the high oxygen content of the air. Excellent chemical and infrared absorption methods are available for determining carbon dioxide in the 300 ppm range of concentration. These methods can be used for continuous automatic monitoring of the gas on the basis of instantaneous or integrated recording of the concentration.

Individual leaves, plants, or field plots may be studied by enclosing the vegetation in a transparent chamber through which a measured volume of air is passed. The chamber is illuminated by artificial light or sunlight, and the carbon dioxide content of the air is determined as it enters and leaves the chamber. The difference represents carbon dioxide absorbed due to photosynthesis or evolved due to respiration.

If the plants are grown in sand culture, root respiration can be measured by the same methods, thus furnishing complete information on the carbon dioxide uptake of the plant, less respiration losses. Excellent concordance between total net carbon dioxide uptake and yield of dry matter in a sugar beet crop and in alfalfa grown over a two-year period (Tables XLIII and XLIV) has been demonstrated. (See also Tables XXX and XXXI.)

In carrying out exchange studies it is essential to supply an adequate amount of carbon dioxide near all the leaves for photosynthesis. Sufficient air should be passed through the plant chamber, and mixing

TABLE XLIV

Net Assimilation of Carbon Dioxide Expressed as Equivalent Dry Matter, for Four Plots of Alfalfa in Nine Crops during Two-Year Period As Compared with Yield Data[a]

Parameter	Plot 1 Net assim. (kg)	Plot 1 Yield (kg)	Plot 2 Net assim. (kg)	Plot 2 Yield (kg)	Plot 3 Net assim. (kg)	Plot 3 Yield (kg)	Plot 4 Net assim. (kg)	Plot 4 Yield (kg)
1940, 4 crops	3.15	1.62	4.21	2.75	3.98	2.12	5.58	3.76
Winter respiration	−0.27	—	−0.27	—	−0.23	—	−0.23	—
1941, 5 crops	8.13	7.23	9.82	8.60	8.20	6.58	8.17	6.78
Crowns and roots	—	2.86	—	2.72	—	3.25	—	2.90
Total	11.01	11.71	13.76	14.07	11.95	11.95	13.52	13.44
Yield/assimilation	—	106%	—	102%	—	100%	—	99%

[a] From Thomas et al. (349).

in the chamber should be maintained by auxiliary fans if necessary, so that the carbon dioxide available to the leaves does not fall more than about 10–15% below intake concentration. Depletion of 15–20% was found by Heinicke and Hoffman (171) and by Decker (93) to cause a substantial reduction in the rate of photosynthesis. If depletion is much less than about 5%, it may be too small to measure accurately. In the operation of the carbon dioxide autometer the standard error of measuring the difference between two sources is about 0.25%. This means that a depletion of 5% can be measured with a standard error of about 5%. The rate of night respiration is usually about 10–20% of assimilation. It is therefore advisable to reduce correspondingly the airflow through the plant chamber at night in order to measure respiration accurately.

2. Chemical Methods of Analysis for Carbon Dioxide in Air

Absorption of carbon dioxide in a dilute sodium or barium hydroxide solution is rather difficult because the molecule must be hydrated before it can be fixed. This requires a few seconds to go to completion. Special absorbers have been designed. These usually involve a fritted glass plate to form small bubbles and a surface tension depressant such as 1% butyl alcohol in the solution to cause the bubbles to persist for the desired time without danger of foaming out of the vessel. Other absorbers employ large tubes with a large volume of solution or long small tubes through which the liquid and gas are drawn concurrently.

The solutions can be titrated with dilute acid, after the addition of an excess of barium chloride to precipitate carbonate. Thymol blue or phenolphthalein end points are satisfactory. Or the electrical conductivity of the solution can be measured either manually or with a recorder. A sodium hydroxide solution has about twice the conductivity of the equivalent carbonate. This is the basis for the carbon dioxide autometer which samples, analyzes and records the gas from a number of sources simultaneously, permitting the continuous determination and comparison of photosynthesis and respiration on two or more plots. The various chemical analyzers are described by Thomas and Hill (350). The "Darmstadt" conductivity apparatus (177) has been reviewed recently by Stocker and Vieweg (339). Brackett (54) determined carbon dioxide exchange by the electrical conductivity of 0.02 N potassium hydroxide solutions in his photosynthesis studies with wheat.

3. Infrared Methods

Carbon dioxide absorbs infrared radiation at 4.2 μ (also at 9 and 14 μ). McAlister (223) was one of the first to apply infrared to photosynthesis studies. Using spectrophotometric equipment with rock salt optics he passed monochromatic infrared at 4.2–4.3 μ through a 1-meter tube attached to a plant chamber containing a wheat plant. Air from the chamber was circulated rapidly through the tube and returned to the chamber. Changes in the carbon dioxide concentration of the air were shown almost instantaneously by changes in the absorption of the infrared, which permitted following the course of the photosynthesis and respiration in minute detail. Fortunately water vapor does not absorb infrared appreciably at 4.2 μ and therefore does not interfere.

Scarth and Shaw (297) employed a nondispersive system to study photosynthesis. Infrared light was passed through two parallel polished steel tubes, provided with mica windows at each end, to two similar thermopiles beyond the far ends of the tubes. The light source was a quartz U-tube filled with carbon dioxide, wrapped on the outside with resistance wire and heated to about 900–1000°C. A stream of humidified air was split, one portion going through a plant chamber, then through a drier containing magnesium perchlorate and barium chloride, and finally through one infrared tube. The other portion was similarly dried and passed through the other infrared tube. Rapid and sensitive responses to parts per million changes in the carbon dioxide content of the air were obtained. Water vapor would interfere if not removed or otherwise compensated. Scarth used the same equipment with short, appropriately calibrated cells for water vapor, to study transpiration.

In recent years infrared instrumentation has been greatly advanced and a number of excellent instruments are available commercially.

4. *Plant Chambers*

Many types of equipment have been used for assimilation studies ranging from a transparent plastic envelope to hold a single leaf or a bell jar for a single plant, to a demountable greenhouse large enough to enclose an 8-year-old apple tree (172). The latter chamber was built on a platform covering the area under the tree. It was provided with blowers, mixing fans, and refrigeration units to maintain the desired atmospheric conditions for the tree. Other field apparatus in use include (a) a portable chamber about $6 \times 6 \times 5$ feet made of plastic sheets on light metal frame work to cover field plots; (b) glass chambers, $6 \times 6 \times 6$ feet mounted above concrete tanks each containing four tons of silica sand and provided with automatic irrigation equipment. An integrated installation comprising 16 such units has been described (350). Air flow through these large chambers was measured with anemometers or calibrated orifices. Carbon dioxide concentrations were determined by continuous sampling of the air as it entered and left the chamber.

Ashton (16) measured the photosynthesis of single large potted sugar cane plants, the shoots of which were mounted in narrow upright plastic cabinets $72 \times 72 \times 4$ inches, with the pots below. A "batch" process was employed. Each cabinet was flushed out during 1 minute with 1000 cubic feet of air. The blower was then stopped, and the batch of air in the cabinet was continuously mixed and cooled for 14 minutes when it was replaced by fresh air for the next batch exposure. A small sample of the air representing the 13th minute of exposure was analyzed for carbon dioxide in an infrared analyzer. Each cabinet, and also the air supply and CO_2-free air were analyzed every half hour from before sunrise until after sunset each day. Calibration was carried out by mixing known volumes of carbon dioxide and CO_2-free air for infrared analysis.

Musgrave and Moss (250) described photosynthesis measurements on one-thousandth acre corn plots in the field. The cabinets were $5 \times 9 \times 12$ feet high. They employed a plastic film, 0.003 inch thick, which was impermeable to carbon dioxide, and very stable in sunlight. The film was cemented to a framework to make a gas-tight chamber. The latter was mounted on a framed plastic-covered floor having the film wrapped around each corn stalk and cemented to make a gas-tight separation from the soil. Air in the chamber was well circulated by mixing fans and cooled to ambient temperatures by refrigeration units. A 2-ton compressor served two chambers. A stream of air from each chamber was drawn continuously through a manifold near an infrared carbon dioxide analyzer and returned to the chamber. The manifold was connected to the analyzer through a multiple port valve that permitted automatic sampling from a number of sources in turn on any desired schedule. In practice, the analyzer drew a small sample for 3 minutes. The carbon dioxide concentration was recorded and the source of the sample was changed for the next 3 minutes. A convenient schedule gave two analyses of each pair of cabinets and one of the outside air each 15 minutes.

Carbon dioxide from commercial cylinders was metered into the return flow from the manifold to the cabinet under control of the analyzer so as to maintain 300 ppm or any other desired concentration in the chamber. Assimilation was calculated from the amount of carbon dioxide added. To determine respiration at night, the carbon dioxide concentration buildup was measured over a 90-minute period, then the air in the cabinet was automatically replaced by fresh air and the process was repeated.

5. Diurnal Carbon Dioxide Exchange Curves

Two similar alfalfa plots (350–352) were covered with portable plastic-covered cabinets; measured air streams were passed through them, and the intake and outlet carbon dioxide concentrations were monitored continuously for about a month, the automatic chemical analyzer being used. The carbon dioxide absorbed or evolved by the plants was calculated for 64-minute periods. Curves for 4 days of this experiment are presented in Fig. 34.

The two plots gave nearly identical curves even on September 25 and 26 when cloudy weather caused photosynthesis to fluctuate extensively. On September 24 one of the plots was definitely less active than the other from about 11 A.M. until 6 P.M. It was noted that the former plot was rather dry on one side and some of the leaves seemed to droop slightly. A heavy irrigation restored the equality of the two plots on subsequent days. Respiration at night was also nearly identical on the two plots. There was a tendency for the curves to rise as the night advanced, suggesting a reduction of respiration due to gradual lowering of the temperature.

Figure 35 shows the assimilation records for the two cloudy days, September 25 and 26, and also for a cloudless day, placed above the pyrheliometer traces for those days. Three similar plots are represented in each graph. On the clear day the assimilation curve has a

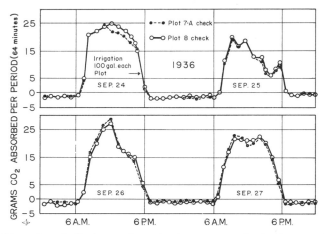

Fig. 34. Diurnal assimilation and respiration on two check plots of alfalfa. One plot showed on September 24 a slight drought effect corrected by irrigation. From Thomas and Hill (352).

shape midway between insolation on a horizontal surface and at normal incidence. The striking effect of the clouds in producing flashes of high intensity light which caused higher-than-normal photosynthesis is illustrated on September 26. It is also of great interest that the exchange curves of three similar plots measured simultaneously are so closely coincident throughout the day, even with large cloud effects.

Figure 36 shows the assimilation curves of three plots in midsummer in relation to the height of the plants, and also of three plots of comparable stage of growth at approximately monthly intervals. In the first group an eightfold increase in height caused a twofold increase in photosynthesis. This is reasonable because side illumination on the 48-inch plants was about equal to top illumination, making total illumination about double that on the 6-inch plants. In the second group the spread of the curves varied with day length, but the

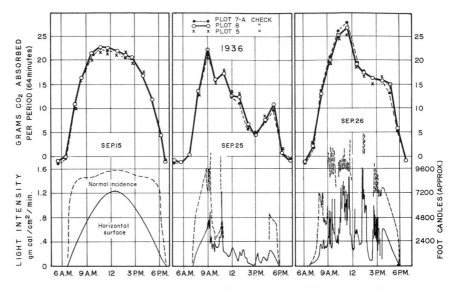

FIG. 35. Relation of photosynthesis to light intensity. Diurnal assimilation curves on 3 days of three similar plots of alfalfa, placed above the corresponding pyrheliometer records. From Thomas and Hill (352).

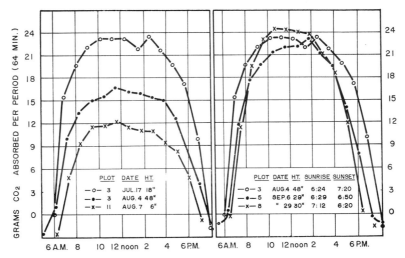

FIG. 36. Diurnal assimilation of similar alfalfa plots at three stages of growth and at three maturity dates. From Thomas and Hill (352).

maximum height occurred late in the season, presumably owing to lower midday temperature and respiration rate.

It may be noted that the midday reduction of photosynthesis described so frequently in older literature (208, 210, 235) has not been observed in many years' operation of this apparatus, except for a few small and questionable dips in the curves. The plots were well supplied with moisture, and leaf temperatures did not exceed ambient temperatures appreciably (maximum about 100°F). However recent observations on citrus indicate serious curtailment of photosynthesis if the ambient temperature reaches 105–110°F.

6. Root Respiration

Root respiration was readily measured on the sand culture plots (350). A small measured volume of air (0.5 cubic foot per minute) was

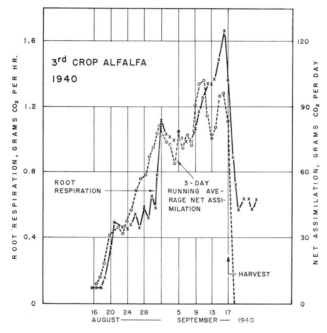

FIG. 37. Daily root respiration of a third-crop 1940 alfalfa plot as compared with the 3-day running-average net assimilation values. From Thomas and Hill (350).

drawn down through the sand into the irrigation manifold. An aliquot of this air was sampled automatically for carbon dioxide, and the solutions were titrated periodically to give one or two average concentration values per day. Figure 37 gives the root respiration in third-crop alfalfa, together with the 3-day running average net assimilation. Starting with low values of both parameters, the two curves rose more or less together until harvest, after which both dropped suddenly to low values again. Evidently the two processes were closely associated, though the root respiration rate was only about 1.4% of the assimilation rate.

Moss, Musgrave, and Lemon (247) determined root respiration of field corn by sampling the gas which diffused from the soil under the floor of the plant chamber. The diurnal pattern showed a maximum at about 4 P.M., and a minimum at about 6 A.M. Dry soil evolved about 70% as much carbon dioxide as moist soil. The average rate of soil respiration was 3.5 pounds of CO_2/acre/hour, representing 5% of the carbon dioxide assimilated by the corn at midday (72 pounds CO_2/acre/hour on a clear day or 50% of assimilation on a rainy day. Probably only part of this carbon dioxide came from the plant roots.

7. Assimilation and Yield Studies

Experiments (350) were carried out in which the gas exchange of sugar beets and a two-year growth period of alfalfa was measured throughout the life of the plantings, including the winter respiration of the alfalfa. The yields of dry matter produced on the sugar beet plots are compared in Table XLIII (p. 148) with the dry matter estimated from the net assimilation. In four experiments, the ratio of the yield to the net assimilation ranged from 97 to 103%.

A similar experiment with four plots of alfalfa gave the data summarized in Table XLIV (p. 149). Again there is excellent agreement between the weight of dry matter from each plot and that estimated from net assimilation. In this case carbon dioxide exchange was measured on four crops in 1940 and on five in 1941. In addition root respiration was determined during the intervening winter months. The ratio of yield to net assimilation ranged from 99 to 106% on the four plots.

Heinicke and Childers (172) studied an 8-year-old apple tree throughout an entire growing season. The tree was enclosed in a glass chamber with a wood floor. Blowers, mixing fans, and refrigeration units provided a measured and conditioned air stream which entered at the top and emerged around the tree trunk below the floor. Carbon dioxide analyses of intake and outlet air streams were made by aspirating sodium hydroxide solutions continuously for the desired number of hours in large absorbers provided with fritted glass bubblers. Gas meters were used to measure the sample volumes and the solutions were titrated as already described. Figure 38 gives the data on a monthly basis, summing up samples taken during three periods each day: 7:30 P.M. to 9:30 A.M., 9:30 A.M. to 2:30 P.M., and 2:30 P.M. to 7:30 P.M. The most rapid assimilation occurred during the midday hours. The 14-hour sampling period represented the excess of the early morning assimilation over night respiration. Root and soil respiration were excluded because the whole chamber was above the ground.

The tree was 3 m high and covered an area of 4 m². It developed most of its leaves between May 20 and June 10. The compensation point was exceeded May 22. Leaf fall began in mid-October and was 90% complete when the experiment was terminated November 17. The tree was then sacrificed and the current year's growth on the tops and roots was separated, dried, and weighed. The yield of dry matter agreed

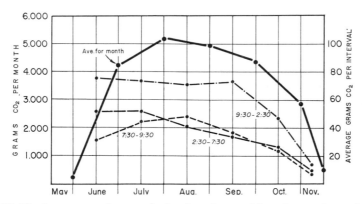

FIG. 38. Total apparent photosynthesis of an 8-year-old apple tree for each month of the growing season and the average daily rates each month from 7:30 P.M. to 9:30 A.M., from 9:30 A.M. to 2:30 P.M., and from 2:30 P.M. to 7:30 P.M. From Heinicke and Childers (172).

closely with total net assimilation, each representing 14 kg, equivalent to 15.5 tons/acre. The yields of sugar beets and second year alfalfa described in Tables XLIII and XLIV were 9.3 and 13.3 tons/acre, respectively, on the sand-culture plots. These yields are compared in Table XLII with maximal yields of other crops recorded in the literature.

8. Aerodynamic Measurement of Photosynthesis

An entirely different method has been suggested for determining carbon dioxide exchange in the field, based on the laws of aerodynamic diffusion.

Absorption of carbon dioxide due to photosynthesis by a field crop would be expected to establish an increasing concentration gradient with height above the plants. Respiration would reverse the gradient. The magnitude of these gradients would depend not only on the rate of absorption or evolution of the gas by the plants, but also on meteorological factors, particularly wind speed gradients near the top of the vegetation. The problem is further complicated by temperature and air density gradients, but these can generally be neglected in the short distances under consideration. Finally the "roughness" of the surface of the vegetation may be important in determining the effective plant height.

The rate of photosynthesis $P(\text{gm/cm}^2/\text{sec})$ is a function of the vertical flux of carbon dioxide, which is proportional to the change of concentration C with height z

$$P = A \frac{dC}{dz}$$

A is the "Austausch" or exchange factor of Schmidt (299).

Inoue *et al.* (186) modified an integrated form of the foregoing equation which had been used successfully in field evaporation-transpiration studies. As applied to carbon dioxide exchange, the equation becomes

$$P = \frac{k^2(U_2 - U_1)(C_1 - C_2)}{\left[\ln\left(\dfrac{z_2 - d}{z_1 - d}\right)\right]^2}$$

U_1 and C_1 are the wind speed and carbon dioxide concentration at height z_1 and U_2 and C_2 are the corresponding values at height z_2. The zero plane displacement or effective height of the plants is d, and k is Karman's constant ($= 0.4$).

The method was tested in a rice (*Oryza sativa*) paddy. The plants were 90 cm high and the effective height was taken as $d = 70$ cm. Wind speeds and carbon dioxide concentrations were measured at 5 elevations between 80 cm and 6 meters.

Small, sensitive cup anemometers were used to measure wind speed. Five air samples were drawn simultaneously from the various levels into 7-liter vinyl bags. The carbon dioxide in 6 liters of each air sample was absorbed in sodium hydroxide and determined by conductivity of the solution. Gradients in the 2-meter space above the plants were used to calculate the assimilation rates for 10-minute periods starting at 1518, 1634, and 1825 hours on August 18, 1956. Values of 14, 4, and -12 mg CO_2/dm^2 leaf surface per hour were found. The sunlight was strong at 1518 and it was nearly sunset at 1825. Photosynthesis in this experiment was not measured independently by other methods, but the values are in fair accord with conventional measurements made at other times.

Huber (182, 183) has measured the gradients of carbon dioxide and also water vapor to determine photosynthesis and transpiration in the field. The infrared method was employed for carbon dioxide determination, and for humidity a method was used based on heat liberation when the moisture was absorbed in sulfuric acid. Illustrative curves and calculations have been published giving reasonable assimilation and respiration data, but, as with the data of Inoue, independent confirmation of the results was not supplied.

Lemon, Musgrave, and associates at Cornell (212, 247, 250) have undertaken a comprehensive study of field corn, in which the measurement of photosynthesis is recognized as a valuable tool for the evaluation of environmental and nutritional factors in relation to crop yield. Their closed chamber gas exchange technique (250) has already been referred to. An investigation of the aerodynamic method was undertaken with the realization "from the onset that such calculations of carbon dioxide exchange are at best semi-quantitative." Nevertheless, the avoidance of artificial conditions associated with a canopy over the plants was sufficient justification for the effort.

The first published paper (212) gives a critical discussion of the method—with models—and defines objectives, requirements, and limitations:

1. The first requirement is a level homogeneous surface of vegetation such as a wheat, corn, or alfalfa field with sufficient upwind "fetch" to ensure equilibrium gradients within a turbulent boundary layer above the plants where velocity and carbon dioxide profile measurements are made. Practical methods of defining the boundary layer with its lower reference plane are discussed.

2. Wind speeds should be moderate and uniform (about 6 to 15 mph). Very low speeds fail to supply the plants with all the carbon dioxide they can utilize, and the assimilation values are therefore low. High speeds give gradients too small to be accurately measured. The infrared analyzer should be sensitive to differences of 1 ppm CO_2 and the anemometers to 0.01 mph (5 cm/sec).

3. With tall crops that wave, such as corn, the height of the reference plane and the "roughness coefficient" may vary with the wind speed, necessitating correction factors in the exchange equation.

4. Carbon dioxide from the soil may need to be corrected for, but no study of this problem is reported.

5. Comparison of photosynthetic rates by the aerodynamic and closed canopy methods gave results that were somewhat higher by the former

method, but of the same order of magnitude. Possibly the differences were real, if the canopy interfered significantly with the supply of light and/or carbon dioxide. The studies are considered to be "encouraging" because of the approximate confirmation by the closed canopy method. They have also "revealed interesting phenomena in the crop growth dependence for carbon dioxide exchange upon the natural air turbulence."

9. Photosynthesis in Water Plants

Large water plants have been preferred objects for the study of photosynthesis since the time of Bonnet (48), who first recorded the evolution of oxygen from submerged leaves under the influence of sunlight. This is due to the ease with which the process can be followed by counting the bubbles of gas emitted. F. F. Blackman and Smith (33) pointed out that this method is applicable with precision only for the intermediate range of temperature, light, and carbon dioxide supply. In more extreme conditions, complications arise that can render the method unreliable. For example, the solubility of oxygen may need consideration, or carbon dioxide may be evolved with the oxygen.

Blackman and Smith circulated boiled tap water, containing a known concentration of carbon dioxide, through a plant chamber at 300 cc per hour. Provision was made for titrating samples of the water for carbonic acid before and after it passed through the chamber; also for analyzing for carbon dioxide all the free gas evolved by the plant. The plant was illuminated by a special gasmantle burner placed at various distances from the leaves. The light passed through a screen of flowing water to remove the infrared wavelengths. The carbon dioxide exchange data obtained by this procedure were evidently very reliable. Each value usually represented an integrated 40- to 60-minute period.

Bose (50) developed an instrument to record each bubble of oxygen issuing from a special outlet on a plant chamber containing a submerged water plant. This outlet had a small amount of nonadhesive oil and a drop of mercury in a constriction in a glass tube. A definite volume of gas (0.2–5.0 mm^2) was emitted each time the drop of mercury moved, depending on the volume of free gas above the plant. Each bubble made an electrical contact which was recorded. This permitted following the assimilation process in great detail and with considerable precision. It might have been desirable to collect the evolved gas for volume measurement and carbon dioxide analysis, but this was not done. Bose carried out many experiments with this equipment on the water plant *Hydrilla verticillata* as reported in his book "Physiology and Photosynthesis."

In general it appears that the aquatic angiosperms respond similarly to land plants with respect to the effects of light, carbon dioxide, and temperature on photosynthesis, as discussed in Section III, A, 1 and 2. For example, Blackman and Smith's original observations of "limiting factors" were made on *Elodea* (*Anacharis*) and *Fontinalis*. Harder (164), E. L. Smith (316) (Fig. 12), and Steemann-Nielsen (327–329) also worked with large submerged plants. Rigid control of the environmental conditions—temperature, light intensity, and carbon dioxide supply was readily possible in these experiments.

10. Effects of Toxic Gases on Photosynthesis

a. Sulfur dioxide. Photosynthetic measurements afford an excellent means of evaluating damage to the leaves of plants by toxic gases and injury due to absorption of toxic substances through the roots. To illustrate the application to a practical case, typical exchange curves for two similar alfalfa plots are given in Fig. 39 covering an 8-day period from August 13 to August 20 (351). The previous crop had been cut on July 27. New growth developed rapidly. In 2 weeks photosynthesis reached its maximum rate, and the total daily assimilation did not increase appreciably thereafter. Fumigation treatments with sulfur dioxide were applied to plot 12 on each of 6 days, starting about 9 A.M. and lasting 4 hours. Concentrations were about 0.57 ppm on August 8, 9, 10, and 0.60–0.66 ppm on August 15, 16, and 17 (Fig. 39). The sulfur dioxide reduced photosynthesis on plot 12 as compared with the control plot 11 from the time of initial application of the gas until about 3 hours after the treatment was stopped. Then the two curves became nearly coincident. The assimilation level was about 76% of the control during the first three treatments and 74% during the second three treatments. On August 11, plot 12 was a little less active than the control, but there was no significant difference in the two plots on August 12–14. Slightly reduced assimilation was also shown from August 18–20 on plot 12 suggesting a little permanent damage. The final yield was reduced about 10% as estimated either from the carbon dioxide exchange or from the weight of the crops. Leaf destruction due to sulfur dioxide was 2.5% of total area.

Evidently the action of the sulfur dioxide was local and for the most part temporary. A small amount of "invisible injury" is suggested by the data. It would have been impossible to draw such a conclusion from yield data alone except by fumigating many plots. A similar experiment with ten 4-hour fumigations at 0.44 ppm on ten different days also caused some temporary reduction in photosynthesis averaging 7–14% of the control when the gas was present. However, total reduction in net assimilation and yield was only about 1% because of a slight post-fumigation stimulation of activity. No more than traces of leaf injury were observed. Concentrations of about 0.3 ppm were applied for about 7 hours each day for a month without interfering with photosynthesis. Above 0.6 ppm the interference became more pronounced

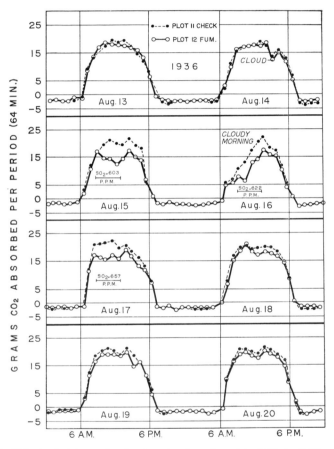

Fig. 39. Effect of 4-hour fumigation treatments with 0.60 to 0.66 ppm sulfur dioxide on three consecutive days on the rate of photosynthesis of alfalfa. From Thomas and Hill (351).

and acute injury was readily produced. Similar experiments have been carried out by Katz *et al.* (196).

b. Hydrogen fluoride. In other experiments, hydrogen fluoride was used as the fumigant (348). Treatment of gladiolus, which is extremely sensitive to hydrogen fluoride, with 1–12 parts per billion (ppb) (or thousand million, pptm), 6–9 hours per day for 2 months gave results shown in Fig. 40. It was observed that the gas was absorbed over the whole surface of the leaves, then was translocated to the tips of the blades. Here it accumulated in sufficient amount to kill the tissue. As the experiment continued, the necrotic area extended progressively farther down the blade with a sharp line of demarcation between the dead and the healthy areas. Figure 40 shows that the rate of assimilation and percentage of green tissue declined together until both had been reduced by about 40–45%. Evidently the green tissue retained its

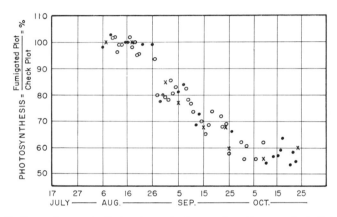

Fig. 40. Daily carbon assimilation by a hydrogen fluoride-treated gladiolus ('Surfside'–1953) plot, expressed as percentage of a control plot, compared with the reduction in functional leaf area due to the fumigation. ●, Days without fumigation; ○, days with fumigation; X, percentage uninjured leaf area on fumigated plot. Thomas (348).

full photosynthetic activity until it became necrotic as a result of translocation which prevented a buildup of fluoride in the green areas, as shown by fluoride analysis.

Figure 41 shows the effect of higher concentrations of hydrogen fluoride on photosynthesis of gladiolus. In this case there was a drastic reduction in assimilation immediately after the fumigation, bearing no relation to leaf destruction. Presumably the fluoride concentrations in the active areas of the leaves exceeded the threshold for interference with photosynthesis. The function gradually returned and eventually reached a level corresponding to the amount of uninjured green tissue that survived the treatment. This is a case of "invisible injury" because the growth of the plant is reduced without the production of corresponding leaf lesions. Similar effects have been observed on other plants.

Tables XLV and XLVI show the results of many experiments on gladiolus and other less sensitive plants with relatively low and relatively high concentrations of hydrogen fluoride. The wide range of concentrations to which the different plants are susceptible is a striking feature of these tables. Cotton, for example, can tolerate about a hundredfold more fluoride than gladiolus.

Recovery from a photosynthesis-inhibiting treatment with hydrogen fluoride is much slower than from an equivalent treatment with sulfur dioxide. The latter is removed from the system by rapid oxidation to sulfate; the former is removed by the slower processes of

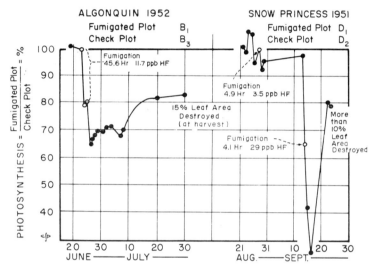

FIG. 41. Effect of relatively high concentrations of hydrogen fluoride on the total daily carbon assimilation of gladiolus. From Thomas (348).

translocation, precipitation, volatilization, and possible conversion to less toxic organic fluorine compounds.

c. Smog. There are two types of urban air pollution that are called "smog" (smoke plus fog): (a) London-type smog due largely to coal smoke containing unburned organic compounds, smoke aerosols, sulfur dioxide and sulfuric acid aerosol, together with more or less fog—a true "smog"; (b) Los Angeles-type or "photochemical" smog resulting from the action of sunlight on mixtures of nitric oxide, nitrogen dioxide, and olefinic hydrocarbons or other organic compounds (without fog) to produce ozone, peroxyacyl nitrates, aldehydes, aerosols, etc. The London-type smog has reducing properties; the Los Angeles type, powerful oxidizing properties.

Both types of smog produce their characteristic lesions on the leaves of plants. They also reduce growth and in some cases increase respiration rates. However, even without the production of leaf lesions there may be interference with photosynthesis and growth. Bleasdale (37) noted that dry weight and tillering of ryegrass, *Lolium perenne*, was reduced by one-third to one-half in the ambient air of Manchester,

TABLE XLV

Effect of Fumigation with Relatively Low Concentrations of Hydrogen Fluoride on the Photosynthesis of Plants[a]

| | HF Fumigation | | | | |
| | | Concentration | | Photo-synthesis reduction[c] (%) | Leaf area destroyed (%) |
Plant	Duration[b] (hours)	Max. (ppb)	Avg. (ppb)		
Fruit trees	183	9.0	2.6	14	10
Fruit trees	39	15.0	4.8	2	<5
Fruit trees	8	4.0	3.9	0	0
Gladiolus					
'Surfside'	53	11.3	6.5	15	15
'Surfside'	44	9.8	6.4	28	30
'Surfside'	205	13.2	4.5	45	44
'Surfside'	128	33.4	5.7	20	22
'Alladin'	30	5.1	3.9	0	3
'Algonquin'	111	7.1	4.3	2	3
'Commander Koehl'	42	10.2	5.5	0	3
Mixed grain	10	200	145	0	Trace
Cotton	77	692	237	0	Trace
Cotton	138	33	17	0	0

[a] From Thomas (348).
[b] Fumigations 4–9 hours per day.
[c] 100 % minus fumigated plot/check plot (as per cent).

England, as compared with water-scrubbed air. The effect was attributed to the removal of sulfur dioxide which was present at about 0.1 ppm, but it seems more probable that other unidentified phytotoxicants were primarily involved. At 0.1 ppm sulfur dioxide does not reduce photosynthesis measurably (351).

Studies on injury to vegetation by photochemical smog (354) have shown a close relationship between leaf destruction and reduction of photosynthetic activity, but until recently "invisible leaf injury" has not been demonstrated by measured changes in assimilation rates.

However, the observations of Koritz and Went (206) indicated clearly that smog could reduce growth rates without necessarily causing visible leaf injury. Preliminary experiments at Riverside, California, with bearing orange (*Citrus sinensis*) trees under plastic canopies have compared the rates of photosynthesis of two similar trees, one supplied with ambient air, the other with carbon-filtered air. The oxidant

TABLE XLVI

EFFECT OF FUMIGATION WITH RELATIVELY HIGH CONCENTRATIONS OF HYDROGEN FLUORIDE ON THE PHOTOSYNTHESIS OF PLANTS[a]

	HF Fumigation			Decrease in photosynthesis (fumigated plot/check plot)				
				Temporary[c]				Leaf area destroyed (%)
		Concentration					Permanent (%)	
Plant	Duration[b] (hours)	Max. (ppb)	Avg. (ppb)	Duration (days)	Average (%)	Time lost (days)		
Fruit trees	4	118	52	18	11	2.0	28[d]	50[e]
Fruit trees	13	22	18	25	15	3.8	35	22[f]
Gladiolus								
'Snow Princess'	4	40	29	9	29	2.6	18	>10
'Algonquin'	46		12	21	12	2.5	17	15
'Surfside'	53	11	7	25	8	2.0	15	15
Young barley	7	68	37	7	15	1.1	1[g]	10
Alfalfa	10	1020	630	14	15	2.1	2	1
Cotton	20	1100	740	4	6	0.2	0	0

[a] From Thomas (348).
[b] All fumigations except on 'Algonquin' gladiolus were applied 4–7 hours per day.
[c] Temporary reduction below level of permanent injury.
[d] Some new growth of leaves reduced percentage loss.
[e] Approximate value 10 days after fumigation and before new growth.
[f] Estimated before leaf fall.
[g] Rapid new growth restored activity.

level of the atmosphere, as indicated by the potassium iodide recorder, usually reached about 0.15 ppm by 3 P.M. each day. Carbon dioxide assimilation by the tree receiving carbon-filtered air increased steadily for about 30 days, then leveled off at a rate corresponding to about 12% above that of the tree receiving ambient air. When the treatments of the two trees were reversed, the photosynthetic rates were also gradually reversed and to about the same extent in the opposite direction as before the reversal of the treatments. There was no significant visible leaf injury due to the smog.

11. Effects of Sprays and Diseases on Photosynthesis

a. *Auxins and herbicides.* Bean (*Phaseolus vulgaris*) plants (134) were sprayed liberally with 0.01% solutions of p-chlorophenoxyacetic acid (CPA), indoleacetic acid (IAA), indolebutyric acid (IBA), 2,4-dichlorophenoxyacetic acids (2,4-D), and naphthoxyacetic acid (NAA). Inhibition of photosynthesis during the next 2–4 days was as follows: CPA, 90%; IAA, 70%; IBA and 2,4-D, 20%. NAA increased photosynthesis 10%. Respiration was increased to 140% by CPA on the first day and declined to 80% on the third day. With IAA, respiration was 100, 120, and 80% and with 2,4-D, 80, 120, and 110% on the first 3 days; with IBA, 100 and 60% on 2 days; and with NAA, 110, 110, 100, and 100% on 4 days.

Wedding *et al.* (381) showed that variations in inhibition of photosynthesis by 2,4-D in the leaves of navel orange or in *Chlorella pyrenoidosa* were due to variations in the concentration of the undissociated molecules, whether produced by acidifying a solution of 2,4-D in the range from pH 8 to 3 or by adding extra 2,4-D to solutions buffered to a definite pH. The inhibition which was proportional to the concentration of the undissociated molecules of 2,4-D reached 50% with orange at 14×10^{-5} M and 100% with *Chlorella* at 5×10^{-5} M. Wedding *et al.* (382) sprayed citrus trees with petroleum oils according to the usual practice. Leaf disks were taken at intervals for measurement of carbon dioxide exchange in the Warburg apparatus. Photosynthesis showed a reduced rate lasting up to 2 months. Respiration was also reduced, but to a lesser extent.

b. *Plant poisons.* Wassink and associates (375) studied the effects of various poisons in lowering the light-assimilation curves, Table XLVII. Ethylurethane, potassium cyanide, hydroxylamine, and pure oxygen were used with *Chlorella* and the diatom *Nitzschia*. Similar data were obtained with purple bacteria. A general lowering of the photosynthesis was produced which was relatively somewhat greater at 6 klux than at 1 klux except with hydroxylamine. Cyanide, oxygen, and hydroxylamine affect principally the dark reactions and have little effect on the light reactions. Ethylurethane affects both dark and light reactions (see Table XXIII).

c. *Diseases.* P. C. Owen (266, 267) found that tobacco mosaic virus lowers the rate of photosynthesis of tobacco leaves. Net assimilation rates (N.A.R.) of whole tobacco plants were measured after infecting the healthy leaves by rubbing them with infected leaves or a purified virus inoculum. A trend toward reduction of growth rate and leaf size was noted in the infected plants during the first week. After 3 weeks, when visible symptoms appeared, dry weights and leaf areas were lowered 25%. There was a 16% reduction of N.A.R. in the third week, but it was not statistically significant. Direct measurements of photosynthesis showed a reduction of 14–20% in the first hour after virus inoculation, and 25–28% in the second hour. Dark respiration was increased by these treatments. Evidently the virus produced a general effect on cell metabolism.

TABLE XLVII

EFFECTS OF POISONS IN REDUCING THE ASSIMILATION OF *Chlorella* AND *Nitzschia*[a]

Plant	Compound	Conc. (%)	Reduction of assimilation at		
			1 Klux (%)	3 Klux (%)	6 Klux (%)
Nitzschia	Ethylurethane	1.5	16	20	26
Nitzschia	Ethylurethane	2.0	27	36	47
Nitzschia	Ethylurethane	2.5	40	56	68
Chlorella	Ethylurethane	2.5	45	47	—
Nitzschia	Cyanide	0.003	43	54	60
Chlorella	Cyanide	0.0015	12	56	—
Chlorella	Oxygen	100	36	47	—
Chlorella	Hydroxylamine	0.008	40	40	40

[a] From Wassink *et al.* (375).

An interesting indirect effect may be mentioned. O. Owen, Small, and Williams (265) observed that when they enriched the atmosphere with carbon dioxide to increase the growth of tomatoes, there was a marked increase in the development of the root disease *Colletotrichum atramentarium* in the soil.

VI. Algal Culture

A. OBJECTIVES

The relatively high efficiency of photosynthesis in many algae, together with a number of other desirable features connected with their culture, has made it worthwhile to initiate the study of the biological feasibility and the economics of large-scale commercial production of these plants for food or other uses. Among advantages that can be listed for algal culture are the following:

1. The plants are simple and devoid of parts, such as stems, roots, and leaves, that may be useless as food. The whole plant can be used. There is thus no waste of photosynthetic energy.

2. Cultivation of algae is possible in many places where agriculture is impracticable, for example, where the water supply or the soil is inadequate or the climate is unfavorable for agriculture.

3. Environmental conditions can be modified more readily for algae than for higher plants. Optimum light absorption can be attained by considering the best depth and density of the cell suspension. Much information is available on nutrient solutions and the limitations of temperature.

4. Continuous harvesting of the crop is possible throughout the year.

5. By varying the composition of the nutrient solution, the composition of the plants can be varied at will, making them rich in either carbohydrates, proteins, or lipids.

6. Two other closely related subjects are the culture of nitrogen-fixing algae to improve soil fertility and the growing of algae in sewage to help in disposing of the waste, at the same time producing useful food or feed.

Specifically it is desired to know whether large-scale installations, covering acres of area and with the inevitable large capital investment and operating costs, can compete economically with conventional agriculture. There are some situations in which the production of food is paramount, as during wars and in overpopulated or famine-threatened areas. In submarines and space travel, there is the added possibility of using algae for air purification. In these cases the economic considerations might not apply. This serves to heighten interest in the fundamental chemical, physiological, and engineering aspects of the problem. The Carnegie Institution of Washington, D.C. initiated work and sponsored projects in this field after World War II. Interest was also manifested in England, Holland, Germany, Japan, Israel, and other countries. In 1953 the Carnegie Institution (69) issued a publication containing 22 invited papers from various countries describing many phases of the subject, including pilot plant studies. Further consideration was given to algal culture at the World Symposium on Applied Solar Energy in Phoenix, Arizona, 1955 (344), also in a review of mass culture of algae by Tamiya in 1957 (345). Other reviews on algae include those by Ketchum (201) and Pirson (271).

B. Growth of Algae

1. Materials

The different algae vary considerably in their growth rates and their response to the environment. Much effort has been directed toward finding high-yielding plants. Promising material includes a number of varieties of *Chlorella* and *Scenedesmus*. The diatom *Nitzschia palea* is also recommended. These generally exhibit optimum growth at about 25°C. However, high-yielding varieties have now been isolated with optimum growth at higher temperatures (35–40°C). These should be useful in hot climates. The blue-green algae *Anabaena* and *Tolypothrix* have strong nitrogen-fixing properties. Their inoculation in rice fields (377) has resulted in marked increases in yield of the crop. Algae that can grow well in a symbiotic relationship with the bacteria of sewage include *Chlorella, Scenedesmus, Chlamydomonas,* and *Euglena.* The objective of finding pure strains is contrasted with the philosophy of Jorgensen and Convit (191), in Venezuela, who work with a mixture of maximum heterogeneity, even including bacteria, protozoa, and ciliates in their "plankton." This material is made up into a broth which supplements effectively the diet of patients in a leprosarium.

2. Effect of Light

Algae are shade plants that are readily injured by strong sunlight. Little success has attended efforts to develop heliophilic varieties. However, it has been shown by experiments with flashing light that the organisms can utilize efficiently high intensity flashes when there are intervening dark periods. Since any large-scale culture of algae must depend upon sunlight, special geometry is required for its most effective utilization. Evidently a rather thin dense suspension of the cells is desirable which will absorb practically all the light. Further, there should be sufficient turbulent mixing, so that the individual cells will be given, alternately, strong illumination in the surface followed by shading in the interior of the suspension. On the other hand, dilute, deep, well-stirred suspensions are also recommended (236, 251, 253).

3. Studies at Carnegie Institution

The Carnegie Institution workers (91) tested four types of equipment exposed outdoors in the sun and a fifth type artificially lighted:

1. Five-gallon bottles, each containing about 3 gallons of suspension aspirated with 5% carbon dioxide.

2. Vertical tapered sedimentation tubes 12.5–15.5 cm in diameter and 67 cm long, with 5% carbon dioxide bubbling through an annular tube in the center of the liquid column. The cells settled out at night and each morning a sludge of 22–68 times the density of supernatant suspension, depending on the light intensity and the ambient temperature, was drawn off at the bottom.

TABLE XLVIII

Production of *Chlorella* in Laboratory Experiments in Sunlight Using about 5% Carbon Dioxide

Type apparatus	Season	Diameter or depth (cm)	Conc. suspension, dry wt. (gm/l)	Yield, dry wt. (gm/m²/day)	Investigator
United States					
5-Gal bottles with 3 gal solution	Summer	25	0.02 to 0.31	16	Carnegie Inst.[b]
5-Gal bottles with 3 gal solution	Summer	25	1.5	8	Carnegie Inst.[b]
Cylinder, 14 × 67 cm (2 gal)	Feb.–Mar.	13 to 15	0.75	8	Carnegie Inst.[b]
Cylinder, 14 × 67 cm (2 gal)	June–Aug.	13 to 15	0.50	15.7	Carnegie Inst.[b]
Rocking tray, 2.65 m²	Summer	0.26 to 0.56	0.3 to 3.0	3 to 7	Carnegie Inst.[b]
Rocking tray, 2.65 m²	Summer, clear days	1.7 ± 1.0	2.7 to 11.5	8.2	Carnegie Inst.[b]
Plastic tubing	Summer	0.8	1.7 to 18	4.5 to 11.7	Carnegie Inst.[b]
Glass tubing	Nov., 30° day, 20° night, 7000 fc,[a] 25°C	1.2	1.5 to 8	10 to 18	Carnegie Inst.[b]
Cylinder with rotor		0.6	40	2.5 to 43	Carnegie Inst.[b]
Plastic tube	Aug.–Sept.	7	0.3 to 0.5	9 to 11	A. D. Little, Inc. (216)
Pilot plant	Nov.–Dec.	7	0.1 to 1.5	1 to 13	A. D. Little, Inc. (216)
Germany					
Trough, 0.7 × 9 m	July–Aug.	9 to 15	0.21 to 0.27	4.5 to 6	Gummert *et al.* (158)
Trough, 0.7 × 9 m	Sept.–Oct.	9 to 15	1.0	3.3	Gummert *et al.* (158)
Holland					
Kolle flasks	July–Aug.	2.4 to 3.0	0.5 to 1.3	10 to 34	Wassink *et al.* (372)
Test tubes	Sept.–Oct.	2.4 to 3.0	0.3 to 0.5	5 to 10	Wassink *et al.* (372)
Japan					
Trough, 1 × 15 m	Feb.	3	1.0	1.7 to 4.1	Tamiya *et al.* (346)
Pond	Winter	5 to 10	4 to 5	2 to 3	Tamiya *et al.* (346)
Pond	Summer	5 to 10	4 to 5	15 to 18	Tamiya *et al.* (346)
England					
Lucite tank	Summer 25°C	5	0.3	15	Geoghegan (141)

[a] Six 300-watt reflector spot lamps.

[b] From Burlew (69) and Davis *et al.* (91).

171

3. A 4×8 foot horizontal tray of 2.65 m² area. A shallow layer of liquid was supplied with 5% carbon dioxide under a thin transparent plastic cover and kept agitated by rocking the tray.

4. A horizontal grid of 8-mm (i.d.) glass or Tygon tubing, 40 feet long. The suspension was pumped through this tubing to and from a separatory funnel that served both as settling chamber and as a place for introducing 5% carbon dioxide gas bubbles into the liquid stream. A sludge was removed from the funnel each morning after permitting overnight settling.

5. A plastic cylinder, 7.5 cm in diameter nearly filled with a stainless steel rotor except for an annular space 6 mm wide. A dense algal suspension (4% dry weight) occupied the annular space. The cells were aspirated with 5% CO_2 and illuminated with about 7000 fc. The rotor was turned at different rates to give controlled degrees of stirring.

Table XLVIII summarizes the results of these experiments together with data by other investigators. Maximum yields were obtained in the summer, when the skies were clear, with concentrations of cells sufficient to absorb practically all the light. The smaller the container or thickness of absorbing layer, the greater the cell concentration. In the large bottles and tubes the optimum concentration was about 0.03–0.5 gm/liter; on the rocking tray with a 1.7-cm layer of culture it was about 3–12 gm/liter; in the 0.8 cm tube, 2–18 gm/liter; and in the rotor-stirred cylinder, it was 40 gm/liter. Yields in the sunlight up to 16–18 gm/m²/day were obtained in the first four types of apparatus. The fifth type gave yields of 25 gm/m²/12 hours at 75 klux with bubble stirring only. With mechanical stirring, yields up to 43 gm/m²/12 hours were obtained. In flat flasks in sunlight, yields up to 28–34 gm/m²/day were obtained in Holland (372) and Japan (344–346). With unfavorable light or temperature conditions the yields were cut down drastically. The Perspex (Lucite) tank in England (141) which was 130 cm high by 45 cm long by 10 cm broad and held 60 liters, gave yields of 0.3 gm/liter/day, equivalent to 15 gm/m²/day, assuming illumination was on both large surfaces. Temperature was maintained at 25°C by circulating water through a glass coil.

4. Effects of Temperature

The effect of temperature of the medium on the growth of *Chlorella* in thermostated suspensions of various densities in sunlight is shown in Table XLIX (91). Half-inch glass tubes with upturned ends were immersed in a slightly inclined position just below the surface of the water bath. In the first experiment, air containing 5% carbon dioxide

TABLE XLIX

EFFECT OF TEMPERATURE AND CELL DENSITY ON THE GROWTH OF *Chlorella*
IN SUNLIGHT[a]

			Yield (gm/m²/day)							
Expt. no.	Dura-tion (days)	Conc. of inoculum (gm/l)	Continuous					25°D[b] 20°N	30°D 20°N	35°D 20°N
			15°	20°	25°	30°	35°			
1	6	0.43	2.5	—	5.7	—	0.3	—	—	—
1	6	1.45	3.3	—	10.5	—	2.1	—	—	—
1	6	7.5	3.6	—	6.8	—	1.0	—	—	—
2	5	1.45	—	12.7	13.4	8.5	—	15.6	17.8	—
3	4	1.45	—	—	15.3	—	3.8	17.0	—	10.7

[a] From Milner *et al.* (243).
[b] D = day; N = night.

was bubbled slowly through the tubes. Three cell densities were employed simultaneously at 15°, 25°, and 35°C. At 15° the yield was small and the densest suspension gave the highest yield by a small mar-

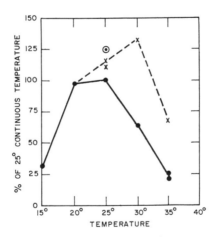

FIG. 42. Growth of *Chlorella* in sunlight at various temperatures, relative to the growth at 25°C. Solid line; continuously maintained temperatures; broken line; day temperatures as shown, night temperature 20°C in each case. From Burlew (69).

gin. At 25° the yield was two to three times that at 15° and the intermediate suspension was best. This culture was just dense enough to absorb practically all the incident sunlight. At 35° the yields were reduced by factors of 4–20 from the 25° values, but the intermediate suspension was still the most effective.

In the second and third experiments growing conditions were better than in the first, judging by the yields of the 25° cultures. The latter were increased 15–30% by operating the system at 20° at night when the day temperatures were 25°. The yield at 30°-day–20°-night was double the 30° continuous value. The yield at 35°-day–20°-night was

TABLE L

GROWTH OF ALGAE IN CONTROLLED LABORATORY EXPERIMENTS: CULTURES ASPIRATED
IN ARTIFICIAL LIGHT WITH 5% CARBON DIOXIDE

Chlorella	Light (klux)	Temp. (°C)	Growth (gm/m²/12 hours)	Efficiency light conversion (%)	Reference
C. pyrenoidosa	75	25°	25[a]	4	Burlew (69)
	75	25°	43[b]	7	Davis et al. (91)
C. ellipsoidea	50	25°	32[a]	8	Tamiya (345)
	50	15°	14	3	
	50	7°	0[c]	0[c]	
	25	25°	23	11	
	25	15°	11	5	
	25	7°	1[c]	0.5[c]	
	10	25°	14	17	
	10	15°	8	9	
	10	7°	3	3	
	2	25°	4	24	
	2	15°	4	22	
	2	7°	1	8	
C. Strain A	30	32°	9.7[b]	5	Wassink et al. (372)
	4.1	24°	3.9[b]	13	
	3.4	32°	2.6[b]	11	
C. vulgaris	14	25°	7.7[a]	20	Geoghegan (141)

[a] Bubbling 5% CO_2.
[b] Bubbling and stirring.
[c] Cells bleached.

reduced below the 25° level, but it was still three times the yield at 35° continuously applied. These yields, expressed as percentage of yields of the applicable 25° continuous cultures, are shown graphically in Fig. 42.

In Table L (345) are listed the yields of four strains of *Chlorella* and the efficiencies of light utilization at 25° with different intensities of artificial light. Yields at lower and higher temperatures are also listed. The yields decrease, and the efficiencies increase, with decreasing light. At the highest light intensity, with type 5 apparatus, the yield and efficiency were significantly higher in the well-stirred suspension than in the suspension stirred only by bubbles. The highest yield, 43 gm/m²/12 hours was obtained with this equipment. This supports the reasoning that turbulent mixing of a dense suspension offers a promising approach to high yields in bright sunlight.

5. Flashing Light

Kok (203, 205) used flashing light to study the theoretical basis of this approach. He concluded that with *Chlorella* the exposure of the individual cells to full sunlight should not exceed 4 milliseconds and that the cells should be dark about 10 times as long as they are illuminated. A shorter dark period is insufficient to complete the "dark" reaction after the flash. A longer period would permit respiration to reduce the yield. Evidently this balance poses difficult engineering problems, the solution of which will doubtless require some compromises.

Halldal and French (161) have suggested a simple apparatus for producing crossed gradients of light and temperature in order to evaluate the optimum conditions for growth of different algae. The latter are grown on agar mounted on a foot-square aluminum plate 2 cm thick having a light gradient in one direction and a temperature gradient in the other. A pH gradient could be arranged in one direction if desired. The apparatus should facilitate the isolation and selection of suitable algae for mass culture or other special purposes.

C. Theoretical Yields

The maximum light flux from the sun and sky is about 10,000 fc, or 108 klux. Only about 40% of this light is photosynthetically active. The conversion factor for this active portion of sunlight is about:

$$1 \text{ lux} = 4 \text{ ergs/cm}^2/\text{sec}$$
$$10,000 \text{ fc} = 108 \text{ klux} = 4500 \text{ kcal/m}^2/12 \text{ hours}$$

If the dry cells contain 50% carbon, and have a heat of combustion of 5.8 kcal per gram, 10,000 fc of sunlight should yield a maximum of about 770 gm dry cells/m²/12 hours, assuming all the light is utilized in photosynthesis.

In another calculation, considering only the photosynthetically active portion of sunlight:

$$1 \text{ lux} = 1.2 \times 10^{12} \text{ quanta/cm}^2/\text{sec}$$
$$10,000 \text{ fc} = 108 \text{ klux} = 1.3 \times 10^{17} \text{ quanta/cm}^2/\text{sec}$$

If the efficiency of photosynthesis is 3 quanta per molecule of carbon dioxide reduced (370), the maximum reduction due to 10,000 fc would be:

$$0.43 \times 10^{17} \text{ mole } CO_2/cm^2/sec$$
or
$$1.8 \times 10^{25} \text{ mole } CO_2/m^2/12 \text{ hours}$$

This represents 30-gm equivalents of carbon or about 720 gram dry cells/m^2/12 hours. If the efficiency of photosynthesis is 9 quanta per CO_2 molecule (229), this value would be 240 gm dry cells/m^2/12 hours.

In practice, insolation varies during the day and with the seasons because of the changing angle of incidence and cloudiness. Table II gives the average daily insolation by months near Boston and also in the arid Southwest at El Paso and Phoenix, as measured by the United States Weather Bureau. Photosynthetically active radiation at these locations represents about one-fourth to one-half the 4500 kcal/m^2/12 hours calculated for a continuous light flux of 10,000 fc. Even so, the average annual theoretical yields on the 9-quanta hypothesis should exceed 100 gm dry cells/m^2/day, in the Southwest and reach 63 gm near Boston (100 gm/m^2/day is equivalent to 160 tons/acre/year). These theoretical yields are much higher than any that have been attained experimentally at high light intensity. The 43 gm/m^2/12 hours yield at 75 klux (Table L) represents an actual conversion efficiency of 7.7% based on heat of combustion and 26% on the 9-quanta assumption. It seems likely that the attainment of higher efficiencies at sunlight intensities will require more complex equipment plus mechanical energy and perhaps also heliophilic varieties of algae. This might be practicable in submarines or space ships. Further, continuous illumination does not appear to injure the cell suspensions, so that it may be possible to obtain yields for 24 hours per day instead of for shorter periods.

D. Culture Techniques

1. Nutrient Solutions

The composition and concentration of the nutrient solution are important factors affecting the yield of algae, particularly with respect to nitrogen, iron, phosphorus, and trace elements. Macroelements are employed in about the same concentrations as in conventional plant nutrient cultures. Adjustments may be made to control contamination. For example, a low calcium medium permitted good growth of *Chlorella*, at the same time eliminating the blue green algae (158). Adequate levels of soluble iron and phosphorus can be maintained by using the chelating agent ethylenediamine tetraacetate (EDTA), or nitroso—R salt. Control of nitrogen levels and the accompanying pH level may be accomplished by the proper ratio of nitrate and ammonium salt. Urea has definite advantages over other forms of fixed nitrogen. It can be used in higher equivalent concentrations to give higher yields. It does not cause appreciable fluctuations of pH and does not support the growth of culture contaminants. Control of contamination of the cultures by rotifers and protozoans is accomplished by the addition of 1–3 ppm of pentachlorophenyl acetate or 2,4-dinitro-6-cyclohexyl phenyl acetate.

2. The "R" Value as Index of Chemical Composition

The chemical composition of algal cells can vary over a wide range according to the environment in which they are grown. Spoehr and Milner suggested (322) an index for the chemical composition of the cells based on the elementary analysis to indicate the degree of reduction of the system:

$$R = \frac{(2.66 \cdot \%C + 7.94 \cdot \%H - \%O)100}{400}$$
$$= 0.67C + 1.99H - 0.250$$

where 2.66 is the ratio of O_2 to C in carbon dioxide, 7.94 is the ratio of O to H_2 in water, and 400% is the summation for the fully reduced

TABLE LI

Constituents of *Chlorella* Calculated from R Values[a]

R Value	Protein (%)	Carbohydrate (%)	Lipid (%)
38	58.0	37.5	4.5
42	50.0	32.2	17.7
50	28.3	26.2	45.5
56	15.7	19.0	65.3
63	8.7	5.7	85.6

[a] From Spoehr and Milner (322).

compound, methane. On this scale, the R value for carbon dioxide is zero and for methane, 100. It is about 28 for carbohydrates, 42 for proteins, and 67 for lipids. The approximate composition of the mixture of these compounds in the cells can be determined from the R value by the solution of simple algebraic equations after estimating the protein from the percentage of nitrogen times 6.25. Satisfactory agreement has been found for the composition estimated by this method and by chemical analysis.

Table LI shows different chemical compositions of the cells and their R values. There is a striking inverse relation between the proteins and carbohydrates on the one hand and lipids on the other, when the R value increases.

3. Nitrogen Supply

Control of the nitrogen level is important not only to control the growth rate of the culture, but also to influence the chemical composition of the cells. High nitrogen is essential for rapid growth, also for high protein levels. Concentrations of about 0.002 to 0.02 M are

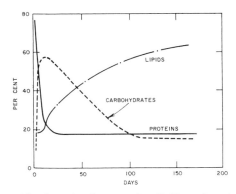

FIG. 43. Changes with time in the protein, lipid, and carbohydrate content of *Chlorella*, Strain A. Temperature 25°C; light 0.03 cal/cm²/min; initial nitrate concentration 2.5 mmoles. From Van Oorschot (361).

desirable, but with urea, concentrations above 0.1 M can be used. Cells grown in these solutions, in bright light, may contain 50–60% protein, 30–40% carbohydrate, and 5–20% lipids. The yields are about 30% higher with ammonium salts or urea than with nitrates.

With lower levels of nitrogen, approaching nitrogen deficiency, the protein is replaced in the cells by carbohydrates and lipids. The effect of nitrogen deficiency is complex. This is illustrated by an experiment of Van Oorschot (361) (Fig. 43). Starting with a culture containing 0.0025 M nitrogen, growth was rapid and the cells were rich in protein during the first few days. As the nitrogen was depleted, the protein level fell rapidly, and the carbohydrate level rose rapidly during the first week. Then the carbohydrate level began to fall and the lipid level rose steadily, until after 2–4 months the lipids predominated in the system.

Spoehr and Milner (322) found that the R value never exceeded 47 when the nitrogen level was greater than 0.001 M. At 0.0004 M nitrogen the R value reached 57. Growth and the efficiency of utilization of light were nevertheless very poor under these conditions. Evidently the production of high-fat cells will involve difficult technical problems. Of these the optimum supply of nitrogen appears to be most critical. When growth is rapid with high nitrogen, protoplasm is synthesized instead of reserve materials. When growth is retarded by low nitrogen the reverse reactions occur. However, the different species of algae vary a great deal in their responses to changes in nitrogen level.

E. Nitrogen-Fixing Algae

In view of the importance of nitrogen in the growth of plants, algae capable of both photosynthesis and nitrogen fixation should have great value. The blue-green algae (Myxophyceae) have these properties, but their observed growth rates have been too slow to be useful. During World War II, Watanabe (377) made an extensive search for more active strains of nitrogen fixers. He found a *Tolypothrix tenuis* in Borneo capable of producing 4 gm dry matter/m²/12 hours, utilizing atmospheric nitrogen alone. This represents 0.24 gm N/m²/12 hours.

Watanabe started large-scale tests with *Tolypothrix* in rice fields in Japan in 1951. He prepared the algae in quantity, at first by pilot plant methods described later, then by growing the algae on a porous andesitic pumice in the form of fine gravel, 2–4 mm diameter. Sterilization of nutrient medium was necessary. The algae grew on the moist gravel in a vinyl tube to which carbon dioxide-enriched air was supplied. When the dry weight of the cells reached about 1.5% of the gravel, the material was stored in air-tight vinyl bags until used. Applications of 130–650 pounds per acre of the moist pumice were made in the summer to the rice fields, after treatments to control pests and also the pH of the water. The *Tolypothrix* grew well in about 70% of the treated fields. In these fields the yields increased progressively in 3–5 successive years, giving finally an increase of about 20%. Nitrogen-rich fields with yields of 73 bushels per acre gave about the same percentage response as nitrogen-poor fields with yields of 42 bushels. More recently the algae have been preserved for 2 years by absorption onto magnesium phosphate (<50 mesh) with addition of a little human serum as a protective agent (376).

Anabaena cylindrica was studied by Allen and Arnon (2). The nutrient solution finally selected contained Na, 4 mM; K, 4 mM; Mg, 1 mM; Ca, 0.5 mM; plus equivalent Cl, SO$_4$, and PO$_4$ and the usual trace elements including molybdenum. Calcium and molybdenum were particularly important. The former at 20 ppm gave double the yield of 1 ppm and 100-fold the yield of zero calcium. Molybdenum at 0.01 ppm raised the yield from 0.3 to 5.8 gm/liter of dry cells.

With the atmosphere as the sole source of nitrogen, *Anabaena* grew a little more rapidly than with nitrate. It grew almost as rapidly as an active strain of *Scenedesmus*. In a 6-day period of continuous illumination with 16 klux, *Anabaena* produced 88, and *Scenedesmus* 105, gm/m². Maximum daily rates were *Anabaena* 26 and *Scenedesmus* 36 gm/m²/24 hours. A yield of 26 gm/m²/24 hours represents 1.8 gm/N/m²/24 hours, equivalent to nearly 3 tons of fixed nitrogen per acre per year. Actual yield would be about 75–80% of the 24-hour yield if illuminated only 13 hours per day at 16 klux. Field tests with this promising species have not been reported.

F. ALGAL CULTURE IN SEWAGE

Disposal of sewage presents a serious economic problem because of its large amount and the necessity of avoiding contamination of rivers, lakes, and ground water. It has been suggested that the growing of algae in the sewage oxidation beds would reduce the biological oxygen demand (B.O.D.) and produce an economically valuable crop of algae to offset the cost of the sewage disposal. Gotaas and Oswald (148, 149) concluded that near San Francisco the inoculated sewage should remain in the oxidation beds at a depth of 6–20 cm for 1.5–4 days. They estimated that such a pond handling East Bay sewage at Richmond, California, properly designed and operated, would reduce the B.O.D. by 70–80% and produce 30–35 tons of algae per acre per year. Day and Tucker (92) found that sewage effluent served effectively as irrigation water for pastures.

G. PILOT PLANT STUDIES

1. In Germany

Pilot plant installations for algal culture have been built in Germany, the United States, Israel, and Japan. The German installation (158) employed four troughs, 9 m long, 0.7 m wide, and 9–15 cm deep with pump and centrifuge. Early tests gave yields up to 6–7 gm/m²/day.

2. In the United States

In the United States, following the laboratory studies by Carnegie Institution and Stanford Research Institute, Arthur D. Little, Inc. (216) built a unique pilot plant, employing a thin polyethylene tube with walls 0.1 mm thick and a circumference of 2.5 m. The tube

was mounted on the roof of the laboratory, on a pair of parallel plat-
forms 21 m long, connected at the ends. The tube described 180° bends
at each end and was placed on the tables to form a closed loop. The
liquid feed and drain were close together near the center of a long run.
The depth of the liquid was about 7 cm and its velocity about 10–15
cm/sec. Facilities were provided for handling the solutions effectively,
even cooling them if necessary, and harvesting the cells. Yields of
9–11 gm/m²/day were obtained in summer. With practicable modifications, yields of 20 gm/m²/day were forecast. The pilot plant was
operated long enough to demonstrate its potentialities, which were
considered promising though the fragility of the plastic envelope and
a tendency for the algae to adhere to it, left much to be desired. The
cost of production in an installation of 100 acres was estimated at
$0.25 per pound of dry cells.

3. In Japan

The first pilot installation in Japan (346) was a trough 15 m long,
1 m wide, and 20 cm deep. The culture was either covered with thin
polyethylene sheeting or operated in the open. It was aspirated in
the trough with 2–9% carbon dioxide, supplied by burning oil. It
was concluded that an open system was more convenient and cheaper
to operate than a closed system. Yields of 2–4 gm/m²/day in winter and
15–18 gm/m²/day in summer were obtained.

Later a circular "pond" type unit, 5 m in diameter and 20 cm deep
was built. It was half-filled with 2000 liters of culture solution. The
bottom of the pond was swept intermittently by pipes with jets
projecting backward. The pipes were pivoted in the center of the
pond, and rotated slowly as liquid was forced out of the jets. Carbon
dioxide-enriched air was introduced into the pipes at the outlet of the
pumps. Advantages of this system include (a) low energy requirement
for operation, (b) prevention of precipitation of the algae on the bottom
of the pond, (c) no cooling required, and (d) easy mixing of the
whole culture. Disadvantages include regulation of water volume after
a heavy rain, and occasional removal of accumulated dust.

It was planned to expand the system by constructing a row of over-
lapping ponds, 20 m in diameter, swept by a series of synchronized
rotating pipes. An installation of 80 multiple ponds, each with 20 components, plus 20 triple component units and 20 single units, together with
a central installation for servicing the crop continuously, was planned
for a 100-acre farm. The cultured area would be 82 acres. One acre of
this farm was in operation in 1958. It was expected that by using
a thermophilic variety of *Chlorella* in the hottest part of the year,

the average annual yield would be 12.5 gm/m²/day. This would produce 9100 pounds per day of dry algae on 82 acres at a cost of about $0.26 per pound. Actual yield (193) was 14 tons/acre/year or 8.6 gm/m²/day in 1958.

4. In Israel

Evenari and Mayer (124) in Israel proposed a sophisticated method of illuminating culture tanks at least 1 m in depth so that sunlight would be more or less uniformly available to all the cells at an intensity of about 400 fc. The method would employ glass tubes with closed lower ends extending to the bottom of the tank. Light-collecting lenses above the tank and reflectors inside the tubes would permit direct illumination of an area in the tank 11 times greater than the surface of the tank. In this way the sunlight would be supplied at optimum intensity to all the cells in the tank.

However, this system was not set up. Instead, tanks $2 \times 1 \times 1$ meters were constructed having a transparent south wall and inside walls painted white (236). Total volume of the solution was about 2000 liters. An effective circulating type stirrer was employed. Carbon dioxide was supplied on a batch basis: 300 liters of pure gas were aspirated through porous candles over a 1-hour period, once a day. The cells were harvested daily by supercentrifuge from half the suspension and the clear solution was returned to the tank. The cell concentration ranged from 40 to 100 mg/liter (dry basis) at harvest and 20 to 50 mg/liter after harvest. The illuminated surface was 3.5 m². In the strong sunlight of Israel, average yields of 16 gm/m²/day with a range from 10 to 32 gm/m²/day were obtained over a period of one year with several different species of *Chlorella*.

The results were considered to be promising, but further studies would be necessary before the system could be competitive with conventional agriculture. Particularly, harvesting by supercentrifuge was uneconomic because of the large power requirement.

The work of Myers and Graham (253) supports the Israeli system of deep dilute suspensions diffusely illuminated. They found maximum yields at a cell density of about 130 mg/liter with surface lighting at 0.5 cal/cm²/min. Yield was equivalent to about 35 gm/m²/12 hours, which was comparable with the yield of the dense, rotor-stirred suspension recorded in Table XLVIII. They also used a long plastic cone extending to the bottom of a deep culture. The directly illuminated surface was thus increased about tenfold. Light intensity equivalent to sunlight was applied continuously. Enough additional light reached the cells to double the yield as compared with surface illumination alone. Pyramids might be better diffusers than cones. A thermophilic *Chlorella* gave a 15% higher yield than one with a 25° optimum growth temperature.

References

1. Ahmed, S., and Evans, H. G. Response of soybean plants to cobalt in the absence of supplied nitrogen. *Plant Physiol.* **35**, xxix–v (1960).
2. Allen, M. B., and Arnon, D. I. Studies on nitrogen fixing blue-green algae. I. Growth and nitrogen fixation by *Anabaena cylindrica*. *Plant Physiol.* **30**, 366–372 (1955).
3. Allmendinger, D. F., Kenworthy, A. L., and Overholser, E. L. The carbon dioxide intake of apple leaves as affected by reducing the available soil moisture to different levels. *Proc. Am. Soc. Hort. Sci.* **42**, 133–140 (1943).
4. Alvim, P. de T. Studies on the mechanism of stomatal behavior. *Am. J. Botany* **36**, 781–791 (1949).
5. Anderson, A. J. Molybdenum deficiencies in legumes in Australia. *Soil Sci.* **81**, 173–182 (1956).
6. Arnon, D. I. Copper enzymes in isolated chloroplasts. Polyphenol oxidase in Beta vulgaris. *Plant Physiol.* **24**, 1–15 (1949).
7. Arnon, D. I. Cell-free photosynthesis and the energy conversion process. *In* "Light and Life" (W. D. McElroy and B. Glass, eds.), pp. 489–569. Johns Hopkins Press, Baltimore, Maryland, 1961.
8. Arnon, D. I., and Hoagland, D. R. Crop production in artificial culture solutions and in soils with special reference to factors influencing yields and absorption of inorganic nutrients. *Soil Sci.* **50**, 463–485 (1940).
9. Arnon, D. I., and Stout, P. R. Molybdenum as an essential element for higher plants. *Plant Physiol.* **14**, 599–602 (1939).
10. Arnon, D. I., and Wessel, G. Vanadium as an essential element for green plants. *Nature* **172**, 1039–1040 (1953).
11. Arnon, D. I., and Whatley, F. R. Is chloride a co-enzyme of photosynthesis? *Science* **110**, 554–556 (1949).
12. Arnon, D. I., Whatley, F. R., and Allen, M. B. Assimilatory power in photosynthesis. *Science* **127**, 1026–1034 (1958).
13. Aronoff, S. Photosynthesis. *Botan. Rev.* **23**, 65–107 (1957).
14. Aronoff, S., and Mackinney, G. Pyrrole derivatives and iron chlorosis in plants. *Plant Physiol.* **18**, 713–715 (1943).
15. Arthur, J. M., and Stewart, W. D. Relative growth and dry weight production of plant tissue under mazda, neon, sodium, and mercury vapor lamps. *Contrib. Boyce Thompson Inst.* **7**, 119–130 (1935).
16. Ashton, F. M. Effects of a series of cycles of alternating low and high soil water contents on the rate of apparent photosynthesis in sugar cane. *Plant Physiol.* **31**, 266–274 (1956).
17. Aykroydt, W. R. "Three Philosophers (Lavoisier, Priestley, Cavendish)." Heineman, London, 1935.
18. Baeyer, A. von. Über die Wasserentziehung und ihre Bedeutung für das Pflanzenleben und die Gährung. *Ber. deut. chem. Ges.* **3**, 63–78 (1870).
19. Baker, J. E., Gauch, H. G., and Dugger, W. M., Jr. Effects of boron on the water relations of higher plants. *Plant Physiol.* **31**, 89–94 (1956).
20. Baly, E. C. C. "Photosynthesis." Methuen, London, 1940.
21. Barker, H. A. Photosynthesis in diatoms, *Arch. Mikrobiol.* **6**, 141–156 (1935).

22. Barthélemy, A. Du rôle que joue la cuticule dans la respiration des plantes. *Ann. sci. nat.: Botan. et biol. vegetale* [5] **9**, 287–297 (1868).

23. Barthélemy, A. Du rôle des stomates et de la respiration cuticulaire. *Compt. rend.* **84**, 663–666 (1877).

24. Basford, R. E., and Huennekins, F. M. Studies on thiols II. *J. Am. Chem. Soc.* **77**, 3878–3882 (1955).

25. Baur, E., and Gloor, K. Photolytic formation of formaldehyde in the eosin group. *Helv. Chim. Acta* **20**, 970–974 (1937).

26. Baur, E., Gloor, K., and Kunzler, H. Photolysis of carbonic acid. *Helv. Chim. Acta* **21**, 1038–1053 (1938).

27. Baur, E., and Niggli, F. Assimilation of carbon dioxide in vitro. *Helv. Chim. Acta* **26**, 251–254, 994–995 (1943).

28. Benedict, F. G. The composition of the atmosphere with special reference to its oxygen content. *Carnegie Inst. Wash. Publ.* **166** (1913).

29. Berzelius, J. J. Über die gelbe Farbe der Blätter im Herbst. *Ann. Chem.* **21**, 257–262 (1837); **27**, 296 (1838).

30. Blackman, F. F. Optima and limiting factors. *Ann. Botany (London)* **19**, 281–295 (1905).

31. Blackman, F. F. Experimental researches on vegetable assimilation and respiration. I. On a new method for investigating the carbonic acid exchanges of plants. II. On the paths of gaseous exchanges between aerial leaves and the atmosphere. *Phil. Trans. Roy. Soc.* **B186**, 485–502, 503–562 (1895).

32. Blackman, F. F., and Matthaei, G. L. C. Experimental researches on vegetable assimilation and respiration. IV. A quantitative study of carbon dioxide assimilation and leaf temperature in natural illumination. *Proc. Roy. Soc. (London)* **B76**, 402–460 (1905).

33. Blackman, F. F., and Smith, A. M. Experimental researches on vegetable assimilation and respiration. VIII. A new method for estimating the gaseous exchanges of submerged plants. IX. On assimilation in submerged water plants and its relation to the concentration of carbon dioxide and other factors. *Proc. Roy. Soc. (London)* **B83**, 374–388, 389–412 (1910).

34. Blackman, G. E., and Black, J. N. Physiological and ecological studies in the analysis of plant environment. XI. A further assessment of the influence of shading on the growth of different species in the vegetative phase. XII. The role of the light factor in limiting growth. *Ann. Botany (London)* [N.S.] **23**, 51–63, 131–145 (1959).

35. Blackman, G. E., and Wilson, G. L. Physiological and ecological studies in the analysis of plant environment. VI. The constancy for different species of a logarithmic relationship between the assimilation rate and light intensity and its ecological significance. IX. Adaptive changes in vegetative growth and development of *Helianthus annuus* induced by an alteration of light level. *Ann. Botany (London)* [N.S.] **15**, 63–94 (1951); **18**, 71–94 (1954).

36. Blackman, V. H. The compound interest law and plant growth. *Ann. Botany (London)* **33**, 353–360 (1919).

37. Bleasdale, J. K. A. Atmospheric pollution and plant growth. *Nature* **169**, 376 (1952).

38. Boawn, L. C., and Viets, F. G., Jr. Zinc deficiency of alfalfa in Washington. *Agron. J.* **44**, 276 (1952).

39. Boehm, J. Stärkebildung in den Blättern von *Sedum spectabile* Boreau. *Botan. Centr.* **37**, 193–201 (1889).

40. Böhning, R. H. Time course of photosynthesis in apple leaves exposed to continuous illumination. *Plant Physiol.* **24,** 222–240 (1949).
41. Böhning, R. H., and Burnside, C. A. The effect of light intensity on rate of apparent photosynthesis in leaves of sun and shade plants. *Am. J. Botany* **43,** 557–561 (1956).
42. Bolas, B. D., and Henderson, F. Y. The effect of increased atmospheric carbon dioxide on the growth of plants. *Ann. Botany (London)* **42,** 509–523 (1928).
43. Bolle-Jones, E. W. The interrelationships of iron and potassium in the potato plant. *Plant Soil* **6,** 129–173 (1955).
44. Bömer, A., and Rintelen, P. Carbon dioxide treatment of the open field. *Z. Pflanzenernaehr. Dueng. Bodenk.* **B12,** 49–85 (1933).
45. Bonde, E. K. The influence of carbon dioxide concentration upon the rate of photosynthesis of *Sinapis alba. Physiol. Plantarum* **5,** 298–304 (1952).
46. Bonde, E. K. The effect of various cycles of light and darkness on the growth of tomato and cocklebur plants. *Physiol. Plantarum* **8,** 913–923 (1955); **9,** 5139 (1956).
47. Bonner, W. D., Jr. Soluble oxidases and their functions. *Ann. Rev. Plant Physiol.* **8,** 427–452 (1957).
48. Bonnet. "Recherches sur l'usage des feuilles dans les plantes." 1754.
49. Bonnier, G., and Mangin, L. Recherches sur l'action chlorophyllienne séparée de la respiration. *Ann. sci. nat.: Botan. et biol. vegetale* [7] **3,** 1–44 (1886).
50. Bose, J. C. "The Physiology of Photosynthesis." Longmans, Green, New York, 1924.
51. Boussingault, J. B. "Agronomie, chimie, agricole, et physiologie," Vol. 3, pp. 266–379; Vol. 4, pp. 359–397; Vol. 5, pp. 1–32. Paris, 1864, 1868, 1874.
52. Boussingault, J. B. De la végétation dans l'obscurité. *Ann. sci. nat.: Botan. et biol. vegetale* [5] **1,** 314–324 (1864).
53. Boussingault, J. B. Étude sur les fonctions des feuilles. *Compt. Rend.* **61,** 493–506, 605–613, 657–665 (1865); *Ann. sci. nat.: Botan. et biol. vegetale* [5], 331–343 (1869).
54. Brackett, F. S. Light intensity and carbon dioxide concentration as factors in photosynthesis of wheat. *Cold Spring Harbor Symp. Quant. Biol.* **3,** 117–123 (1935).
55. Briggs, G. E. Experimental researches on vegetable assimilation and respiration. XVI. The characteristics of subnormal photosynthetic activity resulting from deficiency of nutrient salts. *Proc. Roy. Soc. (London)* **B94,** 20–35 (1922).
56. Brown, A. H. Effects of light on respiration using isotopically enriched oxygen. *Am. J. Botany* **40,** 719–729 (1953).
57. Brown, A. H., and Webster, G. C. Influence of light on the rate of respiration of the blue-green alga anabaena. *Am. J. Botany* **40,** 753–758 (1953).
58. Brown, A. H., and Whittingham, C. P. Identification of the carbon dioxide burst in chlorella using the recording mass spectrometer. *Plant Physiol.* **30,** 231–237 (1955).
59. Brown, H. T., and Escombe, F. Static diffusion of gases and liquids in relation to the assimilation of carbon and translocation in plants. *Phil. Trans. Roy. Soc.* **B193,** 223–291 (1900).
60. Brown, H. T., and Escombe, F. The influence of varying amounts of carbon dioxide in the air on the photosynthetic process of leaves and on the mode of growth on plants. *Proc. Roy. Soc. (London)* **B70,** 397–413 (1902).
61. Brown, H. T., and Escombe, F. Researches on some of the physiological processes

of green leaves with special reference to the interchange of energy between the leaf and its surroundings. *Proc. Roy. Soc. (London)* **B76**, 29–111 (1905).

62. Brown, H. T., and Escombe, F. On the variations in the amount of carbon dioxide in the air at Kew during the years 1898–1901. *Proc. Roy. Soc. (London)* **B76**, 118–121 (1905).

63. Brown, J. C. Effect of dominance of a metallic system requiring iron or copper on development of lime induced chlorosis. *Plant Physiol.* **28**, 495–502 (1953).

64. Brown, J. C., and Holmes, R. S. Iron, the limiting element in a chlorosis. Part 1. Availability and utilization of iron dependent upon nutrition and plant species. *Plant Physiol.* **30**, 451–457 (1955).

65. Brown, J. C., and Holmes, R. S. Iron supply and interacting factors related to lime-induced chlorosis. *Soil Sci.* **82**, 507–519 (1956).

66. Brown, J. C., Holmes, R. S., and Specht, A. W. II. Copper-phosphorus induced chlorosis dependent upon plant species. *Plant Physiol.* **30**, 457–462 (1955).

67. Brown, J. C., Holmes, R. S., and Tiffin, L. O. Hypotheses concerning iron chlorosis. *Soil Sci. Soc. Am. Proc.* **23**, 231–234 (1959).

68. Brun, W. A. Simultaneous and continuous recordings of photosynthesis and transpiration from the upper and lower surfaces of intact banana leaves. *Plant Physiol.* **35**, Suppl. viii (1960).

68a. Buch, K. Carbon dioxide equilibrium and carbon dioxide exchange between the atmosphere and the sea in the North Atlantic Ocean. *Acta Acad. Aboensis, Math. Phys.* **11**, Nos. 9, 12 (1939).

69. Burlew, J. S., ed. "Algal Culture: From Laboratory to Pilot Plant," No. 600. Carnegie Inst. Wash. Publ., Washington, D.C., 1953.

70. Burns, G. R. Photosynthesis in various portions of the spectrum. *Plant Physiol.* **8**, 247–262 (1933).

71. Burns, G. R. Photosynthesis and the absorption spectra of plant pigments. I. *Am. J. Botany* **24**, 257–265 (1937); II. *Am. J. Botany* **25**, 166–174 (1938).

72. Burnside, C. A., and Böhning, R. H. The effect of prolonged shading on the light saturation curves in sun plants. *Plant Physiol.* **32**, 61–63 (1957).

73. Byers, H. R. *In* "The Earth as a Planet" (G. P. Kuiper, ed.), Chapter 7, p. 297. Univ. of Chicago Press, Chicago, Illinois, 1954.

74. Callendar, G. S. Variations of the amount of carbon dioxide in different air currents. *Quart. J. Roy. Meteorol. Soc.* **66**, 395–400 (1940).

75. Calvin, M. The photosynthetic carbon cycle. *J. Chem. Soc.* pp. 1895–1915 (1956).

76. Carpenter, T. M. The constancy of the atmosphere with respect to carbon dioxide and oxygen contents. *J. Am. Chem. Soc.* **59**, 358–360 (1937).

77. Cerighelli, R. Influence de l'anhydride carbonique sur le développement des plantes—emploi de CO_2 comme engrais atmosphérique. *Ann. sci. argon.* **38**, 68–75 (1921).

78. Chapman, H. W., Gleason, L. S., and Loomis, W. E. The carbon dioxide content of field air. *Plant Physiol.* **29**, 500–503 (1954).

79. Chapman, H. W., and Loomis, W. E. Photosynthesis in the potato under field conditions. *Plant Physiol.* **28**, 703–716 (1953).

80. Chesnokov, V. A. Aerial fertilization of plants with carbon dioxide. *Vestn. Leningr. Univ., Ser. Biol.* No. 1, 154–159 (1947).

81. Clarke, F. W. The date of geochemistry. *U.S., Geol. Surv., Bull.* **491** (1911).

82. Clendenning, K. A., and Gorham, P. R. Photochemical activity of isolated chloroplasts in relation to their source and previous history. *Can. J. Res.* **28**, 114–139 (1950).

83. Craig, F. N., and Trelease, S. F. Photosynthesis of chlorella in heavy water. *Am. J. Botany* **24**, 232–242 (1937).

84. Crocker, W. Sulfur deficiency in soils. *Soil Sci.* **6**, 149–155 (1945).

85. Crowther, F. Studies in growth analysis of the cotton plant under irrigation in the Sudan. I. The effect of different combinations of nitrogen applications and water supply. *Ann. Botany (London)* **48**, 877–913 (1934).

86. Cummings, M. B., and Jones, C. H. The aerial fertilization of plants with carbon dioxide. *Vermont, Univ., Agr. Expt. Sta., Bull.* **211**, 56 (1918).

87. Cushing, D. H. Production of carbon in the sea. *Nature* **179**, 876 (1957).

88. Darwin, F. Observations on stomata. *Phil. Trans. Roy. Soc.* **B190**, 531, 621 (1898).

89. Dastur, R. H. The relation between water content and photosynthesis. *Ann. Botany (London)* **39**, 769–786 (1925).

90. Dastur, R. H., and Desai, B. L. The relation between water content, chlorophyll content and rate of photosynthesis in some tropical plants at different temperatures. *Ann. Botany (London)* **47**, 69–88 (1933).

91. Davis, E. A., Dedrick, J., French, C. S., Milner, H. W., Myers, J., Smith, J. H. C., and Spoehr, H. A. Laboratory experiments on chlorella culture. *In* "Algal Culture: From Laboratory to Pilot Plant" (J. S. Burlew, ed.), No. 600, pp. 105–153. Carnegie Inst. Wash. Publ., Washington, D.C., 1953.

92. Day, A. D., and Tucker, T. C. Production of small grains pasture forage using sewage effluent as a source of irrigation water and plant nutrients. *Agron. J.* **51**, 569–572 (1959).

93. Decker, J. P. Effect of air supply on apparent photosynthesis. *Plant Physiol.* **22**, 561–571 (1947).

94. Decker, J. P. Effect of temperature on photosynthesis and respiration in red and loblolly pines. *Plant Physiol.* **19**, 679–688 (1944).

95. Decker, J. P. Further evidence of increased carbon dioxide production accompanying photosynthesis. *J. Solar Energy Sci. Eng.* **1**, 30–33 (1957).

96. Decker, J. P. Some effects of temperature and carbon dioxide concentration on photosynthesis of Mimulus. *Plant Physiol.* **34**, 103–106 (1959).

97. Decker, J. P. A rapid post illumination deceleration of respiration in green leaves. *Plant Physiol.* **30**, 82–84 (1955).

98. Decker, J. P., and Tio, M. A. Photosynthetic surges in coffee seedlings. *J. Agr. Univ. Puerto Rico* **43**, 50–55 (1959).

99. Demoussy, E. Sur la végétation dans des atmosphères riches en acide carbonique. *Compt. rend.* **139**, 883–885 (1904).

100. Deuber, C. Can a pyrrole be substituted for iron in the growth of plants? *Am. J. Botany* **13**, 276–285 (1926).

101. Dhar, N. R., and Ram, A. Photoreduction of carbonic acid, bicarbonates and carbonates to formaldehyde. *Nature* **129**, 205 (1932).

102. Dhar, N. R., and Ram, A. Presence of formaldehyde in rain and dew and its formation by photooxidation of organic compounds and the problem of carbon assimilation. *J. Indian Chem. Soc.* **10**, 287–298 (1933).

103. Dugger, W. M., Jr. The permeability of non-stomate leaf epidermis to carbon dioxide. *Plant Physiol.* **27**, 489–499 (1952).

104. Dugger, W. M., Jr., Humphreys, T. E., and Calhoun, B. The influence of boron on starch phosphorylase and its significance in translocation of sugars in plants. *Plant Physiol.* **32**, 364–370 (1957).

105. Dutrochet, R. J. H. Sur les organes aérifères de végétaux. *Ann. Sci. Nat.* [1] **25**, 212 (1832)

106. Dutrochet, R. J. H. "Mémoires pour servir à l'histoire anatomique et physiologique des végétaux et des animaux." Bruxelles, 1837.
107. Eaton, F. M. Deficiency, toxicity and accumulation of boron in plants. *J. Agr. Res.* **69**, 237–277 (1944).
108. Eaton, F. M., and Ergle, D. R. Effects of shade and partial defoliation on carbohydrate levels and growth, fruiting and fiber properties of cotton plants. *Plant Physiol.* **29**, 39–49 (1954).
109. Eaton, S. V. Effects of potassium on growth and metabolism of sunflower plants. *Botan. Gaz.* **114**, 165–180 (1952).
110. Ebermayer, D. "Die Beschaffenheit der Waldluft und die Bedeutung der atmosphärischen Kohlensäure für die Waldvegetation." Enke, Stuttgart, 1885.
111. Eckerson, S. F. The number and size of stomata. *Botan. Gaz.* **46**, 221–224 (1908).
112. Eckerson, S. H. Influence of phosphorus deficiency on metabolism of the tomato (Lycopersicon esculentum Mill). *Contrib. Boyce Thompson Inst.* **3**, 197–217 (1931).
113. Eckstein, O. Effect of potash manuring on the production of organic matter. *Plant Physiol.* **14**, 113–128 (1939).
114. Egle, K., and Schenk, W. Der Tagesgang des Kohlendioxydgehalts der Luft und das CO_2-Gefälle in Gewächschäusern. *Beitr. Biol. Pflanz.* **31**, 155–177 (1955).
115. Eltinge, E. T. Effect of manganese deficiency upon the histology of Lycopersicum esculentum. *Plant Physiol.* **16**, 189–195 (1941).
116. Emerson, R. Photosynthesis as a function of light intensity and of temperature with different concentrations of chlorophyll. *J. Gen. Physiol.* **12**, 623–639 (1929).
117. Emerson, R., and Arnold, W. The photochemical reaction in photosynthesis. *J. Gen. Physiol.* **16**, 191–205 (1932).
118. Emerson, R., and Green, L. Nature of the Blackman reaction in photosynthesis. *Plant Physiol.* **12**, 537–545 (1937).
119. Emerson, R., and Green, L. Effect of hydrogen-ion concentration on chlorella photosynthesis. *Plant Physiol.* **13**, 157–168 (1938).
120. Emerson, R., and Green, L. Manometric measurements of photosynthesis in the marine alga gigartina. *J. Gen. Physiol.* **17**, 817–842 (1934).
121. Emerson, R., and Lewis, C. M. The dependence of the quantum yield of chlorella photosynthesis on wave length of light. *Am. J. Botany* **30**, 165–178 (1943).
122. Evans, G. C. Ecological studies on the rain forest of Southern Nigeria. II. The atmospheric environmental conditions. *J. Ecol.* **27**, 436–442 (1939).
123. Evans, H. J. Role of molybdenum in plant nutrition. *Soil Sci.* **81**, 199–208 (1956).
124. Evenari, M., and Mayer, A. Experiments on culture of algae in Israel. *In* "Algal Culture: From Laboratory to Pilot Plant" (J. S. Burlew, ed.), No. 600, pp. 197–203. Carnegie Inst. Wash. Publ., Washington, D.C., 1953.
125. Eyster, C., Brown, T. E., Tanner, H. A., and Hood, S. L. Manganese requirement with respect to growth, Hill reaction and photosynthesis. *Plant Physiol.* **33**, 235–244 (1958).
126. Fischer, E. G., Walker, D. R., Boynton, D., and Kwong, S. S. Studies on the control of magnesium deficiency and its effect on apple trees. *Proc. Am. Soc. Hort. Sci.* **71**, 1–10 (1958).
127. Fischer, Hugo. Die Kohlensäure Ernährung der Pflanzen. *Ber. deut. botan. Ges.* **45**, 331–339 (1927).
128. Fischer, Hans, and Stern, A. Pyrrolfarbstoffe, *In* "Die Chemie des Pyrrols" (H. Fischer and H. Orth, eds.), Vol. 2. Akad. Verlagsges., Leipzig, 1940.

129. Fleischer, W. E. The relation between chlorophyll content and rate of photosynthesis. *J. Gen. Physiol.* **18**, 573–597 (1935).
130. Florentin, D. Composition of the air in the streets of Paris. *Compt. rend.* **185**, 1538–1541 (1927).
131. Foster, J. W. The role of organic substrates in photosynthesis of purple bacteria. *J. Gen. Physiol.* **24**, 123–134 (1940).
132. Franck, J. Participation of respiratory intermediates in the process of photosynthesis as an explanation of abnormally high quantum yields. *Arch. Biochem. Biophys.* **45**, 190–229 (1953).
133. Freeland, R. O. Photosynthesis in relation to stomatal frequency and distribution. *Plant Physiol.* **23**, 595–600 (1948).
134. Freeland, R. O. Effects of growth substances on photosynthesis. *Plant Physiol.* **24**, 621–628 (1949).
135. Freeland, R. O. Effect of age on leaves upon the rate of photosynthesis. *Plant Physiol.* **27**, 685–690 (1952).
136. Gabrielsen, E. L. Photosynthesis in leaves at very low carbon dioxide concentrations. *Nature* **163**, 359–360 (1949).
137. Garner, W. W., and Allard, H. A. Effect of abnormally long and short alternations of light and darkness on growth and development of plants. *J. Agr. Res.* **42**, 629–651 (1931).
138. Garreau. Recherches sur l'absorption et l'exhalation des surfaces aériennes des plantes. *Ann. sci. nat.: Botan. et biol. vegetale* [3] **13**, 321–346 (1850).
139. Gassner, G., and Goeze, G. Über den Einfluss der Kali-ernährung auf die Assimilations-grösse von Weizenblättern. *Ber. deut. botan. Ges.* **50**, 412–482 (1932).
140. Gates, C. T., and Bonner, J. Response of the young tomato plant to a brief period of water shortage. IV. Effect of water stress on the ribonucleic acid metabolism of tomato leaves. *Plant Physiol.* **34**, 49–55 (1959).
141. Geoghegan, M. J. Experiments with chlorella at Jealott's Hill. *In* "Algal Culture: From Laboratory to Pilot Plant" (J. S. Burlew, ed.), No. 600, pp. 182–189. Carnegie Inst. Wash. Publ., Washington, D.C., 1953.
142. Gerretsen, F. C. Manganese in relation to photosynthesis. I. Carbon dioxide assimilation and the typical symptoms of manganese deficiency in oats. II. The redox potentials of illuminated crude chloroplast suspensions. *Plant Soil* **1**, No. 4, 346–358 (1949); **2**, No. 2, 159–193 (1950).
143. Gilbert, S. G., Sell, H. M., and Drosdoff, M. Effect of copper deficiency on nitrogen metabolism and oil synthesis in the tung tree. *Plant Physiol.* **21**, 290–303 (1946).
144. Gist, G. R., and Mott, G. O. Some effects of light intensity, temperature and soil moisture on the growth of alfalfa, red clover, and birdsfoot trefoil seedlings. *Agron. J.* **49**, 33–36 (1957).
145. Glueckauf, E. The composition of atmospheric air. *In* "Compendium of Meteorology" (T. F. Malone, ed.), pp. 3–10. Am. Meteorol. Soc., Boston, Massachusetts, 1951.
146. Glueckauf, E. Carbon dioxide content of atmospheric air. *Nature* **153**, 620–621 (1944).
147. Godlewski, E. Abhängigkeit der Säuerstoff-ausscheidung der Blätter von den Kohlensäure-gehalt der Luft. *Arb. Botan. Inst. Würzburg* **1**, 343–370 (1873).
148. Gotaas, H. B., and Oswald, W. J. Utilization of solar energy for waste reclamation. *Conf. Solar Energy: Sci. Basis, Tucson, Arizona, 1954* (1955).

149. Gotaas, H. B., Oswald, W. J., and Ludwig, H. F. Photosynthetic reclamation of organic wastes. *Sci. Monthly* **79**, 368–378 (1954).

150. Graham, A. "Chemical and Physical Researches, 1829–1866" (collected by Angus Smith). Edinburgh, 1876.

151. Green, J. R. "History of Botany 1860–1900." Oxford Univ. Press (Clarendon), London and New York, 1909.

152. Green, L. F., McCarthy, J. F., and King, C. G. Inhibition of respiration and photosynthesis in *Chlorella pyrenoidosa* by organic compounds that inhibit copper catalysis. *J. Biol. Chem.* **128**, 447–453 (1939).

153. Greenfield, S. S. Inhibitory effects of inorganic compounds on the photosynthesis in chlorella. *Am. J. Botany* **29**, 121–131 (1942).

154. Gregory, F. G. The effect of climatic conditions on the growth of barley. *Ann. Botany (London)* **40**, 1–26 (1926).

155. Gregory, F. G., and Richards, F. J. Physiological studies in plant nutrition. I. The effect of manurial deficiency on the respiration and assimilation rate in barley. *Ann. Botany (London)* **43**, 119–161 (1929).

156. Grew, N. "The Anatomy of Plants, with an Idea of a Philosophical History of Plants." London, 1682.

157. Grinfel'd, E. G. Nutrition of plants with carbon dioxide assimilated through the roots. *Dokl. Akad. Nauk SSSR* **97**, 919–992 (1954).

158. Gummert, F., Meffert, M. E., and Stratmann, H. Nonsterile large-scale culture of chlorella in greenhouse and open air. *In* "Algal Culture: From Laboratory to Pilot Plant" (J. S. Burlew, ed.), No. 600, pp. 166–176. Carnegie Inst. Wash. Publ., Washington, D.C., 1953.

159. Haldane, J. B. S. Carbon dioxide content of atmospheric air. *Nature* **137**, 575 (1936).

160. Hales, Stephen. "Statical Essays." London, 1731.

161. Halldal, P., and French, C. S. Algal growth in crossed gradients of light intensity and temperature. *Plant Physiol.* **33**, 249–252 (1958).

162. Hansen, A. Geschichte der Assimilation und Chlorophyllfunction. *Arb. botan. Inst. Würzburg* **2**, 537–626 (1882).

163. Hansen, A. Der Chlorophyllfarbstoff. *Arb. botan. Inst. Würzburg* **3**, 123–146 (1884).

164. Harder, R. Kritische Versuche zu Blackman's Theorie der "begrenzender Factoren" bei Kohlensäure Assimilation. *Jahrb. wiss. Botan.* **60**, 531–571 (1921).

165. Harms, H. Beziehungen zwischen Stomataweite, Lichtstärke, und Lichtfarbe. *Planta* **25**, 155–193 (1936).

166. Harvey-Gibson, R. J. Pioneer investigators of photosynthesis. *New Phytologist* **13**, 191–205 (1914).

167. Heath, O. V. S. Studies in stomatal behavior, V. Part 1. The role of carbon dioxide in the light responses of stomata. *J. Exptl. Botany* **1**, 29–62 (1950).

168. Heath, O. V. S. Control of stomatal movement by a reduction in the normal carbon dioxide content of the air. *Nature* **161**, 179–181 (1948).

169. Heath, O. V. S., and Gregory, F. G. The constancy of the mean net assimilation rate and its ecological importance. *Ann. Botany (London)* [N.S.] **2**, 811–817 (1938).

170. Heath, O. V. S., and Milthorpe, F. L. Studies in stomatal behavior. V. The role of CO_2 in the light response of stomata. Part II. *J. Exptl. Botany* **1**, 227–243 (1950).

170a. Heath, O. V. S., and Orchard, B. Studies on stomatal behavior. VII. Effects of anaerobic conditions upon stomatal movement—a test of Williams hypothesis of stomatal mechanism. *J. Exptl. Botany* **7**, 313–325 (1956).

171. Heinicke, A. J., and Hoffman, M. B. The rate of photosynthesis of apple leaves under natural conditions. *Cornell Univ., Agr. Expt. Sta. Bull.* **577**, 1–32 (1933).

172. Heinicke, A. J., and Childers, N. F. The daily rate of photosynthesis during the growing season of 1935, of a young apple tree of bearing age. *Cornell Univ., Agr. Expt. Sta. Mem.* **201**, 3–52 (1937).

173. Hewitt, E. J. Metal interrelationships in plant nutrition. *J. Exptl. Botany* **5**, 110–118 (1954).

174. Hiesey, W. M. Growth and development of species and hybrids of Poa under controlled temperatures. *Am. J. Botany* **40**, 205–221 (1953).

175. Hiesey. W. M., and Milner, H. W. Physiology of climatic races I and II. *Plant Physiol.* **33**, xxv–xxvi (1958).

176. Hill, R., and Hartree, E. F. Hematin compounds in plants. *Ann. Rev. Plant Physiol.* **4**, 115–150 (1953).

177. Holdheide, W., Huber, B., and Stocker, O. Eine Feldmethode zur Bestimmung der momentanen Assimilationsgrosse von Landpflanzen. *Ber. deut. botan. Ges.* **54**, 168–198 (1936).

178. Holt, A. "A Life of Joseph Priestley." Oxford Univ. Press, London and New York, 1931.

179. Hoover, W. H. The dependence of carbon dioxide assimilation in a higher plant on the wave length of the radiation. *Smithsonian Inst. Misc. Collections* **95**, No. 21 (1937).

180. Hoover, W. H., Johnston, E. S., and Brackett, F. S. Carbon dioxide assimilation in a higher plant. *Smithsonian Inst. Misc. Collections* **87**, No. 16, 1–19 (1933).

181. Hoppe-Seyler, F. Über das Chlorophyll der Pflanzen. *Z. physiol. Chem.* **3**, 338–350 (1879); **4**, 193–203 (1880); **5**, 75–78 (1881).

182. Huber, B. Recording gaseous exchange under field conditions. *In* "The Physiology of Forest Trees" (K. V. Thimann, ed.). Ronald Press, New York, 1958.

183. Huber, B., and Miller, R. Methods for studying transpiration and measuring the water vapor content of a moving air stream. *Ber. Deut. Botan. Ges.* **67**, 323–325 (1954).

184. Ingen-Houz, J. "Experiments upon Vegetables, Discovering Their Great Power of Purifying Common Air in the Sunshine and of Injuring in the Shade and Night." London, 1779.

185. Ingen-Housz, J. "An Essay on the Food of Plants and Renovation of Soils." London, 1796.

186. Inoue, E., Tani, N., Imai, K., and Isobe, S. The aerodynamic measurement of photosynthesis over a wheat field. II. Over a nursery of rice plants. *J. Agr. Meteorol. Tokyo* **13**, 121–125 (1958); **14**, 45–53 (1958).

187. Jacobson, L. Iron in the leaves and chloroplasts of some plants in relation to their chlorophyll content. *Plant Physiol.* **20**, 233–245 (1945).

188. Johnston, E. S. Aerial fertilization of wheat plants with carbon dioxide. *Smithsonian Inst. Misc. Collections* **94**, No. 15 (1935).

189. Johnston, J. A., and Brown, A. H. The effect of light on the oxygen metabolism of the photosynthetic bacterium *Rhodospirillum rubrum*. *Plant Physiol.* **29**, 177–182 (1954).

190. Jones, L. H., Shepardson, W. B., and Peters, C. A. The function of manganese in the assimilation of nitrates. *Plant Physiol.* **24**, 300–306 (1949).

191. Jorgensen, J., and Convit, J. Cultivation of complexes of algae with other fresh water microorganisms in the tropics. *In* "Algal Culture: From Laboratory to Pilot Plant" (J. S. Burlew, ed.), No. 600, pp. 190–196. Carnegie Inst. Wash. Publ., Washington, D.C., 1953.

192. Kamen, M. D., and Barker, H. A. Inadequacies in present knowledge of the relation between photosynthesis and the O^{18} content of atmospheric oxygen. *Proc. Natl. Acad. Sci. U.S.* **31**, 8–15 (1945).

193. Kanazawa, T., Fujita, C., Yuhara, T., and Sasa, T. Mass culture of unicellular algae using open circulation method. *J. Gen. Microbiol* **4**, 1–7 (1958).

194. Katunskii, V. M. Changes in the photosynthetic activity of plants during their growth and development as related to the problem of carbon dioxide fertilization. *Bull. Acad. Sci. USSR, Ser. Biol.* pp. 85–102 (1939).

195. Katunskii, V. M. The intensity of photosynthesis as a basic index of the CO_2 utilization of plants. *Acad. Sci. USSR Inst. Plant Physiol.* pp. 61–89 (1941).

196. Katz, M., Ledingham, G. A., and Harris, A. E. Carbon dioxide assimilation and respiration on alfalfa under influence of sulfur dioxide. *In* "Effect of sulfur dioxide on vegetation," Chapter XV, pp. 393–428. Natl. Res. Council Can., Ottawa, 1939.

197. Keilin, D., and Mann, T. Carbonic anhydrase purification and nature of the enzyme. *Biochem. J.* **34**, 1163–1176 (1940).

198. Kelley, W. C., Somers, G. F., and Ellis, G. H. The effect of boron on the growth and carotene content of carrots. *Proc. Am. Soc. Hort. Sci.* **59**, 352–360 (1952).

199. Kennedy, S. R., Jr. The influence of magnesium deficiency, chlorophyll concentration and heat treatments on the rate of photosynthesis of chlorella. *Am. J. Botany* **27**, 68–73 (1940).

200. Kessler, E. The role of manganese in the oxygen evolving system of photosynthesis. *Arch. Biochem. Biophys.* **59**, 527–529 (1955).

201. Ketchum, B. H. Mineral nutrition of phytoplankton. *Ann. Rev. Plant Physiol.* **5**, 55–74 (1954).

202. Kluiver, A. J. "Chemical Activities of Microorganisms." Oxford Univ. Press, London and New York, 1931.

203. Kok, B. Experiments on photosynthesis of chlorella in flashing light. *In* "Algal Culture: From Laboratory to Pilot Plant" (J. S. Burlew, ed.), No. 600, pp. 63–75. Carnegie Inst. Wash. Publ., Washington, D.C., 1953.

204. Kok, B. Inhibition of photosynthesis by intense light. *Biochim. Biophys. Acta* **21**, 234–244 (1956).

205. Kok, B. Photosynthesis in flashing light. *Biochim. Biophys. Acta* **21**, 245–258 (1956).

206. Koritz, H. G., and Went, F. W. The physiological action of smog on plants. Initial growth and transpiration studies. *Plant Physiol.* **28**, 50–62 (1953).

207. Kostytschew, S. P. Studien über Photosynthese. I. Das Verhältnis CO_2/O_2 bei den Kohlensäure Assimilation. *Ber. deut. botan. Ges.* **39**, 319–328 (1921).

208. Kostytschew, S. P., Bazyrina, K., and Tschesnokow, W. Untersuchungen über die Photosynthesis der Laubblätter unter natürlichen Verhältnissen. *Planta* **5**, 696–724 (1928).

209. Kramer, P. J., and Decker, J. P. Relation between light intensity and rate of photosynthesis of loblolly pine and certain hardwoods. *Plant Physiol.* **19**, 350–354 (1944).

210. Kursanov, A. L. Über den Einfluss der Kohlenhydrate auf den Tagesverlauf der Photosynthese. *Planta* **20**, 535–538 (1933).

210a. Kursanov, A. L. Circulation des matières organiques dans la plante et action du système radiculaire. *In* "Essais de botaniques, I," pp. 142–153. Acad. Sci. U.S.S.R., 1954.

211. Legendre, R. The amount of carbon dioxide in sea air. *Compt. rend.* **143**, 526 (1907).

212. Lemon, E. R. Photosynthesis under field conditions. II. An aerodynamic method for determining the turbulent carbon dioxide exchange between the atmosphere and a corn field. *Agron. J.* **52**, 697–703 (1960).

213. Liebig, J. "Chemistry in Its Application to Agriculture and Physiology," 3rd ed. London and Philadelphia, 1843.

214. Linsbauer, K. Beiträge zur Kenntnis der Spaltoffnungsbewegungen. *Flora* (*Jena*) [N.F.] **9**, 100–143 (1816).

215. Linstead, R. P., Braude, E. A., and Timmons, C. J. A reported photosynthesis in vitro. *Nature* **166**, 557 (1950).

216. Little, A. D., Inc. Pilot plant studies in the production of chorella. *In* "Algal Culture: From Laboratory to Pilot Plant" (J. S. Burlew, ed.), No. 600, pp. 235–272. Carnegie Inst. Wash. Publ., Washington, D.C., 1953.

217. Lloyd, F. E. The physiology of stomata. *Carnegie Inst. Wash. Publ.* **82**, 1–142 (1908).

218. Loftfield, J. V. G. Behavior of stomata. *Carnegie Inst. Wash. Publ.* **314** (1921).

219. Loustalot, A. J. Influence of soil moisture conditions on apparent photosynthesis and transpiration of pecan leaves. *J. Agr. Res.* **71**, 519–532 (1945).

220. Loustalot, A. J., Burrows, F. W., Gilbert, S. G., and Nason, A. Effect of copper and zinc deficiencies on the photosynthetic activity of the foliage of young tung trees. *Plant Physiol.* **20**, 283–288 (1945).

221. Loustalot, A. J., Gilbert, S. G., and Drosdoff, M. The effect of nitrogen and potassium levels in tung seedlings on growth, apparent photosynthesis and carbohydrate composition. *Plant Physiol.* **25**, 394–412 (1950).

222. Lundegardh, H. Fertilization of cucumbers and strawberries with carbon dioxide gas. *Kgl. Landtbruks-Akad. Handl. Tidskr.* **69**, 1044–1056 (1930).

223. McAlister, E. D. Spectrographic method for determining the carbon dioxide exchange between an organism and its surroundings. *Plant Physiol.* **12**, 213–215 (1937).

224. McAlister, E. D. Time course of photosynthesis for a higher plant. *Smithsonian Inst. Misc. Collections* **95**, No. 24, 1–17 (1937).

224a. McAlister, E. D., and Myers, J. Time course of photosynthesis and fluorescence measured simultaneously. *Smithsonian Inst. Misc. Collections* **99**, No. 6 (1940).

225. McElroy, W. D., and Nason, A. Mechanism of action of micronutrient elements in enzyme systems. *Ann. Rev. Plant Physiol.* **5**, 1–30 (1954).

226. McLean, R. C. Studies in the ecology of tropical rain forest: with special reference to the forests of south Brazil. *J. Ecol.* **7**, 5–54, 121–172 (1919).

227. Malpighi, Marcello. "Anatomia Plantarum Idea." Bologna, 1674.

228. Mangin, L. Sur la perméabilité de l'épidermis des feuilles pour les gaz. *Compt. rend.* **106**, 771–774 (1888).

229. Manning, W. M., Stauffer, J. F., Duggar, B. M., and Daniels, F. Quantum efficiency of photosynthesis in chlorella. *J. Am. Chem. Soc.* **60**, 266–274 (1938).

230. Maquenne, L., and Demoussy, E. "Échanges gaseux des plantes vert avec l'atmosphère." Paris, 1913.

231. Maquenne, L., and Demoussy, E. Sur la valeur des coefficients chlorophylliens et leur rapports avec les quotients respiratoires réels. *Compt. rend.* **156**, 506–512 (1913).

232. Maskell, E. J. Field observations on starch production in the leaves of the potato. *Ann. Botany (London)* **41**, 327–344 (1927).

233. Maskell, E. J. Experimental researches on vegetable assimilation and respiration. XVII. The diurnal rhythm of assimilation in leaves of cherry laurel at "limiting" concentrations of carbon dioxide. XVIII. The relation between stomatal opening and assimilation. A critical study of assimilation rates and porometer rates in cherry laurel. *Proc. Roy. Soc. (London)* **B102**, 467–487, 488–533 (1928).

234. Mason, B. "Principles of Geochemistry." Wiley, New York, 1952.

235. Maximov, N. A., and Maximov, T. A. K. Schwankungen in Verlauf der Photosynthese. *Ber. deut. botan. Ges.* **46**, 383–391 (1928).

236. Mayer, A. M., Eisenberg, A., and Evenari, M. Studies on deep mass culture in Israel. *Sci. Monthly* **83**, 198–203 (1956).

237. Mayer, J. R. "Die Organische Bewegung in ihren Zusammenhung mit dem Stoffwechsel." Heilbronn, 1845; also *in* "Mechanik der Wärme," p. 54. Cotta Stuttgart, 1874.

238. Merget, A. Sur les échanges gaseux entre les plantes et l'atmosphère. Résponse aux observations et critiques de M. Barthélemy. *Compt. rend.* **84**, 376–379, 957–959 (1877); **86**, 1492–1495 (1878).

239. Meyer, E. F. H. "Geschichte der Botanik." Koenigsberg, 1854.

240. Miller, E. S., and Burr, G. O. Carbon dioxide balance at high light intensities. *Plant Physiol.* **10**, 93–114 (1933).

241. Miller, L. P. Effect of manganese deficiency on the growth and sugar content of plants. *Am. J. Botany* **20**, 621–631 (1933).

242. Millikin, C. R. Antagonism between molybdenum and certain heavy metals in plant nutrition. *Nature* **161**, 528 (1948).

243. Milner, H. W., Hiesey, W. M., and Nobs, M. A. Physiology of climatic races. *Yearbook Carnegie Inst. Wash.* 346–350 (1958).

243a. Milner, H. W., and Hiesey, W. M. Photosynthesis in climatic races of *Mimulus*. I. Effect of light intensity and temperatures on rate. *Plant Physiol.* **39**, 208–213, (1964).

244. Mitchell, J. W. Effect of atmospheric humidity on rate of carbon fixation by plants. *Botan. Gaz.* **98**, 87–104 (1936).

245. Mituya, A., Nyunoya, T., and Tamiya, H. Prepilot plant experiments on algal mass culture. *In* "Algal Culture: From Laboratory to Pilot Plant" (J. S. Burlew, ed.), No. 600, pp. 273–284. Carnegie Inst. Wash. Publ., Washington, D.C., 1953.

246. Moore, W. E., and Duggar, B. M. Quantum efficiency of photosynthesis in chlorella. *In* "Photosynthesis in Plants" (J. Franck and W. E. Loomis, eds.), pp. 239–250. Iowa State Univ. Press, Ames, Iowa, 1949.

247. Moss, D., Musgrave, R. B., and Lemon, E. R. Photosynthesis under field conditions. III. Some effects of light, carbon dioxide, temperature and soil moisture on photosynthesis, respiration and transpiration of corn. *Crop. Sci.* **1**, 83–87 (1961).

248. Mulder, E. G. Molybdenum in relation to growth of higher plants and microorganisms. *Plant Soil* **5**, 368–415 (1954).

249. Muntz, A., and Laine, E. Proportion of carbon dioxide in the air of Antarctic regions. *Compt. rend.* **153**, 1116–1119 (1912).

250. Musgrave, R. B., and Moss, D. Photosynthesis under field conditions. I. A portable closed system for determining the rate of photosynthesis and respiration of corn. *Crop. Sci.* **1**, 37–41 (1961).

251. Myers, J. Growth characteristics of algae in relation to the problems of mass culture. In "Algal Culture: From Laboratory to Pilot Plant" (J. S. Burlew, ed.), No. 600, pp. 37–54. Carnegie Inst. Wash. Publ., Washington, D.C., 1953.

252. Myers, J., and Burr, G. O. Studies on photosynthesis. Some effects of light of high intensity on chlorella. *J. Gen. Physiol.* **24**, 45–67 (1940).

253. Myers, J., and Graham, J. Mass culture af algae. II. Yield as a function of cell concentration under continuous sunlight irradiance. III. Light diffusers: High vs. low temperature chlorellas. *Plant Physiol.* **34**, 345–352 (1959); **36**, 342–346, (1961).

254. Myers, J., and Johnston J. A. Carbon and nitrogen balance of chlorella during growth. *Plant Physiol.* **24**, 111–119 (1949).

255. National Geographic Society. Observations made in the highest stratosphere flight. U.S. Air Corps. *Nature* **141**, 270–274 (1938).

256. Nichiporovitch, A. A., and Strogonova, L. E. Photosynthesis and problems of crop yield. *Agrochimica* **2**, 26–53 (1957).

257. Nicholas, D. J. D. The function of trace metals in the nitrogen metabolism of plants. *Ann. Botany (London)* [N.S.] **21**, 587–598 (1957).

258. Nicholas, D. J. D., and Thomas, W. D. E. Some effects of heavy metals on plants grown in soil culture, Part 1. The effect of cobalt on fertilizer and soil phosphate uptakes and the iron and cobalt status of tomato. *Plant Soil* **5**, 67–80 (1953).

259. Nixon, R. W., and Wedding, R. T. Age of date leaves in relation to photosynthesis. *Proc. Am. Soc. Hort. Sci.* **67**, 256–259 (1956).

260. Noddack, W., and Kopp, C. Assimilation of carbon dioxide by green plants. IV. Assimilation and temperature. *Z. physik. Chem.* **A187**, 79–102 (1940).

261. Nutman, F. J. Studies of physiology of *Coffea arabica*. I. Photosynthesis of coffee leaves under natural conditions. *Ann. Botany (London)* [N.S.] **1**, 353–367 (1937).

262. Oddo, B., and Pollacci, G. Influenza del nucleo pirrolico nella formazione della clorofilla. *Gazz. chim. ital.* **50**, 54–70 (1920).

263. Oserkowsky, J. Quantitative relation between chlorophyll and iron in green and chlorotic pear leaves. *Plant Physiol.* **8**, 49–68 (1933).

264. Osterlind, S. Inorganic carbon sources of green algae. I. Growth experiments with Scenedesmus quadricauda and Chlorella pyrenoidosa. *Physiol. Plantarum* **3**, 353–360 (1950).

265. Owen, O., Small, T., and Williams, P. H. Carbon dioxide in relation to glasshouse crops. Part III. The effect of enriched atmospheres on tomatoes and cucumbers. *Ann. Appl. Biol.* **13**, 560–577 (1926); also **10**, 312–325 (1923).

266. Owen, P. C. Effects of infection with tobacco mosaic virus on the photosynthesis of tobacco leaves. *Ann. Appl. Biol.* **45**, 456–461 (1957).

267. Owen, P. C. The respiration of tobacco leaves after systemic infection with tobacco-mosaic virus. *Ann. Appl. Bot.* **43**, 114–126, 265–272 (1955).

268. Owen, P. C. The growth of sugar beet under different water regimes. *J. Agr. Sci.* **51**, 133–136 (1958).

269. Paetz, K. W. Untersuchungen über die Zuzammenhänge zwischen Stomataren-offnungweite und bekannter Intensitäten bestimmter Spectral Bezerke. *Planta* **10**, 611–665 (1930).

270. Pfeffer, W. "The Physiology of Plants" (Transl. by A. J. Ewart), Vol. 1. Oxford Univ. Press (Clarendon), London and New York, 1900.

271. Pirson, A. Functional aspects in mineral nutrition of green plants. *Ann. Rev. Plant Physiol.* **6**, 71–114 (1955).

272. Pollacci, G. Einfluss des pyrrolischen Kerns auf die Bildung der Chlorophylle. *Ber. deut. botan. Ges.* **53**, 540–542 (1935).

273. Porter, C. W., and Ramsperger, H. C. The action of ultraviolet light upon carbon dioxide and water. *J. Am. Chem. Soc.* **47**, 79–82 (1925).

274. Pratt, R., and Trelease, S. F. Influence of deuterium oxide on photosynthesis in flashing and continuous light. *Am. J. Botany* **25**, 133–139 (1938).

275. Price, N. O., Linkous, W. N., and Engel, R. W. Minor element content of forage plants and soils. *J. Agr. Food Chem.* **3**, 226–229 (1955).

276. Priestley, J. "Experiments and Observations on Different Kinds of Air," J. Johnson, London, 1774, 1776, 1779, 1781; also, Observations on different kinds of air. *Phil. Trans. Roy Soc.* **62**, 147–264 (1772).

277. Quinlan-Watson, F. The effect of zinc deficiency on the aldolase activity in the leaves of oats and clover. *Biochem. J.* **53**, 457–460 (1953).

278. Rabideau, G. S., French, C. S., and Holt, A. S. The absorption and reflection spectra of leaves, chloroplast suspensions and chloroplast fragments as measured by the Ulbright sphere. *Am. J. Botany* **33**, 769–777 (1946).

279. Rabinowitch, E. I. "Photosynthesis and Related Processes," Vol. 1, pp. 1–599; Vol. II, pp. 600–1205; Vol. II, Part 2, pp. 1211–2088. Wiley (Interscience), New York, 1945, 1951, 1956.

280. Reed, H. S. Effects of zinc deficiency on phosphate metabolism of the tomato plant. *Am. J. Botany* **33**, 778–784 (1946).

281. Reed, H. S. "Jan Ingen-Housz, Plant Physiologist (With a History of the Discovery of Photosynthesis)," p. 102. Chronica Botanica, Waltham, Massachusetts, 1949.

282. Reisenauer, H. M. Molybdenum content of alfalfa in relation to deficiency symptoms and response to molybdenum fertilization. *Soil Sci.* **81**, 237–242 (1956).

283. Reuther, W., and Burrows, F. W. The effect of manganese sulfate on the photosynthetic activity of frenched tung foliage. *Proc. Am. Soc. Hort. Sci.* **40**, 73–76 (1942).

284. Reuther, W., Embleton, T. W., and Jones, W. W. Mineral nutrition of tree crops. *Ann. Rev. Plant Physiol.* **9**, 175–206 (1958).

285. Reville, R., and Suess, H. E. Carbon dioxide exchange between atmosphere and ocean and the question of increases of atmospheric CO_2 during past decades. *Tellus* **9**, 18–27 (1957).

286. Rhykerd, C. L., Langton, R., and Peterson, J. B. Effect of light treatment on the relative uptake of labeled carbon dioxide by legume seedlings. *Agron. J.* **51**, 7–9 (1959).

287. Richards, F. J., and Berner, E., Jr. Physiological studies in plant nutrition. XVII. A general survey of the free amino-acids of barley leaves as affected by mineral nutrition with special reference to potassium supply. *Ann. Botany (London)* **18**, 15–33 (1954).

288. Riedel, F. Effect of carbon dioxide on the carbon balance of plants. *Landwirtsch. Fortschr.* **8**, 196–200 (1933).

289. Riedel, F. The use of purified smoke gas as carbon dioxide fertilizer for plants. *Gesundh.-Ing.* **53**, 257–261 (1930).

290. Riley, G. A. Carbon metabolism and photosynthetic efficiency. *Am.. Scientist* **32**, 132–134 (1944).

291. Riley, G. A. Plankton studies. II. The western North Atlantic. *J. Marine Res. (Sears Found. Marine Res.)* **1**, 335–343 (1938); **2**, 145–162 (1939).

292. Ruben, S., Randall, M., Kamen, M. D., and Hyde, J. C. Heavy oxygen O^{18} as a tracer in the study of photosynthesis. *J. Am. Chem. Soc.* **63**, 877–879 (1941).

293. Ruck, H. C., and Bolas, B. D. The effect of manganese on the assimilation and respiration rate of isolated rooted leaves. *Ann. Botany (London)* [N.S.] **18**, 267–298 (1954).

294. Ryther, J. H. Potential productivity of the sea. *Science* **130**, 602–608 (1959).

295. Sachs, J. von, "History of Botany, 1530–1860" (Transl. by H. E. F. Garnsey and I. B. Balfour), p. 568. Oxford Univ. Press (Clarendon), London and New York, 1890.

296. Saussure, N. T. de, "Recherches chimiques sur la végétation." Paris, 1804.

297. Scarth, G. W., and Shaw, M. Stomatal movement and photosynthesis in Pelargonium. I. Effects of light and carbon dioxide. II. Effects of water deficit and of chloroform. *Plant Physiol.* **26**, 207–225, 581–597 (1951).

298. Scharrer, K., and Jung, J. Optimal micronutrient rates in water culture experiments with corn and field beans. *Landwirtsch. Forsch.* **10**, 3–8, (1957).

299. Schmidt, W. "Der Massenaustausch in freier Luft und verwandte Erscheinungen." Hamburg, 1925.

300. Schneider, G. W., and Childers, W. F. Influence of soil moisture on photosynthesis, respiration and transpiration of apple leaves. *Plant Physiol.* **16**, 565–583 (1941).

301. Schroeder, H. Die jährliche Gesamtproduktion der Grunden Pflanzendecke der Erde. *Naturwissenschaften* **7**, 8–29, 976 (1919).

302. Schwabe, W. W. Physiological studies in plant nutrition. XVI. The mineral nutrition of bracken. Part II. The effect of phosphorus and potassium supply on total dry weights, leaf areas, net assimilation rates, starch and water contents, etc., in the sporophyte. *Ann. Botany (London)* [N.S.] **17**, 225–262 (1953).

303. Senebier, J. "Mémoires physico-chimiques, sur l'influence de la lumière solaire pour modifier les êtres des trois règnes de la nature et surtout ceux de règne végétale," 3 vols. Geneva, 1783.

304. Senebier, J. "Expériences sur l'action de la lumière solaire dans la végétale." Geneva, 1788.

305. Senebier, J. "Physiologie Végétable," 5 vols. Geneva, 1800.

306. Shaw, M., and Maclachlan, G. A. The physiology of stomata. I. Carbon dioxide fixation in guard cells. *Can. J. Botany* **32**, 784–794 (1954).

307. Shaw, M., and Maclachlan, G. A. Chlorophyll content and carbon dioxide uptake of stomatal cells. *Nature* **173**, 29–30 (1954).

308. Shimizu, T., and Tsuno, Y. Studies on yield forecasts. Photosynthesis of wheat and naked barley under field conditions. *Proc. Crop. Sci. Soc. Japan* **26**, 100–102 (1957).

309. Shirley, H. L. Light as an ecological factor and its measurement. *Botan. Rev.* **1**, 355–382 (1935).

310. Sierp, H. Untersuchungen über die Offnungs-bewegungen der Stomata in verschiedenen Spektralbereichen. *Flora (Jena)* **128**, 269–285 (1933).

311. Sievers, E. G. World sources of mineral fuels. *Gas Age-Record* **51**, 757–761 (1923).

312. Singh, B. H., and Lal, K. N. Investigations of the effect of age on assimilation of leaves. *Ann. Botany (London)* **49**, 291–307 (1935).

313. Sisler, E. C., Dugger, W. M., Jr., and Gauch, H. G. The role of boron in the translocation of organic compounds in plants. *Plant Physiol.* **31**, 11–17 (1956).

314. Skok, J., and McIlrath, W. J. Distribution of boron in cells of dicotyledenous plants In relation to growth. *Plant Physiol.* **33**, 120–131 (1958).

315. Smith, E. L. Limiting factors in photosynthesis: Light and carbon dioxide. *J. Gen. Physiol.* **22**, 21–35 (1938).

316. Smith, E. L. Photosynthesis in relation to light and carbon dioxide. *Proc. Natl. Acad. Sci. U.S.* **22**, 504–511 (1936).

317. Smith, J. H. C. Concurrency of carbohydrate formation and carbon dioxide absorption during photosynthesis in sunflower leaves. *Plant Physiol.* **19**, 394–403 (1944).

318. Smith, J. H. C. Molecular equivalence of carbohydrates to carbon dioxide in photosynthesis. *Plant Physiol.* **18**, 207–223 (1943).

319. Smith, P. F., Reuther, W., and Specht, A. W. Mineral composition of chlorotic orange leaves and some observations on the relation of sample preparation technic to the interpretation of results. *Plant Physiol.* **25**, 496–506 (1950).

320. Somers, I. I., and Shive, J. W. The iron-manganese relation in plant metabolism. *Plant Physiol.* **17**, 582–602 (1942).

321. Spoehr, H. A. "Photosynthesis." Chem. Catalog Co., New York, 1926.

322. Spoehr, H. A., and Milner, H. W. The chemical composition of chlorella; effect of environmental conditions. *Plant Physiol.* **24**, 120–149 (1949).

323. Spruit, C. J. P., and Kok, B. Simultaneous observation of oxygen and carbon dioxide exchange during non-steady state photosynthesis. *Biochim. Biophys. Acta* **19**, 417–424 (1956).

324. Stahl, E. Einige Versuche über Transpiration und Assimilation. *Botan. Ztg.* **52**, 117–146 (1894).

325. Stalfelt, M. G. Licht und Temperaturhemmung in der Kohlensäure-Assimilation. *Planta* **30**, 384–421 (1939).

326. Steemann-Nielsen, E. Carbon dioxide concentration, respiration during photosynthesis, and maximum quantum yield of photosynthesis. *Physiol. Plantarum* **6**, 316–332 (1953).

327. Steemann-Nielsen, E. Detrimental effects of high light intensities on the photosynthetic mechanism. *Physiol. Plantarum* **5**, 334–344 (1952).

328. Steemann-Nielsen, E. Experimental carbon dioxide curves in photosynthesis. *Physiol. Plantarum* **5**, 145–159 (1952).

329. Steemann-Nielsen, E. The persistence of aquatic plants to extreme pH values. *Physiol. Plantarum* **5**, 211–217 (1952).

330. Steemann-Nielsen, E. The use of radioactive carbon (C^{14}) for measuring organic production in the sea. *J. conseil, Conseil perm. intern. exploration mer* **18**, 117–140 (1952).

331. Steemann-Nielsen, E. The interaction of photosynthesis and respiration and its importance for the determination of C^{14} discrimination in photosynthesis. *Physiol. Plantarum* **8**, 945–953 (1955).

332. Steemann-Nielsen, E. Light and organic production in the sea. *Rappt. Proces-Verbaux Reunions, Conseil perm. intern. exploration mer* **144**, 141–148 (1958).

333. Steemann-Nielsen, E. Organic production in the oceans. *J. conseil, Conseil perm. intern. exploration mer* **19**, 309–328 (1954).

334. Steemann-Nielsen, E. Production of organic matter in the sea. *Nature* **169**, 956–957 (1952).

335. Stefan, J. Über die Verdampfung aus einem kreisformig oder elliptisch begrenzten Becken. *Sitzber. Akad. Wiss. Wien., Math.-naturw. Kl., Abt. II* **83**, 943–954 (1881).

336. Stiles, W. "Photosynthesis, the Assimilation of Carbon by Green Plants." Longmans, Green, New York, 1925.

337. Stocker, O. Über die Assimilation Bedingungen in Tropischen Regenwald. *Ber. deut. botan. Ges.* **49**, 267–273 (1931).

338. Stocker, O. Assimilation und Atmung Westjavanischer Tropenbaume. *Planta* **24**, 402–445 (1935).

339. Stocker, O., and Vieweg, G. H. Die Darmstädten Apparatur zur Momentanmessung der Photosynthese unter ökologischen Bedingungen. *Ber. deut. botan. Ges.* **73**, 198–208 (1960).

340. Stolwijk, J. A. J. Photoperiodic and formative effects of various wavelength regions in *Cosmos bipinnatus, Spinacia oleracea, Sinapis alba* and *Pisum sativum. I. Koninkl. Ned. Akad. Wetenschap., Proc.* **C55**, 489–502 (1952).

341. Stolwijk, J. A. J. Wavelength dependence of photomorphogenesis in plants. *Mededel. Landbouwhogeschool Wageningen* **54**, 181–224 (1954).

342. Stolwijk, J. A. J., and Thimann, K. V. Uptake of carbon dioxide and bicarbonate by roots and its influence on growth. *Plant Physiol.* **32**, 513–520 (1957).

343. Stoy, V. Action of different light qualities on simultaneous photosynthesis and nitrate assimilation in wheat leaves. *Physiol. Plantarum* **8**, 963–986 (1955).

344. Tamiya, H. Growing chlorella for food and feed. *Proc. World Symp. on Appl. Solar Energy, Phoenix, Ariz., 1955,* pp. 231–241 (1956).

345. Tamiya, H. Mass culture of algae. *Ann. Rev. Plant Physiol.* **8**, 309–334 (1957).

346. Tamiya, H., Hase, E., Shibata, K., Mituya, A., Iwamura, T., Niehei, T., and Sasa, T. Kinetics of growth of chlorella with special reference to its dependence on quantity of available light and on temperature. *In* "Algal Culture: From Laboratory to Pilot Plant" (J. S. Burlew, ed.), No. 600, pp. 204–232. Carnegie Inst. Wash. Publ., Washington, D.C., 1953.

347. Thomas, M. D., Effect of ecological factors on photosynthesis. *Ann. Rev. Plant Physiol.* **6**, 135–156 (1955).

348. Thomas, M. D. Effects of air pollution on plants. *World Health Organ., Geneva, Monograph Ser.* **46**, 233–278 (1961).

348a. Thomas, M. D., Hendricks, R. H., Collin, T. R., and Hill, G. R. The utilization of sulfate and sulfur dioxide for the sulfur nutrition of alfalfa. *Plant Physiol.* **18**, 345–371 (1943).

349. Thomas, M. D., Hendricks, R. H., and Hill, G. R. Apparent equilibrium between photosynthesis and respiration in an unrenewed atmosphere. *Plant Physiol.* **19**, 370–376 (1944).

349a. Thomas, M. D., and Hill, G. R. Unpublished data, 1933–1935.

350. Thomas, M. D., and Hill, G. R. Photosynthesis under field conditions. *In* "Photosynthesis in Plants" (J. Franck and W. E. Loomis, eds.), Chapter 2, pp. 19–52. Iowa State Univ. Press, Ames, Iowa, 1949.

351. Thomas, M. D., and Hill, G. R. Relation of sulfur dioxide in the atmosphere to photosynthesis and respiration of alfalfa. *Plant Physiol.* **12**, 309–383 (1937).

352. Thomas, M. D., and Hill, G. R. The continuous measurement of photosynthesis, respiration and transpiration of alfalfa and wheat growing under field conditions. *Plant Physiol.* **12**, 285–307 (1937).

353. Thorne, G. N., and Watson, D. J. The effect on yield and leaf area of wheat of applying nitrogen as a top dressing in April or in sprays at ear emergence. *J. Agr. Sci.* **46**, 449–456 (1955).

353a. Thorpe, T. E. On the amount of carbonic acid contained in the air above the Irish Sea. *Manchester, Lit. Phil. Soc., Proc.* **V**, 33–41 (1866).

354. Todd, G. W. Effect of ozone and ozonated 1-hexene on respiration and photosynthesis of leaves. *Plant Physiol.* **33**, 416–420 (1958).

355. Tonzig, S. Researches on the daily variations of the CO_2 content in the air available to higher plants during different periods of the year. *Nuovo giorn. botan. ital.* [N.S.] **57**, 583–618 (1950).

356. Turner, J. S., Todd, M., and Brittain, E. G. The inhibition of photosynthesis by oxygen. *Australian J. Biol. Sci.* **9**, 494–510 (1956).

357. Upchurch, R. P., Petersen, M. L., and Hagan, R. M. The effect of soil moisture content on the rate of photosynthesis and respiration in Ladino clover. *Plant Physiol.* **30**, 297–303 (1955).

358. Van der Honert, T. H. Carbon dioxide and limiting factors. *Rec. trav. botan. ned.* **27**, 149–286 (1930).

358a. Van der Paauw, F. The influence of temperature on the respiration and carbon dioxide assimilation of some green algae. *Planta* **22**, 396–403 (1934).

359. van der Veen, R. Induction phenomena in photosynthesis. *Physiol. Plantarum* **2**, 217–234 (1949).

360. Van Niel, C. B. The comparative biochemistry of photosynthesis. *In* "Photosynthesis in Plants" (J. Franck and W. E. Loomis, eds.), pp. 437–495. Iowa State Univ. Press, Ames, Iowa, 1949.

361. Van Oorschot, J. P. L. Conversion of light energy in algal culture. *Mededel. Landbouwhogeschool Wageningen* **55**, 225–276 (1955).

362. Verdeil, F. *Compt. rend.* **33**, 689 (1851).

363. Verduin, J. Diffusion through multiperforate septa. *In* "Photosynthesis in Plants" (J. Franck and W. E. Loomis, eds.), pp. 95–112. Iowa State Univ. Press, Ames, Iowa, 1949.

364. Verduin, J. A table of photosynthetic rates under optimal near-natural conditions. *Am. J. Botany* **40**, 675–679 (1953).

365. Verduin, J., and Loomis, W. E. Absorption of carbon dioxide by maize. *Plant Physiol.* **19**, 278–293 (1944).

366. Vernadskii, V. I. "Geochemie in ausgewählten Kapiteln (Ocerki Geochimii)" (Transl. by E. Kordes). Akad. Verlagsges., Leipzig, 1930.

367. Viets, F. G., Jr. Zinc deficiency of corn and beans on newly irrigated soils in central Washington. *Agron. J.* **43**, 150–151 (1951).

368. Vines, S. H. "Lectures on the Physiology of Plants." Cambridge Univ. Press, London and New York, 1886.

369. Waller, J. C. The katharometer as an instrument for measuring the output and intake of carbon dioxide by leaves. *New Phytologist* **25**, 109–118 (1926).

370. Warburg, O. Photosynthesis. *Science* **128**, 68–73 (1958).

371. Warington, K. Some interrelationships between manganese, molybdenum and vanadium in the nutrition of soybeans, flax and oats. *Ann. Appl. Biol.* **38**, 624–641 (1951).

372. Wassink, E. C., Kok, B., and van Oorschot, J. P. L. Efficiency of light-energy conversion in chlorella cultures compared with higher plants. *In* "Algal Culture: From Laboratory to Pilot Plant" (J. S. Burlew, ed.), No. 600, pp. 55–62. Carnegie Inst. Wash. Publ., Washington, D.C., 1953.

373. Wassink, E. C., and Stolwijk, J. A. J. Effects of light of narrow spectral regions on growth and development of plants. I. *Koninkl. Ned. Akad. Wetenschap., Proc.* **C55**, 471–488 (1952).

374. Wassink, E. C., and Stolwijk, J. A. J. Effects of light quality on plant growth. *Ann. Rev. Plant Physiol.* **7**, 373–400 (1956).

375. Wassink, E. C., Vermeulen, D., Reman, G. H., and Katz, E. On the relation between fluorescence and assimilation in photosynthesizing cells. *Enzymologia* **5**, 100–109 (1938).

376. Watanabe, A. Preservation of blue green algae in viable state. *J. Gen. Appl. Microbiol.* (*Tokyo*) **5**, 153–157 (1959).

377. Watanabe, A., Nishigaki, S., and Konishi, C. Effect of nitrogen-fixing blue-green algae on the growth of rice plants. *Nature* **168**, 748–749 (1951).

378. Watson, D. J. Comparative physiological studies on the growth of field crops. I. Variation of net assimilation rate and leaf area between species and varieties and within and between years. *Ann. Botany* (*London*) [N.S.] **11**, 41–76 (1947).

379. Watson, D. J. Leaf growth in relation to crop yield. *Proc. 3rd Easter School in Agr. Sci. Univ., Nottingham, 1955* (1956).

380. Watson, D. J. The dependence of net assimilation rate on leaf area index. *Ann. Botany* (*London*) [N.S.] **22**, 37–54 (1958).

380a. Watson, D. J., and Baptiste, E. C. D. A comparative physiological study of sugar beet and mangold with respect to growth and sugar accumulation. I. Growth analysis of the crop in the field. *Ann. Botany* (*London*) [N.S.] **2**, 437–480 (1938).

381. Wedding, R. T., Erickson, L. C., and Brannaman, B. L. Effect of 2-4-dichloro-phenoxyacetic acid on photosynthesis and respiration. *Plant Physiol.* **29**, 64–69 (1954).

382. Wedding, R. T., Riehl, L. A., and Rhoads, W. A. Effect of petroleum oil spray on photosynthesis and respiration in citrus leaves. *Plant Physiol.* **27**, 269–278 (1952).

383. Weintraub, R. L. Radiation and plant respiration. *Botan. Rev.* **10**, 383–459 (1944).

384. Went, F. W. "The Experimental Control of Plant Growth," Vol. 17. Chronica Botanica, Waltham, Massachusetts, 1957.

385. Went, F. W. Plant growth under controlled conditions. III. Correlations between various physiological processes and growth in the tomato plant. *Am. J. Botany* **31**, 597–618 (1944).

386. Whittingham, C. P. The chemical mechanism of photosynthesis. *Botan. Rev.* **18**, 245–290 (1952).

387. Whittingham, C. P. Rate of photosynthesis and concentration of carbon dioxide in chlorella. *Nature* **170**, 1017–1018 (1952).

388. Wiederspahn, F. E. Controlled manganese deficiency of apple. *Proc. Am. Soc. Hort. Sci.* **69**, 17–20 (1957).

389. Wiesner, J. Versuche über den Ausgleich des Gasdruckes in den Geweben der Pflanzen. *Sitzber. Akad. Wiss. Wien, Math.-naturw. Kl., Abt. I* **79**, 369–408 (1879).

390. Wiesner, J., and Molisch, H. Untersuchungen über Gasbewegung in der Pflanzen. *Sitzber. Akad. Wiss. Wien, Math-naturw. Kl., Abt. I* **98**, 670–713 (1889).

391. Wiesner, J., "Jan Ingen-Housz: Sein Leben und sein Wirken als Naturforscher und Arzt." Wien, 1905.

392. Williams, R. F. Physiology of plant growth with special references to the concept of net assimilation rate. *Ann. Botany* (*London*) [N.S.] **10**, 41–72 (1946).

393. Williams, W. T. A new theory of the mechanism of stomata movement. *J. Exptl. Botany* **5**, 343–352 (1954).

394. Wilson, C. C. Effect of some environmental factors on the movements of the guard cells. *Plant Physiol.* **23**, 5–37 (1948).

395. Wilson, J. W., and Wadsworth, R. M. The effect of wind speed on assimilation rate. A reassessment. *Ann. Botany* (*London*) [N.S.] **22**, 285–290 (1958).

396. Willstätter, R. Chlorophyll. *J. Am. Chem. Soc.* **37**, 323–345 (1915).
397. Willstätter, R., and Stoll, A. "Investigations on Chlorophyll: Methods and Results" (Transl. by F. M. Schertz and A. R. Merz). Science Press, Lancaster, Pennsylvania, 1928.
398. Willstätter, R., and Stoll, A. "Untersuchungen über die Assimilation der Kohlensäure." Springer, Berlin, 1918.
399. Willstätter, R., and Stoll, A. "Untersuchungen über das Chlorophyll." Springer, Berlin, 1913.
400. Woodward, R. B., and 17 co-authors. Total synthesis of chlorophyll. *J. Am. Chem. Soc.* **82**, 3800–3802 (1960).

CHAPTER TWO

Micrometeorology and the Physiology of Plants in Their Natural Environment[1]

Edgar Lemon[2]

I. Introduction 203
II. Radiation Exchange at the Earth's Surface. 204
III. Exchange of Mass and Momentum at the Earth's Surface 206
IV. Airflow and Turbulent Exchange above the Vegetation Canopy . . . 209
V. Airflow and Turbulent Exchange within the Vegetation Canopy . . . 214
VI. The Diurnal Cycle of Microclimate 216
VII. The Energy Balance 220
References 225

I. Introduction

Increasing understanding of the micrometeorological processes that control the natural physical environment in which plants live is providing a sounder basis for predicting plant response, particularly in plant communities of economic importance. New micrometeorological concepts and methods of measurement are becoming available too for measuring physiological response under natural field conditions. The value of these tools lies in the ability to measure such phenomena as the water vapor, carbon dioxide, and heat exchange taking place in the field over very short periods of time without disturbing the natural environment in any way. Thus we can relate, for example, photosynthesis and transpiration, to other immediate, short-term, climatic factors of the field environment. The understanding of natural processes as they operate in the field environment permits a sounder basis for interpreting results gained in the laboratory or growth chamber under controlled conditions. Indeed, the time is now imminent to use these methods in the study of population dynamics and the interaction between the environment and the genetic constitution of plants.

[1] Contribution from the Northeast Branch of the Soil and Water Conservation Research Division, ARS, U.S. Department of Agriculture, cooperating with the New York State Agricultural Experiment Station at Cornell University, and Meteorology Department, U.S. Army Electronic Proving Ground, Fort Huachuca, Arizona.
[2] Research Soil Scientist, U.S. Department of Agriculture and Professor in Agronomy, Cornell University, Ithaca, New York.

It is the purpose here to present briefly some of this micrometeorological background to plant physiologists. No attempt is made to give a complete review of the subject, which has attracted increasing attention in recent years, but rather to discuss certain of the simpler principles, the consideration of which amplify the conventional treatment of such topics as photosynthesis (Chapter 1), respiration (Chapter 3), and water relations (Chapter 7 in Volume II).

II. Radiation Exchange at the Earth's Surface

The physics of radiation transfer through the atmosphere to the earth's surface is well understood and will not be dealt with here. Reference may be made to Sutton (31). The exchange of radiation within plant communities is not so well understood, however. This arises from the dynamic complexity of the geometric array of plant surfaces, and their reflective and transmissive properties.

The radiant energy available at plant surfaces that is converted into other forms of energy is the difference between incoming and outgoing radiation (net absorption). For the simplest plant communities where the vegetation is dense and uniformly distributed net absorption can be expressed as an equation. Here we can approximate the net radiation absorption in uniform vegetation layers of thickness, $h - z$, by

$$R_{na} = (1 - \alpha_1)R_v[1 - e^{-k_1(h-z)}] + (1 - \alpha_2)R_{ir}[1 - e^{-k_2(h-z)}]$$
$$+ [D_t + U_t - B_{tu} - B_{td}][1 - e^{-k_3(h-z)}] \quad (1)$$

where:

R_{na} = net radiation absorbed, all wavelengths

R_v = downward visible solar radiation (0.3–0.7 μ)

R_{ir} = downward near infrared solar radiation (0.7–3.0 μ)

D_t = downward thermal radiation (>3.0 μ)

U_t = upward thermal radiation at the bottom of the vetegation layer

B_{tu} = thermal radiation emitted upward within the vegetation layer

B_{td} = thermal radiation emitted downward within the vegetation layer

k_1, k_2, k_3 = absorption coefficients

α_1, α_2 = reflection coefficients

h = height of vegetation surface at top of plant canopy

z = height above ground surface.

Each of the three right-hand terms is a net balance for each wavelength concerned. For a corn (*Zea mays*) crop (15) the following representative values have been evaluated for the reflection and absorption coefficients: $\alpha_1 = 0.07$, $\alpha_2 = 0.29$; $k_1 = 0.01$, $k_2 = 0.006$, and $k_3 = 0.008$.

In most micrometeorological studies it is not necessary to measure these various wavelength components separately since there are commerically available instruments that measure net radiation flux directly. Figure 1 gives the distribution of measured net radiation flux

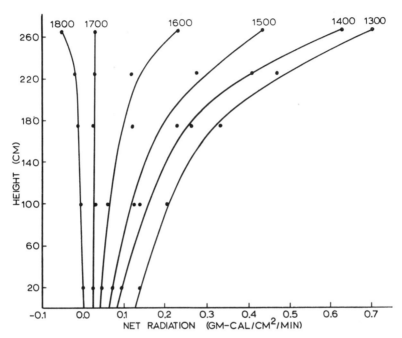

Fig. 1. Representative profiles of net radiation flux in a dense stand of corn approximately 300 cm high. Hour of the day is indicated at the top of each profile. Positive values indicate downward flux, and negative values indicate upward flux. After Allen *et al.* (2).

in the sample corn crop to which reference was made earlier (2). From the figure it is easy to see that during the daylight hours a goodly portion of the net radiation is absorbed in the upper layers of vegetation. It should be pointed out, however, that considerable net radiation is available at the ground surface too, even in dense vegetation. This arises from the relatively high transmission of near infrared solar radiation by vegetation (17, 41).

Net radiation is negative early and late in the day and during the night. This net loss of radiation to the sky usually reaches a maximum shortly after sunset.

In plant communities where the vegetation is not dense or uniform the description of the radiation regime becomes extremely complicated and will not be taken up here. Reference may be made to Aizenshatat (1), Denmead *et al.* (7), Saeki (28), Tanner *et al.* (35), Uchijima (37), and Waggoner and Reifsnyder (40), to mention a few.

III. Exchange of Mass and Momentum at the Earth's Surface

Airflow processes that control the exchange of mass (such as gases and finely divided solids and liquids) play an important role in creating the climate of plant communities through the transfer of, for example, sensible heat, water vapor, and carbon dioxide. Winds impinging upon plant surfaces are slowed by drag or friction so that momentum is also transferred between plants and the atmosphere. These transfers are of course, coupled to physiological processes.

The equations for the transfer of mass and momentum exchanged with the earth's surface have the form of a diffusion equation:

$$f = K(dq/dz) \tag{2}$$

where f is the flux density of the entity in question on a per unit area per unit time basis (flux downward considered as positive); K is the diffusivity coefficient; and dq/dz is the vertical gradient of the entity in question. The quantitative value of K is usually several orders of magnitude larger than molecular diffusivity values. Although relatively slow molecular diffusion plays an important role very near surfaces, perhaps within the first few millimeters, "eddy diffusion," due to turbulent motion, is a rapid transfer mechanism operating over deep atmospheric layers. It is through large values of K, for example, that carbon dioxide concentration is maintained and is not rapidly depleted during daylight by the photosynthetic process. Tremendous volumes of air, hundreds of meters in depth, have to be mixed to bring sufficient carbon dioxide to the surface to maintain a concentration near 300 parts per million.

Within the first few meters of the earth's surface turbulent motion is generated by two processes, the frictional drag of the wind impinging upon the rough elements composing the surface and the buoyant motion caused by change in density when air is heated by a warmer surface. Frictional drag gives rise to "forced convection," which is dominant when the wind blows and there is no appreciable temperature gradient from warmer surfaces. If winds are low and there is a large temperature gradient from underlying warmer surfaces, then buoyancy becomes important too, giving rise to "free convection." Vegetated surfaces that are freely transpiring (and therefore are relatively cool) seldom develop appreciable temperature gradients during the day except under low winds.

One indication of turbulent motion is the rapid fluctuations of the physical properties of the air such as wind speed, temperature, and concentrations of water vapor and carbon dioxide. Figure 2 is a recorder trace of the carbon dioxide gradient over an active corn crop. Instrument response time was only sufficient to record changes in concentration associated with the larger "eddies" in the turbulence regime. Were an instrument fast enough to record all the fluctuations due to all eddy sizes in the turbulent system, such fluctuations as recorded in

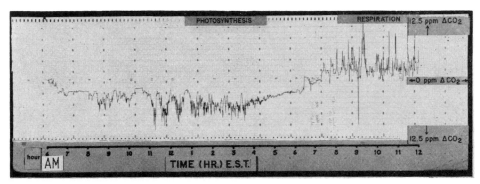

Fig. 2. Recorder trace of carbon dioxide concentration gradients, ΔCO_2, over a rapidly growing field of corn. Differential is measured between two heights over the corn (see Fig. 5). Negative values indicate flux downward (net photosynthesis), and positive values indicate flux upward (net respiration). During daylight hours relatively long-term fluctuations are due to changing cloudiness, i.e., 10–12 A.M., whereas short-term fluctuations are due to turbulence, i.e., 1–4 P.M., when skies were relatively clear. Nighttime fluctuations are due entirely to turbulence. Gradients presented here were used to calculate exchange rates presented in Fig. 6 employing the "aerodynamic" method. Ellis Hollow, New York, July 26, 1961.

the figure would appear on a time scale of fractions of seconds. Since the physical properties of the air are influenced by the same eddies, the instantaneous fluctuations of all these properties are in consort and are strongly correlated as well might be expected. More will be said about this later.

Let us now look at Fig. 3. Here is presented a typical mean wind profile above and within a vegetated layer. It will be noticed that horizontal wind velocity, u, decreases with decreasing height, z, above the ground, going to zero at the ground. Since horizontal momentum of a unit volume of air is ρu (ρ is the mass per unit volume), evidently the vegetation and ground are extracting momentum, due to surface drag, so that there is a vertical transfer of horizontal momentum to

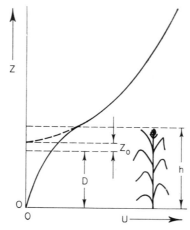

FIG. 3. Schematic representation of wind velocity distribution, u, with height, z, above the ground in a plant community of height h. Wind velocity distribution is logarithmic with height above plants and exponential with height within the plant community. Logarithmic profile extrapolation downward to zero wind speed $(u = 0)$ is indicated by dashed curve with intercept at $D + z_0$ (see text).

the vegetation and ground. Thus we can write an equation in the form of (2) for the vertical transfer of momentum:

$$\tau = \rho \text{Km}(\mathrm{d}u/\mathrm{d}z) \tag{3}$$

where τ is the momentum flux density or the so-called shearing stress and is expressed on a per unit area per unit time basis. Km is the eddy diffusivity coefficient, sometimes called the eddy viscosity.

For the transfer of other entities of interest here we have:

$$E = \text{Kw}(\mathrm{d}e/\mathrm{d}z) \tag{4}$$
$$P = \text{Kc}(\mathrm{d}C/\mathrm{d}z) \tag{5}$$
$$H = C_\text{p}\rho\text{Kh}(\mathrm{d}T/\mathrm{d}z) \tag{6}$$

where E, P, and H are the flux densities for water vapor, carbon dioxide, and sensible heat and Kw, Kc, and Kh are the eddy diffusivities for the same respective entities. The symbols e and C are the masses of water vapor and carbon dioxide per unit volume of air, T is the temperature, and C_p the specific heat of the air.

As mentioned above, the instantaneous fluctuations of physical properties of the air are in consort indicative of simultaneous transport of mass and momentum by the same eddies in the turbulent regime. While turbulent motion is chaotic and always changing, and thus diffusivity values are always changing, it is not surprising that the instantaneous values of the diffusivities are closely related. Under steep temperature gradients it is believed that Kh will be somewhat different from the other diffusivity coefficients. The similarity of these coefficients, assuming Km = Kw = Kh = Kc, however, provides some useful tools for measuring fluxes at the surface. For example, if one has an independent measure of evapotranspiration, E from a weighing lysimeter, then by measuring the vertical gradients of water vapor, Δe, and carbon dioxide, ΔC, over the same interval of distance above the vegetation, Δz, it is simple to calculate the carbon dioxide flux thus:

$$P = E(\Delta C/\Delta e) \tag{7}$$

Monteith and Szeicz (18) have used this principle to evaluate CO_2 exchange rates over a sugar beet (*Beta vulgaris*) field.

IV. Airflow and Turbulent Exchange above the Vegetation Canopy

From the previous discussion it should not be surprising that the physical properties of the air are distributed in similarly shaped mean profiles in the boundary layer above vegetation (26). It was discovered early that in this region somewhat above the roughness elements, the physical properties of the air, under near isothermal conditions, are distributed in logarithmic profiles similar to those found in fluid flow in "rough pipes" under conditions of fully developed frictional turbulence. It was found that the fluid velocity gradients in pipes and over earth surfaces under full frictional turbulence could be expressed by

$$du/dz = u_*/kz \tag{8}$$

where $u_* = (\tau/\rho)^{1/2}$ and $k = 0.4$ (von Karman's constant). Upon integrating (8) we have the empirical logarithmic law which defines the wind speed, u_z, at level, z, above the ground

$$u_z = (u_*/k) \ln (z/z_0) \tag{9}$$

where z_0 is the "roughness length." The parameter, z_0, is a constant of integration and is used to define the roughness characteristics of a given surface. The roughness length, z_0, is found by extrapolating graphically, from measured velocities, the logarithmic profile above the roughness elements downward to where the wind speed theoretically should be zero. Inspection of Fig. 3 reveals, however, that where appreciable vegetation grows on the surface the logarithmic profile distributions of wind (and the other physical properties) are displaced upward from the ground surface. It thus becomes necessary to evaluate this displacement. The value of the displacement, D, must then be subtracted from the height values z measured from the ground surface. Thus equation (9) now becomes:

$$u_z = (u_*/k) \ln (z - D)/z_0 \tag{10}$$

Methods used for evaluating D and z_0 and their shortcomings are taken up elsewhere (6, 14, 36). It suffices to say here that temperature gradients and elastic properties of vegetation as well as vegetation growth all combine to make D and z_0 complicated parameters characterizing vegetated surfaces (11, 30). As mentioned earlier, however, temperature gradients (except when wind speeds are low) are relatively small over vegetation so that plant geometric and elastic properties primarily control D and z_0. Figure 4 presents mean values of D and z_0 for a corn crop as a function of height of the crop. These

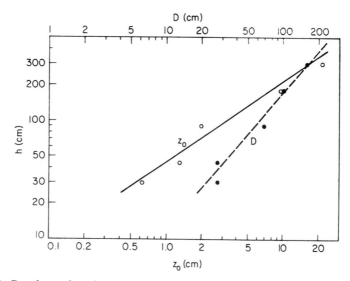

FIG. 4. Roughness length, z_0, and reference plane displacement, D, as a function of height of corn crop, h, during the growing season, 1961, at Ellis Hollow, New York.

values are for near isothermal conditions and wind speeds of 100–400 cm/sec at the 1-meter level above the crop surface.

Returning now to the similarity principle where Km = Kh = Ke = Kc, we can substitute Km for any of the other diffusivity coefficients in Eqs. (4–6). For carbon dioxide exchange, for example, we can substitute Km for Kc and integrate between two levels above the vegetation $z_1 - D$ and $z_2 - D$, obtaining:

$$P = \frac{k^2(u_2 - u_1)(C_1 - C_2)}{[\ln (z_2 - D)/(z_1 - D)]^2} \tag{11}$$

remembering that Km = $(\tau/\rho)/(du/dz)$. Thus we see that by measuring the gradients of wind, Δu, and carbon dioxide, ΔC, over the same height intervals, $z_2 - z_1$, (see Fig. 5), one is able to calculate from equation (11) carbon dioxide exchange rates, provided D has been evaluated from wind profile measurements. We have already mentioned that this presents serious problems and that temperature gradients must be small. The problem of temperature gradients must be considered because equations (8–11) are based upon turbulence created through surface drag or "forced convection." Under conditions of appreciable temperature gradients, or "free convection," equation (11) will underestimate the exchange rate since the added effect of buoyant turbulence will not have been taken into account. By the

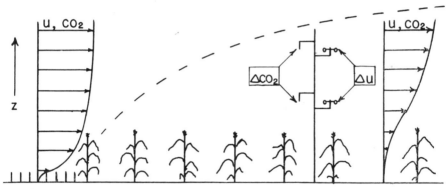

FIG. 5. Schematic representation of wind and carbon dioxide profiles over two different types of vegetation. Dashed curve indicates boundary layer development over taller vegetation when wind is blowing from left to right (vertical scale of boundary layer greatly exaggerated). Mast with sampling devices indicates correct position to sample for carbon dioxide gradients, ΔCO_2, and wind gradients, Δu, above the vegetation but within the boundary layer.

same token temperature inversions over cold surfaces at night damp out turbulent motion, causing Eq. (11) to overestimate exchange rates. Several means for taking into consideration the influence of temperature gradients have been proposed (4, 16, 23, 29, 32).

Perhaps the simplest concept of thermal influence is that of the Richardson number (Ri). This nondimensional number is a quantitative ratio of thermal turbulence effects compared to frictional turbulence effects having the form of

$$\text{Ri} = \frac{g \, dT/dz}{T' \left(\dfrac{du}{dz}\right)^2} \tag{12}$$

where g is the constant of gravity and T' is the absolute temperature.
Equation (12) can be approximately defined by

$$\text{Ri} = [gz/T'][(T_2 - T_1)/(u_2 - u_1)^2] \tag{13}$$

and is applicable to the mean height z between z_1 and z_2 where ΔT and Δu are measured.

Equation (10) can be modified by introducing the Richardson number ratio to take into account thermal effects thus:

$$u_z = [u_*/k \ln (z - D)/z_0][1 - \sigma \text{ Ri}]^{-\frac{1}{2}} \tag{14}$$

where σ is an empirical constant to be determined from a given set of observations depending upon how the mean Richardson number is defined. For example, for observations at Rothamsted, Monteith (20) gives a value of $\sigma = 10$.

Equation (11) can now take the form

$$P = \frac{k^2(u_2 - u_1)(C_1 - C_2)}{[\ln (z_2 - D)/(z_1 - D)]^2} (1 - \sigma \text{ Ri}). \tag{15}$$

Inspection of equation (13) reveals that the Richardson number corrections become less important the smaller the value of z, the mean height above the ground where ΔT and Δu are measured. Thus it is wise to make gradient measurements as close to the surface as possible but still far enough above the vegetation to be within the log profile region.

It should be pointed out again, however, that over wet vegetation, for conditions other than low winds, the thermal effects are small. Under such conditions the Richardson number has absolute values not much larger than 0.03–0.05 so that corrections are small.

Another problem associated with the use of the log profile distribution of physical properties of the air above rough surfaces must be considered. This is the problem of "fetch" and the establishment of equilibrium profiles over the surface in question. Reference should now be made to Fig. 5. The dashed line schematically indicates the growth of the boundary layer downwind from the leading or upwind edge of a uniform vegetated surface. Within this boundary layer above the vegetation, equilibrium logarithmic profiles are established, representing the uniform vegetation underneath. Outside the boundary layer the profiles are no longer logarithmic or representative of the surface below, but are complex representations of conditions upwind. Therefore, the use of equations (8–15) depend upon making measurements within an adequate boundary layer. This can best be assured by having as long a fetch as possible from measurement site to the leading edge and by making measurements close to the surface. A test for adequate fetch is the satisfactory measurement of logarithmic profiles above the vegetation.

Inoue *et al.* (10), Lemon (14), and Monteith (19) have utilized the log profile method (or aerodynamic method) to evaluate net carbon dioxide exchange rates of wheat (*Triticum*), corn (*Zea*), a mixed grass sward (*Phleum pratense* and *Festuca elatior*), and beans (*Vicia faba* [= *V. vulgaris*]). Independent checks on the method are difficult to obtain. An example is given in Fig. 6. Here are plotted net carbon

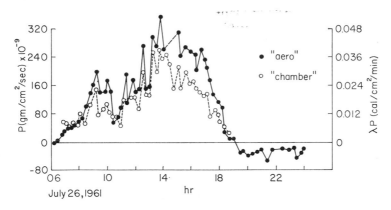

Fig. 6. Comparison of net carbon dioxide exchange rates, *P*, in a corn field determined by "aerodynamic" and closed plastic chamber techniques. Exchange rates also expressed in photochemical energy units, λ*P*. July 26, 1961, Ellis Hollow, New York. From unpublished work of R. B. Musgrave and E. Lemon (1961).

dioxide exchange rates calculated from wind and carbon dioxide gradients obtained by the author over a corn field. These are compared to exchange rates obtained simultaneously from a closed plastic chamber placed over a representative patch of corn in the same field.[3] Close inspection reveals that the two methods agree fairly well with some interesting exceptions: (a) during peak exchange rates until 1400, the aerodynamic calculations significantly exceed the chamber rates; and (b) after 1400, the chamber rates were generally lower. Speculation on the reasons for the differences centers not so much upon errors in the methods, as on differences in the environment of the corn plants.

[3] Appreciation is expressed to Dr. R. B. Musgrave and his former graduate student, Dr. D. N. Baker, for making these data available. The method used by these colleagues has been reported [see Musgrave and Moss (21) and Baker (2a)].

One possible explanation of (a) is a difference in the supply of carbon dioxide. During peak periods under bright sunlight there appeared to be a positive correlation between aerodynamic calculated exchange rates and the wind velocity (14, 15). However, air movement (with CO_2 held at 300 ppm) within the closed chamber was maintained at a uniformly high level by mechanical mixing with fans. A possible explanation of (b) could be a little less favorable moisture regime for the corn in the closed chamber. This effect would most likely show up in the afternoon, although no other evidence is available to indicate that there was any moisture stress. In any event, the two methods agree remarkably well, the differences demonstrating some of the difficulties encountered in comparing plant response under natural conditions and under partially controlled conditions, even assuming there are no errors in the methods. No such agreement could be expected, however, when equation (11) is used where appreciable temperature gradients are experienced without some means of accounting for buoyancy.

V. Airflow and Turbulent Exchange within the Vegetation Canopy

Despite extensive knowledge about profile characteristics and exchange properties of the turbulent airstream immediately above the earth's surface in a well-developed boundary layer, little is known about such phenomena within the canopy of vegetation. This is understandable since the vegetation presents surfaces of sources and sinks distributed in complex geometric array. Recently, however, some empirical models have been proposed for the simplest canopies of relatively uniform distribution of leaves such as are found in corn and wheat fields (5, 12, 22, 33).

It now appears that, in simple uniform canopies at least, the wind profile from the top of the canopy down to at least half-way to the ground, can be represented by an exponential form

$$u_z = u_h \exp\left[-\gamma(1 - z/h)\right] \tag{16}$$

where u_h is the wind speed at the top of the canopy and γ is a velocity attenuation coefficient. By referring to Fig. 3 one can see that the exponential form of the wind profile in the canopy gradually blends into the log profile form near the top of the canopy. This is also true near the base of the canopy since the wind speed has to go to zero at the ground surface.

It might appear that the mean wind speed profiles both above the canopy and within the canopy are expressed by empirical mathematical forms that have little to do with any theory of turbulence. This is not the case, however. Both forms can be arrived at through the mixing-length hypothesis attributed to Prandtl. The mixing length l can be defined by referring back to equation (3) where

$$\tau/\rho = \mathrm{Km}(du/dz) = l^2(du/dz)^2 = u_*^2 \tag{17}$$

(u_* is often called friction velocity). The usual understanding of the mixing length is as follows. An "eddy" breaks away from an original level z and carries across momentum (ρu) appropriate to the mean velocity at level z to a new level, $z + l$, where the eddy mixes again with the mean flow at level $z + l$. The creation and absorption of eddies in the mean motion at various levels create velocity fluctuations. The mixing length concept infers that the quantity l thus introduced is a unique length scale which characterizes the transfer properties of the turbulent motion at a given level.

Above rough surfaces the mixing length increases linearly with height

$$l = kz \tag{18}$$

where k is the von Karman constant introduced earlier. The real success of the mixing length hypothesis lies in the fact that k is indeed a constant, independent of scale of flow. In other words, it is equally applicable to the atmospheric surface boundary layer as to liquid flow in rough pipes.

In the Cionco model (5) for canopy air flow, the mixing length is considered constant with depth into the canopy [also assumed by Inoue et al. (12)]. They reason as follows: Through Eq. (17) we see that $\mathrm{Km} = u_* l$. Above the vegetation u_* is nearly constant with height while Km and l increase linearly. However, down in the canopy, changes in Km should largely be due to changes in u_* since the restrictive action of leaves and stems will cause the mixing length to be more nearly constant within the vegetation canopy. This assumption appears applicable to simple canopies because the wind profile equation derived (Eq. 16) agrees with the experimental data. The diffusivity equation now becomes:

$$K(z) = K(h) \exp \left[-\gamma(1 - z/h)\right] \tag{19}$$

The evaluation of the diffusivity coefficients within the vegetation canopy poses some real problems. Penman and Long (25) report several approaches, one of which is the energy balance to be discussed later. Figure 7 presents a normalized diffusivity profile in red clover (*Trifolium pratense*) obtained by the author using the energy balance approach.[4] The slope of the line gives a value for $\gamma = 2.5$. Inoue *et al.* (12) report $\gamma = 4.25$ for a pine (*Pinus radiata*) forest, and Uchijima and Wright (39) found $\gamma = 2.8$ for a corn field.

Some aerodynamic methods of evaluating diffusivities within the canopy have been attempted. These have been based upon statistical correlation of velocity fluctuations of the eddies themselves (30, 39).

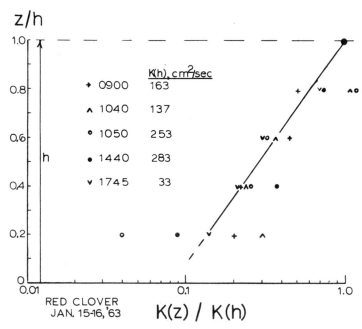

FIG. 7. Relative diffusivity, $(K(z)/K(h))$, as a function of relative height, z/h, in a red clover community. Experimental points are for given hours of the day and absolute diffusivity values are for the top of the crop of height, h, at specified time. The constructed line conforms to Equation (19) where the attenuation coefficient, $\gamma = 2.5$ (the slope). Palmerston North, New Zealand, January 15–16, 1963. From unpublished work of R. W. Brougham and E. Lemon (1963).

The eddy-correlation techniques (8) hold much promise as theory and instrumentation develop.

VI. The Diurnal Cycle of Microclimate

We have devoted considerable attention to the complex subject of turbulent diffusion near the earth's surface because it plays such an im-

[4] Unpublished data obtained in cooperation with Dr. Ray Brougham, DSIR, Palmerston North, New Zealand, January 15–16, 1963. The theory and experimental procedure can be found in a paper by Begg *et al.* (3).

portant role in controlling the microclimate where plants live. The manner in which wind interacts with plants through such parameters as "roughness," z_0, and the "velocity attenuation coefficient," γ, dictates in no small way, the temperature, the water vapor, and the carbon dioxide regime where the plants live. These "aerodynamic" properties of plants are governed by the geometric array and the elastic properties of the leaves and the stems, as has been mentioned before. Now we shall look at the microclimate of a simple plant community over a diurnal cycle and then relate these microclimate factors to dynamic exchange phenomena.

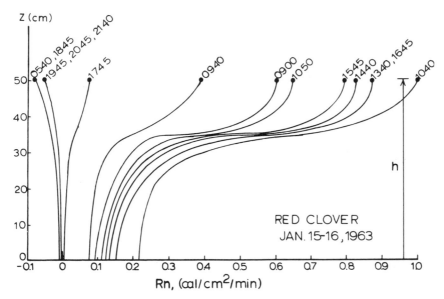

Fig. 8. Net radiation flux profiles in red clover crop of height h. Time of measurement is indicated at top of profiles. Palmerston North, New Zealand, January 15–16, 1963. From unpublished work of R. W. Brougham and E. Lemon (1963).

Figures 8–10 present mean profiles of various microclimate components in a red clover community over a diurnal cycle. They were taken during fair summer weather with moderate winds and temperature.[4] We have already pointed out the pertinent points concerning the distribution of net radiation in vegetation. The differences between net radiation flux profiles in Figs. 1 and 8 are due to leaf area distribution differences, as one can appreciate. The corn plant community had a total leaf area index, LAI (i.e., area of top surface of leaf laminae:ground area), of 4.2 relatively evenly distributed over a height of 300 cm. The red clover had a LAI of 5.8 heavily concen-

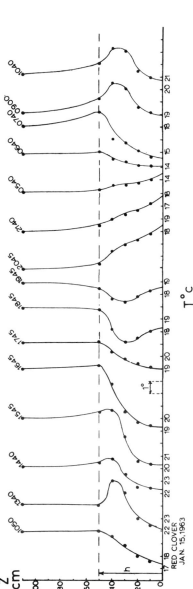

FIG. 9. Air temperature profiles in red clover crop of height h. Measurement times are indicated at the top of each profile. The temperature at the base of each profile is indicated in degrees centigrade. Palmerston North, New Zealand, January 15–16, 1963. From unpublished work of R. W. Brougham and E. Lemon (1963).

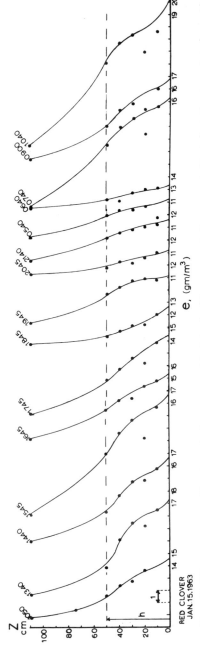

FIG. 10. Water vapor profiles in red clover crop of height h. Measurement times are indicated at the top of each profile. The water vapor concentration at the base of each profile is indicated in grams per cubic meter of air. Palmerston North, New Zealand, January 15–16, 1963. From unpublished work of R. W. Brougham and E. Lemon (1963).

trated just below the canopy top which had a height of 50 cm. The leaf area distribution in the red clover can be found in Fig. 16, and reference to that figure now will assist in understanding all the profiles in Figs. 8–11.

Turning now to the temperature profiles in Fig. 9 one can see first that the profiles above the surface are logarithmic and the gradients very small. The very small daytime gradients here are understandable. Since the clover was a vigorously growing crop with no deficiency of water to impair transpiration, a very large portion of the net radiant energy was expended toward latent heat exchange and very little toward sensible heat exchange. Small gradients both during the day and night can also be attributed to moderate winds. Down in the canopy, gradients were much steeper and complex. Typical day and night profiles are the 1040 and 1945 ones, respectively, where there are distinct zones of heating and cooling giving maximum and minimum temperatures. The other profiles are transitions toward these typical ones, although the 2045, 2140, and 0540 profiles are difficult to explain and might be suspect. It is obvious that more sampling points are needed at both the top and the bottom of the canopy to define more clearly the abrupt changes that take place there. More open canopies, as one might expect, produce much smaller gradients due to greater turbulent mixing. Vegetation that is deficient in water will produce, on the other hand, much larger temperature gradients during the daytime, quite often with the maximum at the soil surface (3).

Figure 10 presents the water vapor profiles for the same periods. Again it is worthwhile to point out the logarithmic characteristic of the above-canopy profiles. Gradients are larger, however, particularly in the daytime with the large loss of transpired water. Typical day and night profiles are the 1050 and 2140 ones. Notice that there is always a gradient favoring water loss upward; thus, dew formation at night is fed by water vapor from the soil below rather than from the atmosphere above the crop. During the day the steepest gradients in the canopy are near the top where transpiration is most active. Again sampling points near the soil and canopy top are needed to define more clearly the abrupt changes that are common at these levels.

Turning now to Fig. 11 we have the carbon dioxide gradients. These profiles are the best ones obtained. Their complexity arises from there being two sources of carbon dioxide during the day when photosynthesis is taking place. Carbon dioxide is then coming from the soil (and the net respiration of the heavily shaded lower leaves of the clover) as well as from the atmosphere above the canopy. During the night carbon dioxide is only moving upward. It should be pointed out that the minimum point in the profile moves downward in daylight during the morning hours, then upward as the afternoon progresses. Absolute concentrations were not measured in this case, but experience has shown that under conditions of moderate wind the absolute concentrations above the crop would vary from a high at night of 350 ppm to a daytime low of, say, 275 ppm, not a very wide variation. The carbon dioxide concentration units in the figure are about double that expressed on a parts per million basis. One division is approximately

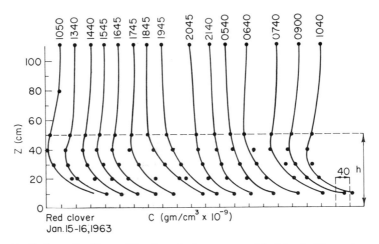

Fig. 11. Carbon dioxide profiles in red clover crop of height h. Time of measurement is indicated at the top of each profile. Experimental points are relative for a given profile only. Absolute values are not indicated for the horizontal axis; the curves have been displaced for clarity. Units of concentration are milligrams per cubic meter. Palmerston North, New Zealand, January 15–16, 1963. From unpublished work of R. W. Brougham and E. Lemon (1963).

equal to 20 ppm. The gradients near the base of the crop must be extremely steep. More sampling points are needed to define this region more clearly.

VII. The Energy Balance

Before taking up the dynamic exchange associated with microclimate profiles, it is necessary to introduce the concept of the energy balance or the so-called heat budget. This concept is based upon the energy conservation law and provides a very useful tool for measuring mass exchange without resorting to difficult aerodynamic methods. Tanner (34) has had remarkable success in proving the reliability of the energy balance method to measure evapotranspiration from vegetated surfaces.

Penman (24) incorporates the principle into his evapotranspiration formula. Rider (27) and Lemon et al. (13) have used it in conjunction with the logarithmic profile aerodynamic method to evaluate the transpiration from vegetation of differing wetness. Most recently Begg et al. (3) have used the energy balance principle to define spatially the transpiration source strength within the canopy of a bulrush millet crop (*Pennisetum typhoides* Stapf & C. E. Hubbard). In this principle

$$\text{Rn} - LE - H - S - \lambda P - B = 0 \qquad (20)$$

where

    Rn = the net radiation flux (upward flux minus downward flux)
    LE = the latent heat exchanged with the atmosphere
    H = the sensible heat exchanged with the atmosphere
    S = sensible heat exchanged with soil storage
    L = thermal conversion factor for latent heat
    λ = thermal conversion factor for fixation of carbon dioxide
    λP = photochemical energy gained or lost from plant storage
    B = sensible heat gained or lost from storage in the vegetation.

For evaluating such a large component in the energy balance as evapotranspiration from vegetation it is possible to ignore such small terms as P and B. Then utilizing the similarity principle where

$$H/LE = (\rho C_p/L)(\Delta T/\Delta e) = \beta \text{ (called the Bowen ratio)} \qquad (21)$$

we can write equation (20) thus:

$$LE = (\text{Rn} - S)/(1 + \beta) = (\text{Rn} - S)/[1 + (\rho C_p/L)(\Delta T/\Delta e)] \quad (22)$$

Thus, by making relatively simple measurements of net radiation flux, Rn, and the gradients of temperature, ΔT, and water vapor, Δe, over the vegetation, and soil heat exchange in the soil, S, it is possible to calculate the latent heat exchange without resorting to any aerodynamics other than accepting the assumption that Kw = Kh.

Also, by accepting the principle of similarity it is possible to utilize the heat budget to evaluate the diffusivity coefficient within the plant canopy at level z (38):

$$K(z) = \frac{\text{Rn}(z) - S - B(z)}{L(de/dz) + C_p\rho(dT/dz) + \lambda(dC/dz)} \qquad (23)$$

Knowing $K(z)$ and the other gradients at z, one can then evaluate the various fluxes at level z through the use of equations (4–6). A good example of this application can be found in Figs. 12 and 13 taken from

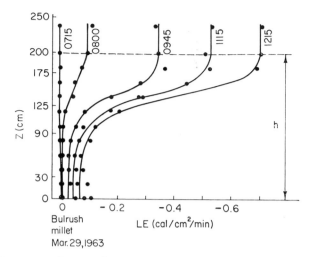

FIG. 12. Representative profiles of latent heat flux at indicated hours in a crop of bulrush millet of height h. Negative flux signifies evapotranspiration and positive flux condensation as dew. Katherine, N.T. Australia, March 29, 1963. After Begg *et al.* (3).

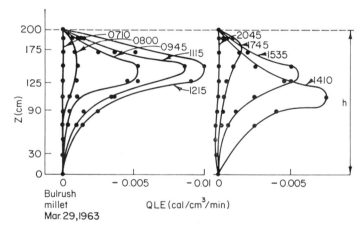

FIG. 13. Source and sink intensity of latent heat exchange, at indicated hours, in a crop of bulrush millet of height h. Negative values signify transpiration and positive values, dew formation. Katherine, N.T. Australia, March 29, 1963. After Begg *et al.* (3).

the work of Begg *et al.* (3). Figure 12 gives some representative latent heat flux profiles in bulrush millet. Now if one takes the slopes of these profiles for the whole day, a distribution of the source intensity of latent heat, (transpiration) can be found as plotted in Fig. 13. The change in shape of the profiles in the afternoon was caused by early

afternoon stomatal closure in the upper leaves with subsequent recovery later in the afternoon.

The exchange of water vapor (latent heat) in the bulrush millet is
thus an example of the dynamics of micrometeorological processes coupling together the microclimate (water vapor profile, for one) and a
physiological function, transpiration.

Let us now return to the red clover and again demonstrate this
coupling between micrometeorology, the microclimate and, physiological function. This time our example will be centered on photosynthesis

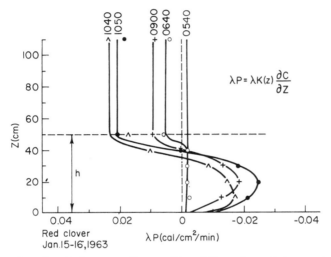

Fig. 14. Photochemical energy flux profiles (CO_2 exchange) in a red clover crop
taken during the morning hours. Negative flux indicates CO_2 movement upward
and positive flux indicates CO_2 movement downward. Palmerston North, New Zealand, January 15–16, 1963. From unpublished work of R. W. Brougham and
E. Lemon (1963).

and respiration. Through the use of equation (23) (see Fig. 7) and
then equation (5) (utilizing the profiles in Figs. 8–11, and knowing
soil heat storage), we have calculated first fluxes of photochemical
energy (carbon dioxide exchange). These are found in Figs. 14–15.
Negative fluxes indicate carbon dioxide moving upward and positive
fluxes show downward movement. Figure 16 is the source and sink
intensity distribution of photochemical energy (net photosynthesis and
net respiration). The profiles were obtained by taking slopes of profiles
in Figs. 14–15. Negative values indicate net respiration and positive
values, net photosynthesis. The leaf area distribution is plotted here
for reference.

It is not the purpose here to analyze the results, but to demonstrate the useful coupling between the fields of micrometeorology and plant physiology. Nonetheless a few points of interest could be observed. One can see that from early morning onward, the compensation point moved deeper into the canopy and then returned toward nightfall. Also, as the day progressed, both net respiration in the lower portion and net photosynthesis at the upper portion of the canopy waxed and waned at both levels. Net photosynthesis, however, outstripped net respiration so that there was a net gain for growth.

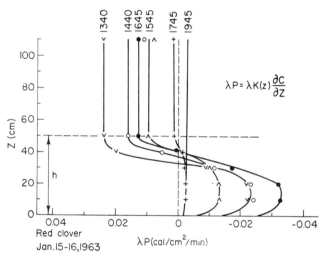

FIG. 15. Photochemical energy flux profiles (CO_2 exchange) in a red clover crop of height h taken during the afternoon hours. Negative flux indicates CO_2 movement upward and positive flux indicates CO_2 movement downward. Palmerston North, New Zealand, January 15–16, 1963. From unpublished work of R. W. Brougham and E. Lemon (1963).

As stated in the introduction, the purpose of this chapter has been to introduce briefly some micrometeorological background to the physiologist. The review has been brief and the author has of necessity drawn heavily upon his own work. Very little has been said about micrometeorological processes close to leaf or other surfaces. We simply do not know much about what happens near surfaces in the natural environment of plant communities. Furthermore, generalization from isolated plants or isolated leaves under controlled conditions and understanding of field results depend upon understanding transfer processes in the field. Micrometeorology is only now beginning to elucidate the gross phenomena of airflow and exchange within the simpler plant com-

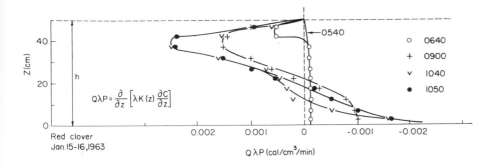

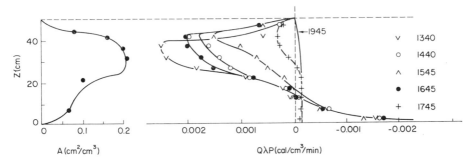

Fig. 16. Source and sink intensity distribution of photochemical energy for indicated time in a red clover crop of height h. Leaf area distribution, A, also presented for reference. Negative values of photochemical energy signify net respiration and positive values of photochemical energy, net photosynthesis. Palmerston North, New Zealand, January 15–16, 1963. From unpublished work of R. W. Brougham and E. Lemon (1963).

munities. The next step will logically be toward the finer structure near surfaces, however. Some speculations on these matters have been taken up by Gaastra (9), Inoue (11), Monteith (20), and Tanner (36).

References

1. Aizenshatat, B. A. Modern problems on the meteorology of the air layer near the ground. *In* "Sovremennye Problemy Meteorologii Prizemnozo Sloia Bozdukha—Cbornik Slatei," collection of articles, pp. 67–131. Gidrometeoizdat, Leningrad, 1958.
2. Allen, L. H., Yocum, C. S., and Lemon, E. R. Photosynthesis under field conditions VII. Radiant energy exchanges within a corn crop canopy and implications in water use efficiency. *Agron. J.* **56,** 253–259 (1964).
2a. Baker, D. N. "The effects of air temperature on the rate of photosynthesis in corn," Ph. D. thesis, Cornell University, Ithaca, New York, 1962.

3. Begg, J. E., Bierhuizen, J. F., Lemon, E. R., Misra, D. K., Slatyer, R. O., and Stern, W. R. Diurnal energy and water exchanges in bulrush millet in an area of high solar radiation. *Agr. Meteorol., Intern. J.* **1**, 294–312 (1964).

4. Businger, J. A. A generalization of the mixing-length concept. *J. Meteorol.* **16**, 516–523 (1959).

5. Cionco, R. M., Ohmstede, W. D., and Appleby, J. F. A model for air flow in an ideal vegetative canopy. Meteorol. Res. Notes No. 5. U.S. Army Electronic Research and Development Activity, Ft. Huachuca, Arizona (1963).

6. Covey, W. A method for the computation of logarithmic wind profile parameters and their standard errors. *U.S. Dept. Agr., Prod. Res. Rept.* **72**, 28–33 (1963).

7. Denmead, O. T., Fritschen, L. J., and Shaw, R. H. Spatial distribution of net radiation in a corn field. *Agron. J.* **54**, 505–510 (1962).

8. Dyer, A. J. Measurements of evaporation and heat transfer in the lower atmosphere by an automatic eddy correlation technique. *Quart. J. Roy. Meteorol. Soc.* **87**, 401–412 (1961).

9. Gaastra, P. Climate control of photosynthesis and respiration. *In* "Environmental Control of Plant Growth" (L. T. Evans, ed.), pp. 113–140. Academic Press, New York, 1963.

10. Inoue, E., Tani, N., Imai, K., and Isobe, S. The aerodynamic measurement of photosynthesis over the wheat field. *J. Agr. Meteorol. (Tokyo)* **13**, 121–125 (1958).

11. Inoue, E. The environment of plant surfaces. *In* "Environmental Control of Plant Growth" (L. T. Evans, ed.), pp. 23–32. Academic Press, New York, 1963.

12. Inoue, E., Lemon, E. R., and Denmead, O. T. A preliminary investigation into the airflow within crop canopies (1964) (in prep.).

13. Lemon, E. R., Glaser, A. H., and Satterwhite, L. E. Some aspects of the relationship of soil, plant and meteorological factors to evapotranspiration. *Soil Sci. Soc. Am. Proc.* **21**, 464–468 (1957).

14. Lemon, E. R. Photosynthesis under Field Conditions. II. An aerodynamic method for determining the turbulent carbon dioxide exchange between the atmosphere and a corn field. *Agron. J.* **52**, 697–703 (1960).

15. Lemon, E. R. Energy and water balance of plant communities. *In* "Environmental Control of Plant Growth" (L. T. Evans, ed.), pp. 55–78. Academic Press, New York, 1963.

16. Lettau, H. H., ed. Computation of heat budget constituents of the earth/air interface. *In* "Exploring the Atmosphere's First Mile," Vol. I, pp. 305–328. Pergamon Press, New York, 1957.

17. Monteith, J. L. The reflection of short-wave radiation by vegetation. *Quart. J. Roy. Meteorol. Soc.* **85**, 386–392 (1959).

18. Monteith, J. L., and Szeicz, G. The carbon dioxide flux over a field of sugar beets. *Quart. J. Roy. Meteorol. Soc.* **86**, 205–214 (1960).

19. Monteith, J. L. Measurement and interpretation of carbon dioxide fluxes in the field. *Neth. J. of Agr. Sci.* **10**, 334–346 (1962).

20. Monteith, J. L. Gas exchange in plant communities. *In* "Environmental Control of Plant Growth" (L. T. Evans, ed.), pp. 95–112. Academic Press, New York, 1963.

21. Musgrave, R. B., and Moss, D. N. Photosynthesis under field conditions. III. Some effects of light, carbon dioxide, temperature and soil moisture on photosynthesis, respiration, and transpiration of corn. *Crop Sci.* **1**, 83–87 (1961).

22. Ordway, D. E., Ritter, A., Spence, D. A., and Tan, H. S. Effects of turbulence

and photosynthesis on CO_2 profiles in the lower atmosphere. *U.S. Dept. Agr.*, *Prod. Res. Rept.* **72**, 3–6 (1963).

23. Panofsky, H. A., Blackadar, A. K., and McVehil. The diabatic wind profile. *Quart. J. Roy Meteorol. Soc.* **86**, 390–398 (1960).

24. Penman, H. L. Natural evaporation from open water, bare soil, and grass. *Proc. Roy. Soc.* **A193**, 120–146 (1948).

25. Penman, H. L., and Long, I. F. Weather in wheat: an essay in micrometeorology. *Quart. J. Roy. Meteorol. Soc.* **86**, 16–50 (1960).

26. Rider, N. E. Eddy diffusion of momentum, water vapor and heat near the ground. *Phil. Trans. Roy. Soc. London* **A246**, 481–501 (1954).

27. Rider, N. E. Water losses from various land surfaces. *Quart. J. Roy. Meteorol. Soc.* **83**, 181–193 (1957); discussion **84**, 190–193 (1958).

28. Saeki, T. Light relations in plant communities. *In* "Environmental Control of Plant Growth" (L. T. Evans, ed.), pp. 79–94. Academic Press, New York, 1963.

29. Sheppard, P. A. Transfer across the earth's surface and through the air above." *Quart. J. Roy. Meteorol. Soc.* **84**, 205–224 (1958).

30. Stoller, J., and Lemon, E. R. Turbulent transfer characteristics of the airstream in and above the vegetative canopies at the earth's surface. *U.S. Dept. Agr.*, *Prod. Res. Rept.* **72**, 34–46 (1963).

31. Sutton, O. G. "Micrometeorology," pp. 158–189. McGraw-Hill, New York, 1953.

32. Swinbank, W. C. Wind profile in thermally stratified flow. *Nature* **186**, 463–464 (1960).

33. Tan, H. S., and Ling, S. C. A study of atmospheric turbulence and canopy flow. *U.S. Dept. Agr., Prod. Res. Rept.* **72**, 13–27 (1963).

34. Tanner, C. B. Energy balance approach to evapotranspiration from crops. *Soil Sci. Soc. Am. Proc.* **24**, 1–9 (1960).

35. Tanner, C. B., Peterson, A. E., and Love, J. R. Radiant energy exchange in a corn field. *Agron. J.* **52**, 373–379 (1960).

36. Tanner, C. B. Energy relations in plant communities. *In* "Environmental Control of Plant Growth" (L. T. Evans, ed.), pp. 141–148. Academic Press, New York, 1963.

37. Uchijima, Z. On characteristics of heat balance of water layer under paddy plant cover. *Bull. Natl. Inst. Agr. Sci.* **A8**, 243–265 (1961).

38. Uchijima, Z. Studies on the microclimate within plant communities. (1) On the turbulent transfer coefficient within plant layer. *J. Agr. Meteorol. (Tokyo)* **18**, 1–9 (1962).

39. Uchijima, Z., and Wright, J. L. An experimental study of air flow in a corn-plant-air layer. *Bull. Natl. Inst. Agr. Sci.* **A11**, 19–66 (1964).

40. Waggoner, P. E., and Reifsnyder, W. E. Difference between net radiation and water use caused by radiation from the soil surface. *Soil Sci.* **91**, 246–250 (1961).

41. Yocum, C. S., Allen, L. H., and Lemon, E. R. Photosynthesis under field conditions VI. Solar radiation balance and photosynthesis efficiency. *Agron. J.* **56**, 249–253 (1964).

PREAMBLE TO CHAPTER 3, PARTS 1 AND 2

In Chapter 4 of Volume IA the subject of respiration was considered at the cellular and subcellular level. There the attention was concentrated upon the reactions and mechanisms by which the energy stored by photosynthesis is made available for useful purposes, in a controlled and sensitively regulated way, in the form of energy-rich phosphorylated compounds. In Chapter 3 attention is to be directed to the respiration of plants as seen at the level of the organ and of the organism. Nevertheless continued reference will be needed to the intermediary reactions of respiration. Therefore, this chapter should also be linked with others on carbohydrate metabolism (Chapter 5, Volume IVB), on organic acid metabolism (Chapter 6, Volume IVB), and on fat metabolism (Chapter 7, Volume IVB) as well as with certain features of Chapter 4 of Volume IVA on the metabolism of nitrogenous compounds.

In Part 1 several systems are selected for more intensive treatment to illustrate how respiration varies with the external conditions during development and also how its pace is brought under regulatory control. Because massive plant storage organs present special problems and have furnished particular kinds of information which require special forms of interpretation their respiration receives separate treatment in Part 2.

The intimate involvement of respiration with other physiological processes, previously noted in Chapter 4 of Volume II, is again made apparent in Part 1. It will be seen that the process of respiration, which produces energy-rich compounds for use in growth and development, is in turn regulated as the phosphorylated compounds and reduced pyridine nucleotides are drawn off as their energy is applied to useful purposes. Thus respiration is not an entirely independent, self-regulating vital function, but it is in turn regulated. To grow and develop, organisms need to respire, but they respire at rates and in ways that are somewhat conditioned by their growth and development.

CHAPTER THREE

Part One

The Respiration of Plants and Their Organs

E. W. Yemm

I. Introduction 231
II. The Respiratory Processes and Their Regulation 232
  A. The Catabolic Reactions of Respiration 233
  B. Oxidation and Oxidative Phosphorylation 235
  C. Anaerobic Metabolism 236
  D. Respiration and Cellular Organization 239
  E. The Rate of Respiration and Its Regulation 241
III. Respiration of Germinating Seeds 246
  A. The Rate of Respiration during Germination 247
  B. Respiration and Seedling Development 252
  C. The Respiratory Quotient of Germinating Seeds 258
  D. External Conditions and the Respiration of Seedlings 261
IV. The Respiration of Leaves. 269
  A. Respiration and the Development of Young Leaves 269
  B. Respiration of Mature Leaves 274
  C. Senescent Leaves. 284
V. The Respiration of Roots 291
  A. Respiration and Root Development 291
  B. Respiration and Salt Uptake in Roots 296
VI. Conclusion 303
  References 304

I. Introduction

In the first volume of this treatise an account of cellular respiration was given with particular emphasis on the biochemical mechanisms and their relation to the organized enzymatic systems of cells. It was made clear that a salient feature of respiratory processes is the aerobic oxidation of organic constituents with the release of energy in forms that can be utilized in the metabolism of living cells. Normally, gaseous exchanges, the uptake of oxygen and the production of carbon dioxide, are external features of plant respiration, but in their relationship with cellular physiology they can be regarded as part of the mechanisms whereby reactive intermediaries and energy are transformed in the maintenance of living structure and in the biosynthetic processes underlying growth and development.

231

Our main concern in this chapter will be to consider the respiration of plants and of plant organs as an integral part of their physiological functions and development. From this standpoint experimental data on plant respiration, generally based on measurements of gaseous exchanges, are recorded in an extensive and scattered literature, but valuable monographs have been prepared in recent years by Stiles and Leach (98), by James (50), and by Beevers (6). "The Encyclopedia of Plant Physiology," edited by Ruhland (85), provides in two volumes a comprehensive review and bibliography of many aspects of the subject. A more selective treatment will be attempted here; an account is given of respiratory metabolism in plants with particular regard to the physiology of their organs, especially their growth and development. Chief attention is given to those systems for which extensive data have been collected, since here it is possible to analyze in some detail the relationship between respiration and other physiological activities of plants and their organs.

In the consideration of respiration at this level of organization, several additional problems arise in the analysis of experimental data. Ideally we should like to relate changes of respiratory rate and metabolic activity directly to the cellular mechanisms involved, but this can rarely be attained. Higher plants and their organs consist of closely integrated cellular systems in which functional and structural differentiation of the cells is often an important feature. To the complexities of intracellular organization must be added interaction between different cells and tissues. From this point of view the plant and its organs must be regarded as a highly coordinated but heterogeneous system in which the different component cells may respond differently to external and internal variables; changes of respiratory rate may depend not only upon the control mechanisms that operate within cells, but also on the relative number of cells that are so affected. Nevertheless, it should be clearly recognized that a better knowledge of respiration at a molecular and cellular level is an essential step toward an understanding of the significance of respiration in the physiology and metabolism of the plant as a whole.

II. The Respiratory Processes and Their Regulation

It is appropriate here to trace in broad outline the chief sequences of catabolic changes with which we are concerned in cellular respiration. Much of the evidence that is now available concerning the biochemical reactions and their mechanism has been reviewed by Goddard and Bonner (38) in Volume IA and need not be considered in detail. Other relevant topics will be taken up in this volume in chapters by

Gibbs on carbohydrate metabolism (Chapter 5) and by Beevers, Butt, and Stiller on organic acid metabolism (Chapter 6).

A. THE CATABOLIC REACTIONS OF RESPIRATION

For the present purpose it is convenient to recognize three main phases in the catabolic reactions that sustain cellular respiration. The first phase can be regarded as a mobilization of reserve substances; it generally involves the breakdown of relatively insoluble constituents, such as polysaccharides, fats, and proteins, to their structural units by means of enzymatic hydrolysis or phosphorolysis. In the second phase, these primary products undergo partial oxidation, commonly with the release of carbon dioxide and the formation of organic acids. A final phase of oxidation by way of the tricarboxylic acid cycle is now clearly indicated as the main pathway of plant respiration, and there is much evidence that the multienzyme systems of mitochondria provide the chief cellular mechanism for the transformations of energy and the production of carbon dioxide.

In most plants, carbohydrates are the main source of respiratory substrate, and the enzymatic reactions that may operate in their breakdown have been reviewed in detail by Goddard and Bonner (38) and by Beevers (6); reference may also be made to Gibbs (Chapter 5, Volume IVB) and to Beevers et al. (Chapter 6, Volume IVB). An outline is given in Diagram I of the catabolic sequence indicating mobilization to hexose or hexose phosphates followed by glycolysis or direct oxidation by the pentose phosphate cycle leading by way of pyruvate to the tricarboxylic acid cycle.

Under some circumstances fats or proteins may provide a substantial part of the respiratory substrate. Fats are particularly important in some germinating seeds and are probably mobilized by breakdown to glycerol and fatty acids. There is evidence that glycerol can be converted to pyruvate, and fatty acids may be degraded by successive β-oxidations to acetyl coenzyme A and thus enter the tricarboxylic acid or glyoxalate cycles (cf. Chapter 6). Protein catabolism appears to contribute to the respiratory substrate, especially under somewhat exceptional circumstances of carbohydrate starvation and in some germinating seeds. It is probably initiated by the action of proteolytic enzymes producing a variety of amino acids, which then undergo oxidative deamination or transamination to α-keto acids. Some of these, such as pyruvate, oxaloacetate and α-ketoglutarate, may be drawn directly into the tricarboxylic acid cycle. It is along these different but convergent pathways that carbohydrates, fats, and proteins probably make their chief contribution to the respiratory substrate of plants.

With regard to the release of energy, a central feature of the various oxidations in glycolysis and in the tricarboxylic acid cycle is the enzymatic transfer of hydrogen or electrons activated by specific dehydrogenases. In most of these reactions di- or triphosphopyridine

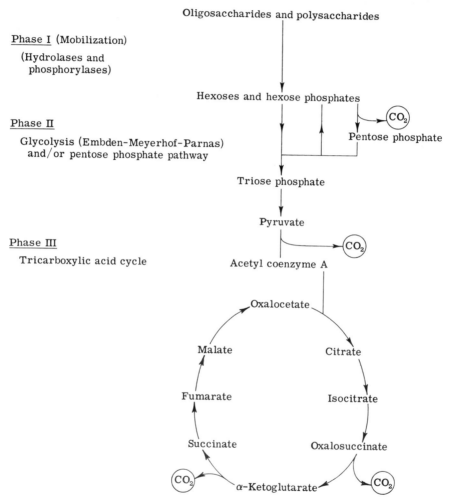

Diagram I. The breakdown of carbohydrate in respiration. (The involvement of glyoxalate in organic acid and fat metabolism is discussed in Chapter 6.)

nucleotides (DPN or TPN) act as hydrogen acceptors; it can be estimated that over three-quarters of the free energy available from the oxidation of hexose is transformed in this way with a smaller fraction (about one-tenth) passing more directly to the cytochrome system. An additional part of the free energy is conserved in anhydrous substrate compounds, usually acyl phosphates, which can be transformed to adenosine triphosphate (ATP), so that less than 10% of the total free energy is dissipated in the substrate reactions directly involved in the breakdown of hexose in respiration. Thus it can be seen that the energy exchanges in the catabolic reactions outlined in phases 2 and 3 of Diagram I may be effectively channeled into the production of reduced pyridine nucleotides and adenosine triphosphate or other "energy rich" compounds.

B. Oxidation and Oxidative Phosphorylation

Under normal aerobic conditions a continuous absorption of oxygen can be demonstrated during the active metabolism of plants. With very few exceptions, the substrates and hydrogen acceptors of the dehydrogenase systems are not readily autoxidized; instead a series of respiratory carriers brings about the transfer of hydrogen or electrons ultimately by means of oxidase action to molecular oxygen. Some uncertainty still exists about the nature of the respiratory carriers in plant cells. The balance of evidence, recently reviewed by Goddard and Bonner (38) and by Beevers (6), indicates that cytochrome oxidase mediates a major part of plant respiration. The general pathway of hydrogen or electrons over various carriers may be outlined as follows: substrate → DPN → flavoprotein → cytochome b → cytochrome c → cytochrome a + a_3 → O_2, with succinate entering the sequence as a hydrogen donor at flavoprotein. Several other oxidases, such as ascorbic acid oxidase and polyphenol oxidase, occur widely in plants, but there is no generally accepted evidence that they have an important role in normal respiration.

The flow of electrons or hydrogen from substrates to oxygen along the increasing oxidation-reduction potential formed by the carrier system constitutes the chief release of energy in respiration. An important advance in our understanding of its significance in cellular metabolism became possible as a result of the demonstration that the synthesis of "energy-rich" phosphate compounds can be coupled with the redox system of the respiratory carriers. Although as yet relatively little is known in detail about the mechanism of the coupling reactions, there are clear indications that phosphorylation occurs at three of the chief redox steps between DPNH and oxygen, as indicated in Diagram II.

Measurements of phosphorylation relative to oxygen uptake in mitochondria indicate that under favorable conditions some 60% of the free energy released in the oxidation of DPNH may be conserved by the formation of anhydrous pyrophosphate bonds of ATP. In efficiently coupled mitochondrial systems such as these, it has been shown that the rate of oxygen uptake is sharply limited by the availability of phosphate and of ADP; high rates of oxidation can be maintained by an adequate supply of phosphate and suitable phosphate acceptors. On the reasonable assumption that a similar and efficient coupling between oxidation and phosphorylation occurs *in vivo*, it is widely held that the level of phosphorylation may exert a controlling influence on the rate of cellular respiration. Some valuable support for this view has been gained from the study of a number of cell poisons that uncouple phosphorylation from oxidation. There is little doubt that the formation of ATP is one of the key reactions in the energy economy

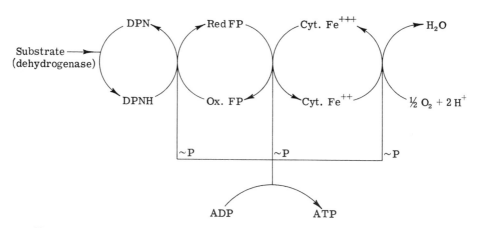

DIAGRAM II. Phosphorylation at three redox steps between DPNH and oxygen. ~P denotes the production of "energy-rich" phosphate groups; FP denotes flavoprotein.

of plant cells, but a better understanding of its regulatory action on cellular respiration will be possible only when more is known of the enzymatic mechanisms which control oxidative phosphorylation.

C. ANAEROBIC METABOLISM

Most of the early theories of the mechanisms of plant respiration were founded on the concept of a close connection between respiration and alcoholic fermentation. This concept was prompted by the observation that many plant tissues continue to produce carbon dioxide when oxygen is excluded and that alcohol often accumulates in appreciable quantities under these conditions. The theory of a common pathway in respiration and fermentation, first put forward by Pfeffer, has been progressively developed by Kostychev and Blackman (see James, 50). It can now be supported by much evidence; many of the enzymes and intermediaries of the Embden-Meyerhof-Parnas sequence of glycolytic reactions have been shown to occur widely in plants (for a summary see Beevers, 6) and, in absence of oxygen, a nearly equimolar production of carbon dioxide and ethanol has been demonstrated in some tissues (50). It is probable, therefore, that in some plants, as in yeast, under anaerobic conditions pyruvate is predominantly decarboxylated to acetaldehyde which is then reduced to ethanol, DPN/DPNH acting as a redox couple between triosephosphate dehydrogenase and alcohol dehydrogenase. When oxygen is admitted, pyruvate is progressively drawn into oxidative mechanisms of the tricarboxylic acid cycle and the redox couples operate with the respiratory carrier system, so that above the so-called extinction point no alcohol is formed.

The existence of an alternative mechanism of carbohydrate breakdown by way of the pentose phosphate cycle has been clearly indicated during the past ten years. All the enzymes and postulated intermediates of the pentose phosphate cycle are known to be present in plants, and some evidence has been obtained by means of C^{14}-labeled intermediates and specifically labeled glucose. A valuable review of our present knowledge of the pentose phosphate pathway and its significance in plant respiration has been given by Gibbs (36) and by Beevers (6); also see Chapters 5 and 6, Volume IVB. There is still room for uncertainty about the contribution made by this alternative mechanism to the normal respiration of plants chiefly owing to the difficulty of interpreting the results of labeling experiments and the possibility that pentose is utilized in biosynthetic reactions. It is of interest that the oxidative steps in the pentose phosphate pathway are dependent upon TPN, not upon DPN as in the glycolytic system (6).

In the development of the common path theory much attention has been focused on the so-called Pasteur effect, which was originally described in studies of the influence of oxygen on alcoholic fermentation and the growth of yeast. Two major effects of oxygen were demonstrated by Pasteur; first the progressive suppression of alcoholic fermentation, when the availability of oxygen is increased, and secondly, the much greater utilization of sugar for assimilation and growth of yeast under aerobic than under anaerobic conditions. The second of these two attributes has been variously interpreted. Many plant physiologists, following the lead of Blackman (12, 13), have emphasized the conservation of carbohydrate under aerobic conditions as an important feature of the Pasteur effect. Resynthesis of carbohydrate from glycolysis products by oxidative anabolism, as suggested by Blackman, has often been regarded as a major part of respiratory metabolism in plants

Studies of the Pasteur effect in plant tissues have usually depended on measurements of the rate of carbon dioxide production in air and in nitrogen. If there is a simple changeover from complete aerobic oxidation of hexose to alcoholic fermentation, without any adjustment in the rate of hexose breakdown, then the rate of carbon dioxide production in anaerobic conditions would be only one-third of the rate in air, as indicated in the over-all equations:

$$C_6H_{12}O_6 + 6\ O_2 \rightarrow 6\ CO_2 + 6\ H_2O$$
$$C_6H_{12}O_6 \rightarrow 2\ CO_2 + 2\ C_2H_5OH$$

In a number of plant organs (102) it has been observed that the initial rate of carbon dioxide production in the absence of oxygen is considerably greater than the corresponding rate in air; but in only a few cases have sufficient data been obtained on the loss of carbohydrates and the accumulation of ethanol, or other end products, to allow a satisfactory definition of the catabolic mechanism. However, the available evidence clearly indicates that the rate of carbohydrate breakdown to carbon dioxide is often lower under aerobic than under anaerobic conditions, at least in mature nongrowing plant tissues. On the other hand, no support has been forthcoming for the suggestion that oxidative anabolism of respiratory intermediates to carbohydrate plays an important part in the conservation of carbohydrate. The experiments of James and Slater (52) and of Neal and Beevers (73) show that C^{14}-labeled intermediates, particularly pyruvate, are not readily reconverted to hexose in a variety of plant materials for which evidence of a Pasteur effect exists.

When plants are deprived of oxygen, it is evident that both the catabolic reactions and oxidation-reduction mechanisms are drastically affected. If alcoholic fermentation replaces respiration, less than one-tenth of the total free energy of hexose is made available. In particular, oxidative phosphorylations coupled with the electron transport system are eliminated, so that the production of "energy-rich" phosphate compounds is restricted to the substrate phosphorylations of glycolysis. It can be estimated that the complete breakdown of 1 mole of hexose in respiration could give rise to 36–40 moles of ATP, whereas the comparable yield under anaerobic conditions would be 2 moles of ATP. Moreover, the production of important metabolites, such as the organic acids of the tricarboxylic acid cycle, is closely dependent upon aerobic oxidation. In the absence of oxygen, most higher plants are able to survive for a limited time, but many of the biosynthetic reactions required for their nutrition and growth depend directly upon respiratory metabolism. As will be discussed in a later section, a more comprehensive interpretation of the Pasteur effect can now be offered in terms of the cellular mechanisms that regulate the breakdown of carbohydrates and link respiration with endergonic processes.

D. RESPIRATION AND CELLULAR ORGANIZATION

In recent years the rapid development of research into both the ultrastructure of plant cells and the activities of cell organelles has provided much additional information concerning the localization and organization of different phases of respiratory metabolism. The important functions of mitochondria in the oxidation of pyruvate by way of the tricarboxylic acid cycle are firmly established; indeed, studies of isolated mitochondria have provided decisive evidence on some of the mechanisms of oxidation and their coupling with phosphorylation. Despite the hazards of the separation of cell fractions by differential centrifugation, an impressive number of fairly pure preparations of mitochondria have been made from a wide variety of plants and plant organs (for a summary, see Beevers, 6). Under favorable conditions these preparations promote the oxidation of pyruvate together with acids of the tricarboxylic cycle, and the oxidation is coupled with the transformation of inorganic phosphate and ADP into ATP. Moreover, the activity of the mitochondria in oxygen uptake, relative to that of the intact cells, indicates that it represents a major part of cell respiration. From a functional point of view, it is evident that mitochondria can be regarded as highly organized multienzyme systems in which the activity depends upon the precise spatial arrangement of different components. The organelle is bounded by a differentially permeable membrane which appears to be a double layered lipoprotein structure, the inner layer being folded so that it extends into the central matrix to form characteristic cristae. There is evidence that the dehydrogenase enzymes operating in the tricarboxylic acid cycle are relatively loosely bound on the membrane or in the matrix, whereas the electron transport system is firmly held in the structure of the membrane. It is of interest that oxidative phosphorylation is particularly dependent upon the intact structure of mitochondria; damage to the structure during preparation or experimental treatment readily leads to a breakdown of the linkage between oxidation and phosphorylation. Besides the enzymes directly involved in the tricarboxylic acid cycle, other enzymatic systems appear to be present in plant mitochondria, including glutamic acid dehydrogenase, transaminase, and the enzymes that bring about the β-oxidation of fatty acids to acetyl CoA. It is probable that these latter systems play an important part in the respiratory catabolism of proteins and fats.

The enzymes that promote glycolysis and the pontose phosphate cycle are mainly present in the soluble fraction which remains after subjecting tissue homogenates to prolonged centrifugation at high speeds. It is, therefore, generally assumed that these enzymes occur in less highly organized sites of the cytoplasm which are readily dispersed into solution. However, the ease with which proteins may be leached from particulate cell structures, such as chloroplasts and nuclei, during separation and fractionation, indicates the need for caution in this direction. The same reservation must be made with regard to soluble oxidase systems commonly found in plants; recent data indicate that at least one of them, ascorbic acid oxidase, may be structurally held in the outer cytoplasmic membrane closely associated with the cell wall (Honda, 47). To what extent the soluble oxidases operate in the oxidation-reduction systems of the glycolytic and the pentose phosphate pathways is at present uncertain; but despite careful examination no evidence has been found that phosphorylation can be coupled with oxidative systems other than those operating with cytochrome oxidase. However, the cytoplasmic membranes including the endoplasmic reticulum clearly offer scope for structural organization which may be very readily lost when cells are disrupted.

In mature plant cells with their conspicuous central vacuoles, it is very probable that a major part of the soluble metabolites, such as sugars, organic acids, and amino acids, are relatively inaccessible to the enzyme of the cytoplasm. The rate at which different substrates are mobilized and drawn into respiratory processes may therefore be decisively influenced by the permeability of the tonoplast and other membranes within the cell or by mechanisms which actively transfer solutes from one organelle to another. Similar conditions probably affect the utilization of insoluble reserve such as polysaccharides. Starch, for example, is generally present in amyloplasts in storage tissues, and its availability will primarily depend on hydrolytic or phosphorolytic activity in these specialized structures.

In his early studies of the respiratory rate in plants, Blackman clearly recognized the importance of organized structure; he used the term "organization resistance" to denote the restraints imposed by cellular structure on the accessibility of enzymes and respiratory substrates. The nature of this organization can now be visualized with greater definition in terms of the differential permeability of intracellular membranes and its effect on the distribution of enzymes and metabolites in different parts of the cell and its organelles. Spatial organization therefore must be a decisive factor not only in preventing reactions, but also in promoting them. The rapid decline (often after a brief increase) of respiratory activity when the protoplasmic structure is destroyed is well known. In all probability, it results from the breakdown of the specific arrangement of enzymatic systems concerned with reaction sequences such as glycolysis and oxidation, as well as the destruction of membranes that control the distribution of coenzymes and other reactants. As already mentioned, the coupling of oxidation and phosphorylation in mitochondria is particularly fragile and is dependent upon the intact structure of the lipoprotein membrane of these organelles. The close spacing of reaction centers of enzymes may well be of great importance in promoting chain reactions that do not readily occur in unorganized solutions.

The existence of separate metabolic pools, which do not rapidly equilibrate within plant cells, has been clearly indicated in recent years by means of isotopically labeled metabolites. But as yet little is known of the mechanisms that control the movement of metabolites and intermediaries of energy exchanges, such as pyridine nucleotides and ATP, between different cellular compartments. It is here that some of the most challenging problems of respiratory metabolism await experimental attack at a cellular level.

E. The Rate of Respiration and Its Regulation

Nearly all the experimental data on the rate of respiration of plant organs have been obtained by measurements of gaseous exchanges under conditions in which photosynthetic activity is precluded. In most of the early work carbon dioxide was measured, but with the development of convenient manometric methods oxygen uptake has been widely used as a measure of respiration. The introduction of sensitive and automatic gasometric methods, such as infrared analysis of carbon dioxide and paramagnetic techniques for oxygen, has been particularly useful for the study of whole plants and relatively bulky plant organs. Of equal if not greater importance has been the development of biophysical and biochemical methods which allow a deeper penetration into the changes of substrates and enzymatic systems of cells and their organelles.

For purposes of comparison, the rate of respiration of plant tissues has been expressed in a variety of ways, including per unit cell, fresh weight, dry weight, surface area, and protein content. As discussed by James (50), none of these provides a very satisfactory basis for comparing widely different organs and tissues. A cell basis has some important advantages, but in studies of respiration in relation to growth and development it is often valuable to express the data per unit plant or well defined plant organ.

The rate of respiration is influenced by many external conditions such as temperature, oxygen, and carbon dioxide concentration; the experimental study of these factors forms an important branch of plant physiology. Until fairly recently, it was generally considered that the protoplast could be regarded as a catalytic complex in which the rate of respiration was primarily determined by the availability of substrates and the intrinsic activity of the enzymatic systems. There is now strong evidence that the rate of cellular respiration is subject to control from several different directions and that the catabolic reactions may be closely regulated in relation to biosynthetic and other processes involved in assimilation and growth. A valuable review of these regulatory mechanisms at a biochemical and cellular level has been given in a symposium of the Ciba Foundation (Wolstenholme and O'Connor, 113).

In plant respiration the operation of regulatory mechanisms was first clearly indicated in studies of anaerobic metabolism and later in the investigation of cell poisons, such as 2,4-dinitrophenol (DNP), which uncouple phosphorylation from cellular oxidations. From these two directions, stress has been placed upon the key positions of inorganic phosphate, ADP, and ATP in glycolytic and in oxidative reactions. Both inorganic phosphate and ADP are required in the glycolytic sequence leading to pyruvate and in the oxidative phosphorylations coupled with electron transport in mitochondria. Consequently, metabolic processes which promote the utilization of ATP with the formation of inorganic phosphate and ADP would be expected to increase the rate of glycolysis and cellular oxidations; conversely, high levels of phosphorylation, associated with a low turnover of ATP, would restrain both these phases of respiration.

Under anaerobic conditions, the major source of ATP by oxidative phosphorylation is eliminated, so that an increase in the availability of phosphate and ADP could account for an acceleration of carbohydrate metabolism by glycolysis. This aspect of anaerobic metabolism in plants has been reviewed by Turner (102), who fully discusses its relation to the Pasteur effect and to the theory of oxidative anabolism developed by Blackman in his pioneer work on plant respiration. The uncoupling action of DNP on oxidative phosphorylation is well established from experimental work on animal and plant mitochondria; here again increases in the rate of oxygen uptake can be interpreted in terms of the greater availability of ADP and phosphate as a result of the suppression of oxidative phosphorylation. The action of DNP on plant tissues has been recently reviewed by Beevers (6); it is evident that many of the effects of the poison on respiratory metabolism, including increases in the rate of carbohydrate breakdown and oxygen uptake together with the inhibition of biosynthetic reactions, are consistent with the regulation of cellular respiration by means of phosphorylation mechanisms.

In normal aerobic growth of plants many of the endergonic reactions, whereby complex cellular constituents are formed from simple precursors, are directly or indirectly dependent upon phosphorylation. The enzymatic synthesis of amides, peptides, polysaccharides, and lipids have been shown to require high energy phosphate compounds, such as ATP, in order to drive the reactions, usually by way of the formation of phosphorylated intermediates or nucleotide compounds. A high rate of turnover of phosphate and ATP would therefore be expected to occur under conditions of active synthesis of cell constituents, thus promoting high rates of glycolysis and oxidation.

Although the phosphorylation mechanisms seem to represent one of the chief control points in cellular respiration, others no doubt operate. For example, pyridine nucleotides are specific reactants not only in glycolysis and respiratory oxidation, but also in many important anabolic reactions, such as the formation of amino acids by reductive amination and the biosynthesis of lipids. In plants, these direct links between catabolic and anabolic reactions at an oxidation-reduction level readily occur under aerobic conditions. They compete successfully with the hydrogen and electron transport systems of respiration and form an alternative mechanism for the oxidation of reduced pyridine nucleotides. It seems probable that the intervention of such effective electron acceptor reactions alters the oxidation-reduction conditions in the cell and may thus promote dehydrogenase activity in glycolysis and in other stages of cell respiration. Moreover the diversion of electrons at a high level of oxidation-reduction energy from the normal carrier systems would preclude oxidative phosphorylations, normally associated with aerobic conditions, and thus reinforce the regulatory mechanism operating through phosphorylations.

The mobilization and activation of respiratory substrates in the first phase of the catabolic sequence is probably another important factor controlling the rate of respiration in plants. Enzymatic interconversion of phosphorylated sugars may play an important part as a preliminary step, but sugars and polysaccharides may not be equally or readily accessible to the glycolytic system. As already noted, starch is often localized in amyloplasts in storage tissues, and its availability will depend upon hydrolytic or phosphorolytic activity in these structures.

In view of the heterogeneity of the cells and tissues of many plant organs, it is not surprising that simple relationships between substrate concentration and respiratory rate have rarely been observed.

There is strong evidence that in growing plant tissues many of the intermediaries of carbohydrate catabolism may be diverted to synthetic reactions in which new cell materials are formed. Long suspected,

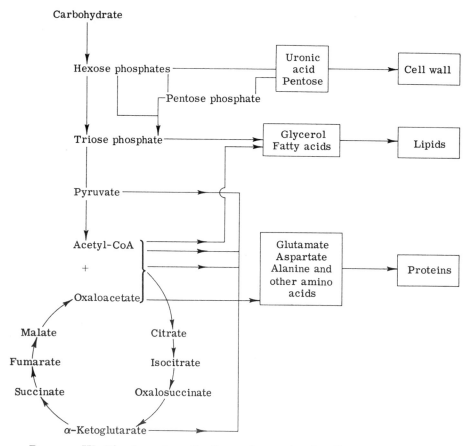

DIAGRAM III. The formation of cell constituents by the diversion of respiratory intermediaries. (See also Chapters 5 and 6 for relations to other cell constituents.)

this diversion of intermediates is now substantiated by many experimental data, including the preparation of carbon balance sheets and the exploration of metabolic pathways by means of isotopically labeled sugars and other metabolites. From these results it is possible to recognize some of the chief points at which withdrawal of intermediaries takes place. Pyruvate, acetate units, and some of the acids of the tricarboxylic acid cycle are indicated as important starting points for biosynthetic reactions. A diagrammatic outline (Diagram III) is given to show the probable relation between the respiratory intermediates and the synthesis of some of the chief structural components of plant cells.

Precise information is lacking on many of the quantitative aspects of the above scheme, but over-all estimates on a number of developing plant tissues indicate that more than half of the respiratory intermediates of carbohydrate metabolism may be drawn into biosynthetic reactions. Both the rate and the catabolic mechanisms of respiration are likely to be affected under these conditions; cyclical processes, such as the tricarboxylic acid cycle, are particularly sensitive to the withdrawal of intermediates, so that partial or modified operation of the cycle may result. It is very probable that the diversion of respiratory intermediates to biosynthesis, itself dependent on energy transfer by phosphorylation and oxidation-reduction, forms part of a closely integrated system regulating the breakdown of carbohydrates and cellular respiration in plants. Changes of nutritional or other physiological conditions would be expected to lead to a readjustment between catabolic and anabolic reactions, a readjustment in which different parts of the control mechanisms may play a decisive part.

The problem of the organization of the concurrent reactions of respiratory catabolism and of anabolism in plant cells is by no means solved. Although many of the reactions of glycolysis and of the tricarboxylic acid cycle are reversible, the free energy relations at certain steps clearly suggest that they do not make a major contribution to biosynthesis under normal physiological conditions; as already mentioned, experiments with C^{14}-labeled intermediaries indicate that, in general, carbohydrates are not readily formed from exogenously supplied organic acids by oxidative anabolism. Indeed, the trend of most recent evidence clearly suggests that the main biosynthetic (anabolic) reactions do not result from a simple reversal of catabolic mechanisms. For example, the biosynthesis of fatty acids does not readily proceed by a reversal of the β-oxidation system found in plant mitochondria, but appears to depend upon a separate reductive mechanism mediated by TPNH (cf. Chapter 7, Volume IVB). Similarly, the formation of anhydrous bonds of proteins and polysaccharides often proceeds by the intervention of nucleoside triphosphates (ATP and UTP) to form nucleotide compounds, as a primary step in the synthesis. The suggestion has been made that the conflicting requirements for anabolism and growth, as distinct from catabolic reactions, may be met by cellular organization (27a). In mitochondria the predominantly catabolic mechanisms of the tricarboxylic acid cycle are closely geared to the production of DPNH and of ATP by oxidative phosphorylation. On the other hand, reductive syntheses mediated by TPNH and ATP may occur in other parts of the cytoplasm. Such an organization might well account in part for the dual systems of DPN and TPN specific dehydrogenases and for the two pathways of carbohydrate breakdown, only one of which operates with TPN. However, it is difficult to reconcile this suggestion with the synthesis of amino acids by reductive amination in which the formation of glutamic acid occupies a key position. Glutamic acid dehydrogenase, the only enzyme known to promote this synthesis in plants, occurs in mitochondria and is specific for DPN. This and other evidence strongly suggests that the biosynthesis of some amino acids is very closely coupled to the catabolic reactions in mitochondria.

The transfer of energy required for the biosynthesis of anhydrides plays an essential part in the formation of the macromolecular constituents of living cells. Many of these mechanisms are as yet incompletely understood, but it is becoming clear that phosphorylated nucleotides are commonly the key reactants. There is strong evidence that ATP promotes the synthesis of amides and peptides from amino acids by the primary formation of amino acid-AMP compounds (cf. Chapter 4, this volume). Uridine triphosphate seems to play a similar role in the formation of UDP derivatives of hexose and pentose sugars as a preliminary step in their condensation to form some of the chief oligosaccharides and polysaccharides in plant cells (cf. Chapter 5, Volume IVB). More recently cytidine diphosphate compounds have been indicated as key reactants in the metabolism of phospholipids (cf. Chapter 7, Volume IVB). In the course of these transfer reactions *in vitro* much of the free energy available in the "energy-rich" phosphate compounds is lost; it can be estimated that about two-thirds of the free energy of ATP is dissipated in the formation of amide or peptide bonds. However, in living cells the synthesis of highly specific and organized macromolecular structures is achieved and the formation of nucleotide intermediates may play a decisive part not only in transmitting energy, but also in imparting specificity and control to the biosynthetic system. It is here that the thermodynamic concept of negative entropy—"nature feeds on negative entropy," see Schrödinger (86)—is relevant to an understanding of the energy exchanges of respiration and their significance in the maintenance and replication of the highly organized structure of living cells.

III. Respiration of Germinating Seeds

Much of the pioneer physiological work on plant respiration was carried out on germinating seeds and young seedlings. Typically, the rate of respiration of dormant seeds is very low, but during the first phases of germination there is a rapid and prolonged rise in the rate of oxygen uptake and carbon dioxide production. In early studies the important physiological effects of temperature, oxygen tension, moisture content, and other conditions on respiration were demonstrated, while measurements of the respiratory quotient ($CO_2:O_2$) were used to explore the nature of the respiratory substrate during the course of germination. Critical reviews of much of this work have been given in monographs by Stiles and Leach (98) and by James (50).

In more recent years the respiratory metabolism of young seedlings has been studied in greater detail with the object of gaining a better understanding of the relationship between respiration and the development of the young seedling. At first, the metabolism of the embryo depends directly upon food reserves stored mainly within the cotyledon or endosperm. Mobilization of these reserves and their utilization in respiratory and anabolic reactions which sustain the early development of the embryo are central problems in the physiology of seed germination.

Dormant seeds commonly remain viable for many years, and much sterile controversy has centered round the relation between respiration and the retention of vitality. Sensitive methods of measurement have indeed shown that a slow production of carbon dioxide and uptake of oxygen often persists in dormancy, but it is by no means certain that this is always so, particularly in seeds of very low water content. The close dependence of respiration on water content has often been demonstrated in seeds and in cereal grains. At all events, it seems clear that respiration in dormant seeds bears little resemblance to the processes brought into play when active metabolism is restored during germination. It is possible that slow gaseous exchanges during dormancy result from decarboxylations and oxidations distinct from those of normal cell respiration. For example, there is some indirect evidence that a slow oxidation of sulfhydryl compounds occurs in dormant pea (*Pisum sativum*) seeds (Spragg and Yemm, 89). Gradual and irreversible inactivation of sulfhydryl and other enzymes may well be an important factor in the ultimate loss of viability in seeds after prolonged storage.

A. THE RATE OF RESPIRATION DURING GERMINATION

When seeds germinate under favorable conditions of aeration, temperature, and water supply, the resumption of respiration is commonly associated with relatively rapid and complex changes in the rate of oxygen uptake and carbon dioxide production. These changes have been particularly clearly shown when sensitive methods are used with single seeds (Stiles, 96), but they are also evident if uniform batches of seeds are germinated under closely controlled conditions. An example of the data obtained when pea seeds, a favorite material for study, are germinated in air at 22°C is shown in Fig. 1. Here, as in several other species that have been examined by comparable methods, the course of respiration can be conveniently analyzed into several phases. The first phase is characterized by a rapid rise in the rate of

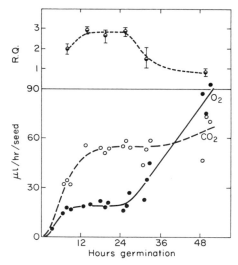

Fig. 1. The respiration of germinating pea seeds. Manometric measurements were made in air at 25°C. Values of the respiratory quotient were calculated from triplicate measurements of O_2 uptake and CO_2 production; mean values are plotted and the range of variation is indicated by vertical lines. From Spragg and Yemm (89).

gaseous exchange, which begins within a few hours of supplying water and extends for about the first 12 hours. This corresponds closely to the rapid uptake of water by the seed and the activation of enzymatic systems, which bring about the mobilization and breakdown of respiratory substrates. In the second phase, the rising rate is interrupted and a period of somewhat variable length, 20–24 hours, follows during which the rate remains fairly constant. The uptake of oxygen is more drastically affected than the production of carbon dioxide, so that the respiratory quotient (R.Q.) rises to a value of 2–3. The third phase begins after about 36 hours, when a conspicuous increase of respiration occurs. Here again, the changes in oxygen uptake are most affected and the R.Q. now falls to a value approaching unity or below. Rupture of the testa by the developing radicle of the embryo occurs at this stage.

Evidence from several directions indicates that a restriction of gaseous diffusion has a decisive influence on the course of respiration in pea seeds during the early stages of germination. Particularly during the second phase, artificial removal of the testa leads to a rapid

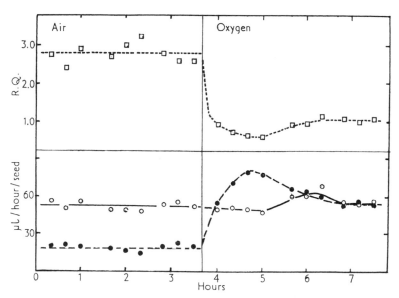

FIG. 2. The respiration of germinating pea seeds in air and in oxygen. The rates of O_2 uptake ($--\bullet--$) and CO_2 production ($--o--$) were measured manometrically in air and in pure oxygen on duplicate samples of seeds which had germinated at 22°C for 18 hours. Changes in the value of the R.Q. are shown in the upper part of the figure. From Spragg and Yemm (89).

increase in the rate of oxygen uptake to a value 3 or 4 times greater than that of the intact seed, but with relatively little change in the rate of carbon dioxide production, so that the R.Q. falls abruptly to a value below 1. Removal of the testa thus reproduces the respiratory effects associated with the transition to phase 3. Very similar changes in respiration are brought about by supplying the intact seeds with pure oxygen instead of air, as illustrated in Fig. 2. On transferring them from air to pure oxygen there is a rapid increase in the rate of oxygen uptake; the R.Q. falls at first to a value of 0.7–0.8 and then rises to a value close to unity after about 2 hours in oxygen. These effects of high oxygen tensions are greatest while the testa is intact during phase 2, but rapidly decline after the emergence of the radicle.

The high values of the respiratory quotient, observed during phase 2, indicate that much of the carbon dioxide produced by the seed at this stage originates in anaerobic metabolism. The production of ethanol by germinating pea seeds in air has often been demonstrated (see, for example, Goksöyr and others, 39). From the data outlined above, it can be estimated that about two-thirds of the carbon dioxide arises from alcoholic fermentation. A powerful Pasteur effect appears to operate to reduce fermentation when oxygen supplies are improved by the splitting of the testa or by exposing the seeds to high oxygen tension, since the rate of carbon dioxide production is relatively little affected under these conditions. Assuming that alcoholic fermentation is replaced by aerobic breakdown of carbohydrates, it can be estimated that the rate of glycolysis is reduced to less than a half when adequate supplies of oxygen are available to the embryo. Other general features of the Pasteur effect have also been demonstrated in pea seedlings since Meeuse (see Goddard and Bonner, 38) has shown that active growth occurs in air, but not in nitrogen. The experiments of Meeuse have further indicated that under some conditions the loss of carbo-hydrate from young seedlings is lower in air than in nitrogen so that a conservation of carbohydrates may occur under aerobic conditions. Thus the chief characteristics of the Pasteur effect can all be recognized in germinating pea seeds. In the natural germination of the seed a transition from partially anaerobic to aerobic conditions may occur in which the Pasteur effect operates to suppress alcoholic fermentation and regulate the breakdown of carbohydrate in respiratory metabolism. With regard to the mechanism of regulation, Rowan and others (84) have obtained some evidence that the level of phosphorylation plays an important part. In developing pea seeds transferred to nitrogen there was a consistent rise in the ratio of ADP:ATP accompanied by an increase of inorganic phosphate in the tissues, a close parallel to the situation in yeast, which was reviewed by Lynen and others (63).

In pea seeds most of the respiratory activities occur at first in the cotyledons; the plumule and radicle contribute less than 1% of the total oxygen uptake after 24 hours' germination, but this rises to 15–20% after 48 hours, when the testa is ruptured (Spragg and Yemm, 89). Starch is the chief reserve of the cotyledons, and many of the glycolytic enzymes which bring about sugar breakdown by the EMP pathway have been demonstrated, as reviewed by Stumpf (99). It is of particular interest that Goksöyr and others (39) have shown that the activity of alcohol dehydrogenase in seeds germinated in air rises rapidly to a maximum during the first 48 hours of germination and then declines, corresponding closely with the phase in which anaerobic metabolism is most active.

There is little doubt that during the early stages of germination of many seeds the limitation of gaseous diffusion by the testa or seed coat has an important influence on respiration. James (50) has reviewed the evidence that a period of anaerobiosis often intervenes in the development of seedlings. There is, however, relatively little quantitative information on the permeability of the seed coat to gases. Brown (17) has shown that moistening causes a marked increase in the rate of diffusion of oxygen and carbon dioxide through the seed coat of *Cucurbita pepo*, but it is evident that the cutinized and lignified wall structures, often found in the testa, may offer relatively high resistance to diffusion. No direct pathway for gaseous diffusion is available through the testa and, within the compact structure of some seeds, intercellular air spaces constitute less than 1.5% of the total volume. Both these circumstances may restrict the gaseous exchanges and so subject the embryo to relatively high concentrations of carbon dioxide as well as low oxygen tensions, thus drastically affecting both the course and rate of respiration.

It is uncertain to what extent the energy exchanges of anaerobic metabolism are effectively harnessed in the development of the embryo. The ability of rice grains to germinate under strictly anaerobic conditions suggests that normal oxygen respiration is not essential, at least in some species, during the early stages of germination. A detailed comparative study of the germination of wheat (*Triticum vulgare*) and rice (*Oryza sativa*) has been made by Taylor (100). At low oxygen tensions and in nitrogen, rice has much greater powers of germination and development than wheat. This is correlated with a higher rate of anaerobic metabolism, in which alcoholic fermentation plays an important part. In rice, as in pea seeds, many attributes of the Pasteur effect are recognizable; the rate of carbon dioxide production is substantially greater in nitrogen than in air, but the growth of the embryo, as indicated by its increase in dry weight, is much more rapid under aerobic conditions. It seems likely that the ability of rice to germinate and to develop under submerged conditions at low oxygen tensions depends in large part upon its enzymatic equipment, which permits a vigorous release of energy by alcoholic fermentation. Such mechanisms are much less highly developed in wheat grains than in rice, and in this respect the different potentialities for respiratory and anaerobic metabolism closely reflect the differences in the biology of the two species.

B. Respiration and Seedling Development

High rates of respiration normally accompany the active development of the young seedling which follows germination. The classical
experiments of Kidd, West, and Briggs (56) showed that the respiratory index, measured as milligrams of carbon dioxide per hour per
gram of dry weight, reached its highest value in the early stages of the
development of sunflower seedlings (*Helianthus annuus*). However, if
for experimental purposes seedlings are germinated in continuous darkness, exhaustion of the limited food reserves intervenes and a gradual
decline of respiration takes place. Some of the most detailed studies of
the drift of respiratory rate in young seedlings grown in the dark have
been made with cereals, particularly with barley (*Hordeum vulgare*).
Here, the chief physiological conditions affecting respiration and its
relationship to the development of the seedling can profitably be
examined.

The drift of carbon dioxide production by barley seedlings germinating in continuous darkness has been described by Barnell (5), James
and James (51), and Forward (35). Under closely controlled conditions the rate increases sharply to a maximum after 6–8 days and then
falls gradually to a low level after about 20 days, followed by a general collapse of the seedlings and attack by saprophytic organisms.
Analytical data for carbohydrates and other reserves of the endosperm
and embryo indicate that the mobilization and exhaustion of these reserves is of primary importance in relation to the respiratory drift. On
this basis, the changes in the rate of carbon dioxide production can
be analyzed into five successive phases of which a diagrammatic summary is given in Fig. 3. An interpretation of the physiological changes
underlying the drift has been discussed by James (50) in terms of
five phases.

Phase 1 is characterized by a rapid acceleration of carbon dioxide
production which may at first be exponential with time, although
oxygen uptake, and consequently the R.Q., often show rapid fluctuations that are very sensitive to conditions of aeration and water supply.
During this time the limited reserves of the embryo itself, sucrose,
raffinose and a small amount of fat, are rapidly broken down and account for a major part of the respiratory substrate as well as the
formation of cellulose and other polysaccharides. In the latter part of
this phase the embryo depends increasingly upon the mobilization and
translocation of reserves from the endosperm.

Phase 2 permits respiration to rise, but more slowly than in the preceding phase. Growth and differentiation of the embryo are supported

by the transfer of reserves, chiefly carbohydrates and proteins from the endosperm. Depletion of these reserves is closely associated with the attainment of a maximum rate of respiration and its subsequent decline.

Phase 3 is characterized by a falling rate of respiration which corresponds to the exhaustion of soluble carbohydrates, chiefly dextrins in the endosperm and sucrose in the embryo. As in phase 2, the respiratory quotient is maintained at a value just below 1, indicating that carbohydrates continue to furnish most of the respiratory substrate.

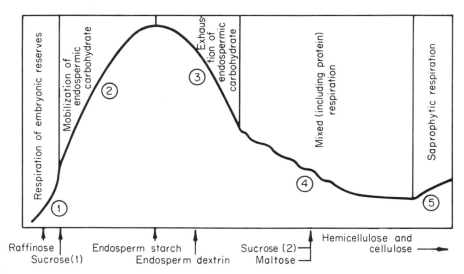

Fig. 3. Diagram to show the chief phases in the respiration drift curves for barley seedlings germinated in continuous darkness. The stages at which different food reserves are exhausted are indicated together with the distinctive features of respiration during phases 1–5. From James (50).

Phase 4 involves a further and somewhat irregular decline in the rate of respiration and the almost complete exhaustion of soluble carbohydrates. At this stage the respiratory quotient begins to fall to a value of about 0.8; this and other evidence indicates that proteins now contribute to the respiratory substrate.

Phase 5 is the terminal phase in which there is a rise in carbon dioxide production, in all probability due to the invasion of saprophytic organisms into the collapsed, disorganized tissues of the seedling.

It is instructive to compare the course of respiration outlined above with that of seedlings grown under more normal conditions in the light. Data were collected by Folkes and others (32) with seedlings

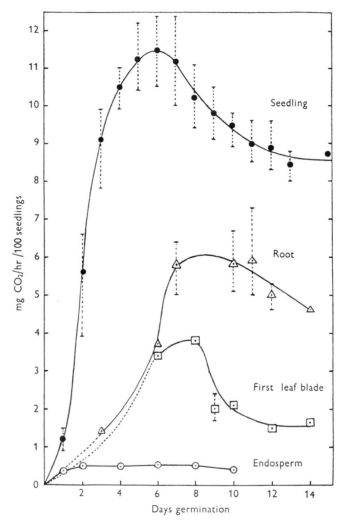

FIG. 4. The respiration of germinating barley seedlings grown with 16 hours' illumination daily at $22 \pm 1°C$. Comparable data for separate parts of the seedlings (endosperm, roots, and leaf blades) are shown. From Folkes *et al.* (32).

subjected to periodic illumination under controlled conditions. The course of respiration for the first 5 to 6 days was very similar to that of seedlings grown in the dark, but the subsequent decline in phases 3 and 4 was partially eliminated. Some of the results of such experiments, which are shown in Fig. 4, further indicate the important effects of starvation on the respiration of etiolated seedlings. Analyses of the seedlings grown in the light show that soluble sugars are maintained at a fairly high level by periodic illumination, no doubt as the result of photosynthetic activity which can be detected about 4 days after germination under these conditions.

However, even when the seedlings are grown in the light the rate of respiration falls appreciably from the maximum value attained at the end of phase 2. There are several lines of evidence which indicate that the supply of nitrogenous nutrients is particularly important at this stage. In the first place, the decline in the rate of respiration coincides with a failing supply of nitrogen from the endosperm of the seedling, and, in samples of grain varying in protein content, there is a close relationship between the level of their respiration and their nitrogen content. Moreover, the exogenous supply of suitable nitrogenous nutrients, such as nitrate or ammonium salts, gives rise to a marked increase of respiration in the seedlings at this phase (cf. Chapter 4, this volume for other relations of respiration and nitrogen compounds).

It is probable that in young seedlings the rate of respiration during the early stages of germination is closely related to the increase of the protoplasmic system in the growing embryo. Tissue proteins are one of the chief components of the protoplast, and their formation *in the embryo* provides an approximate estimate of the formation of new protoplasmic material. An analysis of some of the respiratory data for young barley seedlings along these lines has been made by calculating the rate of carbon dioxide production per unit of tissue protein in the embryo. The results, drawn from experiments with several samples of grain widely different in their nitrogen content, are plotted together in Fig. 5.

Despite the differences in the respiratory rate of the seedlings and in their nitrogen content, it is evident that the data agree closely when respiration is expressed per unit of tissue protein nitrogen. The results indicate that a close relationship exists between the rate of increase of protein and the rate of respiration. However, no simple quantitative relationship between protein nitrogen and respiration, such as that observed by Richards (81) in leaves of barley, has been detected during the early stages of seedling development. It is consistently found that the rate of respiration per unit of protein nitrogen increases to a maximum after about 3 days and subsequently declines to a much lower value. This period of intense respiratory activity during the early phase of germination coincides with the very active assimilation of carbohydrates and nitrogenous reserves from the endosperm. The highest respiratory rate per unit of tissue protein is contemporary with the greatest proportionate increase in dry weight and in total nitrogen content of the embryo.

A parallel study of the metabolism of proteins and amino acids in barley seedlings, grown in the light under similar conditions (Folkes and Yemm, 33, 34), has shown that an extensive interconversion of amino acids accompanies the transfer of reserve proteins from the endosperm to the embryo. The possibility that carbon skeletons of amino

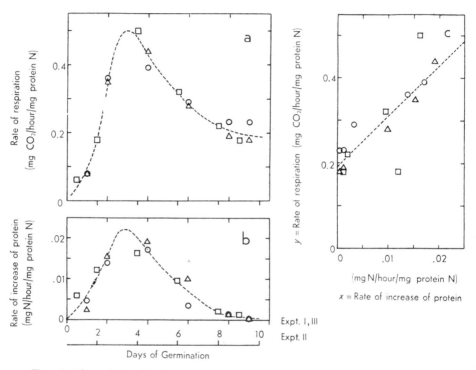

FIG. 5. The relationship between respiration and protein synthesis in developing seedlings of barley. Data of three experiments are included:

Expt.   I   (○)   Low nitrogen content (1.6% N dry wt.)
Expt.   II   (□)   Medium nitrogen content (2.0% N dry wt.)
Expt. III   (△)   High nitrogen content (2.6% N dry wt.)

The rate of respiration and the rate of protein synthesis are expressed per unit of protein in the embryo in figures (a) and (b), respectively. A correlation diagram of respiration rate with rate of protein synthesis is given in figure (c) with a regression line $y = 0.191 + 11.9x$. From Folkes and Yemm (34).

acids make a major contribution to the respiratory substrate in plant tissues has often been suggested (Steward and Street, 94; Steward and Pollard, 93, and for a later discussion of this topic reference may be made to Steward and Durzan in Chapter 4). Such a hypothesis could account for the close relationship between the rate of respiration and protein metabolism in barley seedlings, but other interpretations are possible. As indicated above, the highest rates of respiration coincide with the most rapid synthesis of proteins and other constituents of the embryo, at the expense of the endospermic reserves. It is probable that many of these anabolic processes are closely coupled with cell respiration by way of phosphorylation and oxidation-reduction. Regulatory mechanisms would thus operate to increase glycolysis and respiratory activity in the embryo. After the *de novo* synthesis of protein is virtually completed, the relations of respiration, via phosphorylations, and of protein metabolism would become a function more of protein turnover than of net synthesis.

The experimental data so far obtained in these studies of respiratory metabolism in barley seedlings strongly suggest that carbon dioxide arises chiefly from the breakdown of carbohydrate reserves and that the rate of respiration is regulated by its coupling with the varied endergonic processes, which constitute growth. In this respect protein synthesis occupies an outstanding position; it can be regarded as an approximate measure of the formation of new catalytic systems which promote cell respiration and, at the same time, may make extensive demands upon "energy-rich" phosphate compounds and other respiratory intermediaries. The dependence of respiration both on the *amount* of protein and on the *rate* of its formation in the young embryo, as indicated in Fig. 5, can be accounted for in this way. The relative merits of this hypothesis, compared with that of a protein cycle in which much of the carbon dioxide of respiration arises from the catabolism of amino acids, has been discussed by Yemm and Folkes (118, 119), with regard to these and other data. The concept of a regulation of glycolysis and cell respiration by their close coupling with endergonic processes appears at present to offer a more comprehensive interpretation of the relationship between carbohydrate and protein metabolism in young seedlings.

During the development of barley seedlings, the contribution of different organs to the total respiration of the seedlings changes progressively. Data which give a comparative account of the changes of respiration of the whole seedling and some of its chief parts have been collected by Folkes and others (32) and are illustrated in Fig. 4. As found by Barnell (5) and by Brown (18), the endosperm has a low rate of respiration and makes only a minor contribution to the fluctuations of respiration in the seedlings. A rapid acceleration in the rate of respiration of the embryo is the chief feature of early development, accompanied by the rapid growth of the coleorhiza and roots, followed a little later by that of the coleoptile. After about 6 days' germination,

the roots and the leaf blade dominate the changes, and it is evident
that the decline in respiration in the intact seedling under these conditions is mainly attributable to these organs. The leaf blade in particular shows a sharp decrease in respiration after it is fully expanded.
Here again, there is evidence that the high rate of respiration is closely
associated with the increase of protein content and dry weight of the
expanding leaf. However, it is evident that when the seedlings are
grown under these conditions in the light a drastic change in the metabolism of the leaf occurs as a result of the development of chloroplasts and photosynthetic activity. Measurements of photosynthesis
show that it plays an important part in anabolic processes at an early
stage in the differentiation of the leaf. A substantial part of the protein
of the young leaves is incorporated in the chloroplasts as they develop
in the light, and there are good grounds for supposing that part of the
synthesis of proteins and lipids may be closely linked with photosynthetic activity in chloroplasts (Rhodes and Yemm, 80). The complex
relation between protein synthesis and respiration may in part result
from the extensive synthesis of chloroplastic protein as development
proceeds during the transition to autotrophic nutrition of the seedling.
Photophosphorylation and photoreduction probably play a significant
part in supplementing the respiratory processes and promoting differentiation and development, particularly in photosynthetic organs.

As already indicated, high rates of respiration have often been observed in young developing seedlings (see Stiles and Leach, 98; and
Stiles, 96). In other cereals, as in barley, the endosperm seems to play
a minor part in the activity of the seedling, the developing embryo
being much more active in the uptake of oxygen and the production of
carbon dioxide during germination. However, in dicotyledonous seeds
such as *Brassica hirta (Sinapis alba)*, investigated by Stålfelt (90), and
Pisum sativum (89) the cotyledons respire actively and produce most
of the carbon dioxide during the early stages of seedling development.
The drift of respiration during germination in these seeds usually rises
to a peak during the first few days, similar in some respects to that
observed in cereals. However, there is as yet little detailed information
on the relationship between respiration and growth of the embryo in
these seedlings, and it is clear that the cotyledons play a much more
important part in the fluctuations in the rate of respiration during
germination.

C. The Respiratory Quotient of Germinating Seeds

Much attention has been given to measurements of the respiratory
quotient of seeds and young seedlings as a means of investigating the

nature of the respiratory substrate. Commonly, polysaccharides and fats, either together or alone, are the chief nonnitrogenous reserves of the seed. If it is assumed that complete oxidation of these substrates to carbon dioxide takes place, it would be expected that the value of the R.Q. during carbohydrate breakdown would be 1, whereas fats would give a value close to 0.7. Despite the wealth of information available on this topic (see Stiles and Leach, 98; James, 50), it is often difficult to interpret the data for germinating seeds, in terms of changes in the respiratory substrate. For example, the R.Q. may differ from unity during the breakdown of carbohydrate for a number of reasons. Incomplete oxidation and the accumulation of oxidized products, such as organic acids, often associated with the fixation of carbon dioxide in carboxylation reactions, may lead to quotients considerably lower than 1. On the other hand, low oxygen tensions may give rise to anaerobic release of carbon dioxide and raise the quotient to a value greater than unity. Such complications are particularly important during the early development of young seedlings when gaseous exchanges may be restricted by the seed coat and when carbohydrates and their intermediate products may be extensively diverted to synthetic processes in the young developing seedling.

In many seeds in which carbohydrates are the principal food reserves considerable fluctuations of the R.Q. have often been observed in the early stages of germination. For example, in *Zea mays*, studied by Stiles and Leach (97), it was observed consistently that the R.Q. was at first greater than 1 and then fell to a value of about 0.75 after which it gradually rose to a value close to unity during most of the development of the seedling. Somewhat similar data have been recorded for other cereal grains and for seeds of the Leguminosae, such as *Pisum sativum*, *Lathyrus odoratus* and *Vicia faba*. It is generally considered that these changes result from the preferential utilization of small amounts of fats as substrates in the early stages of germination; particularly in cereals, there is some evidence that fats stored in the embryo itself may be drawn into metabolism before the starch reserves of the endosperm are mobilized. However, it is also possible that the accumulation of products of anaerobic metabolism in the first phase of germination and their subsequent oxidation, when the supplies of oxygen to the tissues improve as a result of the splitting of the seed coat, may be responsible. The data for pea seeds, already briefly described, suggest that fluctuating values of the R.Q. can arise in this way. Detailed measurements of the R.Q. of barley grains in the first stages of germination made by James and James (51) clearly indicate that irregularities in the uptake of oxygen are chiefly responsible for rapid fluctuations of the quotient, which are very sensitive to restriction of diffusion by water films.

Seedlings whose principal reserve is fat, such as those of castor bean (*Ricinus communis*), flax (*Linum usitatissimum*), and sunflower (*Helianthus annuus*), studied by Stiles and Leach (97), generally have R.Q. values that are much below unity. For prolonged periods during the development of the seedling the quotient may be maintained at 0.5 or below. Such low values cannot be attributed simply to the oxidative breakdown of fats. It is evident from analytical data, such as those collected by Murlin (72) for castor bean seeds, that a major part of the fat reserve is converted into carbohydrate. Valuable progress has recently been made in elucidating the biochemical mechanisms underlying the metabolism of fats and its relationship to respiration in plant tissues; this topic is fully considered in other chapters of this volume (Butt and Beevers, Chapter 7, Volume IVB). A rapid conversion of fats to carbohydrates by way of the glyoxalate cycle has been shown to be an important feature of the metabolism of the endosperm of castor bean; the carbohydrate so formed is transferred to the embryo and there sustains synthetic activities and respiration during its growth and development.

The extent to which amino acids and proteins contribute to the main respiratory substrate in young seedlings is at present uncertain. It is probable that the utilization of reserve proteins for the synthesis of tissue protein and other nitrogenous constituents in the growing embryo involves an extensive interconversion of amino acids and that the carbon skeletons of these amino acids may be in part drawn into respiratory metabolism. The effect of such oxidations on the respiratory quotient will depend largely upon the nature of the other nitrogenous compounds which are formed from them. It can be calculated that if amides, such as asparagine, accumulate as an end product of protein catabolism a quotient of 0.7–0.8 would be expected, whereas a value close to unity would result if ammonia were the chief end product, as rarely happens except in acutely starved leaves (115). In lupine (*Lupinus*) seeds with their relatively high protein reserves, an R.Q. of 0.8–0.9 has been recorded over prolonged periods of germination. It is well known (Chibnall, 21) that lupine seedlings usually accumulate considerable quantities of asparagine during germination, and it is possible that in such seeds an appreciable breakdown of amino acids in respiration takes place. For a further discussion of these and other related problems, reference may be made to Chapter 4. The contribution of protein catabolism to respiration may well increase after prolonged germination in the dark, as indicated by the fall in R.Q. observed in highly starved and etiolated seedlings of barley by James and James (51).

It is evident that measurements of the respiratory quotient alone rarely provide a satisfactory identification of the respiratory substrate. Adequate evidence at a biochemical level is available for relatively few seeds, but it is probable that fats and carbohydrates are the chief sources of respiratory substrates during germination. The respiratory metabolism of the intact seedling is probably complicated by the various activities in different parts of the seedling; the metabolism of fats may predominate in some tissues, while concurrently carbohydrates provide the chief substrate in others.

D. EXTERNAL CONDITIONS AND THE RESPIRATION OF SEEDLINGS

Respiratory metabolism in germinating seeds is affected by many environmental conditions; the effects of temperature and aeration have been extensively studied since they often have a decisive influence on germination and the development of the young seedling. Aeration may obviously affect simultaneously the supply of both oxygen and carbon dioxide to the tissues. But despite the important physiological effects of carbon dioxide on germination, there is as yet surprisingly little information on its influence on the respiratory metabolism of seedlings. In fact, most of the measurements of respiration have been made under conditions in which abnormally low external concentrations of carbon dioxide are maintained, compared with those likely to occur under natural conditions in the soil. In this section it is convenient to consider the effects of temperature alongside those of oxygen and carbon dioxide since, as will be shown, there may be important interactions in their effects upon respiration, particularly under conditions of restricted gaseous diffusion during the early stages of germination.

1. Temperature

Two main direct effects of temperature on the rate of respiration can be distinguished. (Other indirect effects of temperature on substrate relations, e.g., the starch:sugar equilibrium, are discussed elsewhere.) At low temperature ranges a primary acceleration of the rate with increasing temperature is observed, but a secondary decline occurs progressively with time in the upper part of the temperature range. It is generally considered that the latter effect is caused by the gradual inactivation of thermolabile enzymes, although it has been subjected to relatively little detailed study. Most attention has been given to the marked acceleration of respiration over the lower part of the temperature range, mainly with the object of analyzing the physicochemical mechanisms. From this standpoint, it is clear that changes of temperature may have a differential effect upon the processes of diffusion whereby gaseous exchanges are maintained and upon the enzymatic reactions that constitute the biochemical mechanisms. Expressed in terms of the van't Hoff temperature coefficient widely used in biological work

$$Q_{10} = \frac{\text{Rate at } (t + 10)°\text{C}}{\text{Rate at } t°\text{C}}$$

diffusion rates fall within the limits 1.2–1.3, while enzymatic reactions, like most thermochemical reactions, commonly have coefficients 2–4 over the physiological range of temperature. It follows that an increase of temperature will accelerate the enzymatic processes of cell respiration to a much greater extent than gaseous diffusion, an effect that is further aggravated by the reduced solubility of gases at higher temperatures.

Most of the early studies of the rate of respiration of seedlings at different temperatures were made by measurements of carbon dioxide production. For example, the extensive data of Kuijper (57) and Fernandes (31) with seedlings of peas (*Pisum sativum*) and wheat (*Triticum vulgare*) show that Q_{10} values of 2–3 generally hold between about 0 and 20°C. Here there is little doubt that the effects of temperature on enzymatic activity predominate. However, there is often a steady decline in the value of the Q_{10} with increasing temperature, and it is probable that this is partly attributable to limitations imposed by the diffusion of oxygen to the tissues. Oota, Fujii, and Sunobe (74) in experiments with *Vigna sesquipedalis* observed a rise in R.Q. associated with the production of ethanol and small amounts of lactic acid at temperatures above 30°C, thus indicating the occurrence of anaerobic metabolism under these conditions. A further complication in these seedlings was that at lower temperatures fat reserves contributed a major part of the respiratory substrate, whereas breakdown of starch predominated at higher temperatures, possibly as a result of thermal displacement of the equilibria involved in the mobilization of these reserves.

An alternative method of analyzing the effects of temperature on the rate of respiration of germinating seeds is by means of the more precise formulation of the Arrhenius equation for thermochemical reactions

$$\log_e \frac{K_2}{K_1} = \frac{E}{R}\left(\frac{1}{T_1} - \frac{1}{T_2}\right)$$

where K_1 and K_2 are velocity constants at absolute temperatures T_1 and T_2, R is the gas constant, and E the energy of activation of the reaction. The activation energy of simple thermochemical reactions, and of some enzymatic reactions over a limited range of temperature (see Vennesland, 104) is effectively constant, and E can be evaluated from the linear relationship when $\log_e K$ is plotted against $1/T$.

Crozier (26) and his collaborators suggested from an analysis of their own and other data, that values of the apparent activation energy could be used to characterize the limiting or "master" reaction in the respiratory processes of some seedlings. Changes in the value of E at "critical" temperatures were attributed to a transition to a different limiting reaction. Crozier's hypothesis has been severely criticized on several grounds. Evaluation of E from the data for small temperature changes indicates that there is nearly always a progressive fall in the value on passing from low to high temperatures with little evidence of a sharp break at a critical temperature. Furthermore, as discussed by Hearon (43), it is improbable that a series of consecutive reactions such as those involved in cell respiration is effectively controlled by a single step. The gradual fall in the apparent energy of activation is attributable in some seeds to the restricted supply of oxygen to the tissues at high temperatures. With pea seeds in the first phases of their germination, Spragg and Yemm (89) have shown that the removal of the testa has a marked effect on the temperature relation of respiration. In the temperature range 4.5–25.5°C the rate of oxygen uptake was much greater for seeds from which the testas had been removed than for intact seeds. In absence of the testa the change of rate with temperature agreed fairly closely with the Arrhenius equation ($E = 17.1$ kcal), as shown in Fig. 6. No such linear relationship was found with intact seeds; there was instead a gradual decline in the increments of rate with increasing temperature, consistent with a progressive limitation of respiration by gaseous diffusion through the testa.

It is not surprising that the analysis of the effects of high temperature on the rate of respiration of germinating seeds in terms of enzymatic inactivation has proved particularly difficult. However, measurements, such as those of Kuijper (57), indicate that a fairly rapid and continuous decline in the rate of carbon dioxide output occurs at temperatures above 40°C. The temperature coefficient of thermal depression calculated between 40 and 50°C has a value often exceeding 4, which would be expected if inactivation of thermolabile enzymes were involved. The available data clearly indicate that the physiological effects of temperature on the respiration of germinating seeds change progressively throughout the temperature range. At low temperatures, activation of thermochemical reactions is of primary importance, but at higher temperatures it is probable that limitation by diffusion of gases and perhaps also of metabolites in the heterogeneous cell system intervenes before an irreversible thermal inactivation of enzymes takes place. The gradual fall of the temperature coefficient

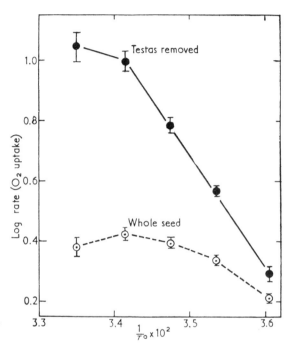

FIG. 6. The effect of temperature on the respiration of pea seeds with or without testas. The seeds were germinated at $22 \pm 1°C$ for 24 hours, and the uptake of oxygen was measured manometrically on duplicate samples at 4.5, 10, 14.8, 20, and 25°C. From the plot of log rate against $1/T$, the line of best fit gave an estimate of activation energy of 17.1 kcal for the seeds from which the testas had been removed. From Spragg and Yemm (89).

and of the apparent heat of activation at higher temperatures may result from a combination of these effects.

2. Oxygen and Carbon Dioxide Concentration and the Rate of Respiration

In germinating seedlings, as in many plant tissues, the rate of oxygen uptake usually shows a hyperbolic relationship with increasing oxygen concentration, such as that found by Forward (35) in barley seedlings cf. Fig. 7. Considerable importance has been attached to the form of this relationship owing to its general resemblance to that of an enzymatic reaction and substrate concentration. The possibility of calculating Michaelis-Menten constants (K_m) to characterize the oxygen affinity of oxidases mediating plant respiration has been indicated by Goddard and Bonner (38). However, there is strong evidence that, in germinating seeds, the form of the relationship is often affected by a restriction of gaseous diffusion. In young seedlings of rice and wheat (see James, 50) during the first phase of germination the increase in oxygen uptake with increasing oxygen concentration is

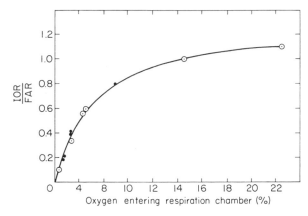

FIG. 7. The relation between respiration and the concentration of oxygen in young barley seedlings. Measurements of oxygen uptake were made in a continuous gas stream at 22.5°C. The rate of uptake in air is taken as 1.0 and other values are expressed as a ratio (IOR:FAR). Circled dots represent replicated observations. After Forward (35).

much less rapid than in older seedlings, an observation indicating a high resistance to oxygen diffusion in the early stages. Similarly, in pea seeds, it was shown by Spragg and Yemm that the form of the curves observed is greatly influenced by removal of the testa as indicated in Fig. 8. Under these circumstances little quantitative significance can be attached to the data with regard to the oxygen affinity of the terminal oxidases of the seedlings.

When carbon dioxide production is used as a measure of respiration, the relationship with oxygen concentration is further complicated. At low oxygen tensions, anaerobic production of carbon dioxide may occur and may be detected by a rise in the respiratory quotient or the accumulation of other end products, such as ethanol and, though more rarely, of lactate. Particular attention has been paid to the estimation

of the oxygen concentration at which anaerobic release of carbon dioxide ceases, the so-called extinction point, as indicated by changes in the
respiratory quotient or by the accumulation of products of anaerobic
metabolism. It is clear that in seedlings, and probably in other plant
tissues, this extinction point has no fixed and determinable value. It
varies widely with temperature and at different stages of germination,
no doubt largely as a result of changes in the accessibility of oxygen
to the cellular mechanisms. For example, the extensive studies of

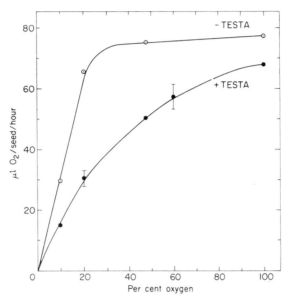

FIG. 8. The effect of oxygen concentration on the respiration of pea seeds. Oxygen
uptake was measured manometrically on seeds germinated for 24 hours in air. The
vertical lines indicate the range of variation in replicated samples of seeds. From
Yemm (unpublished data, 1958).

barley seedlings by Forward (35) indicate that at oxygen concentrations lower than that of air both oxygen uptake and carbon dioxide
production were reduced. As judged by changes in the respiratory quotient, anaerobic production of carbon dioxide was active below 5% in
older seedlings, but at earlier stages of germination an extinction point
of less than 2.5% oxygen was indicated.

The effects of oxygen tension upon carbon dioxide production vary
greatly in different seeds. In some species, such as rice (*Oryza sativa*)
and pea (*Pisum sativum*), the rate of carbon production in pure nitrogen is as high as, or higher than, that in air. Here as already dis-

cussed, there is clear evidence of the occurrence of a Pasteur effect eliminating alcoholic fermentation and regulating the breakdown of carbohydrate at least during the early stages of germination. In other species, detailed comparisons of the rate of carbon dioxide production in air and in nitrogen indicate that the rate is substantially reduced under anaerobic conditions. A systematic study of several different types of seed was made by Leach (58), in which he compared anaerobic and aerobic rates of carbon dioxide production by extrapolation of the carbon dioxide drift curves in air and in nitrogen, following the method developed by Blackman (12). The results, expressed as a ratio of anaerobic to aerobic production of carbon dioxide (NR:OR), gave values of 0.22 to 0.35 for seedlings of sweetpea (*Lathyrus odoratus*), maize (*Zea mays*), and buckwheat (*Fagopyrum esculentum*), in which the principal food reserves are carbohydrates. Since these ratios are often less than the value of 0.33, expected from a simple transition from aerobic oxidation of carbohydrates to alcoholic fermentation, it has been assumed that a Pasteur effect on the conservation of carbohydrate is small or lacking in these seedlings (Turner, 102). However, there are several sources of uncertainty in the interpretation of data based solely on the measurement of carbon dioxide production. The products of anaerobic metabolism in plant tissues rarely consist of carbon dioxide and ethanol in the proportions expected for alcoholic fermentation, but often include appreciable quantities of lactate (102). Moreover, in rapidly growing seedlings under aerobic conditions a major part of the products of carbohydrate breakdown may be diverted to anabolic reactions (see, for example, Kandler, 54; and Betz, 10). In these circumstances measurement of neither carbon dioxide production nor carbohydrate loss can provide unequivocal evidence of the conservation of carbohydrate under aerobic as compared with anaerobic conditions; a much more elaborate balance sheet of metabolic products would be necessary for this purpose.

There is some evidence that in the intense anabolic activities of the growing embryo the respiratory mechanisms are fully exploited by the rapid utilization of "energy rich" phosphate compounds and other metabolites in assimilation and in biosynthetic reactions. Beevers (6) has shown that uncoupling agents, such as 2,4-dinitrophenol, have relatively little effect on the rate of respiration of young seedlings. Thus the restrictive mechanisms operating in mature plant tissues by means of phosphorylation may be relatively ineffective in the actively growing seedling and low NR:OR ratios may arise in this way. However, it is clear the growth and development of the young embryo are largely dependent on aerobic conditions and in many respects offer a close parallel to the Pasteur effect described in relation to the growth of yeast. A broader concept of oxidative assimilation is more helpful in the interpretation of the Pasteur effect in growing plants, rather than that of oxidative anabolism as a mechanism for the conservation of carbohydrate under aerobic conditions.

The inhibitory effect of carbon dioxide on plant respiration was first systematically investigated by Kidd (55) in experiments with germinating seeds of pea (*Pisum sativum*) and white mustard (*Brassica hirta* [*Sinapis alba*]). With concentrations of carbon dioxide ranging from 0 to 80% in the atmosphere, it was shown that both the uptake of oxygen and the release of carbon dioxide declined progressively with increasing concentration. A more rapid fall in carbon dioxide production than in oxygen uptake was observed, so that in white mustard the R.Q. fell from a value of about 0.8 to 0.4–0.5 in 50% of carbon dioxide. Particularly in pea seeds, there was an interaction between oxygen and carbon dioxide concentrations; the inhibitory effects of the latter were enhanced at low oxygen tensions. Such conditions may readily arise within the seed during the early stages of germination, especially at high temperatures, owing to the restriction of gaseous exchanges by the seed coat; they may partly account for the decisive effects of low temperature in promoting germination of some seeds.

The effects of carbon dioxide on the respiration of seeds are of special interest with regard to the physiology of germination. White mustard seeds exposed to conditions of high carbon dioxide concentration together with low oxygen tensions, were shown by Kidd to pass into a dormant state with relatively low rates of respiration. This secondary dormancy could be broken by stripping off the testa or by drying the seed and remoistening it. The possibility was indicated that high concentrations of carbon dioxide and low oxygen tension are important factors in the development of dormancy during the maturation of seeds and perhaps in other compact plant structures subject to restricted diffusion. Although carbon dioxide appears to diffuse across the seed coat more rapidly than oxygen (Brown, 17), it may accumulate in the tissues owing to its production in anaerobic metabolism as well as in respiration.

The mechanism of the inhibitory action of carbon dioxide on respiration has not been extensively studied.[1] However, Ranson and others (79) have shown that in some plant tissues carbon dioxide may act as an inhibitor of succinic acid oxidase, an important enzymatic step in the organic acid cycle. It seems unlikely that this could fully account

---

[1] For some effects of carbon dioxide on the metabolism of potato disks, reference may be made to Chapter 4 of Volume II (Ed.).

for all the effects on seed metabolism since anaerobic production of carbon dioxide is also adversely affected at high concentrations. There is moreover, evidence that low concentrations of carbon dioxide may promote the germination of some seeds, an outstanding example being that of *Trifolium subterraneum*, in which 0.3% of carbon dioxide was effective (Ballard, 4). A reexamination of the metabolism of carbon dioxide in young seedlings by modern methods, which permit the detection of carbon dioxide fixation in carboxylation reactions, may well throw new light on the physiological mechanisms controlling germination and dormancy in seeds.

IV. The Respiration of Leaves

The development of plant organs of limited growth, such as leaves, can be conveniently considered in three stages. In the young leaves, cell division and differentiation leads to expansion of the lamina or leaf blade with its vascular system, mesophyll, and other tissues. The mature, fully grown leaf, primarily engaged in its photosynthetic functions and correlated activities, may persist in different species for a variable period extending from a few weeks to many months. A final stage of senescence can usually be recognized, in which yellowing of the leaf associated with an extensive breakdown of chlorophyll and other leaf constituents precedes leaf fall and death. There is now much evidence that well-marked changes in respiratory metabolism accompany leaf development. One of the chief objectives in this section will be to consider our present knowledge from this standpoint.

Most of the available data depend upon measurements of carbon dioxide production or oxygen uptake of detached leaves in the dark. The great diversity in the morphology and structure of leaves makes comparisons between different species on a dry weight or a fresh weight basis of doubtful value. But for studies of developmental changes, measurements of representative samples of similar leaves at different stages of growth can with advantage be expressed per unit leaf; in this way useful comparisons between respiratory rate and metabolic changes can be made.

A. Respiration and the Development of Young Leaves

Relatively little systematic study has been devoted to the respiration of expanding leaves, but there are clear indications that high rates of carbon dioxide production are associated with this stage of development. For example, Marsh and Goddard (66) in experiments with leaves of carrot (*Daucus carota* var. *sativa*) found that the rate of carbon dioxide production on a fresh weight basis (despite its obvious shortcomings) was nearly three times greater in young expanding leaves than in mature leaves. The first leaf of wheat seedlings was investigated by Roberts (82), who showed that the highest rate of respiration coincided with the later stages of leaf expansion. Similar changes have also been found with leaves of barley (*Hordeum vulgare*) seedlings, as indicated in Fig. 4. As the leaf blade expands owing to continuous cell division and differentiation, the rate of carbon dioxide production per leaf increases rapidly. A maximum rate is reached after about 7 days, and then a sharp decline occurs to a value less than half that of the maximum. This lower rate is maintained in the mature leaf. As already noted, the decline of the rate cannot be attributed to a fall in sugar content in the leaves, since, under the conditions of these experiments, this was kept at a high or rising level owing to photosynthetic activity.

At this point it should be noted that a series of leaves of increasing age taken simultaneously down the stem of the plant cannot be regarded as representative stages in an ontogenetic sequence of individual leaves. Several investigations have shown that during development there is a fairly steady decline in the respiration of the shoot apex, consisting mainly of meristematic tissues of the stem and leaf primordia. In sunflower plants, the respiratory index (milligrams carbon dioxide per gram dry weight per hour) of the stem apex, measured by Kidd, West, and Briggs (56), fell as the plants developed to about one-third of the value found in the seedling and young plants. Similarly, the data of Singh (87) for twelve other species indicate a marked decline in respiration of the shoot apex, particularly in annual plants with a rapid developmental cycle. It follows, therefore, that successive leaves formed on the stem follow a somewhat different metabolic pattern from that of an early stage in their differentiation. The respiratory index of successive leaves of barley plants, grown under a wide range of nutrient conditions, seems to reflect the decline of respiration in the stem apex. Richards (81) has shown that at the time of full expansion in these leaves, the rate of respiration is progressively lower as the plant develops and is closely correlated with the protein content of the leaves. However, the results of Hover and Gustafson (48) with plants of sorghum (*Sorghum vulgare*), sunflower (*Helianthus annuus*) and maize (*Zea mays*) clearly indicate that when the leaves are removed simultaneously from the plant, necessarily at different stages of development, the highest rates of respiration, expressed on a fresh weight basis, are found in the youngest leaves just below the stem apex. Here it seems probable that the high respiratory activity of the young expanding leaves is a predominant factor.

Measurements of the changes of respiration in leaves attached to growing plants have been made in only a few species. Arney (1) studied the drift of carbon dioxide production of leaves of the cultivated strawberry (*Fragaria chiloensis* var. *ananassa*), over a period of several months. A typical record is shown in Fig. 9. The data indicate that the highest rates of carbon dioxide production were observed in the early stages of leaf development. Maximal rates were generally found as the leaf reached its full size, and there was subsequently a fairly rapid decline, often to a rate less than half that of the maximum. In some

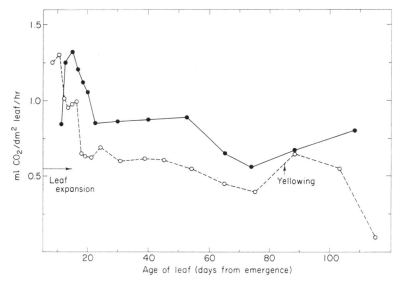

Fig. 9. The respiration of strawberry leaves attached to the plant. Carbon dioxide production was measured in an air stream of 24.5°C. The leaves were periodically sealed into a respiration chamber over a period of 3 months. From Arney (1).

cases a secondary rise of the rate occurred in leaves concurrently with yellowing. In general, these data resemble fairly closely those for barley and wheat, particularly during the phase of active leaf expansion.

More recent studies of the development of barley leaves by Rhodes and Yemm (80) have shown that the high rates of respiration per expanding leaf are closely related to the increase of proteins, nucleic acids, and other constituents (see Fig. 10). The protein content of leaves grown in the light reaches a high value after 8–9 days with relatively little change thereafter. During the active synthesis of protein, RNA attains its highest value in the leaf, but subsequently it decreases sharply to less than half its maximum value, the fall corresponding in time with the decline in protein synthesis and the fall in respiratory activity as the leaf matures. These data are fully consistent with the operation of a close coupling between respiration and anabolic processes in the young leaves and a regulation of cell respiration by mechanisms that depend upon oxidative phosphorylation and other reactions that promote biosynthetic activities. Dinitrophenol has relatively little effect on the respiration of young leaves, as shown by

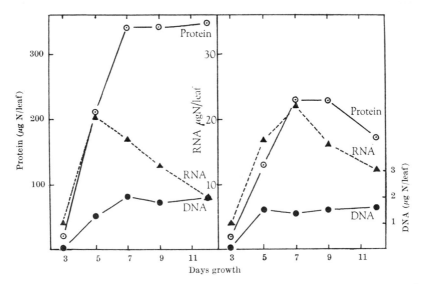

Fig. 10. The biosynthesis of proteins and nucleic acids (RNA and DNA) in developing leaves of barley seedlings grown in the light (about 1000 ft-candles), or in the dark at $22 \pm 1°C$. Data for seedlings grown in the light on the left and in the dark on the right. Samples of about 50 leaves were taken for analysis at different stages of development. From Rhodes and Yemm (80).

MacDonald and DeKock (64) and by Beevers (6), further indicating that the high rate is maintained by a rapid turnover of phosphate and high energy phosphate compounds.

However, it is evident that under normal conditions photosynthetic activity may play an important role at an early stage in the development of young leaves. In the first leaf of barley seedlings grown under normal conditions photosynthetic activity is detectable within a few hours of exposing the leaf to light. Chlorophyll is rapidly formed, and its increase closely parallels that of protein, as the leaf expands. Moreover as shown in Fig. 10, there is a substantially greater synthesis of protein in illuminated than in etiolated leaves and this is to a large extent attributable to the differentiation and development of chloro-plasts. Much circumstantial evidence now indicates that a formation of amino acids and proteins may occur in chloroplasts (Menke, 68); it is probable that photophosphorylation and photoreduction play a significant part, not only in carbon dioxide assimilation, but also in other anabolic reactions in the leaves whereby proteins and lipids are formed, particularly during the differentiation of chloroplasts. To this extent respiration and photosynthesis can be regarded as complemen-tary processes in the physiology of developing leaves. The biochemical

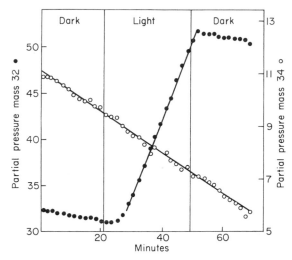

Fig. 11. The respiration of barley leaves in the light and in the dark. A mass spectrometer was used with isotopic oxygen to measure photosynthetically produced oxygen (mass 32) and respiratory uptake of oxygen (mass 34) in a gas mixture $N_2:CO_2:O_2$ 93:5:2. Strips of leaf tissue in a Warburg reaction flask were illuminated at 125 ft-candles and 25°C. From Daly and Brown (27).

mechanisms underlying the two processes have many features in com-mon with regard to both metabolites and the systems of phosphoryla-tion and oxidation-reduction. Interaction between them could, there-fore, occur at many points, but it appears to be limited by their occur-rence in separate cell organelles. The evidence so far gained by means of isotopically labeled oxygen and carbon dioxide indicates that the re-action systems of photosynthesis and respiration are relatively in-dependent of one another in leaves. In experiments with barley leaves using labeled oxygen, Daly and Brown (27) have shown that under some conditions the rate of oxygen uptake continues with little change in the light and in the dark, as indicated in Fig. 11 (cf. Chapter 4,

Volume IA). On the other hand, Weigl, Warrington, and Calvin (110) with C^{14}-labeled carbon dioxide under somewhat similar conditions found that CO_2 production was depressed by strong light, but suggested that this was due to preferential utilization of respiratory carbon dioxide in photosynthesis. These and other observations suggest that the rate of respiration is not drastically affected by illumination, but they do not necessarily imply that the reaction mechanisms are unaffected. In some algae, the data of Brown and Weis (16) indicate that greater interaction between respiratory metabolism and photosynthesis may occur depending upon the intensity of illumination. As yet, very little is known of the conditions that influence movement of metabolites to different cytoplasmic organelles, but it is of interest that phosphorylation, a key reaction in both photosynthesis and respiration, appears to have marked effect on the permeability and other properties of the membranes of both mitochondria and plastids (see Packer and others, 75). Beevers (6) and Walker (109) have reviewed more fully the interrelation between respiration and photosynthesis in plant cells. The rapid transition in young developing leaves from a predominantly respiratory to a photosynthetic metabolism seems to offer a good opportunity for analyzing more fully the interrelationship between these two processes in the development and differentiation of leaves.

B. Respiration of Mature Leaves

Fully grown leaves freshly detached from the plant normally contain plentiful supplies of readily available carbohydrates, such as sucrose and starch. Many measurements of the respiratory quotient indicate that carbohydrates furnish the chief respiratory substrate under these conditions. James (50) has summarized data covering 30 species; all the R.Q. values fall within the limits of 0.94 and 1.17 and most of them are very close to unity. Moreover, when detached leaves are kept in continuous darkness they rapidly lose carbohydrate, and in leaves of several species, including barley, bean, and wheat, analytical data have shown that, at first, the breakdown of carbohydrate accounts for most of the carbon dioxide produced in respiration. On these grounds, it has generally been inferred that the respiration in the tissues is dominated by the catabolism of carbohydrate. On the other hand, there is much evidence that other substrates may readily be broken down in mature leaves and that in some species the catabolism of proteins and organic acids is closely linked to respiration. Early work by Spoehr and McGee also involved amino acids in the respiration of leaves. A simple relationship between the sugar content of leaves and their respiratory rate has rarely been demonstrated. Indeed, the failure to detect a significant correlation between sugar content and respiratory rate of barley leaves led Gregory and Sen (40) to suggest that much of the carbon dioxide originated from the breakdown of protein which is in turn replaced in the operation of a continuous protein cycle. Later experiments of Yemm and Somers (120) with similar leaves strongly suggest that the lack of correlation between respiration and carbohydrate content resulted from the relatively high level of carbohydrate in the leaves investigated by Gregory and Sen. If barley leaves are detached at different times of the day or night, or the plants are darkened for a short preliminary period, it is found that a marked fall in the rate of respiration occurs at low carbohydrate contents. The results of a typical experiment conducted in this way are shown in Fig. 12. At sugar contents exceeding about 20 mg per gram fresh weight, little effect of sugar content on the initial rate of respiration was observed, but at lower values a sharp decrease occurred. Most of the leaves investigated by Gregory and Sen were in the upper part of this range and would, therefore, be relatively insensitive to changes of sugar content.

The form of the relationship between the rate of carbon dioxide production and carbohydrate concentration, shown in Fig. 12, departs widely from that of an enzymatic hyperbola at low sugar concentrations. With barley leaves darkened on the plant in order to reduce their carbohydrate content to a very low level, it is found that a minimal respiratory rate is reached at a rate somewhat less than half that of comparable leaves that are rich in carbohydrate. It seems clear from this and other evidence that at very low sugar levels other substrates are readily drawn into respiration. The R.Q. of leaves with a low carbohydrate content, measured within a few hours of their detachment, is considerably below unity (0.8–0.9), and secondary products of protein catabolism, such as amides, accumulate much more rapidly in them than in leaves of high carbohydrate content (Yemm, 116). Little change of organic acids occurs under these conditions, and in all probability amino acids, formed from the breakdown of tissue proteins, are the chief respiratory substrates at such low levels of carbohydrate.

An important problem in experimental studies of the respiration of detached leaves is the effect of handling, inevitable in the preparation of the tissues for measurements of gaseous exchange in the dark. Audus (2, 3) found a marked temporary rise in the rate of carbon dioxide production of cherry laurel leaves (*Prunus laurocerasus*) when they were subjected to mechanical pressure and bending under carefully controlled conditions, and similar effects have been observed in

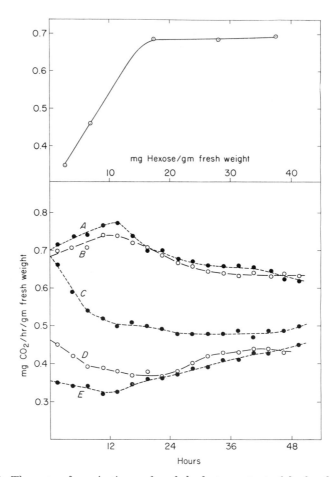

Fig. 12. The rate of respiration and carbohydrate content of barley leaves. Five samples of mature leaves (*A, B, C, D,* and *E*) were collected at different times of the day or night for measurement of carbon dioxide production at 22.5°C and for analysis of total soluble sugars. In the upper part of the figure the initial rate of CO_2 production is plotted against the sugar content of the freshly detached leaves. From unpublished data of Yemm and Somers (120).

leaves of other species. In barley, manipulation of the leaves has been shown to lead to a rise of 5–15% in the rate of carbon dioxide production, and this may persist for 3–6 hours. The cause of this rise is not yet understood. In cherry laurel, the rate of carbon dioxide production under anaerobic conditions is not seriously affected by manipulation, a result suggesting that the oxidative processes are most seriously affected. The possible effects of stomatal changes on the gaseous exchanges of leaves kept for long periods in the dark have not been systematically investigated. As measured by means of a porometer, the stomatal resistance of detached barley leaves in the dark is relatively high, indicating closure of the stomata as would be expected under these conditions. But on the other hand, the rate of respiration of the leaves is not appreciably affected by increasing the oxygen concentration above that of air, so that it is unlikely that the diffusion of oxygen to the leaf cells is seriously impeded. In cherry laurel leaves, the data of Parija (see James, 50) indicate that increasing the oxygen concentration to a value of 33% considerably increases the respiration of the leaves. It is probable that here restriction of diffusion may be of greater importance, and the effects of manipulation may well increase temporarily the supplies of oxygen to the tissues.

The rapid breakdown of sucrose and starch in freshly detached leaves in the dark indicates that they generally provide most of the substrate of respiration. Sucrose, in particular, decreases rapidly at first, but small quantities of both sucrose and starch often persist in highly starved tissues, possibly owing to their inaccessibility in different tissues or cell organelles. In bean (*Vicia faba*) and in barley leaves (Yemm, 114) small quantities of free glucose and fructose temporarily accumulate during the rapid fall in sucrose and starch. Interconversions of carbohydrate, mediated by phosphorylase and phosphokinase enzymes, appear to be involved in the entry of carbohydrates by way of hexose phosphates into catabolic reactions either by glycolysis or by the pentose phosphate pathway. The ease with which different oligosaccharides and polysaccharides are drawn into these interconversions, as well as their accessibility in the cells and tissues of the leaf, probably determine their rate of breakdown. Artificial feeding of starved leaves and of etiolated plants has provided further evidence of the active breakdown of various sugars in respiration. Early experiments conducted in this way have been discussed by James (50). Usually sucrose, glucose, and fructose are about equally effective in raising the rate of respiration; under optimal conditions more than a doubling of the rate has often been recorded together with a rise in the R.Q. toward unity. The use of C^{14}-labeled glucose in feeding experiments has provided more decisive evidence of the origin of the additional carbon dioxide. Vittorio, Krotkov, and Reed (108) supplied uniformly labeled glucose to wheat leaves and observed a substantial increase in the amount of carbon dioxide produced, about 80% of which originated from the glucose which was supplied. To what extent the Embden-Meyerhof-Parnas glycolytic or the pentose phosphate pathways operate in respiration of the mature leaves is at present uncertain. In mature leaves of barley, as distinct from young seedlings, the rate of carbon dioxide production rapidly falls under anaerobic conditions and negligible amounts of ethanol are produced. Extensive damage to these leaves can be detected within a few hours of treatment under anaerobic conditions, and there are some indications that the decarboxylation of glutamate to γ-aminobutyric acid may be a contributory source of carbon dioxide under these conditions. Relatively low molecular ratios of ethanol to carbon dioxide have been recorded in other leaves (James, 50), but the possibility that lactate may be formed anaerobically in these tissues needs further exploration. There has been no satisfactory demonstration, comparable to that in seedlings, of the occurrence and activity of glycolytic enzymes in leaves. Evidence for the participation of the pentose phosphate pathway in leaf respiration is also incomplete. Several of the enzymes involved, such as phosphoribokinase and transketolases have been separated and purified from leaves, but it is possible that they are mainly involved in their photosynthetic activities. Few critical measurements of the respiratory metabolism of specifically labeled glucose by the technique of Bloom and Stetten (14) have been made; the available data (Gibbs and Beevers, 37), suggest that the pentose phosphate pathway is much more active in mature than in young organs.

The tricarboxylic acid cycle appears to be the chief mechanism in the final phase of carbohydrate catabolism in mature leaves. Despite some reports to the contrary, mitochondria have now been shown by electron microscopy to occur widely in mesophyll cells, but it has generally proved difficult to make biochemically active preparations of mitochondria from mature leaves. Special methods of grinding and infiltration with sucrose have yielded preparations with many of the typical properties of mitochondria in the oxidation of organic acids and oxidative phosphorylation (Ducet and Rosenberg, 28). The study of cytochromes in leaves and their relationship to respiration has also proved exceptionally difficult, mainly owing to the presence of other pigments and of special cytochromes in chloroplasts and chloroplast fragments, which cannot readily be separated from mitochondria by the conventional methods of differential centrifugation. Moreover, the treatment of mature leaves with such well known respiratory inhibitors as cyanide, azide, and carbon monoxide has often yielded unexpected and anomalous results. For example, Marsh and Goddard (66) working with mature carrot leaves (*Daucus carota* var. *sativa*) found that cyanide at low concentrations stimulated the rate of oxygen uptake, whereas under comparable conditions the respiration of young leaves was inhibited by more than 50%. Somewhat similar results have been recorded by MacDonald and DeKock (64) for young and old leaves of several other species. Carbon monoxide and azide were also shown to cause only slight inhibition and sometimes to stimulate the respiration of adult and old leaves in contrast to their effects on young leaves.

Several interpretations have been offered to account for these effects; they include drastic changes in the nature of the terminal oxidase as the leaf develops and the presence of excess oxidase, so that the relatively low rate of respiration in adult leaves is determined at some other stage in the respiratory sequence. The balance of evidence, reviewed by Hackett (41) and by Beevers (6), still suggests that cytochrome oxidase is the chief terminal oxidase in the respiration of mature leaves, but that inhibitors, such as carbon monoxide and cyanide, divert the flow of electrons from the normal cytochrome sequence to autoxidizable cytochromes like cytochrome b_7. If this occurs, oxidative phosphorylation would diminish, thus relieving the restraint on early stages of carbohydrate catabolism which may depend on the availability of phosphate and phosphate acceptors. The well-marked difference between the young and old leaves in their sensitivity to the uncoupling action of 2,4-dinitrophenol has already been noted. A general similarity of the effects of cyanide, carbon monoxide, and azide to those of dinitrophenol gives some support to the view that uncoupling of phosphorylation may be an important part of the action of respiratory inhibitors leading to stimulation of respiration in mature and old leaves. The development of effective methods for the separation of active mitochondria from leaves is likely to provide more conclusive evidence on this part of their respiratory mechanisms.

It has often been observed that when detached leaves are kept in the dark their content of organic acids, particularly malic and citric acids, increases. This was first clearly shown by Vickery and his collaborators (106) in leaves of tobacco (*Nicotiana tabacum*) and led to a recognition of the important part played by the tricarboxylic acid cycle in the metabolism of leaves (Chibnall, 21). Somers (88) found somewhat similar changes in mature barley leaves; malate and citrate both increased for about the first 24 hours after leaves rich in carbohydrate had been detached from the plant. The total change in these acids in barley leaves was relatively small compared with their respiratory activities; it represented less than 10% of the total carbon lost as carbon dioxide over the same period and is therefore unlikely to have an important over-all effect on the respiration of the leaves. However, it is probable that the formation of organic acids by carboxylation has an important bearing on the operation of the tricarboxylic acid cycle under some conditions. Withdrawal of dicarboxylic or tricarboxylic acids from the cycle, as intermediaries of biosynthetic reactions, would clearly require replenishment for the continued operation of a catalytic cycle in cell respiration. The fixation of carbon dioxide in the dark is now known to occur commonly in plant tissues; tracer experiments with labeled carbon dioxide have established that in leaves malate and related compounds, such as aspartate, rapidly become labeled. This, and other evidence, strongly suggests that carboxylation of pyruvate, or of a derivative of pyruvate, is an important mechanism of dark fixation of carbon dioxide. These reactions may assume major proportions and play a distinctive role in the respiratory metabolism of some leaves.

Much attention has recently been given to the relationship between respiration and the formation of organic acids in succulent leaves. The occurrence of marked diurnal fluctuations in species of the Crassulaceae and in some cacti has been known for a long time; typically acids accumulate in the tissues at night and decrease during daytime. The fluctuations were first discovered in *Kalanchoe pinnata* (*Bryophyllum calycinum*) and, since they are often found in other species of the family, the term "Crassulacean acid metabolism" (101) is often used for descriptive purposes. A valuable appraisal of our present knowledge of the phenomena, and in particular their relationship to respiratory processes, in these plants has been made by Ranson and Thomas (78) and by Walker (109).

When acids are formed in the dark, the absorption of oxygen continues at a fairly steady rate but relatively little carbon dioxide is released. Thus, very low or negative values of the respiratory quotient are commonly observed during acidification in darkened leaves. Subsequently, as the acid content of the tissues reaches a steady or slowly declining value, R.Q.'s close to unity or above are obtained. Some of the data collected by Thomas and his collaborators are shown in Fig. 13. There is now much evidence drawn from the study of detached leaves of species of *Sedum* and *Kalanchoe* that the fluctuations of acidity are dominated by the formation and disappearance of malic acid; other organic acids, such as citric and isocitric acids, are often present in large amounts, but they show relatively minor changes. A central problem is therefore the biochemical mechanisms underlying the synthesis and breakdown of malate in these tissues and their relationship to respiratory metabolism.

In species of both of the genera listed above, it has been observed that conspicuous losses of carbohydrate, notably starch, accompany the formation of malate, and unequivocal evidence that carbon dioxide fixation plays a major role in acidification has now been gained by means of C^{14}-labeled carbon dioxide. Detailed analytical data, such as those of Vickery (105) on detached leaves of *Kalanchoe pinnata* (*Bryophyllum calycinum*), indicate that the accumulation of malate

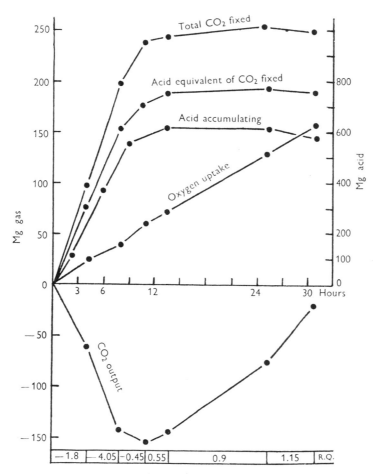

Fig. 13. Respiration and organic acid metabolism in leaves of *Kalanchoe* (*Bryophyllum*). Detached leaves were kept in the dark at 12°C in a gas mixture containing initially 5% CO_2. Measurements of titratable acidity and gas analyses were carried out at intervals. The negative values of the R.Q. during acidification are shown at the bottom of the figure. After Thomas (101).

in the tissues can be largely accounted for by carboxylations involving the products of carbohydrate catabolism. With regard to the relation between acid formation and respiratory metabolism, Thomas (101) suggested that oxygen absorption, which continues during the

acidification, is exclusively involved in oxidative catabolism of carbohydrate and that carbon dioxide produced in this way is drawn into
carboxylation reactions leading to the formation of malate. This
hypothesis can be summarized in over-all equations (1) and (2):

$$C_6H_{12}O_6 + 6\ O_2 \rightarrow 6\ CO_2 + 6\ H_2O \qquad\qquad (1)$$
$$C_6H_{12}O_6 + 2\ CO_2 \rightarrow 2\ C_4H_6O_5 \qquad\qquad (2)$$

Variations in the relative rates of these over-all reactions could readily
account for several important features of acid formation, such as the
fluctuations of the respiratory quotient and the fall in the rate of
carbon dioxide production during acidification.

There are several enzymatic mechanisms widely distributed in plants
which may be directly involved in the synthesis of malate by carbon
dioxide fixation. These systems are summarized as follows:

a. *Malic enzyme*

$$\text{Pyruvate} + CO_2 + TPNH_2 \rightarrow \text{malate} + TPN$$

b. *Phosphoenol pyruvate (PEP) carboxylase–malic dehydrogenase*

$$PEP + CO_2 \rightarrow \text{oxaloacetate} + P_i$$
$$\text{Oxaloacetate} + DPNH_2 \rightarrow \text{malate} + DPN$$

It is possible that phosphoenol pyruvate carboxykinase (Mazelis and
Vennesland, 67) could replace PEP carboxylase in the second reaction
linked with malic dehydrogenase. In a discussion of the participation
of these enzymes in malate synthesis *in vivo*, Walker (109) has indicated the more favorable equilibria in the carboxylase-dehydrogenase
coupled system and its much higher affinity for carbon dioxide. Malic
enzyme, on the other hand, has a relatively low affinity for carbon
dioxide; for effective synthesis it requires high reduction potentials,
maintained by a high $TPNH_2:TPN$ ratio, and rapid removal of malate
from the site of synthesis.

Besides these fairly direct mechanisms of malate synthesis dependent
on carbon dioxide fixation, it seems likely that carboxylations involving
pentose phosphates also operate during acidification. Gibbs and Beevers
(37) using the Bloom and Stettin technique with specifically labeled
glucose found evidence of the pentose phosphate pathway of hexose
breakdown in leaves of *Kalanchoe* (*Bryophyllum*). The interpretation
of their data is particularly difficult in view of the possibility of carbon
dioxide fixation, but they suggest that the glycolytic and pentose phosphate pathways contribute about equally to the catabolism of hexose
in these leaves. The detection of labeled phosphoglyceric acid when
$C^{14}O_2$ is used to trace dark fixation has indicated the possibility that
ribulose diphosphate carboxylase is active under these conditions.

Moreover, in experiments with species of *Kalanchoe*, Bradbeer, Ranson, and Stiller (15) found consistently that malate, formed during dark fixation of $C^{14}O_2$, is labeled only in the carboxyl groups with the label distributed in the ratio of $2:1$ in the C-4 and C-1 positions, respectively. This pattern of labeling suggests that two carboxylation reactions occur; it could result from the formation of phosphoglyceric acid and carboxylation of phosphoenol pyruvate as summarized below:

$$\text{Ribulose diphosphate} + CO_2$$
$$\downarrow$$
$$2 \text{ Phosphoglycerate}$$
$$\downarrow$$
$$2 \text{ Phosphoenol pyruvate} + 2 CO_2$$
$$\downarrow$$
$$2 \text{ Oxaloacetate} + 2 \text{ DPNH}_2$$
$$\downarrow$$
$$2 \text{ Malate} + 2 \text{ DPN}$$

There are, however, several sources of uncertainty in the operation of this sequence of reactions in the synthesis of malate *in vivo*. A continuous production of ribulose diphosphate in the dark is implied together with a supply of reduced pyridine nucleotides presumably from glycolytic or respiratory oxidation-reductions.

It is highly probable that spatial organization of the enzymatic systems and metabolites within the cell has an important influence in regulating carbon dioxide fixation and its coupling with respiratory oxidations. Both the quantity of malate formed and the relative constancy of its labeling indicate that it is mainly held in a storage pool, such as the cell vacuole, remote from the centers of active metabolism. Although there is much indirect evidence that the tricarboxylic acid cycle is active in these leaves, it appears to play only a minor role in the transformation of malate during acidification in the dark. However, the gradual spread of C^{14}-labeling from malate to other dicarboxylic and tricarboxylic acids strongly suggests that the cycle continues during acidification and it may be largely responsible for the fairly steady uptake of oxygen. As already briefly discussed in Section I, C, it is possible that the concurrent breakdown of malate by the DPN-linked oxidation in the tricarboxylic cycle and its synthesis by reductive carboxylation may be achieved by the spatial separation of the different enzyme and coenzyme systems in mitochondria and other cytoplasmic structures. The requirements for effective synthesis of malate are probably met in parts of the cytoplasm remote from the oxidation systems of the mitochondria; here high reduction potentials may be mediated by pyridine nucleotides and the breakdown of carbohydrates by the pentose phosphate pathway or glycolysis. Further information on the distribution of the enzymatic systems in the different cell structures is clearly required, but it is possible that the transfer of malate between these structures and the storage pool has an important effect on acidification and deacidification.

The concentration of carbon dioxide appears to be a decisive physiological factor controlling the rate of malate formation in succulents. Acidification is usually most active at external carbon dioxide concentrations of about 5% and decreases at lower values. It is generally assumed, therefore, that under natural conditions diurnal fluctuations are brought about in this way, since in the light photosynthetic assimilation of carbon dioxide would compete successfully with the carboxylations of dark fixation. The occurrence of acidification in the light has in fact been demonstrated in leaves of *Kalanchoe* when the concentration of carbon dioxide is adequate to saturate the photosynthetic reactions. Very low concentrations of oxygen, on the other hand, retard acidification, and in some species ethanol accumulates under strictly anaerobic conditions with the rapid appearance of damage in the tissues.

Several features of the respiration and metabolism of crassulacean plants suggest that organic acids play a distinctive role in the general physiology and nutrition of these succulents. The plants commonly grow in habitats that are subject to seasonal drought in which water storage and low water loss by transpiration may be of great importance. Under these circumstances, gaseous exchanges would be severely limited, so that the retention of respiratory carbon dioxide by dark fixation, and its subsequent utilization in photosynthetic reactions, may play an important part in their autotrophic nutrition. In some succulents, Loftfield (59) has shown that the stomata close during the day and open at night, thus suggesting that carbon dioxide fixation at night may contribute to photosynthesis during stomatal closure in the day. A reexamination of the water balance in these plants, and particularly the stomatal regulation of gaseous exchanges in relation to acid metabolism, is likely to throw further light upon this aspect of their physiology.

C. Senescent Leaves

A general feature of the metabolism of senescent leaves is the predominance of catabolic reactions in which carbohydrates and proteins are broken down to soluble constituents. Under normal conditions many of these products, such as amino acids and amides, are translocated to other parts of the plant where they may be stored as reserves in seeds or other perennating structures. Yellowing of the leaves, associated with loss of chlorophyll and the breakdown of chloroplastic proteins, is an obvious external symptom of aging, closely related to the decline of normal photosynthetic functions of the leaf. When leaves are detached from the plant, it is generally found that these changes are markedly accelerated; many mature leaves show extensive yellowing and loss of protein within a few days of detachment from the plant. It has often been assumed on rather slender evidence that the metabolic changes in detached leaves can be regarded as an acceleration of the catabolic processes that normally occur in aging leaves. The extensive data now available concerning the changes of respiratory metabolism in detached leaves will be considered mainly from this standpoint.

The drift of carbon dioxide production in detached leaves maintained in continuous darkness provided one of the earliest approaches to the systematic investigation of leaf respiration. F. F. Blackman first studied in detail the respiratory drift of leaves of cherry laurel (*Prunus laurocerasus*) and of nasturtium (*Tropaeolum majus*). Essentially similar methods have subsequently been applied to leaves of several other species, including barley, bean, wheat, and several grasses. In most of these later investigations, particular attention has been given to the changes of carbohydrates, proteins, and other leaf constituents that may be involved in respiratory metabolism in the leaves. A valuable review of much of the data, some of which have not been published elsewhere, has been given by James (50).

Despite wide differences, both in the rate of respiration and in the rapidity with which changes occur in detached leaves, many of the carbon dioxide drift curves have certain features in common. Three examples of the drift curves of contrasting leaf types are illustrated in Figs. 14 and 15. A general account of the chief phases which can be recognized is given below:

Phase 1. A high and sometimes rising rate of respiration is generally observed when leaves are detached after a period of active photosynthesis. As already discussed, the R.Q. is at first close to unity and analytical data for some leaves show that the loss of carbohydrates, chiefly sucrose and starch, accounts for most of the carbon dioxide produced for a period of 12–24 hours. It may be reasonably assumed that the respiratory metabolism in this first phase resembles that of a normal leaf attached to the plant.

Phase 2. Typically, a sharp decline in the rate of respiration occurs sometimes to a value less than half that of the freshly detached leaves. This appears to follow a partial exhaustion of readily available carbohydrates, particularly sucrose. An appreciable fall in the R.Q. to a value of 0.8–0.9 in some leaves indicates that other substrates are drawn into respiratory metabolism at this stage. In several species, including bean, barley, and wheat, there is strong evidence that carbohydrates no longer provide all the respiratory substrate. Secondary products of protein catabolism, such as amides, commonly accumulate in the tissues.

Phase 3. Rapid yellowing of the leaf usually takes place at this stage in adult leaves, and the respiration curves commonly show a well-marked rise and fall in the rate of carbon dioxide production. This is clearly shown in the relatively slow drift curves for leaves of cherry laurel and nasturtium. In other leaves the changes of respiration during yellowing are more variable, but with mature and old leaves of

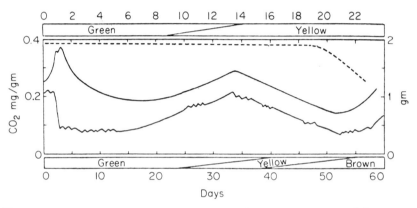

Fig. 14. Respiration drift-curves for *Tropaeolum majus* (upper solid curve and time scale) and *Prunus laurocerasus* (lower solid curve and time scale). The broken line gives the leaf weight for *Tropaeolum* with ordinates on the right. From James (50).

barley and wheat a short rise usually occurs or a fairly steady rate may be maintained. The age of the leaf when it is excised from the plant has an important effect on respiration at this stage. Yellowing is generally delayed and the rate of respiration is maintained at a lower value in young leaves. There is strong evidence, particularly for leaves of barley (Yemm, 114, 115), that carbohydrates provide only a minor part of the respiratory substrate at this stage; the respiratory quotient falls to a value of about 0.8 and there is a rapid accumulation of amides, chiefly asparagine, together with a rise in the level of ammonia in the tissues during the later stages of yellowing. It is very probable that the rapidity and extent of protein catabolism in different types of leaf is a dominant factor influencing the rate of respiration in this phase. At the same time, less readily available sources of carbohydrate may also be drawn into catabolic processes during yellowing. In mature leaves of cherry laurel, cyanogenetic glucosides are rapidly broken down during the secondary rise in the rate of carbon dioxide production, although they contribute only a small part of the total carbon dioxide produced by the leaves.

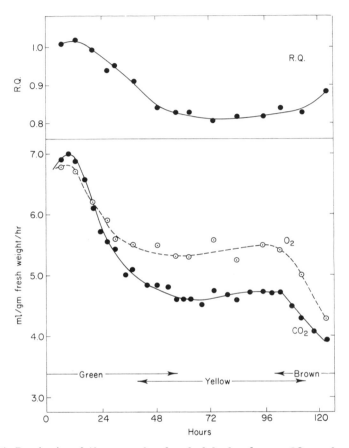

FIG. 15. Respiration drift-curves for detached barley leaves. CO_2 production and O_2 uptake were measured at 25°C in a continuous air stream by gas analysis. Changes in R.Q. are shown in the upper part of the figure and changes in the appearance of the leaves below. From Yemm (unpublished data, 1935).

Phase 4. The transition to this phase is marked by a gradual collapse in cell organization and the degeneration of the tissues. The escape of the cell contents leads to injection of the intercellular spaces with a change in the color of the leaves. It is probable that invasion of the tissues by the saprophytic organisms contributes to the production of carbon dioxide and may be largely responsible for the marked rise in the rate of carbon dioxide production in some leaves during the final stages.

It is very probable that the rapid yellowing and disorganization that occurs in detached leaves differs in several respects from senescence that occurs normally on the growing plant. Respiratory metabolism in particular may be drastically affected when the leaves are kept

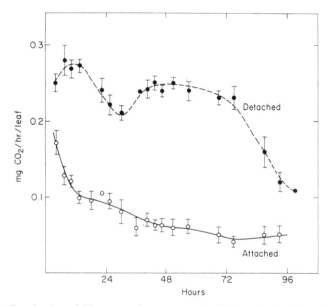

FIG. 16. Respiration drift-curves for attached and detached barley leaves. Measurements of carbon dioxide production were made on replicate samples of ten leaves kept in the dark at $19 \pm 1°C$ in a continuous air stream. The variation between duplicate measurements on samples of attached or detached leaves is indicated by vertical lines. After Yemm (unpublished data, 1938).

continuously in the dark. A rapid depletion of carbohydrate substrates takes place in the starved leaves; at the same time products of protein breakdown accumulate, whereas under normal conditions they could be readily translocated to other parts of the plant. Both these circumstances may lead to a more extensive breakdown of amino acids in respiratory metabolism than would normally take place. Experiments with barley leaves, darkened while still attached to the growing plant, strongly suggest that under these conditions the rate of respiration is greatly reduced, particularly during the phase of yellowing. The results of a typical experiment are shown in Fig. 16. Compared with the detached leaves, those remaining on the plant show a much more prolonged and deep decline in the rate of respiration. This is associated with an extensive loss of protein and other nitrogenous constituents, so that over 70% of the total leaf nitrogen is transferred to other parts of the plant over the period of the experiment. No appreciable accumulation of amino acids and amides could be detected in the attached leaves. Other experiments with leaves when they become yellow and senescent under normal conditions on the growing plant have shown that a rapid breakdown of proteins of the leaves takes place and much of it is transferred to the developing grain (Yemm, 116). The data now available for detached barley leaves indicate that the high rate of respiration observed during Phase 3 of the respiratory drift is dependent upon the accumulation of catabolic products in the leaves, in all probability amino acids formed by the breakdown of chloroplastic and other cytoplasmic proteins. Under the exceptional conditions obtaining in highly starved tissues these products are readily drawn into respiratory metabolism, giving rise to secondary products, such as asparagine and ammonia. It remains to be established whether a high and rising respiratory rate normally accompanies senescence of the leaves on the growing plant. As already noted in Section IV, A, few critical studies have been made of the respiration of attached leaves. The data of Arney (1) on leaves of strawberry have shown that in some cases an appreciable increase in the rate of respiration may occur during yellowing. An example of such a record is included in Fig. 9. But a respiratory rise is by no means always found in old leaves, and its superficial resemblance to that in highly starved detached leaves requires further examination.

The breakdown of protein appears to play an important part in the metabolism of aging leaves, but relatively little is known of the way in which the rate of protein catabolism is controlled in plant tissues. In detached leaves protein breakdown occurs readily both in the dark and in the light. It does not appear to be drastically affected by carbohydrate concentration although the products that accumulate in the tissues depend upon the presence of readily available carbohydrates. Chibnall (22), from experiments on the formation of roots by detached leaves, suggested that hormone effects may be important, and the striking effect of kinetin on the metabolism and yellowing of leaves is a further pointer in this direction (Mothes *et al.*, 71). The possibility of controlling the breakdown of protein in detached leaves by means of hormones may permit a more thorough investigation of the relationship between protein metabolism and respiration in leaves.

It is very probable that the synthesis of the amides, glutamine and asparagine, forms another important link between respiration and the metabolism of carbohydrates and proteins in leaves. The organic acid precursors of the amides, α-ketoglutarate and oxaloacetate, may be readily formed as intermediaries in carbohydrate breakdown, and the biosynthesis of amide bonds may be coupled to oxidative phosphorylation by means of ATP. As barley leaves age on the plant, there is a marked increase in the rate at which glutamine accumulates in the tissues and there is evidence that its formation is closely dependent upon carbohydrate breakdown and respiration (Yemm, 116). The active synthesis of glutamine in aging leaves appears to be part of the normal mechanism whereby tissue proteins are broken down and translocated from the leaves to other parts of the developing plant. Asparagine, on the other hand, is formed in large amounts only under the somewhat exceptional conditions of rapid protein catabolism in highly starved detached leaves. The close relationship between these changes and respiratory metabolism is clearly indicated by the fact that under anaerobic conditions both the breakdown of protein and the synthesis of the amides in the leaves is strongly inhibited.

The suggestion has often been made that the proteins of leaves and other plant tissues are maintained in a continuous dynamic equilibrium, so that a gain or loss of protein results from a change in the balance between synthesis and breakdown. Some valuable support for this view has been gained by the use of isotopic nitrogen (N^{15}) as a metabolic tracer. The methods first developed by Hevesy and others (44) and by Vickery and others (107) have been applied to detached leaves of runner bean (*Phaseolus coccineus* [*P. multiflorus*]) by Chibnall and Wiltshire (23) and to barley leaves by Yemm (117). These experiments have shown that even under conditions in which a net loss of protein occurs in detached and yellowing leaves there is an appreciable incorporation of the isotope into the tissue proteins when N^{15} is supplied as a labeled ammonium salt. These and other data have often been regarded as support for the hypothesis of a continuous protein cycle in leaves in which some of the respiratory carbon dioxide arises from the oxidation of the carbon skeletons of amino acids. There is, however, no decisive evidence that the incorporation of N^{15} into proteins necessarily involves their extensive breakdown and resynthesis;[2] more limited transfer reactions such as transamidation and trans-

---

[2] It was the incorporation of substantial amounts of N^{15} into protein without an equivalent total gain that required, or suggested, the ideas of breakdown and resynthesis. The work of Bidwell *et al.* (see Chapter 4) seems to require that the carbon of the protein of tobacco leaves is turned over at rates that range from 0.2 to 2% per hour. (Ed.)

peptidation could readily account for the labeling of proteins under these conditions. As already indicated, the intensive breakdown of amino acids in the respiration of detached leaves may be largely due to the special conditions of acute starvation together with the accumulation of the products of protein catabolism in the tissues. A gradual decline of biosynthetic activity often associated with high rates of respiration in aging plant tissues may well result from the breakdown of cellular organization and the failure of the regulatory mechanisms dependent upon oxidative phosphorylation.

V. The Respiration of Roots

Two main objectives have prompted the investigation of respiration in roots and in root tissues. In the first place, attention has been focused chiefly on the relationship between respiration and salt uptake; Steward and Sutcliffe (95) have fully reviewed the extensive data on this subject in Volume II of the treatise, so that only certain aspects of assimilation in roots call for further discussion here. A second objective of studies of respiration in roots has been mainly concerned with metabolic changes underlying cell division and differentiation in the growing point. For this purpose the somewhat simpler organization of the growing point of the root than that of the shoot has several advantages. The subterminal meristem and successive zones of expanding and differentiating cells can be readily separated by cutting short segments from the growing tip. Useful information has been gained in this way concerning the respiratory and other metabolic changes that accompany expansion and differentiation of the cells.

A. Respiration and Root Development

It was shown by Machlis (65) that a well-marked gradient of respiration rate occurred along the unbranched roots of barley. With fairly long (10 mm) segments, he found that the rate of oxygen uptake was several times higher in the apical segment than in the older segments taken farther back along the root. A fairly steady rate of respiration was observed in mature segments at distances greater than about 5 cm behind the growing point. Several other investigations have provided a more detailed analysis of the respiratory gradient along the root by examining shorter segments. The experiments of Berry and Norris (7, 8) with onion (*Allium cepa*) root tips are of particular interest, since they studied by manometric methods the influence of oxygen concentration (5–100%) and temperature (15–35°C) on the rate of respiration of three segments taken at 0–5 mm, 5–10 mm, and 10–15 mm from the tip. The highest rates of oxygen uptake on a dry

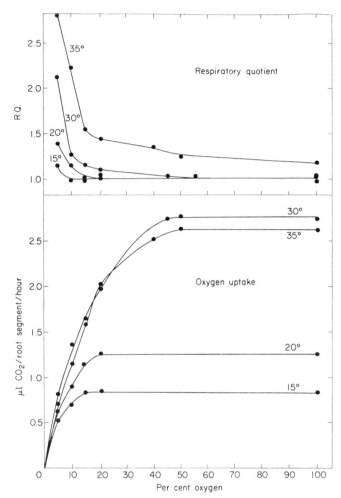

Fɪɢ. 17. The influence of temperature and oxygen concentration on the respiration of onion root tips. The apical segment (0–5 mm) was cut from 2-day-old roots for manometric measurements of oxygen uptake and carbon dioxide production. From Berry and Norris (7).

weight basis were found in the apical segment. In this segment there was strong evidence that at higher temperatures, 30–35°C, the rate of diffusion of oxygen to the tissues limited the rate of respiration in air. As Fig. 17 shows, the highest rate at 35°C was not attained till 50% oxygen was supplied and at lower values high respiratory quotients indicated that anaerobic metabolism occurred. On the other hand, the roots were oxygen saturated at about 20% oxygen at temperatures of

15–20°C.[3] From the data obtained at different temperatures, calculation of the apparent energy of activation from the Arrhenius equation gave a value of 4300 calories, agreeing closely with that for diffusion-limited reactions (Moelwyn-Hughes, 70). Other evidence reviewed by Betz (11) suggests that anaerobic metabolism commonly occurs in plant meristems; high values of the respiratory quotient and the production of ethanol in the meristematic tissues in air have been reported for both root and shoot systems. Respiratory metabolism in the compact root meristem devoid of intercellular spaces appears to be particularly sensitive to limitation by oxygen tension, and this may well occur under conditions of reduced oxygen supply in the soil. The occurrence of a quiescent region in the central part of the root meristem (Clowes, 24), where the cells do not divide and have a relatively low level of biosynthetic activity, may result from the limited supplies of oxygen reaching this part.

Some of the detailed investigations of the growing point of roots have included estimates of the number of cells in the different zones, usually made by means of cell counts on macerated segments of tissue. In this way it has been possible to calculate respiratory rates per cell and to compare these rates with metabolic changes that occur in the cells as they expand and differentiate. An example of respiratory data for roots of *Zea mays* treated in this way has been given by Goddard and Bonner (38) in Volume IA, page 215 of this treatise. Roots of seedlings of *Pisum sativum* were also investigated by Brown and Broadbent (19) and by Betz (9, 10), and Jensen (53) studied in detail the changes of protein and nucleic acids in cells of the meristematic and elongating zones of roots of *Vicia faba* and *Allium cepa*. Some of the data available are illustrated in Fig. 18 (see also Fig. 27 in Chapter 4 of this volume).

Although there are some differences between the roots of different species, the following salient features can generally be recognized. The small undifferentiated cells of the meristem, in which mitosis is most

---

[3] The form of the pO_2/respiration curve is frequently similar to that shown in Fig. 17. However, examples occur in which oxygen saturation is achieved at concentrations as low as water in equilibrium with a few per cent of oxygen in the gas phase (e.g., certain excised roots, including barley, potato, etc.), as high as 20% of oxygen for many tissue slices, and examples are known in which the oxygen concentration to achieve saturation of the system may need to approach 100% (*Azotobacter*); moreover it is a function of the biological activity of the material as, for example, in dormant versus nondormant buds. Therefore, the determination of respiratory intensity is not merely a simple function of the diffusivity of oxygen into complex aggregates of cells, but it also reflects certain innate qualities which determine their response to oxygen supply. (Ed.)

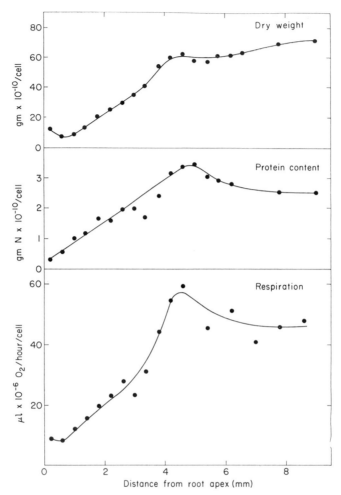

FIG. 18. Cellular respiration and development in the growing point of pea roots. Sections (0.4 mm) were cut from apices of young roots for manometric estimation of oxygen uptake and for estimation of protein-nitrogen and dry weight. Cell counts were made on the different segments after maceration in 5% chromic acid. All data are expressed per average cell. From Brown and Broadbent (19).

active, have a relatively *low* rate of respiration on a cell basis, but as the cells expand by vacuolation in the elongating zone, there is a marked rise in the rate per cell, often to a value 5–10 times greater than that of the meristematic cells. At the same time an increase of proteins and of nucleic acids, both RNA and DNA, occurs, and there is little doubt that an extensive synthesis of specific proteins takes place at this stage, associated with marked changes in the enzymatic activities in the cells. Beyond the region of elongation, when most of the cells have reached their maximum size, respiration is maintained at a high and fairly constant level as differentiation of cortical and stelar tissues continues. But as the cells mature, a gradual decline in respiration is commonly found with a fall in their protein content.

The data for developing roots provide a further example of the close relationship between anabolic activity and respiratory metabolism in plants. In general, the biosynthesis of proteins runs parallel to the increase of respiration and may be an important component in the regulatory mechanisms, such as those already discussed. However, a more detailed analysis of the relation between protein content and respiratory rate indicates discrepancies between the data even for the same species. With roots of *Pisum sativum* Brown and Broadbent (19) found that the rate of respiration per unit of protein is at a maximum in the meristematic zone of the root tip. With similar material Betz (9), on the other hand, recorded much higher rates on a protein basis in the differentiating and expanding zones than in the growing point. It seems likely that differences in the conditions during measurement of the respiration of the segments are chiefly responsible for the discrepancy. Brown and Broadbent used much narrower segments (0.4 mm), and damage to the tissues in cutting may be important. In addition, the oxygen supplies especially to the meristematic cells may be greatly increased in thin slices, and in this regard it is significant that the rates of respiration recorded by Brown and Broadbent are much higher than those of Betz for the meristematic zone.

Prominent changes in the ultrastructure of the differentiating cells of roots have been detected by electron microscopy. An increase in the size and complexity of mitochondria is of particular interest with regard to the rise in respiratory rate of expanding cells. Hanson and his collaborators (42, 60) have shown that an extension of the mitochondrial membranes associated with the development of cristae occurs in differentiating cells although some difficulties have arisen in distinguishing between mitochondria and proplastids of the meristematic cells. Whaley, Mollenhauer, and Kephart (111) have also described the changes in ultrastructure that accompany cell development in roots and have attempted to relate these changes to the rise in respiratory activity. It seems probable that the abundance of mitochondria increases relative to other cytoplasmic structures, and this may be an important feature contributory to the increase of respiratory activity. But improved facilities for gaseous exchanges, as a result of the expansion of the cells and the formation of intercellular spaces, may also be involved.

Autoradiographic methods have now been applied to the study of the metabolism of proteins and nucleic acids in the growing point of the root; Clowes (24) has recently reviewed the data of his own and other experiments. The incorporation of tritiated amino acids, such as leucine and phenylalanine, into proteins, and of tritiated nucleotides into nucleic acids is fully consistent with an active formation of these constituents in the cells during their expansion and differentiation. The interdependence of respiratory metabolism and biosynthetic activity in the developing tissues of the root is again indicated.

B. RESPIRATION AND SALT UPTAKE IN ROOTS

It is well established that the uptake and accumulation of salts in roots and in other plant tissues is dependent upon active metabolism. The pioneer work of Steward (91) with disks of plant tissue and of Hoagland (45) with excised roots demonstrated the important role of cell respiration in the mechanisms whereby relatively high concentrations of salts are often maintained in plant cells when they are supplied with dilute salt solutions. A critical review of our present knowledge in this field and its experimental basis has been given by Steward and Sutcliffe (95) and need not be considered in detail here. In general, a close relationship between respiration and salt uptake has been indicated in several ways. Oxygen concentration has often been shown to affect both respiration and salt uptake in a similar way; many respiratory inhibitors also inhibit salt uptake; a marked increase of respiration is sometimes associated with salt uptake.

With regard to the cellular mechanisms which couple salt uptake and respiration, there is as yet no decisive evidence. Indeed the relative importance of the outer protoplasmic membrane and of the tonoplast of the vacuole is still uncertain. However, most of the hypotheses that have been developed in order to account for the movement of ions across the cell membranes are based on the concept of a carrier system in which ions are combined in an organic complex and thus traverse the membrane. For example, Lundegårdh (61) suggested that such a mechanism may be directly coupled to the electron transport system involving the cytochromes. This theory, which suggests that the outward flux of electrons promotes the accumulation of anions, has been progressively developed and elaborated (see Lundegårdh, 62). Apart from the evidence which now indicates that the cytochrome system is localized in the mitochondria of plant cells rather than in the outer cell membrane, it is difficult to reconcile Lundegårdh's theory with the action of inhibitors which affect differentially the electron transport system and salt uptake. Robertson and others (83) have shown that 2,4-dinitrophenol may increase oxygen uptake and, therefore, presumably electron transport by the cytochrome system, but at the same time strongly inhibits salt uptake. Other properties of a carrier system have been emphasized by Epstein and his collaborators (29, 30), having regard to the selective uptake of cations. They suggest that competition for specific sites is an important part of the mechanism of uptake of cations; some of the kinetic features of competitive enzyme-substrate reactions can be recognized, comparable to the so-called permease systems which have been considered to play an important part in the transport of nutrients across the lipoprotein membranes of the cells of microorganisms (see for example, Mitchell, 69). At least two somewhat distinct processes appear to be involved in the uptake of cations by plant cells; first a relatively unspecific and reversible ion exchange, probably associated with the cell wall, followed by a selective and irreversible uptake dependent upon metabolic processes.

A broader hypothesis of the relationship between salt uptake and respiration has been developed by Steward and his collaborators (92, 95). Particular emphasis is placed upon the dependence of salt uptake on the capacity of the cells for growth and protein synthesis; binding of ions on specific sites formed in the growing cells and their accumulation during vacuolation and expansion are regarded as salient features of salt uptake. From this standpoint, salt accumulation forms an integral part of the assimilatory and biosynthetic processes dependent on cell respiration in developing cells. It is very probable that when essential nutrients are assimilated their effects upon the anabolic reactions of the cells are of major importance in regulating respiratory metabolism, rather than the operation of specific carrier mechanisms. All the available evidence indicates that these latter represent only a relatively small part of the total energy exchanges of respiration during salt accumulation in plant cells.

Experiments with excised roots have often shown that salt uptake is most active in young developing tissues close to the growing point. Prevot and Steward (77) found with young barley roots that the greatest accumulation of bromide occurred within 1 cm of the root apex. More detailed studies of the growing point of pea roots were made by Brown and Cartwright (20), who showed that a maximum accumulation of potassium took place in the zone of rapid cell expansion and differentiation just behind the meristem. Somewhat similar observations were summarized by Steward and Millar (92) with young roots of *Narcissus pseudo-narcissus*, by means of radioactive cesium (Cs^{137}), as indicated in Volume II, page 361. As already discussed, it is in this zone of rapid cell expansion that high rates of cell respiration appear to be closely coupled with an active synthesis of proteins, nucleic acids, and other cell constituents. The localization of rapid salt uptake in this part of the root gives support to the view that salt accumulation is intimately related to growth in plant cells. It seems probable that the active accumulation of salts in the growing zone of the root and their subsequent release as the cells become differentiated and mature, may be an important part of the normal mechanisms whereby the dilute soil solution is progressively exploited as a source of mineral nutrients by the root systems of growing plants.

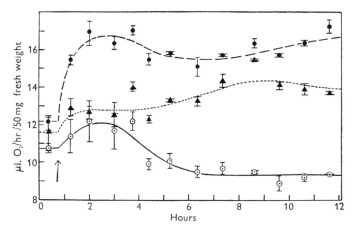

Fig. 19. Respiration of barley roots during the assimilation of nitrate and ammonium salts. Manometric measurements were made on roots of young seedlings in air at 23°C. The time of treatment with different solutions is shown by arrows. — ⊙ — Culture solution only; ··· ▲ ··· culture solution + sodium nitrate (6 mM); ─ ─ ● ─ ─ culture solution + ammonium chloride (6 mM). The difference between duplicate samples is indicated by vertical lines. After Willis and Yemm (112).

High rates of respiration in roots during the uptake of metabolically active nutrients such as nitrates, phosphate, sulfate, and ammonium ions may be closely dependent upon the anabolic reactions by which the nutrients are assimilated by the cells. Hoagland (46) showed that ammonium salts greatly increased the respiration of roots after they had been grown for a short period under conditions of a limited supply of nitrogen. (Other similar effects of nitrate and phosphate on the respiration and metabolism of tissue slices are described in Chapter 4, Volume II.) The respiratory and metabolic changes in excised barley roots during the assimilation of nitrates, nitrites, and ammonium salts have been reinvestigated by Willis and Yemm (112,

121). A marked increase in the rate of respiration accompanies the assimilation of nitrogen, whether it is supplied as an anion or a cation as shown in Fig. 19. This effect of nitrate and ammonium on respiration has frequently been observed in the past (cf. Chapter 4, Volume II). With both nitrate and nitrite, there is a sharp rise in the value of the respiratory quotient (Fig. 20) which may be attributed to a diversion of electrons from the normal carrier system, so that the reduction of nitrate or nitrite in part replaces that of oxygen as a final electron accepter. No appreciable changes of organic acids could be detected

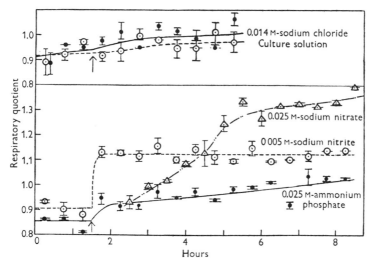

FIG. 20. The respiration quotient of barley roots during the assimilation of nitrate, nitrite, and ammonia. Manometric data for young roots in air at 25°C. The upper part of the figure shows controls, treated with culture solution or sodium chloride. The time of addition of nutrients is shown by arrows. The variation between measurements on replicates is indicated. After Willis and Yemm (112).

in the root tissues under these conditions, indicating that carbon dioxide fixation or the breakdown of organic acids are unlikely to play an important part in the respiratory changes. It is noteworthy that in comparable roots the occurrence of salt respiration induced by ions such as sodium and chloride could not be detected unless the roots were grown under conditions of acute mineral deficiency in solutions containing only calcium sulfate.

When barley roots assimilate nitrogen, an extensive synthesis of glutamine takes place especially when nitrite or ammonium salts are supplied. In some experiments nearly 80% of the nitrogen taken up by the roots could be recovered as glutamine under these conditions.

Experiments in which isotopically labeled ammonium salts were used have shown that a primary synthesis of the amino and amide groups of glutamine occurred, as indicated by the rapid incorporation of N^{15}. After a short period of assimilation of labeled ammonia, the N^{15} abundance, particularly in the amide group of glutamine, closely approached that of the ammonia supplied to the roots, as shown in Fig.

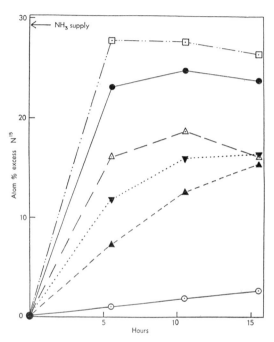

Fig. 21. The assimilation of N^{15}-labeled ammonium phosphate by barley roots. Excised roots of young seedlings were supplied with ammonium phosphate containing 29.3 atom per cent excess N^{15}. The isotopic abundance in soluble and insoluble constituents of the roots is shown as follows: — • •— □ — • •— ammonia; —— ● —— glutamine amide N; — — ▲ — — asparagine amide N; · · · · ▼ · · · · total α-amino N; — — △ — — total soluble N; —— ⊙ —— insoluble N. From Yemm and Willis (121).

21. A rapid loss of carbohydrate, chiefly sucrose, accompanies the synthesis of glutamine and the high rate of respiration; this loss is fully adequate to account for both the production of carbon dioxide and the carbon skeletons of the amides formed in the tissues.

    The cellular mechanisms that give rise to high rates of carbohydrate and respiratory metabolism during the assimilation of inorganic forms of nitrogen have been fully discussed by Yemm and Willis (121).

It seems clear that the enzymatic system which brings about the synthesis of glutamic acid and its amide are closely coupled with respiratory metabolism in several ways. In all probability glutamic acid dehydrogenase operates in the formation of glutamate from α-keto-glutarate and ammonia by reductive amination mediated by pyridine nucleotides, while the synthesis of the amide bond is brought about by glutamine synthetase and depends upon phosphorylation involving ATP. Thus the oxidation-reduction and phosphorylation mechanisms involved in the synthesis of glutamine as well as the supply of its carbon skeleton may be closely linked with the breakdown of carbohydrates and with cell respiration. All the evidence is consistent with the view that carbohydrate and respiratory metabolism in the root tissues are regulated mainly by their close coupling with biosynthetic activity dependent upon phosphorylation, oxidation-reduction, and the diversion of respiratory intermediates. Some further indications that the use of phosphorylated compounds in the assimilation of nitrogen in barley roots may regulate respiration has been gained from a study of the effects of 2,4-dinitrophenol. At low concentrations (10^{-4} to $10^{-5}M$) it is found that dinitrophenol greatly increases the rate of respiration of the roots, sometimes by a factor of 2. This increase is much smaller in presence of ammonia, but the synthesis of glutamine is strongly inhibited at concentrations at which a high rate of respiration is maintained, as shown in Fig. 22. These data strongly suggest that the rate of turnover of ATP is one of the important control points in the respiratory metabolism of the roots which can be released either by uncoupling with dinitrophenol or by the active assimilation of nitrogen.

On the other hand, there are no indications that a continuous protein cycle operates in the direct control of respiratory metabolism in barley roots. Relatively little net synthesis of protein occurs in excised roots when they are supplied with nitrate, nitrite, or ammonium salts, although an appreciable incorporation of nitrogen into proteins can be detected when isotopically labeled ammonia is supplied (see Fig. 21). The rate of incorporation of N^{15} into the proteins continues at a fairly constant rate for a period of 15–24 hours and shows no relationship with the rapid fluctuations of respiration during this time. Similarly no evidence of an active protein cycle could be obtained in a more thorough examination of the incorporation of N^{15} and the synthesis of proteins in roots of intact barley seedlings by Cocking and Yemm (25).

Another way in which respiratory metabolism of cells may be linked with salt uptake emerged from the conditions which promote an unbalanced absorption of anions or cations (cf. Chapter 4, Volume II). In work along these lines, Ulrich (103) has shown that when barley

roots take up an excess of mobile cations, such as potassium from dilute solutions of potassium sulfate, there are marked changes in their
respiration. A fall in the respiratory quotient is accompanied by an
increase of organic acids in the tissues. Conversely, under circumstances
in which an excess of anions, such as chloride, accumulate, a sharp
rise in the R.Q. takes place together with a fall in the organic acid

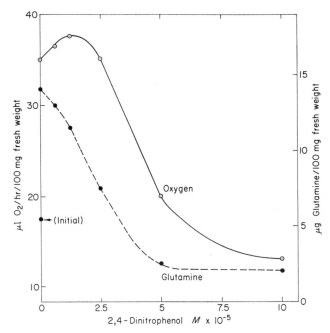

FIG. 22. The effect of 2,4-dinitrophenol on respiration and the synthesis of
glutamine in barley roots. Excised roots were treated with DNP in presence of
ammonium phosphate (0.0025 M), and oxygen uptake was measured manometrically
for a period of 3 hours. Estimation of the glutamine content of the roots were made
at the beginning and at the end of the experiment. From Willis and Yemm (unpublished data, 1957).

content of the roots. Organic acid metabolism may therefore assist in
preserving neutrality during an absorption of an excess of anions or
cations. Subsequent work, in which C^{14}-labeled carbon dioxide was
used by Poel (76) and by Jacobson (49) to measure carbon dioxide
fixation, has demonstrated that the fixation is greatly affected by the
ionic composition of the solution in which the roots are maintained.
When an excess of cation absorption occurs there is a substantial increase in carbon dioxide fixation. Fluctuations of malic acid largely

account for the changes of organic acid, and most of the label was re-covered in the carboxyl groups of the acid.

These data strongly suggest that organic acid metabolism and the fixation of carbon dioxide, or bicarbonate ions, may represent an important link between respiration and salt uptake by roots under some conditions.

VI. Conclusion

In this account of plant respiration attention has been focused particularly on the changes in respiratory metabolism that accompany growth and development in plants. In general, it can be recognized that high rates of respiration are found in young growing regions where anabolic processes are closely linked with the formation of proteins and other cell constituents. There is now much direct and indirect evidence which indicates that endogenous regulatory mechanisms play an important part in the control of respiratory metabolism. High rates of respiration in young growing organs and tissues are in all probability maintained by the extensive diversion of energy and metabolites into biosynthetic reactions by way of oxidation-reduction and phosphorylation. From this standpoint, oxidative assimilation and anabolism may be regarded as closely coupled with, and as important parts of, respiratory metabolism.

The relatively low rates of respiration generally found in mature, fully grown plant organs appear to represent a residual or limited part of the full respiratory potential of the growing cells. It has sometimes been suggested that this residual respiration is of little metabolic significance, but is an inevitable consequence of enzymatic reactions proceeding with a net loss of free energy. Indeed, much of the energy may be ultimately transformed to heat and, under normal circumstances, is rapidly lost by radiation. On the other hand, it is evident that living cells are maintained out of equilibrium with their environment; within the cell, metabolites and multienzyme systems are retained in a highly organized and heterogeneous structure upon which the orderly processes of metabolism depend. Much evidence, gained by means of isotopically labeled metabolites, now indicates that proteins and other structural components of nongrowing cells are held in continuous dynamic equilibria and that appreciable biosynthetic activities persist in mature plant organs. It seems clear therefore, that respiration must make a significant contribution to the energy required for maintaining the organized structure of living cells, although the extent of this contribution is uncertain.

Typically, the growth of higher plants depends upon aerobic

metabolism, although many plants and their organs survive in absence of oxygen. Anaerobic breakdown of carbohydrate, despite its energetic inefficiency, appears to play a significant part in the metabolism of some germinating seeds and possibly in other compact structures with a limited access to oxygen. However, it is clear that respiration with its close coupling to oxidative assimilation is much more effective than anaerobic metabolism in promoting anabolic reactions in higher plants, as in yeast and other microorganisms. Much of the uncertainty with regard to the Pasteur effect in plants has arisen in studies of mature, nongrowing organs and tissues. The broader concepts of respiratory regulation and oxidative assimilation, rather than those of oxidative anabolism and resynthesis of carbohydrate, seem to provide more useful guidance for further studies of the relationship between anaerobic metabolism and respiration in plants.

Our present knowledge of respiratory metabolism in plants is in many respects incomplete both at cellular and at higher levels of organization. Many of the catabolic pathways have been effectively explored by modern biochemical and biophysical methods, but relatively little can yet be said in quantitative terms of their over-all contribution to the respiratory processes. A deeper understanding of the significance of respiration in the metabolism and energy economy of plants will require quantitative information, not only of the catabolic mechanisms, but also of the anabolic systems with which they may be coupled. The unique and versatile biosynthetic mechanisms of autotrophic plants depend primarily upon their respiratory and photosynthetic metabolism. Much common ground between these metabolic systems has been revealed in the oxidation-reduction, phosphorylation, and substrate reactions of plant cells. A better understanding of the relationship between respiratory and photosynthetic processes is, therefore, an important objective in the study of plant metabolism.

REFERENCES

1. Arney, S. E. The respiration of strawberry leaves attached to the plant. *New Phytologist* **46**, 68–96 (1947).
2. Audus, L. J. Mechanical stimulation and respiration rate in the cherry laurel. *New Phytologist* **34**, 386–402 (1935).
3. Audus, L. J. Mechanical stimulation and respiration in the green leaf. III. The effect of stimulation on the rate of fermentation. *New Phytologist* **39**, 65–74 (1940).
4. Ballard, L. A. T. Studies of dormancy in the seeds of subterranean clover (*Trifolium subterraneum* L.). I. Breaking of dormancy by carbon dioxide and by activated charcoal. *Australian J. Biol. Sci.* **11**, 246–260 (1958).
5. Barnell, H. R. Analytic studies in plant respiration. VII. Aerobic respiration

in barley seedlings and its relation to growth and carbohydrate supply. *Proc. Roy. Soc. (London)* **B123**, 312–342 (1937).

6. Beevers, H. "Respiratory Metabolism in Plants." Row, Peterson, Evanston, Illinois, 1961.

7. Berry, L. J., and Norris, W. E. Studies of onion root respiration. I. Velocity of oxygen consumption in different segments of roots at different temperatures as a function of partial pressure of oxygen. *Biochim. Biophys. Acta* **3**, 593–606 (1949).

8. Berry, L. J., and Norris, W. E. Studies of onion root respiration. II. The effect of temperature on the apparent diffusion coefficient in different segments of the root tip. *Biochim. Biophys. Acta* **3**, 607–614 (1949).

9. Betz, A. Zur Atmung wachsender Wurzelspitzen. *Planta* **46**, 381–402 (1955).

10. Betz, A. Zur Atmung wachsender Wurzelspitzen. III. Das Verhalten in Stickstoff und hochprozentiger Sauerstoffatmosphäre; die Pasteursche Reaktion. *Planta* **50**, 122–143 (1957).

11. Betz, A. Die aerobe Gärung in aktiven Meristemen höherer Pflanzen. *In* "Handbuch der Pflanzenphysiologie" (W. Ruhland, ed.), Vol. XII, Part 2. Springer, Berlin, 1960.

12. Blackman, F. F. Analytic studies in plant respiration. III. Formulation of a catalytic system for the respiration of apples and its relation to oxygen. *Proc. Roy. Soc. (London)* **B103**, 491–523 (1928).

13. Blackman, F. F. "Analytical Studies in Plant Respiration." Cambridge Univ. Press, London and New York, 1954.

14. Bloom, B., and Stetten, D. Pathways of glucose catabolism. *J. Am. Chem. Soc.* **75**, 5446 (1953).

15. Bradbeer, J. W., Ranson, S. L., and Stiller, M. L. Malate synthesis in crassulacean leaves. I. The distribution of C^{14} in malate in leaves exposed to $C^{14}O_2$ in the dark. *Plant Physiol.* **33**, 66–70 (1958).

16. Brown, A. H., and Weis, D. S. Relation between respiration and photosynthesis in the green alga, *Ankistrodesmus braunii*. *Plant Physiol.* **34**, 224–234 (1959).

17. Brown, R. An experimental study of the permeability to gases of the seed coat membranes of *Cucurbita pepo*. *Ann. Botany (London)* [N.S.] **4**, 379–395 (1940).

18. Brown, R. Studies in germination and seedling growth. I. The water content, gaseous exchange and dry weight of attached and isolated embryos of barley. *Ann. Botany (London)* [N.S.] **7**, 93–113 (1943).

19. Brown, R., and Broadbent, D. The development of cells in the growing zones of the root. *J. Exptl. Botany* **1**, 249–263 (1950).

20. Brown, R., and Cartwright, P. M. The absorption of potassium by cells in the apex of the root. *J. Exptl. Botany* **4**, 197–221 (1953).

21. Chibnall, A. C. "Protein Metabolism in the Plant." Yale Univ. Press, New Haven, Connecticut, 1939.

22. Chibnall, A. C. Protein metabolism in rooted runner-bean leaves. *New Phytologist* **53**, 31–37 (1954).

23. Chibnall, A. C., and Wiltshire, G. H. A study with isotopic nitrogen of protein metabolism in detached runner-bean leaves. *New Phytologist* **53**, 38–43 (1954).

24. Clowes, F. A. L. "Apical Meristems." Blackwell, Oxford, 1961.

25. Cocking, E. C., and Yemm, E. W. Synthesis of amino acids and proteins in barley seedlings. *New Phytologist* **60**, 103–116 (1960).

26. Crozier, W. J. On biological oxidations as a function of temperature. *J. Gen. Physiol.* **7**, 189–216 (1924).

27. Daly, J. M., and Brown, A. H. The *in vivo* demonstration of cytochrome oxidase in leaves of higher plants. *Arch. Biochem. Biophys.* **52**, 380–387 (1954).

27a. Davies, D. D. "Intermediary Metabolism in Plants." Cambridge Univ. Press, London and New York, 1961.

28. Ducet, G., and Rosenberg, A. J. Leaf respiration. *Ann. Rev. Plant Physiol.* **13**, 171–200 (1962).

29. Epstein, E., and Hagen, C. E. A kinetic study of the absorption of alkali cations by barley roots. *Plant Physiol.* **27**, 457–474 (1952).

30. Epstein, E., and Leggett, J. E. The absorption of alkaline earth cations by barley roots—kinetics and mechanism. *Am. J. Botany.* **41**, 785–791 (1954).

31. Fernandes, D. S. Aerobe und anaerobe Atmung bei Keimlingen von *Pisum sativum. Rec. trav. botan. neerl.* **20**, 107–256 (1923).

32. Folkes, B. F., Willis, A. J., and Yemm, E. W. The respiration of barley plants. VII. The metabolism of nitrogen and respiration of seedlings. *New Phytologist* **51**, 317–341 (1952).

33. Folkes, B. F., and Yemm, E. W. The amino acid content of the proteins of barley grains. *Biochem. J.* **62**, 4–11 (1956).

34. Folkes, B. F., and Yemm, E. W. The respiration of barley plants. X. Respiration and the metabolism of amino-acids and proteins in germinating grain. *New Phytologist* **57**, 106–131 (1958).

35. Forward, D. F. The respiration of barley seedlings in relation to oxygen supply. I, II. *New Phytologist* **50**, 297–356 (1951).

36. Gibbs, M. Metabolism of carbon compounds. *Ann. Rev. Plant Physiol.* **10**, 329–378 (1959).

37. Gibbs, M., and Beevers, H. Glucose dissimilation in the higher plant. Effects of age of tissue. *Plant Physiol.* **30**, 343–347 (1955).

38. Goddard, D. R., and Bonner, W. D. Cellular respiration. *In* "Plant Physiology" (F. C. Steward, ed.), Vol. IA. Academic Press, New York, 1960.

39. Goksöyr, J., Boeri, E. and Bonnichsen, R. K. The variation of ADH and catalase activity during the germination of the green pea (*Pisum sativum*). *Acta Chem. Scand.* **7**, 657–662 (1953).

40. Gregory, F. G., and Sen, P. K. Physiological studies in plant nutrition. VI. The relation of respiration rate to the carbohydrate and nitrogen metabolism of the barley leaf as determined by nitrogen and potassium deficiency. *Ann. Botany (London)* [N.S.] **1**, 521–562 (1937).

41. Hackett, D. P. Respiratory mechanisms in higher plants. *Ann. Rev. Plant Physiol.* **10**, 113–146 (1959).

42. Hanson, J. B. Physiological and biochemical aspects of root development. *In* "Recent Advances in Botany: from lectures and symposia presented to the 9th Intern. Congr. Bot., Montreal 1959," pp. 795–800. University of Toronto Press, Toronto, 1961.

43. Hearon, J. Z. Rate behaviour of metabolic systems. *Physiol. Rev.* **32**, 499–523 (1952).

44. Hevesy, G., Linderstrøm-Lang, K., Keston, A. S., and Olsen, K. The exchange of nitrogen in the leaves of sunflower. *Compt. Rend. Trav. Lab. Carlsberg* **23**, 213 (1940).

45. Hoagland, D. R. Salt accumulation by plant cells, with special reference to metabolism and experiments on barley roots. *Cold Spring Harbor Symp. Quant. Biol.* **8,** 181–193 (1940).
46. Hoagland, D. R. "Lectures on the Inorganic Nutrition of Plants." Chronica Botanica, Waltham, Massachusetts, 1944.
47. Honda, S. Ascorbic acid oxidase in barley roots. *Plant Physiol.* **30,** 174–181 (1955).
48. Hover, J. M., and Gustafson, F. G. Rate of plant respiration as related to age. *J. Gen. Physiol.* **10,** 33 (1926).
49. Jacobson, L. Carbon dioxide fixation and ion absorption in barley roots. *Plant Physiol.* **30,** 264–269 (1955).
50. James, W. O. "Plant Respiration." Oxford Univ. Press (Clarendon), London and New York, 1953.
51. James, W. O., and James, A. L. The respiration of barley germinating in the dark. *New Phytologist* **39,** 145–176 (1940).
52. James, W. O., and Slater, W. G. The aerobic utilization of pyruvate in plant tissues. *Proc. Roy. Soc. (London)* **B150,** 192–198 (1959).
53. Jensen, W. A. The nuclei acid and protein content of root tip cells of *Vicia faba* and *Allium cepa. Exptl. Cell Res.* **14,** 575–583 (1958).
54. Kandler, O. Über den synthetischen Wirkungsgrad *in vitro* kultivierter Embryonen, Wurzeln and Sprosse. *Z. Naturforsch.* **8b,** 109–117 (1953).
55. Kidd, F. The controlling influence of carbon dioxide. III. The retarding effect of carbon dioxide on respiration. *Proc. Roy. Soc. (London)* **B89** 136–156 (1915).
56. Kidd, F., West, C., and Briggs, G. E. A quantitative analysis of the growth of *Helianthus annuus.* Part I. The respiration of the plant and of its parts throughout the life cycle. *Proc. Roy. Soc. (London)* **B92,** 368–384 (1921).
57. Kuijper, J. Über den Einfluss der Temperatur auf die Atmung der höheren Pflanzen. *Rec. trav. botan. neerl.* **7,** 131–240 (1910).
58. Leach, W. Researches on plant respiration. IV. The relation between the respiration in air and in nitrogen of certain seeds during germination. (b) Seeds in which carbohydrates constitute the chief food reserve. *Proc. Roy. Soc. (London)* **B119,** 507–521 (1936).
59. Loftfield, J. V. G. The behaviour of stomata. *Carnegie Inst. Wash. Publ.* **314** 1–104 (1921).
60. Lund, H. A., Vatter, A. E., and Hanson, J. B. Biochemical and cytological changes accompanying growth and differentiation in the roots of *Zea mays. J. Biophys. Biochem. Cytol.* **4,** 87–98 (1958).
61. Lundegårdh, H. Investigations as to the absorption and accumulation of inorganic ions. *Kgl. Lantbrucks-Hogskol. Ann.* **8,** 234–404 (1940).
62. Lundegårdh, H. Anion Respiration. *In* "Handbuch der Pflanzenphysiologie" (W. Ruhland, ed.), Vol. XII, Part 2. Springer, Berlin, 1960.
63. Lynen, F., Hartman, G., Netter, K. F., and Shuegraf, A. *Ciba Found. Symp., Regulation Cell Metab.* pp. 256–273. Little, Brown, Boston, Massachusetts, 1959.
64. MacDonald, I. R., and DeKock, P. C. The stimulation of leaf respiration by respiratory inhibitors. *Physiol. Plantarum* **11,** 464–477 (1958).
65. Machlis, L. The respiratory gradient in barley roots. *Am. J. Botany* **31,** 281–282 (1944).
66. Marsh, P. B., and Goddard, D. R. Respiration and fermentation in the carrot, *Daucus carota.* I. Respiration. *Am. J. Botany* **26,** 724–728 (1939).

67. Mazelis, M., and Vennesland, B. Carbon dioxide fixation into oxalacetate in higher plants. *Plant Physiol.* **32**, 591–600 (1957).

68. Menke, W. The structure and chemistry of plastids. *Ann. Rev. Plant Physiol.* **13**, 27–44 (1962).

69. Mitchell, P. Transport of phosphate across the osmotic barrier of Micrococcus pyrogenes—specificity and kinetics. *J. Gen. Microbiol.* **11**, 73–82 (1954).

70. Moelwyn-Hughes, E. A. "The Kinetics of Reactions in Solution," 2nd ed. Oxford Univ. Press (Clarendon), London and New York, 1947.

71. Mothes, K., Englebrecht, L., and Kulayevo, O. Über die Wirkung des Kinetins auf Stickstoffverteilung und Eiweisssynthese in isolierten Blättern. *Flora (Jena)* **147**, 445–464 (1959).

72. Murlin, J. R. The conversion of fat to carbohydrate in the germinating castor bean. I. The respiratory mechanism. *J. Gen. Physiol.* **17**, 283–302 (1933).

73. Neal, G. E., and Beevers, H. Pyruvate utilization in castor bean endosperm and other tissues. *Biochem. J.* **74**, 409–416 (1960).

74. Oota, Y., Fujii, R., and Sunobe, Y. Studies on the connection between sucrose formation and respiration in germinating bean cotyledons. *Physiol. Plantarum* **9**, 38–50 (1956).

75. Packer, L., Marchant, R. H., and Mukohata, Y. Coupling of energy-linked functions in mitochondria and chloroplasts to the control of membrane structure. *In* "Energy-Linked Functions of Mitochondria" (B. Chance, ed.), pp. 51–71. Academic Press, New York, 1963.

76. Poel, L. W. Carbon dioxide fixation by barley roots. *J. Exptl. Botany* **4**, 157–163 (1953).

77. Prevot, P., and Steward, F. C. Salient features of the root system relative to the problem of salt absorption. *Plant Physiol.* **11**, 509–534 (1936).

78. Ranson, S. L., and Thomas, M. Crassulacean acid metabolism. *Ann. Rev. Plant Physiol.* **11**, 81–110 (1960).

79. Ranson, S. L., Walker, D. A., and Clarke, L. D. The inhibition of succinic oxidase by high CO_2 concentrations. *Biochem. J.* **66**, 57P (1957).

80. Rhodes, M. J. C., and Yemm, E. W. Development of chloroplasts and the synthesis of proteins in leaves. *Nature* **200**, 1077–1080 (1963).

81. Richards, F. J. Physiological studies in plant nutrition. VIII. The relation of respiration rate to the carbohydrate and nitrogen metabolism of the barley leaf as determined by phosphorus and potassium supply. *Ann. Botany (London)* [N.S.] **2**, 491–534 (1938).

82. Roberts, D. W. A. Physiological and biochemical studies in plant metabolism. VI. The effect of ontogeny on the physiological heterogeneity in the first leaf of wheat. *Can. J. Botany* **30**, 558–570 (1952).

83. Robertson, R. N., Wilkins, M. J., and Weeks, D. C. Studies in the metabolism of plant cells. IX. The effects of 2,4-dinitrophenol on salt accumulation and salt respiration. *Australian J. Sci. Res.* **B4**, 248–264 (1951).

84. Rowan, K. S., Seaman, D. E., and Turner, J. S. Phosphorylation as a possible factor in the Pasteur effect in plants. *Nature* **177**, 333–334 (1956).

85. Ruhland, W., ed. "Handbuch der Pflanzenphysiologie," Vol. XII, Parts 1 and 2. Springer, Berlin, 1960.

86. Schrödinger, E. "What is Life?" Cambridge Univ. Press, London and New York, 1945.

87. Singh, B. N. The correlation between life duration and respiratory phenomena. *Proc. Indian Acad. Sci.* **2**, 387–402 (1935).

88. Somers, G, F Organic acids in barley leaves. B.Sc. thesis, Univ. of Oxford, 1939.
89. Spragg, S. P., and Yemm, E. W. Respiratory mechanisms and the changes of glutathione and ascorbic acid in germinating peas. *J. Exptl. Botany* **10**, 409–425 (1959).
90. Stålfelt, M. G. Die 'grosse Period' der Sauerstoffaufnahme. *Biol. Zentr.* **46**, 1–11 (1926).
91. Steward, F. C. The absorption and accumulation of solutes by living plant cells. I. Experimental conditions which determine salt absorption by storage tissue. *Protoplasma* **15**, 29–58 (1932).
92. Steward, F. C., and Millar, F. K. Salt accumulation in plants. A reconsideration of the role of growth and metabolism. *Symp. Soc. Exptl. Biol.* **8**, 367–406 (1954).
93. Steward, F. C., and Pollard, J. K. Nitrogen metabolism in plants: ten years in retrospect. *Ann. Rev. Plant. Physiol.* **8**, 65–114 (1957).
94. Steward, F. C., and Street, H. E. The nitrogenous constituents of plants. *Ann. Rev. Biochem.* **26**, 471–502 (1947).
95. Steward, F. C., and Sutcliffe, J. F. Plants in relation to inorganic salts. *In* "Plant Physiology" (F. C. Steward, ed.), Vol. 2, pp. 253–465. Academic Press, New York, 1959.
96. Stiles, W. Respiration in seed germination and seedling development. *In* "Handbuch der Pflanzenphysiologie" (W. Ruhland, ed.), Vol. XII, Part 2, pp. 465–492. Springer, Berlin, 1960.
97. Stiles, W., and Leach, W. Researches on plant respiration. II. Variations in the respiratory quotient during germination of seeds with different food reserves. *Proc. Roy. Soc. (London)* **B113**, 405–428 (1933).
98. Stiles, W., and Leach, W. "Respiration in Plants," 3rd ed. Methuen, London, 1952.
99. Stumpf, P. K. Glycolytic enzymes in higher plants. *Ann. Rev. Plant Physiol.* **3**, 17–34 (1952).
100. Taylor, D. L. Influence of oxygen tension on respiration, fermentation and growth in wheat and rice. *Am. J. Botany* **29**, 721–738 (1942).
101. Thomas, M. Carbon dioxide fixation and acid synthesis in crassulacean acid metabolism. *Symp. Soc. Exptl. Biol.* **5**, 72–93 (1951).
102. Turner, J. S. Fermentation in higher plants; its relation to respiration; the Pasteur effect. *In* "Handbuch der Pflanzenphysiologie" (W. Ruhland, ed.), Vol. XII, Part 2, pp. 42–87. Springer, Berlin, 1960.
103. Ulrich, A. Metabolism of non-volatile organic acids in excised barley roots as related to cation-anion balance during salt accumulation. *Am. J. Botany* **28**, 526–537 (1941).
104. Vennesland, B. Protein, enzymes, and the mechanism of enzyme action. *In* "Plant Physiology" (F. C. Steward, ed.), Vol. IA, pp. 131–205. Academic Press, New York, 1960.
105. Vickery, H. B. The effect of temperature on the behaviour of malic acid and starch in leaves of *Bryophyllum calycinum* cultured in darkness. *Plant Physiol.* **29**, 385–392 (1954).
106. Vickery, H. B., Pucher, G. W., Wakeman, A. J., and Leavenworth, C. S. Chemical investigations of the tobacco plant. VI. Chemical changes that occur in leaves during culture in light and in darkness. *Conn. Agr. Expt. Sta., New Haven, Bull.* **399** (1937).

107. Vickery, H. B., Pucher, G. W., Schoenheimer, R., and Rittenberg, O. The assimilation of ammonia nitrogen by the tobacco plant: a preliminary study with isotopic nitrogen. *J. Biol. Chem.* **135**, 531–539 (1940).

108. Vittorio, P. V., Krotkov, G., and Reed, G. B. Absorption and utilization of glucose-C^{14} by detached wheat leaves. *Can. J. Botany* **33**, 275–280 (1955).

109. Walker, D. A. Pyruvate carboxylation and plant metabolism. *Biol. Rev. Cambridge Phil. Soc.* **37**, 215–256 (1962).

110. Weigl, J. W., Warrington, P. M., and Calvin, M. The relation of photosynthesis to respiration. *J. Am. Chem. Soc.* **73**, 5058–5063 (1951).

111. Whaley, W. G., Mollenhauser, H. N., and Kephart, J. E. Ultrastructure, physiological and biochemical changes with the development of the cells of the maize root. *Proc. 9th Intern. Congr. Botan., 1959* Vol. 2, p. 429. Univ. of Toronto Press, Toronto, 1959.

112. Willis, A. J., and Yemm, E. W. The respiration of barley plants. VIII. Nitrogen assimilation and the respiration of the root system. *New Phytologist* **54**, 163–181 (1955).

113. Wolstenholme, G. E. W., and O'Connor, C. M., eds. *Ciba Found. Symp., Regulation Cell Metabolism.* Little, Brown, Boston, Massachusetts, 1929.

114. Yemm, E. W. Respiration of barley plants. II. Carbohydrate concentration and carbon dioxide production. *Proc. Roy. Soc. (London)* **B117**, 504–525 (1935).

115. Yemm, E. W. Respiration of barley plants. III. Protein catabolism in starving leaves. *Proc. Roy. Soc. (London)* **B123**, 243–273 (1937).

116. Yemm, E. W. Glutamine in the metabolism of barley plants. *New Phytologist* **48**, 315 (1949).

117. Yemm, E. W. The metabolism of senescent leaves. *Ciba Found. Colloq. Aging* **2**, 202–210 (1956).

118. Yemm, E. W., and Folkes, B. F. The regulation of respiration during the assimilation of nitrogen in *Torulopsis utilis. Biochem. J.* **57**, 495–508 (1954).

119. Yemm, E. W., and Folkes, B. F. The metabolism of amino acid and proteins in plants. *Ann. Rev. Plant Physiol.* **9**, 245–280 (1958).

120. Yemm, E. W., and Somers, G. F. Unpublished data (1939).

121. Yemm, E. W., and Willis, A. J. The respiration of barley plants. IX. The metabolism of roots during the assimilation of nitrogen. *New Phytologist* **55**, 229–252 (1956).

CHAPTER THREE

Part Two

The Respiration of Bulky Organs

DOROTHY F. FORWARD

I. Introduction 311
II. The Ontogenetic Pattern of Respiration in Fleshy Fruits 312
    A. The Course of Respiration 312
    B. The Role of Ethylene in the Climacteric 318
    C. Metabolic Mechanisms Associated with the Climacteric 320
III. The Respiration of Fleshy Roots and Tubers 324
IV. Respiration and the Internal Atmosphere of Bulky Organs 328
    A. Oxygen and Carbon Dioxide in the Internal Atmosphere of Fruits . 329
    B. The Relation between Respiration and the Internal Atmosphere . . 334
V. Oxygen Dependence 343
    A. Diffusion of Oxygen 344
    B. Terminal Oxidases 346
    C. The Pace-Setting Reaction 348
VI. Respiration and Fermentation 350
VII. The Relation of Respiration to Other Metabolism; the Role of Oxygen . 360
    A. Blackman's Schema 360
    B. Interpretation in Terms of Modern Knowledge 361
VIII. Regulation of Respiration Imposed by Organization 368
    References 369

I. Introduction

The organization of higher plants is sufficiently loosely integrated to permit certain organs to be detached from the plant and maintain a life of their own for a matter of days, weeks, or even months. Their life in isolation is different from that in communication with the rest of the plant, but in many organs the progress of change is slow and may be followed easily. This is true of most fleshy storage organs, which, on the whole, are durable and only slowly exhaust the potential with which they begin their separate existence. There has long been an interest in the mechanisms that control the relatively low rates of respiration characteristic of such organs.

The study of their respiration has been oriented in large measure by ideas put forth early in this century by Blackman (30). He introduced an experimental approach that has been widely exploited, and developed a philosophy of interpretation of rates of respiration that has had a far-reaching influence. From his early observations of rates of carbon dioxide production by apples (*Malus sylvestris*) in air, nitrogen, and other oxygen mixtures, Blackman derived a theory of the dependence of plant tissues on oxygen. This theory, although it has not been accepted in its entirety, provided a challenge that has had a profound effect on the character and direction of research on plant respiration, and it contained the germs of more modern theories of the control of metabolism *in vivo*. For example, he set out clearly the idea that in living organisms respiration, like all other metabolic processes, is dependent on organization of the protoplasm, which controls the access of substrate to catalyst, and which may change spontaneously as an organ or organism ages. He also recognized that full comprehension of the control of respiration *in vivo* cannot be derived from consideration of this process in isolation, but requires in addition a knowledge of its linkage to other metabolic processes.

No attempt will be made to review all the extensive work that has been done on the respiration of storage organs. Emphasis will be placed on the special problems associated with massive organs and on the contribution that the study of their respiration can make to our over-all knowledge of the regulation of respiratory processes in living organisms. The reader is referred to reviews dealing especially with the physiology of fruits (24, 25, 27, 87, 128, 129) and to other reviews that will be mentioned in connection with specific subjects.

II. The Ontogenetic Pattern of Respiration in Fleshy Fruits

A. The Course of Respiration

The salient feature of the pattern of respiration in many ripening fruits is a spontaneous rise and fall in intensity. Blackman (30) observed this in 1920 and recognized it as an index to fundamental changes in metabolic balance that are inherent in the aging of the organ. This feature is shown in Fig. 1, which represents the rate of carbon dioxide production by individual apples that fell into three groups ripening at different rates. The realization that a fruit is a dynamic system, continuously undergoing spontaneous changes, and that individuals may vary in their rate of change meant that isolated observations are of little value unless the physiological "age" of the organ is known. Blackman established the course of the spontaneous drift by continuous observation of respiration rates over long periods of time, and referred manipulated deviations to this "ontogenetic drift." This is a basic feature of a great deal of experimental study of the respiration of fruits.

Following the initial experiments carried out by Parija and Blackman, the respiration of apples in storage was studied extensively by Kidd and West, and the annual reports of the Food Investigation Board of Great Britain from 1921 onward attest to the large contribution made by these investigators and their associates to our knowledge of the physiology of fruits. Kidd and West (79, 80) showed that the pattern of respiration in stored apples is qualitatively the same but quantitatively different at different temperatures. This is illustrated in Fig. 2. At 22.5°C a major increase in rate of respiration

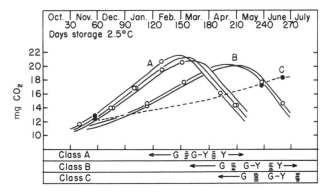

Fig. 1. The rising and falling rates of respiration (milligrams CO_2/100 gm/3 hours) of 'Bramley's Seedling' apples ripening in cold storage. Each point represents the respiration rate at 22.2°C of an apple taken from storage at the time indicated. The color changes are shown below, from green (G) to green-yellow (GY) to yellow (Y), for three groups of apples ripening at different rates. From Blackman (30).

occurs very rapidly, indeed quite dramatically, and they labeled it the "climacteric rise," a term that is now familiar. The temperature coefficient (Q_{10}) for intensity of respiration at corresponding points in the ontogenetic drift is about 2, but is much higher for the rate at which the drift progresses at different temperatures. At 2.5°C the climacteric rise occupied about 25 weeks, but at 22.5°C only a few days.

First observed in apples, a climacteric rise in carbon dioxide output has since been found to occur in many kinds of fleshy fruit during ripening. Not in all fruits, however, is this phenomenon obligatorily associated with ripening. Biale (25), who has made extensive contributions to the study of ripening fruits, made a tentative classification of fruits into "climacteric" and "nonclimacteric," including in the second category citrus fruits, cucurbits, fig (*Ficus carica*), pineapple (*Ananas comosus*), and some berries. But representatives of most of these groups may pass through a climacteric under certain conditions. Lemons (*Citrus limon*) respond to high oxygen concentration in the atmosphere (26) or high carbon dioxide added to air (141). Cantaloupe melon (*Cucumis melo*) undergoes a climacteric at 20°C if picked at an early enough stage and responds to ethylene as "climacteric" fruits do (92, 98). Strawberries (*Fragaria* × *ananassa*) probably undergo a climacteric phase while still ripening on the vine (116). These and other observations make it questionable whether any fleshy

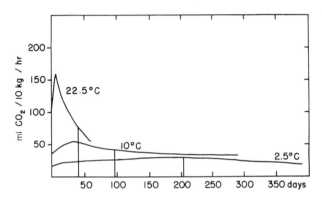

Fig. 2. Respiration rates of 'Bramley's Seedling' apples held at three different temperatures after harvest. The respiration rate was measured at the storage temperature. Vertical lines mark the end of the average duration of life of the apples at each temperature. From Kidd and West (80).

fruits are proof against a climacteric rise if the right conditions are found.

Just as some fruits classed as "nonclimacteric" may be induced to display this feature, so those classed as "climacteric" may have their respiratory pattern modified by external conditions. Early work by Kidd and West (see reports of the Food Investigation Board from 1925 onward) showed that the concentrations of oxygen and carbon dioxide in the storage atmosphere have a pronounced influence on ripening of apples and pears (*Pyrus communis*). Oxygen mixtures richer than air tend to hasten the climacteric, those poorer than air to delay it, and sometimes to depress it. The climacteric is an aerobic phenomenon and fails to appear in the complete absence of oxygen. High carbon dioxide concentrations (5 or 10%) tend to delay and depress the climacteric in apples and pears, and a combination of low oxygen and high carbon dioxide to eliminate it in banana (*Musa acuminata*) (141). The success of commercially used "gas storage" rests on finding the combination that will result in ripening at the desired rate without producing objectionable side effects.

The timing, magnitude, and form of the climacteric are influenced by many other factors in the environment which cannot be detailed here. One of these, ethylene, will be discussed later.

For bulky organs the continuous measurement of carbon dioxide output was until recently more easily accomplished than that of oxygen uptake, and most of the information about the climacteric refers to the production of carbon dioxide. But extra uptake of oxygen is also involved (see, for example, 23, 82, 141). Biale (23) found a respiratory quotient (R.Q.) below 1.0 throughout the ripening of avocado (*Persea*) with no tendency to increase during the climacteric; and throughout the climacteric which is induced in lemons under high oxygen atmospheres the R.Q. remained well below unity (26). The reports for apples are variable. Hackney (57, 58) found an R.Q. close to 1.0 throughout the ripening of 'Granny Smith' and 'Delicious' apples. Kidd and West (82) recorded an increase from 1.0 to 1.25 during the climacteric rise of 'Bramley's Seedling' apples at 22.5°C, but a decrease from 1.24 to 1.16 at 10°C. The reasons for the discrepancy are not apparent, but it is clear that the climacteric rise involves an increase in aerobic respiration and does not merely represent extra carbon dioxide from some other source.

The climacteric rise is best known in fruits that have been detached from the plant, but it also occurs in apples allowed to ripen on the tree (72, 84, 112) and in other fruits attached to the plant, for example, tomato (*Lycopersicon esculentum*) (43) and strawberry (116). Pearson and Robertson (103) found that 'Granny Smith' apples ripening on the tree showed no decline in carbon dioxide output after the climacteric had been reached, and it appears that the climacteric rise is characteristic of ripening fruit whereas the postclimacteric decline may be associated with isolation from the plant. The avocado appears to be an exception (25). It does not display a climacteric rise until after it is harvested, even if picking is delayed for months after maturity is reached.

Ripening in storage is of course the penultimate stage of the ontogeny of the fruit, which begins with fertilization of the ovule and ends with the "death" of the fruit. The early stages have not been so extensively studied as the senescent phase, but the respiratory history of apples has been recorded in several studies, the earliest by Kidd and West (79, 83). They followed the respiration of apples developing on the tree from an average weight of 0.7 gm to full green maturity

(see Fig. 3). During this period of 130 days the rate of production
of carbon dioxide per unit weight fell to about 5% of the initial rate,
and most of the decline occurred during the phase of cell division
in the first 3 weeks. This steep initial decline could be in large part,
but not entirely, attributed to the addition of nonnitrogenous matter
to the cells, for the ratio of respiration to total nitrogen during this
period fell to 40% of its initial value. It seems clear that reference
to a weight basis is at least a major part of the reason for the initial
drop in respiration rate. When Pearson and Robertson (103) estimated

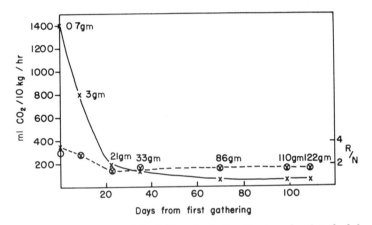

Fig. 3. Respiration rates at 12°C of 'Bramley's Seedling' apples detached from the
tree at various times during their development and maturity. The experimental
period extends from just after petal fall until the apples were full grown but still
green. The average fresh weight of the fruit is shown at each sampling time.
Respiration rate per unit fresh weight is shown by a solid line. The broken line
shows the rate per unit of nitrogen content (R/N, milliliters CO_2/100 mg N/hour).
Drawn from the data of Kidd and West (83).

carbon dioxide production per *cell* by apples developing on the tree
they found a slow but steady increase from the inception of the fruit
until a more or less sharp acceleration announced the climacteric rise
(see Fig. 4). The initial rise is of course partly because the cells
are growing at this time and adding to their complement of enzymatic
protein.

The interpretation of "respiration rates" is always dependent on
the basis of reference, and the selection of an appropriate one is always
a problem. In dealing with mature fruits the problem is not so acute
because the cells are fully grown, but in young fruits it can produce
the sharp contrast that is evident in Figs. 3 and 4. Several investigators

have referred respiration in apples to the content of protein nitrogen, and Kidd and West (86) and Hulme (68, 70, 71) have called attention to a constant proportionality during a large portion of the life of the fruit. Since protein nitrogen was proportional to total nitrogen in the apples observed by Kidd and West (Fig. 3), it may be said that the ratio of respiration to protein nitrogen was constant after the phase of early growth, including cell division, was completed. But it was somewhat higher in the very young growing apple. Robertson and Turner (112) found that protein nitrogen per cell was proportional to the cell surface rather than to cell volume, as if the expanding

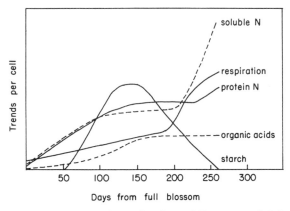

Fig. 4. Generalized curves based on the data of Pearson and Robertson (103) to show the changes in respiration rate and the content per cell of starch, protein nitrogen, soluble nitrogen, and organic acids in 'Granny Smith' apples throughout their development on the tree, from their inception to the postclimacteric stage. The climacteric rise in respiration rate began at 190 days after full blossom. From Pearson and Robertson (104).

cell manufactured enough protein to maintain a layer of protoplasm of constant thickness. When they referred respiration to cell surface they found an initial decline within 3 weeks of about 50%; a later increase occurred at the time of the climacteric. The content of protein nitrogen, or the cell surface, provides a better index to the amount of enzymatic machinery available in a young fruit than its fresh weight. It appears that throughout a large part of the existence of an apple fruit its respiration is proportional to the amount of catalytic machinery, but that during its early growth and again when the climacteric intervenes it can produce more carbon dioxide per unit of protein.

Most of our information refers the rate of respiration to fresh weight and, on this basis, the entire history of the respiration of an apple consists of the sharp fall during growth and the steady low rate of the mature green fruit followed by a climacteric rise, and, if the fruit is picked and stored, a subsequent decline lasting until physiological breakdown or invasion by microorganisms occurs.

B. THE ROLE OF ETHYLENE IN THE CLIMACTERIC

Intensity of respiration is, of course, not the only thing that changes during ripening, and since the earliest studies it has been recognized that many chemical and physical changes associated with "ripening" are initiated at about the time of the climacteric. These will be discussed only as they may bear on the interpretation of the respiratory pattern. Recent summaries by Biale (25, 27) describe many of them.

It has been known for many years that ripening fruits produce an emanation that hastens the ripening of any green fruits that come under its influence, and ethylene is recognized as the effective agent. Ethylene is of widespread occurrence in the internal atmosphere of fruits, and its role as a ripening agent has been the subject of extensive inquiry. Burg (33) has provided a useful review of a large body of literature on ethylene.

Burg reports ethylene to be present in the internal atmosphere of 29 species of fruit that have been examined, and there are no proved exceptions among fleshy fruits. If ethylene is added in low concentration to the atmosphere around green fruits it accelerates ripening, under appropriate conditions of temperature and oxygen supply. In fruits that normally have a climacteric pattern it advances the onset of the rise in respiratory intensity. "Nonclimacteric" fruits, such as lemon, respond with a steep but ephemeral increase in respiration rate, and ripening is accelerated.

For a number of years it has been debated whether ethylene causes the climacteric rise in respiration together with ripening, or whether it is only a by-product of changes that occur in this phase of ontogeny. A part of the difficulty in making a decision lay in the detection and measurement of the small amounts of ethylene that produce physiological effects. The major increase in ethylene content that is relatively easily detected occurs, not before, but near the peak of the climacteric, and this has been the main basis for questioning its role as the causative agent.

The advent of gas chromatography has made it possible to measure concentrations below as well as above the threshold of physiological activity. It is now possible to detect 0.005 ppm in 5 ml of gas (34) and to measure the rate of production of ethylene in 2 or 3 fruits.

Estimates of the minimum internal concentration that is physiologically active range from 0.025 to 1.0 ppm, and 0.1 ppm appears to be a reasonable figure for a threshold concentration (28, 33). In bananas Burg and Burg (34) found that the concentration of ethylene remained close to 0.1 ppm until the day before the climacteric rise. During this day it increased two- to threefold, passing through the threshold value somewhere between 0.1 and 1.0 ppm. It exceeded 1.5 ppm when an increase in respiration rate was first detected. Later during the climacteric rise the internal concentration rose to 6 ppm.

While the major increase in concentration occurs when respiration rate is already rising, the critical question is whether a sufficient concentration has been established at the onset of the climacteric to permit the possibility that ethylene is a causal agent rather than a product of the climacteric rise. In the banana this appears to be established. In the cantaloupe also (92) arrival at an effective concentration just precedes, or coincides with, the first increase in respiration rate. The same appears to be true for a number of fruits of the climacteric type that contain ethylene in physiologically active quantity, usually well over 0.1 ppm, at the onset of the climacteric.

Certain tropical fruits, including mangoes (*Mangifera indica*) (34), contain less than 0.1 ppm ethylene when the climacteric is initiated, and only later does a burst of ethylene occur. There is not yet enough information to establish whether these are sensitive to lower concentrations than other fruits, although there are indications that this is so.

In fruits that normally ripen without undergoing a climacteric ethylene can be detected, but possibly it does not rise above a threshold concentration under the conditions that prevail. Many of these can be induced to undergo a climacteric.

The ability to estimate small quantities of ethylene has made it more, not less, probable that this substance triggers the climacteric rise when its production by the fruit has built up a supracritical concentration in the internal atmosphere. Indeed, remaining doubts center around the possibility that quasi-effective concentrations may sometimes exist in the fruit well before the climacteric begins. The problem of initiation of the climacteric is thrust back to the factors that determine when the rate of production of ethylene will increase severalfold and why a certain concentration triggers a major change in the rate of respiration. Even the site of primary action is not yet clear, although the more rapid transfer of electrons in the mitochondria is the first recognizable sign. It is perhaps significant that Barker and Solomos (16) report that ethylene produces an increase in concentration of fructose 1,6-diphosphate along with an increased rate of respiration.

C. METABOLIC MECHANISMS ASSOCIATED WITH THE CLIMACTERIC

Much research has gone into attempts to explain the pattern of respiration in ripening fruits. Only the more salient features can be discussed here.

The earliest analysis, by Blackman, was in terms of a decrease in "organization resistance" causing respiration rate to rise and "starvation" causing it to fall, the ultimate control depending on "the rate of production of the effective substrate of respiration." Blackman used these terms in a very general sense, to mean on the one hand controls of flux imposed by protoplasmic organization, and on the other hand the progressive expenditure of reserves and the inevitable running down of the system in isolation when no new ingredients can be added. "All along the chain the drift is determined by balance of production and consumption. The controlling factors are the activity of the catalytic components of the system and the amounts of the substrates available." Many efforts have been made to show more specifically the cause of the climacteric and, although a mass of information has been accumulated, the complete story has not yet been told.

Analysis of changes in the metabolism of major carbohydrates and acid constituents revealed no simple relation of respiration rate to the concentrations of any of these, and we now know that in very few plant tissues can such a relation be found. Indeed Blackman had already postulated that the effective substrate for respiration is not a stable sugar but a derivative present in low concentration and dependent on the balance of production and consumption. Haynes and Archbold (61) found that in some apples physiological breakdown occurred when the rate of inversion of sucrose fell below the rate of consumption of sugar in respiration, but this proved not to be invariably true. Kidd and West (80) found that, whether an apple aged rapidly at 22.5°C or slowly at 2.5°C, the total carbon lost in respiration before physiological breakdown was the same. So there is a relation between the survival of the physiological state and the amount of the original store of food that is consumed, but it is not necessarily a direct one.

One constituent that bears a constant relation to the rate of respiration over a considerable part of the total life span of an apple is protein. The relation of respiration to protein has been extensively studied by Hulme (68–71), who confirmed that the ratio of respiration rate to protein content is relatively high during the early growing phase of an apple, then falls to a steady value which endures until the onset of the climacteric, when it rises again. The total nitrogen content of a detached apple remains constant throughout its storage life, but during the climacteric protein accumulates at the expense of asparagine and then of amino acids (67), and this trend may continue after the peak of respiration rate is reached. Apples remaining on the tree import nitrogenous substances during the climacteric, and protein accumulates (see Fig. 4).

An interrelation between respiration and protein synthesis has long been a subject of inquiry and has been discussed frequently by Steward (117, 118, 120). Not only do syntheses require energy, transferred via high-energy phosphate compounds, but the utilization of these compounds may allow respiration to proceed more rapidly. It is well known that the oxidative capacity of mitochondria isolated from numerous species of plants can be controlled by the provision of phosphate acceptors, and that the rate of respiration of plant tissues in some physiological states is increased in the presence of uncouplers, such as dinitrophenol, of the linked reactions of oxidation and phosphorylation.

Studies of the control of respiration during the climacteric in ripening fruits soon became focused on regulation by linked phosphorylation. Ripening is an endergonic process, and a green tomato fruit is prevented from ripening by the injection of dinitrophenol (96). Almost simultaneously Millerd et al. (99) and Pearson and Robertson (104) observed that dinitrophenol does not have the same effect on pre- and postclimacteric tissue. In appropriate concentration it causes increased uptake of oxygen by slices cut from avocado and apple fruits in the preclimacteric condition, but not by slices taken at the climacteric peak. This is shown, for apple tissues, in Fig. 5.

The explanations offered were different. Millerd et al. attributed it to the development of some natural uncoupler during the climacteric so that postclimacteric respiration was released from control by phosphorylation. Pearson and Robertson associated it with increased synthesis of protein during the climacteric resulting in increased ADP:ATP turnover. They found that in the first 50 days of growth of an apple as well as in the climacteric phase, the addition of dinitrophenol had little or no effect on the oxygen uptake of slices (Fig. 5). Only in the intervening phases of cell expansion and preclimacteric maturity did it stimulate respiration. They concluded that only in this intermediate period is respiration limited by the supply of ADP because the utilization of mitochondrial ATP is relatively low, while in both the early and late phases, more rapid synthetic activity releases ADP which no longer limits respiration.

Numerous attempts have been made to determine whether phosphorylation is actually uncoupled from oxidation during the climacteric. These have usually made use of "mitochondrial" fractions of homogenates made from pre- and postclimacteric fruit. Results have been somewhat conflicting, possibly because of the difficulty of making reliable mitochondrial preparations from plant tissues. But Romani and Biale (113) reported that "mitochondria" prepared from avocado fruits at the climacteric peak esterified phosphate. Wiskich *et al.* (139) confirmed this and demonstrated some degree of uncoupling by

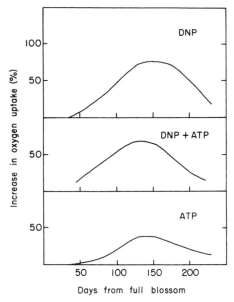

FIG. 5. The effect of 2,4-dinitrophenol (DNP, 0.5 mg/liter) and adenosine triphosphate (ATP, 3 mg/ml) on the rate of oxygen uptake by slices of 'Granny Smith' apples at different stages of development. Curves show the percentage increase in oxygen uptake in the presence of DNP, ATP, and both together. From Pearson and Robertson (104).

dinitrophenol. It proved to be easier to prepare mitochondria showing ADP control from postclimacteric than from preclimacteric fruit, and Howard and Yamaguchi (64) had the same experience with pepper (*Capsicum annuum*) fruits.

The difficulty of interpreting the properties of "mitochondrial" preparations in terms of the intact fruit have led to studies of the incorporation of P^{32} by whole fruits or slices. Biale and Young (27) report that P^{32} applied to slices of avocado fruit in both pre- and postclimacteric phases was incorporated into esters of sugars, nucleotides, 3-carbon acids, and other substances, and the rate of incorporation was higher in the slices from postclimacteric fruit. Phosphorylation did not, then, decrease during the climacteric, but increased. Dinitrophenol decreased incorporation into esters with nucleotides and sugars in both stages, so coupled phosphorylations occurred in both stages. But additional P^{32} was incorporated, at the climacteric peak, into an unidentified compound, by a process that was not inhibited by dinitrophenol.

Marks et al. (96) injected P^{32}-labeled phosphate into whole tomato fruits and found a high degree of esterification both before the climacteric and at its peak. But the incorporation decreased steadily and vanished in late senescence. Jones et al. (78) also report a low P:O ratio for the mitochondria of very senescent apples. Natural uncoupling apparently does occur, but probably not until after the climacteric rise in respiration is past.

The alternative hypothesis, that ATP turnover increases at the climacteric and makes ADP available in a concentration that does not limit electron transport, remains as a possible explanation. Not enough direct evidence is available as yet to establish it positively as a primary cause of the climacteric rise in respiration rate.

Although a great deal of attention is paid to the role of oxidation-coupled phosphorylation of ADP in controlling the rate of respiration, it appears that only in the mature preclimacteric phase of fruit development is this role important. In the very young fruit and after the climacteric some other factor must set the pace of oxidation. Systems are known where the activity of hexokinase appears to be the limiting factor, and Chance and Hess (41) has proposed that in intact cells the rate of respiration may be controlled by a barrier to the release of ATP from the mitochondria to the sites of glucose phosphorylation. Wiskich et al. (139) found that hexokinase was a more efficient stimulator of oxygen uptake by mitochondrial preparations from avocado than was ADP. A somewhat analogous situation has been revealed in ripening banana, where Barker and Solomos (16) found a sevenfold increase in the concentration of fructose 1,6-diphosphate accompanying a fivefold climacteric increase in carbon dioxide output. An approximately hyperbolic relation held between carbon dioxide output and the concentration of fructose 1,6-diphosphate, and they suggest that the activation of phosphofructokinase is responsible for the climacteric rise. Hess (63) postulates that in ascites cells phosphofructokinase activity is controlled by limited diffusion of ATP from the mitochondria, and Barker et al. (7) suggest that glycolytic enzymes are contained in organelles relatively impermeable to mitochondrial ATP. If these suggestions should be correct, a climacteric rise in respiration rate would result from a lowering of the barrier to access of ATP to the sites of glycolysis. Perhaps we have come full circle to Blackman's original concept of "organization resistance" and the "rate of production of the effective substrate of respiration," which can now be stated in more specific biochemical terms.

Still other explanations of the climacteric have been offered. The protein synthesized at this time could represent the synthesis of enzymes as suggested by Hulme (71), and certain enzymes have been shown to increase in activity at the climacteric. These include pyruvic carboxylase in banana (123) and pyruvic carboxylase and malic enzyme in apple, particularly in the peel (73). Hulme *et al.* attribute the climacteric in apple to synthesis of these enzymes in the peel followed by a secondary effect in the pulp. They cite the increase of R.Q. in apples during the climacteric as evidence for a source of decarboxylation other than those occurring as integral parts of the tricarboxylic acid cycle. It must be recalled, however, that in some fruits, and even in some apples, the respiratory quotient does not increase during the climacteric.

More than one mechanism, of course, may be involved. It is obvious that further work is necessary before we shall understand fully the causes underlying the ontogenetic drift of respiration in ripening fruit. When we do understand it, we may have a better perception of the meaning of biological aging in general.

III. The Respiration of Fleshy Roots and Tubers

Fleshy organs biologically different from fruits are those that store reserves for future growth during a dormant phase in a vegetative cycle. The more massive of these are to be found among roots and tubers. Some observations have been made on the respiration of roots, but the only underground storage organ that has received anything like the attention devoted to fruits is the potato (*Solanum tuberosum*) tuber. This is largely the outcome of extensive investigations by Barker (5–15) and Burton (35–40). Slices of a variety of storage organs have been much studied, but their respiration may be qualitatively as well as quantitatively different from that of the intact organ. By following the changes that develop gradually in slices, some indications of the nature of the respiration and metabolism of cells has been gained, but our concern here is with observations on intact organs.

Roots and tubers, so far as their respiratory history is known, do not exhibit the ontogenetic pattern that is characteristic of fruits, where the climacteric is associated with ripening and development is determinate. Storage roots and tubers contain latent meristems where shoots may develop, but there is very little information about the relation between respiration of the storage tissue and the breaking of dormancy. Stout (122) associated a high respiration rate of sugar beets (*Beta vulgaris*) with low temperature induction of bolting, but did not eliminate the possibility that the high intensity of respiration was, as in some other storage organs, a result of the low temperature treatment irrespective of a change in morphogenetic state.

Rather more is known of ontogenetic trends in the potato tuber. The rate of respiration declines as the tuber matures on the vine; when the vine dies or the tuber is detached, the rate falls rapidly to the low value that is characteristic of the dormant stage in storage (3, 50). Burton (38) and Whitehead (137) report that when sprouts begin to grow the rate of respiration rises. But this need not represent an increase in respiratory intensity of the tuber tissue. Tubers in ventilated storage lose water slowly while dormant, but more rapidly when sprouts increase the evaporating surface, and the intensity of respiration per unit weight of the sprouts is very high. When Forward (50) referred carbon dioxide output of the tuber only (with sprouts removed) to original fresh weight, the rate measured at 22°C remained remarkably constant over a period of eight months of storage at 8°C, although sprouting began after about three months. Tubers with sprouts representing 2–3% of the total weight respired at a rate 50–100% higher than that of dormant tubers, but this proved to be largely because of the very intense respiration of the sprouts, which was about 30 times that of the tuber tissue, and there was no indication that respiration of the storage tissue changed with the breaking of dormancy.

Oxygen uptake has, on occasion, been measured along with carbon dioxide output. Platenius (106, 107) found the respiratory quotient (R.Q.) of carrots (*Daucus carota* var. *sativa*) to lie between 0.9 and 1.1 in air and in mixtures containing as little as 3–5% oxygen. The R.Q. of potato tubers has been found by a number of investigators to be close to unity when measured at 10°–24°C, even where the rate of carbon dioxide output is high in sweetened potatoes. But departures from 1.0 may occur. Platenius found a value of 0.5 at 0.5°C. Burton (39) made a systematic study of R.Q. during the development and storage of two varieties of potato. Tubers detached from healthy vines had an R.Q. = 1; but, particularly with the variety 'Majestic,' a disturbance occurred when the vines died. In one season the R.Q. fell below 1, in another it rose above 1, returning to unity during the dormant stage in storage. With 'King Edward' most values were close to unity until sprouting began. In sprouted tubers of both varieties R.Q.'s well above 1.0 were observed when carbon dioxide output increased. The reason for the high R.Q. is not known. No lactic acid or ethanol accumulates in the storage tissue of sprouted 'Sebago' potatoes in air (51a). Nor do sprouted tubers always have a high R.Q. Singh and Mathur (115) found the ratio to be close to unity in immature, mature, and ripe potatoes at harvest. It declined gradually to 0.9 in storage and returned to 1.0 when the tubers sprouted. Evidently the R.Q. is variable with different varieties under different conditions. On many occasions it has proved to be close to unity.

The respiration rates of bulky roots tend to be somewhat higher, of the potato tuber lower, than that of fleshy fruits. The temperature coefficient (Q_{10}) of respiration in fleshy storage organs of several species (2, 77, 106) is of the same order of magnitude, 3–1.7 in different temperature ranges, as that of many less bulky tissues (51). But certain kinds, of which the potato tuber is best known, respond to prolonged exposure to low temperature in a special way. It has long been known that temperature has a profound effect on the carbohydrate and respiratory metabolism of potatoes. Müller-Thurgau as long ago as 1882 discovered that the respiration rate gradually rises when tubers are held at temperatures below about 8°C, and attributed this to the increase in sugar content that also occurs. At a higher temperature after several weeks at low temperature, respiration rate is initially very high and declines as the tuber desweetens. Similar observations have been made with sweet potato (*Ipomoea batatas*) (4, 60). Both are starch-storing organs, and changes in sugar and starch content are, in a general sense, reciprocal.

Barker (5, 6) made an intensive study of the relation between respiration and concentration of sugars in 'King Edward' potatoes at temperatures between −1°C and +15°C. Total sugar, about 0.2% of the fresh weight at 10°C, rises slightly at temperatures below 8° and much more below 3°C, to reach 6.7% of the fresh weight at −1°C. Sucrose, glucose, and fructose are the sugars that accumulate, but the proportion of sucrose to hexose differs greatly at different temperatures. At 10°C sucrose is only about one-fifth of the total sugar, but it increases more than hexose at low temperature, so that at −1°C it represents about one-third, and at +1°C one-half to two-thirds, of the total sugar. Barker found that the rate of respiration during either sweetening or desweetening showed a hyperbolic relation to the concentration of sucrose, but not to that of the other sugars. The relation, illustrated in Fig. 6, suggests that the concentration of sucrose controls the rate of an enzymatic reaction which determines the rate of respiration. The underlying reason for the dependence on sucrose concentration is not known, and it is rare in plants, so far as our knowledge extends.

But sucrose concentration is not the only factor dependent on long exposure to low temperature that affects the rate of respiration. Barker (5) also discovered that at very low temperatures a depressant of respiration accumulates in the tuber in parallel with the accumulation of sugar. Eventually this may distort the relation between respiration and sucrose concentration even at 1°C, but the effectiveness of the depressant develops much more rapidly at higher temperature. Barker interpreted his observations to mean that an inhibiting substance accumulates at low temperature, but that a second process is required

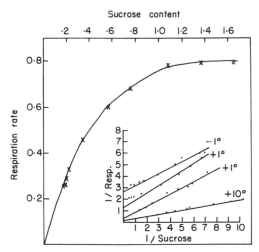

Fig. 6. The relation between respiration rate and sucrose content of 'King Edward' potato tubers during sweetening at −1°C. *Inset:* The reciprocal of respiration rate plotted against the reciprocal of the sugar concentration in potatoes sweetening at −1°C and +1°C, and desweetening at 10°C. For the sake of clarity, the top graph has been raised by two units and the second graph by one unit. From Barker (6).

before the substance can actually inhibit respiration, and this proceeds faster at a higher temperature. An example is shown in Fig. 7. Tubers held for 10 days at 1°C, then for 1 day at 15°C, now respire at a lower rate at 1°C than comparable tubers remaining always at the lower temperature. During the short exposure to 15°C no change in sucrose concentration occurred, so the result of the higher temperature was to accelerate the process that renders effective the inhibitory action of the substance previously accumulated at 1°C. When the depressant action was effective the relation between respiration rate and sucrose concentration was distorted; when it was not interfering, a hyperbolic relation was manifest. Neither the inhibitory substance nor the mechanism that renders it effective has been identified in biochemical terms.

The investigations of Barker and others make it plain that the effect of low temperature on respiration of the potato tuber is complex, and other aspects of metabolism are affected. Temperature is a convenient factor to manipulate, but the response requires close scrutiny. With due regard to complexities, however, a sensitive organ like the potato tuber can be useful in elucidating the metabolic controls of respiration.

Other aspects of the respiration of storage organs will be discussed in succeeding sections. From the foregoing it is apparent that they

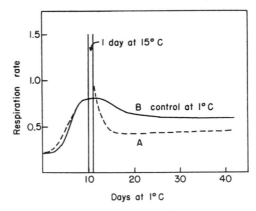

FIG. 7. The rate of respiration of 'King Edward' potato tubers at 1°C showing the development of inhibition, during 1 day at 15°C, by a substance previously accumulated at 1°C. Curve A, after 10 days at 1°C then 1 day at 15°C; curve B, control tubers continuously at 1°C. From Barker (5).

exhibit differences from fruits in their respiratory behavior. These are of interest because, in spite of biological differences, the underground organs have some structural features in common with fleshy fruits. They are bulky organs bounded by a relatively impervious layer of periderm or cuticle, and present physical problems common to all massive organs.

IV. Respiration and the Internal Atmosphere of Bulky Organs

In most plants individual cells come into contact, over a part of their surface, with a gaseous phase that exists in the intercellular spaces and comprises the internal atmosphere of a plant or organ. This is not necessarily the same as the atmosphere external to the plant, and in a fleshy organ it may be radically different. Most of these have an outer skin or rind that is relatively resistant to the movement of gases and provides the chief barrier between the individual cells and the external atmosphere.

Oxygen and carbon dioxide are constituents of the internal atmosphere that are obviously related to respiration, but others may be important as well. Ripening fruits, for instance, certainly develop a number of volatile substances that contribute to their flavor. One of these, ethylene, has a well known effect on the climacteric rise in respiration that has already been discussed. Ethylene also increases, at least temporarily, the carbon dioxide output of organs that do not show a climacteric, including potato tubers of low sugar content (65). Some fleshy organs contain ethanol and acetaldehyde, albeit in small quantity under aerobic conditions (46, 66, 124, 127). Laties (89) has proposed that an endogenous volatile substance, probably acetaldehyde, depresses respiration in whole potato tubers and is released when slices are cut. Apples and oranges (*Citrus sinensis*), however, absorb and metabolize acetaldehyde introduced into the external atmosphere, and apples respond by increased carbon dioxide production with no change in oxygen uptake. Oranges show increased oxygen uptake and carbon dioxide output (48, 127), and there is no direct evidence that the normal content of acetaldehyde depresses respiration. It is not proposed to deal further here with the effects on respiration of volatile substances that occur in low concentration in the internal atmosphere. The subject is discussed briefly in a review by Pentzer and Heinze (105) and also by Laties (89).

Attention will be directed to oxygen and carbon dioxide and to the problems associated with the fact that the cells inside massive organs may be exposed to very different concentrations of these gases from those existing in the external atmosphere. These problems involve such questions as whether the rate of emission of carbon dioxide is a true measure of respiration rate and what effects elevated carbon dioxide and depressed oxygen concentrations have on respiration. In relation to carbon dioxide concentration most attention has been given to interpreting carbon dioxide emission in transitional states when external conditions are changed. In relation to oxygen concentration interest has been mainly in the question whether an oxygen deficiency caused by slow diffusion limits the rate of respiration in bulky organs.

A. Oxygen and Carbon Dioxide in the Internal Atmosphere of Fruits

Carbon dioxide in high concentration in the external atmosphere depresses respiration (see Part 1 of this chapter, p. 268), and the intensity of respiration in many plant organs is dependent on oxygen concentration over a wide range. It is therefore important to know the concentrations of these gases inside the organ. This was realized many years ago by Magness (93) and by Kidd and West (81) among others. The most influential work was that of Wardlaw and Leonard and their associates, who worked in Trinidad with tropical fruits. In a series of papers (131–135) they published the results of a systematic

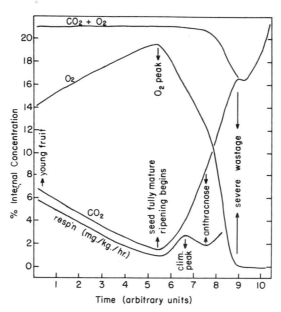

FIG. 8. Generalized diagram of the drifts in respiration rate and the percentage concentration of oxygen and carbon dioxide in the internal atmosphere of the papaw (*Carica papaya*) fruit during development, ripening, and senescence. The rate of respiration was initially 290, at the commencement of ripening 35, and at the climacteric peak 140 mg/kg/hour. "Anthracnose" marks the stage of senescence when fungal lesions develop and provide entry for secondary infections. From Wardlaw and Leonard (132).

study of the internal atmosphere of papaw (*Carica papaya*) and banana (*Musa acuminata*) in relation to their development and ripening.

The papaw, a massive fruit weighing several kilograms when fully developed, with a large internal cavity, provides an easily accessible internal atmosphere which is in communication with that of the intercellular spaces in the flesh. The whole is separated from the exterior by a rind that is relatively impervious to gases. Wardlaw and Leonard sampled the gas in the cavity and determined the oxygen and carbon dioxide concentration at successive stages of development and ripening of individual fruits.

A generalized diagram of the trends is shown in Fig. 8. In very young fruits weighing 27 gm, the carbon dioxide concentration was about 6%. It was less in older fruits, declining in parallel with the rate of respiration to a minimum value which endured until the fruits were full grown and weighed more than 1100 gm. In mature green fruits the internal atmosphere had 1.5–4% carbon dioxide in different individuals and 17–19% oxygen. Such fruits, picked and stored (26.7°C), were observed as they ripened.

The beginning of ripening of the papaw is marked by a substantial climacteric rise in respiration rate and, commencing somewhat earlier

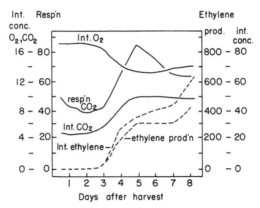

FIG. 9. Changes in respiration rate and percentage concentration of oxygen and carbon dioxide in the internal atmosphere of a cantaloupe (*Cucumis melo*) fruit during ripening and senescence. The production of ethylene and its internal concentration are also shown. The units are: internal concentration of oxygen and carbon dioxide, per cent; respiration, milligrams CO_2/kg/hr; ethylene production, μl/kg/day; internal concentration of ethylene, ppm. Temperature 20°C. From Lyons *et al.* (92).

but proceeding in parallel fashion, an increase in internal concentration of carbon dioxide with a corresponding decrease in oxygen. By the time the fruit was fully yellow, oxygen had fallen to 13–16% and carbon dioxide had risen to 5–9%. The changes in composition of the internal atmosphere did not, however, stop at the climacteric peak of respiration, but continued their trend until a late stage in senescence of the fruit when the carbon dioxide concentration reached 16–21% and oxygen fell to zero or close to it. By this time severe wastage of the fruit had occurred. The continued trend of oxygen depletion and carbon dioxide accumulation after the climacteric peak, occurring when respiration rate is falling, was attributed to an increase in the resistance to movement of gases through the flesh of the fruit. It is only in senescence that this occurs. Until the climacteric is passed the permeability of papaw tissue to carbon dioxide is roughly constant throughout the development of the fruit.

Other fruits show essentially similar trends up to the time of the climacteric peak. Mature green fruits of cantaloupe (*Cucumis melo*) (92), avocado (*Persea*) (21), banana (135), and apple (55, 56, 126), have all been found to have roughly 17–20% oxygen and 1% to 3 or 4% carbon dioxide in the internal atmosphere. This condition changes but slightly until about the time of the climacteric rise in respiration, when a drastic decline in oxygen content occurs. By the time the climacteric peak is reached fruits of the various kinds contain about

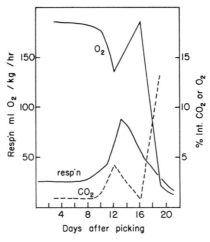

FIG. 10. Changes in respiration rate and percentage concentration of oxygen and carbon dioxide in the internal atmosphere of (Hass) avocado fruits during ripening and senescence at 15°C. From Ben-Yehoshua *et al.* (21).

10–15% oxygen in the internal gas phase, or occasionally less, and usually about 5–10% carbon dioxide. In the phase of senescence after the climacteric peak, however, fruits differ greatly. In 'Granny Smith' apples (55, 57) the oxygen concentration continues to fall, as in papaw, but carbon dioxide does not rise above about 5%. In cantaloupe (92), as shown in Fig. 9, and avocado at 20°C (21) the composition of the internal atmosphere changes but little after the climacteric peak of respiration. But in avocado at 15°C (Fig. 10) and banana at 29.4°C (135), the oxygen content recovers rapidly to 16–18% as the rate of respiration falls away rapidly, and carbon dioxide concentration declines toward its preclimacteric value. This is but a temporary recovery, and sooner or later the trend is again reversed. The final decline in oxygen concentration is precipitous when breakdown of tissue occurs, and carbon dioxide content rises steeply.

These differences among fruits that appear in the later stages of senescence are related in some fruits at least to the time when a breakdown of protoplasmic structure results in marked changes in the physical system. Up to the time of the climacteric all fruits that have been examined show similar relations between the internal concentrations of oxygen and carbon dioxide and the rate of respiration. That is to say, when oxygen is being consumed and carbon dioxide produced rapidly, the concentration of the former is low and of the latter is high. And when the rates of consumption and production fall, the trend is reversed. In papaw, apple, and cantaloupe, and in avocado held at 20°C, this relation breaks down at about the time of the climacteric peak, and oxygen concentration does not increase as the rate of consumption falls. But in the avocado at 15°C and in the banana the breakdown of the relation is delayed, and immediately after the climacteric peak, as the rate of consumption of oxygen declines, its concentration in the internal atmosphere rises. Only at a later stage does oxygen disappear from the atmosphere inside the fruit although the rate of its consumption is low. This is the stage at which resistance to gas movement increases markedly.

Where resistance to diffusion of oxygen and carbon dioxide has been estimated, it is found to be very low in young growing fruits and then to reach a stable low value which endures until the climacteric peak of respiration is reached (81, 85, 132). Thereafter, sooner or later, it increases in most fruits (21, 81, 132, 135). Until this occurs the resistance to gas movement is chiefly in the skin and the internal atmosphere is nearly uniform throughout the interior (126, 135). During the final phase the skin may possibly become more impervious (55), but in avocado the increase in resistance is in the flesh and is apparently the result of clogging of intercellular spaces by fluid exuding from degenerating cells (21). For other fruits as well there is evidence that the restriction of movement of gases that develops in late senescence is associated with degeneration of cells in the flesh.

The composition of the internal atmosphere can, of course, be modified by artificially increasing the resistance to gas diffusion. Petroleum jelly applied to the surface of a papaw fruit causes the carbon dioxide in the cavity to rise from less than 2% to 15% in green fruit, or to 23% in mature fruit within 24 hours. Apples and other fruits respond similarly (56, 126), and the oxygen concentration falls to low values. It is probable that in these fruits gas exchange with the external atmosphere occurs over a large part of the surface. Tomatoes, on the other hand (42), may be waxed in all parts but the stem end without affecting the output of carbon dioxide, which occurs mainly through the stem scar and the immediately adjacent region. 'Delicious' apples (58) continue to allow ready exchange of oxygen and carbon dioxide throughout their storage life, and some fruits of this variety remain open at the calyx end. Fruits differ, therefore, in the permeability of the periderm to gases, and their morphology affects the internal atmosphere. The striking thing is not that fruits differ in the post-climacteric phase, but rather that so many kinds conform in being well ventilated until the climacteric occurs.

B. THE RELATION BETWEEN RESPIRATION AND THE INTERNAL ATMOSPHERE

Although a good deal of information on the respiration of storage organs has been accumulated, and a lesser but substantial amount on their internal atmosphere, ideas about the relation between the two have not been fully clarified. The problems that have attracted most attention have already been indicated, and these may now be examined more closely.

The atmosphere inside fleshy organs differs from that outside, and we may ask whether the respiratory activity of the component cells is the cause or the consequence of this difference. We have already seen that during the development and ripening of fruits, at least until the climacteric is past, the natural changes in the internal atmosphere are in the direction to be expected if they are the result of changes in the rate of respiration. During the climacteric they are in the opposite sense to that required if they were the cause rather than the result of changes in respiratory intensity. In the final phase of ontogeny when resistance to gas movement increases, oxygen may be depleted to the extent that it must limit respiration. This is a complex phase. Some or all of the carbon dioxide may be derived from fermentation in some fruits and we are largely ignorant about the point at which irreversible physiological breakdown begins. It is of more interest that, until the organ reaches this phase of its senescence, it appears on the face of it probable that the rate of respiration is the cause rather than the consequence of the modified internal atmosphere.

Before going further we should clarify what we mean by the rate of respiration. Strictly it is the rate at which certain chemical processes occur within a cell. For our purposes interest lies in the rate at which oxygen is consumed and carbon dioxide is produced in these reactions, and the first question that concerns us is whether this can be measured by observing the rate at which these gases are exchanged between the surrounding atmosphere and the whole organ. Since the rate of carbon dioxide output has been most frequently used for this purpose, it will be considered first.

1. Carbon Dioxide

The rate of evolution of carbon dioxide by a fruit is directly determined by the concentration gradient between the internal and external atmospheres and the resistance offered by the periderm. The system involved is:

combined $CO_2 \rightleftharpoons$ dissolved CO_2 in cell $\rightleftharpoons$
gaseous CO_2 in intercellular spaces $\rightleftharpoons$ atmospheric CO_2

In a constant external atmosphere changes in rate of liberation of carbon dioxide may result from a change in rate of metabolism of the tissue, or a change in its capacity to retain the gas in solution, or a change in the resistance offered by the external layers of tissue to egress of the gas. During that long portion of its life when the permeability of a fruit to carbon dioxide is roughly constant, the ontogenetic changes in carbon dioxide emission depend largely on changes in concentration gradient determined by production and retention of the gas in the tissues. The dissolved carbon dioxide parallels the concentration in the gaseous phase, so these must equilibrate readily. In steady states, therefore, within this ontogenetic phase the rate of emission of carbon dioxide represents the rate of its production.

When, however, new conditions are rapidly imposed, a new equilibrium must be established within the tissue, and this requires a finite time which may be long or short according to the magnitude of the change and the physical system involved. Wardlaw (131) focused attention on this fact in relation to the transitional release of carbon dioxide when the composition of the external atmosphere is suddenly changed.

When high concentrations of carbon dioxide (up to 15%) are allowed to accumulate in the atmosphere surrounding a papaw fruit (131), this results in high internal concentration (about 18%). And when the chamber enclosing the fruit is recharged with normal air, an excess of carbon dioxide is liberated that is not metabolic in origin but results from a sharp increase in the differential between internal and external concentrations. Carbon dioxide already present in the internal atmosphere diffuses out, and the concentration in the cavity falls rapidly. A transitional excess of carbon dioxide is also emitted when nitrogen gas is introduced into a chamber enclosing the fruit. Wardlaw (131) attributed this to a change in the capacity of the tissue to retain carbon dioxide, caused by the removal of oxygen, and a consequent release of dissolved gas. He extended his interpretation to cover all observations of increased rates of emission when bulky fruits are transferred from air to nitrogen. Rather wide adoption of this interpretation warrants a close look at the experimental data.

In fact, the transitional outbursts of carbon dioxide following this manipulation were artifacts of the experimental conditions. A fruit was sealed into a chamber and its rate of carbon dioxide production was measured for an unspecified time, during which carbon dioxide accumulated in the chamber.[1] Wardlaw assumed that when he replaced the atmosphere in the chamber by nitrogen gas he changed only the oxygen concentration. Actually he transferred the fruit from "air" containing an elevated concentration of carbon dioxide to nitrogen containing none. This resulted in an increase in gradient of carbon dioxide concentration from the inside to the outside of the fruit. The respiration record showed a sudden small outburst of carbon dioxide (about 25 ml from a fruit 4 liters in volume) lasting for about half an hour. That this resulted from a change in carbon dioxide gradient rather than a change in the retentive capacity of the tissue in the absence of oxygen, is indicated by other data: (a) In the same experiment carbon dioxide gradually accumulated in the nitrogen-charged chamber until it reached 4.5%, and the chamber was then recharged with nitrogen, reducing the external carbon dioxide concentration to zero. This resulted in liberation from the fruit of excess carbon dioxide amounting to about 100 ml over a period of 2 hours. Here the tissue was not suddenly deprived of oxygen. It had been under anaerobic conditions for 5 hours. (b) When, in other experiments, normal air replaced air charged with 10% or 15% carbon dioxide as a result of respiration, the transitional outburst involved about 200 and 475 ml, on two occasions, from a 2.8 liter fruit with a cavity 950 ml in volume. Here the tissues, far from being deprived of oxygen, received a sudden access of oxygen (its internal concentration rose from 8½% to 15½% in one instance and from 3% to 16% in the other).

When these data are considered it is clear that Wardlaw demonstrated that a sudden increase in the concentration gradient of carbon dioxide across the flesh and periderm of the papaw fruit leads to the temporary emission of this gas at a rate higher than that of the steady state emission determined by the rate of its production by the cells. The impression remains that he demonstrated that a deficiency of oxygen decreases the capacity of tissue to retain carbon dioxide, and that the release of bound carbon dioxide explains the increase in rate

[1] In 2½ hours a concentration approaching 3% would develop under the conditions stated.

of its emission whenever an organ is placed in an atmosphere of nitrogen. Actually none of Wardlaw's experiments were mounted in such a way as to yield decisive information about this.

Boswell and Whiting (32) equated the excess carbon dioxide liberated on transfer of potato tubers from air to nitrogen with the decrease in amount of bound carbon dioxide in the tubers under anaerobic conditions. But the tissues used for determination of bound carbon dioxide were thin slices immersed in 5 ml fluid in Warburg vessels, a system that is radically different from the intact tuber. Judging by Burton's (36, 37) data for potatoes, one would expect the dissolved carbon dioxide in a whole tuber at the temperature used by Boswell and Whiting to be very much smaller than the total transitional excess that he recorded. Potato tubers may show an increase, decrease, or no change in initial carbon dioxide output when transferred from air to nitrogen after different times in storage (50) although the composition of their internal atmosphere is stable over long periods of storage life (35, 36).

The experiments of Wardlaw and of Boswell and Whiting are regarded as two classical demonstrations that excess carbon dioxide emitted when an organ is transferred from air to nitrogen has its origin in the inability of anaerobic tissue to retain as much of this gas as aerobic tissue. It is therefore important to realize that they did not, in fact, demonstrate this. This is not to say that carbon dioxide already present in the tissues may not be released when an organ becomes anaerobic. For instance, Wager (130) reported that the total carbon dioxide content of green peas (*Pisum sativum*) decreased when they were transferred to nitrogen from air, and that the difference could account for the transitional excess of carbon dioxide output over the basic level. But it cannot be assumed that this is always so. The rate of production of carbon dioxide by peas decreased immediately in nitrogen to about half its value in air, so the shift in solution equilibrium was toward lower content of carbon dioxide. But where the basic rate of production is unchanged or increased on transfer to nitrogen (30, 50) the shift in equilibrium, if any, would be in the direction of higher carbon dioxide content in the tissue and its interim result would be retention, not release, of the gas. It has not been shown that under these conditions anaerobiosis causes a reduction in "bound" carbon dioxide.

A transient outburst of carbon dioxide frequently appears when bulky organs are suddenly transferred to a higher temperature, and this is sometimes attributed to the released of carbon dioxide from solution, because the solubility of gas decreases as temperature rises. This intepretation neglects the fact that the rate of production of carbon dioxide by the cells also increases at higher temperature and leads to a higher concentration in the intercellular spaces. Leonard (91) set out useful information on the differential solubility of carbon dioxide and oxygen under conditions likely to be encountered in organisms. His tables show that the amount of carbon dioxide dissolving in 100 ml of water varies more with its concentration in the gas phase than with temperature in the appropriate range. He states that, for the conditions observed in papaw fruit, an increase in temperature from 25° to 35°C, would lead to an increase, not a decrease, in the amount of dissolved carbon dioxide.

It has been observed for a number of fleshy organs, for example, the apple (126) and the potato (36, 44), that the concentration of carbon dioxide in the internal atmosphere rises with temperature, and Burton found that in potato tubers dissolved and "bound" carbon dioxide increased with temperature above 10°C. Readjustment to satisfy the new physical conditions when the temperature is increased would therefore require retention of more carbon dioxide in the liquid phase rather than its liberation (44). This and other effects of a sudden change of temperature have been discussed by Forward (51).

Particularly in massive organs it is important to know how the physical system responds when any sudden change in conditions is imposed, and the relation of the rate of emission of carbon dioxide to its rate of production in transitional states poses special problems. But there is no good reason to suppose that in steady states the rate of evolution is not a satisfactory index of the rate of production even in fleshy fruits.

It remains to inquire whether the elevated carbon dioxide concentration in the internal atmosphere of massive organs normally limits their rate of respiration. Denny (44), Ben-Yehoshua et al. (21), among others, believe that it may. It is well known that high carbon dioxide in the external atmosphere may depress the respiration of fruit, but the effect is dependent on the species and the ontogenetic stage. The respiration of lemons may even be increased by the addition of 5 or 10% carbon dioxide to the storage atmosphere (141). Further information is required, but at present it appears probable that the respiration of fruit is under the control of other mechanisms.

2. Oxygen

The system involved in oxygen movement is of course like that for carbon dioxide with the direction reversed:

$$\text{atmospheric } O_2 \rightleftharpoons O_2 \text{ in intercellular spaces} \rightleftharpoons O_2 \text{ in solution} \rightleftharpoons \text{combined } O_2$$
$$(1) \qquad\qquad (2) \qquad\qquad (3)$$

Oxygen will diffuse into the intercellular spaces at a rate directly determined by the concentration gradient across the periderm (or cuticle) and by the resistance offered by this layer of tissue. Its internal concentration will depend on the rate of its removal in steps (2) and (3) as well as on the rate of diffusion from the exterior. The question that has received most attention is whether the internal concentration of oxygen sets a limit on the rate of respiration. If it does, is the effective barrier at step (1), (2), or (3)?

It may appear that in a massive organ enclosed by a relatively impervious skin the first step, diffusion to the surface of individual cells, especially those near the center of the tissue mass, would probably set a limitation on the rate of respiration of the intact organ, which is known to be much lower than that of slices of the same tissue. The available evidence does not, however, lend any very solid support to this supposition. While the internal atmosphere of fleshy fruits contains less oxygen than that in the surrounding air, the differential measured in a variety of organs is not so great as to lead one to expect any very marked effect on respiration rate.

Up to the time of the climacteric the fruits that have been examined have more than 15% oxygen in their internal atmosphere, and their respiration rate is at a minimum. During the climacteric the concentration of oxygen falls to 10% or sometimes less. It seems as if the internal oxygen concentration is the result, not the cause, of the activity of the chemical processes of respiratory metabolism.

In the ripening papaw (132), the banana (135), and the avocado (21), there appears to be no deficiency of oxygen until late senescence. In apples developing on the tree the same is true (57), while in apples ripening in storage the respiration rate can sometimes, but not always, be increased by substituting an external atmosphere of pure oxygen for air (30, 57). This is a usual test for oxygen deficiency. It greatly increases the internal oxygen concentration, and so demonstrates whether the equilibrium concentration with air outside was sufficient to support the maximum rate of respiration of which the cells were otherwise capable. It does not, however, reveal the cause of the deficiency if one exists.

When the resistance to diffusion of oxygen into apples is artificially increased by an impervious coating such as oil the respiration rate falls in proportion to the internal concentration, but the correlation does not always hold with uncoated fruits undergoing ontogenetic changes in rate of respiration and oxygen content (126). There is in fact, no good evidence that, before a late stage of ontogeny, the diffusion of oxygen through the skin of a fruit sets a limit on its rate of respiration, although the chief barrier to diffusion from the exterior to the surfaces of the cells lies in the skin.

Questions concerning oxygen deficiency have probably been more thoroughly examined for the potato tuber than for any other organ. Here the expectation that oxygen supply in the interior limits its consumption is at its height. A tuber is a compact organ with only about 1% of its total volume in intercellular space, and an early estimate of the resistance of its tissue to the diffusion of oxygen led Stiles and Dent (121) to believe that a potato tuber must be anaerobic except

TABLE I

GASEOUS AND DISSOLVED CARBON DIOXIDE AND OXYGEN IN POTATO TUBERS AT DIFFERENT TEMPERATURES[a]

	Carbon dioxide		Oxygen		
Temperature (°C)	In intercellular spaces (%)	Free in solution (ml/ml sap)	In intercellular spaces (%)	Free in solution (ml/ml sap)	Equil. conc.[b] in intercellular spaces (%)
5	2–2.5	0.025–0.03	18–18.5	0.007	19.4
10	2	0.02	18	0.006	15.8
15	4–4.5	0.035–0.04	17–18	0.006	17.0
20	5	0.04	16	0.0045	14.3
25	7–12	0.05–0.08	10–13	0.0025	11.5

[a] Data of Burton (35, 37).

[b] The concentration of oxygen in the intercellular spaces that would be in equilibrium with the oxygen in solution.

in a thin superficial layer. But when the composition of the internal atmosphere was determined, this proved to be a false conclusion. Burton (35) has made the most thorough study of the oxygen relations of the potato tuber, which shows no climacteric rise, but maintains a low rate of respiration and a stable internal atmosphere for months at all temperatures above those that cause sweetening. The chief barrier to diffusion is in the skin, and in the interior even the restricted intercellular passages are sufficient to provide for virtually uniform distribution of oxygen throughout the gas phase. Its concentration is above 16% in tubers held in air at temperatures between 5° and 20°C.

The composition varies with storage temperature, as shown in Table I. Not only are the intercellular spaces well aerated, but the oxygen in solution is to all intents and purposes in equilibrium with the gaseous phase, so there appears to be no barrier to dispersal through the liquid phase. The solubility of oxygen is low, and the diffusion coefficient of oxygen through water is several orders of magnitude lower than that through the gas phase; and it is conceivable that the main barrier to diffusion would be within the cells themselves. But Burton calculated that at the rate of oxygen consumption observed in potato tubers the oxygen pressure gradient from the surface to the center of a spherical cell 100 μ in diameter would be of the order of 10^{-5} atmosphere, which is negligible. Ducet and Rosenberg (45) has discussed the relation of cell size to diffusion of oxygen to the sites of the oxidative enzymes, presumably in the mitochondria. In the vacuolate plant cell, however, size is irrelevant, since the protoplasm with its inclusions is spread in a thin superficial layer. Furthermore, it streams in cyclosis, and in all probability the mean diffusion path for oxygen in the liquid phase is extremely short.

When oxygen is removed by combination with an enzyme the amount dissolved in the cell sap is decreased, and some will move in from the gaseous phase to restore equilibrium. This steepens the gradient between the internal and external atmosphere and increases the rate of diffusion across the periderm. The question is whether this last is adequate to supply the amount required by the respiratory machinery operating at a rate set by other factors, or whether the rate of diffusion of oxygen limits the rate at which the machinery can operate. Burton's data show that at temperatures up to 25°C the rate of diffusion does not limit the respiration rate, during dormancy or after sprouting begins, of several varieties of potato stored in air.

This conclusion is supported by Woolley's (140) measurements of the diffusion of argon (A^{37}) through cylinders of potato tissue at 24°C (and also of oxygen at 2°C). Woolley also estimated that oxygen diffusion would not limit respiration (at 24°C) in pieces of tuber whose diameter was less than 10 cm.

Burton confirmed his conclusion by showing that at 10° and 25°C the rate of respiration was not increased in an atmosphere of pure oxygen although the internal concentration of oxygen increased fivefold.

It is the common finding of those who have observed respiration rates of bulky organs at different temperatures that the temperature coefficient (Q_{10}) is of the order of 2 to 3. This suggests metabolic control rather than control by diffusion, which has a low Q_{10}. Woolley found the apparent diffusion coefficient of argon (A^{37}) through cylinders of potato tuber to be the same at 15.6 and 23.9°C.

The evidence is impressive that in the intact tuber the rate of respiration in air is not controlled by diffusion of oxygen from the exterior, and there is no positive evidence that this is a controlling factor in any fleshy organ before an advanced stage of senescence is reached.

3. Respiration of Whole Organs and Slices in Relation to Diffusion of Oxygen

Before other aspects of respiratory control are discussed, brief mention should be made of the relation between oxygen uptake by whole organs and by slices of tissue cut from them. Thin slices when freshly cut frequently respire at a much higher rate (five- to tenfold) than equal masses of the parent organ, and it was formerly assumed that this is because oxygen gains freer access to the bulk of the cells. But there is little justification for this assumption in the absence of evidence that it is diffusion of oxygen into whole organs that sets the pace of their respiration. When slices are cut and their individual surfaces are exposed to moist air or an aerated liquid medium, the superficial layers enter upon a new and radically different course of metabolic development that involves a further increase in the intensity of respiration. Many aspects of metabolism are affected and, although oxygen is necessary for the development of the divergent metabolism, other controlling factors have been implicated (see, for example, Volume II, Chapter 4 of this treatise; also Laties, 89). This is not the place to review the growing body of information about the respiration and concomitant metabolism of slices of storage organs. For our present purpose the point of interest is that improved access of oxygen to the cells is not now regarded as the effective reason for the higher respiration rate of slices. Evidence relating to this conclusion is discussed at some length by Laties (87–89). A recently published case in point may be cited (17). Slices from preclimacteric bananas respire at 5 times the rate of whole fruits, but slices taken from fruit at the climacteric peak respire at only one-third the rate in whole fruits, which is now equal to that of slices from preclimacteric fruit.

The internal atmosphere of cut tissues is not necessarily richer in oxygen and poorer in carbon dioxide than that of intact organs although the periderm is normally the chief barrier to oxygen movement. When avocado fruits are peeled at the time of picking, a climacteric rise in respiration is induced, and if they are peeled in the postclimacteric stage the rate of oxygen uptake drops (21). The effect is primarily an acceleration of the ontogenetic drift. Physical changes in the superficial layers of cells produce a new tough rind and the internal concentration of oxygen and carbon dioxide are very nearly the same, at corresponding stages of the respiratory drift, as that in unpeeled fruits.

Observations made by Woolley (140) showed that when cylinders are cut from potato tubers most intercellular spaces in a superficial layer a few microns thick become clogged with fluid, and this layer appears to offer the major resistance to discussion of argon (A^{37}) through the cylinder. Air can be easily drawn through the unwetted surface of a tuber under a low pressure differential (0.1 atmosphere), while a higher differential (0.7–2 atmospheres) is required to draw it through cylinders of the pulp.

The physical system is different when slices of tissue are used for measurement of oxygen uptake by conventional Warburg methods. The slices are immersed in a fluid medium, and the gaseous phase in the intercellular spaces communicates with that in the flask only through the fluid. Even when this is shaken to equilibrate its oxygen content with that in the gas above it, unstirred layers at the tissue surfaces may effectively increase the length of the diffusion path of oxygen in the liquid phase (cf. 102).

Laties (87–89) presents an array of evidence that neither in potato slices nor in the whole tuber does diffusion of oxygen limit the rate of respiration.

V. Oxygen Dependence

Although the rate of respiration of storage organs surrounded by air frequently appears not to be limited by access of oxygen to the cells in the interior, the respiration of many such organs is dependent on oxygen concentration in the atmosphere over a broad range, and particularly at a partial pressure much below that in air. Figure 11 shows four examples of the relation between respiration and the partial pressure of oxygen in the gas supplied. It will be evident from foregoing sections that the relation might be quantitatively different for other species or for the same organs in different ontogenetic states. But the general form is similar for a wide variety of organs. It approaches that of a rectangular hyperbola, and therefore resembles the relation between substrate concentration and rate of reaction in an enzyme-catalyzed system according to the formulation of Michaelis and Menten. Such oxygen dependence curves are usually interpreted as representing the effect of the concentration of oxygen as a reactant in the final step in the chain of transport of electrons, i.e., from the oxidase to oxygen, and this carries the implication that this final step limits the rate of respiration.

This interpretation leads at once to difficulties, because cytochrome oxidase, the one oxidase that is firmly established as a universal mediator of the normal respiration of cells, has a very high affinity for oxygen, and should be saturated at a partial pressure below 0.01 atmosphere.

Three kinds of explanation of this apparent anomaly have been offered at one time or another:

1. Respiration is catalyzed by cytochrome oxidase, but diffusion of oxygen through plant tissues is slow, and the actual concentration of oxygen in the mitochondria is very much lower than that in the

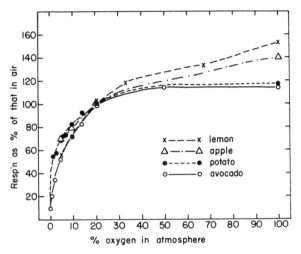

FIG. 11. The relation between respiration rate and percentage concentration of oxygen in the external atmosphere. Drawn from the data of Biale (23) for avocado, Biale and Young (26) for lemon, Blackman (30) for apple, and Mapson and Burton (95) for potato.

external atmosphere. Consequently even in air the enzyme may not be saturated with oxygen.

2. A substantial part of the respiration in organs that show oxygen dependence is not mediated by cytochrome oxidase, but rather by an enzyme with a much lower affinity for oxygen.

3. In the intact organ the rate of respiration is not limited by the rate of the final transfer of electrons to oxygen, but by some earlier step in the integrated metabolism that supplies reduced oxidase and hydrogen ions. A corollary of this interpretation is that the presence of oxygen affects the pace-setting reaction directly or indirectly.

A. DIFFUSION OF OXYGEN

The first explanation, limitation by slow oxygen diffusion, continues to express the popular view, although experimental evidence against it exists for a variety of organs, and especially for the potato tuber which is a classical example of an ill-ventilated organ. To the evidence presented in the foregoing section we may add that the form of the curve relating oxygen uptake and external concentration is not that to be expected if diffusion were the limiting factor. Diffusion is directly proportional to the concentration gradient of the diffusing substance, and Burton (40) has, in fact, shown that the diffusion of oxygen into tubers previously freed of oxygen is linearly, not hyperbolically, related to the external partial pressure. And Laties (89) has shown that the relation of oxygen uptake by potato slices to their thickness

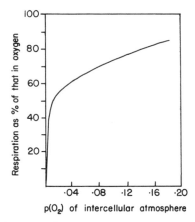

FIG. 12. The relation between the rate of oxygen uptake and the partial pressure (in atmospheres) of oxygen in the intercellular spaces of 'King Edward' potato tubers. From Mapson and Burton (95).

does not conform to that between oxygen diffusion and thickness of the slices. Furthermore Mapson and Burton (95) have shown that the oxygen uptake by potato tubers increases with partial pressure of oxygen up to 0.2 atmosphere *in the intercellular spaces* (see Fig. 12).

There remains the possibility that diffusion within the cell limits the rate of oxygen utilization. Our knowledge of oxygen movement within the cell is negligible. We know that in a vacuolate cell the diffusion path to the mitochondria from the cell surface is extremely short, and that the cytochromes are incorporated in the membrane of the mitochondria. This would lead us to expect that, once at the cell surface, oxygen would gain rapid access to the oxidase. Mention has already been made of Burton's (35) demonstration that dissolved oxygen in a potato tuber is in virtual equilibrium with the gaseous oxygen in the intercellular spaces, and of his calculation that diffusion through the liquid phase to the interior of a cell would cause only negligible lowering of concentration along a 50 μ path. It is not known, however, whether some architectural feature of the cell may interpose a barrier to diffusion into the mitochondrial membrane.

Instances are known where diffusion of oxygen appears to limit its uptake, and their infrequent occurrence serves to emphasize that the respiration sites in most plant tissues are accessible to oxygen. The intact testa of some seeds is highly resistant to the diffusion of oxygen, and fermentation accompanies respiration in early stages of germination (see Part 1 of this chapter, p. 249). But when the seed coat is ruptured, fermentation ceases and there is no further indication of limited oxygen access. In the meristematic tip of a root, intercellular spaces are absent and respiration rate is high. At moderately high temperatures and low oxygen pressures, respiration has a low temperature coefficient ($Q_{10} = 1.0\text{--}1.5$) and aerobic fermentation appears. At lower temperature and higher oxygen pressure, respiration has a high Q_{10} (2.0), and no fermentation occurs. Under the first set of circumstances diffusion of oxygen may well limit the rate of oxygen uptake, while under the second set it probably does not [see for example, Berry and Norris (22) and Forward (51)]. Plants readily ferment excess substrate when access of oxygen is limited and the absence of fermentation is in itself an indication that diffusion of oxygen is not setting the pace for its uptake. It is only in exceptional cases that respiration of the organs of higher plants is accompanied by fermentation unless the external oxygen concentration is reduced to such an extent that the respiration rate is much below that in air. And it is only in exceptional cases that their respiration rate has a temperature coefficient appreciably below 2.0 (51). It is probable that very generally their rate of respiration is not limited by oxygen diffusion.

B. TERMINAL OXIDASES

It has been repeatedly suggested that oxidases other than cytochrome a_3 play a major role in mediating the respiration of higher plants. This is not only because of oxygen dependence, but also because inhibitors of cytochrome oxidase do not always inhibit the respiration of higher plant tissues, and because these are known to contain other active oxidases. Cytochrome b_7, peroxidase, phenol oxidase, and ascorbic acid oxidase are widely distributed, in addition to glycolic oxidase, which is confined to green tissues. Cytochrome b_7 is abundant in some tissues, especially those whose respiration is cyanide resistant. It is oxidized by a route that by-passes cytochrome a_3 and may be autoxidizable. While no link between peroxidase and respiration has been revealed, systems have been discovered that would provide for the cyclic oxidation and reduction of phenol oxidase and ascorbic acid oxidase. Potentially, both could serve as catalysts for continuous electron transfer to oxygen. Both have relatively low affinity for oxygen, of such magnitude that if they supplemented cytochrome oxidase, the combination could explain the oxygen dependence that is observed.

No agreement has yet been reached about the role of these so-called soluble oxidases in the normal respiration of plants. James (75) and Mapson (94) have reviewed arguments in favor of the participation of ascorbic acid oxidase. Hackett (54) and Beevers (19), for example, have presented the case for cytochromes as the unique carriers of electrons in normal respiration.

Certainly some plant tissues in some physiological states exhibit resistance to known inhibitors of cytochrome oxidase, and some are more sensitive to inhibitors of copper enzymes. On this behavior rests a large part of the argument that a copper-containing enzyme such as ascorbic acid oxidase or phenol oxidase is mediating some part of the respiration. The objections to evidence based on inhibitors are: (a) No inhibitor can be guaranteed to be absolutely specific in its reaction, and in organized tissue it may have unknown effects. For example, cyanide, which inhibits all heavy metal oxidases, and which inhibits oxygen uptake by freshly cut potato slices, increases the rate of oxygen uptake by intact tubers (59). This is apparently because it also increases the concentration of sucrose, which in potato tubers is correlated with respiration rate. (b) If the rate of respiration is not limited by the final transfer of electrons by the oxidase, an inhibitor may impair the activity of the enzyme without having a corresponding effect on the rate of respiration. In many tissues (see Hackett, 54) there is a large excess of cytochrome oxidase over that required to maintain observed rates of oxygen uptake and a substantial proportion of it could be inactivated without affecting the rate of respiration.

Mapson and Burton (95) found evidence that the same is not true in small potato tubers. Figure 12 represents a two-phase relation between respiration and partial pressure of oxygen in the intercellular atmosphere, which is interpreted as the outcome of mixed oxygen uptake by two terminal oxidase systems. The apparent half-saturation point for over-all oxygen uptake is at 0.01 atmosphere or less. Respiration rate rises rapidly from zero to half-maximal at about 0.01 atmosphere of oxygen in contact with the cells and then much more slowly up to 1 atmosphere. These data could be explained by assuming the participation of two oxidases, one with an oxygen affinity $(1/K_m)$ of

2.5×10^6. This corresponds approximately to the affinity of cytochrome a_3 for oxygen, and would account for about 70% of the respiration in air and 50% in pure oxygen. Cytochromes a, c, and b were observed spectroscopically. At a partial pressure of 0.01 atmosphere oxygen, cytochromes a and c were virtually all oxidized. If, at oxygen concentrations higher than 1% this high affinity system operated at full capacity and a second system with an oxygen affinity of 6.5×10^3 were added, the observed curve of oxygen uptake would be reproduced. The hypothetical second oxidase would correspond in oxygen affinity to ascorbic acid oxidase or phenol oxidase. No ascorbic acid oxidase could be found in the tubers. Phenol oxidase is present in potatoes and has been proposed by Boswell and Whiting (31) and others as a mediator of respiration. In no tissue, however, has it been unequivocally established as fulfilling this function. An alternative possibility is suggested below (p. 367).

One difficulty in conceiving a role for low-affinity oxidases is that oxygen would be snapped up by the universally occurring high-affinity cytochromes unless these were insufficient in amount, or unless oxygen were deterred from access to them. A more serious obstacle to the acceptance of either phenol oxidase or ascorbic acid oxidase as playing an important role in respiration is that no system other than a cytochrome chain has been clearly shown to be linked to the generation of ATP in respiration. It has been suggested (see Marré, 97) that ascorbic acid oxidase may be associated with oxidative phosphorylation, but until it is clear that extramitochondrial oxidations are linked with an energy-transferring mechanism they cannot be seriously considered as major respiratory catalysts.

On the other hand there is clear evidence [see for example, Simon (114) and Wiskich and Bonner(138)] that some alternative to cytochrome a_3 operates in the mitochondria of some plants, including the storage root of sweet potato. But there is not yet enough information to identify the alternate pathway. The weight of evidence at present is that respiration yielding energy through ATP formation occurs in the mitochondria and is mediated by some variant of the cytochrome system of catalysts. Other oxidases are undoubtedly present, but their role is yet to be determined.

C. THE PACE-SETTING REACTION

Evidence for the control of respiration rate by some step other than the final transfer of electrons and hydrogen ions to oxygen is now familiar. Spectroscopic evidence is that the reduction of cytochromes is a slower process than their oxidation and that the pace of electron transfer is determined earlier in the chain. Many plant tissues have oxidative capacity well beyond that used in normal respiration and can be stimulated by various means to respire at a higher rate. One means that is frequently effective is the addition of a phosporylation uncoupler. It has already been mentioned that apple and avocado fruits in the preclimacteric phase respond to dinitrophenol by a marked increase in oxygen uptake, and in this phase presumably the rate of respiration is kept low by the relatively slow pace at which phosphorylation occurs. In the young growing phase and at the climacteric peak this is no longer true although the best evidence is that oxidation is linked to phosphorylation in these stages.

One conclusion that can be drawn from this example is that the control of respiration is not always the same. The organism must have more than one kind of regulatory mechanism at its disposal, but there is no really convincing evidence that, except at external oxygen concentrations well below that in air, the concentration of oxygen as a reactant in the terminal step in oxidation is an effective one.

Much of our knowledge of the mechanism of respiratory control within the electron transport chain is derived from studies of isolated mitochondria. While we may presume that the same mechanisms operate in organized tissue, their operation will necessarily be regulated by the milieu in which the mitochrondia exist.

The usual criterion for successful preparation of mitochondria is respiratory control by ADP. If mitochondrial oxidation is also under the control of a phosphate acceptor *in vivo*, then we must look for the means of regulation of the availability of the acceptor. Our knowledge of such regulation is decidedly meager, but not entirely lacking. ADP may arise outside the mitochondria by the utilization of ATP in endergonic reactions. One such reaction is the phorphorylation of sugar. For ascites cells Chance and Hess (41) have proposed a feedback control in which the limiting step is diffusion of ATP, formed during respiration, from the mitochondria to the sites of glucose phosphorylation. Other syntheses, including protein synthesis, occur in specific regions of the cell outside the mitochondria and are also dependent on the availability of ATP. Some of these, in addition to releasing ADP, may be responsible for removal of inhibiting substances. Wiskich and Bonner (138), for example, suggest that removal of oxaloacetate, which inhibits succinic dehydrogenase, depends on a high ATP level. In all probability the effective regulation is not always the same, nor does it always act through regulation of phosphorylation.

Little attention has yet been paid to the implication that, in tissues showing oxygen dependence, if the pace-setting step for oxygen uptake is not the final transfer of electrons to oxygen, then oxygen must have an additional role beyond that of a reactant in the terminal step, i.e., it must affect the pace-setting reaction directly or indirectly. There is evidence of a secondary effect of oxygen on metabolic balance in the gradual approach of respiration rate to a new steady state when an apple is transferred from air to high or low oxygen (30). It requires 45 hours at 22°C for adjustment to a new level of drift paralleling that in air (cf. Fig. 16). This implies that oxygen has some regulatory role beyond the immediate acceptance of electrons and hydrogen ions.

At present we are largely ignorant of the secondary effects of oxygen in the higher plants, whch are strict aerobes. Goldfine and Block (52) have called attention to certain specific syntheses that require the presence of molecular oxygen, and there may well be other reactions still unknown that are regulated by oxygen supply in organized tissue. Chance and Hess (41) suggested that removal of oxygen removes a barrier to diffusion of ATP to the sites of glucose phosphorylation in intact ascites cells and, by implication, that oxygen is required for maintenance of the barrier. Protoplasmic organization breaks down in the absence of oxygen, and it is reasonable to assume that in organized tissue oxygen may control the convergence of substrate and catalyst. The whole subject of secondary effects of oxygen on the metabolism of aerobes is largely unexplored.

It is obvious that the dependence of plant organs on the concentration of oxygen in the atmosphere has not yet been fully explained. Of the three solutions offered, the third holds the greatest potential for providing an explanation.

VI. Respiration and Fermentation

One approach to the possible metabolic roles of oxygen is through a comparison of aerobic and anaerobic metabolism. Storage organs can be exploited for this purpose because they survive long enough without oxygen to allow metabolic adjustments to occur.

Higher plants readily ferment sugar in the absence of oxygen, and also in its presence when for any reason oxidation of pyruvate by the usual channels is inadequate. By far the most general products are ethanol and carbon dioxide together with traces of acetaldehyde. Theoretically, equimolar amounts of alcohol and carbon dioxide should result from the fermentation of glucose, and this condition is met, at least approximately, in some storage organs, such as apple fruits and carrot roots in certain physiological states. The potato tuber is classically an exception. It may produce little or no alcohol in the absence of oxygen while the production of carbon dioxide is considerable. Initially, on transfer of a tuber to nitrogen, the rate of carbon dioxide production may be above or below that in air. The usual drift on continued exposure to nitrogen is downward to an adjusted rate parallel to that in air, but lower. The form of the drift, however, is variable, as is the adjusted ratio to air. These depend on time in storage and on temperature (11–13, 50) among other factors. Figure

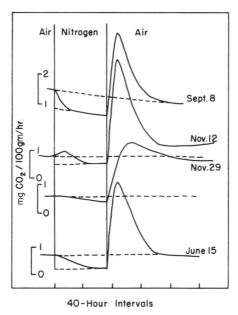

40-Hour Intervals

Fig. 13. Aerobic and anaerobic production of carbon dioxide at 22°C by 'Sebago' potato tubers at harvest (Sept. 8) or after various periods of storage (Nov. 12, Nov. 29, June 15). Solid lines show four types of response to transfer of the tuber from air to nitrogen, and back to air. Broken lines show the drift in tubers held in air throughout, or extrapolation of the adjusted anaerobic rate to the time of transfer to nitrogen. From Forward (50).

13 illustrates the spontaneous changes in form of drift that develop as tubers age in storage.

The anaerobic metabolism of potato tubers is complex and has not been completely analyzed. Barker (11, 12) reported that lactic acid may accumulate instead of, or along with, alcohol, and this has been confirmed by Forward (50a). Figure 14 shows the rate of production of carbon dioxide, lactic acid, and alcohol by 'King Edward' potatoes held in nitrogen at 10°C. In these tubers lactic acid began to accumulate immediately and was succeeded by a period of alcohol accumulation which was delayed for 8–10 days. For 20 days the carbon dioxide emitted exceeded the amount that would correspond to the alcohol produced. The source of this extra carbon dioxide has not yet been determined. It is too great in quantity to be dissolved or bound carbon dioxide released in the absence of oxygen (9), and its appearance indicates the occurrence of a reaction other than the reduction of pyruvate to lactic acid.

The pattern of fermentation is modified by temperature. At 1°C potatoes sweeten and develop a high rate of respiration in air, and when sweetened tubers are transferred to nitrogen, the production of carbon dioxide drops away rapidly, lactic acid is formed, and alcohol

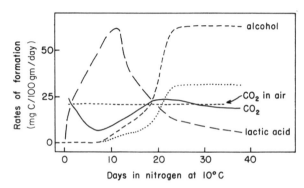

FIG. 14. The rates of production of carbon dioxide, lactic acid, and alcohol by 'King Edward' potatoes in nitrogen at 10°C. The dotted line shows the theoretical carbon dioxide production corresponding to ethanol formation. From Barker and El Saifi (13).

appears after a lag of only one day (11). Alcoholic fermentation accompanies lactic acid production instead of succeeding it. Barker associates the early appearance of alcohol with high sugar content of the tubers. But this is not the only cause of a change in fermentative metabolism, for at 25° and 35°C other patterns of production of carbon dioxide, alcohol, and lactic acid appear (51a).

When air is restored major readjustments in metabolism occur, at least as complex as those associated with anaerobiosis. These have been observed with 'King Edward' potatoes at 10° and 1°C by Barker (8, 9, 11, 14, 15) and with 'Sebago' at 22°C by the author. The pattern differs at different temperatures, but the basic responses are similar (Fig. 15). An outburst of carbon dioxide production precedes a steady drift equal to or higher than that of tubers remaining always in air. An equally dramatic increase in the content of sugars, especially sucrose, follows at varying time intervals. At 10°C sucrose content begins at once to subside to the level characteristic of aerobic tubers, while at 1°C and (at certain storage ages) at 22°C, it may remain high temporarily or indefinitely. Lactic acid disappears and cannot all be accounted for as carbon dioxide. Keto acids, first pyruvic and then α-ketoglutaric, increase sharply and then diminish during the carbon dioxide outburst. Barker has interpreted this as an indication that lactic acid is reoxidized to pyruvic, which traverses the tricarboxylic acid cycle, yielding carbon dioxide from pyruvic, oxalosuccinic,

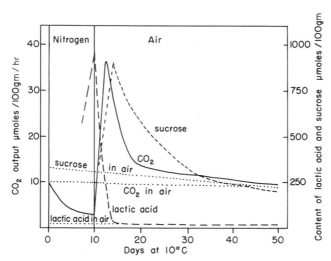

FIG. 15. Changes in the rate of carbon dioxide production and in the content of lactic acid and sucrose in 'King Edward' potatoes in nitrogen and in air after nitrogen at 10°C. From Barker and Mapson (11).

and α-ketoglutaric acids. Laties (90), however, presents evidence that the tricarboxylic acid cycle is not active in slices freshly cut from potato tubers, but is blocked between citric and α-ketoglutaric acids. Since Barker (10) has shown that under high oxygen pressure a block develops at the same point in the cycle, the possibility arises that a period of anaerobiosis removes the block and temporarily allows freer operation of the cycle during the "after effect." Not enough information is yet available for a decision on this point, but it is clear that the presence or absence of oxygen may have important secondary effects on the balance of metabolic processes, even to causing qualitative differences in the pathways traversed.

In apple fruits the situation is simpler. Anaerobic fermentation is essentially alcoholic, and no "after effect" occurs when air follows nitrogen. With such an organ the output of carbon dioxide can be used with more confidence as an index to respiratory and fermentative metabolism. Blackman's (30) analysis of the carbon dioxide output

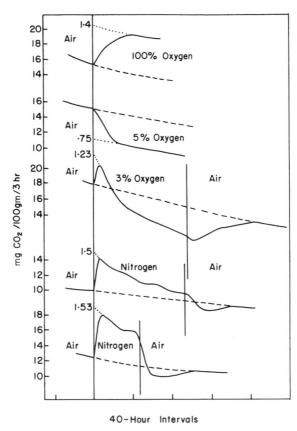

Fig. 16. The carbon dioxide output of 'Bramley's Seedling' apples in air, in nitrogen, and in different mixtures of oxygen and nitrogen. Solid lines show continuous records of carbon dioxide production in air, then in the experimental gas, then in air. Broken lines show the drifts in air only. Vertical lines mark the times of transfer. From Blackman (30).

of apples exploited this approach. He transferred individual apples from air to nitrogen gas and to atmospheres containing from 3% to 100% oxygen and made continuous records of their carbon dioxide production. Examples are shown in Fig. 16. In atmospheres with 5% oxygen or more, the rate of output gradually assumed a new level parallel to the airline drift and bearing the relation to it that is shown in Fig. 17. In nitrogen or 3% oxygen, the output rose rapidly and then declined toward the air line. There is no question of the extra carbon dioxide being simply the result of the release of bound carbon dioxide, since the rate of output remained high for many hours. Failure to reach an adjusted level parallel to the air line means that there is no constant relation between the rate of carbon dioxide output in air and that in nitrogen. Consequently, if a ratio is to be used as a test for the conservation of carbon in air the only significant one is that of the initial rate in nitrogen to the final rate in air.

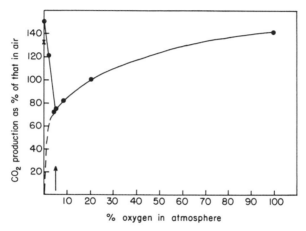

FIG. 17. The respiration of 'Bramley's Seedling' apples in relation to oxygen supply. Above 5% oxygen in the external atmosphere, oxygen uptake and carbon dioxide output are equal; below 5% the oxygen uptake is shown by a broken line and the carbon dioxide output by a solid line. The difference represents anaerobically produced carbon dioxide. The arrow marks the extinction point. Drawn from the data of Blackman (30) and Watson (136).

Blackman, presuming that the initial rising rate reflected physical adjustments of oxygen and carbon dioxide content of the tissue, extrapolated the declining drift to the time of transfer from air to obtain the ratios that are incorporated in Fig. 17 for 3% and 0% oxygen.

It is apparent that the output of carbon dioxide reached a minimum at about 5% oxygen and increased both below and above this concentration. Watson (136) later showed that this minimum carbon dioxide production coincides with the lowest external oxygen concentration at which no fermentation occurs, i.e., the extinction point. At higher concentrations oxygen uptake and carbon dioxide output increase together; at lower concentrations oxygen uptake decreases to reach zero in nitrogen, while fermentation increases to reach a maximum in nitrogen.

The concentration of oxygen that exists in air has no special physiological significance, and the appropriate indication of carbon conservation is not the ratio of carbon dioxide output in nitrogen to that in air, but rather to that at the extinction point. This has not been determined for many tissues, but where it has, no instance has been encountered in which the ratio is as low as 1:3. A few ratios based on the extinction point are cited by Thomas *et al.* (125).

The products of fermentation in apples are ethanol and carbon dioxide; so, for every carbon atom that appears as carbon dioxide, two are, ideally, retained in the tissue as alcohol. The carbon loss in fermentation, in terms of carbon would therefore be three times the carbon dioxide production.[2] The carbon loss in aerobic respiration is, however, all represented by the carbon dioxide. In the apples that were placed in nitrogen, the carbon loss, estimated in this way, was initially 4.5 times that in air at the moment of transfer (cf. Fig. 16).

There are two possible explanations. Either the rate of carbon loss indicates the rate of sugar breakdown which rises rapidly to 4.5 times its rate in air, or carbon that is conserved in air cannot be used in the same way in the absence of oxygen, and intermediate carbon compounds are fermented. Blackman regarded the second alternative as more probable. In the first place it seems unlikely, *prima facie,* that the removal of oxygen would increase the rate of sugar breakdown fourfold and at once cause it to decline steadily. In the second place, it is difficult to understand why the rate of sugar breakdown should be lower in 5% oxygen than in air, but higher in 3% oxygen or nitrogen than in air. Blackman's interpretation is illustrated in Fig. 18, where the upper lines represent the hypothetical rate of sugar breakdown declining slowly in air, dropping away more steeply in nitrogen, and returning to the original drift when air is restored. The difference between carbon dioxide production and sugar breakdown in nitrogen represents the carbon of alcohol, that between carbon dioxide production and sugar breakdown in air represents carbon incorporated into other tissue components which can be formed only in the presence of oxygen.

The alternative interpretation of Fig. 16 is that the rate of sugar breakdown is higher in nitrogen than in air, and this is the more generally accepted view. There is very little pertinent information

[2] The actual production of alcohol in apples is sufficiently close to the ideal to validate this assumption for the purposes of the analysis that Blackman made (47).

about higher plants on which to base a decision. Higher carbon loss in nitrogen than in air has been observed with a variety of tissues, but this merely shows that the phenomenon is not a unique property of apples. Only rarely have measurements of sugar loss been made. Neal and Girton (101) report that sugar is consumed more rapidly by corn (*Zea mays*) root tips in nitrogen than when they are in air. They found that in nitrogen the net loss of sugar was equivalent to the carbon dioxide plus ethanol produced; in air it was equivalent to carbon dioxide production alone. This presents a simple picture

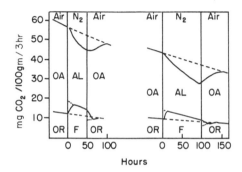

Fig. 18. Hypothetical construction of the glycolysis line for apples in air, then nitrogen, then air. The lower graphs show the rate of production of carbon dioxide (*OR* in air, F in nitrogen). The upper graphs represent the hypothetical rate of glycolysis. In nitrogen this is $3 \times F$ (i.e., the carbon loss) extrapolated to the time of transfer from air. In air the glycolysis line bears a constant relation to the carbon dioxide production and is made to coincide with carbon loss in nitrogen at the time of transfer from air. In nitrogen the difference between the carbon dioxide line and the glycolysis line represents the carbon retained in ethanol (*AL*). In air the difference is assigned to the carbon conserved in oxidative anabolism (*OA*). From Blackman (30).

of inhibition of sugar breakdown by oxygen and the oxidation of all the glycolytic product in air. These authors point out that the apparent equivalence of substrate loss and product gain in nitrogen is deceiving, and one can surmise that the same is true in air. It is difficult to believe that in a root tip containing actively dividing and elongating cells the only fate of sugar is to be lost as carbon dioxide. However, the net loss of sugar was, under the conditions of these experiments, greater in nitrogen than in air. Fidler (49) found the loss of carbohydrate by apples greater in nitrogen than in air after long periods of exposure. Measured loss over a 2-day period was less in nitrogen than in air. But his method did not show significant differences for exposures up to a week or more, and longer exposures probably entail secondary effects of anaerobiosis. Platenius (107) measured sugar loss in shelled peas at different oxygen pressures, and found that the net loss decreased along with respiration rate, from air down to the extinction point (3% oxygen). The sugar loss was the same at 2% and 3% oxygen, but unfortunately he did not determine it in the complete absence of oxygen, nor has anyone explored the whole range of oxygen concentration from that in air down to zero. Obviously the evidence is inadequate and, at best, the net loss of sugar does not necessarily represent the rate of sugar breakdown. But there is no positive evidence from analyses of this kind that in the presence of oxygen sugar breakdown is greater than in its absence, although there is evidence that it is enhanced as the oxygen concentration rises above the extinction point.

The whole question of the relation between aerobic and anaerobic metabolism is obscured by ambiguity in interpretation of the Pasteur effect. It has already been mentioned by Yemm in Part 1 of this chapter that Pasteur in his studies on beer discovered that in air yeast produced less alcohol and more yeast per unit of sugar consumed than it did in the absence of oxygen. In other words fermentation was replaced by respiration and synthesis of cell material. It would seem reasonable, as James (74) has stated, to regard this as the "Pasteur effect." But in the succeeding years various aspects of the relation between aerobic and anaerobic metabolism have arrested the attention of biochemists and physiologists, and have been associated with the Pasteur effect. To most biochemists today the term carries the implication that sugar catabolism is inhibited by oxidative processes. There was no such implication in Pasteur's treatment of the phenomenon, and since, as we now know, the pathway to protein synthesis lies through the tricarboxylic acid cycle, it is hard to see how inhibition of sugar breakdown could lead to greater synthesis of protein under aerobic conditions. Plant physiologists generally associate the Pasteur effect with conservation of carbon in the presence of oxygen, a conception that is nearer the original, but the assumption is usually implicit in their interpretation of experimental data that the breakdown of sugar is inhibited.

In spite of many attempts the Pasteur effect as it is known to the biochemist has not been fully explained, and to Chance and Hess (41) this suggests that "the consequences of initiation or cessation of oxygen metabolism are manifold and not easy to interpret." Various hypotheses have been discussed recently by Aisenberg (1). It seems highly probable that the ADP–ATP system is involved. Addition of a phosphorylation uncoupler such as 2,4-dinitrophenol (DNP) increases the rate of aerobic fermentation in many tissues. In yeast, for instance, the total carbon loss rises in the presence of DNP to the anaerobic level, and this is usually interpreted as release of the inhibition imposed by oxidative phosphorylation. With higher plant tissues, however, the effects of DNP cannot always be explained in the same way. Beevers (18) found, with a variety of tissues, that carbon loss in air with DNP present rose well above the anaerobic level. He attributed this to a direct stimulation of glycolysis by DNP, but with one tissue at least, corn coleoptile, DNP did not stimulate glycolysis in the absence of oxygen, and with the uncoupler present in both aerobic and anaerobic states, carbon loss was much greater in air than in nitrogen. It could be because sugar breakdown was greater in air, and intermediates that were blocked by DNP from entering synthetic reactions were diverted to carbon dioxide production by respiration and fermentation.

With disks of tobacco (*Nicotiana tabacum*) leaves in an atmosphere of oxygen, Porter and Runeckles (108) found that DNP in a concentration that increased respiration did not increase the utilization of labeled sugar and the loss of starch. The extra carbon dioxide produced in respiration was evidently derived from a noncarbohydrate source. Their observations were not consistent with the view that DNP increases glycolysis, but were consistent with the assumption that the rate of glycolysis is controlled by the phosphorylation of sugar, which requires phosphate bond energy supplied in the presence of oxygen.

Evidence concerning the effect of oxidative phosphorylation on glycolysis in plant tissues remains inadequate, but the above examples indicate that hypotheses that may serve to explain the behavior of systems reconstituted *in vitro*, or of the cells of yeast or of tumors that normally exhibit aerobic fermentation, may not be adequate for the strictly aerobic tissues of higher plants. In a reconstituted system of supernatant plus mitochondria, ADP and inorganic phosphate may well be depleted by oxidative phosphorylation, but in an organism that is utilizing ATP they are restored. If availability of phosphate acceptor controls glycolysis a higher rate of ATP turnover would be expected to cause a higher rate of glycolysis. If the control point is the production of hexose diphosphate, then it becomes necessary to explain why the vastly greater production of ATP in oxidative than in glycolytic phosphorylation should not increase rather than decrease glycolysis. Sequestration of ATP in the mitochondria (41) or in a glycolyzing organelle (7) has been proposed. The former would, of course, restrict ATP at all extramitochondrial sites of endergonic reactions, and would interpose a difficulty in explaining why protein synthesis and growth are associated with active respiration. An alternative hypothesis is that the rate of sugar utilization is not actually lower in organized tissues in the presence of oxygen, but that some of the products are diverted to other reactions and do not appear as carbon dioxide. Respiration is not an isolated process, but is interlocked with other reaction sequences, and these should not be neglected in considering the relation between the aerobic and anaerobic metabolism of an organ.

VII. The Relation of Respiration to Other Metabolism; the Role of Oxygen

Interactions between respiratory and other metabolism have been discussed at some length by Yemm in Part 1 of this chapter. The bearing of some of them on an interpretation of respiratory control will be considered here. A comprehensive dynamic framework for a theory of the relation of respiration to the metabolic system as a whole was constructed by Blackman (30) to explain the production of carbon dioxide by apple fruits. Since it remains the only proposal of the same scope, its basic provisions will be set out, and then it will be considered in relation to more recent developments.

A. Blackman's Schema

The essential features may be summarized as follows:

1. The purely oxidative aspects of respiration involving the flux of hydrogen and oxygen cannot be divorced from the flux of carbon.

2. Oxygen has more than one effect on respiratory metabolism. Not only is it a reactant in oxidation, but it also affects the balance of processes in the metabolism of carbon.

3. The effective substrate for respiration is a labile derivative of sugar, and the rate of its production determines the rate of glycolysis.[3]

4. The products of glycolysis are removed as rapidly as they are formed.

5. Their oxidation is accompanied by diversion of carbon into syntheses which utilize energy rendered available in conjunction with oxidative respiration.

6. In the absence of oxygen the whole product of glycolysis is fermented, both that which would be oxidized and that which would go into tissue components in fully aerated tissue.

7. The rate of production of the labile substrate for glycolysis increases with the supply of free oxygen in the tissues.

[3] Glycolysis is used here to refer to the breakdown of sugar to intermediates that may serve as substrate for fermentation, oxidation, or other reactions. It does not include the final steps of fermentation.

8. Respiration occurs within a complex metabolic framework, and the whole is regulated by the organization of the protoplast.

These principles were derived from analysis of continuous records of carbon dioxide output by apples, examples of which are shown in Fig. 16. From the gradual approach to a new adjusted level when an apple is transferred from air to another supercritical oxygen concentration, Blackman concluded that oxygen has a secondary role in the metabolism of the apple, in which it is not consumed as a reactant but affects metabolic balance. From the initial relatively rapid increase in carbon loss when an apple is transferred to a subcritical oxygen concentration, including nitrogen, he concluded that in the absence of free oxygen in the tissue a qualitative change occurs in the fate of carbon. Since carbon loss declines at a measurable pace in nitrogen, he concluded that the initial change was not in the rate of glycolysis, and since synthetic reactions requiring energy provided by respiration could not proceed without it he proposed that the initial change was a diversion of intermediates of sugar breakdown, away from endergonic reactions to fermentation. Because carbon loss increases with oxygen concentration above the extinction point and the products of glycolysis do not accumulate, he proposed that free oxygen in the tissues increases the rate of glycolysis which, at any concentration above the extinction point is higher, not lower, than in nitrogen or at any oxygen concentration below the extinction point.

All these postulations are the outcome of dynamic considerations. Blackman speculated to some extent about the actual substances and mechanisms involved, but none of the principles stated is dependent on the participation of any specific substance or metabolic pathway.

B. INTERPRETATION IN TERMS OF MODERN KNOWLEDGE

Since this scheme was proposed, great advances have been made in knowledge of the biochemical pathways involved in respiratory metabolism, and of interconnections among metabolic pathways. It is now clear, for instance, that the labile derivatives of sugars that serve as respiratory substrate are phosphate esters of hexoses, and there is evidence that in some instances the rate of glycolysis is controlled by the reactions catalyzed by phosphohexokinase and hexokinase. It has been established that plant tissues very generally use the Embden-Meyerhof pathway of sugar breakdown, but in addition it now seems highly probable that storage organs, among other tissues of higher plants, use the pentose phosphate pathway. It has not yet been possible to estimate the proportion of glucose that enters this cycle, but it is evidently exploited more by plants than by animals (see a discussion

by Beevers, 19). It is not, however, entirely independent of the glycolytic pathway. Ap Rees and Beevers (110, 111) found, with tissues
that use the pentose pathway to some extent, that specifically labeled
carbon atoms of glucose are retained in detectable quantity only in
components derived from the oxidation of pyruvate or in carbohydrates.
This suggests that the pentose pathway may act as a shunt route
from glucose phosphate to phosphoglyceraldehyde and so to pyruvic
acid. It is not known to what extent it acts as a cycle and to what
extent as a shunt.

Pyruvic acid may be alternatively fermented or oxidized by known
pathways. The tricarboxylic acid cycle operates in many plant tissues
and affords connections with amino acid and protein metabolism. These
and other linkages between intermediary respiratory metabolism and
synthetic reactions are illustrated in Diagram III in Part 1 of this chapter (p. 244). We may add that there is now direct evidence that under
aerobic conditions the breakdown of sugar provides carbon both for
carbon dioxide and for derivatives of sugar that remain in the tissues.
When glucose or pyruvate labeled with C^{14} is fed to a variety of
tissues (100, 110, 111), not all the labeled carbon absorbed and metabolized appears as $C^{14}O_2$. For example, when carrot slices are fed uniformly labeled glucose (110), only about a quarter of the activity absorbed appears as $C^{14}O_2$ within 24 hours. Of the activity remaining
in the tissue, a substantial fraction is in compounds (organic acids,
amino acids, proteins, etc.) whose appearance implies the breakdown
of sugar. Specific labeling of individual carbon atoms of glucose or
pyruvate showed that those carbon atoms which would enter the tricarboxylic acid cycle are mainly the ones that are retained in the
tissue. At the same time it indicated that some of the sugar is degraded
via the Embden-Meyerhof pathway, some via the pentose pathway.

The important studies of protein synthesis by Steward and his associates have established that carbon from sugar is incorporated in protein.
Steward et al. (118), working with cultured explants of carrot, found
much more label in the protein amino acids when radioactive glucose
was supplied than when glutamine or γ-aminobutyric acid was the
substrate. Evidence from trends in relative specific activities indicated
that the carbon skeleton of protein is acquired most readily from derivatives of glucose in intracellular compartments segregated from the
free amino acids, and that some of the carbon dioxide evolved had
passed through protein.

The general concept of a "protein cycle" in plants was first clearly
enunciated by Gregory and Sen (53). According to their scheme, the
cycle is fed by newly synthesized amino acids and it provides, via

protein breakdown, the carbon skeletons of amino acids which are converted to carbon dioxide in respiration. Direct evidence for the existence of such a cycle was first presented by Steward *et al.* (118), who showed that the proteins of cultured carrot explants are in a state of turnover, and that the rate of turnover is roughly proportional to the rate of growth of the tissues. Unequivocal demonstration that such turnover does provide a substantial amount of the respiratory substrate of the tissues comes from a new series of experiments by Bidwell *et al.* (29). In these experiments radioactive substrates, frequently glucose, were used to label either the proteins or the soluble compounds, and the passage of carbon through various pools and fractions was determined from the analyses in time of the total and specific activities of the compounds. It was shown that protein turnover varied from 0.5 to 50% of the total protein per day, and the contribution of protein breakdown products to respiration ranged from 7% of the total carbon dioxide produced in nongrowing, slowly metabolizing tissues to 27% in actively growing, rapidly metabolizing tissues.[4]

It may be regarded as established that, at least in some nonphotosynthesizing tissues, under aerobic conditions the respiration of sugar provides not only energy, but also carbon skeletons for synthetic reactions.

The linkage of energy transfer with respiration has been assigned to the use of energy released during the transport of electrons to synthesize ATP. Control of the rate of electron transfer by the linked phosphorylation has been established for mitochondria and for many tissues, although in some physiological states the turnover of ATP appears to be rapid enough to remove any limitation by phosphate acceptor or inorganic phosphate.

Finally, no one would now seriously dispute the claim that the ultimate regulation of metabolism *in vivo* lies in protoplasmic organization.

This account of modern developments has stressed certain features that are related to Blackman's theory, and it will be clear that some provisions of his theory were predictions that are now established as common knowledge. But two postulates have not been generally accepted. These are "oxidative anabolism" and "oxygen augmentation" of sugar breakdown.

The term oxidative anabolism was avoided in outlining the essential features of Blackman's theory because, like the "Pasteur effect" it has acquired a connotation not necessarily inherent in the original conception. Experiments on the carbon dioxide output of apples indi-

[4] The author is indebted to R. G. S. Bidwell for this account of the protein cycle.

cated that fermentation in the absence of oxygen was not entirely replaced by the carbon dioxide of respiration when oxygen was supplied. Since no known intermediate of sugar breakdown accumulated, Blackman concluded that the missing carbon was "somehow worked back into the system" (30). The amount of carbon remaining in the tissue he labeled the carbon of oxidative anabolism. This has come to be interpreted as resynthesis of carbohydrate from the products of glycolysis, and since there is little evidence that the majority of tissues, including slices of apple (76), can synthesize carbohydrates in major quantity from added pyruvate, many plant physiologists reject the concept of oxidative anabolism. Some tissues (20, 109) can synthesize sugar from malic acid via phosphoenol pyruvic acid, but this is apparently not a major function except in specialized tissues.

Although Blackman speculated about the fate of the carbon retained in the tissue and thought it likely that it is "anabolized back into the carbohydrate region of the system" so that the same carbon atom might circulate several times before it was thrown out as carbon dioxide, he was careful to point out that "while we shall speak of OA as a substance, as far as evidence goes it is only a magnitude . . . *OA is only the amount of carbon by which produced* CO_2 *falls short of carbon glycolysis* (30).[5] This broader concept would include not only the reappearance of a carbohydrate or an ester by any pathway, but also the diversion of carbon to any metabolic product that retains it in the tissue, provided only that its formation depends on breakdown of sugar and is linked with oxidative respiration.

The chief reserves in plant tissues are carbohydrates and, for example, an apple separated from the tree must ultimately draw upon them for the carbon incorporated in organic substances synthesized after the fruit is picked. No evidence has appeared that glycolysis is reversible in such tissues, but this is not the course predicted by Blackman, who postulated that it is irreversible. Blackman conceived of all sugar breakdown as glycolysis and did not propose a specific pathway for the cycling of carbon. But we now know that the pentose cycle operating in the presence of oxygen may provide for the recycling of hexose phosphate. In the complete operation of this cycle

6 hexose phosphate $+$ 12 TPN $\rightarrow$ 5 hexose phosphate $+$
$$6 \ CO_2 + P_i + 12 \ TPNH + 12 \ H^+$$

If the reduced TPN is oxidized via the normal electron transport channels:

6 hexose phosphate $+ \ 6 \ O_2 + 6 \ H_2O \rightarrow$ 5 hexose phosphate $+$
$$6 \ CO_2 + 12 \ H_2O + P_i$$

[5] Italics introduced by the present author.

But the evidence of Ap Rees and Beevers (110) is that the carbon from glucose which is retained in the tissues is detected only in the normal products of pyruvate metabolism or in carbohydrates, so it is likely that some of the triose phosphate that is formed in the cycle is diverted to pyruvate and goes through the tricarboxylic acid cycle. If this pathway is followed, 4 out of 6 hexose phosphate molecules would reappear as hexose phosphate. It is not, of course, known to what extent this pathway participates in the respiration of any one tissue, and recycled hexose phosphate may again enter the cycle or go by the Embden-Meyerhof pathway. Not all the carbon of pyruvate supplied by ap Rees and Beevers to tissue slices appeared as carbon dioxide within 24 hours, but some of those carbon atoms that enter the tricarboxylic acid cycle were retained in the tissue, so the diversion of carbon at this stage is also established. Oxidative anabolism, defined as the difference between the carbon of sugar catabolized and that lost as carbon dioxide, might then be composed partly of carbon recycled to hexose phosphate in the pentose cycle, partly diverted to tissue constituents including protein, where some may be retained and some may contribute to the measured carbon dioxide if a protein cycle is operating.

In tissue where respiration involves oxygen consumption and carbon dioxide production by both the pentose and glycolytic pathways together with the tricarboxylic acid cycle and is accompanied both by recycling of hexose phosphate via the pentose cycle and diversion of carbon, temporarily or permanently, to proteins or other metabolites derived from intermediates, then the actual rate of entry of hexose into catabolic progressions in aerated tissue may greatly exceed the rate of loss of carbon as carbon dioxide. When oxygen is withheld the pentose cycle would cease to operate and synthetic reactions depending on the product of oxidative phosphorylation would be prevented. If there is competition between the glycolytic and the oxidative enzymes for hexose phosphate, the elimination of the pentose pathway in the absence of oxygen might increase the rate of operation of the Embden-Meyerhof pathway and increase the production of carbon dioxide by glycolysis and fermentation, although at the same time the total catabolism of hexose is decreased. Evidence that plant tissues utilize the pentose pathway must affect the interpretation of both the Pasteur effect and oxidative anabolism. Plant physiologists would perhaps do well to drop both terms and concentrate their attention on the metabolic pattern. They may eventually arrive at an interpretation that embraces the principles originally conceived by both Pasteur and Blackman.

There remains the question whether, in the whole organism, the rate of breakdown of sugar, either via the Embden-Meyerhof pathway or via the pentose pathway, is higher or lower in the presence of oxygen than in its absence. This has already been discussed in the section on aerobic and anaerobic metabolism. Blackman's postulation that in an intact organ it is higher rests primarily on the observation that, in apples, as the concentration of oxygen is increased above the extinction point of fermentation the rate of carbon loss increases, and some secondary effect of free oxygen gradually changes the metabolic balance when the oxygen concentration is suddenly changed. He concluded that the secondary effect is dual: (a) it makes possible reactions that cycle carbon or retain it in tissue components, and (b) it increases the conversion of stable sugar to the labile substrate of glycolysis.

This runs counter to the idea that inhibition of glycolysis by respiration coupled to phosphorylation is a universal phenomenon. Current theories assign the control to availability of ADP, inorganic phosphate, or ATP at one point or another in the reaction chain. There is nothing illogical in the supposition that where the turnover of the system $ADP + P_i \rightleftharpoons ATP$ is rapid any one of the constituents might be more readily available to the glycolytic enzymes, and it is inhibition, not augmentation, of glycolysis in the presence of oxygen that requires a special explanation. Taking into account alternative explanations of observed phenomena, there is no decisive evidence that in the normal metabolism of higher plants the actual rate of involvement of sugar in catabolic processes is decreased by respiration or colligate reaction systems. If it were it would be difficult to account, for instance, for active incorporation of the carbon atoms of glucose in protein in actively respiring cells (118), or the association of growth with a high rate of respiration.

Whatever the terminology used, if we apply broadly the ideas originally promulgated by Blackman, we can reconcile some of the apparently conflicting evidence about the effect of oxygen on the respiration of plant organs.

In the apple, as well as in some other organs, carbon loss in air is lower than in nitrogen, but higher than in intermediate concentrations of oxygen, and is minimal at the extinction point of fermentation. Oxygen therefore appears, on the surface, to be both inhibiting glycolysis below the extinction point and stimulating it at higher concentrations. It is more logical to suppose that oxygen has only one or the other effect on glycolysis. Inhibition alone could not explain increased carbon loss above the extinction point, but augmentation of sugar breakdown together with diversion of intermediates to fermentation below the extinction point could account for the whole pattern of carbon dioxide output. Recognition of the extinction point rather than air as the significant reference point for comparison of anaerobic and aerobic metabolism is essential to understanding the oxygen relations of higher plants where the rate of respiration is usually much lower at the minimum oxygen concentration that eliminates fermentation than it is in air.

The dependence of respiration rate on atmospheric oxygen concentration over a wide range requires explanation. It has not been established by decisive evidence that in air the oxygen concentration in the mitochondrial membrane is so low that cytochrome oxidase is not saturated, nor that oxidases with low affinity for oxygen mediate any substantial proportion of normal respiration. Many plant tissues can be stimulated to higher rates of respiration without increase of oxygen supply, and fruits undergo their normal climacteric rise while the oxygen concentration in the intercellular spaces decreases. Products of glycolysis do not normally accumulate, and where fermentation is not occurring it must be presumed that the rate of breakdown of sugar is geared to the rate of respiration. Augmentation of the rate of sugar breakdown by oxygen would resolve the dilemma, if the extinction point represents the concentration at which oxygen ceases to control the rate of respiration as a reactant, and free oxygen in the tissues begins to exert some indirect effect on the rate, and on the disposition of products, of sugar breakdown. Blackman predicted that a break might appear in the curve of oxygen uptake plotted against oxygen pressure, such as that observed by Mapson and Burton (Fig. 12) with potatoes. Above the break, the rate of oxygen uptake might be determined by some reaction other than the transfer of electrons to oxygen, if oxygen were to exert an effect upon it directly or indirectly. Such a reaction might be oxidative phosphorylation or one of the steps in the production of reduced substrate.

There is a good deal of evidence that the rate of phosphorylation frequently does control the rate of oxygen uptake. Where ATP turnover is high this control is not manifested and the respiration rate is high, possibly because glycolysis is not restricted. High respiration rates depending on high synthetic activity, high ATP turnover, and high rates of sugar breakdown in the presence of oxygen are consistent with the general theory and would also account for the observed retention in the tissue of some of the carbon from sugar under aerobic conditions.

The conservation of carbon by maintaining it in stable sugars would render it useless to the organism. Conservation by diverting it to tissue products would explain Pasteur's original observation that oxygen redirects metabolism from fermentation to respiration and synthesis of cellular material. In growing tissue, carbon is incorporated in new cellular matter. In nongrowing tissue, turnover is implied. While turnover of protein does occur (29, 62, 118), there is as yet relatively little information about its extent, and none for bulky quiescent organs. Even in so unlikely an organ as a postclimacteric senescent fruit, however, net protein synthesis occurs and phosphorylation is active. But more information is needed about the activity of metabolism in nongrowing tissue. Until this is available there can be no well-founded conception of the traffic of carbon in such tissue.

The foregoing scheme offers an integrated picture of respiratory metabolism in the setting of the whole metabolism of an organ, based on the principles originally formulated by Blackman but expressed in terms of more recent knowledge. It remains the only hypothesis of equal scope. Not all its postulates can be regarded as experimentally established, but it satisfies many observations on the respiration of organized tissue and offers an explanation for some otherwise puzzling phenomena. How near the truth it lies can be determined only by future expansion of knowledge, and the greatest need for enlightenment concerns the role of oxygen in the metabolism of aerobic organisms.

VIII. Regulation of Respiration Imposed by Organization

As we learn more of the biochemical mechanisms at the disposal of plants, the challenges of the future lie more and more in attempts to find the mechanisms that determine the balance and direction of metabolic processes.

The cell is the unit of function, and much of the organization that regulates the interplay of substrate and catalyst, of energy production and utilization, of source and sink, of the organism and its environment, resides in the architecture of the protoplast; it is here that we must look for the machinery that controls the normal course of respiration of an organ. But the relation of an individual cell to its fellows also influences the course of its metabolism. This was dramatically demonstrated by Steward (119) when he showed that cells that break loose from an explant of carrot phloem will differentiate to produce an embryo and a new carrot plant, while cells remaining attached to the mass continue to divide without differentiation. Each cell must have had the potentialities of a whole organism, but those incorporated in a tissue were restrained by their position from expressing them. While the difference is one of morphogenetic development, this undoubtedly depends on differences in metabolic balance within the cell.

Within the realm of respiration a good illustration of the importance of the milieu of a cell is to be found in the consequences of cutting slices from a potato tuber. The slice promptly begins to develop respiration, and in fact a whole pattern of metabolism, that is profoundly different quantitatively and qualitatively from that of the same mass of cells when within the tuber (cf. this treatise Vol. II, Chapter 4, Section II, C).

Again, the respiration of the whole organ is affected by its relation to the entire plant. The respiration rate of a potato tuber falls rapidly as soon as it is detached from the vine, although it contains reserves that will keep it alive for months. There are many familiar examples of the influence of one part of a plant on the intensity and direction of metabolism in another part.

Investigation of the mechanism of these controls of higher order is still in its infancy and probably must await a better understanding of controls within the cell. But their existence should not be forgotten. Potential mechanisms can be revealed by studies of cells or cell fractions, but caution must be exercised in inferences about their operation in the whole organism, where they may be overridden by intercellular or organismal controls.

The study of whole organs provides a frame of reference for the operation of individual chemical or physical mechanisms, and the bulky organs are useful because they have a relatively high degree of self sufficiency and can survive for a long time. They present special problems, but when due attention is given to these, they can be used to illuminate some of the fundamental problems of respiratory control in the organism.

REFERENCES

1. Aisenberg, A. C. "The Glycolysis and Respiration of Tumors." Academic Press, New York, 1961.
2. Appleman, C. O., and Brown, R. G. Relation of anaerobic to aerobic respiration in some storage organs with special reference to the Pasteur effect in higher plants. Am. J. Botany 33, 170–181 (1946).
3. Appleman, C. O., and Miller, E. V. A chemical and physiological study of maturity in potatoes. J. Agr. Res. 33, 569–577 (1926).
4. Appleman, C. O., and Smith, G. L. Effect of previous cold storage on the respiration of vegetables at higher temperatures. J. Agr. Res. 53, 557–580 (1936).
5. Barker, J. Analytic studies in plant respirations. IV and V. The relation of the respiration of potatoes to the concentration of sugars and to the accumulation of a depressant at low temperatures. Part. I. The effect of temperature history on the respiration/sugar relation. Part II. The form of normal respiration/sugar relation and the mechanism of depression. Proc. Roy. Soc. (London) B112, 316–358 (1933).
6. Barker, J. Analytic studies in plant respiration. VI. The relation of the respiration of potatoes to the concentration of sugars and to the accumulation of a depressant at low temperatures. Part III. The relation of the respiration

to the concentration of sucrose. *Proc. Roy. Soc. (London)* **B119**, 453–473 (1936).

7. Barker, J., Khan, M. A. A., and Solomos, T. Mechanism of the Pasteur effect. *Nature* **201**, 1126–1127 (1964).

8. Barker, J., and Mapson, L. W. Studies in the respiratory and carbohydrate metabolism of plant tissues. V. Experimental studies of the formation of carbon dioxide and of the changes in lactic acid, sucrose and in certain fractions of keto-acids in potato tubers in air following anaerobic conditions. *Proc. Roy. Soc. (London)* **B141**, 321–337 (1953).

9. Barker, J., and Mapson, L. W. Studies in the respiratory and carbohydrate metabolism of plant tissues. VI. Analysis of the interrelationships between the rate of carbon dioxide production changes and the changes in the content of lactic acid, sucrose and of certain fractions of keto-acids in potato tubers in air following anaerobic conditions. *Proc. Roy. Soc. (London)* **B141**, 338–362 (1953).

10. Barker, J., and Mapson, L. W. Studies in the respiratory and carbohydrate metabolism of plant tissues. VII. Experimental studies with potato tubers of an inhibition of the respiration and of a "block" in the tricarboxylic acid cycle induced by "oxygen poisoning." *Proc. Roy. Soc. (London)* **B143**, 523–549 (1955).

11. Barker, J., and Mapson, L. W. Studies in the respiratory and carbohydrate metabolism of plant tissues. XII. Further studies of the formation of CO_2 and the changes in lactate, alcohol, sucrose, pyruvate and α-ketoglutarate in potato tubers in nitrogen and in air following anaerobic conditions. *Proc. Roy. Soc. (London)* **B157**, 383–402 (1963).

12. Barker, J., and El Saifi, A. F. Studies in the respiratory and carbohydrate metabolism of plant tissues. 1. Experimental studies of the formation of carbon dioxide, lactic acid, and other products in potato tubers under anaerobic conditions. *Proc. Roy. Soc. (London)* **B140**, 362–385 (1952).

13. Barker, J., and El Saifi, A. F. Studies in the respiratory and carbohydrate metabolism of plant tissues. II. Interrelationship between the rates of production of carbon dioxide, of lactic acid and of alcohol in potato tubers under anaerobic conditions. *Proc. Roy. Soc. (London)* **B140**, 385–403 (1952).

14. Barker, J., and El Saifi, A. F. Studies in the respiratory and carbohydrate metabolism of plant tissues. III. Experimental studies of the formation of carbon dioxide and of the changes in lactic acid and other products in potato tubers in air following anaerobic conditions. *Proc. Roy. Soc. (London)* **B140**, 508–522 (1953).

15. Barker, J., and El Saifi, A. F. Studies in the respiratory and carbohydrate metabolism of plant tissues. IV. The relation between the rate of carbon dioxide production in potato tubers in air following anaerobic conditions, and the accompanying changes in lactic acid content and sugar concentration. *Proc. Roy. Soc. (London)* **B140**, 522–555 (1953).

16. Barker, J., and Solomos, T. Mechanism of the "climacteric" rise in respiration of banana fruits. *Nature* **196**, 189 (1962).

17. Baur, J. R., and Workman, M. Relationship between cell permeability and respiration in ripening banana fruit tissue. *Plant Physiol.* **39**, 540–543 (1964).

18. Beevers, H. 2,4-Dinitrophenol and plant respiration. *Am. J. Botany* **40**, 91–96 (1953).

19. Beevers, H. "Respiratory Metabolism in Plants." Harper & Row, New York, 1961.

20. Benedict, C. R., and Beevers, H. Formation of sucrose from malate in germinating castor beans. II. Reaction sequence from phosphoenol-pyruvate to sucrose. *Plant Physiol.* **37**, 176–178 (1962).

21. Ben-Yehoshua, S., Robertson, R. N., and Biale, J. B. Respiration and internal atmosphere of avocado fruit. *Plant Physiol.* **38**, 194–201 (1963).

22. Berry, L. J., and Norris, W. E. Studies of onion root respiration. I. Velocity of oxygen consumption in different segments of root at different temperatures. *Biochim. Biophys. Acta* **3**, 593–606 (1949).

23. Biale, J. B. Effect of oxygen concentration on respiration of the Fuerte avocado fruit. *Am. J. Botany* **33**, 363–373 (1946).

24. Biale, J. B. Post harvest physiology and biochemistry of fruits. *Ann. Rev. Plant Physiol.* **1**, 183–206 (1950).

25. Biale, J. B. Respiration of fruits. *In* "Handbuch der Pflanzenphysiologie" (W. Ruhland, ed.), Vol. XII, Part 2, pp. 536–592. Springer, Berlin, 1960.

26. Biale, J. B., and Young, R. E. Critical oxygen concentrations for the respiration of lemons. *Am. J. Botany* **34**, 301–309 (1947).

27. Biale, J. B., and Young, R. E. The biochemistry of fruit maturation. *Endeavour* **21**, 164–174 (1962).

28. Biale, J. B., Young, R. E., and Olmstead, A. J. Fruit respiration and ethylene production. *Plant Physiol.* **29**, 168–174 (1954).

29. Bidwell, R. G. S., Barr, R. A., and Steward, F. C. Protein synthesis and turnover in cultured plant tissue: sources of carbon for synthesis and the fate of the protein breakdown products. *Nature* **203**, 367–373 (1964).

30. Blackman, F. F. "Analytic Studies in Plant Respiration." Cambridge Univ. Press, London and New York, 1954.

31. Boswell, J. G., and Whiting, G. C. A study of the polyphenol oxidase system in potato tubers. *Ann. Botany (London)* [N.S.] **2**, 847–863 (1938).

32. Boswell, J. G., and Whiting, G. C. Observations on the anaerobic respiration of potato tubers. *Ann. Botany (London)* [N.S.] **4**, 257–268 (1940).

33. Burg, S. P. The physiology of ethylene formation. *Ann. Rev. Plant Physiol.* **13**, 280–302 (1962).

34. Burg, S. P., and Burg, E. A. Role of ethylene in fruit ripening. *Plant Physiol.* **37**, 179–189 (1962).

35. Burton, W. G. Studies on the dormancy and sprouting of potatoes. 1. The oxygen content of the potato tuber. *New Phytologist* **49**, 121–134 (1950).

36. Burton, W. G. Studies on the dormancy and sprouting of potatoes. II. The carbon dioxide content of the potato tuber. *New Phytologist* **50**, 237–296 (1951).

37. Burton, W. G. The effect of the concentrations of carbon dioxide and oxygen in the storage atmosphere upon the sprouting of potatoes at 10°C. *European Potato J.* **1**, 47–57 (1958).

38. Burton, W. G. The physiology of the potato: problems and present status. *Proc. 1st Triennial Conf. European Assoc. Potato Res., 1960* (W. H. de Jong), pp. 79–117. Wageningen, 1961.

39. Burton, W. G. The effect of stage of maturity and storage on the respiratory quotient of potato tubers. *European Potato J.* **6**, 268–270 (1963).

40. Burton, W. G. The permeability to oxygen of the periderm of the potato tuber. *J. Exptl. Botany* (in press).

41. Chance, B., and Hess, B. Spectroscopic evidence of metabolic control. *Science* **129**, 700–708 (1959).

42. Clendenning, K. A. Studies of the tomato in relation to its storage. II. The effect of altered internal atmosphere upon the respiratory and ripening behaviour of tomato fruits stored at 12.5°C. *Can. J. Res.* **C19**, 500–518 (1941).

43. Clendenning, K. A. The respiratory and ripening behaviour of the tomato fruit on the plant. *Can. J. Res.* **C20**, 197–203 (1942).
44. Denny, F. E. Accumulation of carbon dioxide in potato tuber tissue under conditions for the continuous removal of the exhaled gas. *Contrib. Boyce Thompson Inst.* **14**, 315–324 (1946).
45. Ducet, G., and Rosenberg, A. J. Leaf respiration. *Ann. Rev. Plant Physiol.* **13**, 171–200 (1962).
46. Fidler, J. C. Studies in zymasis. IV. The occurrence of zymasic products in apples during senescence. *Biochem. J.* **27**, 1614–1621 (1933).
47. Fidler, J. C. Studies in zymasis. V. Seasonal fluctuations in zymasis and in carbon dioxide/alcohol number ratios in apples in the absence of oxygen. *Biochem. J.* **27**, 1622–1628 (1933).
48. Fidler, J. C. The role of acetaldehyde in the catabolism of carbohydrate. *Gt. Brit., Dept. Sci. Ind. Res., Food Invest. Bd. Rept., 1937* pp. 124–126 (1938).
49. Fidler, J. C. A comparison of the aerobic and anaerobic respiration of apples. *J. Exptl. Botany* **2**, 41–64 (1951).
50. Forward, D. F. The response of potato tubers to a period of anaerobiosis. I. Drifts in carbon dioxide production. *Can. J. Botany* **31**, 33–62 (1953).
50a. Forward, D. F. Unpublished data (1955).
51. Forward, D. F. Effect of temperature on respiration. *In* "Handbuch der Pflanzenphysiologie" (W. Ruhland, ed.), Vol. XII, Part 2, pp. 234–258. Springer, Berlin, 1960.
51a. Forward, D. F. Unpublished data (1964).
52. Goldfine, H., and Block, K. Oxygen and biosynthetic reactions. *In* "Control Mechanisms in Respiration and Fermentation" (B. Wright, ed.), pp. 81–103. Ronald Press, New York, 1963.
53. Gregory, F. G., and Sen, P. K. Physiological studies in plant nutrition. VI. The relation of respiration rate to the carbohydrate and nitrogen metabolism of the barley leaf as determined by nitrogen and potassium deficiency. *Ann. Botany (London)* [N.S.] **1**, 521–561 (1937).
54. Hackett, D. P. Respiratory mechanisms in higher plants. *Ann. Rev. Plant Physiol.* **10**, 113–146 (1959).
55. Hackney, F. M. V. Studies in the metabolism of apples. II. The respiratory metabolism of Granny Smith apples of commercial maturity after various periods of cool storage. *Proc. Linnean Soc. N. S. Wales* **68**, 33–47 (1943).
56. Hackney, F. M. V. Studies in the metabolism of apples. III. Preliminary investigations on the effect of an artificial coating on the respiratory metabolism of Granny Smith apples after various periods of cool storage. *Proc. Linnean Soc. N. S. Wales* **68**, 48–56 (1943).
57. Hackney, F. M. V. Studies in the metabolism of apples. IV. Further studies in the respiratory metabolism of Granny Smith apples, with special reference to the importance of oxygen supply. *Proc. Linnean Soc. N. S. Wales* **69**, 91–107 (1944).
58. Hackney, F. M. V. Studies in the metabolism of apples. V. The respiratory metabolism of Delicious apples of commercial maturity after various periods of cool storage. *Proc. Linnean Soc. N. S. Wales* **69**, 108–119 (1944).
59. Hanes, C. S., and Barker, J. The physiological action of cyanide. I. The effects of cyanide on the respiration and sugar content of the potato at 15°C. *Proc. Roy. Soc. (London)* **B108**, 95–118 (1931).
60. Hasselbring, H., and Hawkins, L. A. Respiration experiments with sweet potatoes. *J. Agr. Res.* **5**, 509–517 (1915).

61. Haynes, D., and Archbold, H. K. Chemical work on fruit. *Gt. Brit., Dept. Sci Ind. Res., Food Invest., Bd. Rept.*, 1927 pp. 100–114 (1928).
62. Hellebust, J.A., and Bidwell, R. G. S. Protein turnover in wheat and snapdragon leaves: preliminary investigations. *Can. J. Botany* **41**, 969–983 (1963).
63. Hess, B. Control of metabolic rates. *In* "Control Mechanisms in Respiration and Fermentation" (B. Wright, ed.), pp. 333–350. Ronald Press, New York, 1963.
64. Howard, F. D., and Yamaguchi, M. Respiration and the oxidative activity of particulate fractions from developing pepper fruits (*Capsicum annuum* L.). *Plant Physiol.* **32**, 418–428 (1957).
65. Huelin, F. E. The effect of ethylene on the metabolism of potatoes. *Gt. Brit., Dept. Sci. Ind. Res., Food Invest., Bd. Rept.*, 1931 pp. 90–92 (1932).
66. Hulme, A. C. The acetaldehyde and ethyl alcohol contents of apples during storage. *Gt. Brit., Dept. Sci. Ind. Res., Food Invest., Bd. Rept.*, 1933 pp. 70–73 (1934).
67. Hulme, A. C. The metabolism of nitrogen in apple fruits. *Gt. Brit., Dept. Sci. Ind. Res., Food Invest., Bd. Rept.*, 1936 pp. 126–131 (1937).
68. Hulme, A. C. A relation between protein content and rate of respiration in the cell of the apple. *Gt. Brit., Dept. Sci. Ind. Res., Food Invest. Bd. Rept.*, 1938 pp. 127–128 (1939).
69. Hulme, A. C. The nitrogen content of the apple fruit in relation to its growth and respiration. *Gt. Brit., Dept. Sci. Ind. Res., Food Invest. Bd. Rept., 1939* pp. 55–57 (1949).
70. Hulme, A. C. The relation between the rate of respiration of an apple fruit and its content of protein. I. The value of this relation immediately after picking. *J. Hort. Sci.* **26**, 118–124 (1951).
71. Hulme, A. C. Studies in the nitrogen metabolism of apple fruits. The climacteric rise in respiration in relation to changes in the equilibrium between protein synthesis and breakdown. *J. Exptl. Botany* **5**, 159–177 (1954).
72. Hulme, A. C. Studies on the maturity of apples. Respiration progress curves for Cox's Orange Pippin apples for a number of consecutive seasons. *J. Hort. Sci.* **29**, 142–149 (1954).
73. Hulme, A. C., Jones, J. D., and Wooltorton, L. S. C. The respiration climacteric in apple fruits. *Proc. Roy. Soc. (London)* **B158**, 514–535 (1963).
74. James, W. O. "Plant Respiration." Oxford Univ. Press (Clarendon), London and New York, 1953.
75. James, W. O. Reaction patterns in the respiration of the higher plants. *Advan. Enzymol.* **18**, 281–318 (1957).
76. James, W. O., and Slater, W. G. The aerobic utilization of pyruvate in plant tissues. *Proc. Roy. Soc. (London)* **B150**, 192–198 (1959).
77. Johnstone, G. R. Physiological study of two varieties of *Ipomoea batatas*. *Botan, Gaz.* **80**, 145–167 (1925).
78. Jones, J. D., Hulme, A. C., and Wooltorton, L. S. C. Mitochondrial preparations from the fruit of the apple. II. Oxidative phosphorylation. *Phytochemistry* **3**, 201–212 (1964).
79. Kidd, F., and West, C. The course of respiratory activity throughout the life of an apple. *Gt. Brit., Dept. Sci. Ind. Res., Food Invest., Bd. Rept., 1924* pp. 27–34 (1925).
80. Kidd, F., and West, C. Physiology of fruit. I. Changes in the respiratory activity of apples during their senescence at different temperatures. *Proc. Roy. Soc. (London)* **B106**, 93–109 (1930).

81. Kidd, F., and West, C. The internal atmosphere of apples in gas-storage. *Gt. Brit., Dept. Sci. Ind. Res., Food Invest., Bd. Rept., 1934* pp. 110–111 (1935).

82. Kidd, F., and West, C. The uptake of oxygen by apples. *Gt. Brit., Dept. Sci. Ind. Res., Food Invest. Bd. Rept.,* pp. 102–108 (1938).

83. Kidd, F., and West, C. Respiratory activity and duration of life of apples gathered at different stages of development and subsequently maintained at a constant temperature. *Plant Physiol.* **20,** 467–504 (1945).

84. Kidd, F., and West, C. The climacteric in early and late varieties of apple on the tree. *Gt. Brit., Dept. Sci. Ind. Res., Food Invest., Bd. Rept., 1939* pp. 57–59 (1949).

85. Kidd, F., and West, C. Resistance of the skin of the apple fruit to gaseous exchange. *Gt. Brit. Dept. Sci. Ind. Res., Food Invest., Bd. Rept., 1939* pp. 59–64. (1949).

86. Kidd, F., West, G., and Hulme, A. C. Rise in insoluble ("protein") fraction of the total nitrogen during the climacteric in apples and pears, and effect on this phenomenon of retarding the climacteric by carbon dioxide or stimulating the climacteric by ethylene. *Gt. Brit., Dept. Sci. Ind. Res., Food Invest., Bd. Rept., 1938* pp. 119–125 (1939).

87. Laties, G. G. Respiration and cellular work and the regulation of the respiration rate in plants. *In* "Survey of Biological Progress" (B. Glass, ed.), Vol. 3, pp. 215–299. Academic Press, New York, 1957.

88. Laties, G. G. Controlling influence of thickness on development and type of respiration activity in potato slices. *Plant Physiol.* **37,** 679–690 (1962).

89. Laties, G. G. Control of respiratory quality and magnitude during development. *In* "Control Mechanisms in Respiration and Fermentation" (B. Wright, ed.), pp. 129–156. Ronald Press, New York, 1963.

90. Laties, G. G. The onset of tricarboxylic acid cycle activity with aging in potato slices. *Plant Physiol.* **39,** 654–663 (1964).

91. Leonard, E. R. Studies in tropical fruits. VI. A preliminary consideration of the solubility of gases in relation to respiration. *Ann. Botany (London)* [N.S.] **3,** 825–843 (1939).

92. Lyons, J. M., McGlasson, W. B., and Pratt, H. K. Ethylene production, respiration and internal gas concentrations in cantaloupe fruits at various stages of maturity. *Plant Physiol.* **37,** 31–36 (1962).

93. Magness, J. R. Composition of gases in intercellular spaces of apples and potatoes. *Botan. Gaz.* **70,** 308–316 (1920).

94. Mapson, L. W. Metabolism of ascorbic acid in plants. I. Function. *Ann. Rev. Plant Physiol.* **9,** 119–150 (1958).

95. Mapson, L. W., and Burton, W. G. The terminal oxidases of the potato tuber. *Biochem. J.* **82,** 19–25 (1962).

96. Marks, J. P., Bernlohr, R., and Varner, J. E. Esterification of phosphate in ripening fruit. *Plant Physiol.* **32,** 259–262 (1957).

97. Marré, E. Phosphorylation in higher plants. *Ann. Rev. Plant Physiol.* **12,** 195–218 (1961).

98. McGlasson, W. B., and Pratt, H. K. Effects of ethylene on cantaloupe fruits harvested at various ages. *Plant Physiol.* **39,** 120–127 (1964).

99. Millerd, A., Bonner, J., and Biale, J. B. The climacteric rise in fruit respiration as controlled by phosphorylative coupling. *Plant Physiol.* **28,** 621–631 (1953).

100. Neal, G. E., and Beevers, H. Pyruvate utilization in castor bean endosperm and other tissues. *Biochem. J.* **74,** 409–416 (1960).

101. Neal, M. J., and Girton, R. E. The Pasteur effect in maize. *Am. J. Botany* **42,** 733–737 (1955).
102. Ohmura, T., and Howell, R. W. Inhibitory effect of water on oxygen consumption by plant materials. *Plant Physiol.* **35,** 184–188 (1960).
103. Pearson, J. A., and Robertson, R. N. The physiology of growth in apple fruits. IV. Seasonal variation in cell size, nitrogen metabolism and respiration in developing Granny Smith apple fruits. *Australian J. Biol. Sci.* **6,** 1–20 (1953).
104. Pearson, J. A., and Robertson, R. N. The physiology of growth in apple fruits. VI. The control of respiration rate and synthesis. *Australian J. Biol. Sci.* **7,** 1–17 (1954).
105. Pentzer, W. T., and Heinze, P. H. Postharvest physiology of fruits and vegetables. *Ann. Rev. Plant Physiol.* **5,** 205–224 (1954).
106. Platenius, H. Effect of temperature on the respiration rate and the respiratory quotient of some vegetables. *Plant Physiol.* **17,** 179–197 (1942).
107. Platenius, H. Effect of oxygen concentration on the respiration of some vegetables. *Plant Physiol.* **18,** 671–684 (1943).
108. Porter, H. K., and Runeckles, V. C. Some effects of 2,4-dinitrophenol on the metabolism of tobacco leaf discs. *Biochim. Biophys. Acta* **20,** 100–108 (1956).
109. Ranson, S. L., and Thomas. M. Crassulacean acid metabolism. *Ann. Rev. Plant Physiol.* **11,** 81–110 (1960).
110. Ap Rees, T., and Beevers, H. Pathways, of glucose dissimilation in carrot slices. *Plant Physiol.* **35,** 830–838 (1960).
111. Ap Rees, T., and Beevers, H. Pentose phosphate pathway as a major component of induced respiration of carrot and potato slices. *Plant Physiol.* **35,** 839–847 (1960).
112. Robertson, R. N., and Turner, J. S. The physiology of growth of apple fruits II. Respiratory and other metabolic activities as functions of cell number and cell size in fruit development. *Australian J. Sci. Res.* **B4,** 92–107 (1951).
113. Romani, R. S., and Biale, J. B. Metabolic processes in cytoplasmic particles of the avocado fruit. IV. Ripening and the supernatant portion. *Plant Physiol.* **32,** 662–668 (1957).
114. Simon, E. W. Respiration rate and mitochondrial oxidase activity in *Arum* spadix. *J. Exptl. Botany* **10,** 125–133 (1959).
115. Singh, B. N., and Mathur, P. B. Studies in potato storage. III. Respiration of potato tubers during storage. *Ann. Appl. Biol.* **25,** 79–87 (1938).
116. Smith, W. H. The cooling and storage of strawberries. *Gt. Brit., Dept. Sci. Ind. Res., Food Invest., Bd. Rept., 1936* pp. 155–159 (1937).
117. Steward, F. C., and Bidwell, R. G. S. Nitrogen metabolism, respiration, and growth of cultured plant tissue. *J. Exptl. Botany* **9,** 285–305 (1958).
118. Steward, F. C., Bidwell, R. G. S., and Yemm, E. W. Nitrogen metabolism, respiration, and growth of cultured plant tissue. *J. Exptl. Botany* **9,** 11–51 (1958).
119. Steward, F. C., Mapes, M. O., Kent, A. E., and Holsten, R. D. Growth and development of cultured plant cells. *Science* **143,** 20–27 (1964).
120. Steward, F. C., Stout, R. R., and Preston, C. The balance sheet of metabolites for potato discs showing the effect of salts and dissolved oxygen on metabolism at 23°C. *Plant Physiol.* **15,** 409–447 (1940).
121. Stiles, W., and Dent, K. W. Researches on plant respiration. VI. The respiration in air and in nitrogen of thin slices of storage tissues. *Ann. Botany (London)* [N.S.] **11,** 1–34 (1947).
122. Stout, M. Relation of oxidation-potential, respiration, and catalase activity to

induction of reproductive development in sugar beets. *Botan. Gaz.* **110**, 438–449 (1949).

123. Tager, J. M., and Biale, J. Carboxylase and aldolase activity in the ripening banana. *Physiol. Plantarum* **10**, 79–85 (1957).

124. Thomas, M. The controlling influence of carbon dioxide. V. A quantitative study of the production of ethyl alcohol and acetaldehyde by cells of the higher plants in relation to concentration of oxygen and carbon dioxide. *Biochem. J.* **19**, 927–947 (1925).

125. Thomas, M., Ranson, S. L., and Richardson, J. A. "Plant Physiology." Churchill, London, 1956.

126. Trout, S. A., Hall, E. G., Robertson, R. N., Hackney, F. M. V., and Sykes, S. M. Studies in the metabolism of apples. I. Preliminary investigations on internal gas composition and its relation to changes in stored apples. *Australian J. Exptl. Biol. Med. Sci.* **20**, 219–231 (1942).

127. Trout, S. A., and Kidd, F. The role of acetaldehyde in the respiration of plants. *Gt. Brit., Dept. Sci. Ind. Res., Food Invest., Bd. Rept., 1931* pp. 78–85 (1932).

128. Ulrich, R. Postharvest physiology of fruits. *Ann. Rev. Plant Physiol.* **9**, 385–416 (1958).

129. Varner, J. E. Biochemistry of senescence. *Ann. Rev. Plant Physiol.* **12**, 245–264 (1961).

130. Wager, H. G. The effect of anaerobiosis on acids of the tricarboxylic acid cycle in peas. *J. Exptl. Botany* **12**, 34–46 (1961).

131. Wardlaw, C. W. Studies in tropical fruits. II. Observations on internal gas concentrations in fruit. *Ann. Botany (London)* **50**, 655–676 (1936).

132. Wardlaw, C. W., and Leonard, E. R. Studies in tropical fruits. I. Preliminary observations on some aspects of development, ripening and senescence with special reference to respiration. *Ann. Botany (London)* **50**, 621–653 (1936).

133. Wardlaw, C. W., and Leonard, E. R. Studies in tropical fruits. III. Preliminary observations on pneumatic pressures in fruits. *Ann. Botany (London)* [N.S.] **2**, 301–315 (1938).

134. Wardlaw, C. W., and Leonard, E. R. Studies in tropical fruits. IV. Methods in the investigation of respiration with special reference to the banana. *Ann. Botany (London)* [N.S.] **3**, 27–42 (1939).

135. Wardlaw, C. W., and Leonard, E. R. Studies in tropical fruit. IX. The respiration of bananas during ripening at tropical temperatures. *Ann. Botany (London)* [N.S.] **4**, 269–315 (1940).

136. Watson, D. J. A study of the respiration of apples, with particular reference to oxygen consumption in respiration, and to the effect of oxygen concentration Ph.D. thesis, University of Cambridge, 1932.

137. Whitehead, T. The physiology of potato leaf-roll. I. On the respiration of healthy and leaf-roll infected tubers. *Ann. Appl. Biol.* **21**, 48–77 (1934).

138. Wiskich, J. T., and Bonner, W. D. Preparation and properties of sweet potato mitochondria. *Plant Physiol.* **38**, 594–604 (1963).

139. Wiskich, J. T., Young, R. E., and Biale, J. B. Metabolic processes in cytoplasmic particles of the avocado fruit. IV. Controlled oxidations and coupled phosphorylations. *Plant Physiol.* **39**, 312–322 1964).

140. Woolley, J. T. Potato tuber respiration and ventilation. *Plant Physiol.* **37**, 793–798 (1962).

141. Young, R. E., Romani, R. J., and Biale, J. B. Carbon dioxide effects on fruit respiration. II. Response of avocados, bananas and lemons. *Plant Physiol.* **37**, 416–422 (1962).

PREAMBLE TO CHAPTER 4

Carbohydrates, fats, and proteins have been recognized, ever since cell physiology began, as classes of compounds in protoplasm which fit it for its role as the "physical basis of life." Chapter 5 (on carbohydrate metabolism), Chapter 6 (on organic acid metabolism), and Chapter 7 (on fatty acid and fat metabolism) deal primarily with molecules and metabolites which do not contain nitrogen. Chapter 4 in its several parts is concerned with those metabolites which do contain nitrogen.

Because plants have the distinguishing characteristic that they can elaborate all their essential nitrogenous compounds from inorganic sources of nitrogen, nitrate, or ammonia, their nitrogen metabolism is of particular significance in the economy of nature. Recent work has exposed a far greater range of simple nitrogen compounds in plants than was hitherto believed to exist. Since the nitrogen metabolism of plants which culminates in the synthesis of proteins and nucleic acids, deals with compounds that are of especial importance in the interpretation of life and of the mechanism of heredity, it has special importance to plant physiologists and to botanists.

Because this is the only chapter devoted specifically to nitrogenous compounds and their metabolism, it is divided into three parts. Although to see each part as an entity some topics necessarily recur. These parts (1) stress historical perspectives, (2) the nature of the nitrogenous compounds which are now known to occur (which is summarized in a series of Appendices), and the metabolic and physiological significance of their reactions, and (3) proteins and protein metabolism.

Although here presented as a distinct area of plant metabolism the problems of nitrogen should not be treated in isolation, for they impinge upon carbohydrate and organic acid metabolism (Chapters 5 and 6), which furnish the carbon frameworks on which nitrogen compounds are built; upon photosynthesis (Chapter 1), because the conditions under which that process occurs are conducive also to protein synthesis for the intermediary metabolites of green cells in the light play a role in the synthesis of nitrogen compounds; and upon respiration, because energy-rich compounds from respiration furnish energy for the endergonic reactions of nitrogen metabolism and also because some of the end products of nitrogen metabolism may be reabsorbed into the respiratory cycle of plants.

The organic phosphorylated compounds of plants have been aptly described as the "energy currency" of organisms; this being so, their nitrogen metabolism comprises the metabolic "market place" where many of the energy-requiring transactions are negotiated.

CHAPTER FOUR

Metabolism of Nitrogenous Compounds[1,2]

F. C. STEWARD[3] AND D. J. DURZAN[3,4]

Part 1. Nitrogen in Nature and in Plants: Historical Perspective

I. Introduction 381
   A. Compounds of C, N, P, and S in the Chemistry of Organisms . . 381
   B. The Utilization of Nitrogen: Nitrogen Metabolism and Other
      Physiological Functions 384
   C. Plants and Animals, Some Comparisons and Contrasts 385
   D. Nitrogen Autotrophic and Heterotrophic Cells 387
   E. Origin of Complex Nitrogenous Compounds and of Life 388
II. The Determination of the Forms of Nitrogen That Occur in Plants . . 389
   A. Analysis of Total Nitrogen 389
   B. The Total Soluble Nitrogen: Its Analysis and Determination . . . 391
   C. Separation of the Soluble Nitrogen into Its Components 393
   D. Analysis of Functional Organic Nitrogenous Groups 396
   E. Protein Nitrogen: Its Analysis and Determination 397

[1] This chapter is based on the work of the senior author's laboratory over many years. It draws upon research carried out with numerous collaborators, some of which is documented in theses that may have been published only in part. To acknowledge all this except in this general way is not feasible, nor is it possible to acknowledge specifically all the sources of support upon which original work was based. Nevertheless, this synthesis of knowledge would not have been possible without all this help and experience.

[2] The total number of sources consulted for the text of this chapter was very large. To cite all these in the bibliography is obviously not feasible. Therefore, the following plan has been adopted. Where it is necessary to cite the actual source of a publication this will be done in the usual way for the treatise; e.g., Vickery (564). However, there are many examples in which it will suffice to mention in the text an author without giving the full reference in the bibliography, usually because this can be found in the review sources which have been cited or because the reference in question would be of historical interest only and should be easily traced. In these cases the name and date only will be given in the text: e.g., Schulze (1886).

Observations that are recorded in the tables are cited there by author and reference only, without titles; wherever a text reference is made to the occurrence of such a substance, which is also cited in the tables, the reference is not duplicated in the bibliography. Since the material cited to record the occurrence in plants of the free amino acids and other nitrogen compounds, the protein amino acids, and the keto acids that are related to nitrogen compounds, is extensive, it is presented in the form of Appendixes I, II, and III to this chapter.

[3] Laboratory of Cell Physiology, Growth and Development, Cornell University.
[4] Present address: Department of Forestry, Chalk River, Ontario, Canada.

379

III. Trends in the Understanding of Nitrogen Metabolism 401
  A. Nitrogen in Agriculture and Plant Nutrition 401
  B. Exogenous and Endogenous Sources of Nitrogen for Plant Metabolism 407
  C. Types of Systems Investigated for Their Nitrogen Metabolism . . 418
  D. Recognized Classes of Soluble Nitrogenous Compounds 424
  E. Concepts of Nitrogen Metabolism 436

Part 2. Free Nitrogen Compounds: Their Nature, Reactions, and Determination

IV. The Range of Soluble Nitrogen Compounds Now Found in Plants . . 451
  A. The Application of Chromatography 452
  B. Amino Acids Other Than the α-Amino Acids of Protein 457
  C. Amino Compounds That Are Decarboxylation Products 457
  D. Hydroxyamino Acids. 459
  E. Imino Acids and the Origin of New Nitrogen Compounds by
    Cyclization 466
  F. Substitution in Ring Compounds 470
  G. Other Amides and Substituents of Common Amino Acids or Amides 472
  H. Other New Amino Acids and Ring Configurations 473
  I. γ-Glutamyl Peptides in Plants 474
  J. Metabolic Cycles in Relation to the Occurrence of Newly Discovered
    Amino Acids 474
V. Factors That Affect the Complement of Soluble Nitrogen Compounds in
  Plants 480
  A. Divisions of the Plant Kingdom, Taxonomic and Genetic
    Considerations 480
  B. Some Developmental Considerations 491
  C. Some Effects of Nutrition and Environment 511
  D. Some General Interpretations of the Response of Nitrogen Metabolism
    to Environmental Factors 520
  E. Some Observations on Nitrogen Metabolism in Normal versus
    Pathological Situations 528
  F. The Effect on Nitrogen Metabolism of Exogenous Substances and
    Stimuli to Growth 531

Part 3. Proteins and Protein Metabolism: Regulatory Control

VI. The Complement of Protein Nitrogen in Plants 533
  A. Early Work on Fractionation of Protein Nitrogen: the Characteriza-
    tion of Proteins 533
  B. Proteins and Enzymes: Enzymatic versus Structural Protein and
    Metabolic versus Nonmetabolic Protein 538
  C. Seeds and Storage Protein 541
  D. Leaf Protein, Chloroplast and Nonchloroplast Protein 542
  E. Proteins in Growing Points and in Relation to Development . . . 544
VII. Ribonucleic Acids, Nucleoproteins, and Virus Protein 548
  A. Kinds of Ribonucleic Acid 549
  B. Summary of RNA's and Their Helical Structure 550
  C. Controls over the Activity of RNA and DNA in Chromosomes: the
    Role of Histones 552
  D. Normal and Virus Proteins 553

VIII. The Mechanism of Protein Synthesis in Plants 555
    A. The Problem Introduced 555
    B. Sites of Protein Synthesis in the Cell 560
    C. The Presentation of Nitrogen and Energy to the Synthesizing Surface:
       Modern Ideas on Protein Synthesis 563
    D. Current Concepts of Protein Synthesis 565
    E. The Template and Coding Problems in Protein Synthesis 570
    F. The Genetic Control of Protein Synthesis 573
    G. Protein Synthesis and Morphogenesis 576
    H. Protein Metabolism in Higher Plants in Relation to Modern Views
       on Protein Synthesis 577
  IX. The Regulation of Nitrogen Metabolism and of Protein Synthesis *in vivo* 579
    A. The Regulatory Control of Protein Metabolism 579
    B. Protein Metabolism in Relation to Respiration 583
    C. Effects Due to Light and the Role of Chloroplasts 598
    D. Nitrogen Compounds in Relation to Translocation and Storage . . 601
  X. Epilogue 602
References 604
Appendix I. Amino Acids and Amides Found in the Protein of Plants and
Their Free Occurrence 637
Appendix II. Amino, Imino Acids and Their Derivatives That Occur Free in
Plants 644
Appendix III. Keto Acids That May Occur in Plants, Their Amino Acid
Analogs and Their Metabolic Roles 680

PART 1

NITROGEN IN NATURE AND IN PLANTS: HISTORICAL PERSPECTIVE

I. Introduction

A. COMPOUNDS OF C, N, P, AND S IN THE CHEMISTRY OF ORGANISMS

Out of the list of ten essential elements (C, H, O, N, P, S, K, Ca, Mg, Fe), which were the first to be recognized as necessary for all higher plants, six (C, H, O, N, P, S) are light, nonmetallic elements. Three of these (C, H, O), which are furnished by carbon dioxide and water, respectively, contribute most to the structures and substances that are involved in nature. However, it is as difficult to conceive of the properties of protoplasm without nitrogen as it would be to conceive of organic chemistry and biochemistry without the properties of carbon. Carbon is characterized by its unique position at the center of the second period of the periodic table and owes its ability to combine with itself, forming indefinitely repeating chains and rings, to properties of the atom which are consistent with this position. As quoted by Wald (593), Lewis had earlier (1923) commented that the ability to form multiple bonds is almost entirely, if not entirely, confined to elements of the first period of eight elements and essentially to carbon, nitrogen, and oxygen. This, together with the fact that they are the smallest elements that achieve stable electronic configurations by adding 1, 2, 3, and 4 electrons, accounts for their predominant role in the chemistry of organisms (comprising 99% of their living parts).

Nitrogen—only slightly heavier than the carbon atom and in an adjacent position in the periodic table—has quite different properties. These properties include a wider range of valence states in compounds which may be used in the metabolism of plants. The combining power of nitrogen may be 3 as in ammonia, 5 as in ammonium chloride, but nitrogen may exist in oxidation or reduction states which range from $+7$ to -3 (see Table VII, Volume III, Chapter 3).

Although compounds in which two nitrogen atoms are combined together are fairly common, compounds in which three nitrogen atoms are so combined are rare, and they are usually very unstable. However, when nitrogen enters into combination with carbon and oxygen it forms an integral part of the complex repeating structures that are important in the molecular architecture, which is the basis of so much biological structure and function. The examples here are the proteins, built up from amino acids, and the nucleic acids, built up from nucleotides which contain both nitrogen and phosphorus. Whereas the fragments (amino acids, nucleotides) from which these macromolecules are constructed were recognized much earlier, the intimate knowledge of the role of these macromolecules (proteins, nucleic acids) in cells is much more recent.

The role of sulfur in the physiology of plants is comparable to that of phosphorus, except that this role is played on a smaller scale. However, like phosphorus, sulfur enters into important nitrogenous compounds which are functionally important, e.g., in soluble and structural protein and in enzymes. Sulfur also occurs in organic combinations, as in coenzymes, which mediate fundamental biochemical reactions, e.g., biotin, α-lipoic acid, thiamine pyrophosphate.

Sulfur, like nitrogen, has a range of valence states ($+6$ to -2). One of the peculiarities of sulfur compounds is that they tend to take part in certain regulatory functions of general metabolism, e.g., glutathione has a major role in oxidation-reduction, in certain biosynthetic pathways, and in cell division. Indeed, the fundamental role of coenzyme A in carbohydrate and fat metabolism now constitutes a special chapter in biochemistry. Also, when sulfur is in combination with nitrogen, e.g., as in methionine, it provides key substances that can act as methyl donors, as in the formation of lignin or the methylated bases of ribonucleic acids.

Minerals, built by many elements (Si, Al, O, . . .) combined by ionic bonds, may achieve elaborate repeating patterns in space (cf. Chapter 1, Volume III); these have maximum stability and very low energy levels, they are not readily subject to change and they are "very dead." Compounds of C, H, and O in the polysaccharides (e.g., starch and cellulose) also form stable, elaborate repeating patterns in space, and though these have variety by utilizing the properties of the asymmetric carbon atom bound by covalent bonds, they do not have the range of properties nor contain the information out of which protoplasmic organization is built. It is with the insertion of nitrogen into these repeating molecular patterns in space that a greater variety of building blocks is furnished, and the linkages to combine them also become more varied; thus structural complexity becomes evident at different levels of organization.

The polypeptide chains which constitute the proteins have complexity of structure at different levels (cf. Vennesland, Chapter 2, Volume IA). This appears first at the level of the over-all amino acid analysis, and in the sequence of peptide-bonded residues along the chain. Second, complexity appears in the actual geometric arrangement of individual protein chains, which is determined by the backbone polypeptide configuration and is stabilized by interactions of the amino acid side chains. A third order of complexity appears as the chains are folded and contorted in space. The latter includes the spatial relations between neighboring segments of the same or different polypeptide chains. Fourth, the term quaternary structure has now come into use to denote aggregated subunits, e.g., the four polypeptide chains of hemoglobin [Perutz *et al.* (389, 390)].

From a chemical point of view, there is essentially an equal probability that any amino acid could be peptide-bonded with any other; consequently in a polymer of 100 amino acid residues built of approximately 20 different amino acids, there will be a possible 20^{100} different arrangements. Therefore, only a few of the possible proteins have been selected for use in cells from a very large number of possibilities.

Thus the properties of the nitrogen atom help to convey diversity, the lability, and the inherent reactivity that fits the proteins as carbon compounds for their special role in nature. The molecules out of which the form of animate protoplasm (as the physical basis of life) is built are more labile, more asymmetrical, have higher energy content, greater order (therefore lower entropy) than obtains in even the most elaborate mineral. With the still further entry of phosphorus and sulfur (from the third period of the periodic table) into macromolecules like nucleoproteins, which are considered to be the physical basis of heredity and of specificity, and also into molecules which have "high energy content," these considerations may be carried even further (593).

The prevailing view until recent times was that of all the cellular constituents only proteins were complex enough to qualify as hereditary material. The work with bacteria and phages has now provided (220, 432) convincing evidence that nucleic acids can play this role. Some question still remains, however, whether proteins, other than nucleoprotein, are also represented among the hereditary determinants of cells. There is no evidence in direct support of such a possibility, but neither is there any decisive evidence against it.

B. The Utilization of Nitrogen: Nitrogen Metabolism and Other Physiological Functions

The proper subject of this chapter is the way in which plants and cells elaborate from nitrate, or even elementary nitrogen, compounds as complex as proteins and nucleic acids and nucleoprotein; it also includes the consideration of most other nitrogen compounds which occur in plants, their intermediary metabolism, and the relationships between nitrogen metabolism and other physiological functions.

One cannot legitimately separate the course of nitrogen metabolism from that of carbohydrate metabolism nor from the growth and organization of cells. In the broadest terms carbohydrate metabolism (cf. Chapter 5 of this volume and Chapter 4 of Volume IB) can be considered to begin with the transfer of hydrogen from water to carbon dioxide and the elaboration of a great variety of reduced carbon compounds, some of which, through the alternating cycles of anabolism and catabolism, may be reoxidized with the eventual transfer of hydrogen back to oxygen in the re-formation of water, as in terminal oxidation. Nitrogen metabolism begins with the reduction of nitrate, or elementary nitrogen, and the entry of reduced nitrogen compounds into organic combination with appropriate intermediates of carbohydrate metabolism (e.g., keto acids). From the soluble organic nitrogen compounds so formed such complex molecules as proteins are built up, but these, both in the dynamic system of the cell and in the nitrogen cycle in nature, may need to be recycled. In the re-use of nitrogen for synthesis, carbon fragments may be released and respired. The events of protein turnover and recycling of nitrogen in cells require the complex cellular oganization so that catabolism and anabolism may proceed side by side but in different compartments of the same cell. Through the nitrogen cycle in nature the organic products of one organism, or life cycle, are rendered again usable by the nitrifying bacteria for the needs of another. Thus the compounds of nitrogen are maintained in, and are essentially a part of, a continuously dynamic state, which is a feature of the organization of living cells and organisms.

Ideas of the dynamic state in organisms require that at least some protein is synthesized only to be broken down again. Thus the significance of such nitrogenous compounds in the economy of cells may lie not merely in their existence, but in their formation and re-formation. As the highly ordered macromolecules of proteins, etc., are built up and assume their characteristic shape, order is created out of randomness, new features appear by virtue of these spatial arrangements, the free energy of the system is increased, entropy decreases, and chemical work will be done on the system. Conversely, when these compounds are broken down, the reverse holds, so that work may now be done by the system. Therefore, alternations in the synthesis and breakdown of protein, which so characterizes the dynamic state of living cells, may be a part of the way in which cells do work.

Many points of contact between nitrogen metabolism and other physiological functions exist. Carbohydrate metabolism and organic acid metabolism impinge upon nitrogen metabolism where carbon compounds (e.g., keto acids or intermediates of photosynthesis) act as "ports of entry" for nitrogen into organic combination. Reciprocally, and after protein breakdown, the re-use of the nitrogen for synthesis may release carbon compounds for respiration. The frequent association of protein synthesis in leaves with light and protein breakdown with darkness, involve nitrogen metabolism in light-dark effects. Through the pyridine nucleotides (reduced in the light) many of the biochemical effects of light are mediated by hydrogen transfers and even the phytochrome of Borthwick and Hendricks, which is now believed to be at least one of the pigments that trigger off morphogenetic effects of light (62, 329), is also believed to be a protein. Interpretations of differentiation have frequently invoked an orderly sequence of nitrogen metabolism and protein synthesis during development. And of course the gene-enzyme hypothesis in its later forms (232, 233, 331, 432) credits the hereditary mechanism as acting through gene-determined enzyme reactions.

It is, therefore, difficult to name a basic property of cells in plants that does not, at some point, make contact with their nitrogen metabolism.

C. PLANTS AND ANIMALS, SOME COMPARISONS AND CONTRASTS

There are fundamental distinctions between plants and animals. A letter from Berzelius to Mulder in 1838 states that "protein" is the fundamental substance in nutrition and is synthesized by plants "pour

les herbivores et que ceux-ci fournissent ensuite aux carnassiers" (564).
Even prior to this (28) Beccari had recognized in the gluten of wheat
the source from which herbivores might draw the "animal-principle"
that Mulder was later to call protein. The conversion of inorganic
nitrogen, nitrate or elementary nitrogen, to organic form is essentially
a characteristic of plants, whereas, to build protein, animal cells essentially require already elaborated organic nitrogen compounds. For
example, mature humans require eight essential amino acids, and
certain tissues in culture even require twelve (140). By contrast such
inorganic substances as ammonia and nitrite are toxic to humans. The
animal body is characteristically organized about the gastrointestinal
tract, which ingests comparatively solid particles that are later dissolved;
this was observed directly by Beaumont (1833) through a bullet hole
in the stomach of Alexis St. Martin. By contrast green plants are
organized to receive their essential raw materials in the form of carbon
dioxide and water and to absorb nitrogen in the form of nitrate or
ammonia. However, some situations (e.g., acid bogs), in which the
nitrogen cycle (see Section III, A, 2) fails to restore organic nitrogen
to the usable inorganic form, have fostered the occurrence of plants
that may obtain nitrogen in other ways; examples are the insectivorous
or carnivorous plants, which secrete proteolytic enzymes and absorb
the products of proteolysis [for references to history and to examples
of this behavior see Darwin as quoted by McKee (313, 314)].

Although nitrate is the main ultimate source of nitrogen for plants,
it is rarely accumulated as such. Indeed neither nitrate nor ammonia
is commonly accumulated, although the latter (NH_3) does occur as
salts of organic acids in extremely acid sap plants (427–429). (Many
records of free ammonia in plants are erroneous because of the unsuspected decomposition of unstable amides such as glutamine.) The
entering nitrate is, therefore, rapidly converted by plants to organic
forms. But, having elaborated complex nitrogen compounds from such
simple starting points as nitrate and ammonia, or elementary nitrogen.
plants—unlike animals—do not usually excrete nitrogen compounds in
quantity. (The special case of excretion from roots of leguminous plants
is dealt with in Chapter 5, on nitrogen fixation, in Volume III.)
Whereas in the animal body nitrogenous compounds are constantly
being excreted in some quantity, as in urine, feces, sweat, tears, etc.
(603), such large losses usually do not occur from higher plants, which
convert and re-use their nitrogen; all this, therefore, contributes to
a very different economy of nitrogen in animals and in plants. However, some losses of nitrogen do occur in plants by leaching of foliage
by rain and by the annual leaf drop. Of the "nitrogen-rich storage

and mobile reserves" glutamine is now recognized to be an equally important metabolite in the economy of animals and of plants, but asparagine is still conspicuous as a peculiarly plant product, and it seems to be usable in protein formation only by ruminants among higher animals (307).

Since animals search out their food by mobility, they find the circumstances that are conducive to their nitrogenous nutrition, which are thereby confined to rather narrow limits. Plants, on the other hand, must withstand a wide range of environmental fluctuations, and these, it will be seen, greatly modify their nitrogenous nutrition (see Part 2). Indeed plants have used for their photosynthetic nutrition certain unique nitrogenous compounds (the chlorophylls) whereas the respiratory pigments of plants and animals, which are also nitrogen compounds (the iron cytochromes) are much more universal.

D. NITROGEN AUTOTROPHIC AND HETEROTROPHIC CELLS

Just as some cells are autotrophic for carbon and can utilize completely inorganic sources via photosynthesis, whereas others are heterotrophic, and so require organic carbon compounds elaborated by other cells or other organisms, so in the utilization of nitrogen similar differences obtain. Most green plants are autotrophic for nitrogen in the sense that they can utilize such completely inorganic sources as nitrate or even in special cases elementary nitrogen, yet this assimilation of nitrogen may be essentially confined to certain cells and tissues. Thus many cells of the plant body may, in fact, receive only the already elaborated nitrogenous compounds available to them from the assimilating regions. In other words, whereas the organism as a whole may be regarded as autotrophic for nitrogen, many individual cells or tissues may have a more heterotrophic nitrogenous nutrition.

Turning from the simple sources of plant nutrients to the key role of such elaborated compounds as proteins and nucleic acids, it is a problem to understand why nature often elaborated such complicated molecules to perform what often appear to be such chemically simple tasks (9, 10, 258). A salient example is hemoglobin, even more complicated than myoglobin, and in both cases the essential function is performed by heme iron which acts to promote oxygen transport. In both cases the protein molecule has, and acts by virtue of, a "native" orientation of the coiled molecule in space. However, in some other cases, e.g., catalase, somewhat simpler and smaller molecules than the natural enzyme have proved to be quite effective (cf. Vennesland, Chapter 2, Volume IA, for reference to Wang's work on catalase). These and other prominent examples of relationships between the configuration of complex molecules and physiological function are drawn from the animal body (e.g., as in the structure of insulin, of the pituitary hormones, or of DNA), but the presumption is that equally intimate relationship could be disclosed between the molecular structure of enzymes and their functions in plants. Therefore, we should see in the complex molecules produced in nitrogen metabolism something more than mere end products of a metabolic chain, for they, in turn, create and are a part of the internal environment of the cell in which metabolism proceeds in a controlled and regulated way.

E. Origin of Complex Nitrogenous Compounds and of Life

Since nitrogenous compounds play an essential role in the functioning of all cells, any concept of the origin of life or "biopoiesis" (a word due to Pirie which means the "natural evolution of life out of the inorganic world") must take into account how such compounds were formed initially [for discussions of this topical subject reference may be made to (36, 52, 87, 89, 161, 173, 352, 372, 373, 551, 592, 593)].

The very word protein, by its derivation from "*proteios*" (from the Greek, meaning of the first rank or primary), was designed to stress that these compounds, above all, contributed to the properties of protoplasm. One may note, however, that this suggestion is now attributed to Jöns Jacob Berzelius (1779–1848), who communicated the suggestion in a letter dated July 10, 1838, to the Dutch chemist, Gerardus Johannes Mulder. Mulder first used the term (1838) in the general scientific literature (203, 564), and the word protein was used to designate the complex radical which was thought to be combined with sulfur and phosphorus as they occur in nature. Therefore, ideas upon the origin of life face certain critical problems in the understanding of what Sir F. G. Hopkins has referred to as "the most improbable and most significant event in the history of the universe." How, in fact, did proteins emerge in an otherwise inorganic world? Enzymes, which are fundamental units of cell mechanism, therefore, take an important place during biopoiesis (131). How did the property of self-duplication and the storage of information in molecules, which permitted such complex molecules to be recapitulated, originate *de novo?* (416).

It is now generally conceived that the earth's atmosphere, prior to the abundance of life, was a much more reducing atmosphere than was at first supposed, so that ultraviolet light of short wavelength could penetrate the earth's atmosphere, which was also more transparent than now to radiation (173). Also, such molecules as methane and ammonia were then abundant. It has been suggested—and indeed shown —that through the effects of radiation and electrical discharge on molecules of this sort a variety of organic nitrogenous compounds, such as amino acids, and even nucleic acid precursors, could have been formed; e.g., see Table I, after Calvin (89).

Furthermore amino acids and urea give an unexpected array of compounds by the action of heat (162, 396). An unexpected variety of complex organic molecules can be elaborated and chromatographically detected after irradiation of suitable mixtures of the compounds mentioned (1, 2, 161, 162, 327, 374–376, 439). Therefore, there is much current interest in complex nitrogen compounds that occur in meteorites from outer space. According to modern concepts (416), which are

TABLE I

Yields in Moles ($\times$ 10^5) from Sparking a Mixture of
CH$_4$, NH$_3$, H$_2$O, and H$_2$[a,b]

Glycine	63	Succinic acid	4
Glycolic acid	56	Aspartic acid	0.4
Sarcosine	5	Glutamic acid	0.6
Alanine	34	Iminodiacetic acid	5.5
Lactic acid	31	Iminoacetic-propionic acid	1.5
N-Methylalanine	1	Formic acid	233
α-Amino-n-butyric acid	5	Acetic acid	15
α-Aminoisobutyric acid	0.1	Propionic acid	13
α-Hydroxybutyric acid	5	Urea	2.0
β-Alanine	15	N-Methylurea	1.5

[a] After Calvin (89).
[b] Carbon (710 mg) was added as CH$_4$.

greatly influenced by ideas drawn from virus multiplication and the role of nucleic acids in self-duplicating systems and in protein synthesis, "life may have started with the coupling of nucleic acid polymerization and amino acid polymerization." However, these interesting ideas are not subject to proof at this time.

The earth's original reducing atmosphere was presumably replaced, owing to green plant photosynthesis, by the present oxidizing atmosphere, so that, for all practical purposes, nitrate is now the essential starting point of nitrogen metabolism, and it is in this form that inorganic nitrogen enters into the nitrogen cycle of nature.

II. The Determination of the Forms of Nitrogen That Occur in Plants

A. Analysis of Total Nitrogen

To recognize nitrogenous compounds of plants as such, a means of detecting their nitrogen content was necessary. Fusion with sodium and conversion of the nitrogen of the organic compound to cyanide (except where the nitrogen might be combined with oxygen) permitted the nitrogen to be detected by the sensitive color reactions for cyanide. For quantitative purposes, however, the nitrogen of the organic compound was converted to elementary nitrogen (as in the Dumas method, 1831) or to ammonia (as in the Varrentrapp-Will method, 1841). Indeed, in 1836, Boussingault had already published elementary analyses of several plant proteins which marked a new step in their study.

Johan Kjeldahl's simple and rapid method (1883), which utilizes sulfuric acid digestion to produce ammonia and which is applicable both to pure substances and to crude mixtures, made the total analysis of nitrogen more generally feasible. Bradstreet[4a] (72) and Vickery (562, 563) have reviewed the history of the Kjeldahl determination. Hübsch and Nehring (223) have made a comparison of the Kjeldahl with the Dumas method. The method can be modified to include nitrate nitrogen if this is necessary (227). Nitrate usually occurs in only very low concentrations in plants and thus constitutes a separate quantitative estimation. Steward and Street (502) discussed some of the difficulties then encountered in determining nitrate and its reduction products, including hydroxylamine, in plant extracts. Even for pure proteins, and for simple mixtures of proteins, the nitrogen which is determined by some form of the Kjeldahl method is now almost universally used as the standard determination of total nitrogen. However, despite its wide acceptance and use, the method is still fraught with some difficulties.

Difficulties in the Kjeldahl determination arise chiefly from the digestion. Variables known to influence the speed and completeness of conversion of protein nitrogen into ammonia by sulfuric acid include the chemical form of the nitrogen in the sample. For example, the nitrogen of tryptophan and lysine and histidine are known to be difficult to convert quantitatively to ammonia. The presence in the digest of salts (which raise the boiling point) and the use of oxidizing agents (which influence the digestion time and the completeness of the recovery of nitrogen) have been much discussed in the analytical literature. The use of metallic catalysts, the problem of interfering materials (e.g., ammonium salts), the time of digestion, and the use of reducing or hydrolyzing agents prior to the digestion have all been discussed by Kirk (252).

The advantages of the micro-Kjeldahl method, over the macro or even the semimicro methods, are now generally conceded (252, 253). Indeed, there are forms of the total nitrogen determination which have

[4a] Bradstreet, R. B. "The Kjeldahl Method for Organic Nitrogen" Academic Press, New York, 1965 (in press).

been adapted to such small amounts of material as are used in Cartesian diver studies, e.g., the method of Bruel *et al.* (cf. 253) has a claimed accuracy of ± 0.005 μg and can be applied to samples having as little as 0.1 μg of total nitrogen.

The less accurate and less dependable direct Nesslerization of the Kjeldahl digests has often been employed instead of the standard distillation procedure.

Thus, the analysis of nitrogen-containing compounds frequently resolves itself into the determination of ammonia. (This is true not only of the determination of total nitrogen, but also of urea and of amides.) For minute amounts of ammonia and ammonium compounds, the convenient procedures are modifications of the Conway and Byrne diffusion method (115). Paech and Tracey (381) have outlined many of the methods available for the analysis of nitrogen compounds in plants. However, investigations today are much less dependent than hitherto upon accurate determinations of the total nitrogen in plants, for so many cases where synthesis was earlier in question can now be answered by the use of N^{15}- or even C^{14}-amino acids to detect the change that occurs strictly during the period of observation.

B. The Total Soluble Nitrogen: Its Analysis and Determination

Hartig (1858) observed that when he treated sections with absolute ethanol there occurred the crystallization of a substance (really a variety of compounds) rich in nitrogen which he called "Gleis." He concluded that "The apparently general occurrence of this crystallizable substance in all the young cell tissues, indicates that its solution is the form in which the nitrogenous nutritive substance of plants, formed from the reserve material, is transported from cell to cell" (106).

Extracts of plants, obtained after heat coagulation, may well contain some protein; this emphasizes the need for some caution in interpreting the fraction so obtained. Heat coagulation and plant protein precipitants, e.g., trichloroacetic acid (TCA), have the advantage of quickly inactivating proteolytic enzymes so as to limit artifacts and postmortem changes. There have been few critical studies (e.g., 46, 353) of nonprotein nitrogen after different deproteinizing procedures; although Steward and Preston (498) found that proteins synthesized by potato (*Solanum tuberosum*) disks could be equally well measured as the gain of TCA-precipitable protein or as the gain of alcohol-insoluble nitrogen.

Clean-cut separations of "soluble" and "insoluble" nitrogen (or protein) are not always easy, and a measure of compromise in the methods adopted is necessary.

Umbreit and Wilson (1936) and Orcutt and Wilson (1936) used the term *basic nitrogen* for that which is precipitated by treatment of deproteinized plant extracts with phosphotungstic acid. The amino nitrogen of this precipitate was termed basic amino nitrogen. The determination of nonbasic amino nitrogen has not proved useful, in view of its uncertain composition (514). Hydrolysis of deproteinized plant extracts with acid or proteolytic enzymes can result in an increase of amino nitrogen, which is called peptide nitrogen (522).

Protein precipitants (e.g., picric acid, trichloroacetic acid, sulfosalicylic acid) were reviewed by Rona and Strauss (424). Exhaustive extraction of plants with cold 70% ethanol leaves a residue termed "alcohol insoluble nitrogen," but this should be regarded as a measure of protein only where the alcohol soluble protein can be neglected (this excludes cereal grains that contain gliadins soluble in alcohol).

The use of aqueous alcohol to extract soluble nitrogen compounds (378, 379, 554–557, 567, 575) has been widely practised and, together with the various procedures available for the precipitation of protein from plant extracts, has been discussed by Steward and Street (502).

The most convenient method of extracting the nonprotein nitrogen of plants is by the use of ethanol (70–80%) with precautions (concentrating the neutral or only very weakly acid extract in the cold) to prevent the decomposition of glutamine. Crook (1946) and Crook and Holden (1948) studied the extraction of nitrogenous materials from green leaves of various plant species. MacDougall and de Long (1948) studied nitrogen extractions preparatory to subsequent determination of the lignin content of plants. Mitchell and Hamilton (328), Winterstein (612), Martin and Synge (295), Steward and Thompson (505, 507), Thompson *et al.* (537), Thompson and Morris (535), and Synge (522) have all covered the methods available for the extraction and analysis of free nitrogen compounds from plants. However, Ikawa and Snell (228) now claim that during the evaporation of concentrated syrupy solutions in an acidic medium, glutamic acid may react with serine to give O-(γ-glutamyl serine.

Although Schoenbein is said to have first studied chromatography in 1861, its development as a systematic analytical tool derives from the botanist Michael Tswett in 1906 (634). Tswett was then interested in separating leaf pigments, hence the term chromatography for the methods used to effect a visible separation of the different pigments. Knowledge of the range of free and soluble nitrogenous compounds in plants essentially awaited the application of the two-directional chromatographic methods, particularly on paper (114), although the continued use of the term chromatography for the means now employed to separate the various compounds of a mixture is justified only by its universality.

C. SEPARATION OF THE SOLUBLE NITROGEN INTO ITS COMPONENTS

Greenstein and Winitz (185) have covered the chemistry of the amino acids thoroughly in a useful work of reference.

The two-directional procedure on paper was first used to separate the free nitrogen compounds of plants in the author's (F.C.S.) laboratory and was described by Dent et al. (129). Although paper chromatography separated the nitrogen compounds, certain auxiliary techniques were also needed to detect them. The ninhydrin reagent, Ehrlich's reagent (51, 264, 459), and particularly the use of C^{14}-labeling combined with autoradiography, have all proved effective for this purpose.

The methods of qualitative paper chromatography were not immediately adaptable to the quantitative estimation of amino acids as in a protein hydrolyzate, although Thompson and Steward (540) were able to reproduce and extend by quantitative paper chromatography the partial analyses of certain globulins which had been isolated from cucurbitaceous plants by Smith et al. (Table II). The prior detection of amino acids on paper with ninhydrin, followed by their elution and complete reaction in the test tube, has been most carefully elaborated in the procedures of Hanes (196) and Tigane et al. (541). This method is complicated by the very long prewashing of papers which is recommended. After chromatography traces of ammonia are removed from the paper by a borate buffer and the amino acids are detected by partial reaction with ninhydrin. After elution, the amino acids are allowed to react in a test tube with a ninhydrin reagent, which in this case causes asparagine to react blue like other amino acids although when allowed to react on paper it may often appear brown. It used to be thought that the ninhydrin reagent was specific for α-amino acids, but it is now known that it reacts with a very much wider range of compounds to give substances which on paper are sometimes bluish purple (e.g., alanine, glucosamine); blue (β-alanine), brown (asparagine and some cyclic compounds), or yellow (hydroxyproline). In this versatility lies ninhydrin's chief usefulness. Thompson and Morris (535) have described a method which submits different fractions (e.g., basic, neutral, and acidic amino acids) separately to paper chromatographic analysis. In its various forms, the paper method of separation requires less sample and enables the variety of nonprotein nitrogen constituents to be more easily recognized than by other means. Alternative methods involve the use of ion exchange resins (332, 474) and microbiological assay (513).

As in the case of the amino acids, the earliest methods for the detection of carbamyl and guanidyl compounds involved their actual isolation. Later methods involved the action of enzymes. Urease is

TABLE II

ANALYSIS OF GLOBULINS ISOLATED FROM (a) CUCUMBER AND (b) SQUASH BY VARIOUS PROCEDURES AND AT DIFFERENT PERIODS[a]

Amino acid	Chemical analysis[b] (1947)		Quantitative paper chromatography[c] (1951)	
	(a) Cucumber	(b) Squash	(a) Cucumber	(b) Squash
Cystine	1.1	1.1	1.2	0.8
Aspartic acid	—	—	9.3	6.8
Glutamic acid	—	—	21.4	24.2
Serine	—	—	5.1	5.7
Glycine	—	—	4.9	5.5
Threonine	3.6	2.8	3.4	3.0
Alanine	—	—	5.1	5.7
Histidine	2.3	2.2	—	—
Lysine	2.9	3.0	3.0	4.0
Arginine	15.8	16.2	15.7	15.2
Methionine	2.5	2.3	2.2	2.5
Proline	—	—	4.9	5.4
Valine	7.0	6.5	5.4	5.6
Leucine(s)	14.0	13.5	12.8	13.3
Phenylalanine	6.5	6.8	10.0	8.3
Tryptophan	1.9	1.7	—	—
Tyrosine	4.6	4.4	3.8	3.7

[a] Data stated as grams amino acid per 100 gm protein.

[b] Material isolated by E. L. Smith et al. [J. Biol. Chem. 164, 159–163 (1946)] and supplied by Dr. H. B. Vickery of the Connecticut Experiment Station.

[c] See J. F. Thompson and F. C. Steward [J. Exptl. Botany 3, 170–187 (1952)].

commonly believed to be specific for urea, and it is used to determine urea in the free state and also bound urea after this is released by acid hydrolysis. Tracey [1955, cf. Paech and Tracey (381), p. 155] now questions this specificity because it does not distinguish between different ureides.

Enzymatic procedures for the nitrogen of guanidine compounds are less convenient. Arginine nitrogen can be determined by the combined action of arginase and urease. A less specific enzyme, heteroarginase, will release urea, detectable by urease, from a range of guanidine compounds (255). An incompletely characterized arginase exists in plants.

Ureides and guanidine compounds may be separated on paper by the usual procedures of chromatography, but the reagents in use for their detection on paper are not entirely specific. These are Ehrlich's reagent (p-dimethylaminobenzaldehyde) for urea and ureides and Sakaguchi's reagent for substituted guanidines (533). All imino compounds, including those that do not react with ninhydrin, may be located on the papers by the method of Rydon and Smith (431). The use of ninhydrin may also be extended somewhat to include guanidine compounds by treatment with alkali after ninhydrin (511).

Stein and Moore (473) carried the use of starch columns to a high degree of perfection for quantitative purposes. This method could separate accurately fifteen amino acids from 2–3 mg of material and solved in principle the problem of the complete amino acid analyses of a protein by a single procedure. Ion exchange resins are now used for desalting soluble plant extracts prior to free amino acid analysis, and their use constitutes the best general procedure now available for the analysis of protein hydrolyzates. Since many still unidentified nitrogen compounds occur in plants and quite different substances frequently superimpose on paper chromatograms (511) and on ion exchange columns (332, 464, 632a), any procedure that does not utilize the flexibility and resolving power of a two-directional system is still somewhat limited in its scope. Hence the elegant techniques of Stein and Moore reach their full use in the analysis of protein hydrolyzates, but their application to the wide range of substances found free in plants is more restricted. The combination of qualitative paper chromatography to identify the constituents and the quantitative Stein and Moore amino acid analyzer to determine them, now constitute a valuable procedure for the study of metabolism, and even more so when automatic recording devices to measure the radioactivity of labeled compounds can be used simultaneously. For use in the author's laboratory extensive maps have been prepared to show the behavior of a large number of free nitrogen compounds that occur in nature on the recently adopted automatic amino acid analyzer. The maps avoid mistaken identification by superposition of different substances in the eluates from the columns. Moreover, the relative accuracy of the automated Stein and Moore system has been compared with the best estimates that can be achieved by rigorously standardized quantitative analyses on paper. Unquestionably the former is capable of greater accuracy, but the latter can be used with adequate accuracy to cover a much larger range of samples than is usually feasible with the former, and the Stein and Moore procedure can be used to calibrate the paper method. The flexibility of the paper method, with its use of different solvent systems renders it much more adaptable to the study of the free nitrogen compounds, where so many unexpected substances may occur (data to substantiate these points are to be published elsewhere).

Filter paper ionophoresis of cupric complexes of neutral amino acids and oligopeptides has been described (94), as well as gas-liquid chromatography of amino acid derivatives (237).

Johnson *et al.* (237) even converted the amino acids into volatile *N*-acetyl amino acid *N*-amyl esters and separated a mixture of 35 components in minimum quantities of amino acid (10^{-10} mole) by gas chromatography; however, their quantitative application still involves difficulties. Separation of peptide derivatives by gas chromatography has been combined with mass spectrometry to determine amino acid sequences (41).

D. Analysis of Functional Organic Nitrogenous Groups

Knowledge of the total nitrogen content told nothing about the nitrogenous groups in which the element is contained. At first no clear distinction was made between amino acids and amino acid amides. Both were referred to indiscriminately as "amides" and their nitrogen as "amido-N." Von Gorup-Besanez (1874) determined "amido-N" by the laborious method of Sachsse-Kormann (1874) based on the interaction between amino groups and nitrous acid ($KNO_2 + HCl$) to give gaseous nitrogen. Chibnall (106) believes that the modern term amino nitrogen did not come into general use until the Van Slyke method appeared. For a long period, primary amino nitrogen ($-NH_2$) needed to be determined by the Van Slyke procedure. The method, first suggested by Hausmann (1899) and technically perfected much later by Van Slyke, depends on reaction with nitrous acid. Anomalous values for Van Slyke amino nitrogen may be given by certain compounds, e.g., by ammonia, urea, and even various peptides that react with nitrous acid to yield nitrogen. Lysine, cystine, glycine (436), and glutamine (570) all give values that deviate from the theoretical. Tyrosine and other phenolic compounds give values in excess of the theoretical when the reaction chamber is exposed to light (163).

The Van Slyke nitrogen method determines the nitrogen of all amino groups not involved in peptide linkages, but, under the acidic conditions required, some peptides may be hydrolyzed. Therefore, the values obtained on crude plant extracts may not measure the free amino nitrogen very accurately.

The gasometric ninhydrin methods of Van Slyke are based on the measurement of released carbon dioxide. This method is specific for free amino acids (though glutathione also reacts) in that it requires the presence, in the free unconjugated state, of both —COOH and a neighboring —NH$_2$ or NH—CH$_2$ group. The interference due to volatile aldehydes can be surmounted (195, 438). For a time this ninhydrin method replaced the nitrous acid method for the determination of the free amino acid nitrogen of plant extracts. There are, however, complications in that 2 moles of carbon dioxide are given by aspartic acid, whereas only 1 mole is liberated from glutamic acid. Also, γ-aminobutyric acid is not determined by these procedures, although β-alanine reacts partially.

Amino nitrogen in plant extracts has been determined by the formol titration, which is applicable to the determination of microgram quantities (253, 514). It is fairly specific for amino groups but will not distinguish between amino acids and other amines. Amide nitrogen (CO·NH$_2$), however, had to be determined by acid hydrolysis of the amide group to ammonia. So long as asparagine (Vauquelin and Robiquet, 1806) and glutamine (e.g., Schulze and Bosshard, 1883) were the only amides known to occur in quantity in plants, they could be distinguished by their respective stability toward acid by a method modified from Boussingault (1850). The total amide nitrogen of plants is readily determined by hydrolysis, using 6 N H$_2$SO$_4$ (570) of a nonprotein nitrogen fraction. The more stable amide, hydrolyzed with normal H$_2$SO$_4$ at 100°C for 3 hours, is calculated as asparagine; the less stable amide, hydrolyzed at pH 6.5 at 100°C for 2 hours by the procedures devised by Vickery et al. (569), is calculated as glutamine (108).

E. PROTEIN NITROGEN: ITS ANALYSIS AND DETERMINATION

Present knowledge of the composition of plant proteins is based on the amino acid analysis of a relatively small number of reserve proteins. Vickery (560, 561) has given a historical account of this problem from the first determinations, in 1871, of the aspartic and glutamic acid content of proteins from lupine (*Lupinus* sp.) and soybean (*Glycine max*), through the classical work of Osborne. The work of Osborne prepared the way for the doctrine of essential amino acids for animal nutrition as enunciated by Osborne and Mendel (1914).

The application by Lowry et al. (283), of color reactions for the analytical determination of protein is primarily due to Folin. The first comprehensive attempt to determine a single amino acid in a series of proteins by a colorimetric method was that of Folin and Denis

(1912). This was a method for tyrosine which depended on the reduction of phenolic reagents to produce a blue color. Folin and Looney (1922) later added colorimetric methods for the determination of tryptophan and of cysteine in protein. Colorimetric methods based on the biuret reaction are also used to determine protein nitrogen (e.g., Baudet and Giddey, 1948; Feinstein, 1949).

In the eighteenth century, however, substances now known to be proteins were found to yield colored products when treated with certain reagents. In 1780 Scheele recorded the formation of "liver of sulfur" from egg white and casein that had been treated with acid; i.e., he detected the presence of sulfur in the proteins by the blackening of lead or silver acetate (cf. 561). Welter in 1779 observed that nitric acid leaves a yellow stain on skin—the reaction that Mulder (1836) named the xanthoproteic reaction. Tiedemann and Gmelin (1826) described a color reaction of enzymatic protein digests with chlorine water which later became known as the tryptophan reaction (Neumeister, 1890). In 1849 Millon described the reaction with mercuric nitrate, and its relationship to tyrosine was known to Kühne (1868). The idea of employing color reactions for amino acid determination is, however, more recent (252).

Since proteins contain about 16% of nitrogen, the protein content of plant material was often estimated by multiplying the total nitrogen content (in per cent) by the factor 6.25 (i.e., 100/16) [cf. Chibnall (106)]. Since the nitrogen content of different purified proteins may vary between 12 and 19%, and since in large part the total organic nitrogen of tissue samples may consist of nonprotein nitrogen stored in a soluble form, the use of the factor 6.25 to convert the total nitrogen of plants to protein may lead to much error.

In fact, the analysis of protein mixtures of unknown composition is only very approximate if an empirical factor is used that does not specifically apply to the system in question. The uncertainty of the nitrogen content of protein arises from the fact that some proteins are rich in glycine or in lysine, arginine, and histidine, whereas others are almost devoid of these amino acids, which have a high content of nitrogen relative to carbon. Also, there is often uncertainty as to what constitutes a perfectly dry, unaltered protein (107).

Boussingault (1836) and Mulder (1836) published what were probably the first ultimate analyses of protein, and Mulder formulated the first hypothesis of protein constitution from what then appeared to be regularities in protein composition (559, 568), work in Liebig's laboratory to characterize proteins developed the view that there were only four such fundamental substances in nature, namely, albumin, fibrin, casein, and gelatin, found both in plants and animals (Liebig, 1842). This view was greatly oversimplified; nevertheless, it influenced thinking on the nature of proteins for half a century.

By 1872 Ritthausen had published a table of amino acid analyses of proteins and was making use of the amino acid composition to differentiate proteins of somewhat similar properties; he was also the first to attempt to establish stoichiometric relations. E. Schulze in Switzerland grasped the significance of this approach and applied it to the composition of the extract obtained from germinated seedlings, which proved to be entirely different in its amino acid composition from that to be expected from the composition of the seed protein if no metabolic process other than hydrolysis had occurred (Schulze, 1880).

A method of hydrolysis using hydrochloric acid and stannous chloride, introduced by Hlasiwetz and Habermann (1873) was used in 1880 by Schulze, to study the seed proteins of *Cucurbita* and of *Lupinus*. The narrow range in the composition of the proteins with respect to their content of carbon, hydrogen, nitrogen, and even sulfur, made clear differentiation of these proteins impossible at that time. [Pure proteins of *Cucurbita* were later isolated and partially analyzed by Smith *et al.* in 1946, 1947, 1948, cf. Smith and Greene (458); the analyses were later supplemented chromatographically by Thompson and Steward (540).] However, throughout the last decade of the nineteenth century there was no other method of characterizing protein, aside from precipitation reactions and coagulation temperature.

Hydrolytic products of proteins have been isolated by the fractional crystallization of the compounds themselves or of their copper, silver, and other salts. Only when one or more of the amino acids occurred free, or after hydrolysis in conspicuous amounts, was their isolation and characterization effected, e.g., arginine by Schulze (1886, 1896) in lupine and the conifers *Picea*, *Pinus*, and *Abies*. A great advance was made when Drechsel (1882) discovered that the protein molecule contained diamino acids as well as monoamino acids, and to Kossel and Kutscher (1900) we owe our chief knowledge concerning their isolation and estimation. In 1901 Kossel's lecture, delivered on June 1 (561), summarized these accomplishments to that date.

Emil Fischer (1901) had introduced the "ester method" for isolating and separating the monoamino acids, which depended upon the fractional distillation *in vacuo* of their esters. The method, though not fully quantitative, enabled one to (a) obtain some 70% of the total products which resulted from protein hydrolysis; (b) show that phenylalanine, serine, and alanine, which were known to occur only in a few, were present in all, proteins; (c) recognize phenylalanine as the principal aromatic constituent, for it often exceeds tyrosine and also occurs even when the latter is absent; (d) demonstrate the presence of two new compounds, proline and hydroxyproline.

In 1902 Fischer and Hofmeister independently advanced the hypothesis that in proteins the α-amino group of one amino acid and the α-carboxyl group of another amino acid are joined by loss of water to form an amide linkage. Although it is now necessary to postulate additional types of covalent bonding (cf. Vennesland, Volume IA of this treatise, 1960) to account for all the physical and chemical properties of proteins, the peptide linkage has remained their main structural feature.

Abderhalden, a pupil of Fischer, devoted much labor to the analysis of proteins and polypeptides and to enzyme action. His analyses [see Plimmer (394)] were of great physiological interest for the purpose of ascertaining whether proteins of similar origin are identical and whether they differ during development. In 1924 Osborne provided analyses of vegetable proteins and came to the conclusion that only about 5–15% remained to be accounted for. Vickery (565) commemorated the life and work of Osborne.

One cannot trace here all the ramifications of technique involved in the chemical determinations of protein hydrolytic products designed to build up a stoichiometric picture of what a plant protein is. Such names as Levine, Van Slyke, Dakin, and Cohn have an honored place in the history of the first part of the twentieth century (561). In retrospect, however, the problems of protein composition and structure could not be solved by the stoichiometric analysis of protein hydrolytic products alone, and the later rapid progress which has been made had to wait upon new methods such as the use of chromatographic procedures, using paper and columns, and of group analysis. Nevertheless, the modern ideas upon protein structure were ushered in by Svenson (1930), who added to the familiar Hofmeister and Fischer polypeptide hypothesis by stating that long molecular chains are flexible and are apt to roll themselves up by virtue of cohesive forces between different parts of the polypeptide chains which thereby attract and adhere to each other. In the 1930's Bergman and Niemann developed the periodicity hypothesis of amino acid arrangement in protein. Today, however, from much work on such proteins as insulin, hemoglobin, myoglobin, and ribonuclease the idea that amino acids fall into definite and characteristic linear sequences along the protein chains is becoming a familiar feature of even semipopular scientific writing (135, 170, 274, 361, 382, 474). Also, by the use of N^{15}, concomitant synthesis and breakdown of nitrogen compounds have now come to be regarded as a major characteristic of life (558, 571). A new approach to the analysis of protein hydrolyzates by mass spectrometry has been made possible. Amino esters are sufficiently volatile for analysis, and the identification of molecular fragments formed on electron impact may lead to unequivocal interpretations of their structure (40, 213).

The publication of many reviews on protein synthesis (cf. Part 3) indicates the rapid progress in a currently very active field where, until recently, great satisfaction was derived from the mere finding that an amino acid could be shown to find its way into protein. Thus, the current state of knowledge has rapidly turned from over-all protein nitrogen determination to mechanisms of protein synthesis and to such newer concepts as those involving ribosomes, messenger RNA, and coding systems.

III. Trends in the Understanding of Nitrogen Metabolism

The way our knowledge of nitrogen in plant nutrition and metabolism has developed can now be summarized.

A. NITROGEN IN AGRICULTURE AND PLANT NUTRITION

The early history of nitrogen metabolism is a part of the history of inorganic plant nutrition as this is related to agriculture. Reference may here be made to the relevant chapters of Volume III of this treatise. Various empirical practices in agriculture developed over the years to restore or maintain fertility in soils subjected to intensive cultivation. The early eighteenth century crop nutrition, theory and practice, has recently been reviewed from this standpoint (172).

1. Nineteenth Century Ideas and Developments

The trend of events may be summarized by the state of knowledge at the midpoint of the nineteenth century, at its end, at the first quarter of the twentieth century, and again midway through the twentieth century.

Priestley, to whom physiologists were indebted for their knowledge of the atmosphere, thought in 1779 that plants could feed on and absorb nitrogen gas although he had no means to confirm this notion. The development of better eudiometric methods for analyzing gaseous mixtures led de Saussure (1804) to make several important statements that helped to lay the foundations of modern plant physiology (348).

De Saussure (1804, p. 242) stated in his well-known "Recherches Chimiques sur la Végétation": "Experiment proves that most plants assimilate no elementary nitrogen whatever. Yet nitrogen is an essential part of plants. It is generally found in wood, in plant extracts, and in the green coloring matter of plants." "The superior quality of fertilizers of animal origin in comparison to those of plant origin apparently is due to the higher content of nitrogen in the former." Later (1842) he maintained that ammonia does *not* serve directly as a plant nutrient, but that it benefits the plants by dissolving soil humus which does contain nitrogen.

Sir Humphrey Davy in 1819 anticipated Liebig to some extent by stressing the importance of ammonia as a source of nitrogen and that a known solution containing nitrogen, which was used for watering barley (*Hordeum vulgare*) plants, could "furnish azote to form albumen or gluten in plants that contain them; but the nitrous salts are too valuable for other purposes to be used as manures" [(610), p. 26].

In the first half of the nineteenth century ideas of vital force were to be doomed by Wöhler's synthesis of urea; the known value of legumes in crop rotation was to be attributed by Boussingault to their ability to use nitrogen of the air and, by the scorn of Liebig, the humus theory fell into disrepute; the organic matter of soil could be regarded by Liebig as but a clumsy way of making ammonia available to plants. Thus, according to Russell (430), the following general ideas were held by 1856.

Nonleguminous plants were known to require a supply of nitrogenous compounds, nitrate and ammonium salts being almost equally effective. Without an adequate supply of nitrogen no increases of growth are obtained even when ash constituents are added. The amount of ammonia obtainable from the atmosphere is insufficient for the needs of crops. Leguminous plants behave abnormally in that they require no nitrogenous manure and yet they contain large quantities of nitrogen as well as enriching the soil with this element.

Great advances in soil microbiology were to be made in the sixties and seventies by Winogradsky, Schloesing, and Müntz, among others, to show that nitrification was a bacterial process and that nitrogen compounds rapidly change to nitrates in the soil so that, whatever compound is supplied as manure, plants receive their nitrogen in effect only as nitrate. This closed the discussion as to the nitrogenous food of nonleguminous plants, but the situation with respect to legumes was clarified only later.

In 1888, Hellriegel and Wilfarth concluded that nitrogen fixation was due to the activity of a "soil ferment" present and active in the root nodules of leguminous plants. This "organized ferment" was subsequently isolated in pure culture by Beijerinck in the same year and was known as *Rhizobium radicicola* (later *Rhizobium leguminosarum*).

Hellriegel, Wilfarth, Bertholet, Laurent, and Beijerinck, among others, ended another controversy by concluding that leguminous plants alone, like nonleguminous plants, have no power of assimilating gaseous nitrogen, but rather it is the associated soil bacteria that do this.

These events were to be supplemented later in the nineteenth century by a clearer understanding of the nitrogen cycle in nature (cf. Fig. 1) and of the role of the nitrifying organisms that restore

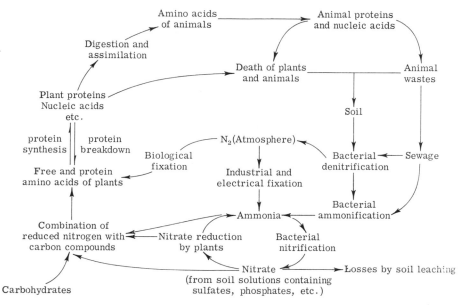

FIG. 1. The nitrogen cycle in nature.

organic nitrogen compounds to the condition in which the nitrogen may again be used as a plant nutrient, either as ammonium or as nitrate.

Prianishnikov (403a) stated that 'towards the beginning of this century exact data were obtained proving the possibility of the assimilation of ammonia by plants without preceding nitrification.' The recognition of both nitrate and ammonia as being directly available to plants was known only to a limited circle of specialists; for the majority of agronomists continued to think that availability of ammonia depended upon prior nitrification. The general state of agronomic knowledge on the North American continent is shown by Johnson (238). In a general commentary upon constituents of the atmosphere that affect the life of plants, ammonia, as its carbonate, was regarded as absorbed by foliage probably at all times; whereas nitrous and nitric acid, which when united to ammonia dissolve in water, were thought to be absorbed only through the roots.

2. Nitrogen Fixation and the Nitrogen Cycle

One of the many ways of portraying the relationships of nitrogen and its turnover in nature is shown in Fig. 1. The paramount role of the microorganisms is to be seen at several points, such as: ammonification of plant and animal residues followed by the processes of nitrification; denitrification causing loss to the atmosphere; nitrogen fixation by free and symbiotic organisms. In spite of the heavy applications of inorganic or other nitrogenous fertilizers to the soil, the principal mechanisms for restoring the nitrogen removed by plant nutrition involve the bacteria. Furthermore, the principal losses are those incurred under Western urban living in which much biological nitrogenous waste is eventually destined for the sea. As early as 1936 Lipman and Conybeare (cf. 610) estimated that there was a steady drain on the nitrogenous reserves of the soils of the United States and that if this trend were not modified its "half-life" might be as short as 142 years, but the most available nitrogen for plants in the soil could have an even shorter half-life. For later comparisons of the withdrawal and resupply of nitrogen on a world and regional scale, reference may also be made to Schmalfuss (435), Munson and Doll (346), and Jacob (234).

Redfield (411) now approaches this problem from the standpoint of the "turnover" in the sea of the various elements important to life. Redfield estimates the number of atoms relative to phosphorus that are turned over in the sea by plankton; as the figures show, 16 nitrogen and 106 carbons are "turned over" for each phosphorus atom that enters the biochemical cycle. To bring these changes about, 276 atoms of elementary oxygen are combined in respiration. The changes in the composition of sea water which occur are attributed solely to the synthesis and decomposition of organic matter. For every atom of phosphorus (and its accompaniment of nitrogen and carbon) that enters and leaves the cycle, 235 atoms of oxygen are consumed. In the average composition of sea water neither carbon (as carbonate) nor oxygen seem to be limiting, whereas phosphorus and nitrogen are present in limiting quantities (cf. Table III).

Thus Redfield concludes that these cyclical biochemical events have reached their present steady state and are controlled by the solubility of phosphate in the sea and by the oxygen content of the atmosphere, which in turn are regulated by the pace of photosynthesis and the removal of carbon dioxide from the atmosphere. Thus the nitrogen cycle now operates at a rate determined by the two parameters mentioned, so that limitations are set in this way to the turnover of nitrogen by all living organisms.

3. Nitrogen Fertilizers and the Food Resources of Man

By the turn of the twentieth century, food economists were predicting, on the basis of their known supplies of nitrates available for fertilizers, that the world was nearing its capacity to produce such important crops as wheat. They predicted that by the 1930's drastic shifts would have to be made in the diet of the wheat-eating people. However, new synthetic processes for the chemical fixation of atmospheric nitrogen soon discredited these predictions. Whereas, according to Jacob (234),

TABLE III

ATOMIC RATIOS OF ELEMENTS IN THE BIOCHEMICAL CYCLE IN NATURE
RELATIVE TO PHOSPHORUS: ANALYSIS OF PLANKTON AND SEA
WATER RELATIVE TO PHOSPHORUS[a]

Parameter	P	N	C	O
Plankton	1	16	106	—
Changes in sea water	1	15	105	—
Average composition of sea water	1	15	1000	200–300

[a] After Redfield (411a); see also (411).

natural sources of fertilizers supplied in the U.S.A., 90% of the commercially used fertilizers in 1900, they supplied only 2% in 1956. Again, 8 million tons of nitrogenous fertilizer were used in the United States in 1937, but by 1957 some 22 million tons were used. The increase was due to either ammonium nitrate or anhydrous ammonia while the use of sodium nitrate remained relatively steady.

With the recognition of the participation of soil microorganisms in nitrogen transformations, efforts were soon directed to measuring the effectiveness of amino acids as nitrogen sources for the growth of higher plants under sterile conditions (Lutz, 1898; Molliard, 1910; Hutchinson and Miller, 1911; and Brigham, 1917). More recent investigators, of which only a few are cited (15, 178, 355, 576, 584, 595, 624) indicate that the results depend upon the species and the amino acids used (cf. Quastel, Chapter 6, Volume III of this treatise).

In fact the growth of such new high-yielding crops as hybrid corn (*Zea mays*) can now be maintained unlimited by nitrogen, which is furnished by artificial means; in the outcome, there is a heavy over-production in the United States of both corn and wheat (*Triticum* spp.). Thus there is no sign yet that agricultural production may become limited by lack of available nitrogen. Furthermore, such agricultural practices as "clean cultivation" versus "grassing-down," as in the nutrition of fruit trees, came to be well understood in terms of their bearing upon the nitrogenous economy of the trees in question (99).

Eastern and tropical agriculture, however, present very different situations to these in Europe or the U.S.A. The ratio of arable land to units of the population and the comparative lack of the needed nitrogen as fertilizer, aggravated by outmoded cultural practices, all make the Eastern agricultural economy one that is "nitrogen and protein poor." In 1963, W. Slater, (*Nature* 199, 1225–1226) drew attention to the problems of feeding an increased world population by the end of the century. However, as man moves into outer space he will again encounter limitations due to his inability to furnish organic nitrogen for his needs (i.e., lacking the equivalent of the arable land that supports each man's needs, and the nitrogen cycle to reconvert waste products for his use). In this respect man's nitrogenous nutrition will probably be harder to adapt to conditions in space than his carbohydrate and oxygen requirements.

Also the occurrence of unexpected nitrogen compounds in common food plants now raises the question of their role in human and animal nutrition. The possibility that S-methylmethionine found in *Brassica* leaves and other plants may have nutritionally beneficial effects has been noted by McRorie *et al.* (1954). Examples of nutritionally injurious effects attributable to other natural compounds are: leucenol in *Leucaena glauca* (Yoshida, 1944), the *Lathyrus* factor (Schilling and Strong, 1954), β-nitropropionic acid[4b] in *Indigofera* (336). The discovery of γ-glutamyltyrosine and γ-glutamylphenylalanine in soybean meal may have a bearing on the toxicity of soybean meal. Borchers has found (59) that the toxicity of raw meal is reduced by tyrosine but not by phenylalanine.

The large array of ninhydrin-reactive compounds found in human urine implies that many of these constituents when ingested may be excreted rather than metabolized. In fact, some substances have been first recognized to occur in urine and then their natural occurrence in the food material has been detected. For example, an amino acid, later identified as 5-hydroxypipecolic acid, was first seen by Gartler and Dobzhansky (176) in human urine and associated with the use of the edible date (*Phoenix dactylifera*) in certain diets before the chemical

---

[4b] Other nitro compounds are known in plants; cf. *J. Org. Chem.* 24, 2070–2071 (1959); *Chem. Comm.*, p. 75 (1965).

identity was known (188). The urinary excretion of 5-hydroxyindole-acetic acid has been related to banana (*Musa*) feeding (8) and then to the detection of its precursor in the fruit.

B. Exogenous and Endogenous Sources of Nitrogen for Plant Metabolism

For all practical purposes, nitrate is the ultimate inorganic source of nitrogen for plants. For a number of microorganisms, nitrate may be the major source of nitrogen, and the biological processes by which it is absorbed and converted to ammonia, or to the amino acid level, for ultimate use in protein synthesis is known as "nitrate assimilation" and "ammonification." Nitrate may be used by certain microorganisms under aerobic or practically anaerobic conditions as a terminal electron acceptor in place of oxygen. The energy-yielding reactions under

TABLE IV
SOME HIGH VALUES FOR THE NITRATE CONTENT OF HIGHER PLANTS[a]

Plant material	Mmole of nitrate in 1 gm dry matter	Author
Tomato, leaves	1.28	Eisenmenger, 1933
Peach, stem	0.88	Davidson and Shive, 1934
Peach, roots	1.80	Davidson and Shive, 1934
Amaranthus, aerial parts	0.83	Olson and Whitehead, 1940
Oats, young plants	1.6	Sessions and Shive, 1933
Wheat, young leaves	0.75	Burstöm, 1943

[a] After Burström (82).

these conditions are necessary for growth and have been described as "nitrate respiration" or "true dissimilatory nitrate reduction."

Nevertheless, nitrate is rarely accumulated as such in the internal aqueous fluids [cf. Table IV, after Burström (82); cf. also Chapter 4, Volume II]. The reason is that the nitrate is rapidly reduced and enters into organic combination. The actual pathway of reduction was suggested by Meyer and Schulze as early as 1894 to be $NO_3^- \rightarrow NO_2^- \rightarrow H_2N_2O_2 \rightarrow NH_2 \cdot OH \rightarrow NH_3$. The oxidation-reduction change in the state of the nitrogen atom is a feature of the use of nitrate.

Sessions and Shive (1933) stated that in oats (*Avena sativa*), nitrate nitrogen may form up to one-third of the total amount of nitrogen, and Davidson and Shive (1934) found that half the amount of nitrogen in peach (*Prunus persica*) roots consisted of nitrate. But, this work was done prior to the general acceptance of the role of molybdenum in nitrate reduction, and it is now open to the interpretation that the plants in question were deficient enough in molybdenum to accumulate nitrate.

Hamy (1945) has discussed the amount of nitrate in plant extracts. Nitrite, the most obvious first reduction product of nitrate, is also rarely present in quantity, and when it does so appear, largely as a result of some nutritional disorder, it may be toxic to stock which browse upon such foliage (118, 606). Nitrate reduction via, perhaps, hyponitrous acid and hydroxylamine has been considered to operate principally in roots and in nongreen organs. The energy for this reduction is supplied by aerobic carbohydrate respiration (106, 349, 350, 358, 359). The presence of nitrous acid in proximity to amino acids would normally be expected to cause loss of nitrogen gas, and in fact it has been suggested that this occurs to some extent in *Narcissus* leaves when nitrate was being converted into amino compounds (386, 573). Vickery *et al.* (573) have detected losses of nitrogen in developing leaves of *Narcissus* and attributed this to the formation of nitrite by oxidation of ammonia resulting from amino acid deamination. Losses of organic nitrogen by grasses was observed by Chibnall and independently by Curtis (1944). Under specific climate conditions and rich nitrogen supply, pure glutamine exuded from the secretory organs at the tips of leaves of grasses.

Pearsall and Ewing (387) early presented evidence that in plants supplied with abundant nitrates, amino acids tend to accumulate and influence the metabolism of plants in such a way that acid production is reduced with a consequent reduction in the hydrogen-ion concentration of the sap. Both the high amino acid content and the higher pH value allow a greater swelling of the protoplasmic colloids, and to this feature the higher water content and reduced transpiration of the high nitrogen plants were then attributed.

Conditions for maximum absorption of nitrate and ammonium nitrogen may lead to their greater utilization (110, 357) and greater absorption of ammonium salts produces more of the soluble nitrogen fractions (asparagine, amino acids, etc.) and thus causes the carbohydrate reserves to be depleted. The latter effect is regarded as the chief cause of the reaction of clover (*Trifolium*) to ammonium sulfate (49).

The most important reduced form of nitrogen that can be utilized directly by plants is unquestionably ammonium, although this again rarely accumulates as such. It was called by Prianishnikov the "alpha and omega of nitrogen metabolism." Ruhland and Wetzel (427–429) distinguished between what they called "amide plants" and "ammonium plants," the latter being mainly plants with very acid saps which can form ammonium salts with malic and oxalic acids, e.g., leaves of *Begonia semperflorens* which have a pH of 1.5 and contain much ammonia when cultured in the dark. Kultzscher, who worked in Ruhland's laboratory (1932), also emphasized the same distinction in *Oxalis deppei* and *Pelargonium*, but Schwab (440) with more accurate analyses showed that "acid or ammonia plants" also produce high amide concentrations. Table V illustrates the data for *Oxalis deppei*.

There is a measure of contrast between most angiosperms and bacteria or fungi with respect to their affinity for ammonium versus nitrate as the main source of nitrogen. The latter, often grown in carbohydrate-rich media, tolerate ammonia, and the media that are prescribed for their best growth often contain organic ammonium salts in relatively high concentrations. (An example is the early use of ammonium tartrate, etc., in Pasteur's culture solution for fungi.) By contrast, most

TABLE V
pH and Analysis of Sap of *Oxalis deppei*[a]

Sample	pH	NH$_3$	CONH$_2$	CONH$_2$ / NH$_3$
Leaves	1.3	13.7	50.0	3.6
Roots	4.5	5.0	93.9	19.0

[a] Data after Kultzscher (1932); cf. Chibnall (106).

angiosperms preferentially use nitrate. It is also noteworthy that plants frequently absorb ammonia better at the more alkaline reactions (optimal at pH 6.0), whereas uptake of nitrate is facilitated by more acid reactions (optimal at pH 4.5), as suggested by Tiedjens (1933). Subsequent workers (110, 357) have in general confirmed the pH optima for use of nitrate. Tidmore (1933) emphasized that such experiments must not imply that ammonium salts are invariably nontoxic since toxicity becomes evident at lower phosphate concentrations.

Although their particular significance for plants cannot be stated in detail at this point the types of biochemical reactions that may (a) utilize ammonia (Table VI) and (b) yield ammonia (Table VII) have been conveniently summarized by Takahashi *et al.* (526), to whom reference may be made for the individual citations.

On the basis of dry weight per unit of absorbed nitrogen, nitrates are still the most effective source, as Pardo (1933) pointed out. The increased respiration of roots (214) which can be obtained with nitrates, and of storage tissues (498), may also promote the absorption of other

TABLE VI
Reactions That Utilize Ammonia[a,b]

Enzyme	Reaction
Glutamic dehydrogenase (DPN, TPN)	L-Glutamate $+ DPN^+ + H_2O \rightleftharpoons$ α-ketoglutarate $+ DPNH + H^+ + NH_3$
Alanine dehydrogenase	L-Alanine $+ DPN^+ + H_2O \rightleftharpoons$ pyruvate $+ DPNH + H^+ + NH_3$
Aspartase	L-Aspartic acid $\rightleftharpoons$ fumaric acid $+ NH_3$
Glutamine synthetase	L-Glutamic acid $+ ATP + NH_3 \rightleftharpoons$ glutamine $+ ADP + P_i$
Asparagine synthetase	L-Aspartic acid $+ ATP + NH_3 \rightleftharpoons$ asparagine $+ ADP + P_i$
Carbamyl phosphate synthetase	$CO_2 + NH_3 + 2ATP \rightarrow$ carbamyl phosphate $+ 2ADP + P_i$ (animal)
	$CO_2 + NH_3 + ATP \rightleftharpoons$ carbamyl phosphate $+ ADP$ (bacteria)
Cytidine triphosphate synthetase	$UTP + NH_3 + ATP \rightarrow CTP + ADP + P_i$
Guanylate synthetase	Xanthosine 5′-phosphate $+ NH_3 + ATP \rightarrow$ guanosine 5′-phosphate $+ ADP + PP$
DPN synthetase	Desamido-DPN $+ NH_3 + ATP \rightarrow DPN + AMP + PP$
D- and L-Amino acid oxidase	D-(L-)-Amino acid $+ E$-FAD $\rightleftharpoons E$-FADH$_2 +$ imino acid
	Imino acid $+ H_2O \rightleftharpoons$ keto acid $+ NH_3$

[a] After Takahashi et al. (526).
[b] DPN, DPNH, TPN, TPNH are, respectively, in new terminology: NAD, NADH, NADP, NADPH.

410

TABLE VII

REACTIONS THAT YIELD FREE AMMONIA[a]

Enzyme	Reaction
Hydrolyzing reactions	
Adenine deaminase	Adenine $+ H_2O \rightarrow$ hypoxanthine $+ NH_3$
Guanine deaminase (guanase)	Guanine $+ H_2O \rightarrow$ xanthine $+ NH_3$
Cytosine deaminase	Cytosine $+ H_2O \rightarrow$ uracil $+ NH_3$
Adenosine deaminase (nonspecific adenosine deaminase)	Adenosine $+ H_2O \rightarrow$ inosine $+ NH_3$
	Adenosine compound $+ H_2O \rightarrow$ inosine compound $+ NH_3$
Cytidine deaminase	Cytidine $+ H_2O \rightarrow$ uridine $+ NH_3$
5'-Adenylic acid deaminase	5'-Adenylic acid $+ H_2O \rightarrow$ inosinic acid $+ NH_3$
Adenosine diphosphate deaminase	$ADP + H_2O \rightarrow IDP + NH_3$
Arginine desimidase	Arginine $+ H_2O \rightarrow$ citrulline $+ NH_3$
Citrullinase	Citrulline $+ H_2O \rightarrow$ ornithine $+ NH_3$
Histidine deaminase	Histidine $+ H_2O \rightarrow$ urocanic acid $+ NH_3$
Pterin deaminase	Pterin $+ H_2O \rightarrow$ deaminopterin $+ NH_3$
Urease	Urea $+ H_2O \rightarrow CO_2 + NH_3$
β-Ureidopropionase	β-Ureidopropionate $\rightarrow \beta$-alanine $+ CO_2 + NH_3$
L-Ureidosuccinase	L-Ureidosuccinate $\rightarrow$ L-aspartic acid $+ CO_2 + NH_3$
Oxidative formation of ammonia	
L-Amino acid oxidase	L-Amino acid $+ O_2 + H_2O \rightarrow$ corresponding α-keto acid $+ H_2O_2 + NH_3$
D-Amino acid oxidase	D-Amino acid $+ O_2 + H_2O \rightarrow$ corresponding α-keto acid $+ H_2O_2 + NH_3$
Glycine oxidase	Glycine $+ \frac{1}{2}O_2 \rightarrow$ glyoxylate $+ NH_3$
D-Aspartic acid oxidase	D-Aspartic acid $+ O_2 + H_2O \rightarrow$ oxalacetic acid $+ H_2O_2 + NH_3$
D-Aspartic-D-glutamic acid oxidase	D-Aspartic acid $+ O_2 + H_2O \rightarrow$ malonate $+ CO_2 + NH_3$
	D-Glutamic acid $+ O_2 + H_2O \rightarrow \alpha$-ketoglutaric acid $+ NH_3$
Desulfurase and dehydrolase	
Cysteine desulfurase	Cysteine $\rightarrow H_2S + NH_3 +$ pyruvate
Homoserine dehydrase	L-Homoserine $\rightarrow \alpha$-ketobutyrate $+ NH_3$
L-Serine dehydrase	L-Serine $\rightarrow$ pyruvic acid $+ NH_3$
D-Serine dehydrase (D-threonine dehydrase possibly same protein)	D-Serine $\rightarrow$ pyruvic acid $+ NH_3$ (D-Threonine $\rightarrow \alpha$-ketobutyrate $+ NH_3$)
L-Threonine dehydrase	L-Threonine $\rightarrow \alpha$-ketobutyrate $+ NH_3$

TABLE VII (*Continued*)

Enzyme	Reaction
Amidases	
Glutaminase	Glutamine → glutamic acid + NH_3
Asparaginase	Asparagine → aspartic acid + NH_3
Amino acid amidase	Amide → amino acid + NH_3
Aminoimidazolase	4-Aminoimidazole → glycine derivative + NH_3
AMP-amidase	$AMP-NH_2$ → AMP + NH_3
D-Glutamic acid oxidase	D-Glutamic acid + O_2 + H_2O → α-ketoglutaric acid + NH_3
L-Alanine dehydrogenase	L-Alanine + DPN^+ + H_2O → pyruvate + DPNH + H^+ + NH_3
L-Glutamic acid dehydrogenase	L-Glutamate + DPN^+ + (TPN^+) → α-ketoglutarate + DPNH (TPNH) + H^+ + NH_3
L-Cysteinesulfinate dehydrogenase	L-Cysteine sulfinate + DPN^+ → β-sulfinyl pyruvate + NH_3 + DPNH + H^+
Amine oxidases	
Monoamine oxidase	Monoamine + O_2 + H_2O → corresponding aldehyde + NH_3 + H_2O_2
Diamine oxidase	Diamine + O_2 + H_2O → corresponding aldehyde + NH_3 + H_2O_2
Reductive formation of ammonia	
Hydroxylamine reductase	NH_2OH + cytochrome$_{red}$ → NH_3 + cytochrome$_{ox}$ NH_2OH + TPNH(DPNH) + H^+ → NH_3 + $TPN^+(DPN)^+$

[a] After Takahashi *et al.* (526).

ions (e.g., potassium). The ability of nitrate, as a mobile anion, to promote cation absorption (cf. Chapter 4, Volume II) could indirectly affect the growth of the plant in ways which ammonium salts cannot duplicate.

Nitrate reductase has been known for many years as a powerful reducing system that is widely dispersed in plants [Eckerson, 1931; Nason and Takahashi (350)]. In 1924, Eckerson (143) grew tomato (*Lycopersicon esculentum*) plants in sand cultures that lacked nitrogen. Positive tests for nitrates, nitrites, ammonia, and amino acids could not be obtained although the plants still contained an abundance of carbohydrate. When calcium nitrate was added to the sand, nitrate ions could be detected within 24 hours in all parts of the plant. Nitrites were detected 36 hours later, and at the end of 2 days, the quantity of nitrite, ammonium ion and asparagine had increased. Amino acids tended to increase in quantity for 3 weeks after the addition of nitrate while the carbohydrate content of the cells decreased. The nature of the detailed mechanism has only recently been better understood.

It is this area of nitrogen metabolism that has produced evidence of, and reasons for, the essentiality of molybdenum for plants, since a molybdenum-containing protein is now known to be the functional enzyme of nitrate reduction. (For a discussion of work in this field reference may be made to Chapter 4, by Nason and McElroy, in Volume III.)

The effects of age, concentration, and external pH have been studied by Prianishnikov (1934) in relation to the relative absorption of nitrate and ammonium nitrogen from ammonium nitrate by etiolated seedlings (cf. 403a). In young high-carbohydrate seedlings the absorption of ammonium exceeds that of nitrate, but with prolonged "starvation" this relation may be reversed and the "physiological acidity" of this salt may give place to an apparent "physiological alkalinity." Age, increased concentration and an acid reaction, all tend to increase nitrate absorption relative to ammonium. In extreme cases of plants in which the synthetic mechanism is impaired with age, nitrate reduction may still continue and ammonium may actually be released to the external solution. Sideris *et al.* (448, 449) and Street *et al.* (516) found that nitrate plants, in contrast to ammonium plants, usually have comparatively high contents of inorganic and of complex organic forms of nitrogen, such as polypeptides.

However, some plants are also able to utilize atmospheric nitrogen, and by the new well-known symbioses this becomes a means of furnishing nitrogen even to certain angiosperms. The property of nitrogen fixation (cf. Virtanen and Miettinen, Chapter 5 in Volume III), is possessed by both aerobic and anaerobic free living organisms and is also exemplified by symbiotic associations between the nitrogen-fixing organism and a far wider range of vascular plants than was at first thought to be the case. Thus, plants can chemically utilize nitrogen in all its forms, ranging from the lowest reduction state of the nitrogen $(NH_3, -3)$ to one of its more fully oxidized levels in the form of nitrate $(NO_3, +5)$. The possible chemical forms in which nitrogen might exist in plants and their oxidation levels are indicated in Volume III (page 382).

When biological nitrogen fixation and photosynthesis go on in different parts of the same organism, as in nodulated leguminous plants, a considerable degree of interdependence may exist (610, 611). When these processes take place in the same cells, e.g., blue-green algae, close biochemical relations can exist between them. The suggestion arises that in blue-green algae and photosynthetic bacteria the hydrogen donors, carbon skeletons, and energy required in nitrogen fixation are supplied more or less directly by photosynthesis [Fogg and Than-Tun, 1958 (153), and earlier references there cited].

Fogg and Than-Tun (153) have studied the kinetic relationships between the photosynthetic assimilation of carbon and the assimilation of nitrogen into the cell material of *Anabaena cylindrica* Lemm. At any given temperature there was a close correlation between rates of the two processes over a wide range of variation in other factors. Both carbon and nitrogen assimilation were found to be inhibited by relatively low concentrations of carbon dioxide. The results of experiments using N^{15} (288) suggested that the nitrogen first appears in blue-green algae as glutamic acid, which is known to be readily derived from the phosphoglyceric acid produced in photosynthesis. Linko *et al.* (275) found that during short periods of photosynthesis and under conditions that probably allowed nitrogen fixation, up to 20% of the carbon fixed by *Nostoc muscorum* appeared as citrulline. In *Alnus* roots, exposed for a few hours to an atmosphere containing excess N^{15}, citrulline usually had the highest atom per cent excess of the tracer after glutamic acid; this suggested that it is formed directly from ammonium produced by fixation of nitrogen. It appears likely that, in blue-green algae, glutamic acid and citrulline are indeed the first organic products of nitrogen fixation and that their formation is closely linked with photosynthesis (153).

The fixation of nitrogen is carried on by a wide range of microorganisms, aerobic and anaerobic, photosynthetic and nonphotosynthetic (630). Numerous hypotheses have been outlined regarding the mechanisms (216, 309, 403), but none has yet gained complete acceptance.

Much attention has been devoted to the intermediate steps in the conversion of atmospheric nitrogen to organic forms. Important consequences would follow if, by chemical means, the biological process could easily be simulated, and atmospheric nitrogen be made chemically available for agriculture in very different ways from the well-known Haber processes. However, the course of biological nitrogen fixation which stems from nitrogen in the air may well follow similar routes to the normal course of nitrogen metabolism that starts from ammonia or nitrate.

Thus, at different times virtually all the conceivable reduction products of nitrate have been invoked as principal intermediary forms of nitrogen prior to their entry into organic combination with carbon; but of these reduced forms of nitrogen the three most widely mentioned have been ammonia, hydroxylamine, and more recently hydrazine. Among the forms of nitrogen available for the use of plants, though not naturally occurring, the substance cyanamid deserves special mention because of its wide use as an artificial nitrogenous fertilizer. Both cyanamid and liquid ammonia are now applied widely under agricultural conditions; urea also is widely used in the form of over-head sprays or by direct application to the soil.

The source of nitrogen supplied may affect the subsequent course of the nitrogen metabolism. Plants supplied with urea do not normally store it as such, but often convert it rapidly to glutamine. By contrast, in animals urea is said to be formed more directly from the amide group of glutamine. Freiberg and Payne (1957), Freiberg et al. (168), and Boynton (1954) have dealt with urea as a nutrient applied to the leaves of various plants. Reinbothe and Mothes (413) have also evaluated the recent evidence of urea and its derivatives in plants. Urea supplied to apple leaves appears mainly as glutamine (Boynton et al., 1953). Steward and Pollard (493) stressed that there may be something unique about the incorporation of carbon from urea which distinguishes it from the incorporation of carbon dioxide. In aerated potato disks, urea led to more glutamine than was obtained by the use of ammonium salts, and the C^{14} of the urea that was furnished appeared in the glutamine. The incorporation of urea carbon and nitrogen into guanine by *Chlorella* and *Scenedesmus* has also been demonstrated by Ellner and Steers (1955).

A common idea is that urea may be an alternative source of ammonia, the carbon being lost as carbon dioxide. However, urea may not be merely broken down to carbon dioxide and ammonia, as by the enzyme urease, for it may be fixed in some plants or organs in which urease is not abundant, or from which it is even absent. Urea is an extremely versatile compound when it is able to react with other substances. In this laboratory it has been found that, when autoclaved with the normal organic and inorganic constituents of a plant nutrient solution, C^{14}-urea gives rise to a great variety of compounds and even radiation will produce a similar range of new substances. Moreover a urea-related substance that now looms as a possible important intermediary of nitrogen metabolism is carbamyl phosphate

$$\left(\begin{array}{c} \quad\nearrow NH_2 \\ CO \\ \quad\searrow OPO_3H \end{array} \right)$$

which is known to be a donor of the carbamyl group [cf. Reichard, 1959; Cohen and Sallach (113), Reinbothe and Mothes (413)] but is also a substance that may have other far-reaching implications.

A study of the mechanism of CO_2 fixation in the biosynthesis of carbamyl phosphate has been reported by Jones and Spector (239). Thiouracil apparently has an effect on the *de novo* synthesis of carbamyl phosphate synthetase during the morphogenesis of tadpoles (113).

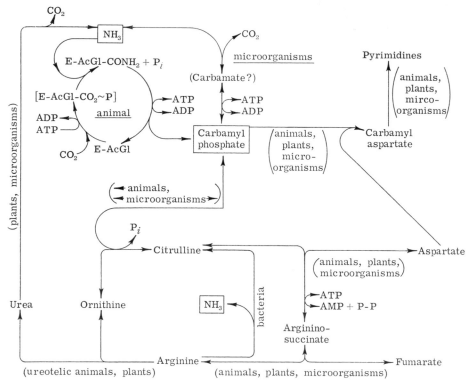

Fig. 2. Metabolic interrelationship of carbamyl phosphate, the urea cycle, and pyrimidine synthesis. AcGl, acetyl glutamate; E-AcGl, enzyme-bound acetyl glutamate. After Cohen and Sallach (113).

There are good indications that the Krebs-Henseleit (260) ornithine cycle (cf. Fig. 2) operates in some plants so that the cycle may constitute a method of urea synthesis. The rarity of reports of the requisite enzymes associated with the interconversions of the cycle has led to the belief that the ornithine cycle is mainly associated with amino acid synthesis and is only secondarily a source of urea. Urea can also arise from aerobic breakdown of purines.

The efficacy of urea as a nitrogen source, as in the special nutrition of dividing carrot (*Daucus carota* var. *sativa*) cells in tissue culture, has not been sufficiently explained (445). Until recently, urea has been thought of more as a product of catabolism than as an intermediate or precursor for synthesis, though the synthesis of allantoin from glyoxylic acid and urea, and the reciprocal relations of allantoin and glutamine, have been noted [Tracey, 1955; in Paech and Tracey (381); Reinbothe and Mothes (413); see also Fig. 3]. Shantz and Steward (445) suggest that urea nitrogen may intervene in the reactions of carbamyl groups, particularly when the cell-division growth factors create a heavy demand for the fragments necessary to build nucleic acids. This links

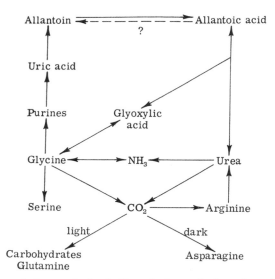

FIG. 3. Scheme of the synthesis of allantoin from glycine, glyoxylic acid, or urea. After Mothes (342).

urea with the synthesis of important pyrimidines which might otherwise be limiting in cells [cf. Böttger (65) pp. 791 *et seq.*].

It would be inappropriate to conclude even a short reference to exogenous sources of nitrogen to growing plants without reference to anomalous nutritional situations. The first concerns situations in which the normal processes of nitrification are impaired, as in acid bogs, and in which plants have resorted to carnivorous or insectivorous habits for obtaining their nitrogen.

The biology of these situations in which plants resort to the insectivorous habit attracted many observers even prior to the work of Charles Darwin, who wrote a book on insectivorous plants (1875). Most of their attention was directed to the elaborate trapping machinery [as in *Drosera, Nepenthes, Dionaea,* etc.; cf. Schmucker and Linnemann

(437)] rather than to the biochemistry which these anomalous methods of nutrition involve. The presence of protein digesting enzymes, such as papain in such plants as *Carica papaya* and the bromelin of the pineapple (*Ananas comosus*), has long been known. Early attempts were made to study the enzymes involved in protein digestions by insectivorous plants (314). However, the insectivorous habit has not attracted the attention it deserves since the development of modern techniques for the study of intermediary nitrogen metabolism.

The idea that plants normally may secrete organic nitrogen from their roots into the rhizosphere has become generally appreciated from the study of nitrogen fixation. In the chapters by Virtanen and Miettinen on nitrogen fixation (Chapter 5, Volume III) and by Quastel on the role of microorganisms in soil (Chapter 6, Volume III), one may see that some direct absorption even of nucleic acids or amino acids by plants from soil is a factor in their nutrition. Moreover, the techniques of biochemical genetics may now render some plants, that normally subsist on simple exogenous forms of nitrogen, dependent upon the exogenous supply of particular nitrogen compounds, for which they lack the means of synthesis. Thus, mutants have become known which require an exogenous supply of such amino acids as lysine and arginine for growth whereas the wild form of the organism, in the case of *Neurospora*, can readily manufacture this substance in its ordinary metabolism (29). Such terms as "auxotrophs" and "prototrophs" have been used to describe mutant strains with different degrees of ability to manufacture their necessary nitrogenous metabolites from simple sources and have led to the consideration of the problem of gene action in single cells. Reference should also be made here to such symbiotic situations as those which exist in lichens, in which the green organism, i.e., the alga, may furnish a carbohydrate and the associated fungus provides much of the ability to metabolize nitrogen.

Recent literature on the culture of angiosperm cells, tissues, and organs under aseptic conditions is replete with examples in which much better growth is obtained by furnishing nitrogen in some other form than nitrate [e.g., casein hydrolyzate (497)].

C. TYPES OF SYSTEMS INVESTIGATED FOR THEIR NITROGEN METABOLISM

Early preoccupations with the maintenance of fertility in soils led naturally to the first techniques of studying the relations of plants to nitrogen by the use of field plots and the testing of conditions necessary to maintain the growth of crops without recourse to fallowing. The balance sheets of Boussingault (cf. 238, 430) need no reemphasis here.

Nor need one stress again the attempts of Liebig in Germany and, more successfully, of Lawes and Gilbert in Britain to maintain fertility in soils by the use of artificial (or mineral) fertilizers. In fact by the end of the nineteenth century stress was being laid on the routine application of appropriate mixtures of fertilizers to make available the elements nitrogen, phosphorus, and potassium. So much so that the NPK content of a fertilizer has become a widely familiar prescription to maintain fertility and yield. This point had been reached when, in 1893, Gilbert made his well known lecture tour of the United States to communicate this new knowledge to the New World. The study of the nutrition of seedlings with the rise of the water and sand culture techniques also developed in the nineteenth century; indeed de Saussure was already a pioneer in this respect in his well-known book published in 1804.

Having passed from agricultural crops in the field to the growth of whole plants in a water or sand culture there is an easy transition to the study of nitrogen metabolism in seeds and to the changes that take place on their germination. From Boussingault's work (1864, 1868), to quote Chibnall (106), "we gain our first insight into the quantitative aspects of chemical changes which take place during germination" (*loc. cit.*, p. 43). In one of the first metabolic balance sheets with germinated seedlings Boussingault showed by elemental analysis that, whereas there were extensive changes in the carbon, hydrogen, and oxygen (i.e., carbohydrate) during germination (more being lost in the dark than in the light), the seedlings nevertheless retained all their nitrogen.

From this type of material, concrete ideas on nitrogen metabolism were derived. The role of cotyledons in the nutrition of the embryo, as they store nitrogen prior to germination and release it to the growing regions after germination, directed attention to prominent sites in the plant body in which nitrogenous reserves were mobilized for later use. This type of experimental material, together with the study of the superimposed effects of light and darkness upon both growth and nitrogen metabolism of the seedling provided the basis for much of the classical work in this field. Indeed, the white (*Lupinus albus*) and yellow (*L. luteus*) lupine and seedlings of the grasses (e.g., barley) were prominent experimental materials for studies of this sort (106, 314).

Quite early in the history of nitrogen metabolism the study of buds and other organs of perennation, such as bulbs and tubers, made a contribution. Borodin (1878) studied the developing or dormant buds of *Alnus glutinosa, Quercus robur* (*Q. pedunculata*), and *Spiraea sorbifolia* and concluded that only a limited number of nitrogen-free substances (perhaps only glucose) can, *together with asparagine*, regenerate protein. In addition, he implicated cell growth and respiration, linked to the continuous breakdown of protein in all living plant parts, to form asparagine in the presence of adequate supplies of carbohydrate. Asparagine would thus accumulate in the absence of carbohydrate. Hartig's (1885, 1888) experiments implicated the stem of a tree as a storehouse of such reserve substances. Trees after a heavy seed year contained only traces of nitrogen and granular starch in the wood and rind. He concluded that the recurrence of seed years depended on the times required by the tree to refill its storage organs, i.e., the stem. In pine (*Pinus* spp.) or spruce (*Picea* spp.) this might take some five to seven years (83).

The chemical changes that occur in starving leaves presented a system on which much was done. Despite observations on the formation of protein in the light and its breakdown in the dark, respectively, these provided no clear picture of the mechanism that controls the protein level. The striking but rapid loss of protein in detached leaves, led to the suggestion by Chibnall (106), supported by Lugg (284), that the synthesis of protein is ultimately controlled by hormones coming from other parts of the plant. In particular, the leaves of grasses, or monocotyledonous plants, show breakdown very rapidly after they are excised [Yemm's data (627–629); pp. 217, 218 of Chibnall (106)].

Late in the nineteenth century, in which details of the nitrogen cycle were being worked out, the science of microbiology was in its infancy. The role of bacteria in the conversion of organic nitrogen to ammonia, of ammonia to nitrite, and of nitrite to nitrate needs no re-emphasis here (cf. Fig. 1). However, the great stimulus that was to come from microbiology in the study of nitrogen metabolism required the development of pure culture techniques and the recognition that microorganisms, cultivated with appropriate sources of carbohydrate, represented very convenient systems on which to study not only nitrogen metabolism, but a variety of other physiological problems. Thus in the present century much has been learned about nitrogen metabolism by reference to bacterial metabolism, or to the intermediary metabolism of *Neurospora* (152). However it was other, and even later, trends in enzymology and biochemical genetics that made the work on microorganisms rewarding.

After the isolation, by Sumner (1926), of the enzyme urease as a crystalline protein, the general recognition of enzymes as proteins came slowly but surely; this endowed nitrogen metabolism and protein synthesis with a particular significance. The rise of genetics and the subsequent gene-enzyme and related hypotheses focused attention upon the genes as the means of determining metabolism via the enzyme proteins that they regulate, and this stimulated much work on sequences of metabolism and on the genetic determination of specific intermediary steps (cf. 29, 232, 233, 331).

The selection of particular traits well suited to an intensive analysis of gene action included, first, the choice of a microorganism instead of a multitissued plant or animal; second, the selection of auxotrophs, blocked at a particular step in some biosynthetic pathway; and third, the selection of mutant strains containing altered enzymes that differed from the wild type in a manner that could account for the phenotypic difference between strains. To date, about a dozen systems have been found in which the mode of gene action involves an influence of specific proteins (293, 432). Such organisms as *Escherichia coli* (419) and *Neurospora* (152) had and still have much to commend them in this field of work.

The work on gene-controlled protein enzyme synthesis had two further consequences in relation to nitrogen metabolism. The focusing of attention upon the self-reproducibility of the gene and its ability to carry the message which permits a given protein to be produced, led naturally to the consideration of virus synthesis as a special, but related, case. Thus, we may know more about the nitrogen metabolism involved in virus synthesis through investigation of such systems as that of tobacco mosaic virus than we do about the metabolism of nitrogen that culminates in the proteins needed for ordinary cytoplasmic metabolic activities. Evidence against protein as a hereditary material comes from experiments with synthetic plant viruses, in which the nature of the protein "coat" seems to have no apparent genetic effect upon the progeny of the viruses (165, 450, 549, 550).

Inevitably, however, as knowledge of nitrogen metabolism and of protein synthesis grew, a new type of experimental system came into prominence, namely, the cell-free system on which attempts are now being made to emulate the synthetic nitrogen metabolism of the cell, particularly in protein synthesis, without its complete organization. Therefore, many workers in the field of nitrogen metabolism now study such isolated organelles as mitochondria or even smaller particles, which are variously described as microsomes or ribonucleic acid granules, etc., and which may be considered to be the "physical basis of metabolism." Enzymes that activate individual amino acids are now being actively investigated (cf. Part 3). All experimental systems, which range from the whole plant in its habitat in the field to the isolated granules obtained from cells and even to single enzymes, have their place, but the questions posed, or the answers expected, are largely determined by the kind of system that is selected for test.

There are, of course, some features that all plants have in common. Ability to use simple forms of nitrogen has been recognized as a distinctively plant characteristic, whereas dependence upon previously elaborated nitrogenous compounds is a distinctively animal feature. Although comparative biochemistry has tended to stress similarities between organisms, much is also to be learned from their differences. In fact, in any actual case, one is apt to be concerned not only with nitrogen metabolism in general but with the special metabolism of some particular organism or organ. Work on such classical objects as the white and yellow lupine has already been mentioned. Later studies of a somewhat comprehensive nature came to be made on the *Narcissus* bulb with its ability to store and later to use nitrogenous reserves (360).

The growing regions of the shoot and the root have been studied in more recent times. An earlier concept was that the cells in the meristem proper were the most active in protein synthesis, and it was earlier thought that, as cells grew by enlargement, their emphasis shifted to other aspects of metabolism. It now appears that protein synthesis of one sort or another may continue throughout the main course of the growth of cells, whether this is by division or by enlargement. Brown (78) now identifies the point of maximum protein synthesis as being somewhat behind the tip of the root.

The shoot meristem particularly has been studied to detect the kind of compounds from which protein is elaborated and, by the combined use of the chromatographic techniques and critical morphological sampling, it has been possible to reveal the compounds that are present in the growing apices of a dicotyledonous plant (white lupin) and of a fern (*Adiantum*) (509, 510).

The classical investigations on the nitrogenous economy of the tobacco (*Nicotiana tabacum*) plant, in which nitrogen metabolism came to be integrated with carbohydrate and organic aicd metabolism, are described in the papers by Vickery and his associates [cf. (566) (and references there cited) and also see Shmuk (446)].

The economic importance of the nitrogen-fixing organism *Rhizobium* and its legume hosts has given to this system a very prominent place in the history of nitrogen metabolism. In fact, as reference to the work entitled *Nitrogen Fixation* by Wilson (610, 611) will show, much of the existing knowledge of nitrogen metabolism in this period could be described by reference to the special problem of the fixation of atmospheric nitrogen by plants. This involved the basic assumption that much of the subsequent metabolism is common to the nitrogen-fixing system and to the plants that utilize the more conventional nitrate sources.

Nevertheless, it is somewhat surprising to note how relatively few comprehensive studies have been made of the nitrogen metabolism of a particular plant species, clone, or variety; so much so that the largest single volume specifically devoted to nitrogen metabolism of plants (cf. 340) did not find it necessary to include an index of plant names. Although not primarily concerned with nitrogen metabolism per se, nevertheless some rather comprehensive information on nitrogen metabolism does accrue from the studies made on the cotton (*Gossypium*) plant by Mason and Phillis (304, 305) and Maskell and Mason (299–303), particularly with reference to the problem of translocation of nitrogen from one part of the plant body to another. On two plants (*Mentha piperita* and *Musa acuminata*) some rather comprehensive information has been obtained by the techniques of chromatography and in the author's laboratory [(486–488); cf. Chapter 2, Volume III].

One should also recognize that the study of photosynthesis, which was greatly facilitated by work on *Chlorella* and *Scenedesmus* and *Rhodospirillum*, has contributed to a comprehensive knowledge of their nitrogen metabolism inasmuch as the path of carbon in photosynthesis integrates closely with the path of carbon in amino acid and protein metabolism (88).

More recently the use of plant tissue cultures has furnished experimental material for the critical study of the metabolism of cells in the rapidly proliferating state, as compared to cells in the resting, more quiescent state, characteristic of the storage organ. In particular, explanted tissue from carrot root, potato tuber, or similar organs can be exposed to a nutrient solution with or without the growth factors, which will make them grow by cell division. These growth factors are prominently those that are encountered in such nutrient fluids for immature embryos as coconut milk or the homologous fluids obtained from immature fruits of *Juglans*, *Aesculus*, and *Zea* (499). The significance here is that the nitrogenous composition of the tissue, as it is explanted from the organ of its origin, changes promptly as the cells pass into the dividing state. After this growth induction has occurred, the intermediary metabolism of the cells can be studied in relation to different exogenous sources of nitrogen under aseptic conditions while growth and protein synthesis are in progress.

The recent research on amino acid metabolism and uptake by higher plants has also utilized tissue cultures [(43–45, 120, 497, 624), and earlier papers cited]. The transition from the quiescent to the growing state involves changes in both the total amount and the relative composition of the soluble nitrogen pool (508), and it is also evident that the protein fraction is affected in both amount and kind.

D. Recognized Classes of Soluble Nitrogenous Compounds

The history of nitrogen metabolism may be seen in terms of the substances which were thought to be involved.

1. Amino Acids and Amides

The first free amino compound in plants (1802) may have been observed in the sap of *Asparagus officinalis* in the form of cubic crystals by Delaville. In 1806, when only about 36 chemical elements were known, Vauquelin and Robiquet proved that the new readily crystallizable neutral substance contained nitrogen. Street and Trease (517) have traced the history of the substance asparagine.

The study of asparagine was first made possible by its ability to crystallize in the form of "Gleis" when plant preparations were treated with alcohol (Hartig, 1858).

Wollaston (1810) isolated cystine (cysteic oxide) from urinary calculi although he did not recognize the relationship of cystine to the protein molecule, nor did he determine that it contained sulfur. Cystine was not, in fact, isolated from a plant protein till 1912.

Glycine was actually the first amino acid to be isolated from a protein hydrolyzate. Vickery and Schmidt (574) and Meister (320), both thought that Braconnot, who used sulfuric acid hydrolysis was unaware of the presence of nitrogen in glycine. However, Barrales (25) quoted the original work to show that Braconnot (1820) in fact noted that his newly isolated crystals of "sucre de gelatin" (*loc. cit.* p. 114): "Distilée il a donné un léger sublimé blanc et un produit ammoniacal, d'où il résulte qu'il contient de l'azote." Abderhalden (1903) found 3.8% of glycine in the hydrolyzate of the plant protein edestin. Braconnot's observation therefore marks the beginning of our knowledge of the amino acid units of which protein molecules are composed.

The detection of asparagine and the use of Hartig's alcohol method focused attention on one of the two principal amides that are still important in any comprehensive survey of plant metabolism. The emphasis on the semiamide of glutamic acid, namely glutamine, came much later (Schulze and Urich, 1877; also Schulze and Bosshard, 1883).

Glutamine was recognized by Schulze and Urich (1877) in beet (*Beta vulgaris*) roots and again in 1877 with Barbieri, using gourd seedlings. Hlasiwetz and Habermann (1873) had suspected its presence as an amide in protein, but it was not until 1932 that the presence of amides in protein came to be generally accepted (121).

After it became known that proteins on hydrolysis gave amino acids, there was a search for the amino acids of protein hydrolysis occurring in the free state in plants, the implicit assumption being that the amino acids should precede the formation of protein. The search for the amino acids deemed necessary to make protein, as they occur in the free state in plants, occupied the long and productive lives of Schulze and of Prianishnikov. Surprisingly, however, the list arduously compiled by Schulze (1906) consisted only of the following substances: lysine, arginine, proline, valine, leucine, isoleucine, phenylalanine, tyrosine, tryptophan, and histidine. In 1939 Chibnall added the amino acids glycine, alanine, and serine.

Since some of the amino acids commonly present in the structure of proteins were yet to be discovered, e.g., methionine (cf. Mueller, 1923) and, at about the same time, threonine (McCoy et al., 1925), these amino acids could not have been found free by Schulze or Prianishnikov. In fact, prior to the use of chromatography, there was some evidence of the free occurrence in plants of such other amino acids as glutamic and aspartic acid, although Schulze and Prianishnikov had tended to regard them only as products of amide hydrolysis whenever they were noticed to occur free. The progress made possible by the modern techniques of chromatography in its various forms is recorded in a comprehensive tabulation (see Appendix I).

However, attention has also been given in different periods to other classes of nitrogenous substance which are categorized below.

2. The Simpler Nitrogen Bases

The simple nitrogen bases have been much investigated because of their relation to alkaloids on the one hand and to such important complex compounds as the phosphatides, or lipids, on the other. This is described in such older works as that of Barger on the Simpler Natural Bases (1914), or the more recent works by Guggenheim (193) entitled "Die biogenen Amine." These simpler nitrogen bases range from substances like choline, many betaines, some amines like putrescine, and many simple nitrogenous bases, which are commonly part of alkaloid synthesis or metabolism (Table VIII). Choline sulfate formation mediated by pyridine nucleotides in microorganisms and higher plants (Zea mays) has received recent attention (364, 609).

Mention should also be made of knowledge gained from studies on "alcoholic fermentation" and putrefaction giving rise to ptomaines or natural amines. Amino acids were found to be the mother substances of the by-products (fusel oil) of alcoholic fermentation and of the products of putrefaction. A description of these compounds is given also in Harden's "Alcoholic Fermentation" and more recently (1961) in the Pfizer *Handbook of Microbial Metabolites*. Kamienski (243, 244) has separated 36 primary, secondary, and tertiary amines and diamines from plants by paper chromatography in a survey which included 220 species of flowering plants and mosses.

The ubiquity of the natural bases establishes their interest for the general physiologist, and they have been variously reviewed by McKee (312, 314), Steward and Street (502), and Guggenheim (193). The

TABLE VIII
SOME SIMPLE NITROGEN BASES IN PLANTS

$[HOCH_2CH_2N(CH_3)_3]^+OH^-$	Choline, a substituted N base
$(CH_3)_3\overset{+}{N}CH_2COOH$ $\overset{\mid}{OH^-}$	Betaine, glycylbetaine
$H_2NCH_2(CH_2)_2CH(NH_2)COOH$	Ornithine, a diamino acid
$H_2NCH_2(CH_2)_2CH_2NH_2$	Putrescine, a diamine

earlier neglect of the simpler natural bases by plant physiologists was due to their relatively low concentration in cells and to the lack of a good method for their investigation.

The heterogeneous class of nitrogen compounds known as alkaloids, which may account for a considerable amount of the nitrogen metabolized by plants, will not be discussed in detail in this chapter except where their metabolism seems to impinge upon protein metabolism or respiratory metabolism. It has, however, been dealt with fully in the writings of the following: Mothes and Romeike (345), Mothes (341), Manske and Holmes (292), Sokolov (463). Battersby (27) has reviewed the biosynthesis of alkaloids. Indeed, one of the paradoxes of nitrogen metabolism is that plants often form many chemically complex compounds that seem to have little to do with the main stream of nitrogen (i.e., protein) metabolism. The factors in evolution which brought this situation about are obviously complex.

3. Urea and Ureides

One of the first nitrogenous compounds to be recognized was the animal catabolic product urea, as investigated by Rouelle in 1773. The belief was current for a time that organic compounds could arise only through operation of a "vital force" inherent in the living cell. Berzelius even held that the chemical synthesis of organic substances was beyond the realm of possibility. Thus Frederich Wöhler's (1779–1884) synthesis of urea from ammonium cyanate refuted the then currently postulated "vital force," and, in a letter to Berzelius, he stated, "I must tell you that I can prepare urea without requiring a kidney or an animal, either man or dog." Through the classical period of Pfeffer, Schulze and Prianishnikov, references to urea in plants were infrequent and probably resulted from the accepted role of urea in the animal, which has, however, little counterpart in plants. Observations of the occurrence of urea in plants were made; they were summarized by Verschaffelt (1911) and more recently by Tracey [cf. Paech and Tracey (381)] and Reinbothe and Mothes (413).

Tracey (381) recognized that this was a much neglected field, despite the widespread occurrence of urea, ureides, and their enzymes in 29 then known families of angiosperms.

However, it was not until 1912 that urea was found to occur free in plants when Fosse (1912, 1913) isolated dixanthyl urea from plants of the Compositae, Cucurbitaceae, Umbelliferae, and Solanaceae. Later he grew maize aseptically and demonstrated urea in various parts of that organism (1913). Brunel and Capelle (1947) have reviewed the importance and distribution of ureides in plants (381, 413).

Even so, for a long time urea was not thought to occur in significant quantity in plants, although the enzyme which converts it to carbon dioxide and ammonia (Sumner, 1926) was isolated as a storage protein from jack bean (*Canavalia ensiformis*). It is only recently that urea and ureides have been detected more frequently and in some quantity in plants.

Fosse (1939) considered urea to be an end product of purine metabolism, not an initial source of nitrogen or even a general metabolite. Between 1927 and 1932 Klein and Tauböck used urease to demonstrate free urea and ureides. Urea was most conspicuous in the early stages of germination, in the etiolated plant and in immature seeds. These authors recognized (1932) that this urea could originate from arginine rather than by oxidation of purines, but Damodaran and Venkatesan (1948) found considerably more urea than could be accounted for by the breakdown of arginine. Reifer and Melville (412) believe that a urea precursor exists in grasses and that this forms urea when the foliage is dried. When urea was furnished to the roots or leaves it was rapidly converted to glutamine: this was later shown for a number of plants using C^{14}-urea. The study of urea metabolism took a new turn from the use of urea labeled with C^{14} and N^{15}. Reinbothe and Mothes (413) have summarized the recent work on urea metabolism.

The Krebs and Henseleit (260) theory of urea formation via the ornithine cycle was developed with reference to animals. However, Mothes (1961) described the different ways that urea and ureides can be formed in plants, and thence introduce their carbon and nitrogen atoms into the general metabolism of amino acids. Genetically determined differences and the physiological aging of the organ in question play a decisive role in the metabolism of these compounds. The sequence of biochemical reactions for the biosynthesis of arginine was first established by study of the nutritional requirements of *Neurospora* mutants [Srb and Horowitz (471)].

Such substances as allantoin and allantoic acid are now known to be prominent in some plants and in some situations notably in the xylem sap, and these compounds and their metabolic relationships have been extensively investigated and written upon (54, 381, 413). There is now reason to believe that compounds of this type may be among the main mobile forms of nitrogen in certain plants.

Discussion of urea in plants is incomplete without reference to its possible role in the metabolism of the purines and pyrimidines (cf. Section V). There are two possible routes of allantoin synthesis in plants, one directly from simpler molecules via condensation with urea, or a second via aerobic purine breakdown (Fig. 3). However, the main route of allantoin formation appears to be via the aerobic purine (e.g., adenine) degradation [cf. Reinbothe and Mothes (413), and references there cited].

Recent work by Cook[5] on the yeast *Candida flareri* deals with the use of urea. It is believed that condensation with an unidentified carbon fragment leads to hydantoic acid

$$
\begin{array}{ccc}
HN & \!\!\!\!\!\!\!\! & CO \\
| & & | \\
OC^* & & CH_2 \\
& N & \\
& H &
\end{array}
$$

Hydantoin

which equilibrates with hydantoin.

$$
\begin{array}{ccc}
& NH_2 & \\
& | & \\
OC^* & & COOH \\
| & & | \\
NH & \!\!\!\!\!\! & CH_2
\end{array}
$$

Hydantoic acid

[5] Cook, A. R. Studies on the utilisation of urea by *Candida flareri*. Ph.D. thesis, University of Liverpool, England, 1963. Cook, A. R., and Boulter, D. Utilization of urea by *Candida flareri*. Proc. Biochem. Soc. *Biochem. J.* **88,** 69p (1963).

Kinetic studies support these early products. The hydantoic acid is then believed to lose glycine, which is consumed in purine and ribo-flavine synthesis, leaving a residue as carbamyl phosphate. The C^{14}-carbon of the urea passes to carbamyl phosphate and then may meet any of several fates. It may pass as carbon dioxide into the Krebs cycle; as ureidosuccinic acid (carbamyl aspartate) into pyrimidines or into the ornithine cycle.

4. Guanidino Compounds

There is a wide range of guanidino compounds in plants. Schulze studied the occurrence of guanidine in plant organs as early as 1892. Arginine, which was discovered by Schulze (1886) in lupine and conifers (1896), is now known to be widespread in the free state [Mothes and Reinbothe (413)] and also is an important constituent of protein and of histones (391) [e.g., prior to germination 94% of the arginine is combined in seed protein of *Pinus pinea*, as in the work cited by Chibnall (106)]. The peptide arginylglutamine was found in the freshwater alga *Cladophora* (291). Galegin (isoamylene guanidine) occurs in *Galega* (20). Reuter (415) administered $C^{14}O_2$ to ex-cised *Galega officinalis* shoots to show that galegin is synthesized in the shoot. When arginine with the amidine group labeled with C^{14} was fed to the seedlings, leaves, flowers, and fruits, most of the radioactivity was lost as CO_2 although some radiocarbon was incorporated into galegin. Canavanine, which is a guanidine derivative of homoserine, is present in *Canavalia ensiformis*, *Glycine max* (*G. hispida*), and

TABLE IX
STRUCTURE OF SOME GUANIDINO COMPOUNDS

$H_2NC(=NH)NH_2$	Guanidine
$H_2NC(=NH)NHCH=C(CH_3)_2$	Galegine
$H_2NC(=NH)-NH(CH_2)_3COOH$	γ-Guanidinobutyric acid
$H_2NC(=NH)NHO(CH_2)_2CH(NH_2)COOH$	Canavanine
$H_2NC(=NH)NH(CH_2)_4NH_2$	Agmatine
$H_2NC(=NH)NH(CH_2)_3CH(NH_2)COOH$	Arginine

Colutea arborescens (see Table IX). Bell (30), Birdsong *et al.* (42), and Tschiersch (547) found canavanine to be more generally distributed than has previously been assumed. Streptidine (a compound of inositol and guanidine) is part of the streptomycin molecule (388). *Streptomyces* also produces γ-guanidinobutyramide by an unusual oxidation of arginine (531).

Using paper chromatographic methods for substituted guanidine derivatives [for references see Steward and Pollard (494)] evidence has been furnished that guanidino compounds other than arginine occur in castor beans (*Ricinus communis*). These compounds include agmatine, γ-guanidinobutyric acid, and two unidentified substances. Morel (334) has also noted that winter-stored artichoke (*Helianthus tuberosus*) tubers are rich in arginine, but in the tissue cultured from these, arginine disappeared, protein increased, and seven unidentified, Sakaguchi-positive substances appeared. One of these is now identified by co-chromatography as γ-guanidinobutyric acid, and the other is thought to be arginic acid.

Canavanine is known to be a competitive inhibitor of arginine in *Neurospora* metabolism (218). Canavanine inhibition in plants has been demonstrated with *Avena* coleoptiles (Bonner, 1949), peas (*Pisum sativum*) minus their cotyledons (Fries, 1954), carrot tissue cultures [Steward et al. (497)], and maize embryos (Wright and Srb, 1950). Canavanine inhibition of *Ginkgo biloba* pollen was only slightly reversed by arginine (Tulecke, 1960), whereas growth inhibition of isolated *Pinus serotina* roots was reversed by arginine, a reversal indicating a competitive inhibition of arginine utilization (23).

Mothes (342), Reinbothe and Mothes (413), and Steward and Pollard (494) have summarized the limited information concerning canavanine biosynthesis. Presumably canavanine arises via transamidination from arginine [Fig. 23 of Mothes (342), p. 1804]. Transamidinase, partially purified from kidney, has been shown to catalyze the transfer of the amidine group from arginine, canavanine, and guanidinoacetic acid to ornithine, canaline, and glycine in each possible combination of donor and acceptor [Walker, 1956, 1957; Ratner and Rochovansky (409)]. Although there is a formal resemblance between transamidination and transcarbamylation, there is no evidence that an analogous amidine derivative is formed (413).

In jack bean, not only is there canavanase or arginase present, but also an enzyme able to condense canavanine and fumaric acid to canavaninosuccinic acid (Walker, 1956). The condensation product has not been reported in the jack bean itself. Walker postulated (1956) that an enzyme-amidine complex (active urea) is involved in transamidination. In *Pseudomonas* canavanine can yield hydroxyguanidine and homoserine (Kalyanckar et al., 1958), and in resting cells of *Streptococcus faecalis*, guanidine and serine can be formed (Kihara et al., 1957). Canavanine can also yield ammonia and O-ureidohomoserine [Kihara and Snell, 1957; for references see (413, 494)]. The structures of these compounds are shown in Table X. Thoai and Olomuki (531,

532) have studied the properties of arginine decarboxyoxidase from *Streptomyces*. The reactions carried out by the enzyme are shown below, after Thoai and Olomuki (531, 532):

arginine → γ-guanidinobutyramide + CO_2 + H_2O
canavanine → β-guanidinoxypropionamide + CO_2 + H_2O
homoarginine → δ-guanidinovaleramide + CO_2 + H_2O

Since guanidine and some of its related compounds do occur free in plants, the extension of chromatographic methods for their detection should be profitable. Agmatine (1-amino-4-guanidinobutane) has been chromatographically detected in potassium-deficient barley leaves (460).

TABLE X

THE STRUCTURE OF COMPOUNDS RELATED TO METABOLISM
OF GUANIDINO DERIVATIVES

$HOOCCH(CH_2)_2ONHC(=NH)NHCH(COOH)CH_2COOH$	Canavaninosuccinic acid
$HONHC(=NH)NH_2$	Hydroxyguanidine
$NH_2C(=NH)NHCH_2COOH$	Guanidinoacetic acid
$H_2NOCH_2CH_2CH(NH_2)COOH$	Canaline
$H_2N(C=O)NHO(CH_2)_2CH_2COOH$	Ureidohomoserine
$HO(CH_2)CH(NH_2)COOH$	Homoserine

5. Purines and Pyrimidines

The great importance now attached to the pentose nucleic acids in modern biology and the modern knowledge of the chemical composition of important coenzymes give to the natural occurrence of purine and pyrimidine bases a particular interest. With due regard to the importance of these substances, the occasions on which their isolation or detection in plants in the free state has been demonstrated may seem small. Reference may be made to Böttger (65) for a summary of purines and pyrimidines in plants. Appendix II refers to uracylalanine (willardiine, a compound that has as yet undeveloped implications). However, the synthesis, occurrence, and metabolism of this important class of compounds will not be dealt with fully in this chapter, for it belongs more properly to the discussion of the catalytic systems of which they form important parts, or of the more complicated compounds into which they enter.

Pyrimidine analogs (5-fluorouracil) can replace a site in the growing point normally occupied by the pyrimidine in ribonucleic acid (RNA), and thus inhibit photoperiodic induction in cocklebur (*Xanthium* sp.) The inhibition can be counteracted by orotic acid. Bonner and Zeevaart (58) have thus claimed that deoxyribonucleic acid (DNA) multiplication is not essential for floral induction and that it is RNA synthesis which is here required. Thiouracil has been used by Heslop-Harrison (211) in a similar way to inhibit the flowering of *Cannabis sativa*. Both uracil and orotic acid have been implicated in the metabolism of β-alanine in pine embryos (24). The nucleotide composition of ribonucleic acid from vegetative and flowering cocklebur-shoot tips is given in Table XI after Ross (425). Very little difference was observed.

TABLE XI

The Nucleotide Composition of Vegetative and Flowering Cocklebur[a]

Nucleotide	Vegetative cocklebur (%)	Flowering cocklebur (%)
Adenylic acid	23.4 ± 0.7	23.5 ± 0.9
Cytidylic acid	23.9 ± 0.9	23.5 ± 1.1
Guanylic acid	30.7 ± 1.1	31.5 ± 1.0
Uridylic acid	21.9 ± 0.8	21.6 ± 0.8
6-Amino/6-oxo	0.90	0.89
Purine/pyrimidine	1.18	1.22
Total RNA (% fresh weight)	0.49	0.45

[a] After Ross (425).

Of special interest has been the purine analog known as kinetin (6-furfurylaminopurine). Other substituted aminopurines have been shown to have varying degrees of cell-division promoting qualities (499) and in conjunction with purine and pyrimidine derivatives have been credited with flower induction in isolated buds cultivated *in vitro* (97).

Henderson *et al.* (208) have postulated that one mechanism of action of kinetin and its analogs is the regulation of the function of xanthine oxidase.

In recent years the use of tritium- and C^{14}-labeled bases has been particularly effective in the demonstration of DNA synthesis and self-duplication during meiosis and mitosis in such plants as *Lilium* and *Allium*.

Ureido compounds (e.g., ureidosuccinic acid) can be the principal product of thermal reaction of malic acid and urea under the conditions of hot springs (162). Schramm *et al.* (439) have evidence that similar reactions can go farther and yield nucleic acids similar to those isolated from living tissue; consequentially they implicate these reactions in the origin of life. Catalytic and information control of the polymerization of bases to produce a polynucleotide has been demonstrated in the nonenzymatic polymerization of uridine phosphate in the presence of polyadenylic acid.

6. Nucleosides, Nucleotides, and Nucleic Acids

The nucleic acids were discovered when the chemistry of the proteins was little understood and when everything incapable of being otherwise rationally explained was attributed to the proteins as the somewhat mysterious agents of all physiological complexity. Nucleosides and nucleotides were within living memory regarded as complex, ill-understood, but potentially important compounds.

Although yeast nucleic acid was first obtained in 1889 by Altmann, the first chemical studies on this substance were made by Kossel (1893). Until 1902, all the knowledge of plant nucleic acid had been obtained upon yeast nucleic acid, and, in that year the contribution of Osborne and Harris on "tritico-nucleic acid" from wheat embryos appeared (1902). Having about 750 gm of material, they used a variety of methods which revealed that guanine and adenine could be produced in equivalent quantities, and, like yeast nucleic acid, gave uracil and pentose. In the mother liquor from which Osborne and Harris obtained uracil, Wheeler and Johnson (1902 or 1903) found cytosine.

Loew (279), who was concerned about the general value of certain mineral salts stated: "It is clear that the division and multiplication of cells depends on the development of the nucleus, and that upon the continuous formation of nucleoproteids. Hence, when phosphoric acid is present in but small amount, only a corresponding amount of nucleoproteid can be produced." Also, "The observation that the total mass of protein in seeds is increased by an increased supply of phosphoric acid would also be easily understood on the basis of the hypothesis of Strasburger and Schmitz that the nuclei are manufacturers of protein matter. This hypothesis is highly probable and in fact has been confirmed by Hofer in the case of enzymes, which must, partially at least, be considered as a class of proteins." Prianishnikov (403a) noted that Loew introduced a number of totally incredible hypotheses related to the polymerization of aspartic aldehyde during the synthesis of proteins and that this created so much distrust of his work among scientists that even his valuable ideas, such as those in the seemingly prophetic passages quoted above, were not fully appreciated.

The close resemblance of tritico-nucleic acid to yeast nucleic acid was noted by Osborne and Harris (1902). By 1914 Jones (242) stated, "We come to understand quite clearly that there are but two nucleic acids in nature, one obtainable from the nuclei of animal cells and the other from the nuclei of plant cells." Calvery (85) and Calvery and Remsen (86) isolated guanylic and uridylic acid from wheat seedlings.

In 1929, Heulin, employing a colorimetric method for estimating nucleic acid nitrogen, found that in wheat the nucleic acid N:total coagulable N, excluding nucleic acid N, decreases with age. Pearsall (384) also stated that if this is true for *Beta* leaves, his results would in part reflect a decline in the nuclear cytoplasmic ratio with leaf age. Albaum and Ogur (1947) and Albaum *et al.* (1950) isolated diadenosinetetraphosphoric acid and adenosine triphosphate from plants (3). Euler *et al.* (1949) and Euler and Hahn (148) have investigated the nucleoprotein and nucleic acid content of pollen and leaves. Aspartyl and glutamyl derivatives of ADP have been reported (200), as well as activated compounds of amino acids with adenosine monophosphate (127).

A number of nucleotide–peptide complexes have been identified in which an amino acid is bound by an anhydride link (between the carbonyl and the phosphate group) to a nucleotide. Davies and Harris (1960), cf. Davies *et al.* (124), have identified many such compounds in yeast extracts and have shown that they consist of several types, i.e., simple nucleotides linked to a single amino acid or amino acids linked to polynucleotide chains of variable lengths. An example is arginyl-alanylarginylalanyl-5'-uridylate. In addition di- or trinucleotides linked to single amino acids or to peptides have been isolated; e.g., alanyl [3'-(5'adenyl)]-5'-uridylate. Canellakis (90) lists other examples. Nucleotide peptides have been identified in RNA (404).

Nevertheless, through the remarkable advances to knowledge in the intervening years, such symbols as RNA and DNA have now come to have a widely appreciated significance. It should not, therefore, be regarded as any failure to recognize the importance of these compounds that they are not here discussed along with nitrogen metabolism. Except insofar as the pentose nucleic acids contribute as templates to the mechanism of protein synthesis and as carriers of nitrogen compounds as in soluble RNA (sRNA) or messenger RNA (mRNA), it seems more appropriate that they should be related to the discussion of genetics and development, rather than to the normal course of intermediary nitrogen metabolism with which this chapter is primarily concerned (95). Davidson (122), Grunberg-Manago (192), and Warren (596) have covered the metabolic pathways related to nucleotides and nucleosides in animals and bacteria, and Burns (80) has outlined the role of these components in synchronization of cell division, while Böttger (64), Canellakis (90), Albaum (3), Brawerman and Shapiro (76) have reviewed some aspects of nucleic acid metabolism in plants.

7. Cyanogenetic, Isothiocyanogenetic Glucosides and Nitriles

These compounds occur free in plants, but because they seem remote from the amino acid protein metabolism of plants, or even from the intermediary metabolism of carbon which is concerned with respiration, they will not be discussed here. Nevertheless, reference may be made to Dillemann (130) and Kjaer (253a). Some recent work on the use of cyanide by plants and its relation to asparagine, etc., is relevant to this chapter.

Blumenthal-Goldschmidt et al.[6] have noted that hydrocyanic acid is incorporated into asparagine by seedlings that contain relatively large amounts of cyanoglucosides [flax (*Linum usitatissimum*), sorghum (*Sorghum vulgare*), and white clover (*Trifolium repens*)] and in seedlings that contain small amounts, if any, of organic nitriles. The carbon

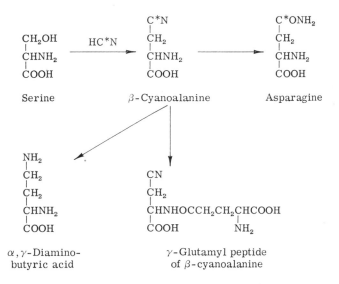

is incorporated in relatively large amounts into the amide carbon of asparagine in seedlings of six species. They suggest that asparagine can be synthesized by a condensation of a C_1 unit with a C_3 moiety. The C_1 unit may be hydrocyanic acid or a related compound, such as formamide. Experiments with serine-C^{14} support this hypothesis. The reaction sequence may be serine, β-cyanoalanine, asparagine as shown above. Moreover, Ressler[7] has shown that β-cyano-L-alanine is the

[6] Blumenthal-Goldschmidt, S., Butler, G. W., and Conn, E. E. *Nature* **197**, 718–719 (1963).

[7] Ressler, C. *J. Biol. Chem.* **237**, 733–735 (1962).

likely intermediate in the conversion of asparagine to α,γ-diamino-butyric acid by dehydration of the amide group of asparagine. α,γ-Diaminobutyric acid may also be formed by the reduction of the cyano group of β-cyanoalanine. Structurally, β-cyanoalanine is also the parent amino acid which, by biological decarboxylation, could form the lathyrus factor, β-aminopropionitrile, or the γ-glutamyl peptide (Appendix II).

Vetch (*Vicia sativa*) when treated with cyanide (private communication, Dr. L. Fowden) did not give β-cyanoalanine or asparagine, but produced a strong unknown substance chromatographically near serine. Apparently this substance is the γ-glutamyl peptide of β-cyanoalanine, which gives on hydrolysis aspartic acid which contains C^{14} from $HC^{14}N$ and cold glutamic acid. This γ-glutamyl peptide is now regarded as the precursor by decarboxylation of the lathyrus factor.

E. Concepts of Nitrogen Metabolism

Having listed the classes of nitrogen compounds that traditionally are found in plants, we may now turn to the major schools of thought and to the general concepts of nitrogen metabolism which they represent. This important topic need be dealt with in outline only, because reference may be made to such works as those of Chibnall (106), of Wilson (610), and of Prianishnikov (403a); to the detailed writings of Vickery (559–561), and Vickery and Schmidt (574); and to the many summaries that have been written from time to time, such as those of Nightingale (358, 359), Pearsall (385), Wood (617), McKee (311–314), Mothes, who edited Volume 8 of the Ruhland Encyclopedia of Plant Physiology (1958), Steward *et al.* (494, 502, 507), Virtanen (578), and Webster (599, 600).

1. Concepts Prior to Knowledge of Transamination and the Carboxylic Acid Cycle

The first unified concept of nitrogen metabolism regarded ammonia as the main ultimate source of raw material from which all nitrogenous metabolism stems. Liebig became preoccupied with ammonia and enunciated the hypothesis that the organic components of soil were of comparatively little significance, apart from their role of furnishing nitrogen for higher plant nutrition. One may note, however, that Liebig probably failed by his obsession with ammonia to discover allantoin and allantoic acid in maple (*Acer* sp.) sap. He noted in 1834 that copious amounts of ammonia were obtained by the treatment of evaporated maple sap with lime. To avoid any suspicion of contamination of the maple sap, "vessels which hung upon the trees in order to collect the juice were watched with greater attention on account of the suspicion that some evil-disposed persons had introduced urine into them, but still a large quantity of ammonia was again found in the form of neutral salts" (271). In the light of more recent work, this ammonia really came from allantoin and allantoic acid, which occurs in the maple sap (53, 342, 344).

The strategic place to be reserved in higher plant metabolism for the amides asparagine and glutamine can be easily summarized. They represent nitrogen-rich reserves which, together with arginine in some organisms, are the form in which much soluble nitrogen may be stored against its later use and to which much nitrogen is returned after protein has been broken down. At different points in time the relation of these substances to protein metabolism was expressed in different ways, and this may be conveniently illustrated by the schemes of Schulze and later of Prianishnikov (cf. Fig. 4).

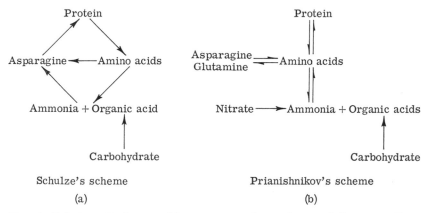

Fig. 4. Relation of nitrogen-rich reserves and protein metabolism according to Schulze and to Prianishnikov.

Schulze's observations brought him into contradiction with Pfeffer's theory that asparagine formed in the cotyledons of seedlings is a primary product of protein decomposition and is transported to the growing points. In 1888 Schulze introduced a number of hypotheses to explain the observed discrepancy between the primary (hydrolytic) and the secondary decomposition (with the predominance of asparagine) of protein. In one of Prianishnikov's concepts (403a), he suggested that in the secondary formation of protein in the stems of seedlings a major role is played by leucine, aminovaleric acid, and others, whereas, the utilization of asparagine in the process of protein synthesis encounters difficulties so that the asparagine accumulates as a product left over from the course of repeated decomposition and synthesis of proteins.

Prianishnikov's scheme was an obvious adaptation to place the amino acids needed for protein synthesis along the direct route to and from protein and also to recognize that the carbon needed to convert nitrate or ammonia to protein had its ultimate origin in carbohydrate via organic acids. Studies started in 1909 (403a), conducted by a number of Russian investigators, established that the decisive factor in the synthesis of asparagine from exogenous ammonia by plants is the content of carbohydrate, but not light per se. Light is beneficial to the synthesis of proteins, but not of asparagine.[8]

By contrast, and to focus attention on the march of events since Schulze and Prianishnikov, a modern metabolic scheme of ammonia utilization may be presented in advance of the discussion of all the evidence [see Fig. 5, from Cohen and Brown (112)]. However, in view of the uncertainty, the route from aspartic acid to asparagine there shown should be regarded as tentative.

Once comprehensive schemes purporting to portray the course of nitrogen metabolism appeared, ideas of the physiological factors that regulate protein metabolism were not slow to emerge.

The original postulate of Knoop and Oesterlin (1925) concerning the reductive formation of amino acids from α-keto acids and ammonia was well founded chemically and strongly supported by animal experiments. At that time, because no enzyme system was known that could cause reductive amination, Knoop's theory was reduced to a hypothesis. However, Quastel and Woolf (1926) and Woolf (1929) discovered that resting cells of *Escherichia coli* reversibly synthesized L-aspartic acid from fumarate and ammonium chloride. Virtanen and Tarnanen (586) claim to be the first to have synthesized an amino acid with a cell-free enzyme solution. The distribution of the enzyme aspartase covered many but not all microorganisms and higher plants; thus it did not give rise to a general explanation of amino acid synthesis from ammonia.

Knoop's concept of amino acid synthesis was realized with the work of Euler *et al.* (1938), who synthesized L-glutamic acid with glutamic acid dehydrogenase. The mechanism of the reaction could not explain the biosynthesis of the then known amino acids, but, at least, two amino acids in which the amino group is formed by incorporation of an inorganic nitrogen compound into the carbon skeleton were established.

---

[8] Recent evidence often links glutamine, rather than asparagine, to light [(cf. 481, 487); also Figs. 29, 41 of this chapter].

There is no good proof of different pathways for amino acid synthesis from nitrate and from ammonia (81). According to our present knowledge (578), it seems that the prominent pathway for incorporation of ammonia into the carbon skeleton to form amino acids is through the biosynthesis of L-glutamic acid. Thorough investigations are still

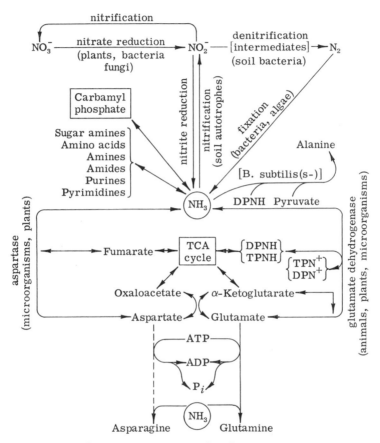

Fig. 5. Composite scheme for formation and utilization of ammonia. After Cohen and Brown (112).

needed to elucidate the biosynthesis of L-alanine and L-aspartic acid, the latter especially in some organisms in which aspartase is to be found.

The importance of the green leaf in the synthesis of protein, because of the availability of carbohydrate in a form suitable for conversion to protein, was also recognized. Chibnall (104, 105) deduced that there was a fall in the protein content of leaves at night which he ascribed

(105) to a diurnal variation in the rate at which a continuous protein synthesis and decomposition took place in bean leaves (106). The predominating role of carbohydrate in protein regulation was given expression in the Paech hypothesis, which is adequately documented by Chibnall (106).

According to Paech's formulation, active carbohydrate (e.g., monoses) and active nitrogen (e.g., ammonia) could not coexist in high concentrations in the cell because there is a limit to the cell's capacity for holding protein; neither could they coexist at very low concentrations because there is then a limitingly small amount of protein to keep the cell alive. The formulation was applicable to ripening seeds of the Gramineae and starving leaves, but it has received criticism from both Chibnall (106) and Wood and Petrie (620). Chibnall (106) postulated that "active carbohydrate" could be replaced by the "necessary series of α-keto acids" and that the α-keto acids could arise in organic acid cycles such as that of Krebs (cf. 261).

Paech also considered the withdrawal of nitrogen from the older leaves of a plant during the intensive growth of the flowering period when the nitrogen requirement could no longer be sufficiently supplied by root intake from the soil. He assumed that growing plant parts must exert some attractive force on the soluble nitrogenous compounds of the older leaves which automatically led to protein decomposition to preserve the constancy of protein nitrogen: total nitrogen value. Chibnall (106) stated that, "In any single organ—within the limits of its cell capacity—the protein content is the resultant of the quantity of monosaccharides and soluble nitrogenous materials ($-NH_2$) present, and that on the part of the whole organism these single systems are linked together, being balanced against each other by means of the translocation stream whose direction and velocity is determined by the force exerted at the growing centers."

Chibnall [(106), p. 86] has summarized what was then regarded as a modern interpretation of protein metabolism in the seedling based on the views of Paech, Schulze, and Prianishnikov:

"(i) On germination, the reserve proteins undergo enzymic digestion to give, as primary products of hydrolysis, asparagine, glutamine and the usual series of amino acids, including aspartic acid and glutamic acid if these be combined as such (i.e., in the unamidised form) in the intact protein molecule. The reserve carbohydrates also undergo enzymic breakdown to give simpler products, including monoses.

(ii) The decomposition of the reserve proteins is controlled by the outflow from the cotyledons or endosperms, as the case may be, of either monoses ("active" carbohydrates) or amino acids, etc. ("active" ammonia), depending on which component is present in the lesser amount (Paech).

(iii) The primary products of reserve protein decomposition are transported to the growing parts, where they become available for protein regeneration.

(iv) Obligatory metabolic processes, which Prianishnikov connected with respiration rather than with growth, bring about a secondary decomposition of certain of the amino acids, which results in the formation of asparagine, glutamine or both these amides.

(v) Protein synthesis in the growing parts takes place at the expense of the amino acids translocated from the reserve organs, and of the acid amides resulting from secondary changes.

(vi) The proteins of the growing parts may have a different amino acid composition from those of the reserve organs. Protein regeneration, therefore, in the absence of external sources of nitrogen, may necessitate an interconversion of amino acids (really a transfer of nitrogen from one nitrogen-free acid to another), and this probably takes place through the agency of asparagine and glutamine."

In 1937 Gregory and Sen (186) discussed the interrelationship of carbohydrate and nitrogen metabolism and re-presented some of the earlier ideas of Spoehr and McGee (470) that respiration was closely linked to nitrogen compounds.

But none of these emerging ideas could gain much permanent stature until the course of intermediary carbohydrate metabolism had been worked out in sufficient detail, and this required developments that are now expressed by the path of carbon through glycolysis and the tricarboxylic acid cycle [Szent-Györgyi, 1936; 1937; Krebs and Johnson, 1937; Krebs and Lowenstein (261)], as these are conventionally portrayed. Some implications of these developments are indicated below.

2. The Tricarboxylic Acid Cycle and Ports of Entry of Nitrogen Into Organic Combination with Carbon

By whatever route the nitrogen source is reduced and whatever the reduction product which finally enters into organic combination, there is a crucial step to be negotiated when nitrogen and carbon compounds combine. This step must be a peculiarly plant property since it is not feasible for most animals to utilize nitrate or ammonia. Modern ideas on this crucial problem required the prior knowledge of the intermediary carbohydrate metabolism which is now represented by glycolysis and the tricarboxylic acid cycle.

Almost as soon as the so-called tricarboxylic acid cycle began to be accepted, that is, about 1937, Chibnall and Vickery saw its significance for the interpretation of nitrogen metabolism of the keto acids, which were destined to play such a key role in the carboxylic acid cycle.

The discovery of transamination in animal tissue at about the same time (Braunstein and Kritzmann, 1937; Braunstein, 1939) also placed the keto acids in a position such that they might receive amino groups from another source and produce both primary and secondary α-amino acids. Subsequently it was shown in Virtanen's laboratory (583) to function also in higher plants. Albaum and Cohen (4) observed a very rapid transamination between glutamic and oxaloacetic acid in oat seedlings. Braunstein (74) has reviewed the transamination of dicarboxylic acids in nitrogen metabolism.

A survey of transaminase activity in various plants has been made by Leonard and Burris (267). The tissues were tested for their ability to catalyze the forward and reverse glutamic-aspartic and glutamic-alanine transamination reactions. Apparently, the transamination rate per unit of tissue decreased with age in the growing plant. Transamination was first believed to function only in systems where L-glutamic acid was the donor or α-ketoglutaric acid the acceptor of the α-amino group. Not only dicarboxylic acids, but also amino groups in other than the α-position can be transferred as shown by the system: γ-aminobutyric acid $+$ α-ketoglutaric acid $\rightleftarrows$ succinic semialdehyde and glutamic acid [for references to this and other unusual transaminations, see Cohen and Sallach (113), p. 47]. Meister et $al.$ (323) have observed the transamination of the α-amino nitrogen of glutamine and asparagine to a number of α-keto acids with the simultaneous splitting of the amide nitrogen to form ammonia. Wilson et $al.$ (608) have used C^{14}-α-ketoglutarate to study transamination in plants. The kinetics and mechanism of action of glutamic-oxaloacetic transaminase of cauliflower ($Brassica$ $oleracea$ var. $botrytis$) have been described by Davies and Ellis (123).

Hydroxylamine has been of great interest in connection with the biosynthesis of "primary" amino acids. The fact that keto acids will combine so readily in solution with hydroxylamine to form oximes, and that these may be readily reduced to amino acids, also suggested these substances to be significant as "ports of entry" for nitrogen when the reduced form of nitrogen was thought to be hydroxylamine. Indeed, even when hydrazine became a prominent contender for this role (cf. 81, 309) the keto acids remained important as the receptors of nitrogen.

The first ideas on this question derived from the readiness with which pyruvic acid could give rise to alanine, oxaloacetic acid to asparatic acid, and α-ketoglutaric acid to glutamic acid, and also from the well-known efficiency of glutamic acid as a donor of amino groups in transamination reactions (577, 578). Since pyruvic acid and oxaloacetic acid are relatively unstable substances, α-ketoglutaric acid was the substance with which tissue could be perfused or which could be injected into the blades of perennial ryegrass (*Lolium perenne*) with consequential increase in the glutamic acid content. Early evidence of this sort was obtained by Chibnall (106).

TABLE XII

Early Evidence of the Formation of Glutamine
from α-Ketoglutaric Acid[a,b]

Hours	4	20
Decrease in α-ketoglutaric acid	17.1	26.2
Calculated equivalent increase in total glutamine N	3.3	5.0
Observed increase in total glutamine N	3.2	5.3

[a] Data after Chibnall [(106) cf. p. 209].

[b] Data given in milligrams per 100 gm fresh weight.

The observation by Virtanen and Laine (583) that roots of legumes fixing nitrogen secreted into the medium a substance which gave asparatic acid on reduction led to the view that the excreted substance was oximinosuccinic acid.

The controversy about the fact of the excretion of nitrogen fixation products from leguminous plant roots into the medium seemed for a while to be capable of resolution on the following lines. In Finland the fact of secretion of nitrogenous substances from pea roots seemed readily to be established, whereas in Wisconsin, by similar means and similar plants, this was much less evident. Furthermore, the work in Finland tended to lead more to aspartic acid via the oxime of oxaloacetic acid, whereas the work in Wisconsin seemed to indicate a path to glutamic acid via ammonia as the product of reduction of nitrogen. Thus, what Virtanen called the "Grundaminosäuren" could be regarded as a function of the conditions under which the host plants were grown, so that these conditions might foster the supply to the nodules by the host plant of α-ketoglutaric acid under the Wisconsin conditions, but a predominance of oxaloacetic acid under Finnish conditions. This difference might not be unreasonable when one recognizes the disparity in the length of day and prevalent summer night temperatures in these two diverse situations. [One may note that length of day does affect the metabolism of mint plants in ways that seem to revolve about effects upon their keto acids; cf. Steward *et al.* (487).] Were this to be regarded as an acceptable hypothesis, one could then visualize a reaction scheme such as that indicated below, in which the nutritional and environmental conditions determine whether the main port of entry is via a 4-carbon keto acid leading to aspartic acid or a 5-carbon keto acid leading to glutamic acid as shown below:

$$NO_3 \rightarrow NO_2 \rightarrow \quad NH_2OH \quad \rightarrow \quad NH_3$$

$$+ \qquad\qquad +$$

oxaloacetic acid $\qquad$ α-ketoglutarate

$$\downarrow \qquad\qquad\qquad \downarrow$$

oximinosuccinic $\qquad$ glutamic acid

$$\downarrow$$

aspartic acid

Thereafter, both the appropriate amides and, by transamination, a series of other amino acids could be derived from these two dicarboxylic amino acids.

As the years passed, the work in Wisconsin on nitrogen fixation seems to have entrenched the view that ammonia, or even hydrazine, is the form by which nitrogen enters and that glutamic acid is the primary product. Meanwhile later work in Finland seems to have conceded that the role of hydroxylamine may have been somewhat overemphasized in comparison with that of hydrazine or ammonia. However, the formation of oximes, and their subsequent reduction as a feasible path of nitrogen into the formation of an α-amino acid need not yet be eliminated as a theoretical possibility. Perhaps the protagonists of one or the other point of view may, in seeking to establish their particular theory, have done some disservice inasmuch as the truth may lie in a balanced combination of *both* routes, in ways that may well be determined by the environmental conditions.

The fact that an exogenous supply of the appropriate intermediates has not resulted in any dramatic confirmation of the hydroxylamine hypothesis need not be disturbing, for the endogenous supply of keto acid and of the reduced nitrogen source may occur under such conditions, or on intercellular sites, that may be very difficult to duplicate from endogenous sources. The widely dispersed presence in plants of a small amount of nitrogen in the form of free nitrite, free hydroxylamine, and hydroxylamine combined with a keto acid as oxime [all of which may be estimated by methods that were reviewed and summarized by Steward and Street (502), Paech and Tracey, eds. (381)] seems to be inescapable.

However, it is now necessary to assume that almost every amino acid has a naturally occurring keto acid analog. To demonstrate this fact certain techniques needed to be developed for the detection of the frequently very unstable keto acids which are present in very small amount. Towers and Steward (545) fixed the keto acid of the tissue in acidic, alcoholic 2,4-dinitrophenylhydrazine, the purpose of which was to stabilize the otherwise reactive keto acids in the form of their 2,4-dinitrophenylhydrazones. The latter compounds were purified by forming the sodium salts of the acids, extracting other material in nonaqueous solvent, and then subjecting the purified phenylhydrazones to hydrogenolysis so that the keto group was replaced

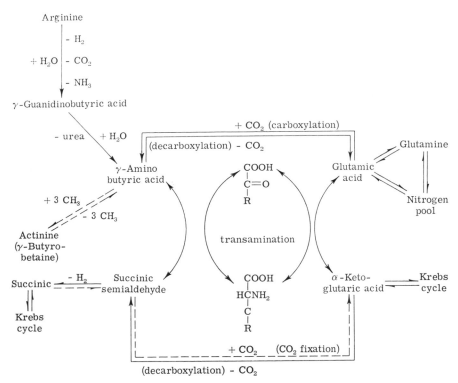

Fig. 6. Some metabolic relations of γ-aminobutyric acid. After Steward and Pollard (493).

by an amino group on the same carbon. By using a different device, essentially the same ends were achieved in Virtanen's laboratory (577, 585) and further modifications of this method have been suggested by Isherwood and Jones (231).

One can now recognize the occurrence in small amounts in plant cells and organs of a far wider range of keto acids than the list of plausible intermediates in the Krebs cycle would alone require. Thus Steward and Pollard (494), who summarized the evidence that had accrued by recognizing free keto acids in the form of their amino acid analogs, were forced to the view that the keto acids were virtually as numerous as the rapidly expanding list of amino acids detectable by chromatography. One or two dramatic examples illustrate this conclusion.

γ-Aminobutyric acid, one of the first prominent amino acids whose wide occurrence in plants was discovered through paper chromatography, was first thought to arise by decarboxylation of glutamic acid. However, when experiments are done with carrot tissue and carbon-labeled glutamine or carbon-labeled γ-aminobutyric acid, it is found that the glutamine only passes back into γ-aminobutyric acid with difficulty, whereas the γ-aminobutyric acid passes in the tissue very readily toward glutamic acid and to glutamine (482, 493). This raises the possibility, therefore, that γ-aminobutyric acid is not merely the decarboxylation end product that it might otherwise seem to be, but that it is a normal intermediate which lies along a route of synthesis, and it may, therefore, be a substance which is produced when a nitrogen-free carbon compound acts as an acceptor of nitrogen or as a "port of entry" for nitrogen. Possible metabolic relations of γ-aminobutyric acid are shown in Fig. 6. Indeed, one might infer from the frequency with which γ-aminobutyric acid is conspicuous in growing tissue cultures that the succinic semialdehyde is its normal precursor in such situations. The presumed reactions through succinic acid and its semialdehyde to γ-aminobutyric acid have been substantiated by enzymatic studies with *Pseudomonas* (17) as shown in the accompanying sequence [after Bachrach (17)].

$$
\begin{array}{ccccccc}
\text{CH}_2\text{NH}_2 & & \text{CH}_2\text{NH}_2 & & \text{CHO} & & \text{COOH} \\
| & \text{NADP} & | & & | & \text{NADH} & | \\
\text{CH}_2 & (\text{TPN}) & \text{CH}_2 & & \text{CH}_2 & (\text{DPNH}) & \text{CH}_2 \\
| & \rightleftarrows & | & \longrightarrow & | & \rightleftarrows & | \\
\text{CH}_2 & \text{NADPH} & \text{CH}_2 & & \text{CH}_2 & \text{NAD} & \text{CH}_2 \\
| & (\text{TPNH}) & | & & | & (\text{DPN}) & | \\
\text{CH}_2\text{OH} & & \text{COOH} & & \text{COOH} & & \text{COOH}
\end{array}
$$

| 4-Amino-butan-1-ol | γ-Amino-butyric acid | Succinic semialdehyde | Succinate |

Dixon and Fowden (132) observed that γ-aminobutyric acid was metabolized with an oxygen uptake and supplied organic acids in intact peanut (*Arachis hypogaea*) mitochondria. Aspartate, alanine, and ammonia were formed using the nitrogen atom of γ-aminobutyrate. Work with tracers suggested that γ-aminobutyrate was converted to succinate. Glutamic decarboxylase was present largely in the non-particulate (soluble) cytoplasm in peanut and pea plants. Dixon and

Fowden concluded that the enzymes responsible for γ-aminobutyrate synthesis resided in different subcellular fractions.[8a]

A further example may be cited. It will be mentioned later that the soluble nitrogen compounds in the tissues and growing point of a fern (*Adiantum pedatum*) contained large amounts of an amino acid (90% of the total soluble nitrogen) which later proved to be γ-methyl-γ-hydroxyglutamic acid (188). When *Adiantum* was subjected, however, to the keto acid technique, it was clear that there existed in the tissue the keto acid which, when converted to the phenylhydrazone and thence to the amino acid, gave the substance now known to be γ-methyl-γ-hydroxyglutamic acid. In this case, therefore, the natural occurrence of the keto acid was inferred before either its own nature or the nature of the amino acid to which it gives rise was established. Pyruvic aldol from the condensation of 2 molecules of pyruvic acid is now the logical keto acid precursor of this relatively new amino acid, which occurs in the tissue in such large amount. From these two examples, one may therefore infer that a far greater range of keto acids than those now known to be implicated in the Krebs cycle may act as "ports of entry" through which reduced nitrogen compounds may enter into combination with carbon to form, eventually, amino acids. This point of view is consistent with published work cited in much greater detail (188).

Meister and Abendschein (322) have synthesized a wide range of keto acids and then tested their conversion to amino acids through an analytical procedure essentially similar to that of Towers *et al.* (546). Rabson (406) and Steward and Pollard (494) also compiled a list of amino acid analogs of the keto acids for which there is some evidence of natural occurrence. A modified form of this list is given in Appendix III.

The role of keto acids bears upon the origin of glutamine and asparagine, respectively. Glutamine may arise directly from glutamic acid by the action of glutamine synethetase, which requires ATP and magnesium (144, 465). The corresponding enzyme reaction from aspartic acid seems not to yield asparagine under physiological conditions. α-Ketoglutaramic acid is known to arise from glutamine by transamination with α-ketosuccinic acid, with the release of asparagine. This reaction sequence (Fig. 7) is a possible alternative route of syn-

[8a] More recent work by Vickery (*J. Biol. Chem.* **238**, 2453–2459 (1963) shows that when glutamate is metabolized by tobacco leaves in the dark with the release of CO_2 the nitrogenous product which accumulates is asparagine not γ-aminobutyric acid.

thesis for asparagine if the α-ketoglutaramic could again act as a "port of entry," for nitrogen glutamine would be reconstituted (also see reference to asparagine from HCN in Section III, E, 7). Despite all this, however, asparagine biosynthesis seems to be one of the important

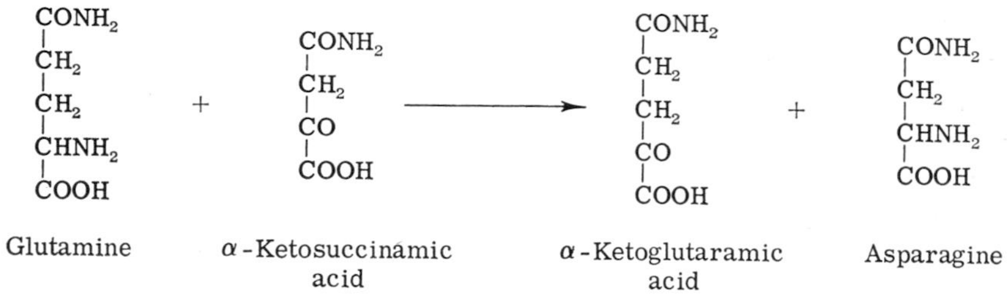

| Glutamine | α-Ketosuccinamic acid | α-Ketoglutaramic acid | Asparagine |

FIG. 7. A possible reaction leading to asparagine.

unsolved problems. Contrasts between glutamine and asparagine have frequently been drawn (see Section V, D), and it is now well known that it is far easier to label glutamine from $C^{14}O_2$ in leaves than it is asparagine [cf. Steward and Bidwell (481)].

3. Some Recent Trends

Amino acids in protein are not merely linked together in the general proportions in which they are to be found in the protein hydrolyzate, but they are also arranged in definite linear order, and this order determines the property of the protein as much as does the total amino acid composition. Although plant proteins have as yet contributed little to this type of sequential amino acid analysis, it seems inescapable that these considerations will apply to plants. Thus the template hypothesis of protein systhesis, the gene-enzyme hypothesis, all require the prior formation of an amino acid in the precise time and place needed for its incorporation into protein in the course of synthesis. Furthermore, there is also need to direct to that point of synthesis the energy-donating molecule, probably in the form of ATP. Lipmann (277) conceived that the energy from a single phosphate bond (ATP ≅ 12,000 cal) might well activate the single peptide link (about 3000 cals) needed to incorporate an amino acid into protein.

Thus, most of the work now being done on the biochemistry of protein synthesis, which is associated with various schools of thought summarized in Part 3 of this chapter, regards protein formation and protein metabolism in this general way. However, much of the supporting evidence for these ideas comes from the area of bacterial metabolism, of virus synthesis, or of the genetic determination of the course of protein metabolism.

It has to be recognized that *in toto* the problem has ramifications which are not immediately obvious from the amino acid hypothesis of protein synthesis in its most simple and direct form. This has been true historically, from the time of Schulze and Prianishnikov and is also apparent from the great variety of compounds, other than the α-amino acids needed to make protein, which are to be found in plants (see Part 2). Present evidence shows that many amino acids essential for protein synthesis are hard to demonstrate free in plants, whereas many others not so needed may frequently occur in quantity. Thus nitrogen metabolism now needs to be examined from the standpoint of the balance between, and the interplay of, both anabolism and catabolism and on the idea that these apparently conflicting phases of metabolism may be able to proceed, side by side, in different compartments or organelles of the same cell. Thus, the role of the amides in the scheme of Prianishnikov, which were quoted above, may be facilitated by their storage in aqueous phases out of the direct line of synthesis or breakdown. A given amino acid, like γ-aminobutyric acid or like γ-methyl-γ-hydroxyglutamic acid, neither of which enters directly into the structure of protein, may nevertheless be important as temporary reservoirs of nitrogen and carbon, which are to be considered in the over-all plan of nitrogen metabolism.

Steward and Millar (492) combined the ideas of Hanes *et al.* (198, 199) and Dounce (136) to visualize that γ-glutamylpeptides could present the amino acids to the nucleic acid surface in such a way that prior phosphorylation becomes unnecessary, the energy for their combination being in the γ-linkage. The template surface was conceived to act as a γ-glutamyl peptidase so that the γ-glutamyl peptides could donate at one and at the same time their amino acid radicals for synthesis and the energy to create the peptide bonds.

Thus there are now two complementary trends. The first is toward a very direct interpretation of how plants convert sugar via carbohydrate and keto acids and nitrate or ammonia to amino acids, and how these are arranged in their appropriate order at the site of synthesis with the source of energy, almost certainly derived from ATP, on a template surface which is regulated by the nucleic acids of the hereditary material: all this contributes to a highly specific formation of proteins with a highly ordered amino acid configuration. Thus prior amino acid synthesis seems essential to the culminating events of nitrogen metabolism in protein synthesis. By contrast, when protein is broken down, the same amino acids do not appear at the site where the products accumulate, for the amino acid products of protein breakdown tend to be reworked so that nitrogen-rich storage reserves in the form of the amides or of arginine, or indeed of certain other compounds, accumulate whenever breakdown predominates over synthesis, or the accumulation of nitrogenous material occurs in excess of its conversion to protein. Thus, in the understanding of the *whole cell*, as distinct from the events at the *immediate site* of synthesis, other considerations need to be invoked as follows.

The economy of plants, as indicated at the outset, permits the re-use of nitrogen. Therefore, whether soluble nitrogen results from prior protein breakdown, or the accumulation of nitrogen-rich reserves in excess of their conversion to protein, it is destined to be re-used in the formation of protein. This means that readily utilizable nitrogen-rich compounds are mobilized at centers of synthesis to be supplemented by carbon frameworks, derived from sugar, and probably in the form of keto acids. Thus the crucial event of protein synthesis occurs when carbon from sugar and nitrogen in a suitable form, are condensed at a template site of synthesis at which amino acids not only are made, but *immediately* incorporated into the protein. Therefore, the amino acids that are the true and final intermediates of protein synthesis *may never mingle with those that occur free and in large amount in the cells*. Thus the immediate source of carbon for the protein that can be recognized may well be carbohydrate. Similarly when the soluble nitrogen compounds are being drawn upon for synthesis, much of the carbon with which previously it had been combined in protein and from which it had been released during breakdown may not necessarily accompany that same nitrogen into the *new* protein which is to be synthesized at a *new* site or in a *new* organ or in a *new* growing region. On the contrary, when nitrogen is withdrawn from the storage and translocatory forms to contribute again to protein synthesis, it may leave behind nitrogen-free carbon frameworks, i.e., carbon skeletons which are to be oxidized as they enter into the normal reactions of catabolism in the form of organic and keto acids.

It is, therefore, essential to combine concepts and biochemical considerations that are dictated by the current knowledge of protein structure and protein synthesis in rather homogeneous systems with those equally important biological considerations that are dictated by the need to balance anabolic and catabolic processes in the interpretation of over-all nitrogen metabolism in more heterogeneous systems. Indeed, one may note here that current knowledge of the structure of even a single parenchyma cell or the cytoplasm of a single meristematic cell, with its highly intricate vesicular system of organelles, of granules, endoplasmic reticulum, etc., provides ample opportunity for the segregation of these two types of function within one cell. (Parenthetically one may note that it is in these more heterogeneous plant systems that most of the world's protein is manufactured *de novo*.) Thus, a major problem now is the need to interpret how anabolism and catabolism may proceed in a coordinated and integrated way, for it is apparent that no single organelle can achieve, in terms of over-all protein synthesis and over-all nitrogen metabolism, what the complete cellular organization can so easily do.

Having, therefore, surveyed the various landmarks in the historical trend of our knowledge of protein metabolism in plants by reference to the types of system investigated, to the substances that have been recognized to be important, and to some of the theoretical concepts by which the course of nitrogen metabolism has been interpreted, certain particular topics must now be discussed in more detail.

PART 2

FREE NITROGEN COMPOUNDS: THEIR NATURE, REACTIONS, AND DETERMINATION

IV. The Range of Soluble Nitrogen Compounds Now Found in Plants[9]

Nitrogen is now known to be dispersed from its initial "ports of entry" through an array of soluble compounds that are produced during the metabolism of plants. These compounds are far more numerous and varied in their structure than follows if their main purpose were only to form α-amino acids for condensation to protein. Group transfer reactions such as transaminations [Braunstein and Kritzman, 1937; Braunstein, 1939; cf. Braunstein (74)], transamidations (113, 126, 281, 518), transpeptidations (197–199), all contribute to this dispersal of nitrogen, but many previously unsuspected metabolic pathways have now been disclosed (112, 113, 184, 526).

So long as the knowledge of nitrogen metabolism remained limited, it was possible to think in broad general terms. Pfeffer, Schulze, and Prianishnikov, regarded carbohydrate as the source of the organic acid with which ammonia entered into organic combination. The concept was outlined by relatively simple general schema (e.g., those of Fig. 4). Prianishnikov regarded ammonia as the central substance of nitrogen metabolism. Prior to the even more recent increase of knowledge, Virtanen could speak of the dicarboxylic acids (glutamic and aspartic acids) as "Grundaminosäuren" of plants. The importance of ammonia is still recognized, but it is less easily regarded as the *sole* pivotal nitrogenous reduction product of nitrogen metabolism. While the key role of the dicarboxylic acids and their semiamides is still apparent, other compounds also play significant roles.

[9] Wherever new compounds are referred to in the text without formulas, reference should be made to Appendixes I and II, where they may be found.

In the study of carbohydrate metabolism, and as knowledge of glycolysis and the carboxylic acid cycle increased, the role of such substances as pyruvate has become submerged in a much more complicated pattern consisting of many phosphate compounds, complexes such as acetyl coenzyme A, etc., and the previously unsuspected intermediaries of the pentose cycle (see Chapter 3, Volume IA). Thus, one should now survey a much wider chemical landscape in order to see its outstanding topographical details. In nitrogen metabolism, also, we have now a much more varied metabolic scene to survey, but it is again important to discern its most salient features. In so doing, the first impression may be one of greater complexity, and that definition may seem to be lost in the detail. However, as new pathways and relationships emerge, new signs of order will inevitably appear.

Understandably, new pivotal substances now attract attention. One such example is carbamyl phosphate, for it has obvious significance as a simple activated molecule that may enter into varied types of condensation. Moreover, it may be synthesized from carbon dioxide and ammonia in the presence of ATP and magnesium (180, 240, 241). Similarly, the substituted glutamic acid, known as γ-methyl-γ-hydroxy-glutamic acid, which was not known until 1955, now emerges as a pivotal molecule from which many lines of synthesis may diverge (493).

Thus, one should now summarize the knowledge of the naturally occurring, simple, free nitrogen compounds that has accrued from the application of chromatography on paper and on various types of columns. A comprehensive tabulation (cf. Appendices I and II) has been compiled to record the occurrence of, and the main evidence for, the recognition of the new compounds that are most important. This tabulation records, as it were, the intimate details of the chemical landscape that has been uncovered; the next task is to treat this mass of information [which has been variously summarized in a number of reviews by Vickery and Schmidt (574), Virtanen (577), Steward and Pollard (494, 496), Fowden (155),[9a] Loomis (280), Webster (599), McKee (314)] in such a way that signs of order and new concepts emerge.

A. THE APPLICATION OF CHROMATOGRAPHY

The first major application of two-directional partition chromatography on paper to an extract of a plant was that published by Dent, Stepka, and Steward (129). An alcoholic extract of the potato tuber,

---

[9a] A more recent article by L. Fowden [*Ann. Rev. Biochem.* **33**, 173–204 (1964)] was published too late to be utilized in the preparation of this chapter.

about which much had been learned by the previous and more arduous methods (515), was used. Steward and Street (501) had already summarized the factors that modify the composition of the alcohol-soluble nitrogen of the potato tuber as measured by the total amino nitrogen (by the Van Slyke determination), the total amide, and the relative content of asparagine and glutamine [as measured by controlled amide hydrolysis under the procedures of Vickery et al. (569, 570), Chibnall and Westall (108)].

It was already known that much of the total nitrogen of the potato tuber was alcohol soluble (approximately two-thirds); much of this consisted of amino acid nitrogen, and, of the amide nitrogen, some was the relatively stable substance asparagine and much consisted of the less stable substance glutamine. The problem was to fractionate the soluble nitrogen compounds and so reveal the identity of the constituent amino acids, and also to be certain that the two then known amides might not be accompanied by a hypothetical "third amide" which had so long been sought. Prior to the development of chromatographic methods, Steward and Street (501) and also Neuburger and Sanger (353) had established the identity of glutamine in the potato tuber and, by the more laborious chemical isolation of that substance from the tubers, Steward and Street had satisfied themselves that all the amide present could be accounted for as asparagine and glutamine. It is noteworthy that the actual sample of glutamine that was isolated later proved to be chromatographically pure and free of γ-aminobutyric acid, which was not known to occur in potato tuber when the isolation was made.

After the development of two-directional paper chromatography as first used by Martin and Synge (294) and Consden, Gordon, and Martin (114), it was logical to apply these procedures to plant extracts. When this was done (129) a landmark in the study of nitrogen metabolism was passed. In a simple set of operations and at one dramatic moment the entire complement of ninhydrin-reactive nitrogen compounds could be seen dispersed upon the paper sheet. The obvious amino acids and the two amides were immediately recognized, and their relative amounts were surmised from the intensity of the ninhydrin color by which they were revealed. Even such amino acids as β-alanine, readily recognizable by its blue color with ninhydrin, and threonine, previously recognized only very rarely in plants, could be seen. However, one conspicuous ninhydrin-reactive spot occurred directly below valine on phenol/collidine-lutidine chromatograms (see Fig. 8). This unexpected substance proved to be γ-aminobutyric acid, which was eventually isolated in the quantity of 4 gm from 13 kg

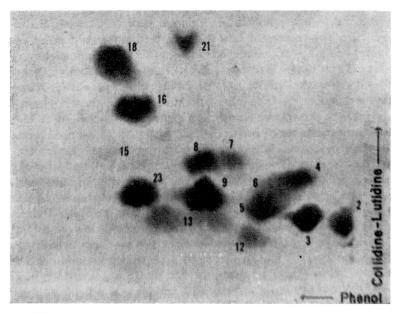

Fig. 8. The ninhydrin-reactive, alcohol-soluble nitrogen compounds of the potato tuber. Reproduction from an original Kodachrome picture (somewhat scarred) of a chromatogram dating back to the period of Dent, Stepka, and Steward (129). The spots are designated by number according to the convention then adopted, namely: *2*, aspartic acid; *3*, glutamic acid; *4*, serine; *5*, glycine; *6*, asparagine; *7*, threonine; *8*, alanine; *9*, glutamine; *12*, lysine; *13*, arginine; *15*, proline; *16*, valine; *18*, leucines; *21*, tyrosine; *22*, β-alanine; *23*, γ-aminobutyric acid; *18*, leucine and isoleucine, not here separated.

of potato tuber (538). From these beginnings, γ-aminobutyric acid is now known to be ubiquitous in plants, and one would now be more surprised to find a plant in which it did not occur than to find another example of its presence. By standardizing these procedures the familiar amino acids of protein, together with such substances as γ-aminobutyric acid, asparagine, and glutamine, may be located on chromatograms and their positions recorded in such maps as those shown in Figs. 9A and 9B (note that 9A is a phenol/butanol-acetic acid/water map and 9B is a phenol/collidine-lutidine map). Although many other free compounds than these occur in plants and their positions have been recorded on similar chromatographic maps (511) their identification by their position alone is unwise (494, 505).

When Steward and Street (502) reviewed the nitrogen metabolism of plants in 1947, the possible role of chromatography was then merely foreshadowed in a footnote. By 1950 Steward and Thompson

(505) were able to report the recognition by these techniques of many expected compounds in many plants and also to show that a large number of hitherto unsuspected substances could be detected in the organs of common plants, many of them food plants. About this time respiration—long considered in terms only of carbohydrate breakdown —was shown to include in the potato tuber a moiety that varied concomitantly with protein synthesis and with the metabolism of amino acids. This was especially true of cells of the potato tuber in well aerated solutions, which also promoted salt accumulation (500).

The general problem was reviewed again, ten years after the first footnote reference to chromatography by Steward and Street (502). Steward and Pollard (494) dealt with the then large number of

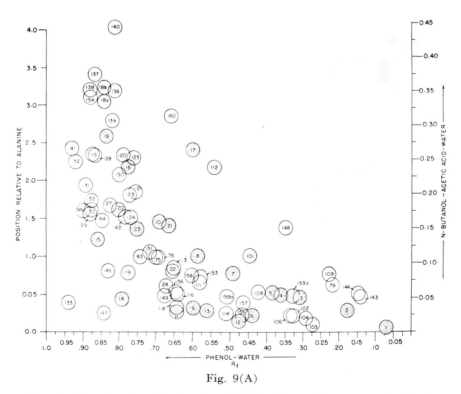

Fig. 9(A)

FIG. 9. Maps showing the location of ninhydrin-reactive compounds. Numbers 1–25: amino acids and amides established as occurring widely, free or in proteins, in plants. Numbers 75, 76, 79, 83, 84: unidentified nitrogen compounds that occur in plants. Numbers 100–160: nitrogen compounds that are either unnatural or not established as occurring widely in plants. (A) Phenol/butanol-acetic acid chromatograms. (B) Phenol/collidine-lutidine chromatograms. After Steward et al. (511).

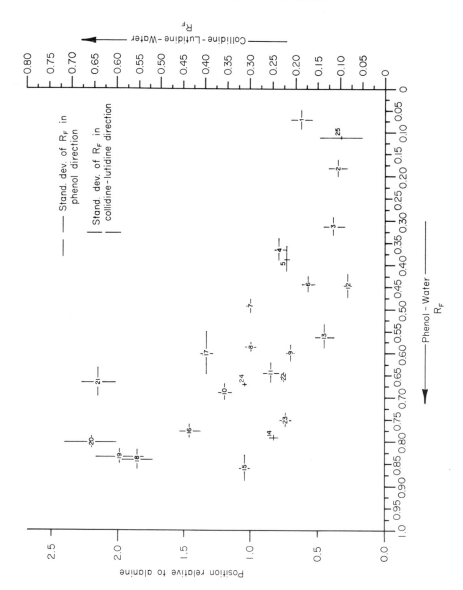

Fig. 9(B)

substances which had either been identified or merely recognized as spots on chromatographic papers; but they also pointed to the occurrence of many keto acid analogs of the nitrogen compounds. The latest comprehensive summaries of the occurrence of nitrogen compounds in plants are those published by Steward et al. (511), Fowden (155, 156), Fowden and Gray (158), Steward and Pollard (496).[10]

It will be apparent, however, that the end is not yet in sight, for a great deal of work is still to be done by combined chromatographic detection, chemical isolation, and synthesis before the full harvest of knowledge has been reaped from the application of these procedures. In fact, since this manuscript was started many new substances have appeared. It is evident, therefore, that the knowledge of the nitrogen metabolism of plants as it existed prior to 1947 now needs to be rewritten in terms of the many new compounds that have been disclosed. These compounds can be summarized in the following categories.

B. Amino Acids Other Than the α-Amino Acids of Protein

An amino group may occur in a natural free amino acid in potentially any place in a carbon chain; β-alanine and γ-aminobutyric acid were the two first examples of this. However, other examples (cf. Appendix II) are now known. These extend the list of amino acids that are of interest to the plant physiologist far beyond those which occur in protein. For example, Evans et al. (151) have tracer evidence that β-aminoisobutyric acid in Iris can form via dehydrothymine and β-ureidoisobutyric acid to link this amino acid with pyrimidine metabolism (Fig. 10, A and B).

C. Amino Compounds That Are Decarboxylation Products

Substances that could be related to amino acids as their corresponding decarboxylation products frequently occur. However, even where this seems to be the case, more rigorous proof may be required that the substance in question really originates in this way.

Thompson et al. (538) gave a list of 28 angiosperms belonging to 14 families and 21 genera in which γ-aminobutyric acid had been recognized. Like α-aminobutyric acid, γ-aminobutyric acid could arise by decarboxylation of glutamic acid, and in some situations both

---

[10] Too recent to be cited in the reference list is an account by Tschiersch and Mothes in "Comparative Biochemistry" (M. Florkin and H. S. Mason, eds.), Vol. 5, Part C, Academic Press, New York, 1963. In this account specific compounds are mentioned together with a brief account of their occurrence and metabolism.

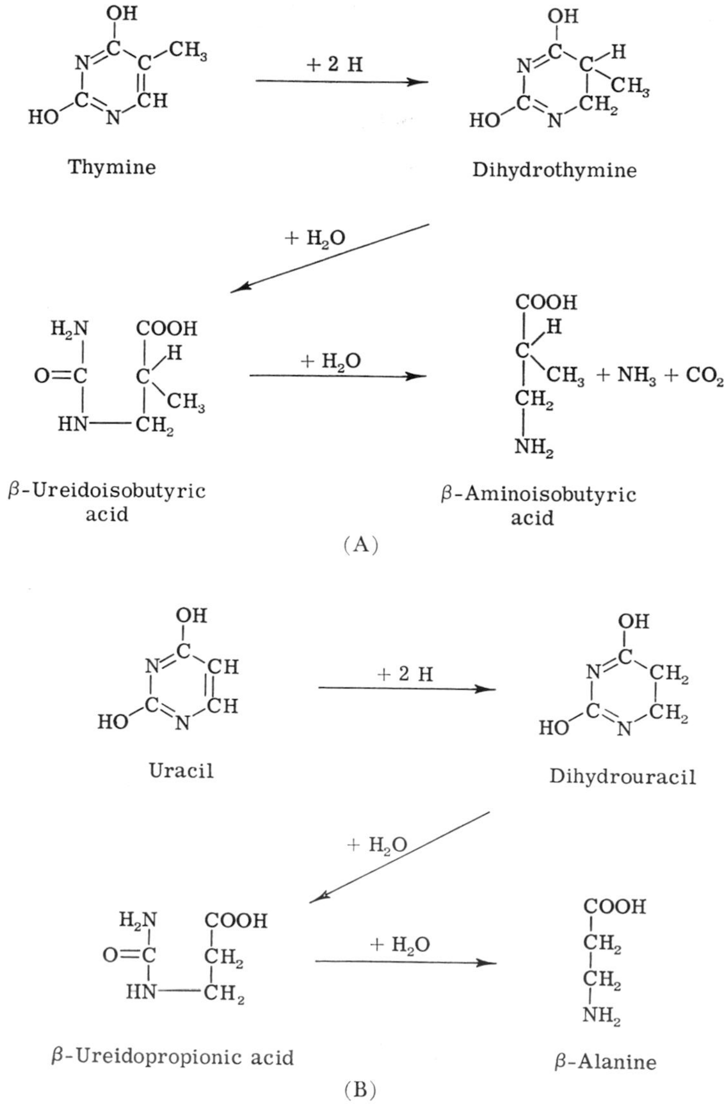

FIG. 10. (A) Degradation of thymine. (B) Degradation of uracil.

decarboxylations are known to occur. Schales *et al.* (433) studied the glutamic acid decarboxylase activity of a number of plants, but if γ-aminobutyric acid does originate in this way there would almost certainly need to be a recarboxylation before the acid could be used in metabolism.

Similarly, β-alanine could arise by decarboxylation of asparatic acid. β-Alanine came into prominence by the work of Virtanen and Laine (583) on nitrogen fixation. This compound is usually present in smaller amount than γ-aminobutyric acid. Putrescine and agmatine, the decarboxylation products of ornithine, and arginine are especially evident in potassium-deficient plants (417, 460). Tyramine, the decarboxylation product of tyrosine, occurs in tobacco under conditions of mineral nutrient deficiency [especially in the tissues lacking molybdenum and boron, where it amounted to about 50% of the soluble nitrogen (485)]. Tyramine has been demonstrated chromatographically by Erspamer and Fulconieri (1952) in barley germ, and by Correale and Cortese (1953, 1954) in *Cytisus* (*Sarothamnus*) *scoparius* and in "*Genista spartium.*" Hydroxytyramine (from dihydroxyphenylalanine) was also detected in the former species. Although taurine, the decarboxylation product of cysteine, often occurs in large amounts in animals, it is not commonly encountered in plants. Therefore, the decarboxylation products of amino acids are widespread in their occurrence in plants, at least in trace, or just detectable, amounts, although some of these substances are, of course, volatile amines (511). Isobutylamine via valine decarboxylase has been reported in *Arum* spadix by Simon (451). However, the possibility that this reaction was due to bacteria on the exposed spadix surface seems not to have been eliminated.

D. HYDROXYAMINO ACIDS

Until recently serine, threonine, and tyrosine (with dihydroxyphenylalanine as an unstable intermediary metabolite) were the outstanding hydroxyamino acids. It will be noted, however, that threonine was not chemically isolated until 1935 by McCoy *et al.* (308), and the free occurrence of alanine and serine in *Medicago sativa*, detected by Vickery (557), had not been frequently verified prior to the advent of chromatography. Not only are these substances (i.e., serine and threonine) regarded as of widespread, free, occurrence, but it is now known that a variety of other hydroxyamino acids also occur, with the hydroxyl group in various positions (cf. Appendix II). Examples will be cited here to show how these discoveries were made and to indicate their probable significance.

The controversial natural occurrence of hydroxyglutamic acids since Dakin (1918, 1919) claimed to have isolated β-hydroxyglutamic acid from proteins has been resolved, although Dakin himself concluded (1941) that the acid was not in fact present in protein. Dent and

Fowler (128) showed chromatographically that Dakin's original isolate contained none of this compound.

When an alcoholic extract of the growing point of a fern (*Adiantum pedatum*) was examined by Steward, Wetmore, Thompson, and Nitsch (510) and Steward, Wetmore, and Pollard (509), it was found that the most outstanding ninhydrin-reactive substance on the chroma-

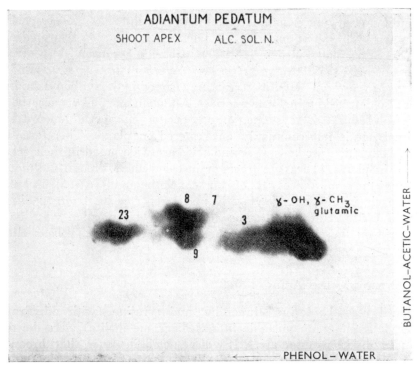

Fig. 11. Chromatogram of the ninhydrin-reactive, alcohol-soluble compounds of the shoot apex of the maidenhair fern (*Adiantum pedatum*). Compounds designated by numbers follow the convention of Fig. 8. After Steward and Pollard (496).

togram moved to a position somewhat close to that of serine (see Fig. 11), but it was quite clear that it was not, in fact, identical with this substance. Subsequent isolation of the compound from *Adiantum* and comparison with a synthetic product, showed that this material was γ-methyl-γ-hydroxyglutamic acid (see Appendix II). This acid accounted for about 90% of the soluble nitrogen in the growing regions of *Adiantum*. The possibility that this substance might be the parent from which γ-methyleneglutamic acid and its amide were derived was noted. A watch was kept for this substance in a survey of the Liliaceae (in which the γ-methylene compounds are prominent) and, in fact, it was found to occur in a number of plants of this family (Fig. 12).

If a substance occurs in quantity in a fern and is widely dispersed in an angiosperm family like the Liliaceae, and if it is also known to occur elsewhere in the plant kingdom, it seems very probable that it plays a more frequent role in intermediary metabolism than might be supposed from its late detection. The synthesis of this substance

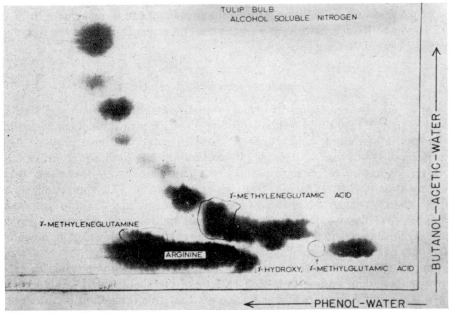

FIG. 12. The ninhydrin-reactive, alcohol-soluble compounds of the tulip bulb. Chromatogram arranged to show the designated γ-glutamyl compounds. After Steward and Pollard (496).

from pyruvic aldol, that is, from its corresponding keto acid, indicates also its probable biogenesis. However, this substance can exist as four diastereoisomers, and it is still not known to what extent nature makes use of all the structural possibilities, or whether—as seems more probable—the naturally occurring substance (496) is confined to one form only. From γ-methyl-γ-hydroxyglutamic acid, a great many metabolic pathways may flow, as indicated in Fig. 13.

In Fig. 14 Maitra and Dekker (289) have outlined the following scheme for γ-hydroxyglutamic acid metabolism in mammalian systems.

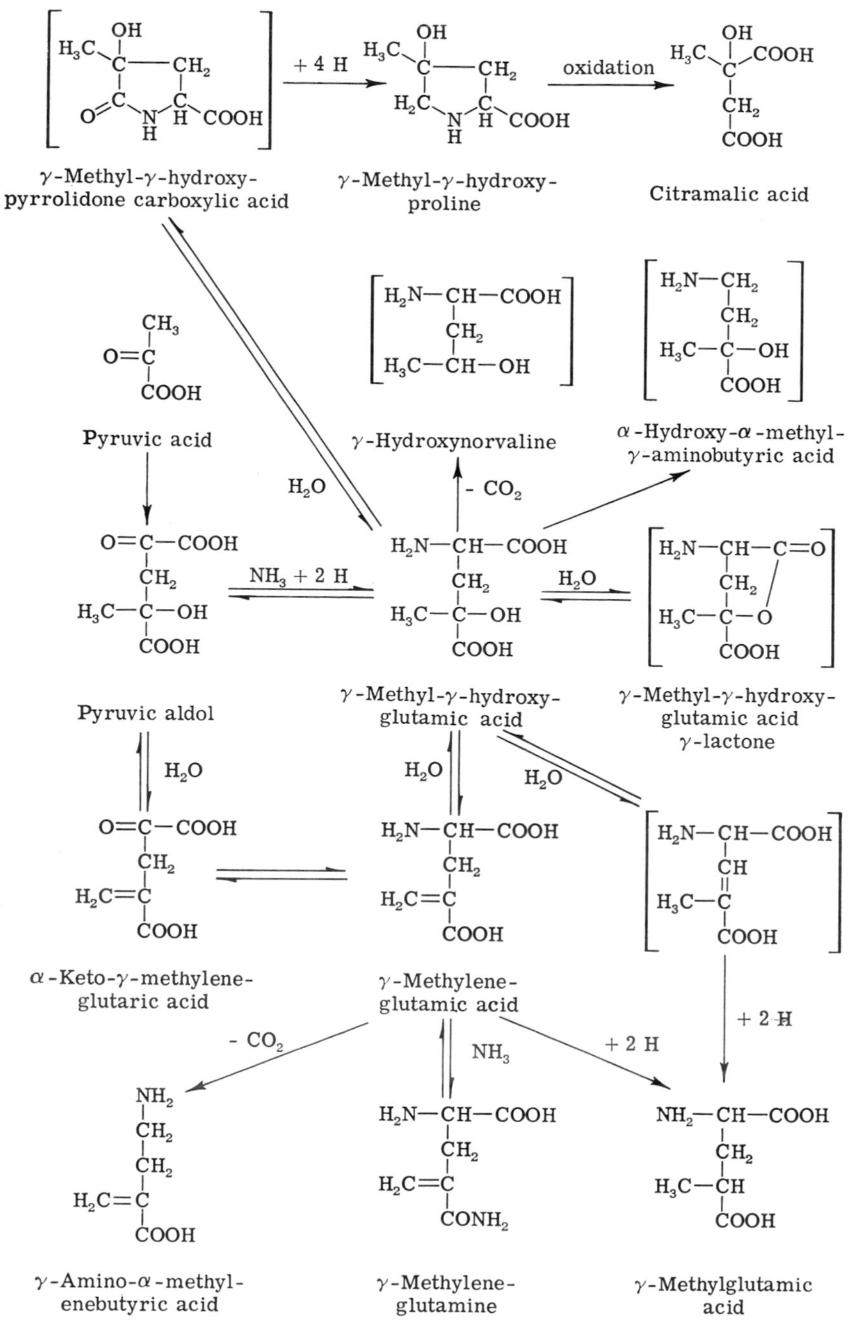

FIG. 13. Some possible relationships of γ-methyl-γ-hydroxyglutamic acid. Substances shown in brackets are hypothetical. After Steward and Pollard (493).

The sequence requires that α-ketoglutamic acid transaminates with γ-hydroxyglutamic.

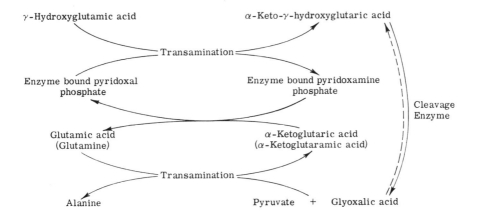

Fig. 14. Scheme [after Maitra and Dekker (289)] to show γ-hydroxyglutamic acid metabolism in mammalian systems by pathways which involve transaminases and the cleavage of α-keto-γ-hydroxyglutaric acid so that alanine and glyoxalic acid are the end products.

Thus glyoxylate and alanine may form from hydroxyglutamate by sequential loss of the amino group, cleavage to pyruvate and glyoxylate and reamination of one of the fragments (125).

A new amino acid, present in *Kalanchoe daigremontiana*, was seen to exist in two forms on paper. This suggested that one form was a lactone and the other a free acid. Finally the acid was isolated and shown to be γ-hydroxyvaline by Pollard *et al.* (398). Again various isomeric possibilities exist. By a combination of an unequivocal organic synthesis (carried out by Sondheimer) and by certain enzymatic reactions (496), it is possible to write the precise structure of the natural substance. The chart shown in Fig. 15 indicates the synthesis and relationships of γ-hydroxyvaline. This then is an example of a widespread problem. When two or more asymmetric centers exist,

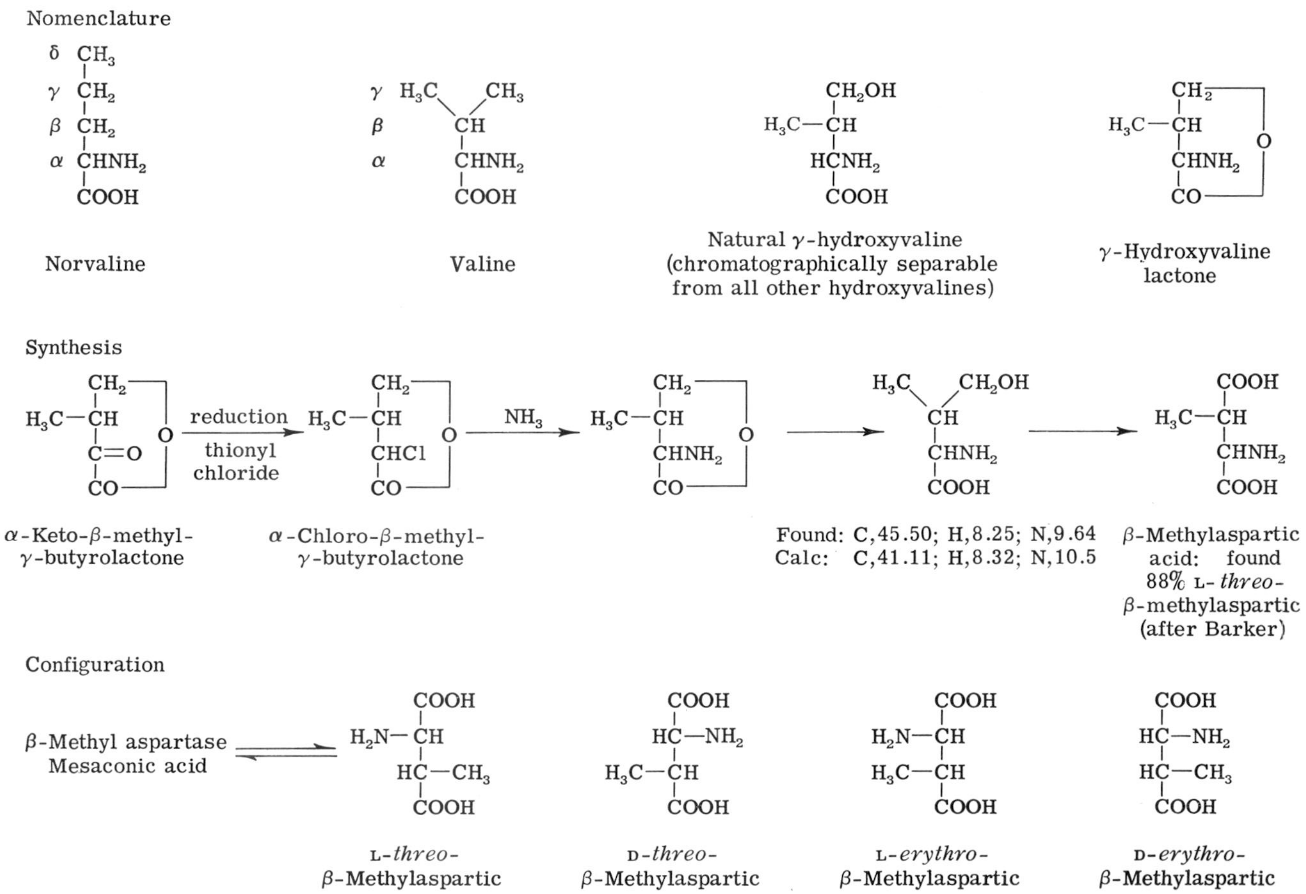

Nomenclature

δ CH₃
γ — CH₂
β — CH₂
α — CHNH₂
    COOH

Norvaline

γ H₃C   CH₃
          \\ /
β   CH
β — CH₂...

$$\begin{array}{ll} \delta & CH_3 \\ \gamma & CH_2 \\ \beta & CH_2 \\ \alpha & CHNH_2 \\ & COOH \end{array}$$

Norvaline

$$\gamma\; H_3C \;\; CH_3$$
$$\beta \quad CH$$
$$\beta \quad CH_2$$
$$\alpha \quad CHNH_2$$
$$\quad COOH$$

Valine

$$\begin{array}{l} CH_2OH \\ H_3C—CH \\ HCNH_2 \\ COOH \end{array}$$

Natural γ-hydroxyvaline
(chromatographically separable
from all other hydroxyvalines)

$$\begin{array}{l} CH_2—O \\ H_3C—CH \quad | \\ \quad CHNH_2 \\ \quad CO \end{array}$$

γ-Hydroxyvaline
lactone

Synthesis

$$\begin{array}{l} CH_2—O \\ H_3C—CH \quad | \\ \quad C=O \\ \quad CO \end{array}$$

α-Keto-β-methyl-
γ-butyrolactone

reduction ⟶
thionyl
chloride

$$\begin{array}{l} CH_2—O \\ H_3C—CH \quad | \\ \quad CHCl \\ \quad CO \end{array}$$

α-Chloro-β-methyl-
γ-butyrolactone

NH₃ ⟶

$$\begin{array}{l} CH_2—O \\ H_3C—CH \quad | \\ \quad CHNH_2 \\ \quad CO \end{array}$$

$$H_3C \;\; CH_2OH$$
$$\quad CH$$
$$\quad CHNH_2$$
$$\quad COOH$$

⟶

$$H_3C \;\; COOH$$
$$\quad CH$$
$$H_3C—CH$$
$$\quad CHNH_2$$
$$\quad COOH$$

Found: C, 45.50; H, 8.25; N, 9.64
Calc: C, 41.11; H, 8.32; N, 10.5

β-Methylaspartic
acid: found
88% L- threo-
β-methylaspartic
(after Barker)

Configuration

β-Methyl aspartase ⇌
Mesaconic acid

L-threo-
β-Methylaspartic

$$\begin{array}{l} COOH \\ H_2N—CH \\ HC—CH_3 \\ COOH \end{array}$$

D-threo-
β-Methylaspartic

$$\begin{array}{l} COOH \\ HC—NH_2 \\ H_3C—CH \\ COOH \end{array}$$

L-erythro-
β-Methylaspartic

$$\begin{array}{l} COOH \\ H_2N—CH \\ H_3C—CH \\ COOH \end{array}$$

D-erythro-
β-Methylaspartic

$$\begin{array}{l} COOH \\ HC—NH_2 \\ HC—CH_3 \\ COOH \end{array}$$

FIG. 15. The structure and synthesis of natural γ-hydroxyvaline and its lactone, and certain criteria used to prove its identity. After Steward and Pollard (496).

the task of identifying a new nitrogenous compound is not completed until the stereoisomerism is fully known. Also the physiology of these compounds will not be fully comprehended until it is known why a given plant gives preference to one isomer or the other.

Largely by the efforts of Virtanen's laboratory, the importance of homoserine, particularly in leguminous plants, is now well recognized (Miettinen *et al.*, 1953; Virtanen *et al.*, 1953; cf. Appendix II), and its *O*-acetyl derivative also occurs [cf. Fig. 16 and Grobbelaar and Steward (190)].

The relationships of homoserine and *O*-acetylhomoserine to other naturally occurring compounds have been summarized [(496), cf. p. 36]. Whereas, Grobbelaar [cf. Steward (479)] noted that homoserine and *O*-acetylhomoserine varied in peas reciprocally with asparagine, Larson and Beevers [*Plant Physiol.* 38, Suppl. vi (1963)] noted that homoserine received carbon from C^{14}-aspartate. This suggests that there is an environmentally determined switch of the metabolism of aspartate in peas which diverts the end products from homoserine to asparagine according to the conditions.

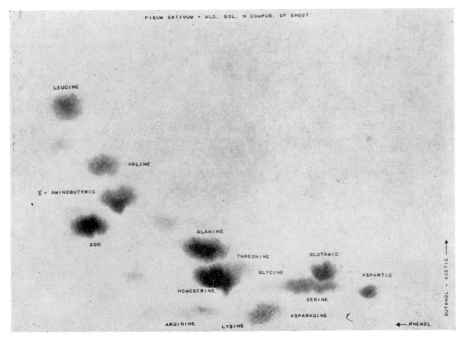

Fig. 16. The ninhydrin-reactive, alcohol-soluble compounds of the pea (*Pisum sativum*). Note the occurrence of homoserine and of an unidentified substance (No. 200) now known to be *O*-acetylhomoserine. After Steward and Pollard (496).

It has long been known that hydroxyproline occurs in certain proteins (e.g., as in collagen, the structural protein of animals) and even when it does occur free in some situations, hydroxyproline does not normally accumulate in green plants. (The occurrence of hydroxyproline in a protein moiety of proliferating cells will be discussed later, see Section V, B). Allohydroxyproline, however, was isolated in bulk from the plant *Santalum album* (407).

The substitution of hydroxyl groups in other cyclical compounds is also known to occur. The prominent example of this is the occurrence of 5- and perhaps 4-hydroxypipecolic acid in a variety of plants. It is more convenient to refer to this substance after referring to another concept which explains the presence of hitherto unrecognized amino (or more strictly imino) acids, namely their origin by cyclization.

E. Imino Acids and the Origin of New Nitrogen Compounds by Cyclization

One of the first new imino acids to be recognized by chromatography is known as pipecolic acid, or piperidine-2-carboxylic acid. This ap-

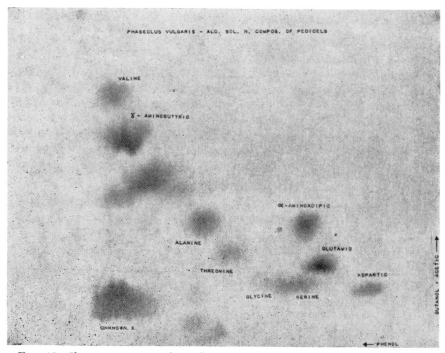

Fig. 17. Chromatogram to show the presence of α-aminoadipic acid, which became radioactive when C¹⁴-lysine was supplied. After Steward and Pollard (496).

peared on paper chromatograms [(505), unknown No. 1] of the ovary of green beans (*Phaseolus vulgaris*), but it quickly proved to be of widespread occurrence (Morrison, 1952; Zacharius *et al.*, 1954; Hulme and Arthington, 1952; for references see Appendix II).

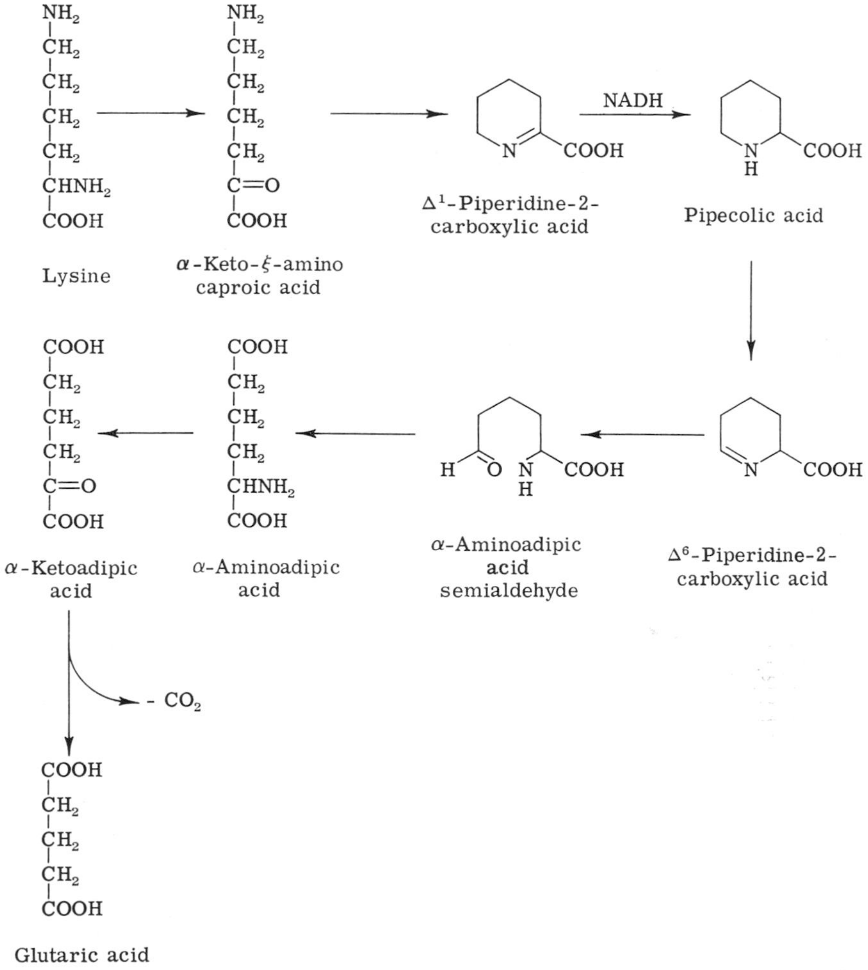

FIG. 18. General pathway of lysine degradation.

Bulk isolation of this imino acid yielded 13.4 gm, isolated from 68 kg of green beans (191). It has also been isolated from coconut milk in gram quantities in this laboratory by Shantz. The suggested origin of this substance was from lysine by the removal of ammonia between the α-amino and ε-amino groups, and there was precedent for this in other organisms. Both in *Neurospora* and in beans this has now been

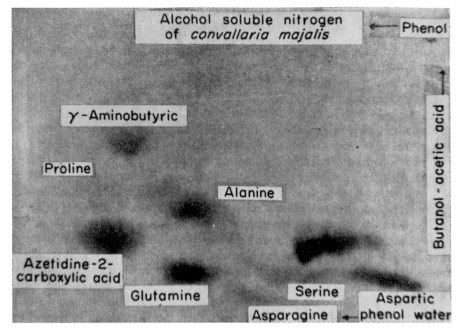

FIG. 19. The ninhydrin-reactive, alcohol-soluble compounds of lily-of-the-valley (*Convallaria majalis*). Note azetidine-2-carboxylic acid (brown). After Steward and Pollard (496).

amply proved by applying C[14]-labeled lysine and finding that it is converted to C[14]-labeled pipecolic acid [(189), and references there cited]. However, the converse reaction, i.e., the reconversion of pipecolic acid to lysine, does not occur, for the pipecolic acid is metabolized, via α-aminoadipic acid, to glutaric acid (see Fig. 18). In fact, when it is realized that this reaction sequence may occur, critical examination of chromatograms, especially prepared to reveal α-aminoadipic acid, shows that this substance is a normal constituent of green beans (Fig. 17). (It will be noted that α-aminoadipic acid also belongs in the category of an α-amino acid that had not previously been thought of as occurring in plants because it is not a normal constituent of protein hydrolyzates.)

The cyclization of lysine to pipecolic acid is suggestive by analogy with the formation of proline from ornithine, but this parallel becomes even more meaningful with the identification of another compound which is analogous to piperidine-2-carboxylic acid (with its 6-membered ring) and to proline (with its 5-membered ring). The new compound is azetidine-2-carboxylic acid, first detected by Fowden on chromatograms (Fig. 19) of the rhizome of lily-of-the-valley (*Con-*

vallaria majalis), later seen in many liliaceous plants (159), and conclusively shown by Fowden (154) to have the unexpected structure of a 4-membered ring with a carboxylic group in the 2-position to the imino group. Analogy with the other sequences of cyclization suggests that α,γ-diaminobutyric acid could lead to azetidine-2-carboxylic acid by somewhat analogous steps. α,γ-Diaminobutyric acid is known to occur in combined form in *Bacillus circulans* (256). However, attempts in this laboratory to find free azetidine carboxylic acid in this organism were unsuccessful. These relationships are shown in the chart at Fig. 20 (494).

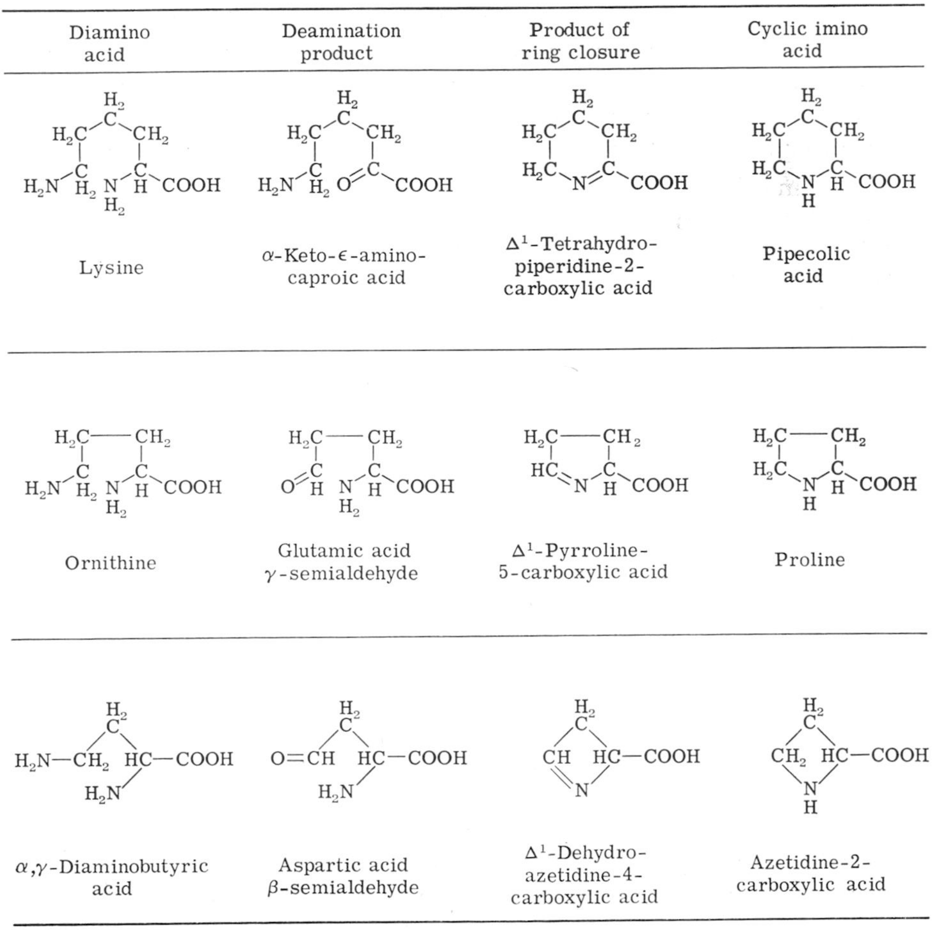

Diamino acid	Deamination product	Product of ring closure	Cyclic imino acid
Lysine	α-Keto-ϵ-amino-caproic acid	Δ^1-Tetrahydro-piperidine-2-carboxylic acid	Pipecolic acid
Ornithine	Glutamic acid γ-semialdehyde	Δ^1-Pyrroline-5-carboxylic acid	Proline
α,γ-Diaminobutyric acid	Aspartic acid β-semialdehyde	Δ^1-Dehydro-azetidine-4-carboxylic acid	Azetidine-2-carboxylic acid

FIG. 20. Metabolic intermediates related to some naturally occurring cyclic amino acids, After Steward and Pollard (494). *Note:* Glutamic acid α-semialdehyde occurs in microorganisms; alternatively α-keto-γ-aminovaleric acid could occur. Alternatively to aspartic acid β-semialdehyde; α-keto-γ-aminobutyric acid could occur.

However, as indicated above, the cyclic imino compounds may also have other substituents in the ring, and this leads to a number of other possibilities, some of which are now well known.

Knowledge of the role of azetidine-2-carboxylic acid as a competitive proline antagonist and protein synthesis inhibitor has been extended by Fowden and Richmond[11] (1963). The seedlings which do not form azetidine-2-carboxylic acid are now found to be suppressed by exogenous application of this substance. Whereas it might have been supposed that the azetidine ring might open, forming, say α-hydroxy-γ-aminobutyric acid, it is now found that the acid enters into combination in polypeptides or proteins because the amino acid activating enzymes of these plants cannot discriminate between proline and the azetidine compound. The larger ring compound, pipecolic acid is, however, too big to fit into the active site on the enzyme-amino acid complex, and so again it is found that it never enters into the structure of protein. The protein structure which contains azetidine instead of proline has, however, such different geometry (since proline normally occupies a key place in the turns of an helical spiral) that this "unnatural" protein proves toxic. Thus azetidine-2-carboxylic acid is a new antimetabolite for an amino acid (proline) where the mode of action seems to be known. But what of the plants like *Convallaria*, which make this substance? These have amino acid-activating enzymes which by actual test do discriminate between proline and the smaller ring of azetidine, and so proteins are protected against this substance even when it is formed in metabolism. Since the azetidine compound does not go into protein it is free to be stored, presumably by vacuoles.

These ideas open up new vistas, not only in the interpretation of the role of some protein amino acid analogs (ethionine for methionine), but also in the interpretation of the role of such compounds as fluoroacetate and fluorocitrate, which may be toxic to plants that do not contain them, but are tolerated by those that do.

F. Substitution in Ring Compounds

Prior to the discovery of piperidine-2-carboxylic acid as a widespread constituent of plants, King *et al.* (251) had isolated an unsaturated imino acid which, because it was obtained from the wood of Rhodesian teak (*Baikiaea plurijuga*), they called baikiain (cf. Appendix II). The possible relationship of this substance to pipecolic acid was immediately obvious, and in fact it was the reduction of the optically active form

[11] Fowden, L., and Richmond, M. H. Replacement of proline by azetidine-2-carboxylic acid during biosynthesis of protein. *Biochim. Biophys. Acta.* **71**, 459–461 (1963).

of baikiain that finally proved the stereo configuration of the (L-) pipecolic acid which was isolated from green beans (188). However, the
presence in the piperidine ring of the double bond in baikiain became
significant when, in some other legumes (e.g., *Gleditsia*) in the work
of Zacharius (in Steward's laboratory) and in *Acacia* species and
Rhapis excelsa (*R. flabelliformis*) [in Virtanen's laboratory (580)], a
substance was both recognized and isolated that proved to be a hy-
droxy-substituted piperidine carboxylic acid. This substance is now
known to be 5-hydroxypipecolic acid, which could arise by addition
of water across the double bond of baikiain. This substance occurs in
the fruit of the edible date (*Phoenix dactylifera*) (188), and is also
quite widespread. It presumably originates as suggested. It has been
claimed that the 4-hydroxy compound also occurs (581), and the 4-
amino compound, which would add ammonia across the double bond
of baikiain, has also been recognized (434).

Substitution of a hydroxyl group in proline had, of course, long been
known, and methylproline has been recognized in apples (Hulme and
Arthington, 1952, 1954). Another imino acid, independently isolated
by both Hulme (1954) and Urbach (1955) from apples (*Malus sylvestris*), gave some trouble in its identification (for these references see
Appendix II). There was need here to distinguish between the alter-
natives (II) and (III) shown below: under one possibility a methyl
and a hydroxyl group would be attached to the same carbon of the
proline ring, and in the other the group OH·CH$_2$ was the substituent
in the ring. This problem was resolved (39) by the proof that the compound isolated from apples had the structure shown at (III)

$$
\begin{array}{ccc}
\text{H}_2\text{C}\!-\!\!-\!\text{CH}_2 & \text{H}_3\text{C} & \\
\text{H}_2\text{C}\diagdown_{\substack{\text{N}\\\text{H}}}\!\!\text{CHCOOH} & \text{HO}\!\!>\!\!\text{C}\!-\!\!-\!\text{CH}_2 & \text{HOH}_2\text{CHC}\!-\!\!-\!\text{CH}_2 \\
& \text{H}_2\text{C}\diagdown_{\substack{\text{N}\\\text{H}}}\!\!\text{CHCOOH} & \text{H}_2\text{C}\diagdown_{\substack{\text{N}\\\text{H}}}\!\!\text{CHCOOH} \\
\text{Proline} & & \\
\text{(I)} & \text{(II)} & \text{(III)}
\end{array}
$$

although its isomer may conceivably exist.

The recent discovery by Gray and Fowden (183) of a 4-methylene-
substituted proline in *Eriobotrya japonica* suggests that the substance
may occur naturally.

Up to the present time, substitution in the azetidine ring seems not to
have been observed. However, it can be seen that cyclization by loss of
ammonia from diamino compounds, with the possible insertion of
various substituents in the resultant ring, gives rise to an array of
possibilities not previously considered as major nitrogenous metabolites

in plants. Hydroxyamino acids may, of course, give rise to cyclic structures as they form their lactones (or lactams), and the possibilities of compounds originating in this way needs also to be considered more widely. In fact, when substituent hydroxyl groups occur in the ring, there is also the possibility that other new compounds may arise by acetylation, as well as by methylation, of the hydroxyl group. This possibility was realized in the case of homoserine, since O-acetylhomoserine occurs in legumes. This compound, isolated by Grobbelaar (187) was distinguished from the N-acetyl compound that might perhaps have seemed to be the more likely one to occur (190).

G. Other Amides and Substituents of Common Amino Acids or Amides

It is interesting to note that the semiamides of aspartic and glutamic acid are so characteristic of plants. Isoglutamine, i.e., with the amino group attached to the same carbon as the amide group, is known as a synthetic substance, but not as a constituent of plants. Also the diamide of glutamic acid does not apparently occur although the diamide of acetylenic carboxylic acid has been reported from actinomycetes (Suzuki *et al.*, cf. Appendix II).

Although many have searched for a third plant amide, and although free urea occurs in bulk too rarely in plants to be so regarded, nevertheless a third amide was destined to appear in the form of a substituted glutamine. This compound appeared as a spot on chromatograms of the stem exudate from cut shoots of the peanut (*Arachis hypogaea*) in Fowden's laboratory and also on chromatograms of the bulb of the tulip (*Tulipa*) in Steward's. Although it was isolated and obtained in pure form independently in the two laboratories, the synthesis and final identification of the substance was accomplished by Done and Fowden (134), Wailes, Whiting, and Fowden (591), who showed unequivocally that the parent substance was γ-methyleneglutamic acid and that it occurred in the plant mainly in the form of its amide γ-methyleneglutamine. Subsequent studies have shown that these compounds are surprisingly unreactive as metabolites, being more stable *in vivo* than many common amino acids (155).

The free acid γ-methyleneglutamic acid, its amide γ-methyleneglutamine, and its reduction product γ-methylglutamic acid could all be demonstrated in the tulip (159). Prior to the discovery of γ-methyl-γ-hydroxyglutamic acid (188), there was an obvious speculation that this substance arose from a carbon progenitor which could be derived by the condensation of two molecules of pyruvic acid (160). C^{14}-Pyruvate permits the incorporation of C^{14} into γ-methyleneglutamate (157). Labeled glutamic acid, glutamine, asparagine, sugars, and sugar phosphates arose in peanut leaves from at least part of the carbon skeleton, including the α-carbon of γ-methyleneglutamate, and it then rapidly entered intermediate metabolites associated with the tricarboxylic acid cycle (157).

A considerable number of natural substituted glutamine compounds, particularly γ-glutamyl compounds, are now known. These developments recall that Hanes *et al.* (197–199) had shown a mechanism for the enzymatic synthesis of γ-glutamyl peptides. Recently a variety of such γ-glutamyl peptides, with various amino acids involved as the substituent group, have been detected by Thompson *et al.* (536). The probable metabolic importance of the variety of glutamine substituents was referred to by Steward and Pollard (493), and by contrast they pointed to the fact that, as yet, rather few substituted asparagine compounds have appeared. It is true that N^4-(2-hydroxyethyl)-L-asparagine has since been isolated and also N^4-ethyl-L-asparagine [(158, 182); for these references see Appendix II]. The concentrations of N^4-ethyl-L-asparagine (159) and asparagine in tulip leaves grown at different temperatures showed a well-defined correlation. An unidentified compound (U131) seemed to be an important constituent of leaves grown at 40°F, while asparagine was predominant in leaves grown at 70°F (182). N^4-Ethyl-L-asparagine is closely related in structure to theanine (see Appendix II). There may well be as yet undiscovered compounds of asparagine to be revealed, even as it is also known that a number of still incompletely identified glutamine compounds occur (395, 496). A compound present in *Phlox* and *Hemerocallis* yields upon hydrolysis hydroxyglutamic acid, ammonia, and an unidentified carbon residue.[11a] This substance seems to be a substituted hydroxyglutamine (work of Pollard and Grobbelaar in this laboratory). The hydroxyglutamic acid also occurs free in *Phlox* (579).

H. Other New Amino Acids and Ring Configurations

The recognition of the azetidine ring configuration has been noted above. In the work of Fowden (1955) still other ring configurations have appeared, notably a ring system isomeric with that which occurs in histidine (158, 365). In fact this substance (β-pyrazol-1-ylalanine) also occurs as its γ-glutamyl peptide (see Appendix II). This ring configuration occurs in substances isolated from *Cucurbita* seeds and it is mentioned here to emphasize that the full potentialities of the occurrence of hitherto unrecognized ring configurations in relatively low molecular weight compounds have not been fully appreciated. The cyclopropane ring is also represented by compounds which contain nitrogen.

[11a] This carbon residue now appears to be a furan derivative.

as evidenced by the occurrence of aminocyclopropanecarboxylic acid in products derived from pears (*Pyrus communis*), as shown by Burroughs (1957) (cf. Appendix II). If one adds also the range of compounds that may exist when one deals with sulfur-containing compounds as well as those that contain nitrogen, then the chemical possibilities are very great indeed. Several such compounds have already appeared (cf. Appendix II). Among the new sulfur-containing amino acids, the following may be mentioned as examples of an obviously incomplete list: cycloalliin (Virtanen and Matikkala, 1959); γ-L-glutamyl-S-methyl-L-cysteine (Zacharius *et al.*, 1959).

I. γ-Glutamyl Peptides in Plants

In addition to free glutamine, triglutamine[11b] (L-pyrrolidinonyl-γ-L-glutaminyl-L-glutamine) has been found in nature by Dekker *et al.* (1949) in the brown alga *Pelvetia fastigiata* (cf. 505), and of course glutamic acid polymers, such as the folic acids, are now well known. Glutathione (γ-glutamylcysteinylglycine) is a classical example of a universally distributed glutamyl peptide.

Since 1958 γ-glutamyl compounds, mostly in plant storage organs, have become numerous (Appendix II). The discovery of α-glutamyl peptides suggests a possible function for transpeptidases to form a series of dipeptides in plants whereas in animals they may aid the hydrolysis of dipeptides in the diet (536). The glutaminotransferases (590) of animals and higher plants transfer γ-glutamyl groups to amines, but not to amino acids, and thus are not implicated in γ-glutamyl peptide metabolism. Also, since several of the naturally occurring plant peptides contain amino acids not generally found in protein (e.g., S-methylcysteine, β-alanine, β-aminoisobutyric acid, and hypoglycin A), their direct role in protein synthesis is unlikely. However, it is still possible that this interesting mechanism could be part of the means by which amino acids are presented to the right place, at, or near, the synthesizing surface; this having been done, all the amino acids could be released from such complexes by a single γ-peptidase.

J. Metabolic Cycles in Relation to the Occurrence of Newly Discovered Amino Acids

The rich array of new amino acids was revealed by the combination of chromatography, the ninhydrin reagent as a color reaction, and radioautography to detect new compounds as spots on a paper or on an X-ray film to show substances for which there was no previous clue. Thus the search was not limited by the preconceived ideas of what

[11b] Trivial name: fastigiatine.

might be there. Nevertheless, it is instructive to look for the substances that might have been anticipated from known metabolic cycles, other than the formation of protein, through amino acid condensation.

The formation of amino acids from the keto acids, of the Krebs cycle, in its original or modified forms, has been mentioned. Free glycine from glyoxylic acid, which has recently assumed a more prominent place than hitherto (261), is not an amino acid that commonly occurs free in plants in quantity. In fact, in many plants it is somewhat difficult to detect. Oxalosuccinic acid might reasonably function as a "port of entry" for nitrogen, in the same way that oxalacetic acid could lead to aspartic, and thus conceivably lead to α-aminotricarbalyllic acid (367).

The ornithine cycle (260)—so prominent in discussions of the metabolism of *Neurospora*, of certain microorganisms, and of animal organs—might have been expected to lead to much more frequent evidence of the occurrence of citrulline and ornithine than has in fact occurred, arginine, of course, being a common nitrogen-rich storage amino acid in plants. Since the main routes for entry of nitrogen into organic combination in higher plants are via the reduction products of nitrate and their eventual combination with α-keto acids, it is perhaps not surprising that the Krebs-Henseleit cycle is not as prominent as it is in liver, the animal organ in which ammonia is detoxified for excretion as urea. Unequivocal evidence for the occurrence of citrulline, ornithine, and argininosuccinic acid is, therefore, not plentiful, although citrulline does occur free in the watermelon (*Citrullus vulgaris*), for which it was originally named. In fact, if these substances do occur free (particularly ornithine) then they probably do so only in small amount because they are very transient, reactive, intermediates in the cycle in question. Barnes and Naylor (1959), cf. Naylor (351), claim to have observed intermediates of the ornithine cycle in pine root cultures, but the requisite enzymes for the interconversion of the intermediates were not readily demonstrable. Apparently Barnes (22) also detected γ-guanidinobutyric acid in ethanol extracts of pine and this can be verified (Durzan, unpublished data). Citrulline was isolated as a product of nitrogen fixation in certain plants (e.g., Miettinen and Virtanen, 1952).

Since the above was written Boulter and Barber (67a) have claimed that they found substantial amounts of citrulline, ornithine, and argininosuccinic acid in the metabolism of germinated seeds of *Vicia faba* L. If substantiated, this represents one of the few cases where *all* the key intermediates of the ornithine cycle have been demonstrated in higher plants. Boulter and Barber (67a) claim to have isolated argininosuccinic acid, although the evidence for this compound and its degradation products (ornithine and aspartic acid) is based on their positions on chromatograms. Evidence is cited to indicate that C^{14}-arginine applied to *Vicia faba* cotyledon tissue yields proline and argininosuccinic acid quickly (30 minutes), and *later* (3 hours) glutamic, ornithine, citrulline, and aspartic acid became radioactive and by 24 hours C^{14} was dispersed through a variety of compounds. The claim is that arginine is metabolized by reversing the ornithine cycle.

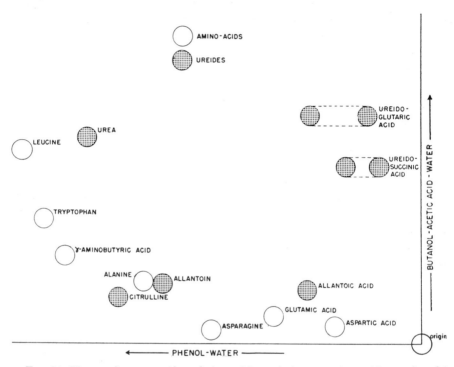

Fig. 21. Urea and some ureides: their position relative to amino acids on phenol/acetic acid-butanol chromatograms. After Steward *et al.* (488).

While this paper is interesting, there are many points to be confirmed before it can be accepted as conclusive evidence that the ornithine cycle works in the manner suggested.

Urea and ureides, detectable by Ehrlich's reagent on chromatograms [cf. Fig. 21, from Steward *et al.* (488)], almost certainly account for more nitrogen and more compounds than is commonly recognized, and the possibilities of deriving metabolically interesting compounds by condensation of urea, or carbamyl phosphate, with other carbon compounds may not yet have been fully exploited. It has been proved that higher plants degrade purines and uric acid to allantoic acid, glyoxylic acid, and urea, but the problem remains whether allantoin and allantoic acid are always formed via the purines (54, 342, 413). Klechkowski and Kretovitch (254) have shown the transamination by pea and wheat seedlings of ornithine, citrulline, and arginine in the presence of added α-ketoglutarate.

Despite the many chromatographic analyses of plants carried out in the past fifteen years, only a minute fraction of the discrete substances that are detectable on chromatograms can be assigned to known compounds. Therefore, the possibilities for new discoveries are still very great. In fact, this laboratory, like others, keeps a register of what are described as unidentified nitrogenous compounds which react on paper with ninhydrin, or other general reagents. These can be clearly recognized by the coordinates of their position on standard two-directional chromatograms. It has even become necessary to keep such maps for particular plant families because, even within the genera and species of one family, the number of such compounds which require to be identified may be very large. The Liliaceae are a case in point. When a survey of the Liliaceae was made by Fowden and Steward (159), there were prior reasons to look for pipecolic acid, hydroxypipecolic acid, γ-methyleneglutamine, γ-methyleneglutamic acid, γ-methylglutamic acid, γ-methyl-γ-hydroxyglutamic acid, and azetidine carboxylic acid, and evidence for the occurrence of these substances was found throughout the family. However, the more conspicuous outcome of the survey was that one had still to recognize at least 53 compounds, detectable by ninhydrin, as occurring in these plants and still to be identified. No doubt, many of these compounds, which can be located on the chromatographic map [(159), cf. Fig. 2] could be isolated if sufficient plant material were to be extracted.

The γ-methylene compounds and their derivatives could be regarded as being particularly prominent in the Leguminosae and in the Liliaceae. The cyclization of lysine to give piperidine derivatives also seemed to be prominent in legumes (cf. Fig. 22). However, all these metabolic pathways, plus the occurrence of the azetidine compounds, may occur even within the storage organs of one plant. An example of this is to be seen in *Saraca indica*, a tropical leguminous plant, the seeds of which reveal the striking array of compounds shown in Fig. 22 from Barrales (25, 496).

It is surprising how some plants seem to emphasize compounds which hitherto were not known, even to the exclusion of the more common nitrogenous compounds that were familiar even to Schulze and Prianishnikov. For example, the seeds of *Baikiaea plurijuga* are rich in pipecolic acid, rich in baikiain, and rich in a substance first called unknown A, but which has since turned out to be 5-hydroxypipecolic acid; however, the amount of such amino acids as alanine, glutamic acid, aspartic acid, glutamine, etc., is comparatively slight. Figure 23 shows a photograph of a chromatogram of the alcohol extract of the seeds of this plant, and for comparison a chromatogram after the nitrogenous compounds of the alcohol soluble nitrogenous compounds were catalytically reduced in such a way that the baikiain was converted to pipecolic acid (Fig. 24).

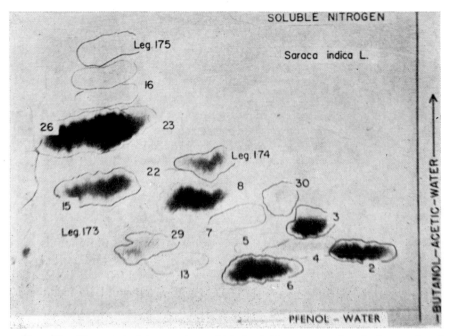

FIG. 22. The ninhydrin-reactive, alcohol-soluble compounds of the seeds of *Saraca indica* L. Spots numbered *2–23* follow the convention of Fig. 9; no. *26* represents pipecolic acid; no. *29* (brown) represents γ-methyleneglutamine; no. *30* (brown) represents γ-methyleneglutamic acid. Spots designated *Leg. 173–175* represent unidentified substances encountered in this and other legumes. From Steward and Pollard (496). After Barrales (25).

Many questions now arise concerning the role of the numerous newly recognized but simple nitrogen compounds. The fact that so many of them appear in storage organs, organs of perennation, etc., may well suggest that they represent end products of metabolism, providing inert storage forms of nitrogen. This does not preclude their participation in small amount in the normal metabolism of more active tissue. Indeed, the occurrence of γ-methyl-γ-hydroxyglutamic acid in the meristem of a fern as well as in trace amounts in the bulb of the

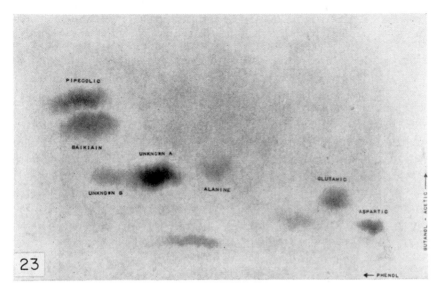

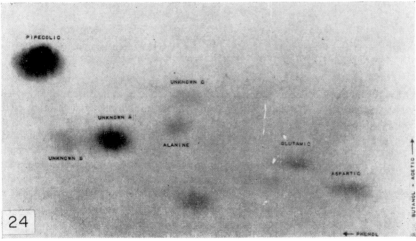

FIGS. 23 and 24. The ninhydrin-reactive, alcohol-soluble compounds of the seeds of *Baikiaea plurijuga* prior to reduction (Fig. 23) and after catalytic hydrogenation (Fig. 24). Note the conversion of baikiain (brown) to pipecolic acid (purple). "Unknown A" (purple) is now known to be 5-hydroxypipecolic acid. After Steward and Pollard (496).

tulip may indicate this. Perhaps some of these substances may also act as antimetabolites, and in fact azetidine carboxylic acid has been shown to be a proline antagonist (497). No hydroxyazetidine-2-carboxylic acid derivative has yet been found, but an analogy with hydroxyproline and hydroxypipecolic acid their eventual discovery might be anticipated (155), and such compounds, like free hydroxyproline, could have marked antimetabolic activity.

The vast array of soluble nitrogen compounds that may occur in plants, indicated above, shows that nitrogen, once it enters into organic combination with carbon, may be rapidly dispersed through many chemical configurations and families of compounds. The next problem that arises is the way in which this array of soluble compounds is affected by the variables that determine the physiology and development of plants.

V. Factors That Affect the Complement of Soluble Nitrogen Compounds in Plants

A. Divisions of the Plant Kingdom, Taxonomic and Genetic Considerations

Since virtually all plants convert inorganic nitrogen to protein, it is not surprising that broad principles of classification do not emerge from the consideration of the composition of the soluble nitrogen pool with respect to the compounds that are most closely linked to protein metabolism. This follows because the normal metabolism of nitrogen to protein is such a fundamental and basic feature that little deviation is permissible.

The distribution of the more recently discovered soluble compounds does not seem to follow any rigid taxonomic lines. It appears that these compounds represent the outcropping, as it were, of underlying veins of metabolism which are more general, though hidden, than was hitherto believed to be the case. The conspicuous compounds that do occur in some plants seem to be more in the nature of biochemical variants, often determined at the gene level within the species, rather than at the level at which broad taxonomic differences are determined [cf. Fowden and Steward on the Liliaceae (159)]. Moreover such compounds seem very prone to modificatons due to nutritional and environmental factors. Some examples that illustrate these points follow.

γ-Aminobutyric acid was the first of the "newer" soluble nitrogen compounds to attract attention; it was first noticed in the potato tuber, then in yeast, and rapidly it was seen to be ubiquitous, even occurring prominently in brain (420). Thus a growing plant that never formed γ-aminobutyric acid would now excite as much attention as one that never formed glutamic acid.

1. Divisions of the Plant Kingdom

More work may yet be usefully done to reveal the compounds in plants which belong to different divisions of the plant kingdom. However, the range and diversity of the compounds encountered within single angiosperm families, within one species and even between cultivars and genetic strains, may deprive such a study of broad taxonomic significance. The lessons to be learned, therefore, are physiological, metabolic, and developmental, not evolutionary or taxonomic. Indeed there is much to be said for the point stressed by Erdtman (147) that the most taxonomically useful compounds may be those that are the least important biologically so that they may exist as somewhat inert metabolic end products.

Many marine algae and higher fungi are replete with still unidentified compounds that appear as spots on ninhydrin-stained chromatograms, but hardly more so than the seeds or other storage organs of many angiosperms. Despite their different evolutionary levels, one set of organisms appears to have as diverse nitrogenous metabolism as the other. Conversely organisms high in the evolutionary scale do not necessarily possess more varied, or a more complex range of, nitrogenous compounds than those lower in the scale.

In survey work of this kind it is important that the chemical identifications should be critical. For example, Lewis and Gonzalves (269, 270) now claim that such compounds as acraine, phenylserine, carnosine, glycocyamine, 2-aminocaprylic acid occur free; in protein of various marine algae, ornithine, homocystine, β-alanine, and γ-aminobutyric acid are said to occur. However, their methods of circular paper chromatography consist at best of single directional chromatograms, and it stretches credibility too far to believe that these compounds could be unequivocally identified by these methods alone. Full use of the selectivity of two-directional systems, of different solvent combinations, may be needed to demonstrate that new compounds do occur in given plants, but there is no substitute for isolation, critical biochemical identification, and eventual synthesis.

2. Cultivars of Economically Important Plants

After the first use of chromatographic methods to determine the constituents of the soluble nitrogen of the potato tuber, analyses were made of extracts from tubers from various cultivars. Briefly it was found that several long-established cultivars had high amino acid content in their soluble nitrogen and an amide moiety which tended to be rich in asparagine. By contrast several later cultivars, many of them developed in the United States and grown at the high levels of nitrogen supply that now prevail, had higher total soluble nitrogen and a higher relative proportion of amide—and this amide was predominantly glutamine (493: see Table 2, pp. 384–385). A long-established English cultivar, viz., 'King Edward,' is adapted to cooler nights and longer days and stores more asparagine, whereas certain other cultivars, developed in the United States and adapted to higher temperatures and perhaps shorter days, emphasize glutamine rather than asparagine (see Table XIII). Lacking a critical study in which all cultivars were grown side by side, it is somewhat difficult to evaluate the role of nutritional and environmental effects on the one hand and the genetic effects on the other. Nevertheless, data of this sort do suggest that progressive selection and breeding may have modified the content of

TABLE XIII

Differences in the Composition of the Soluble Nitrogen
of Various Strains of Potato Tuber[a]

Amino Acid	Sample 1[b] 1951	Sample 2 1951	Sample 3 1955	Sample 4 1955	Sample 5 1955
Aspartic acid	107[c]	78	205	246	168
Glutamic acid	178	134	190	276	431
Serine	66	67	53	42	29
Glycine	28	—	—	—	48
Asparagine	138	1661	2200	2672	1083
Threonine	100	54	43	76	81
Alanine	131	55	68	72	24
Glutamine	3021	1142	858	754	91
Lysine	63	42	30	66	72
Arginine	356	159	55	110	81
Methionine	83	66	—	—	—
Proline	—	25	225	—	—
Valine	244	197	68	141	254
Leucines	93	167	25	24	196
Phenylalanine	138	203	—	—	—
Tyrosine	128	179	Trace	Trace	Trace
γ-Aminobutyric acid	300	244	225	86	240

[a] After Steward and Pollard (493).

[b] Sample 1: cultivar 'Sebago,' Sample 2: cultivar unknown; Sample 3: hybrid of English parentage (from Dr. L. C. Peterson, Cornell Univ., Ithaca, New York; Sample 4: seed size potatoes—presumed 'Katahdin'; Sample 5: cultivar 'Houma,' tubers developed in greenhouse in the spring and under short days. Note the range of variation in glutamine/asparagine from sample $1 > 2 > 3 > 4 > 5$.

[c] Values given as micrograms of amino acid per gram fresh weight.

the soluble nitrogen of the tubers with reference to the proportion of
amide to amino acid and of glutamine to asparagine.

Some hitherto unpublished data of S. O. Trione, working with one
of us, are relevant here. Trione examined potato tubers of different
varieties drawn from a breeding program being carried out by
Dr. Plaisted at Cornell University, so that all the plants were grown
in the same place and in the same year. The alcohol-soluble free
nitrogen compounds were examined as well as the soluble proteins

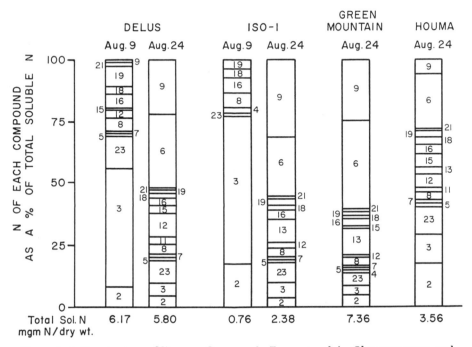

*Numerical Key to Some Nitrogen Compounds Encountered in Chromatograms and
Represented in Histograms*

1, cysteine as cysteic acid   *9*, glutamine   *18*, leucine(s)
2, aspartic acid   *11*, histidine   *19*, phenylalanine
3, glutamic acid   *12*, lysine   *20*, tryptophan
4, serine   *13*, arginine   *21*, tyrosine
5, glycine   *14*, methionine as sulfoxide   *22*, β-alanine
6, asparagine   *15*, proline   *23*, γ-aminobutyric acid
7, threonine   *16*, valine   *24*, hydroxyproline
8, alanine   *17*, methionine as sulfone   *26*, pipecolic acid

Fɪɢ. 25 The soluble nitrogen compounds of potato tubers as affected by variety
and stage of development when grown in one locality. See accompanying numerical
key to compounds. Compiled from the work of Trione with one of us (F.C.S.).

shown by acrylamide gel electrophoresis. The principal genetically determined differences that appeared among the soluble nitrogen compounds seemed to be the proportion of asparagine to glutamine. One genetic class (designated N 1.1-2) was characterized by a ratio of asparagine to glutamine which was almost 1.0; in another (57G-16-17) the ratio was 0.5; and in yet another (NUR-18) this ratio was about 3.0. In different genetic strains the percentage of the total nitrogen of the tubers which is alcohol soluble (i.e., nonprotein) ranged from 71% ('Katahdin') to 59% (N 1.1-2). It was also observed that tubers from different genetic classes gave different protein patterns on acrylamide gel electrophoresis. There is therefore a substantial degree of genetic determination of the nitrogen compounds which are laid down in the potato tuber during its storage period. This is also shown by the data of Fig. 25 (cf. data at August 24 for the different varieties).

It is clear that the conditions which prevail during growth in different localities also have their effect. A variety ('Katahdin') grown in Maine showed a relatively constant content of total amide as the tubers developed (65% of total soluble nitrogen), but the ratio of asparagine to glutamine increased as the season advanced and as the days became shorter and the nights cooler. On Long Island the same variety in the same year was similar early in the growing season, but it diverged markedly later on. Amide nitrogen was reduced to about 16% of the total soluble nitrogen and the rest was mainly arginine (60%) (Table XIV). This condition prevailed in tubers which grew under longer days and at generally higher night temperatures.

TABLE XIV

The Effect of Locality on the Nitrogen Compounds in Potato Tubers (cultivar 'Katahdin') during Their Development

	Tubers grown in Maine			Tubers grown in Long Island	
	July 19/62	Aug. 15/62[a]	Aug. 28/62	July 20[a]	Aug. 7
Total soluble N (mg/gm dry wt.)	1.26	5.65	7.62	6.41	5.66
Total amide N as % of soluble N	62.13	60.80	65.05	56.00	15.73
Asparagine; glutamine	0.36	0.80	1.10	0.65	0.21
Arginine N as % of soluble N	0.0	0.0	8.13	12.2	60.3

[a] The tubers were at a similar stage of development on these dates in the two localities. From work of Trione with one of us (F. C. S.).

Still another approach to the problem of the factors that determine nitrogen metabolism in potato tuber tissue is possible by using potato tuber tissue cultured by the aid of coconut milk and a suitable synergist [2,4-D dichlorophenoxyacetic acid (2,4-D) or naphthaleneacetic acid (NAA)]. There was more amide nitrogen (predominantly asparagine) in the cultured tissue which grew least; there was relatively more glutamine than asparagine in the tissue which grew best (in presence of NAA). Also the tissue that was most effective in growth was rich in γ-aminobutyric acid. Treatments which gave graduated effects on the growth of the tissue were reflected in the amount of soluble (pH 8.3) protein present and in the variety of compounds that it contained (i.e., the number of bands that appeared after gel electrophoresis); each of these parameters increased according to the following series of nutrient treatments: basal medium < basal medium plus coconut milk < basal medium plus coconut milk plus 2,4-D < basal medium plus coconut milk plus NAA.

Obviously there are certain genetically determined potentialities of the potato plant for nitrogen metabolism, but these may be modified, during development, by the effect of different environments and different growth-regulating substances. All these variables must make their impact upon the genetically determined centers of synthesis through some chemically invoked mechanism. A similar study made upon the banana plant also points to genetic, seasonal and developmental, and environmental factors that affect the complement of soluble nitrogen compounds that accumulate in the cells of the fruit.

The edible banana represents a vegetative structure with some unusual features of nitrogen metabolism (namely, the conversion of amide-nitrogen to histidine late in the ontogeny of the fruit). These facts were discovered by chromatographic means. However, in view of the unexpected importance of histidine, which is not common in the free state in large quantity, it was deemed necessary to isolate this compound and identify it unequivocally (486, 488).

Several cultivars of *Musa acuminata* have been examined ['Lacatan,' 'Cavendish,' 'Gros Michel' (Jamaica and British Cameroons)] from this point of view, and the conclusion now is that differences in the nitrogenous composition which are due solely to the cultivars are easily obscured by the even greater differences that may be superimposed by nutrition and environment at the time of the development of the fruit. In fact, the same cultivar ('Gros Michel') grown in Honduras has a different complement of compounds in its pool of soluble nitrogen if the fruit develops predominantly in the spring, when cooler nights prevail, as shown below.

As the parthenocarpic fruit of the banana advances through its development it has a different complement of soluble nitrogen compounds at different stages. The "embryonic" primordium may be rich in soluble nitrogen, but this is mostly amino acid nitrogen (alanine, aspartic and glutamic acid, etc) with little amide nitrogen and virtually no histidine. Very young flowers are rich in amide nitrogen. During growth of the fruit this falls to a low level, but as development continues, the storage products enter as the organ develops until, after the fruit has emerged from the pseudostem, it has a higher soluble nitrogen content and the bulk of this consists of asparagine, glutamine, and histidine. The presence of histidine is a feature of the later stages of development of the fruit and especially of the ripened fruit. In this organ, it is clear that the "winter-grown fruit" becomes asparagine rich earlier in the course of its development than does the "summer-grown fruit," and the histidine content (at the expense of amide, mainly glutamine) also appears earlier in development if the fruit is winter grown. Thus a complex of nutritional and varietal effects is brought to bear upon the complement of soluble nitrogen compounds which are to be found in some economically important plants.

3. Legumes and the Piperidine Compounds

Among the first plants to yield newly identified compounds were the legumes. Pipecolic acid was observed in green beans (*Phaseolus vulgaris*) and in other angiosperms. The derivation of pipecolic acid from lysine and its conversion via α-aminoadipic acid which occurs in peas (*Pisum sativum*), and thence to glutaric acid, seems to occur equally well in *Neurospora* and in leguminous fruits and seeds. Moreover, pipecolic acid frequently accumulates in response to mineral deficiency [e.g., to potassium deficiency in the banana (169)] and to the attack of pathogens [e.g., *Verticillium alboatrum* in mint (*Mentha*) (426)].

The occurrence of baikiain (dehydropipecolic acid), or of 5-hydroxypipecolic acid in other legumes might suggest that distinctive metabolic pathways are peculiar to, or characteristic of, leguminous plants. These ideas are, however, quenched by the equally conspicuous presence of 5-hydroxypipecolic acid in the seeds of edible dates. These examples, however, all suggest various situations in which lysine may be metabolized in different plants so that it does not itself accumulate but gives rise to other compounds. However, some plants (notably certain conifers) seem incapable of utilizing lysine in the manner now familiar in legumes, the banana, and other plants. Thus lysine accumulates as such in potassium-deficient spruce leaves, and the nitrogen compounds that are often conspicuous in conifers are arginine and proline (observations of Durzan on *Picea glauca*). Vogel (587) has studied lysine formation in *Azolla, Lemna, Ginkgo, Agave, Melilotus,* and *Helianthus* to show that lysine-threonine-aspartic acid patterns (revealed by C^{14}-labeling), like those found in bacteria and green algae but unlike those found in higher fungi, exist.

4. Legumes and Liliaceous Plants and Their Nitrogen Compounds

The discovery of γ-substituted glutamyl compounds (γ-methyleneglutamine and γ-methyleneglutamic acid) in either the tulip bulb or in the peanut (*Arachis hypogaea*), referred to earlier (see Section IV, G), might have aroused thoughts that these unusual compounds were restricted or specific in their distribution. However, their occurrence in two such disparate genera discourages any idea that they are characteristic of either the genera or the families in question. The data on the occurrence of nitrogen compounds in such families as the Liliaceae (159) do not lend themselves to interpretation on taxonomic lines; rather they suggest that the potentialities for synthesis of these compounds run throughout the family. Whether a given compound accumulates, or not, seems to be determined at the level of far fewer genes than would be required to cause specific, or genetic, differences. No doubt the biochemical potentialities to synthesize these nitrogen compounds are very widespread, but it requires quite subtle combinations of genes, nutrition, environment, and the type of organ to evoke their accumulation.

Nevertheless, the view that the γ-methylene compounds are to some extent typical of liliaceous plants or the legumes and the substituted piperidines of the legumines has something in its favor. In fact, seeds of some leguminous plants (e.g., *Saraca indica*) show both classes of compounds in the same extract and on the same chromatogram (496). However, any seeming significance of the occurrence of the substituted glutamyl compounds in plants in any particular part of the plant kingdom seems to disappear when one considers that γ-methyl-γ-hydroxyglutamic acid, probably the parent of this class of compounds, occurs in great abundance in the apex of a fern (*Adiantum*), but is likewise known to occur in many liliaceous plants.

Similarly the frequent occurrence of homoserine in legumes and of *O*-acetylhomoserine in *Pisum* testifies to the biochemical versatility of plants in this family, but again this fact seems to have no direct taxonomic significance. It may be noted that some cultivars of *Pisum* respond to day length and to night temperature in such a way that reciprocal relationships appear in their content of asparagine, under some circumstances, contrasted with homoserine, under others: cf. Steward, Pollard, cited from Grobbelaar *et al.* (188).] However, both Liliaceae and Leguminosae, the two families upon which most data have been assembled, agree in one important respect, namely, the great number of recognizable but still unidentified ninhydrin-reactive compounds that appear on chromatograms of extracts from plants in these families (25, 187, 496, 511, 631). Apparently there are still many well traversed metabolic routes in plants that have escaped detection through the preoccupation of plant physiologists with preconceived paths that supposedly lead to protein.

5. *Sulfur Compounds*

O-Choline sulfate has been identified in mycelia and conidiospores of fungi (467, 527, 622), in a genus of lichens (272) and also in red algae (273). Choline phosphate occurs in higher plants (290) in *Zea mays, Hordeum vulgare,* and *Helianthus annuus;* choline sulfate was detected as a major compound in sulfur-deficient roots (364). Wilson (609) and Kun (262) have reviewed the sulfate metabolism of microorganisms.

6. *Genetic Considerations*

Cultivated varieties often differ greatly in their storage of soluble nitrogen compounds, as already noted (pp. 481 *et seg.*) for the relative predominance of glutamine and asparagine in potato tubers (493, 501). This effect is, however, apt to be encountered in other organs and with other constituents of the soluble amino acid pool [cf. data of Steward *et al.* (1960) on banana varieties and of Fowden and Steward (159) on *Tulipa* varieties]. Thus many of the observed differences in composition of plants with respect to their simple nitrogen compounds may be due to varietal, or genetic, factors that may have arisen adventitiously insofar as nitrogen metabolism is concerned, in the course of selective breeding for other characteristics. It is important that this point be appreciated by plant breeders.

Many questions still have only partial answers. If almost every enzymatic step were to be regarded as gene controlled, there should be some relationship between the possible complexity of the metabolism and the total number of genes. Some ferns may have hundreds of small chromosomes; at least one angiosperm (*Haplopappus gracilis*) has a diploid number of only 4 ($n = 2$). However, the latter plant seems no more restricted in its nitrogen metabolism than any plant that has a much larger array of chromosomes. *Neurospora,* which is notably versatile in its nitrogen metabolism, is a haploid, and even *Escherichia coli* has a rich range of nitrogen metabolism (419). Comparisons between haploid and diploid phases of development are few. In the angiosperms this comparison may be made by comparing pollen with sporophytic

TABLE XV

COMPARISON OF THE SOLUBLE NITROGEN FROM
HAPLOID (*N*) AND DIPLOID (2 *N*)
POLLEN AND EMBRYO TISSUE[a]

Amino acid	Pollen, *N*	Dormant embryo, 2 *N*
Aspartic	1.98[b]	2.34[b]
Glutamic	0.81	1.34
Serine	0.96	2.23
Glycine	1.93	2.79
Asparagine	0.16	1.36
Threonine	3.22	0.66
Alanine	5.55	6.31
Glutamine	0.74	1.14
Histidine	0.87	Trace
Lysine	3.18	1.73
Arginine	46.65	71.21
Methionine	0.90	1.60
Proline	25.42	0.34
Valine	0.73	1.18
Leucines	0.82	2.25
Tyrosine	0.66	2.41
β-Alanine	0.13	Trace
γ-Aminobutyrate	4.05	1.09
Hydroxyproline	0.69	—
μgm N/gram fresh wt	2070	1085

[a] Results of D. J. Durzan, The nitrogen metabolism of *Pica glauca* (Moench) Voss, and *Pinus banksiana* L. with reference to nutrition and environment. Ph.D. thesis, Cornell Univ., Ithaca, New York, 1964.

[b] Values for the amino acids given as nitrogen in each compound as a percentage of the total soluble nitrogen.

tissue, and this has been made qualitatively by Virtanen and Kari (582) and also quantitatively by Durzan (Table XV). Although there are some notable differences in the accumulation of certain amino acids in pollen, which do not occur conspicuously in leaves, e.g., hydroxy-proline, the main conclusion usually is that the biochemical potentialities of the haploid tissue are just as great as those of the diploid.

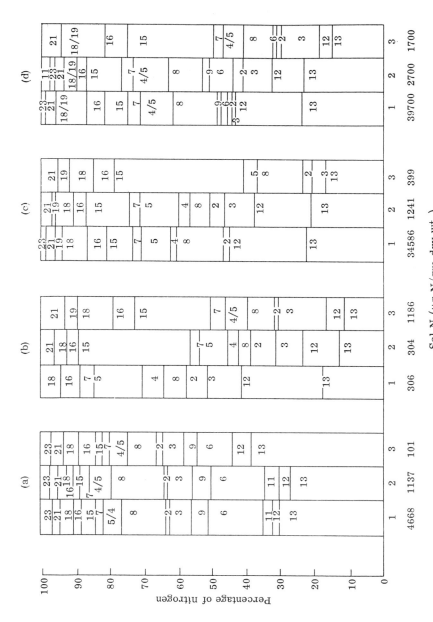

Fig. 26. The amino acid composition of the embryo, scutellum, and endosperm of wheat (cultivar 'Genesee') grain, (a) Alcohol-soluble free amino acid nitrogen fraction. (b) Alcohol-soluble protein nitrogen fraction. (c) Alcohol-insoluble protein nitrogen fraction. (d) Total nitrogen. Columns 1, embryo; 2, scutellum; 3, endosperm (including testa, etc). For numerical key, see Fig. 25. Data after Chang (100).

Pollard's data (395) for *Lycopodium cernuum* show that the sporophyte and gametophyte have nitrogen complements which are qualitatively similar but quantitatively different. Both tissues were characterized
by large quantities of arginine although the gametophyte had nearly
eight times as much total soluble nitrogen (395).

A comparison between a commonly triploid tissue and a diploid one
may be made using the endosperm of the cereal grains and the corresponding embryo. While differences in their composition with respect
to the soluble nitrogen compounds certainly do exist (Fig. 26) these
differences are not obviously attributable to their respective chromosome complements, i.e., the degree of ploidy *per se.*

Momotani (330) has, however, proposed a revised system of classification for the Aceraceae after the re-examination of morphological
characters and analysis of the seed proteins. Generally the protein
patterns accorded with taxonomic affinity.

B. SOME DEVELOPMENTAL CONSIDERATIONS

Although developmental studies which relate nitrogen metabolism
of various organs (leaves, fruits, bulbs, roots, etc.) to their stage of
development have traditionally been made, it is logical to begin this
discussion with the growing point itself.

1. Growing Regions

Generally, plant growing points are not conspicuously rich in stored
soluble nitrogen, for these regions are actively engaged in its conversion to protein. Nevertheless, chromatograms made using extracts from
shoot apices reveal a number of free, soluble compounds. This number
is much greater than Allsop (6) first supposed. The white lupine (a
classical plant in work on nitrogen metabolism) was used in a study
of this kind (509, 510).

The central portion of the shoot apex of lupine was isolated, extracted, and analyzed separately from a series of leaf primordia. In
lupine, the cotyledons (which are, of course, modified leaves) of the
germinating seed are very rich in asparagine under the conditions in
which protein breaks down in the cotyledons and nitrogen is being ex-
ported. Thus asparagine, normally a storage substance, even makes its
appearance in the growing point itself, although the amount present
there falls very rapidly if the cotyledons are removed. However, the
characteristic feature of the soluble nitrogen complement of the shoot
apex of lupine was the presence of all the common amino acids and
the amides, with a tendency for the basic amino acids to be more
prominent in those regions where cell division and nucleic acid syn-

thesis must also be important. Gradations with age were also noted in the composition of the soluble nitrogen fraction from the first few leaf primordia of lupine. Although this has not yet been done, it would be interesting to have a comparable comparison for the proteins which occur in the first few leaves of lupine by the method that has been successfully applied to the soluble proteins of *Neurospora* (101) and certain angiosperms (cf. page 545).

A fern (*Adiantum*) was examined from this point of view (509, 510). Even in the shoot apex of this plant there are many relatively large and somewhat slowly dividing cells. Apparently this fern elaborates its protein during growth by drawing upon a complement of relatively few soluble nitrogen compounds. In this much more restricted pool of soluble nitrogen than that of lupine γ-methyl-γ-hydroxyglutamic acid was curiously and unexpectedly predominant. (Whether exogenously supplied γ-methyl-γ-hydroxyglutamic acid can be equally effective as a nitrogen source for these fern apices is still a moot point.) When the apex of *Adiantum* was even further dissected, so as to distinguish between the outer layer of larger and less meristematic cells (including the apical cell) and the inner zone of smaller, more meristematic, prestelar cells and leaf primordia, the differences in composition of these regions were noted. Although it was necessary to dissect 700 fern apices to obtain enough material for these analyses, the amount of dry matter so available was only 4 mg. Even so, the differences were interpretable on the hypothesis that the outermost layer had the soluble nitrogenous compounds more typical of quiescent, mature, cells than of active meristematic cells, as shown by the following quotation:

"These trends are:
    a. The arginine, and combined arginine-lysine, content of the protein fraction is highest in the leaf primordia (Table 3) and is lower in the more mature, vacuolated cells. In this respect the prismatic layer also has the characteristics of an older tissue. b. In the protein of the old leaves, the ratio of glutamic to aspartic acid was > 1.0 (Table 3). This was also true of the outermost apical layer of the shoot but not of the leaf primordia, nor the young leaves. c. With respect to certain amino acids (valine, alanine), the growing cells of the leaf primordia show a tendency toward a low content in the protein in contrast to the older tissue—including the prismatic layer—whereas other amino acids (serine) show a tendency toward a high content in the protein of the growing cells."

But the most cogent comparison of the nitrogenous composition of actively dividing cells with that of their quiescent, mature, counterparts comes from the analysis of growing and nongrowing cells of tissue cultures. It has long been known that the transition of the cells of potato tuber from the storage state to that of cells in division, as at the surface of a cut slice, was accompanied by a decrease in their total soluble nitrogen [cf. papers by Steward and Preston (498), and references there cited]. However, both carrot and potato explants have now been analyzed for their free and combined nitrogen compounds; the changes are shown in Table XVII (493).

"The preponderance of glutamyl (glutamic acid, glutamine and γ-aminobutyric acid related to them) over aspartyl (aspartic, asparagine β-alanine) compounds is evident in the growing tissue of both the potato and the carrot. This suggests that whereas the special importance of compounds that can give rise to glutamine in the economy of the growing, proliferating, protein-synthesizing cells remains, the importance of compounds that can give rise to asparagine has largely disappeared.

"Returning to the classical case of the lupine, one can see from the soluble nitrogen in different parts of the shoot apex that the relative predominance of asparagine tends to decrease as one approaches the meristematic cells of the central apex and the first leaf primordia. Conversely, the relative importance of compounds related to glutamic acid (including glutamine and γ-aminobutyric acid) over aspartyl compounds (aspartic and asparagine) increases.

"Therefore, although *both* asparagine and glutamine may function as soluble storage reserves in systems in which protein synthesis is at a minimum, or even in which protein breakdown may occur, as, e.g., in the leaf of monocotyledonous plants in the dark, there are striking examples to be found in which glutamine, rather than asparagine, is particularly important in the economy of growing, protein-synthesizing cells. Such ideas are consistent with the view that the nitrogen en route to protein synthesis reaches that goal through association with the glutamine molecule, whether this be through the group-transfer reactions of transamination, transamidation, or transpeptidation."

2. Growing versus Resting Cells

The respiration and protein nitrogen per root cell both increase to a maximum at about that point along the axis of roots (5 mm from the tip for peas) at which the cells are fully extended (78). Thus the great increase of protein and the greatest respiration are not associated, as much as might have been supposed with cells that divide and multiply their self-duplicating structures. When respiration is measured per unit *of protein in the cell*, the protein present in the youngest cells (0.0–0.4 mm from the tip) is found to be more effective in maintaining respiration than is that in the older ones. Clearly, there is initially a rapid phase of growth in cell volume, accompanied by a rapid increase in both protein nitrogen and respiration, followed by a protracted phase of slower development with respect to all three parameters (volume, respiration per cell, and protein-nitrogen content). Figure 27 summarizes the development of root tissue in terms of their known DNA, RNA, and protein metabolism.[11c]

[11c] A recent paper by Steward, Lyndon, and Barber [*Am. J. Botany*, **52**, 155–164 (1965)] traces the changes in proteins of pea roots during development

The shift from the quiescent, resting state of the storage tissue to that of active growth and cell proliferation may be brought about to a limited extent, as in the events of wound healing, or to the maximum extent under the stimulus of such growth regulators as coconut milk plus 2,4-D when they are added as supplements to a culture medium. The shift from the quiescent to the growing state is accompanied by a great and downward change in the ratio of alcohol-soluble nitrogen to alcohol-insoluble (protein) nitrogen. The change in the protein nitro-

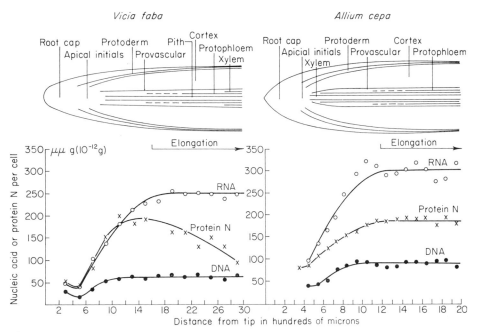

FIG. 27. DNA, RNA, and protein nitrogen content per cell in root apices. After W. A. Jensen [*Exptl. Cell Res.* **14**, 575 (1958)].

gen, however, is not only one of total amount, for there are also changes in the composition of the bulk protein. This has been observed by (a) analysis of the amino acids which comprise the protein in bulk (508); (b) studies of C^{14}-incorporation into the protein and its subsequent turnover (508); and (c) actual observations of the proteins as they are separated by paper (286) or acrylamide gel electrophoresis (489).

Whereas the nitrogen-rich compound asparagine virtually disappears from the rapidly dividing cells of carrot root or potato tuber, glutamine persists, although at a lower level of concentration than in the storage cells. Alanine and γ-aminobutyric acid, however, remain conspicuous elements of the soluble nitrogen of the rapidly growing cells. These effects are shown in the data of Table XVII. A similar example concerns the virtual disappearance of arginine from the winter-stored artichoke tubers when French workers grew the tuber tissue in culture (138, 139).

It has long been known that actively dividing cells of potato tuber formed sugar at the expense of starch [(480), references there cited]. Treatments which promoted growth and protein synthesis depleted this accumulated reserve of sugars, whereas those which retarded growth and protein synthesis conserved it. This has been prominently shown in the effects of oxygen tension; the K:Ca ratio; the effects due to CO_2/HCO_3^-, etc., on the soluble nitrogen of disks of potato tuber tissue in aerated solutions at 23°C. There is not only net loss of carbon from a carbohydrate pool to be considered, but also the cyclical turnover of carbon as it replenishes, from sugars and a suitable soluble nitrogen source, the protein that is being metabolized. The results show that the rate of turnover of protein is also increased by the stimulus of coconut milk that causes the cells to divide. This system has, therefore, been interpreted as follows. The same soluble nitrogen compound (e.g., glutamic acid) may exist simultaneously in different phases in the cell. In one such phase glutamic acid may derive its carbon readily from sugar, and the amino acid so formed may be readily accessible to the site of protein synthesis. In another phase the same acid may come more readily from protein breakdown, and its carbon may become available for respiration, while its nitrogen is re-used to make more protein. This concept implies a marked compartmentation of certain metabolites in cells. Thus certain metabolites, as they are formed, immediately participate in protein synthesis and are not free to mingle with their molecular counterparts that arise in other ways and are stored and metabolized elsewhere in the cells.

In this concept the balance between anabolic and catabolic processes is altered in quiescent and growing cells, and it has been presented in a diagrammatic scheme. Before considering the merits of this scheme the movement of carbon and nitrogen to and from the protein fraction in these cells should be summarized.

Briefly, exogenous C^{14}-labeled sugar penetrates the cell, labels the free sugars and certain metabolites (sugar phosphates) that are immediately derived from carbohydrate. Despite rich labeling of these substances in the cell, the carbon dioxide which emerges is at first and for some time much more weakly labeled. Although the endogenous free amino acid remains weakly labeled, when the exogenous source is sugar, the amino acids of the protein (e.g., glutamic acid) quickly become more strongly labeled, showing that they arise more easily from the sugar of the cell than from the free amino acid pool. In due course, protein which has incorporated C^{14}-amino acid breaks down and labels the free amino acid pool; as the nitrogen is re-used, the C^{14}-labeled carbon frameworks come off as $C^{14}O_2$. When C^{14}-glutamine is supplied exogenously, the endogenous sugars are not labeled, but in spite of this C^{14} readily enters the pathways that promptly yield $C^{14}O_2$. However, C^{14}-asparagine behaves quite differently: it remains surprisingly inert and is neither readily incorporated into the amino acids of protein nor easily respired away. These ideas culminate in suggested inter-locking cycles of respiration and protein metabolism in which glutamic acid and glutamine seem to play a key role. Glutamine is a source of the nitrogen which, together with sugar, forms protein and also represents a nitrogen-rich substance toward which protein breakdown products are reworked before resynthesis takes place (Fig. 28). These concepts are summarized in various papers [(482), and references there cited].

The protein fraction of actively growing cells appears to behave in two ways. One part (metabolically active protein) readily "turns over," so that the growing cell is distinguished not only by a greater rate of total synthesis but by a greater rate of "turnover." It is conceived that these cyclical events keep the cellular machinery turning in ways that allow its work to be done. There is also a part of the protein that does not turn over, and this is most easily distinguished by that part of it which, having incorporated C^{14}-proline, turns it *in situ* into C^{14}-hydrox-yproline. The ratio C^{14}-proline to C^{14}-hydroxyproline for the whole cell approaches asymptotically a figure of 0.7 for both carrot and potato tuber tissue, but this is made up of several fractions in which the ratio may be different (286, 399, 484). In fact, some soluble pro-tein moieties, separable by acrylamide gel electrophoresis, though they contain C^{14}-proline, remain virtually free of hydroxyproline. There-fore, this type of protein is not wholly confined to the cell wall as some have supposed (263), although in the surface of the cytoplasm it may have some special relationship to the formation of the first-formed wall. Hydroxyproline is formed *in situ* from proline after proline has become part of the peptide chains in collagen formation (478), the oxygen being supplied from the air as shown by the use of O^{18} (171). Struc-tural protein, which is so important to the rapidly proliferating plant cells, is conspicuous in a number of examples of cultured cells that proliferate under the stimulus of coconut milk. This signifies that the protein of the cell may be both metabolic (deriving its carbon readily

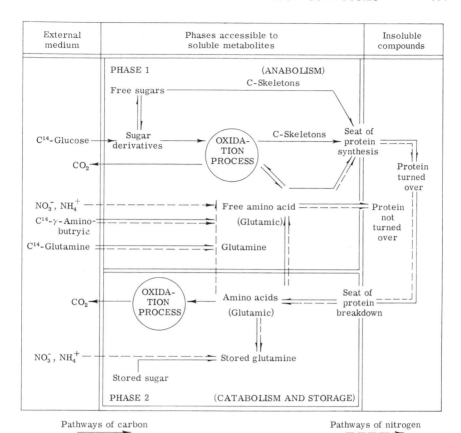

FIG. 28. Diagram to illustrate the main metabolic pathways in cultured carrot tissue explants. Pathways emphasized by the experimental treatments are indicated by heavy lines. After Steward *et al.* (482).

from sugars, and contributing to the soluble nitrogen pool on breakdown) and nonmetabolic [incorporating at least some of its amino acids, e.g., proline, directly, and, having done so, retaining them indefinitely (either as proline or hydroxyproline)]. These ideas of protein synthesis and turnover have been discussed also by Steward and Bidwell (481).[11d]

3. Nitrogen Metabolism of Leaves

Much attention has been given to the nitrogenous composition of leaves, almost more than to that of any other organ (313, 314, 507).

[11d] A later paper on this subject is one by R. G. S. Bidwell, R. A. Barr, and F. C. Steward, *Nature* 203, 367–373 (1964).

Nitrogen metabolism of leaves needs to be discussed with respect to both developmental factors and other variables that affect the composition of leaves. Suffice it to say here that leaves pass sequentially in their development through stages that are reflected by their nitrogenous composition. Prior to the use of chromatographic methods this was investigated by Pearsall (384) for plants such as Swiss chard (*Beta vulgaris* var. *cicla*) and by Gregory and Sen (186) on the cereal grasses, by Nightingale and Robbins (360) on *Narcissus*, and by Vickery *et al.*, on *Narcissus* and *Nicotiana* (572, 573). Of necessity these early studies dealt only with the broad categories of soluble nitrogen that were then determinable.

Systematic observations on the respiration rates of starving, detached, leaves were made by F. F. Blackman early in the present century. Most of this work remained unpublished and became known through the writings of his pupils and through his lectures (cf. 47, 235). Blackman showed that a preliminary fall in leaf respiration was followed by a rise and fall to which the term senescent rise, or senescent hump, has been applied. He considered the senescent rise to be caused by a partial breakdown in cell organization which allows enzymes to reach substrates to which they are not normally accessible [(48), p. 444]. Later the biochemical changes induced by starvation came to be investigated, and from the work of Deleano (1912), of Mothes (337–339), and of Yemm (627, 628) it appeared that the senescent rise occurred because protein was being utilized as a respiratory substrate, a process necessitated by the carbohydrate deficiency attendant upon starvation.

The following processes have been considered to occur simultaneously and to be associated with the climacteric rise in leaf respiration, as summarized by Wood *et al.* (619):

(i) The formation of glutamine (a) from ammonia (derived from oxidation of amino acids) and glutamic acid (derived from the protein molecule) at a rate depending on rate of oxidation of amino acids and (b) from α-ketoglutaric acid and ammonia in the presence of enzyme, the equilibrium position being determined by rate of oxidation of carbohydrates in the system.

(ii) The formation of asparagine from ammonia and aspartic acid derived from protein hydrolysis.

(iii) The production of citric from pyruvic acid (derived from glycolysis of sugars) and oxaloacetic or malic acids, at a rate determined by the sucrose and oxaloacetic acid contents; and its utilization by oxidation through the Krebs cycle or in the formation of glutamic acid.

(iv) The production of oxaloacetic acid from aspartic acid by aspartic acid aminophorase, by oxidation of citric acid, or by oxidation of α-ketoglutaric acid derived from glutamic acid; and production of malic acid in equilibrium with this.

(v) As a result of falling respiration rate α-ketoglutaric acid increases at the expense of glutamic acid. Malic and oxalacetic acids increase by oxidation of α-ketoglutaric acid and further production of asparagine from oxaloacetic acid at the expense of amino acids by action of aminophorase. Possible transamination of glutamic acid and oxalacetic acid to aspartic acid may also occur under the influence of glutamic aminophorase.

(vi) The cessation of protein synthesis due to the exhaustion of sulfur-rich proteins and the breakdown of chloroplast structure resulting in cessation of amino acid production and of accumulation of amides.

(vii) Oxidation of amino acids and amides, probably by L-amino oxidases and which possibly has been continuous, becomes obvious, and ammonia accumulates. Disappearances of citric and malic acids in undetermined ways.

Diurnal cycles of nitrogen metabolism in leaves have been studied for *Acer negundo* (Schulze and Schütz, 1909), *Phaseolus* (104, 105), cotton (299, 300), tobacco (366, 456), and mint (485, 487), and contrasted effects of light and darkness on the formation of asparagine

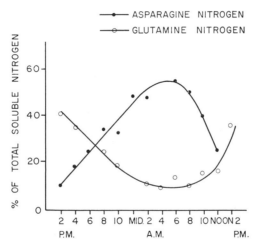

FIG. 29. Diurnal variation in composition of soluble nitrogen of leaves of *Mentha piperita* L. grown under short days. Variation of glutamine and asparagine expressed as percentage of total soluble nitrogen. After Steward *et al.* (487).

(in dark) and glutamine (in light) have been observed for mint (Fig. 29) and tobacco (493). The trend of the work on the biochemistry of mint leaves is to emphasize that asparagine is characteristic of dark metabolism of leaves (especially at high temperatures) and glutamine of light metabolism.

The seasonal cycles of nitrogen metabolism have also been studied for spruce buds on leader shoots grown under natural conditions and under natural light supplemented by a period of continuous light [Durzan and Steward, *Plant Physiol.* **38**, Suppl. vi (1963)]. Essentially the free arginine nitrogen as a percentage of the total soluble nitrogen increases in all parts of the shoot (buds, stem, foliage) during its active growth (Fig. 30). When apical growth ceases arginine nitrogen (now at a maximum) declines concomitantly with an increase in proline nitrogen. However, the decrease of arginine per unit fresh weight is greater than can be accounted for as proline. In winter, shoots are characterized by a lower content of soluble nitrogen which is rich in

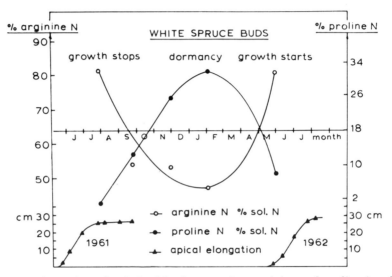

Fig. 30. The reciprocal relationships between free arginine and proline in white spruce in relation to growth and the onset of dormancy. Unpublished data of Durzan and Steward.

proline. In the spring, prior to growth, a flush of arginine-rich soluble nitrogen occurs and the percentage of proline nitrogen declines.

Thus the morphogenetic factors (light versus dark, short days versus long days, high night temperature versus low, nutritional deficiencies, etc.) that control the status of the plant or organ in its growth profoundly alter its nitrogen metabolism. Since these factors operate on the growing regions, this must mean that it is the behavior of growing regions which largely determines what the nitrogen compounds do. This recalls the "triggering role of coconut milk in the nitrogen metabolism of cultured cells."

4. Nitrogen Metabolism and Floral Development in the Tulip Bulb

Although both floral parts and bulbs represent morphologically transformed leaves, each may represent different modifications from the normal metabolism of the leaf as shown by their content of nitrogen compounds. This is conveniently illustrated by the tulip bulb and the flowers that develop within it.

The nitrogen metabolism of the tulip bulb has two characteristic features. First it contains several of the newer compounds which were discovered by chromatography; γ-aminobutyric acid, γ-methylene-glutamic acid and its amide γ-methyl-γ-hydroxyglutamic acid are among these (159). Secondly it has a heavy emphasis on arginine as the nitrogen-rich reserve; arginine not only exists free in quantity in the bulb, but it is also present in the alcohol-insoluble nitrogen to the extent of 47% of that nitrogen in the form of some prolamine-like complex. (Thus about 60% of the total nitrogen of the bulb is in the form of combined arginine.) In this respect the bulb scales are completely different from the tulip leaves to which they correspond, for free arginine is virtually absent from the expanded green leaf, or is present only at a low level.

The floral parts of the tulip are initiated during the period of apparent rest in the bulb. The process is very temperature dependent and its limitations are now well known (brief exposure to 68°F for floral initiation; longer exposure to 48°F for floral development prior to planting). Advantage was taken of this situation to study the changes in the size and composition of the pool of soluble nitrogen compounds concomitantly with these morphological events. Bulbs of the cultivar 'William Pitt' were used; these initiate flowers at 68°F which subsequently develop at 48°F. The sequence of chemical changes which accompanied the initiation and growth of the flowers is shown in Fig. 31 (632).

The heavy emphasis on arginine in the soluble nitrogen of the bulb remained during the period of initiation, although the total amount of soluble nitrogen increased at the expense of the arginine-rich reserve. However, when the floral parts grew, still within the bulb, they acquired their own distinctive pattern of metabolism, which was distinguished by a marked transition from the emphasis on arginine, as the main soluble reserve, to a more familiar composition in which amide nitrogen (mainly glutamine with a smaller amount of asparagine) and alanine were conspicuous. Surprisingly enough the substituted glutamyl compounds (γ-methyleneglutamic acid and its semiamide) did not seem to enter into these changes in any decisive way.

Thus green leaves of tulip in the light, bulb scales, and floral parts, all have their distinctive patterns of soluble nitrogenous compounds. Indeed, later work (Steward and Barber, unpublished) shows that these organs also have characteristic complements of soluble protein as shown by acrylamide gel electrophoresis (see Fig. 48). Since each such structure presumably contains cells with the same genetic information, its biochemical expression must be controlled in some way that also determines the course of morphogenetic development. How these distinctive patterns of nitrogen metabolism are superimposed upon organs which are all transformed leaves represents a problem as difficult as the interpretation of their growth and form. What then is the mechan-

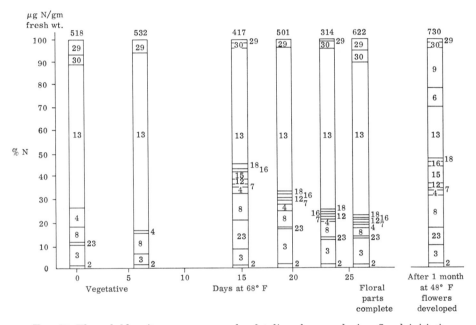

FIG. 31. The soluble nitrogen compounds of tulip: changes during floral initiation and development. Where not shown, compounds nos. 4, 7, 12, 15, 16, 18, 21, and 30: are present in trace amounts. For nos. 2–23, see numerical key accompanying Fig. 25; no. 29, γ-methyleneglutamine; no. 30, γ-methyleneglutamic acid. After Zacharius *et al.* (632).

ism that controls the biochemical expression of the parenchyma cells, which may be, and probably are, genetically totipotent for all the attributes of the tulip plant?

The mobilization of soluble nitrogenous reserves during the growth of storage organs presents another opportunity to see the changes that ensue during development. This problem could be examined with reference to an array of fruits and seeds, to bulbs, rhizomes, tubers, etc., to resting vegetative buds as organs of perennation, but only a few examples must suffice. The great variety of soluble nitrogen compounds to be found in such storage organs has already been commented upon.

5. Nitrogen Metabolism in the Growth and Development of Fruits

The growth of fruits is commonly triggered off by pollination and by the setting of seed. The growth of the fleshy pulp of an apple is stimulated by events in the ovule and by the presence of seed, though the fleshy pulp of the fruit represents a complex fusion of the bases of floral parts with the wall of this inferior ovary. In the edible banana the fruit grows parthenocarpically and has, as it were, its own genetically "built in" stimuli, for neither pollination nor seed formation are

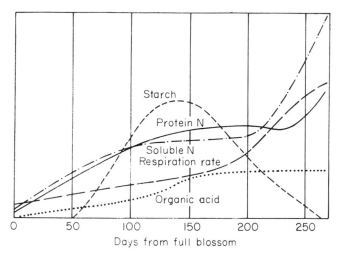

FIG. 32. Trends in starch, protein nitrogen, soluble nitrogen, organic acids, and respiration rate of 'Granny Smith' apples during development on the tree. Reproduced from J. A. Pearson and R. N. Robertson [*Australian J. Biol. Sci.* **7**, 1 (1954)].

necessary to stimulate this growth and mobilization of organic reserves. The regions that grow are here represented by an inner layer of pulp-initiating cells that border upon the wall of the loculus, and in the trilocular ovary itself the cavities become filled by developments that arise from the septa, that separate the carpels, and also from the central axis and placental region (408). Both of these examples of fruit development have been examined in relation to their nitrogen metabolism.

a. Nitrogen metabolism in the growth and development of the apple. After the first rapid phase of growth by cell division is over (30–35 days from the fall of the petals, or until early June under English conditions; 21 days from full bloom under Australian conditions), the accumulation of alcohol soluble nitrogen compounds increases, until it remains steady at about 50% of the total nitrogen, while the fruit expands (224) (cf. Fig. 32). During this period the respiration of the fruit per unit of protein nitrogen is approximately constant (226, 421), and it was also shown that the protein nitrogen per cell increases throughout the phase of cell enlargement in such a manner that the

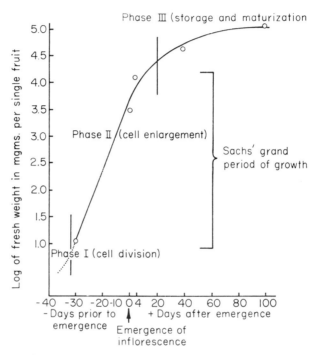

Fig. 33. Growth curve of single 'Gros Michel' fruit. After Steward *et al.* (486).

protein per unit cell surface tends to remain constant. At the climacteric, a crisis in the fruit development is passed, marked by a temporary increase in respiration, and this is accompanied by some increase of protein, seemingly at the expense of asparagine (225, 250). However, beyond this point breakdown of protein is irreversible and the metabolism of soluble nitrogen compounds ensues in a system in which the protoplasmic organization disintegrates progressively.

b. Nitrogen metabolism in the growth and development of the fruit of the banana. The parthenocarpic development of the banana fruit represents a remarkable sequence in development [Steward *et al.* (486, 488); Freiberg and Steward (169); Ram *et al.* (408)]. The vegetative shoot apex is enclosed at about ground level within a hollow sheath, or

pseudostem, formed by the rolled bases of the leaves which this growing point has produced. When the floral transition occurs this growing
point changes and now produces rapidly a series of bracts with axillary
growing regions (previously absent), which form the primordia of
the flowers [Barker and Steward (21)]. These develop as the tip of the
inflorescence grows up the pseudostem until it finally emerges. About

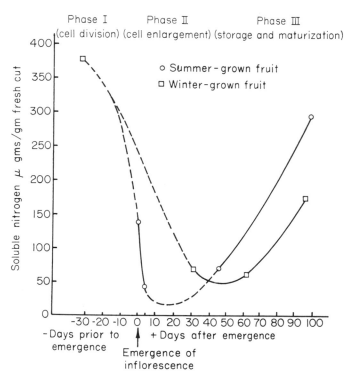

Fig. 34. Change in total soluble nitrogen of 'Gros Michel' fruit during development. After Steward *et al.* (486).

30 days prior to this emergence the whole inflorescence is small (only
a few centimeters in length), but as it moves up the stem individual
"hands" (or rows of young fruits) and "fingers" (individual fruits)
can be seen. When the inflorescence emerges from the pseudostem the
individual fingers in the cultivar 'Gros Michel' may be almost 2–6 gm
in weight, whereas at maturity they may reach some 100 days later a
weight of about 180 gm per fruit and, of course, a mature "stem" of
bananas may weigh 100 pounds or 45 kgm.

Two figures (Figs. 34 and 35) describe the growth of the fruit in time and the changes in the total soluble nitrogen which it may contain. The essential formative growth is over many days prior to emergence, when the individual fruits are still small. Virtually all the cell divisions which form the parts of the fruit are completed by this time, a long phase of growth predominantly by cell enlargement ensues, and thereafter a period of maturation and storage of soluble reserves. The

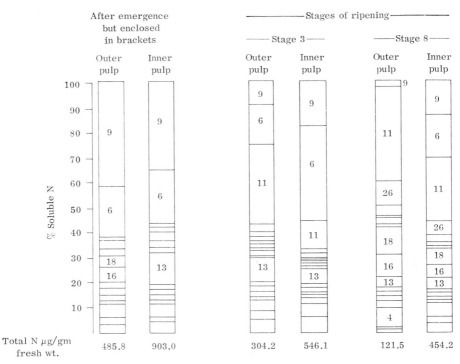

Fig. 35. Alcohol-soluble nitrogen of outer pulp and inner core of banana (cultivar 'Gros Michel'). for numerical key, see Fig. 25; other compounds are identified in the original source from which these data are taken, namely R. A. Barr, The nitrogen compounds of the banana plant: A comprehensive study. Ph.D. thesis, Cornell Univ., Ithaca, New York, 1963.

post harvest ripening and senescence of the fruit and the climacteric represent a special and still further problem.

The figures show that the very young inflorescence (at 30 days prior to emergence) is richly endowed by the leafy shoot and rhizome with soluble nitrogen compounds which are necessary for its growth (nearly 400 μg nitrogen per gram fresh weight). However, the inflorescence and young fruits grow faster than the organic nitrogen reserve can be mobilized, and the soluble nitrogen therefore falls to a very low level (of the order of 25 μg per gram fresh weight) when the fruit has about completed its fastest rate of growth in size [(Sachs' Grand Period of Growth) i.e., at about 20 days after emergence]. Therefore, growth slows down, and mobilization of soluble nitrogen in excess of its immediate requirement occurs, so that storage ensues. The time scale of this development is somewhat different at different seasons of the year since the fruits that develop in the winter do so more slowly (Table XVI). Even at the time of emergence, when the content of soluble

TABLE XVI

SEASONAL EFFECTS IN THE PROPORTION OF ASPARAGINE, GLUTAMINE, AND HISTIDINE PRESENT IN ALCOHOL-SOLUBLE NITROGEN OF BANANA 'GROS MICHEL' FRUIT AT HARVEST[a]

Condition	Summer-grown fruit	Intermediate		Winter-grown fruit
Time of harvest	July	October		December
Days from emergence	95	—	—	115
Amino acid[b]				
Asparagine	35.1	32.1	27.4	26.8
Glutamine	25.6	27.7	22.5	7.4
Total amides	60.7	59.8	49.9	34.2
Histidine	8.2	15.4	21.4	35.0
Amides + histidine	68.9	75.2	71.3	69.2

[a] After Steward et al. (486).
[b] Values given as nitrogen of each amino acid as a percentage of total soluble nitrogen determined.

nitrogen is still declining, it is made up of the usual compounds that seem typical of actively metabolizing cells, with a predominance of amide nitrogen (mostly glutamine), alanine, and γ-aminobutyric acid along with the two dicarboxylic acids. When the soluble nitrogen content reaches its lowest point it is also poorest in its content of glutamine nitrogen, but, as nitrogenous reserves are now being stored in excess of their immediate requirement for growth, the relative composition of the soluble pool which accumulates is quite different from that which prevailed earlier. Asparagine predominates over glutamine, and histidine, not commonly present in quantity in the free state in plants, soon becomes conspicuous. Moreover, the metabolism of the ovary of pistillate and staminate flowers is different (Fig. 36). As would be expected the soluble nitrogen content of the mature pulp is greater than that of the peel. Although the absolute amount of free histidine per unit weight of tissue was greatest in the inner region of the fruit (composed of septa and placental tissue), it was actually a more conspicuous fraction of the total soluble nitrogen of the outer region of pulp (i.e., ovary wall). These trends, which occur during the development of the fruit, become even more significant when they are continued into ripening and senescence after the climacteric of the

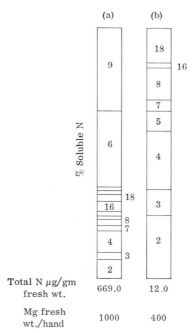

Fig. 36. Alcohol-soluble nitrogen of (a) ovaries (divested of peel) in the female and (b) vestigial ovaries of male flowers of banana (cultivar 'Gros Michel'). For numerical key, see Fig. 25; other compounds are also identified in the original source from which these data are taken, namely R. A. Barr, The nitrogen compounds of the banana plant: A comprehensive study. Ph.D. thesis, Cornell Univ., Ithaca, New York, 1963.

fruit has been passed. Subsequently there is a marked transfer of nitrogen from the previously stored amide nitrogen (glutamine and asparagine) into histidine nitrogen. Apparently this unexpected change occurs without any drastic change in the total amount of protein nitrogen although some slight change in its amino acid analysis does occur.

c. Nitrogen metabolism of pine cones. Distinct differences have been observed also between the female and male cones of *Pinus strobus* (Fig. 37). The larger female cones contain more total soluble nitrogen (mostly amide) than the male cones. The latter, however, are richer

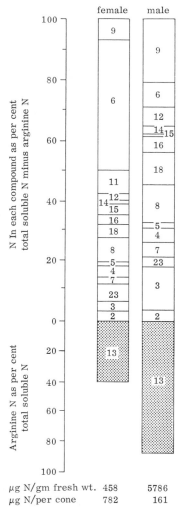

Fig. 37. The soluble nitrogen composition of the male and female cones of *Pinus strobus*. For numerical key, see Fig. 25. After Durzan, Ph.D. thesis, Cornell Univ., Ithaca, New York, 1964.

in arginine and dicarboxylic acid nitrogen and, when compared on a unit fresh weight basis, they contain much more soluble nitrogen than do the female cones. As Fig. 37 shows, the female cone emphasizes asparagine and arginine in its soluble nitrogen whereas in the male cone the emphasis is more upon arginine than amide nitrogen, with glutamine predominating over asparagine.

6. The Regulatory Control of Nitrogen Metabolism during Ontogeny

The reasons for all these diverse changes in nitrogen metabolism during growth and development are obscure for they can be presented only as part of the sequence of ontogenetic events. These examples, of which many more could now be cited, serve, however, to pose one of the outstanding problems of growth and chemical differentation, namely, how are the enzymes and biochemical potentialities that are presumably present in all the diploid cells so controlled as to produce the observed results? Why is it that different but even morphologically homologous organs can be characterized by such different biochemistry? This cannot merely be an expression of the gene-enzyme relationships that serve to explain how genetic constitution can control and transmit metabolic properties. Such genetic ideas explain only what biochemical events it is *feasible* for an organism to carry out; they do not tell us the machinery that determines how far the organism will take advantage of its innate biochemical powers. For this we must look to some extragenetic or "epigenetic" machinery which determines the extent to which given reactions will actually proceed, the extent to which some metabolites will accumulate as subsequent stages in the sequence are slowed down. The underlying mechanism which controls this in terms of biochemical differentiation as seen through nitrogen metabolism must be a part of the larger problem that determines whether leafy organs become true leaves or bracts or modified floral parts or even, by fusion, part of the fruit. Presumably all the cells have the same innate, genetically controlled, potentialities, but in these different internal environments in which they develop they use this information to a quite different extent. Some mechanism "turns on or off" or "modulates" the genes that determine specific biochemical events.

Organs which exhibit a climacteric have a crisis in their development that will be expected to have repercussions upon nitrogen metabolism. A very plausible view is that the climacteric inactivates that part of the nitrogen metabolism which is concerned with the resynthesis of protein from soluble reserves. (The cells of potato tubers stored at $+1°C$ for some time certainly undergo this change. After the climacteric, banana cells will not respond by growth to applied growth-promoting substances as if synthesis is no longer able to be activated.) This being so, an organ in nitrogen balance prior to the climacteric may undergo a change, with accumulation of some products of protein breakdown, after the climacteric. Because this problem is intimately bound up with protein synthesis and breakdown it is referred to again in Part III, Section IX of this chapter.

There is, however, another, almost converse, topic which concerns renewed activity on the part of organs that have become mature, if not senescent. Mothes (342), Mothes and Engelbrecht (343), and Engelbrecht (145) have shown that the local application of such exogenous (even "unnatural") growth factors as kinetin, will cause C^{14}-labeled nitrogenous substrates (e.g., glutamic acid, methionine, glycine, lysine) that are applied elsewhere to the leaf to be mobilized, against the concentration gradient, at the center of kinetin stimulation (Fig. 38). This

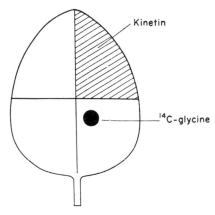

Kinetin

^{14}C-glycine

FIG. 38. Radioautograph of a *Nicotiana rustica* leaf. Application of kinetin to the upper right quadrant and of C^{14}-glycine to the lower right quadrant of the blade surface. After Mothes (342).

type of experiment, like those described with tissue cultures, shows that localized application of growth factors may regulate redistribution and metabolism of the nitrogenous solutes in the cells.

C. SOME EFFECTS OF NUTRITION AND ENVIRONMENT

Any factor that controls growth and development in plants may be expected to have repercussions upon their nitrogen metabolism and hence upon the total amount and relative composition of the stored soluble nitrogen compounds that they contain. As already indicated, most growing regions are so actively turning soluble nitrogen into protein that they do not store soluble nitrogen compounds in their cell vacuoles to high concentrations. However, when the processes of growth subside as in fleshy, resting storage organs, although nitrogenous nutrition continues, the soluble nitrogen content reaches a high level. Conversely, when the resting cells such as those of the potato tuber and the carrot root, or the artichoke tuber, again become highly meristematic through the stimulus of wound healing and the exposure to moist air, or still more due to the growth factors which cause them to proliferate, then the high content of stored soluble nitrogen is reduced to a much lower level (Table XVII). Therefore, the complement of soluble nitrogen to be found in plants will be a function of their previous nutrition and exposure to specified environmental conditions.

TABLE XVII

Comparison of the Nitrogenous Composition of the Alcohol-Soluble Fraction of Mature, Nongrowing, and Tissue-Cultured Plant Tissues[a,b]

	Potato tissue		Carrot tissue	
Amino acid	NonGrowing	Cultured	NonGrowing	Cultured
Aspartic acid	11.22	1.44	25.9	1.2
Glutamic acid	16.95	12.73	42.0	4.5
Serine	8.81	7.28	17.0	1.5
Glycine	5.15	8.00	1.7	0.4
Asparagine	291.5	—	153.3	0.0
Threonine	11.83	4.50	11.8	0.7
Alanine	20.75	21.50	93.0	10.4
Glutamine	579.0	15.72	77.8	0.0
Lysine	12.1	2.62	0.5	0.0
Arginine	114.5	15.46	35.2	0.0
Methionine	7.74	0.0	2.61	0.0
Proline	0.0	0.0	4.2	0.1
Valine	29.1	6.31	8.6	0.9
Leucines	11.15	8.05	7.8	0.7
Phenylalanine	11.71	6.18	13.3	0.0
Tyrosine	9.38	3.66	1.6	0.0
γ-Aminobutyric acid	40.75	88.3	18.4	2.3
Total	1187.0	201.0	515.7	22.7

[a] Data from Steward and Thompson (507).
[b] Values given as micrograms of nitrogen per amino acid per gram fresh weight.

A special case is encountered when plants grow and develop with a *minimal* supply of nitrogen for they then store sugars and organic acids in excess, as shown by the "low salt" barley plants used by Hoagland (214) to study ion uptake (cf. 503). These and other similar questions will not be examined in detail here.

Prior to the use of chromatography, knowledge of the nitrogenous economy of the plant body was necessarily limited to broad classes of substances. Since the intake of inorganic nitrogen was via the roots, a predominantly upward movement of nitrogen to the shoot was recognized and, since the green leaf in the light was a prime organ of protein synthesis, a net intake of nitrogen during the day was evident. Chibnall (104, 105) spoke about the intake of nitrogen to *Phaseolus* leaves by day, and the well known experiments of Maskell and Mason (299, 300) (also see Chapter 5, Volume II) on the cotton (*Gossypium*) plant indicate the general state of knowledge at that time. Whereas Maskell and Mason had been able to correlate carbohydrate movements, predominantly downward, with positive gradients of soluble carbohydrate in the stem and petiole regions (299–303), they were unable to achieve a similarly satisfying result for the nitrogen compounds. Since nitrogen movements were predominantly into the shoot and into the leaf, most of their observed longitudinal gradients of concentration were in the opposite direction (299–303), no doubt because they reflected the gradations in the accumulation of soluble nitrogen compounds in parenchyma cells. Concentrations of solutes accumulated in vacuoles are commonly greater near the apex and decrease basipetally downward. The only category of nitrogen compounds in the cotton plant that seemed to yield concentration gradients that were positively correlated with the observed *upward* movements into the shoot was the so-called "residual-nitrogen"; this represented an indeterminate category of nitrogen, namely, the nitrogen not otherwise determined as amide, amino, or ammonia nitrogen. Since then the role of allantoin and allantoic acid in the xylem of some plants has become more conspicuous (54, 342, 413) and, from Benson's group, Maizel *et al.* (290) have even found that choline as its phosphate is important as a form in which phosphate (and no doubt also nitrogen) is mobile within the plant body. Choline sulfate in the roots of sulfur-deficient maize also suggests its function as a major sulfur reserve (364).

Germinating seeds in the dark, unable to convert their soluble nitrogen to protein as in leaves in the light, accumulate much soluble nitrogen whereas excised and senescent leaves break down protein; all this suggests that light will be implicated in the nitrogen metabolism of growing leaves. Light energy is not, however, directly involved in protein synthesis in the same sense that it is directly involved in photosynthesis, although much carbon from $C^{14}O_2$ does enter rapidly into protein in *Chlorella* (32). The involvement of light in protein synthesis is rather because the green leaf, furnished with the carbon intermediates of photosynthesis and either nitrate or its reduction products, is an effective general milieu for protein synthesis, albeit mainly for the synthesis of chloroplast protein. The newer interpretations of photosynthesis (88, 512, 553), show that the production of ATP by photosynthetic phosphorylation and the formation of reduced pyridine nucleotides (nicotinamide adenine dinucleotides) may permit the energy of ATP and the reducing effect of TPNH (NADPH) to be harnessed to nitrogen metabolism of leaves in the light.

However, one should first describe how light and mineral nutrients affect the soluble nitrogen compounds which occur in green leaves.

1. The Effects of Mineral Nutrients

This topic is to a large extent discussed in Volume III, to which reference may be made (cf. Chapters 3 and 4).

The mineral nutrient condition which is most immediately relevant to nitrogen metabolism was in point of fact, discovered quite late. The

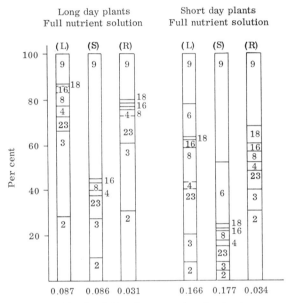

FIG. 39. Relative composition of soluble nitrogen of leaves (L), stems (S), and roots (R) of *Mentha piperita* L. grown under long and under short days in the greenhouse (70°F). Nitrogen of each amino acid is expressed as a percentage of total soluble nitrogen determined by reaction with ninhydrin. For numerical key, see Fig. 25. After Steward *et al.* (487).

conversion of nitrate to its reduced and organic forms presupposes the role of molybdenum, though this may possibly be replaceable in part by vanadium. Thus, if the supply of molybdenum is adequate, the soluble nitrogen complement of green plants in the light should be unlimited by nitrate reduction. Even if molybdenum is limiting, the synthesis of soluble nitrogen compounds may still be feasible if ammonia is the source of nitrogen (cf. Hewitt, Volume III, Chapter 2).

Assuming that reduced nitrogen, either by nitrate reduction or the supply of ammonia, is available and that the carbohydrate is plentifully furnished via photosynthesis or in the culture medium, then the accumulation of soluble nitrogen compounds will be determined by the extent of protein synthesis on the one hand and the synthesis of the soluble compounds in the cells on the other. Almost any mineral deficiency which becomes acute enough to limit growth will retard protein synthesis, so it is a frequent feature that minerally deficient plants accumulate soluble nitrogen compounds, often to very high levels, at

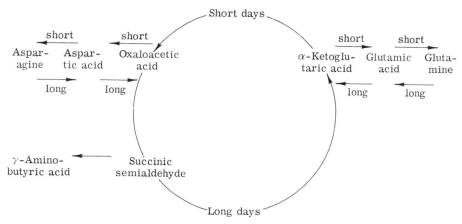

FIG. 40. Diagram of the metabolism of *Mentha piperita* L. grown under long, as contrasted with short, days. After Steward *et al.* (487).

least until the cells succumb to morbid changes which reduce their ability to retain the accumulated compounds.

Lack of sulfur, an essential but small ingredient of protein, has long been known to be associated with the accumulation of soluble nitrogen. This was found in the early work of Eaton (141), but has been verified for other plants: in this laboratory notably for mint, tobacco plants, and pine seedlings.

In mint (*Mentha piperita*) plants grown under long days, the accumulated soluble nitrogen in the leaves takes the form of the nitrogen-rich compounds arginine and glutamine, though under short days there is a much greater amount in the form of asparagine than under long days (485) (cf. Figs. 39 and 40). In fact, when sulfur is resupplied to sulfur-deficient mint plants, so that the plants can recover,

the following events occur. The total soluble nitrogen falls by its conversion to protein and, as soluble nitrogen is metabolized, glutamine
accumulates under long-day conditions and asparagine under short-day
conditions (see Chapter 2, Volume III, p. 316, Fig. 63).

2. Interactions of Light, Mineral Nutrients, and Nitrogen Metabolism

At this point another idea is relevant. The contrasted effects of light
and darkness and the resultant diurnal fluctuations in plants lead to
frequent evidence that light tends to favor glutamine more than asparagine, whereas darkness (or short-day conditions which emphasize long
nights) tends to favor asparagine. In addition to the above example
cited from the study of sulfur deficiency in mint plants, this same effect
has been encountered through the study of diurnal fluctuations in the
leaves of mint plants (485, 487). It was found that the level of asparagine tended to go up at night and down by day, whereas the tendency
of glutamine was to rise by day and fall by night (cf. Fig. 29). These
effects are associated with the known synthesis of protein by day and
its tendency to break down by night. Thus the glutamine was more
associated with conditions conducive to synthesis and the asparagine
with conditions which lead to protein breakdown. The underlying and
more direct cause may turn out to be some consequence of a ready
supply of ATP (and reduced pyridine nucleotides) by day and in the
light and the converse by night. A similar effect has also been noted in
the conversion of C^{14}-proline to other metabolites in tobacco leaves, for
C^{14}-proline readily gave rise to glutamine which was accumulated in
the light (Fig. 41A), and to asparagine which accumulated in the dark
(Fig. 41B).

Therefore, any interpretation of the soluble nitrogen complement of
green leaves must comprehend the interactions between the mineral
nutrients supplied and the conditions of light or darkness under which
the plants were grown. The histograms which show the relative composition of the leaves of mint plants, as they are grown under different
conditions of acute mineral deficiency, show that this composition is
also a function of growth under long or short days. These data were
quoted as Fig. 62 by Hewitt (Chapter 2, Volume III, p. 314) and have
since been published in extenso (487); therefore they need not be recapitulated here. One may, however, refer to the tendency of potassium-deficient mint plants under long-day conditions to accumulate
glutamine and their greater tendency to accumulate asparagine in their
leaves under short-day conditions. Thus the relative composition of the
accumulated soluble nitrogen of the leaves of the plants is a diagnostic
criterion of the mineral deficiency in question, and the distinctive

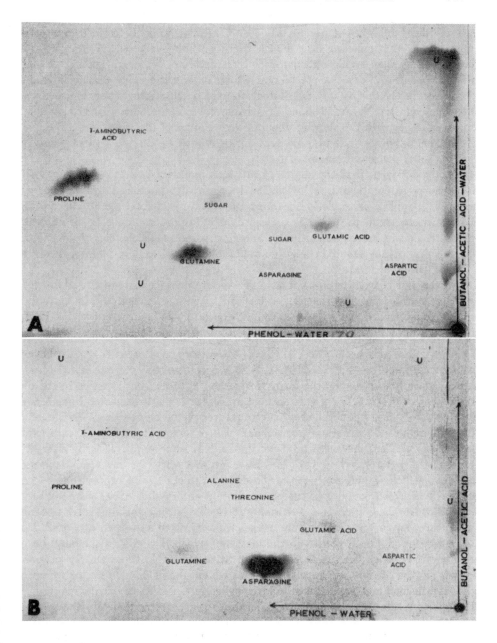

FIG. 41. Radioautographs of the chromatograms of the alcohol-soluble fraction of tobacco leaf disks. C^{14}-proline was supplied for 72 hours (A) in continuous light and (B) in the dark. U designates a radioactive but and unidentified spot. After Steward Bidwell (481) from work with J. K. Pollard.

metabolic responses to the particular mineral nutrient need to be described for both long- and short-day conditions.

3. Nitrogen Metabolism and Nutrition of Conifers

The interaction of mineral nutrition and nitrogen metabolism has been investigated in two species of conifers (Durzan, unpublished data obtained in this laboratory). Jack pine (*Pinus banksiana*) is a very shade intolerant species, utilizes ammonia better than nitrate for growth, and stores more of its soluble nitrogen as arginine from a nitrate than an ammonium source. White spruce (*Picea glauca*) is more tolerant of shade, grows better on nitrate, but stores a higher percentage of its soluble nitrogen as arginine from an ammonia source. The heavy accumulation of arginine nitrogen in these plants reflects their inability to use the soluble-nitrogen they contain in growth and protein synthesis. Figures 42A and B show the analysis of the soluble nitrogen of leaves of these two species as affected by levels of mineral nutrition. Clearly the size and relative composition of the pool of soluble nitrogen compounds is extremely responsive to the mineral nutrition of the trees, especially where this limits (e.g., by low phosphorus or potassium) the use in growth and protein synthesis.

4. Effects Attributable to Night Temperature

These interactions, however, become even more complex when one also takes into account the specific effects of night temperature. It is to be expected that more carbohydrate is consumed by respiration at high temperature, and protein breakdown should be faster. The effects of high night temperatures on the soluble nitrogen compounds have been surprisingly large in the few cases examined. The notable case is again that of the mint plant. This plant behaves, morphogenetically, more nearly like a short-day plant, when grown under long-day conditions and at high night temperatures. The night temperature is seen also to effect the soluble nitrogen compounds which are stored in the leaves (Table XVIII). To quote Steward *et al.* [(487), Part VII, p. 132]: "Under any specified photoperiod the level of soluble nitrogen in both leaves and stems is much greater in plants grown at low night temperature than in comparable organs of plants grown at high night temperature. Most of this extra soluble nitrogen at low night temperature is in the form of amide. Similar, but not as dramatic, results were obtained after only 3 days . . .". The original paper (cf. Tables 2 and 3) showed that the characteristic effect due to night temperature can be seen after as little as 1–6 diurnal cycles.

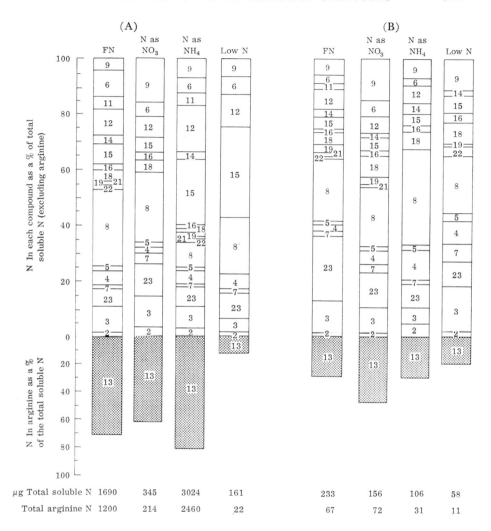

Fig. 42. Relative nitrogen composition of (A) *Picea glauca*, a shade-tolerant conifer, and of (B) *Pinus banksiana*, a very shade-intolerant conifer. For numerical key,
see Fig. 25. F.N. = full nutrient; N as NO_3 means nitrate as sole source of nitrogen.
N as NH_4 means ammonium as sole source of nitrogen. Low N means low level
nitrogen supply. After Durzan, Nitrogen metabolism of *Picea glauca* (Moench)
Voss and *Pinus banksiana* L. with special reference to nutrition and environment.
Ph.D. thesis, Cornell Univ., Ithaca, New York, 1964.

TABLE XVIII

Amino Acids and Amides of *Mentha piperita* Grown 2 Weeks
under Different Photoperiods and Night Temperatures[a,b]

	Short days (8 hours)		Long days (15 hours)	
	50°F	80°F	50°F	80°F
Aspartic acid	52	101	64	54
Glutamic acid	396	167	222	150
γ-Aminobutyric acid	47	19	85	42
Serine	16	tr	—	10
Glycine	11	7	29	—
Threonine	—	—	—	—
Alanine	55	38	44	56
Valine	—	9	—	—
Asparagine	119	tr	82	—
Glutamine	732	152	595	77

[a] After Steward *et al.* (487).

[b] Values given as micrograms of amino acid per gram fresh weight of tissue.

D. Some General Interpretations of the Response of Nitrogen Metabolism to Environmental Factors

1. General Considerations

Is any general synthesis of the environmentally induced effects possible? The following are the considerations against which this problem may be approached:

(i) Nitrate supply usually tends to be conducive to growth and protein formation under optimal nutritional balance; ammonium supply tends to cause soluble nitrogen to accumulate with the consequential need to "detoxify" and convert it to soluble organic forms such as amides and arginine.

(ii) Light and young leaves are conducive to protein synthesis whereas darkness and senescent leaves, especially detached leaves of monocotyledons, are conducive to protein breakdown.

(iii) In the diurnal fluctuations of soluble nitrogen compounds in leaves, there are now several examples in which asparagine is the favored compound by night and glutamine by day although this may not be generally so.

(iv) The overriding factor that determines the formation of soluble nitrogen compounds is accessibility to the carbon acceptors of suitably reduced forms of nitrogen. This may, within limitations, be furnished by ammonia, but it is preferentially supplied from nitrate; thus all conversion to soluble nitrogen compounds from nitrate presupposes a mineral nutrition that is not limited by molybdenum which is necessary (although perhaps partially replaceable by vanadium) for nitrate reduction.

(v) Granted carbohydrate acceptors (keto acids) from the carbohydrates photosynthesized by the leaves, and reduced nitrogen supplied from the roots, etc., the accumulation, or not, of soluble nitrogen compounds will be largely determined by the incidence of protein synthesis. Actively growing regions like growing points, or rapidly proliferating tissue cultures, store only minimal amounts of soluble nitrogen while their more slowly growing. mature or even senescent, counterparts store much more. However, the difference between the nitrogen compounds in growing meristematic cells and storage tissues is not only one of amount, for these nitrogenous moieties differ greatly in their respective contents of such substances as histidine and the amides. A striking case of this has been described for the banana fruit (refs. 86, 488) (cf. Fig. 34).

(vi) Of the elements, other than nitrogen, carbon, hydrogen, and oxygen, which are directly implicated in protein formation, sulfur and phosphorus impinge directly upon nitrogen metabolism. The effects of sulfur are immediate and great. Inability to keep pace by methionine and cysteine formation (which rarely accumulate free in large amount if other nutritional factors and growth conditions are optimal) causes protein formation to be rapidly limited by lack of sulfur. It is under these conditions that other nitrogen-rich reserves accumulate, especially arginine and the amides; when sulfur is resupplied, these compounds are again rapidly metabolized. One may note also that, according to Wood and collaborators (26, 201, 618), the sulfur amino acids of leaves of grasses play a regulatory role in the course of nitrogen metabolism. Apparently the chloroplasts contain 70% of the sulfur in leaves (201).

(vii) There is much evidence that potassium and calcium are implicated in the utilization of the soluble nitrogen compounds of plants, or the effective use of exogenous nitrogenous supplies. Lacking individual and distinctive reasons for the essentiality of potassium, one can nevertheless point to many of its indirect roles; one of the most important of these, which is historically well recognized, is its prominent presence in the growing tips of plants. This was well known even in the mid-nineteenth century, but was also proved by Penston (1931, 1935, 1938). The prevalence of potassium in growing regions is compatible with its obvious involvement in growth and protein synthesis. Both in the metabolism of potato disks and of mint leaves the external conditions of potassium supply are found to favor protein synthesis, and, since calcium often tends to have the reverse effect (possibly because of its contrasted effect on phosphorylation), the relative potassium-calcium nutritional levels prove to be important in the regulation of nitrogen metabolism.

(viii) Against this background almost every mineral nutrient that causes recognizable effects on growth, or leads to claims of the essentiality of the nutrient element in question, results in, or may be presumed to cause, changes in the soluble nitrogen pool. This is evident for an element like manganese, which is intimately concerned with the conversion of the first products of nitrate reduction to protein (490, 491), but effects of one sort or another can be attributed to all the trace elements.

Incidentally, the best way to find the point of impact of the nutrient elements on the nitrogen metabolism is not merely to describe the nitrogen compounds which accumulate when the element in question is grossly deficient. A better method is to establish the symptoms of deficiency, at a reversible level, and then study the changes that ensue with time when the deficient element is re-supplied. It is in this way that the most critically limiting steps may be discerned. For example, the major effects of manganese in tomato leaves were found by Margolis (491) to be:

"(a) Manganese deficiency in the presence of either nitrate or ammonium-nitrogen produced an accumulation of soluble nitrogen compounds which consisted mainly of the amides, the accumulation from ammonium being much greater than from nitrate. (b) A rapid decrease in the level of soluble nitrogen compounds (especially the amides) occurred when manganese was restored to plants which had become manganese deficient while they received adequate nitrate or ammonia. (c) Restoration of manganese and nitrate to plants rendered deficient in both of these nutrients resulted in a rapid formation of soluble nitrogen compounds (mainly the dicarboxylic amino acids and glutamine). This synthesis was most rapid when both manganese and nitrate were furnished. (d) The effects attributable to manganese are, however, complex; they cannot be interpreted in terms of a single and invariable effect, for they interact with factors of environment, nutrition, and development. (e) Manganese seems not to be directly implicated in the reduction of nitrate, nor in its immediate conversion to organic form, but it is conducive to the more rapid build-up of soluble nitrogen compounds, because it governs the supply of carbon acceptors for nitrogen via the Krebs cycle. (f) In addition to the effects of manganese upon the mobilization of carbon at "ports-of-entry" for nitrogen into organic combination, the main seat of the effect of manganese on nitrogen metabolism is on the conversion of soluble products to protein as manganese acts to determine growth and protein synthesis. In this role interactions with the environment are conspicuous."

A schematic representation of interacting effects on nitrogen metabolism insofar as they involve light and darkness, photoperiodicity and night temperature, potassium and calcium on the behavior of mint plants, may be shown below by quoting a series of interpretative diagrams from recent summaries at different periods [(487); Steward *et al.* (485); Fig. 43].

To the evidence already given one may now add the further points that light may be presumed to furnish ATP through photosynthetic phosphorylation (cf. Volume 1B, pp. 212 et seq.), and this could have the same effects as if it resulted from potassium-stimulated oxidative phosphorylation in nongreen cells. Also light may be expected to intervene through furnishing reduced pyridine nucleotides, which are known to be implicated in so many biochemical reactions including those of carboxylation and decarboxylation.

The case, therefore, is overwhelming for the participation in, and the interactions between, a variety of nutritional and environmental factors that determine the balance of soluble and protein nitrogen in plants and the relative composition of the pool of soluble compounds that will be stored under any given circumstances. The problem is to detect in so multifactor and heterogeneous a system a particular step by which it may be limited at a particular time. In fact, the effective control may well be dispersed among the interacting networks of

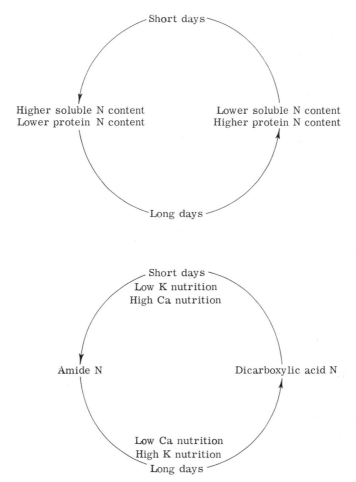

FIG. 43. Schematic representation of interacting effects on nitrogen metabolism. After Steward *et al.* (487).

factors that impinge upon the soluble nitrogen of plants and that are determined by the nutritional and environmental conditions. This is, therefore, a much more complex situation than it seemed necessary to visualize so long as nitrogen metabolism was only the very direct conversion of inorganic nitrogen to amino acids and thence to protein.

2. Role of Keto Acids and the Tricarboxylic Acid Cycle

The ideas that flowed originally from Chibnall and Vickery regarding the keto acids as ports of entry (pyruvic acid leading to alanine; oxaloacetic acid leading to aspartic acid, and α-ketoglutaric acid leading to glutamic acid) still stand with the following additions. Other keto acids, such as glyoxylic acid (leading to glycine) and succinic semi-aldehyde (leading to γ-aminobutyric acid) as well as the keto analog of almost every known amino acid (see Appendix III) are almost certainly involved. However, the roles of photoperiodicity and of night temperature, respectively, in regulating the keto acids that occur in leaves, are suggested by the work that has been done on the mint plant (487).

No doubt all these effects operate indirectly through the consequences of light as a source of phosphorylated compounds (ATP) and of reduced pyridine nucleotides as a source of reducing power, and they are no doubt affected by nutritional factors that operate in ways not yet fully known.

3. The Regulatory Role of Carbon Dioxide

In the diverse metabolic interactions that link carbohydrate and nitrogen metabolism in plants, it is often seen that carbon dioxide may play a regulatory role. This could operate directly through the need for carbon dioxide to combine with the C_3 or C_2 products of glycolysis and so permit them to build up and maintain the level of acids at the complexity of citric-isocitric acids, which are then decarboxylated and dehydrogenated in the tricarboxylic acid cycle. The Krebs cycle, as a line of synthesis rather than a source of energy, should then operate very differently according to the carbon dioxide tension or the bicarbonate supply. It is of interest here that alternative paths of development of certain water molds (*Blastocladiales*) have been found to be "triggered off" by the presence or absence of external bicarbonate, and Cantino (91) and Lovett and Cantino (282) have made a biochemical interpretation of this effect along the following lines. Cantino believes (Fig. 44) that the presence of carbon dioxide could be conducive to the carboxylation of α-ketoglutaric acid, rebuilding isocitric, so that this compound could give glyoxylic and succinic acids. The glyoxalate so produced could then transaminate with alanine to form glycine and thus lead on to consequences upon the nitrogen metabolism which in turn determine the morphogenetic development. The essential point, however, is that this is another example in which carbon dioxide seems strategically placed to play a determining metabolic role which involves nitrogen metabolism, and thus it may exert a profound effect on growth and development. Another example is that of potato tuber disks in well-aerated solutions (see p. 588). The rapid conversion of soluble nitrogen compounds to protein in the potato disks, which is stimulated by low CO_2 tensions and by oxygen saturation, is completely suppressed if CO_2/HCO_3^- is introduced into the solutions; this has been interpreted as an effect mediated by suppressing decarboxylations in the absence of free carbon dioxide and bicarbonate and by fostering carboxylations in their presence.

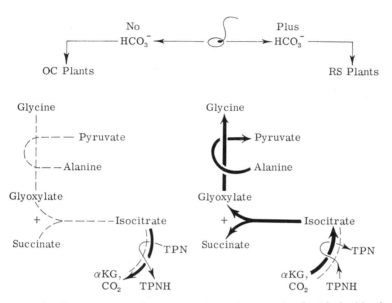

FIG. 44. The shift in a critical sequence of reactions associated with the bicarbonate trigger mechanism for morphogenesis, showing how bicarbonate alters the mechanism of isocitrate. After Cantino (91).

4. Asparagine and C_4 Compounds: Glutamine and C_5 Compounds

This point in metabolism at which nutritional and environmental effects on nitrogen metabolism are mediated is where, at the level of keto acids, it interlocks with carbohydrate metabolism. The factors in question seem to determine whether synthesis shall stem from 5-carbon compounds like α-ketoglutarate, glutamic acid, and glutamine, or from the 4-carbon compounds like asparagine and its immediate precursors. This suggests that carboxylations are fostered by one set of conditions (that emphasize glutamine and the C_5 compounds) and that decarboxylations are fostered by others that lead to asparagine and the C_4 compounds. Actual determinations of the keto acids in mint leaves [Steward *et al.* (487): see work of Rabson] lend support to these ideas, for the α-ketoglutarate content of mint leaves undoubtedly responds to effects of environment in ways that are compatible with its suggested role.

The interpretation is somewhat limited by precise knowledge of how asparagine is made *in vivo* (also see Part 1, III, 2). The appropriate ATP-mediated amidating enzyme to convert glutamic acid, from α-ketoglutaric acid, to glutamine has been well known since the work of Speck (465, 466) and Elliot and Gale (144), and this enzyme functions under physiological conditions. All attempts to demonstrate with equal ease the conversion of oxaloacetate to asparagine have failed. Indeed, in mint plants oxaloacetate is the one keto acid which is difficult to demonstrate under conditions appropriate to its role as an amino acid precursor. In fact, the biochemistry of asparagine is so different from that of glutamine as still to raise questions whether *in its physiological reactions* it should be regarded as an exact homolog of glutamine or whether it may assume a configuration different enough to explain its divergent properties. [This problem still remains because the precise suggestion made by Steward and Thompson (506a) to resolve the dilemma has not been substantiated.]

5. Role of Cellular Organization: Anabolism and Catabolism

In order to interpret, in all their complexity, the situations referred to above, much more needs to be known about the interlocking systems of anabolism and catabolism that represent the metabolic machinery of intact cells and of whole plants.

Beverton and Holt (38) have seen merit in von Bertallanfy's concept that metabolic processes are determined between the active mass which undergoes catabolism and the necessary surface to support anabolism. The anabolism or synthesis is expressed as the product of the rate at which mass is created per unit area (M) and surface (S), where S is the physiologically effective surface through which exchange with the environment takes place. Thus $dw/dt = MS - KW$. M is a coefficient of anabolism, anabolism being proportional to the surface area through which essential exchanges take place; and K is a coefficient of catabolism, catabolism being proportional to the entire bulk (W) of living material. This concept now needs to be given meaning in terms of the conversion of soluble nitrogen to protein in anabolism and its breakdown in catabolism.

Work with purified enzyme systems, even with isolated particulate preparations, lacks the essential features which the whole cellular organization presents. An interesting example is the contrast between the incorporation of C^{14}-proline into the protein of intact cells and of particulate preparations. Whereas both systems will perform the incorporation, only the intact cell converts the incorporated proline into hydroxyproline (521).

In general, one may also say that green plants in the light could so reduce the effective carbon dioxide content of their cells that they might divert the products of glycolysis, by lack of carbon dioxide from the Krebs cycle and so "feed" it, as it were, from the other end by keto acids released when nitrogen from nitrogen-rich reserves forms protein with extra carbon from sugar, or directly from the intermediates of photosynthesis. Such compounds (keto acids) need not, therefore, be carboxylated with external carbon dioxide prior to their use in the respiratory cycle. Under cover of darkness and carbon dioxide the normal entry of C_3 or C_2 carbon fragments from glycolysis into the Krebs cycle may be resumed.

Perhaps the form of the Krebs cycle that may be most implicated is that which now admits of a greater role for glyoxylic acid. For example, the step from α-ketoglutarate to isocitric acid by TPNH (NADPH)-mediated carboxylation is now said to be possible (cf. Chaps. 5, 6, Vol. IVB), and the isocitric may then cleave to form glyoxylic acid and succinic acid. If the former (glyoxylic acid) could with a suitable donor of nitrogen, and perhaps energy, yield asparagine (for which there is no published evidence), this might solve many problems. However, in a recent paper, which is entitled, "Pathways leading to the formation of amino acids and amides in leaves," R. G. S. Bidwell describes one experiment in which glucose-UL-C^{14} was supplied to bean seedlings, and the pattern of labeling in asparagine was determined. He then writes [*Can. J. Botany* 41, 1623–1638 (1963)]:

"The pattern of labelling in glutamic acid and glutamine was as expected, indicating normal entry of glucose carbon into the Krebs cycle, and synthesis of glutamic acid and glutamine after very little cycling. The pattern in asparagine, however, was unexpected; only ten to twenty per cent of the label was present in the C_1 of this amide. Such a labelling pattern could have arisen if, for example, oxalosuccinate were split to form glyoxalate and malate or oxaloacetate [reactions of the glyoxylate cycle]. If the oxalosuccinate had been formed from citrate obtained by normal condensation of acetyl (C^{14}-labelled) with oxaloacetate (unlabelled), this would give rise on cleavage to a molecule of oxaloacetate which was labelled in C_3 and C_4, but not C_1 and C_2. Also, the glyoxalate formed (unlabelled) could unite with acetyl (radioactive) and this would also produce oxaloacetic acid labelled in C_3 and C_4. This in turn would lead to aspartate and asparagine labelled in C_3 and C_4. It does not seem possible to explain the observed labelling pattern in any other way."

All these points are mentioned here, not to introduce unsupported speculation, but to show how much interpretation is limited by precise knowledge of what really occurs *in the intact cell* under the influence of these different environmental stimuli. Suffice it to say, however, that effects due to environment and nutrition all seem to converge upon those areas of metabolism where carbohydrate and nitrogen metabolism meet. Here the keto acids, their reactions and formation, seem to be mediated by photoperiod and by night temperature and play a critically regulatory role.

Some recent, not yet fully published, results of C. P. Whittingham (private communication) relate to the similar work of Tolbert, now published (544a); they concern the link between early products of photosynthesis and nitrogen metabolism and therefore invoke effects due to light. These results are important because they involve glycolic acid, glycine, and serine as primary products that are close to the path of carbon in photosynthesis. Briefly the evidence is as follows. Whittingham has studied the path of carbon from $C^{14}O_2$, using concentrations closer to those with which green cells normally contend than has often been the case. Under these circumstances a higher percentage of the $C^{14}O_2$ than expected appears as glycolate instead of sucrose. Isonicotinyl hydrazide (INH) does not inhibit photosynthesis, but it does cause glycolate to accumulate greatly (up to 20% of $C^{14}O_2$ fixed). Low concentrations of carbon dioxide favor a glycolate, rather than sucrose pathway, and the concentration of carbon dioxide controls the amounts of sucrose and of glycine from $C^{14}O_2$ that may be obtained. Therefore glycine (from glycolic and glyoxylic acids) is as much a primary product of the photosynthetic cycle as is sucrose, for the light-activated acceptor molecule splits (at low concentrations of CO_2) to give glycolic → glyoxylic → glycine → serine → aspartate. The inhibitor (INH) blocks the glycine → serine conversion. Thus the photosynthetic cycle seems to have two exits, both of which may lead to serine; one of these is via phosphoglyceric acid (PGA) (at higher concentrations of CO_2) when the serine is only COOH-labeled; the other is via the decomposition of the acceptor (at lower concentrations of CO_2) when the serine is generally labeled. Only the latter path to serine is blocked by INH. Although these results are very recent, they are important because they bear upon the now somewhat general finding that glyoxylic acid, glycine, and serine are closely connected with photosynthesis.

E. Some Observations on Nitrogen Metabolism in Normal versus Pathological Situations

One might expect nitrogen metabolism of the host plant to be profoundly altered by the presence of a parasite, whether this results in a benign or a pathological situation. Some examples are properly discussed more fully elsewhere. The association of *Rhizobia* with their hosts is the subject of Chapter 5 in Volume III; the changed protein metabolism of the host in response to the presence and multiplication of a virus is now so basic to our understanding of protein synthesis that this topic is usually best dealt with under that heading. The contrasted metabolism of plants in health and disease is also the subject of Chapter 8 Volume IVB. Any remarks here may, therefore, be brief.

Having described the great range of effects that may be superimposed upon the nitrogen metabolism of an angiosperm by the interacting effects of environmental factors and of mineral nutrients, it is a still further complication to introduce yet another heterotrophic organism with its own metabolic requirements and its complex relationships to the host plant. Where this results in such symbiosis as exists between an alga and a fungus as in a lichen, or in the legume-nodule-forming bacteria, one can see that the green partner may mobilize the carbon frameworks (keto acids) and the fungus may play a role in mobilizing the nitrogen donors. In pathological situations, however, the pathogen may compete with the host for essential nitrogenous metabolites, it may secrete toxins or antimetabolites that divert or block the normal metabolism, and in these ways profoundly alter the composition of the soluble nitrogen pool and the balance between soluble nitrogen and protein nitrogen in the host cells. One might have thought, therefore, that a knowledge of the nitrogen metabolism of plants in health and disease could have told much about how the pathogen acts.

A recent advanced treatise on plant pathology, edited by Horsfall and Dimond (219), cites surprisingly little along these lines. The well known example of lycomarasmin, the peptide which acts as the wilt toxin in tomato plants affected by *Fusarium oxysporum* f. *lycopersici* (73, 177); an antimetabolite for methionine which acts as the toxin of wildfire of tobacco (621); the ability to reproduce symptoms of "frenching" in tobacco by application of an imbalanced excess of certain amino acids so that frenching was associated with an increase of isoleucine and other amino acids (475), all are examples in which specific nitrogen compounds seem to be associated with the symptoms of disease (cf. 219). The effect of rust on wheat [(219); cf. Vol. 1, p. 376] seems to be associated with an accumulation of certain free (especially aromatic) amino acids, and in other situations (*loc. cit.* p. 295) the parasite causes some normal nitrogenous constituents to disappear.

Two plant diseases have been examined in this laboratory, using hosts that have been extensively investigated by chromatographic methods. These are the effect of *Verticillium alboatrum* on the metabolism of the peppermint plant and the effect of *Fusarium oxysporum*

f. *cubense* on the nitrogenous constituents of the banana. Admittedly, however, one cannot from studies of healthy and diseased plants in this way immediately point to the place in the metabolism where the pathogen intervenes. This is hardly to be expected, since the effect of many factors in normal development cannot yet be attributed to specific metabolic reactions or metabolites. So far as the evidence goes, however, these two situations yielded the following results.

From a comparison of the soluble nitrogen compounds of healthy and diseased mint (*M. piperita*) and chrysanthemum (*C. morifolium*) plants one can say that the general effects of *Verticillium alboatrum* is to increase the soluble nitrogen and decrease the protein nitrogen moieties (426). In chrysanthemum the soluble compound which seemed most responsive in the diseased plants was proline, but in infected mint plants the total amino acid nitrogen increased several times, and pipecolic acid (not normally present in quantity) became conspicuous. [In fact pipecolic acid was noted by Stepka and Takahashi (477) in a diseased condition of prunes.] But quantitatively the most outstanding shift was from glutamine to asparagine in the stems and leaves of the diseased plants. Again the fungus seems to intervene at that sensitive area of metabolism where the responses due to nutrition and environment are also exerted.

The organism of Panama disease (*Fusarium oxysporum* f. *cubense*) when grown in culture (Freiberg, unpublished data from this laboratory) proves to be rich in arginine, and the "flecked" areas of leaf sheaths, normally symptomatic of affected tissue, are also seen to contain large amounts of free arginine. However, the roots and rhizomes of diseased plants that were examined contained more alanine than their healthy counterparts. Beyond stating again that the presence of the fungus alters the nitrogen metabolism of the host, little can be said at this stage.

Lioret (1957) made a study of the amino acid composition of tumors and normal tissue of *Tragopogon*. He uncovered a new free amino acid in the crown gall tissue of tobacco (*Nicotiana tabacum*), Virginia creeper (*Parthenocissus quinquefolia*), and Jerusalem artichoke (*Helianthus tuberosus*). The amino acid was identified as lysopine (Biemann *et al.*, 1960), the lysine analog of octopine, which is a characteristic component of octopus muscle. No physiological function has yet been assigned to this compound (for references see Appendix II).

From the above, one concludes as follows. The full interpretation of the normal nitrogen metabolism of plants during their growth and development, and as affected by the range of nutrient and environmental conditions to which they may be subjected, presents a difficult task of correlating the effects of many interacting variables affecting many substances and their reactions. Any normal or pathological situation which intervenes drastically enough to alter the growth or morphology of the plant will almost certainly have detectable repercussions on its nitrogen metabolism. Only in rare cases can such observed effects be singled out as the cause of disease rather than as one of its many consequences.

F. The Effect on Nitrogen Metabolism of Exogenous Substances and Stimuli to Growth

Most of the exogenous substances used as herbicidal sprays, defoliants, etc., are somewhat similar in their action to the natural hormones. Hormones were first postulated by Starling (1905). However, Fitting (1909) first used the term hormone in the botanical literature when he described the induction of growth in the styles and ovaries of emasculated tropical orchids by extracts of pollinia. Despite the difficulty in defining the term hormone, many thousands of chemicals have been synthesized and tested for effectiveness on one or more aspects of growth regulation, such as those that have been ascribed to auxins, kinins, gibberellins, etc.

Active compounds are used for the stimulation of root cuttings, prevention of tuber sprouting, induction of parthenocarpic fruit formation, prevention of preharvest fruit drop, and selective weed control; however, the herbicidal use far outweighs all others. It is interesting, to question the effect of these substances on nitrogen metabolism. Comparatively little, however, has been done on these lines.

Christiansen and Thimann (109) reported that the free amino acid content decreases markedly during the growth of stem tissue in water and falls essentially to zero with the presence of auxin. In their view, the synthesis of protein and asparagine paralleled the consumption of the amino acids. Reinhold and Powell (414) analyzed the changes brought about by indoleacetic acid (IAA) on radioactive amino acid uptake by sunflower tissue. After 2 hours, the glutamic acid uptake was increased by 12% over the control, whereas water uptake increased 10% and the use of glutamic acid in respiration increased 52%. On the other hand, Galston and Kaur (175) have implicated effects on the cytoplasmic proteins in the action of auxin.

The effects on nitrogen metabolism of the substances which, in such fluids as coconut milk (or coconut water) induce rapid growth in previously quiescent cells, have been extensively investigated.

The growth induction occurs concomitantly with a rapid decrease in free, alcohol soluble, nitrogen compounds, and there is a simultaneous increase of protein. This has been investigated by the use of aseptic cultures of carrot root, potato tuber, artichoke tuber, etc. The agent which induces growth in explants from these organs is either whole coconut milk (at 5–10% by volume in the medium) or this supplemented by one of many possible synergistic substances. The first such synergist to be discovered was 2,4-D at almost 5 ppm, which stimulated growth in the potato tuber (92, 483).

The decreased soluble nitrogen of the actively growing tissue has, however, a different relative composition from that of the quiescent organ. Notably, amides (especially asparagine) decline and disappear. Often glutamine persists and alanine and γ-aminobutyric acid become relatively more conspicuous. In a case like the artichoke tuber, which is rich in arginine, this also may lead to the disappearance of arginine, which may be replaced by proline. However, the new protein synthesized in these rapidly proliferating cells is richer than the normal in a hydroxyproline-rich protein moiety than is the quiescent, storage tissue. Such tumors as those of crown gall behave toward nitrogen metabolism (protein synthesis) in a way that is reminiscent of the effect of coconut milk on dormant tissue (Steward, Caplin, and Shantz, 1955). Briefly, it has been shown that growth substances that stimulate cell division also stimulate over-all protein synthesis. They stimulate too a metabolically inactive structural protein which contains hydroxyproline, and also the "turnover" of the more metabolically active protein which these cells contain (482).

The experimental application of exogenous substances structurally related to purines and pyrimidines has yielded some information on the following lines:

The retardation of senescence, e.g., the Richmond-Lang effect (418) on isolated tobacco leaves (342, 343) implicates movement of soluble nitrogen compounds to "green" nonsenescent (but noncell dividing) areas on the leaves. The movement of these and other metabolites to regions where kinetin has been applied, is against a concentration gradient.

Wollgiehn (616) has studied the influence of kinetin on the metabolism of nucleic acids and protein of excised leaves of *Nicotiana rustica*. Kinetin inhibits the yellowing and degradation of nucleic acids and proteins in excised leaves. Protein and RNA, but not DNA, are synthesized in the kinetin-treated parts of leaves, and kinetin is more effective in darkness than in light.

The topic of flowering and fruiting properly falls in Volumes V or VI. However, it has been a classical idea that the balance between vegetative and reproductive growth involved nitrogenous compounds and nutrition on the one hand, and carbohydrates and their metabolism on the other. These ideas are traceable to such authorities as Klebs (1918) and Kraus and Kraybill (259). The problem has always been to define these concepts by reference to specific compounds rather than to very general categories. In fact, these ideas fell into disrepute because this could not be done. Chromatography now permits the soluble nitrogen compounds, in their response to flowering stimuli, to be more intimately investigated, and the new techniques of protein separation may well shed new light on the role of specific protein fractions. Since many floral organs and fruits have distinctive patterns of nitrogen metabolism, their induction may well be correlated with changes in the nitrogen compounds to which they have access [Bollard (53, 55)]. Kessler et al. (247, 248) have utilized uracil, xanthine, and caffeine sprays which are active in inducing flowering in fruit trees. Adenine-treated plants increased the DNA:RNA ratio, which paralleled induced tolerance to elevated temperatures.

In summary, while any exogenous substance or stimulus which regulates or modifies growth may be expected to exert a profound influence on nitrogen metabolism, it may not be possible to define easily its first point of contact with the metabolism. Moreover the effect of the substance on growth may not flow causally from its effect on nitrogen metabolism, possible as this may seem to be.

PART 3

PROTEINS AND PROTEIN METABOLISM: REGULATORY CONTROL

VI. The Complement of Protein Nitrogen in Plants

A. EARLY WORK ON FRACTIONATION OF PROTEIN NITROGEN: THE CHARACTERIZATION OF PROTEINS

Knowledge of protein structure has advanced so rapidly since the application of chromatography to protein hydrolyzates, by the use of the so-called "finger printing techniques," and by methods which locate the precise position of amino acids along the protein backbone that it may almost seem unprofitable to refer to the earlier history of our knowledge of plant proteins. On the other hand, it is still necessary to understand how present knowledge of plant proteins emerged.

From the work of Beccari (1747) and Fourcroy (1789) in the eighteenth century, on gluten and albumin, respectively, knowledge advanced in the nineteenth century by such works as those of Boussingault (1836), Mulder (1839), Liebig (1841), and Ritthausen (1872). But the work of Osborne (377) really ushered in the modern knowledge of plant proteins and may be consulted for these references.

Osborne (377) reviewed and extended Ritthausen's work. The proteins that occur in the physiologically active cells and fluids of plants were then little studied, owing to their relatively small quantity and the difficulty of separating them from each other. The majority of seed proteins thus far studied were globulins, which are soluble in neutral saline solutions, from which they were precipitated by dilution or dialysis.

Osborne (377) recognized that the metabolically or physiologically active proteins of plants had different properties from the reserve substances so prominently found in seeds. He commented (*loc. cit.* p. 9) on the amount of nucleic acid which is associated with the physiologically active protein, in contrast to the reserve protein, and showed that the embryo of wheat was virtually free from the proteins that are characteristic of the endosperm. By contrast, the embryo contained a globulin extractable by salt solution (*loc. cit.* p. 115).

Nucleoproteins were first described as existing in vegetable cells by Hoppe-Seyler (1879), who obtained a preparation from yeast which was very similar to those then recently obtained from animal sources. By 1900, the only extensive study of the "nucleoproteins" obtained from plants made since the true character of the nucleic acids was established (242) and the basic properties of the proteins recognized, was that of Osborne and Campbell (1900). These workers obtained a large number of products, which consisted of protein combined with very different proportions of nucleic acid, from the wheat embryo. The aqueous extract of the wheat embryo contains a large proportion of nucleic acid, which was isolated and its composition and properties determined (Osborne and Harris, 1902).

Chibnall's ether method for the separation of total protein of plant leaves yielded three main categories, which were roughly characterized as vacuolar, cytoplasmic, and chloroplast protein. The cytoplasmic proteins (glutelins) were remarkably similar even though the samples were derived from different plants. The only trends noted among the three categories were in their amino acid content; but even this may now be doubted (507).

The precipitation of proteins by deproteinizing agents was one of the most widely employed operations of the biochemical analyst, but probably no such agent separates protein completely from all other constituents of a plant mixture. Hiller and van Slyke (1922) found that 2.5% trichloroacetic acid was superior to other reagents to separate protein from nonprotein nitrogen.

By 1947 (502) the incomplete knowledge of plant proteins was evident from the accepted designation "protein nitrogen" to cover what was still a clearly heterogeneous fraction. This fraction comprised the structural proteins of the cytoplasm and the nucleus and the cytoplasmic inclusions, together with the storage proteins which may be dissolved as globulins in the cell sap or deposited in seeds as crystals or granules. The best known plant proteins were still the storage proteins of seeds of crop plants.

The fractionation of protein has been reviewed in various volumes [e.g., "The Proteins" (354), "Modern Methods of Plant Analysis," Vol. 4 (cf. 380), and Pirie (392, 393); "The Enzymes" (520), "The Nucleic Acids" (102), and by Lugg (284), McCalla (306), Vickery (560), and Anson (11)].

The distinction between classes of proteins, e.g., albumins and globulins, was for a long time made completely on the basis of their solubility in water, salt solutions, and other solvents. The inadequacy of such a classification was pointed out by Block (50), who attempted to introduce chemical characteristics into the classification. It became apparent from electrophoresis (after Tiselius) and ultracentrifuge measurements correlated with separation by neutral salts (e.g., Cohn et al., 1940; Leutscher, 1940; Taylor and Keys, 1943; Gutman et al., 1941; Svensson, 1940; Dole, 1944; Pillemer and Hutchinson, 1945) that neutral salts are incapable of accurately fractionating protein mixtures. Homogeneous protein fractions were prepared by the use of the ultracentrifuge, by electrophoresis, and also by the density gradient methods used in virus preparations.

The first pure crystalline enzyme, urease, was obtained from jack bean and characterized by J. B. Sumner in 1926. The final process was, in Sumner's words, "absurdly simple," but there is no single process for the crystallization of all enzymes or proteins.

The older crystallization methods (e.g., for globulins: Sumner, 1919) as well as that for the enzyme urease (519) paved the way for the application of physical methods to the study of proteins, including the use of X-ray diffraction and of the electron microscope [(625) loc. cit. p. 9; (354)].

The chemical interpretations of protein have advanced by the use of a variety of techniques to protein hydrolytic products. These include countercurrent distribution (Craig, 1950); partition chromatography on columns [Stein and Moore (473, 474), Moore et al. (332), Moore and Stein (333), Spackman et al. (464)]; or on paper [e.g., Consden et al. (114), Thompson and Steward (539), Hanes (196), Tigane et al. (541)]. Other useful procedures may be found by consulting sources such as Glick (181) and Meister (320).

Katz et al. (245) described the separation and analysis of peptides from proteolytic digestion mixtures by a rapid method and introduced features into two-directional chromatography and electrophoresis on large sheets of filter paper with the identification of up to 80 spots. High voltage paper electrophoresis in the analysis of protein digests, as this has been applied to determine the structure of such proteins as hemoglobin (18, 229, 230), has led to great developments in these techniques. Martin (297) has discussed the basic considerations that must be recognized in the design of electrophoretic apparatus.

In the electrophoresis of nucleic acids on freshly prepared silica gel, the gel acts as a molecular sieve that allows ribonucleic acid and peptides to migrate, but restrains deoxyribonucleic acid and proteins [Harris and Davis (202)]. Synge and Youngson (523) have described molecular sieve methods for separations of polypeptides according to their molecular weight.

Starch gel electrophoresis is now in general use and often gives good resolution (461, 462). By these means 16 fractions of basic proteins have been found in the ribosomes of pea seedlings (444). Starch gel electrophoresis has been applied especially well to the separation of proteins of corn (*Zea*) or wheat, where 12 or more distinctive bands have been separated (C.S.I.R.O. Wheat Res. Unit Report 30.9.62).

Keil (246) and Semenza (442) have reviewed different applications of modified celluloses to the preparation of proteins. A high degree of resolution can be expected from materials that combine the advantages of gel filtration with ion exchange, e.g., DEAE-Sephadex (400).

Acrylamide gel is the latest supporting medium for proteins during electrophoretic separations. In some cases, e.g., *Neurospora*, soluble proteins, a higher degree of resolution is obtainable by this method than by others (101). It has been possible to distinguish the protein complement of different closely related species of *Neurospora* and even between mutant strains of the same species of *Neurospora* (Fig. 45). A two-dimensional gel electrophoresis has been used to separate the isozymes of lactic dehydrogenase (410).

Various enzymes and the changes in protein structure by mutation have been investigated by disk electrophoresis and by immunological methods (167, 626). Medvedev (317) has even described the paper electrophoresis of proteins while preserving the pattern of their localization in tissues.

In the period 1955–1962 the interest of biochemists and cytologists in histones was stimulated and much of the work has demonstrated that histones are more numerous, complicated, and metabolically active than had earlier been supposed. The occurrence, preparation, fractionation, and analysis of histones has been well reviewed by Phillips

(391) and, as part of the discussion of wider subjects, by Butler and Davison (84) on deoxyribonucleoprotein, Shooter (447) on deoxyribonucleic acid, and Peacocke (383) on the nucleic acids and nucleoproteins.

The study of ribosomes is a rapidly advancing field of molecular biology owing to the direct participation of ribosomes in protein syn-

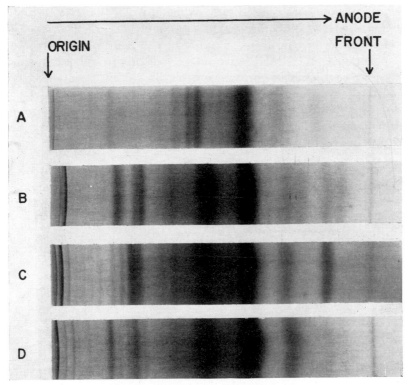

Fig. 45. Electrophoretic separations on acrylamide gels of the soluble proteins of *Neurospora*: protein patterns of (A) *N. intermedia*; (B) *N. crassa* (74A); (C and D) two independently derived "peak" mutants (pk 1A and pk 2A, respectively) of *N. crassa* (74A). From the unpublished work of Srb, Barber, and Steward.

thesis through the role of nucleic acids. Developments in this field have flowed from advances in instrumentation, notably the electron microscope (325, 401, 402, 605) and the ultracentrifuge, in isotope techniques, and in the chemistry of nucleic acids (548).

The most powerful techniques arose from ultracentrifugation methods that enabled the quantitative separation of the particles from the supernatant protein and from other cellular components. The wide application of isotopically labeled amino acids and the basic knowledge about the chemical linkages of RNA (542–544), the notion of helical DNA (416, 598), and the work on virus and its infectious RNA (164, 165), all provided the foundation of what has almost become a "ribosome science" (316, 548, 597).

However, these developments in protein biochemistry do not of themselves necessarily lead to full understanding of the more complex biological phenomena, like growth, which involve the transformation of chemical energy into other forms (mechanical, electric, or osmotic). The chemical or energy transformations in protein metabolism are, in fact, negotiated in cellular structures (ribosomes, reticulum, membranes, etc.), which biochemists prefer to discard as "residues" in their preparation of cell-free systems, so that what is now called protein chemistry is mainly the chemistry of soluble proteins. An exception is the recent observation of Rich *et al.* (1962) that hemoglobin synthesis requires a cluster or aggregation of ribosomes.

B. Proteins and Enzymes: Enzymatic versus Structural Protein and Metabolic versus Nonmetabolic Protein

The number of individual protein molecules of a given estimated molecular weight of 64,000 that could be present in any plant cell is very large indeed (estimated at 10^9 molecules of the order of 64,000 molecular weight compared with about 3×10^{15} water molecules in an active carrot root cell). But how many distinct molecular species this represents cannot be stated with any certainty, and estimates will vary. If every gene has its protein counterpart, then there may be potentially several thousand protein species in any totipotent cell; and if every enzyme requires a distinctive protein moiety, then the known enzyme-catalyzed reactions that any cell might be called upon to carry out might require some 1500 to 2000 distinct enzyme proteins. Against these estimates even the best methods for separating protein seem crude.

The amino acid composition of plant protein is not now regarded as so rigidly determined as used to be supposed. Although the age of daffodil (*Narcissus*) leaves did not greatly affect their amino acid composition (79), there were greater deviations in the storage tissue [cf. Steward and Thompson (507); also note effects due to mineral nutrition, as described in Chapter 2, Volume III, on mint leaves under different mineral nutritional regimes, etc.]

In Part 2 (Section IV) the great increase in knowledge of the nonprotein nitrogenous constituents of plants, which followed upon improved techniques for their separation and recognition, has been described. The same should be true of the all-important bulk protein fraction, except that the methods for the separation and recognition of subfractions are not yet as equal to the task.

The purely structural role of proteins is much less conspicuous in plants than in animals—there is really nothing comparable in plants to collagen and the connective tissues of animals. It is said that the proteins of actively metabolizing tissues, whether of plant or animal origin, all have about the same over-all amino acid composition, and all are, so far as composition is concerned, good proteins for the nutrition of nonruminants (12).

A protein moiety, richer by far in hydroxyproline than plant proteins commonly are, was found in rapidly proliferating cells, especially those induced to do so by treatment with coconut milk. During synthesis in the cell, this protein moiety incorporates C^{14}-proline directly and thereupon converts it *in situ* to hydroxyproline. The ratio of C^{14}-hydroxyproline: C^{14}-proline combined in the whole cell approaches asymptotically a value of about 0.70, but this is made up of a number of electrophoretically recognizable entities which have a graded content of hydroxyproline, so that the ratio ranges from 0.7 to 4.0. Although a large part of the combined hydroxyproline adheres to the larger particles and to the cell debris (including the cell wall), it is erroneous to say [as Lamport and Northcote (263) have] that *all* this material is in the wall. By acrylamide gel electrophoresis (484) a soluble protein preparation from cultured carrot explants was resolved into 7 distinct bands, 6 of which contained C^{14}-hydroxyproline, and the ratio to C^{14}-proline varied from 0.1 to 1.0 (Fig. 46). Because of its metabolic inactivity, this protein moiety seems to fulfill a structural role. If the synthesis of this type of protein is reversibly inhibited, e.g., by exogenous treatment with hydroxyproline, then the growth of the cell is also inhibited even in the presence of coconut milk (497). Therefore, this substance plays some important role, and its initial formation is part of the growth induction response brought about in the cells by coconut milk, which is the prelude to much subsequent morphogenetic development (356, 495). In some slime molds there also seems to be precedent for a contractile protein (Loewy, Ts'o, Nakajima). However, with these two exceptions, there is little evidence of structural protein in plants. Some fungi, of course, have chitinous walls that contain nitrogen in the form of a polymer of N-acetylglucosamine (256).

An important role of plant protein, however, may be its part in the fine structure of plant organelles, including in this category the plasma membranes.

Since Lord Rayleigh discovered the remarkable ability of proteins to spread on the surface of water (as little as 1.0 mg may occupy 1.7 square meters of surface), proteins have been known as the most surface active substances and have entered into all ideas on the fine structure of surface membranes (see Volume IA, Chapter 1; Volume II, Chapter 1). When studied at a water–air interphase, the protein may form an expanded film with its peptide backbone chain lying in the water surface and the side chains disposed around it. Theories

Anode	Protein section No.	Movement relative to front (Rf)	C^{14}-Hydroxy-proline (counts per second)	C^{14}-Proline	Ratio $\frac{C^{14}\text{-Hydroxyproline}}{C^{14}\text{-Proline}}$
	1	100.0	0	2.1	0
	2	91.3	2.3	25.1	0.092
	3	80.3	1.4	15.6	0.090
	4*	70.0	3.5	24.3	0.144
	5	54.8	4.6	20.6	0.223
	6	36.0	0.7	5.8	0.121
	7	19.4	3.1	5.8	0.534
	8	7.8	2.4	4.5	0.533
Cathode	9	2.0	4.5	11.7	0.385

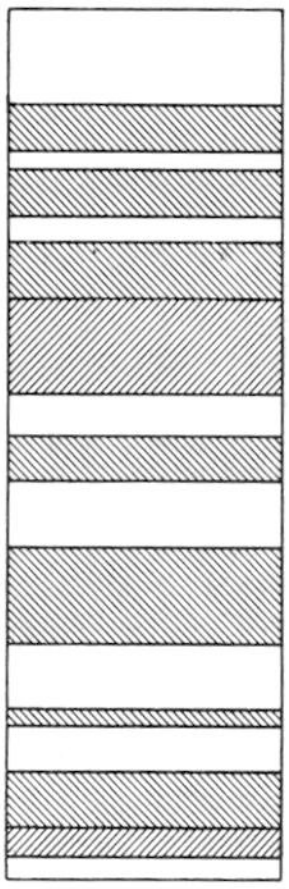

Fig. 46. The electrophoretic separation of the soluble protein of carrot; the absorption of C^{14}-proline into the various proteins and the degree of its conversion to C^{14}-hydroxyproline. Asterisk indicates section that seems to contain two strong but incompletely separated bands that are here treated as one. After Steward and Chang (484).

of membrane structure have drawn freely upon this property of proteins and have pointed out that given amino acid residues would impart acidic or basic properties to given areas of this mosaic-like surface. Protein can exist, however, as in the body of the cytoplasm, in the globular state (molecules of the order of 20–100 A in diameter). Since the normal range of molecular forces is of the order of 4–5 A, the protein film should be in a more reactive state than the globular molecules.

All views of protein synthesis presuppose that it occurs at a surface where the protein is first formed on a template and in the form of a film, the precise arrangement of the amino acids being determined by the nucleic acid surface on which it is supposed to form. Much of the proof of this idea comes from the ability of the nucleic acid part of the virus molecule to convey infectivity and to induce in the host the synthesis of its appropriate protein moiety (549); it is nevertheless being extended to describe protein synthesis generally. However, when one examines the fine structure of such cellular organelles as mitochondria and plastids, it is impossible to escape the view that much of their architecture and function is traceable to the highly organized protein films and lamellae that they contain (325, 401, 402, 605). Moreover, each of these organelles which retains the property of self-duplication is likewise supposed to contain the RNA on which its protein may be formed and may even have the DNA which may give rise to RNA.

Different growth habits may be associated with different protein complements. For example, Hotta (221) has estimated the protein content of *Dryopteris erythrosora* and has found that two-dimensional growth of the gametophyte provided a sharp increase in protein concentration. During one-dimensional (filamentous) growth the protein concentration was gradually lowered. Amino acid antimetabolites, e.g., ethionine, and 5-methyltryptophan, inhibited two dimensional growth with a reversion of "artificial" one-dimensional filamentous) growth. 8-Azaguanine inhibited growth similarly, but the inhibition could be removed by the addition of guanine or the corresponding amino acid. Selective inhibition always was accompanied by an immediate cessation of protein concentration followed by an immediate decrease.

Ribonucleic acid from the "2-dimensional" prothallium possesses a nucleotide composition distinct from that of the filamentous growth. Hotta (221) has tentatively inferred that a special RNA sensitive to 8-azaguanine is closely related to the "two-dimensional" growth of the gametophyte.

C. SEEDS AND STORAGE PROTEIN

Generally plants with fleshy cotyledons have seeds that are richer sources of proteins than are cereal grains, hence they are important protein suppliers for man and beast. Even so plant proteins are commonly deficient in lysine and methionine, and much attention has been given recently to the preparation of a protein diet from plants (such as soybean) which when supplemented with these amino acids, would be fully nutritive for animals and humans (7). Despite this, most of the early knowledge of the composition of plant proteins is based on the analysis of a relatively small number of reserve substances. Vickery (559, 561) gave a historical account of this problem, from the first determinations, in 1871, of the aspartic and glutamic acid content of protein preparations obtained from lupine and soybean and then through the classical work of Osborne, which paved the way for the doctrine of essential amino acids in animal nutrition, as enunciated by Osborne and Mendel in 1914.

Pure proteins (globulins) have been prepared from a number of cotyledonary sources. The various globulins prepared by Sumner (519) from jack bean, and the globulin prepared from *Cucurbita* seeds by Smith in Vickery's laboratory (458), are cases in point. For the composition of seed proteins and what is known about them, references will need to be made to other sources (11, 77, 284). The utilization of this stored protein is a principal event in seed germination, as shown by classical works in this field (cf. 106, 314, 338).

The seed reserve breakdown and concomitant protein synthesis elsewhere in the germinating seeds may be related to the ribonucleic acid metabolism of *Vigna sesquipedalis* (370, 371). Large quantities of RNA are stored in the cotyledons and the presence of the cotyledons attached to the embryo is essential for the RNA breakdown.

D. LEAF PROTEIN, CHLOROPLAST AND NONCHLOROPLAST PROTEIN

The current belief, based on electron microscopy, is that chloroplasts develop from preexisting proplastids of about 1 μ in diameter (146, 217, 604). These bodies are identified by their size and the existence of a double membrane from which lamellae are seen to develop upon exposure of the cells to light. The proplastids are thought to arise from less discrete, small, single-membraned cytoplasmic vesicles, which coalesce to form the larger bodies.

Neither the proplastid nor its precursor have the pigment composition, enzyme activity, or protein-lipid structural elements of the functional chloroplast. According to Wolken (613), photosynthetic activity first becomes measurable after 24 hours of exposure to light and coincides with a significant concentration of photosynthetic cytochrome. Wolken and Gross (615) and Epstein and Schiff (146) claim that photosynthetic oxygen evolution can be measured as early as 6 hours after illumination, which parallels the formation of the first lamellar strand.

Apparently 75% of *Oenothera* leaf protein exists in the chloroplasts, and many of the soluble proteins and enzymes in the leaf are not cytoplasmic but originate mainly in the chloroplast (638). The precise composition of the chloroplast is still incompletely known, but data on isolated chloroplasts from a variety of plants indicates 35–55% protein, 18–37% lipids (mostly phospholipids), and 5–8% inorganic material on a dry weight basis. The pigments make up about 8–20% of the chloroplast. Leaf protein (i.e., largely chloroplast protein), especially of older leaves, is more readily extractable in borate buffers (0.1 M at pH 8.3) than by the use of either phosphate or Tris buffers (638).

In a communication to a symposium on chloroplasts held at a meeting of the American Society of Plant Physiologists, August 1963, Kahn reported on the presence of "phytoferritin," and its resemblance to ferritin in animals has been noted. These phytoferritin bodies are rich in iron, with a protein coat, and under the electron microscope they appear circular with a central core. They are found in chloroplasts from young leaves but are sparse in older ones. Their function is not known.

Total nucleic acid (i.e., both RNA and DNA) contributes from 0.3 to 3.5% of the dry weight of chloroplasts; this is compatible with the self-duplicating character of the chloroplast (614). However, the evidence for DNA in chloroplasts has been histochemical only and is presumed to follow from their self-duplicating nature.

Recently, however, Gibor (work in press)[11b] has demonstrated DNA in the chloroplast from enucleated, aseptic cells of *Acetabularia* and finds approximately 1 mole of DNA to 5000 of protein. Sissakian and Filippovich (1957) using C^{14} and S^{35} claim to have confirmed the participation of the chloroplasts in protein synthesis (453) (cf. Section IX, C).

Bové and Raacke (68) have demonstrated that chloroplasts have amino acid-activating enzymes and conduct a light-dependent incorporation of amino acids into protein (476). In fact, according to Zucker and Stinson (638) many enzymatic functions of leaves are virtually confined to chloroplasts. Although only a small fraction of the total RNA of a green leaf is localized in the chloroplasts, a close relation exists between the total RNA of a green leaf and its capacity for photosynthetic phosphorylation (455), which may in turn be harnessed to protein synthesis.

Claims have been made that isolated chloroplasts incorporate amino acids into their protein (207, 588). Spinach chloroplasts apparently contain 15–40% of the total amino acid-activating enzymes of the cell (206, 207, 588). Heber (207) supplied $C^{14}O_2$ to *Spinacia oleracea* leaves and has concluded that protein synthesis occurs in the chloroplast as soon as photosynthesis is initiated. According to Lyttleton (287) spinach chloroplasts contain ribosome particles that differ somewhat from those in cytoplasm. Rhodes and Yemm (415a) postulate that developing

---

[11b] Since this paragraph was written, the work of Gibor has appeared [*Proc. Natl. Acad. Sci. U.S.* **50**, 1164 (1963)]. Also an important paper in this general area by Gibor and Granick [*Science,* **145**, 890–897 (1964)] should be consulted.

chloroplasts couple photosynthesis directly to protein synthesis, presumably by virtue of photosynthetic phosphorylation.

By contrast, the incorporation of radioactivity into soluble cytoplasmic proteins shows a lag period of at least 2 minutes. Brawerman et al. (75) have shown that during chloroplast formation in resting Euglena cells exposed to light, the protein of the plastid fraction increases rapidly after a considerable lag period. RNA of both plastid and microsome fractions increased without apparent lag prior to the plastid protein rise. The authors suggest that the RNA produced during chloroplast formation represents specific ribosomal material which is necessary for the synthesis of the plastid proteins (76).

Thus it is now generally accepted that plastids contain RNA (454) although there is no agreement on the amount of RNA present in chloroplasts. A number of chemical and physical treatments can inhibit the formation of chloroplasts. These can produce permanent changes in Euglena, but usually only temporary changes in other photosynthetic organisms. Loss of the capacity to produce chloroplasts in Euglena may be due to an interference in the replication of DNA-containing cytoplasmic entities that are responsible for plastid formation (454).

The light-dependent maturation of Euglena proplastids is paralleled by changes in the content and composition of plastid RNA. Chloroplast growth is affected by 5-fluorouracil only during the early period of chloroplast development before most of the chloroplast protein and lipid is synthesized. Thus it has been postulated (454) that the light-dependent growth of chloroplasts requires the synthesis of new ribosomal species and that this synthesis is initiated by light. More recently Böttger (66) has observed that before the regeneration of roots the content of RNA and protein in leaves of Euphorbia pulcherrima decreased in the particle-free fraction and the slowly sedimenting fractions in which microsomes and mitochondria occur. A concomitant increase was observed in the fractions that contained the nuclei and chloroplasts. The changes in the content of RNA and protein were reversed when roots appeared.

E. Proteins in Growing Points and in Relation to Development

What one would really like to know, however, are the answers to the following questions: (a) What is the precise nature of the proteins in a given meristematic region? Are they different in the shoot and root tip? (b) How does the protein complement change during the ontogeny and differentiation of a given cell or tissue? Is it the cause or the result of the differentiation that occurs? (c) What are the distinguishing features of chloroplast protein that may constitute up to 35–45% of the leaf protein, and how does it differ from the proteins of the cytoplasm? (d) Do the factors that determine morphogenesis (length of day, night temperature, etc.) influence the kind of protein that is made in the growing point?

It is idle to pretend that our present state of knowledge of plant proteins is adequate to deal with questions such as those listed above. The more nearly cells approach the meristematic condition (as in those that are in active growth in tissue culture), the more nearly the segments analyzed approach the central meristem, the richer the proteins tend to be in the more basic amino acids (78). A first study by the gel electrophoresis method examined the soluble proteins along the axis of the primary root of pea and compared these with the epicotyl, the plumule, and the cotyledons (cf. Fig. 47, after Lyndon, Barber, and Steward). Although their significance is not yet clear, it is very apparent that there are differences in the electrophoretic pattern which can be associated with the region from which the sample was drawn. Heyes and Brown (212), Robinson and Brown (422, 423), Morgan and Reith (335), and Avers (16) have previously traced sequential changes in the protein complement along the axis of roots by measuring the activity due to certain selected enzymes as examples of identifiable proteins. The evidence furnished by the gel electrophoresis method renders visible and much more apparent the fact that such differences do exist. In fact, when one recalls how firmly ideas of protein synthesis are now oriented toward the gene-template ideas (cf. Section VIII, D–F), it becomes a problem to know how the same genes are responsible during morphogenesis for the synthesis of quite different proteins. This is well illustrated by Fig. 48, which shows the changing pattern of proteins in different organs of the tulip (work of Barber and Steward).

A start has been made on the converse problem. Certain mutant genes determine the form of the colonies of cultured *Neurospora*. It has been shown that these mutant strains differ significantly from the wild type in the banded structure revealed by gel electrophoresis. Moreover, when the gene in question is put into *Neurospora* with different genetic backgrounds, certain characteristic protein bands appear (101). One can, therefore, now hope to correlate the form of these colonies in *Neurospora*, i.e., their morphogenesis, with a genetically determined protein complement. If it can also be shown that the great events of flowering and fruiting are triggered off by the formation of characteristic protein moieties in response to the environmental conditions, one of the most baffling morphogenetic problems may yield to experimental attack. Phytochrome itself, the receptive pigment which according to

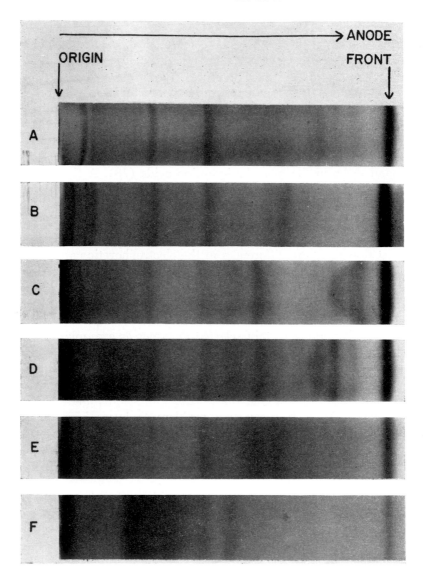

Fig. 47. Electrophoretic separations on acrylamide gels of the soluble protein (pH 8.3) of 3-day-old pea seedlings. (A) Segment, 0–1 mm from the tip of the root; (B) segment, 5–6 mm from the tip of the root; (C) hypocotyl; (D) epicotyl; (E) plumule; (F) cotyledons. Data of Lyndon, Barber, and Steward, *Am. J. Botany* 52, 155–164 (1965).

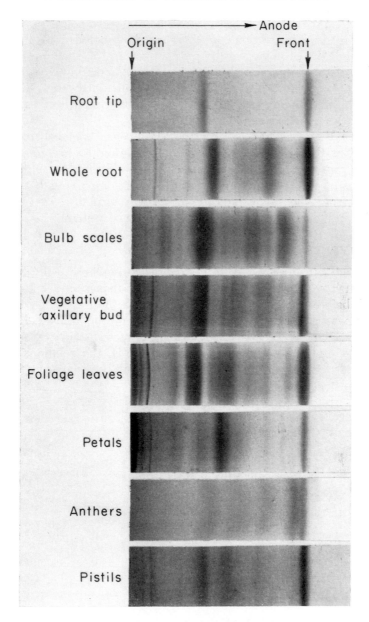

FIG. 48. Electrophoretic separations on acrylamide gels of the soluble proteins (pH 8.3) of *Tulipa*. After Barber and Steward, unpublished data.

Hendricks and Borthwick (210) mediates these responses, is supposed to be a protein. This section can, therefore, be concluded as follows.

Although the proteins are the most important of all the nitrogenous constituents of plants and although much of the distinctive metabolism of plants is directed to their formation, we actually know very little in detail about these important substances in comparison with others that occur in plants (e.g., carbohydrates).

The next great wave of progress in nitrogen metabolism may be expected as the structure, synthesis, and role of the proteins that are actively concerned in growth, metabolism, and development yield to detailed study. The gains thus made should compare with those that accrued from the understanding of protein structure in relation to genetics and to genetically transmitted disease in humans.

VII. Ribonucleic Acids, Nucleoproteins, and Virus Protein

Although nucleic acids were discovered quite early (cf. Section III, D) in such cells as the sperms of fishes, a surprisingly long interval elapsed before their significance in genetics and in self-duplicating systems was generally recognized. Because of the role of deoxyribonucleic (DNA) in the nucleus and of ribonucleic acid (RNA) in the cytoplasm and their part in protein synthesis, the relations of these compounds to nitrogen metabolism should now be discussed.

TABLE XIX

Characteristics of Purified Ribonucleoprotein Particles[a]

Source	Composition (% dry wt.)			Diameter (Å)	$S_{20}°C^b$
	RNA	Protein	P-lipid		
Pea seedling	35	55	4.5	250	75
Pea seedling	45	51	—	—	—
Wheat roots	30	70	—	100	—
Yeast	42	58	—	210	80
Azotobacter vinelandii	—	—	None	200–250	86

[a] For references and other data, see Raacke (405) Table VII.

[b] Svedberg sedimentation units at 20°C.

The bulk of the RNA of the cell is present in ribonucleo-protein particles and ribosomes (76). The composition of ribonucleoprotein particles is given in Table XIX. The need for RNA during protein synthesis in intact roots has been shown by soaking the roots in ribonuclease (69–71). The roots can recover from the ribonuclease inhibition of protein synthesis by the addition of yeast RNA.

The base composition of the total RNA in different microorganisms with widely different DNA compositions is surprisingly uniform (31).

A. KINDS OF RIBONUCLEIC ACID

In cells, mainly of microorganisms, there are now thought to be three categories of RNA:

1. Transfer or Soluble RNA. The transfer or soluble RNA (sRNA) is almost 70% of the total, with a molecular weight of almost 2.5×10^4. During protein synthesis this RNA selectively combines with the amino acid–adenylate enzyme complex in such a way as to transfer the amino acid to the terminal adenylic acid group of the sRNA chain. This amino acid is supposed to be linked through its carboxyl group to the ribose hydroxyl group at the end of the specific RNA chain. The adenosine monophosphate and enzyme so released are then available for rephosphorylation by ATP.

2. Ribosomal RNA. The sRNA, with its amino acid attached, approaches the protein synthesizing template on the ribosome. In this way it brings the amino acid moiety to the template surface, which also consists of RNA (ribosomal RNA has a much higher molecular weight of the order of 1.7×10^5). The structure of the template RNA is such as to attract the sRNA–amino acid complex to a particular site on the surface. The point of the so-called genetic code is to arrange these amino acids in their correct sequence prior to protein synthesis. The point of the sRNA is to act as what Crick (119) calls "an adaptor molecule" which enables the amino acid to find its proper place on the template. This requires specificity in the sRNA; one area finds the specific location on the template, another is specific for the amino acid in question.

3. Messenger RNA. The ribosomal RNA and the transfer or sRNA's do not have precisely the same base ratios as the DNA of the cells from which they were derived; however, a small portion of the total RNA, i.e., messenger RNA (mRNA), has the same base ratio as the DNA of the cell. This small amount of RNA (3% of the total) with a molecular weight of at least 250,000 turns over rapidly (incorporating P^{32}-uracil and 5-fluorouracil) and does seem to have all the properties required of the "messenger" which determines the ribosome template at which protein is formed. In 1962 Tissieres and Watson (cf. 597) showed that mRNA is unstable in cell-free extracts and can be broken down to acid-soluble fragments. Although this concept is widespread mRNA has not been clearly demonstrated in higher plant or mammalian cells (76, 192, 548). However, Loening (278) does claim some very recent evidence for the occurrence of mRNA in peas. An RNA polymerase has been isolated from *Escherichia coli* by Chamberlin

and Berg (98) which makes RNA with a base composition complementary to that of DNA, and it thus resembles mRNA. This enzyme
requires the four ribonucleoside triphosphates, a divalent ion, and DNA.
Both strands of the DNA can act as the primer upon which RNA
may be formed.

B. SUMMARY OF RNA's AND THEIR HELICAL STRUCTURE

Unlike DNA, the RNA occurs in at least three forms, each serving
a different function. The structure of sRNA has turned out to be
another double helix, similar in its geometry to one of the forms of
DNA; it consists, not of two separate chains that run in opposite
directions, but of a single chain coiled back upon itself (389). This
model represents the structure of transfer RNA, which is a single
chain of no more than about 80 nucleotides with a ratio of G:C
and U:A equal to unity. It is conceivable that the sequence of bases
in one-half of each molecule is complementary to that in the other half
and that hydrogen bonds between specific pairs of bases give these
molecules their well-ordered, double, helical configuration. According
to some (361, 607) the double coil of sRNA has three unmatched
bases at the closed end of the coil near the bend (Fig. 49A); these
serve to attract the AA-sRNA complex to the template. One free end
of the chain (Fig. 49A) is occupied by guanine, which may be paired
with cytosine, and the other three bases (two cytosines and one terminal
adenine), which serve to attract the amino acid and form the complex
(see also Figs. 49A and B). Some consideration is being given to ideas
that ascribe part of the specificity to features of the tertiary structures
of the molecule as well as to the terminal bases.[11c]

Ribosomal RNA, when detached from its protein carrier, still appears
to contain some double-helical regions but, in the absence of complementary sequences of bases, these are more disordered and interrupted by nonhelical regions. The protein complement of ribosomes
reveals a complicated picture. Each ribosome most likely contains
twenty or so polypeptide chains with a net basic charge (597).

Ribosomal and transfer RNA each have a distinct base composition
(different from that of the DNA) with little species specificity. Although the concept of a short-lived messenger RNA and of the mechanism of its replication from a DNA primer plays an important role
in current schemes of protein synthesis, its structure is not known.

[11c] Very recently B. Commoner [*Nature* **203**, 486–491 (1964)] questions the hypothesis that "the capability of the living cell for self duplication and inheritance
is *uniquely* determined by the DNA" and he invokes other components of "a
multimolecular system."

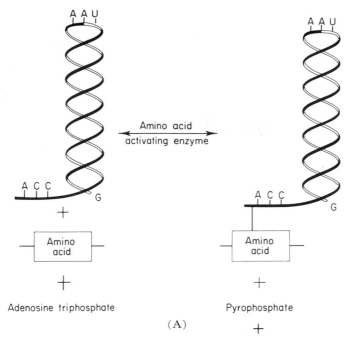

Amino acid
activating enzyme

AAU

ACC

G

+

Amino
acid

+

Adenosine triphosphate

(A)

AAU

ACC

G

+

Amino
acid

+

Pyrophosphate

+

Adenosine monophosphate

ADJACENTLY ADSORBED AMINO
ACID ADAPTER RNA MOLECULES

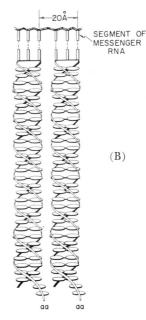

├─20Å─┤

SEGMENT OF
MESSENGER
RNA

(B)

aa   aa

Fig. 49. (A) Transfer RNA or sRNA is a special helical form of RNA that transports amino acids to their proper site in the protein chain. There is at least one transfer RNA for each of the 20 protein amino acids.* All, however, carry the bases ACC where the amino acids attach and G at the opposite end. The attachment requires a specific enzyme and energy supplied by adenosine triphosphate. Unpaired bases in transfer RNA (AAU in the example) may provide the means by which the transfer RNA "recognizes" the place to deposit its amino acid package. From Nirenberg (361).

(B) Transfer RNA, with amino acid attached, acting as an adapter molecule [after Zubay (636)] which is located upon a specific segment of the messenger RNA; in the process of this alignment amino acids are brought into juxtaposition in preparation for their condensation to peptides and protein.

* The base sequence in an alanine transfer RNA from yeast is now known from the work of R. W. Holley et al., Science **147**, 1462–1465 (1965).

Moreover, concerning the concept of a transient, short-lived mRNA, which is derived from microorganisms, there are some reasons to believe [cf. Morton and Raison (336a)] that an mRNA in the more highly organized cells of higher plants, or even of animals [Harris (202a)], might need to be more stable. The point of contact between growth substances and their effect on protein metabolism may need to be sought in terms of their intervention; as seen in Fig. 53B, either they act as "messengers" or they affect the kind of mRNA message that is relayed.

C. Controls over the Activity of RNA and DNA in Chromosomes: the Role of Histones

Histones, which constitute the protein moiety of nucleohistone, are basic proteins rich in the hexone bases (approximately one in four of the amino acid residues may be histidine or arginine). Zubay et al. (637) have shown by electron microscopy and X-ray diffraction that the histone molecules are wound spirally around the DNA of the chromosome, the whole assemblage being held together by ionic bonds. In this structure, the protein moiety very probably possesses the α-helical structure. Data quoted by Huang and Bonner (222) also indicate that the structure of the DNA molecule may be stabilized by histone.[12]

Using higher plants, McLeish (315) showed that nuclear arginine quantitatively parallels that of DNA and that the amount of nuclear arginine, like DNA, is doubled during interphase. The synthesis of arginine and DNA in interphase occurs within the same relatively short period of time. Bonner and Huang (57) have argued that histone may, in some way, inhibit or otherwise depress the ability of the DNA of chromatin to support DNA-dependent RNA synthesis: thus histone would appear as a modifier or regulator of genetic information. This being so, histone could be a means by which genetic information is turned "on or off," as in numerous examples where such controls may operate during growth, development, and morphogenesis.

Gifford and Tepper (179) have related histones to morphogenesis (during flower induction) of Chenopodium and Ipomoea (Pharbitis) species. Shortly after buds receive the flowering message from the leaf and before any visible signs of differentiation, there is a dramatic

---

[12] At an international conference on the biological chemistry of histones held in 1963, the proceedings of which are now in press, these ideas were developed further. Various hypotheses are being proposed to account for the ways in which histones could modify the behavior of DNA [cf. Science 141, 651–654 (1963)].

drop in the histone content of the cells of the bud. This is followed by an equally dramatic increase in the RNA of the same cells.

Thus modern work and thought is implicating the nucleic acids and nucleoproteins in the mechanism of protein synthesis in ways that could not have been anticipated even a short time ago (cf. Section VIII).

D. NORMAL AND VIRUS PROTEINS

Viruses are capable of autocatalytic growth and multiplication in living cells. One tobacco mosaic virus (TMV), which is present in the filtered juice of infected plants, was isolated by Stanley (472) in crystalline form and found to be a nucleoprotein. The molecular weight is about 40 million; the nucleic acid moiety amounts to about 6% (TMV) of the total weight. The crystalline protein is highly infectious and its virus activity parallels the range of pH stability. The protein associated with the RNA appears to be devoid of enzymatic activity. Since this pioneering work, a number of other plant viruses have been isolated and shown to be nucleoproteins (e.g., potato X, cucumber mosaic, tomato bushy stunt). The RNA of turnip yellow mosaic virus can constitute up to 35% of the total weight.

Fraenkel-Conrat and Williams (166) and Fraenkel-Conrat et al. (164) have given the details of the demonstration that the protein viral coats can be removed from their RNA cores in such a way as to leave the RNA uninjured. The RNA alone has infectivity and can produce normal "offspring" with protein coats when put into a susceptible tobacco plant cell. Tsugita and Fraenkel-Conrat (549) have shown that presumably pure viral RNA with some of its bases deaminated by nitrous acid can initiate the formation of viral protein which differs from the original by replacements of one or more amino acid residues.

The bridge between one generation of TMV and another can thus be reduced to the RNA alone, and the transfer of synthetic information is here characteristically *not* dependent upon DNA. Doi and Spiegelman (133) conclude that the incoming viral RNA contains the information for an RNA polymerase, dependent upon a RNA primer, which can then cause the virus to multiply. Since this enzyme must be synthesized *before* replication, the viral RNA must be conserved during the formation of viral protein. Their data lend no support to the otherwise attractive speculation that the RNA viruses might represent "escaped genetic messages of the host." Furthermore, they imply that RNA viruses replace the "DNA to RNA to protein" path of information transfer with one that stems from RNA to RNA to protein. Tsugita

et al. (550), using viral RNA in a ribosome system from *E. coli,*
observed a 75-fold increase in C^{14}-amino acid incorporation into protein
by the preparation, so that the viral RNA here acted as messenger
RNA.[13]

Some light may be shed upon the sources of carbon for viral protein
synthesis by considering the entry of C^{14} from various labeled substrates
into either the general protein of tobacco leaves or into the protein
of newly synthesized tobacco mosaic virus (397). By the appropriate
application of the C^{14}-substrate, prior to the inoculation with TMV,
the carbon of general leaf protein may be labeled and its subsequent
movement into TMV protein may be traced. This has been shown to
occur by Gates (Tables XX–XXI). The C^{14} of arginine in the newly

TABLE XX

The Source of Arginine Carbon for TMV Synthesis in
Illuminated Tobacco Leaves[a]

Specific activity of arginine carbon[b] in leaf protein during TMV synthesis	Specific activity of arginine carbon[b] in newly synthesized TMV	Per cent of arginine carbon in TMV which came from leaf protein previously synthesized
77	42	
51	27	
38	33	ca 50%
30	16	

[a] Leaf protein was previously labeled with arginine.

[b] Data are percentages of the specific activity of C^{14} in the arginine originally
supplied to the leaf. They are taken from L. F. Gates, "The source of amino acids
used for synthesis of tobacco mosaic vires by illuminated tobacco leaves." Ph.D.
thesis, Cornell Univ., Ithaca, New York, 1963. Data are in counts per second per
100 μg of carbon.

TABLE XXI

The Specific Activities of the Carbon of the Amino Acids in Various
Fractions of Noninfected Tobacco Leaves That Received $C^{14}O_2$[a]

Amino acid fraction	Serine	Alanine	Glutamic acid	Glutamine
Amino acids that entered the TMV	76	45	20	—
Ethanol-extracted free amino acids	16	26	17	12
Leaf protein amino acids	10	5	2	—

[a] Data from L. F. Gates, "The source of amino acids used for synthesis of tobacco
mosaic vires by illuminated tobacco leaves." Ph.D. thesis, Cornell Univ., Ithaca,
New York, 1963.

[13] Ochoa *et al.* have summarized their views on viral RNA and its replication
in *Federation Proc.* **23,** 1285–1296 (1964).

synthesized TMV had to come in this way from arginine that was previously present in protein. Hence protein breakdown products were obviously accessible for incorporation into TMV protein.

All this is consistent with the existence of more active pools of amino acids that are more accessible to the synthesis of virus protein than *the main body of free amino acids in the leaf;* these active pools seem to be fed readily via photosynthesis, and they may also be replenished, at least in part, from protein breakdown. Some of the data that support these conclusions are given below.

$C^{14}O_2$ may also contribute carbon to the virus and leaf protein. Bidwell *et al.* now find that C^{14} can enter leaf protein more readily via photosynthesis even than from C^{14} amino acids already present in the leaf (see page 527). Similarly, Gates now finds that such amino acids as serine, alanine, and glutamic acid, which receive their carbon from $C^{14}O_2$ at especially high specific activity, are also readily incorporated into the TMV. In fact, the specific activity of the carbon in these amino acids in the TMV may be greater than that of the same acids as they occur free in the cell, or even as they occur in the leaf protein which is concurrently present in the leaf.

VIII. The Mechanism of Protein Synthesis in Plants

A. THE PROBLEM INTRODUCED

The culmination of any discussion on nitrogen metabolism in plants, animals, or microorganisms would be a complete interpretation of the mechanism of protein biosynthesis and of protein breakdown. This is currently one of the most active areas of biochemical inquiry and speculation. It is in this area of knowledge that biochemistry and genetics have come together, with the evident possibility of interpreting some of the most basic attributes of living systems, namely, their ability for self-duplication and adaption. Therefore, a resumé of current status of this field is necessary as the background against which the problem in plants may be discussed.

There are a number of excellent diagrammatic summaries of current concepts of protein synthesis and to the more detailed citations to the relevant literature, which can be found in the following references: Watson (597), Crick (119), Nirenberg (361), Tsugita and Fraenkel-Conrat (549), Schweet and Bishop (441), T'so (548), Simpson (452), McQuillen (316), Allfrey (5), Ochoa (368), Perutz (389), Raacke (405), Webster (600), Gale (174), Bonner (56), Zamecnik (633), Berg (34), Brachet (71), Steward and Thompson (507), Hoagland (215), Meister (320).

The most important plants in nature for their protein biosynthesis are autotrophic green plants, for these combine the ability to fix energy by photosynthesis and to convert inorganic forms of nitrogen into protein. These organisms bring about that primary linkage of carbohydrates and nitrogenous compounds, without which the animal kingdom would be ineffective. Therefore, it might seem that an understanding of the mechanism of protein metabolism would have come first from those systems that utilize the simplest starting points and which carry out the process in its entirety. This is far from true, for most of the work on protein biosynthesis has been done with systems that are often quite specialized in their protein metabolism and are more restricted than is the autotrophic green plant.

Viruses have furnished special protein, which could be both recognized and purified; moreover, their synthesis is possible only in the environment of the cell or in situations that recapitulate special features of this environment. Nevertheless viruses have contributed greatly to current knowledge and may even have affected disproportionately ideas on the mechanism of primary protein synthesis. Such special proteins as hemoglobin, and physiologically active protein molecules like insulin, have received great attention because they are obtained pure and their physiological functions have both permitted and justified a great deal of attention to these molecules. Thus the gross amino acid composition of proteins has been extended into the understanding of the detailed linear arrangement of amino acids along the polypeptide chain, and this has led, not only to the interpretation of the genetic determination of this arrangement, but to quite specific views upon how the amino acids must be incorporated *de novo* into a system of this sort. Other proteins which are endowed with distinctive properties as enzymes, e.g., ribonuclease, and which can also be purified, yield to similar methods and have furnished similar information. A summary and discussion of this area of knowledge will be found in Anfinsen (9).

These examples do not necessarily represent the most general protein synthesis as it occurs in plants. At the outset the question should be raised whether all protein synthesis must of necessity occur by similar routes and by similar mechanisms? In growing, dividing cells, where self-duplication of their respective units is the chief event, not only is the protein manufactured, but the organization is recapitulated. In later aspects of growth in which the total protein of the cell increases greatly during elongation and maturation, but without the dramatic reduplication of organelles, the chief emphasis is upon increase of substance, not the duplication of complex organization. In storage organs, protein may be deposited as a reserve substance in cells that have ceased to grow either by division or enlargement. Therefore, the following questions arise. Is the large amount of globulins deposited in such storage organs as tubers, bulbs, rhizomes, etc., of necessity synthesized by exactly the same type of machinery that applies to the *de novo* formation of protein ingrowing, dividing cells? Or, when such special proteins as the glutelins, the gliadins, etc., are deposited in the endosperm of cereal grains, must the mechanism again be the same? After the primary synthesis of protein in a nitrogen autotrophic cell, the breakdown products are re-used in some other site or some other cell; is it necessarily true, therefore, that the mechanism of resynthesis must be the same as that which applied *de novo?* Is the synthesis of the very large amount of chloroplast protein necessarily by the same route that leads to the cytoplasmic proteins, or to the proteins of nuclei? It may not yet be possible to answer these questions, but they should, be posed. Nevertheless all speculation upon protein biosynthesis leans very heavily upon knowledge gained on systems that represent only very restricted examples drawn from the whole range of situations in which protein is synthesized in nature.

Concepts of protein structure and protein biosynthesis naturally go back to the polypeptide theory of protein structure and to Emil Fischer's attempt to build such polypeptide chains by chemical condensation of amino acids. Although these attempts at protein synthesis, which regarded synthesis as the reverse of hydrolysis, seem now as ingenuous as the early attempts to synthesize such carbohydrates as starch and glycogen by reversing their hydrolysis, they nevertheless achieved some significant results. Fischer himself built up polypeptides of substantial size although he never achieved anything of a sufficient degree of complexity even to approach the structure of a protein. Wasteneys and Borsook (1930) and Borsook (60, 61) hydrolyzed proteins and attempted to recombine their hydrolytic products; although they did not achieve the reformation of the protein in question, they nevertheless produced materials of some complexity which they termed "plastein." All these early attempts, necessary as they were, were nevertheless doomed to failure because it is now known that protein synthesis, like much carbohydrate synthesis, is not merely the reverse of hydrolysis. In fact Borsook's earlier work, summarized in 1950, demonstrated that the incorporation of labeled amino acids into the proteins of cell free preparations requires energy. Significant progress could not be made until much more became known about the means which living organisms use to apply the chemical energy of respiration at the site where synthetic work is done.

Although it had long been recognized that a principal outlet for the energy of respiration was the synthesis of such energy-requiring compounds as proteins, no significant progress was made until the role of the phosphorus compounds became known. The idea that the important consequence of respiration was to develop a flow of phosphorylated compounds like ATP, which could apply their energy at selected points and behave, as it were, like the currency through which biological energy transactions were negotiated, derived from the respected work of Kalckar and Lipmann (cf. Chapters in Volume IA).

Lipmann (276, 277) suggested that the energy of one phosphate bond, as in ATP, could be donated to produce one peptide linkage. The same idea was exploited by Bergman and Fruton [cf. Meister (320) for references] in the formation of synthetic polypeptides. In fact, the role of ATP and similar compounds that specifically donate the energy derived ultimately from respiration to the formation of particular peptide bonds, permeates all the conceptual work on protein biosynthesis. Much of the later speculation and research is to see how this synthesis can be brought about in ways that produce the highly specific and ordered structures that emerge.

A first step toward modern ideas on protein biosynthesis was the gene-enzyme hypothesis, usually attributed to Beadle and Tatum (29). The mode of action of genes required that they should be able to modify specific biochemical reactions by determining the presence, or absence, of the enzyme necessary to catalyze that reaction. In its earliest form, the hypothesis regarded each gene as leading to the formation and action of one enzyme, and in so doing the link was forged between the gene, as a site on the chromosome, and the determination of some protein of highly specific properties. Among the reactions that lent themselves to genetic study and to association between particular genes and the reaction in question were some that involved the synthesis of amino acids essential to the formation of protein required in growth (152). Thus mutant strains of *Neurospora* could be found which, being unable to manufacture lysine, for example, could grow only if this amino acid were supplied exogenously in the medium. This type of result not only attributed the formation of the particular enzyme needed in the formation of lysine to the presence or absence of a particular gene, but it also gave credence to the view that free exogenous lysine could be used at the site of protein synthesis. This supported the idea that all the amino acids needed in protein synthesis are synthesized by, and exist as free entities in, the cell prior to their condensation. It was, however, pointed out quite early (502, 507) that this latter conclusion did not necessarily follow, for higher plants certainly do not manufacture and store in quantity all the free amino acids in the proportions in which they are required for protein synthesis. Thus one has to regard the gene-enzyme relationship, as it related to the preformation of the amino acids needed for protein synthesis, as determining not the prior accumulation of all the protein amino acids, but their synthesis *at a point close to the protein synthesizing template;* hence the intermediary amino acids of synthesis may not in fact mingle with those that do exist in the free state. If, the gene for the formation of an amino acid is lacking, then this forces the exogenous amino acid to become conspicuous by its necessary exogenous supply. This, however, does not mean that the normal strain which has all the genes necessary for the manufacture of protein from nitrate would necessarily lead to the *free* formation of the amino acid in question.

Later it was shown [cf. Steward *et al.* (482), and references there cited] that the intermediary events of primary protein biosynthesis, which draw carbon from sugar and nitrogen from nitrate, do in fact proceed via intermediates that do not mingle with the soluble amino acids of cells that exist in quantity, as in vacuoles. These results were obtained from cells (carrot) in an active state of division under tissue culture conditions.

The achievements of the gene-enzyme hypothesis, which for the first time forged the link between genetics and metabolism; the knowledge that the chromosomes contained pentose nucleic acids and that these substances were the most likely to constitute the material substance of the genes; and the knowledge that a virus could be interpreted as a protein moiety whose synthesis was stimulated by a nucleic acid moiety, all combined to suggest similarities between genes, enzymes, and viruses—and in fact to suggest that the primordial, living, self-duplicating particle might have had something of the properties, or constitution, of a gene or a virus. The fact that the infective part of a virus is located in the nucleic acid, so that virus and protein can recombine even in unnatural combinations, also gave great impetus to the idea that nucleic acids formed the essential substance upon which, as a template, proteins were formed. This work has been recently summarized (528).

Thus there emerged a philosophy of protein synthesis, rather than a body of evidence that amounted to absolute proof. The essential structure of the genetic material is regarded as being composed of pentose nucleic acids. In some way the pentose nucleic acid holds the key to its own self-duplication and to the formation of structures, surfaces, or templates at which characteristic proteins may be made. Hence the link between the hereditary units, which are genes, and the metabolism through which they act, is determined by the formation of proteins with definite characteristics as enzymes. This general philosophy has received some dramatic confirmations but still has areas that are more properly to be regarded, even yet, as convenient working hypotheses.

The concept of Watson and Crick (598) which showed how the coiled helices which constitute the molecule of DNA might hold the key to their own self-duplication represented one great advance. This idea, coupled with knowledge of the combinations possible in the arrangement of the nitrogen bases along this coiled helical thread, permits explanations of the means by which the biochemical information which is the chemical blueprint of the organism's metabolism is held to reside in the DNA. Since abnormalities in the hereditary material produced by mutant genes result in abnormal proteins, which have now been shown to possess highly specific derangements of their amino acid configuration (as in the work on the abnormal genetically determined hemoglobins) (230) also gave a great impetus to the general theory as outlined. Ingram (230) showed that different human hemoglobins which are inherited as if controlled by a single gene, have identical sequences of amino acids except for a single amino acid which is replaced by another at a given position (19).

Early in the modern development of ideas of protein synthesis, Dounce (136) attempted to formulate the type of reactions that would permit an amino acid to be placed on a condensing surface, or template, in such a way that regularity of sequence could be achieved and, at the same time, make provision for phosphate bond energy to result in the formation of a peptide bond. In fact, Dounce (137) has very recently revised these earlier ideas to fit the observations of Cavalieri and Rosenberg (1961) (cf. 95), who, on the basis of the isolation of four-stranded DNA from bacteria during the nondividing state proposed that the double helix itself, rather than two single strands, represents the unit which is concerned in the replication process.

The current view is that such macromolecular biosynthesis proceeds on templates, complementary or otherwise. The wide acceptance of this view flows from the success of the model proposed by Watson and Crick (598) for replication and from the *in vitro* demonstration of DNA replication by Kornberg (257) (cf. also 37).

B. Sites of Protein Synthesis in the Cell

Being committed to the idea that protein synthesis is associated with RNA in the cell, it is natural to associate any local accumulations of RNA with protein synthesis. Knowledge of the fine structure of cells, as it is revealed by the electron microscope (325, 401, 402, 605), has made rapid advances due to techniques of fixation with potassium permanganate, which now preserve detail in the cytoplasm. Even so, they are particularly applicable to cells which are relatively non-vacuolate, so that for higher plants at least the information derives almost entirely from the densely cytoplasmic cells of the apical meristems. These are, however, cells which are especially active in growth by self-duplication and in concomitant protein synthesis.

In addition to such self-duplicating organelles as nuclei, plastids, mitochondria, and Golgi bodies, the electron microscope has shown that the cytoplasm has a vesicular system of fine threads, or tubules, known as the endoplasmic reticulum. The reticulum ramifies throughout the cytoplasm and may even be closely associated with the properties of membrane surfaces. It also presents channels of communication from the nuclear envelope, which has visible pores through which large molecules could pass, and may be a possible path of communication for such molecules throughout the cytoplasm. Bordering the endoplasmic reticulum, however, abundant minute granules appear in electron micrographs. These granules are supposed to consist of ribonucleoprotein material, and thus to be local centers of protein synthesis (cf. Fig. 50, which shows ribosomes both attached to reticulum and also free in cytoplasm). It is now conceded that much of the particulate matter, from homogenized cells hitherto termed "microsomes," consists of the fragmented reticulum with its adhering minute granules of ribonucleoprotein material (548). Thus, any residual capacity of such microsome particles for protein synthesis may derive from their association with this ribonucleic acid material. Although much of this evidence of fine structure of cytoplasm comes from work on animal cells, plant cells seem to be essentially similar. A close correlation between the RNA content of the microsome fraction pellet and the rate of protein synthesis has been noted by Oota and Osawa (370) and by Martin and Morton (296) for plant material. Such observations lead to the consideration that microsomes, probably because of their high RNA content, play an exceedingly important role in protein synthesis.

One can now visualize that the mitochondrion of plant cells produces the energy which would be locally stored as ATP or some equivalent substance. Meanwhile appropriate sources of nitrogen would be mobilized from the vacuole or the storage regions. One may then expect the synthesis of cytoplasmic protein to occur in association with the ribonucleic acid granules already referred to. Membrane surfaces, upon which the reticulum is condensed and to which ribonucleic acid particles adhere, might then become active areas of synthesis by virtue of these features.

Work by Morton and Raison (336a) deals with the synthesis of storage protein in cells of wheat endosperm. Work on amino acid incorporation in cell-free preparations and on the electron microscopy of the endosperm cells is reported. These authors recognize a special cytoplasmic plastid-like inclusion in the cytoplasm of the endosperm cells which is electron dense and which synthesizes protein. This is termed a "protein-forming plastid" or "proteoplast." This type of syn-

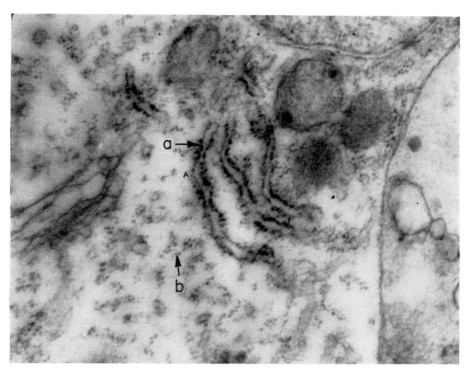

Fig. 50. Cytoplasm of actively growing carrot cells cultured by the aid of coconut milk, showing ribosomal particles both attached, as at *a*, to the surface of the reticulum and free, as at *b* in the cytoplasm. From work of H. W. Israel working with one of us (F.C.S.).

thesis is assumed to occur on ribonucleoproteins (plastid ribosomes) which receive their messenger RNA in relatively stable form. Another novel feature of this work is that it invokes the inositol phosphates (phytols), which are rich in seeds and are supposed to account for certain very dense protein-bound phytate bodies seen in electron micrographs. This is believed to furnish phosphate and to be able to regenerate ATP from ADP and so replace the need, partially or wholly, for glycolysis and the Krebs cycle as the source of ATP. All this suggests that the formation of these storage proteins occurs in different ways from that of the proteins of cells in plant growing regions, or even of bacteria. However, final judgment should be reserved until this work is extended and confirmed.

Studies with microbial and mammalian cells indicate that external amino acids can by-pass a large intracellular pool and be incorporated

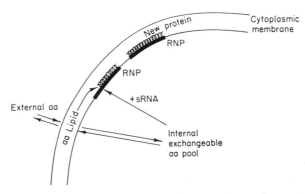

FIG. 51. (1) Protein is synthesized at an interface between a lipoprotein membrane and a ribonucleoprotein film. (2) Amino acid preferentially approaches the interface from the lipoprotein membrane side, either from the outside or inside of the cell. (3) Amino acid from the interior pool could approach from the RNA side utilizing sRNA reactions. RNP = ribonucleoprotein or ribosome; sRNA = soluble ribonucleic acid. Cytoplasmic membrane is to be representative of any cell membrane such as endoplasmic reticulum or nuclear membrane. After Hendler (209).

directly into protein. This and other observations have led Hendler (209) to propose a model for protein synthesis (Fig. 51) which has some points of similarity with certain ideas of Steward, Bidwell, and Yemm (482) for plants.

C. THE PRESENTATION OF NITROGEN AND ENERGY TO THE SYNTHESIZING SURFACE: MODERN IDEAS ON PROTEIN SYNTHESIS

A great deal of biochemical experimentation and a great deal more speculation have been addressed to this question. All the prominent ideas assume that the protein-synthesizing surface receives the appropriate amino acids already prefabricated. These ideas may be true in the sense that the amino acids may be manufactured in the appropriate proportions *closely adjacent to the site of synthesis*. However, they are obviously not true in the sense that the protein amino acids occur free in the right proportions and in the right qualitative mixtures in the cell as a whole. The ultimate source of carbon for the synthesis of protein in plant cells is sugar. The ultimate source of nitrogen is either nitrate or ammonia, which is made available via certain nitrogen-rich compounds like the amides or arginine. Nevertheless one has to assume that amino acids are formed at some point immediately prior to the actual formation of the peptide bonds. All the current speculations upon protein synthesis and protein metabolism begin, therefore, with the preformed amino acids, and try to visualize their arrangement in the appropriate order along the peptide chain and also their condensation to form the peptide bonds, by which successive amino acid residues are united.

There is available a curious mechanism for making a variety of γ-glutamyl peptide bonds, but these are of a kind that do not contribute directly to protein synthesis. This is the mechanism investigated by Hanes and others (197, 199) by which a variety of γ-glutamyl peptides can be made. This is a sequel to the known synthesis of glutamine. Indeed, glutamine can even be regarded as a "glutamyl hydrogen peptide." Glutamine can be synthesized from glutamic acid and ammonia, with energy donated by ATP, and thus energy is incorporated into the amide linkage of glutamine. An attractive earlier hypothesis was to regard the known ease with which glutamine acts as a source of nitrogen for protein synthesis, as being in some way associated with the energy which had been incorporated into its amide group.

Later, when Hanes et al. (197–199) showed that amino acid residues could be moved from a peptide like glutathione and be attached to form γ-glutamyl peptides, leaving a cysteinylglycyl residue to pick up another amino acid residue, this mechanism seemed to have potential and general significance. In short, the cysteinylglycyl moiety of glutathione acts like a carrier which can transfer an amino acid residue onto the glutamyl portion of the eventual γ-glutamyl peptide, and it is conceived that the energy built into the amide linkage is contributory to this end. However, the fact that the peptides so manufactured involve γ-linkages, instead of α-linkages, seems at first sight to rob this mechanism of any relationship to protein synthesis. Indeed, it has to be admitted that no proof yet exists that it is so related to protein synthesis, although it was earlier suggested that the glutamyl molecule might in this way be a means of conducting preformed amino acids up to the protein-synthesizing surface and of carrying with them a measure of energy that is associated in the γ-peptide linkage. Thus, if the template or pentose nucleic acid surface could act as a γ-glutamylpeptidase, it would permit a great variety of amino acids to reach their destination in the form of a single type of carrier molecule. These thoughts, associated with the ideas of Dounce (136) on the phosphorylation of the template surface, seemed earlier (492) to have some merit.

Work of Thompson, Virtanen, and Fowden (see Appendix II) has revealed a variety of glutamyl peptides in plants which potentially correspond to most of the protein amino acids, excluding the basic ones (arginine, lysine, histidine), but also including amino acids *never* found in protein. These observations are still interesting even though they do not readily find a place in current thinking on protein synthesis. Conceivably, this mechanism by which free amino acids are attached to another molecule, may yet serve as a storage device or carrier mechanism to mobilize certain amino acids at cellular sites, or organelles, even though it may not be a direct part of the final synthesis of the peptide or protein. Ito and Strominger (1962) have in fact described a different mechanism of peptide synthesis, which occurs in bacteria. Apparently there is here a stepwise addition of single amino acids, e.g., L-alanine, D-glutamic acid, and L-lysine, which are added successively to a uridine nucleotide, in the formation of long peptides. Thus, in this system, amino acids are successively presented to a uridine nucleotide on which a polypeptide grows.

D. CURRENT CONCEPTS OF PROTEIN SYNTHESIS

1. The ultimate biochemically coded information that determines the metabolism of the organism is carried in the DNA of the nucleus, which is conceived as a double helix of deoxyribose phosphate nucleic acid chains which, on the Watson-Crick model, are wound around each other. While the two chains have the bases arranged in a similar linear sequence, the order is reversed in such a way that they may be held together by hydrogen bonding between cytosine and guanine on the one hand, and by thymine and adenine on the other (cf. Figs. 11a and b of Volume IA, pp. 58–59).

2. The information that is carried by the base sequence on the DNA of the nucleus is transmitted to the RNA of the cytoplasm. It is now claimed that a site for the control of RNA synthesis has been located and mapped on what is conceived to be the chromosome of *E. coli* (4a). Although usually the RNA is regarded as comprising a single-stranded helix, it may be coiled back upon itself and appear as two strands held together by hydrogen bonds. As in the case of DNA, so in the case of RNA, the strand consists of ribose phosphate chains, to which nitrogen bases are attached to each of the sugar units, though in this case the sugar is D-ribose not D-deoxyribose. Adenine, guanine, and cytosine represent three of the bases in RNA, though a fourth is uracil in place

of the thymine of DNA. The linear structure of the DNA imparts to a single strand of RNA (mRNA) a certain amount of information which is represented by the sequence of bases along its length. DNA replication may occur in the manner of Fig. 52. The mRNA finds its way into the cytoplasm, reaches the ribosomes and determines the surface or organization which can then act as the template in synthesis. For the over-all scheme, reference may be made to Fig. 53B, the details of which will be commented upon below.

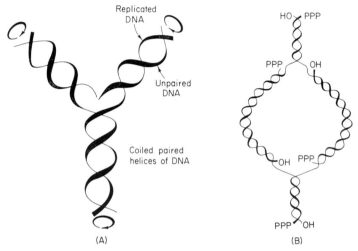

Fig. 52. Schematic replication mechanism for DNA reproduced from Delbrück and Stent *in* "The Chemical Basis of Heredity" (W. D. McElroy and B. Glass, eds.), pp. 699–736. John Hopkins Press, Baltimore, Maryland, 1957. (A) "Speedometer" model of Levinthal and Crane (1956).

(B) Scheme to emphasize the occurrence of unpaired DNA in the middle of a strand, which also replicates, and that there is polarity in the single strands which is determined by their end groups.

3. ATP brings about the necessary linkage to form an amino acid enzyme complex with the liberation of pyrophosphate and AMP. (The latter can be reconverted to ATP only by a reentry into the aerobic respiratory cycle.[13]) The formation of the amino acid–AMP-enzyme complex, termed amino acid activation, is said to be magnesium dependent, and, of course, specific activating enzymes (pH 5 enzymes) are required to catalyze these reactions. The amino acid is then transferred from its amino acid–AMP-enzyme complex to a specific sRNA. thus liberating, for re-use, free AMP and enzyme. Thus, an sRNA-

[13] It has been suggested (cf. page 562) that an inositol phosphate regenerating system for ATP may function in plant seeds.

amino acid is formed, and it is this entity which moves to the template or messenger RNA activated surface (see Fig. 53A). Zubay (636) visualizes adjacent adapter, or sRNA, molecules, which can attach to the mRNA at one end "by unpaired bases" and bring amino acids into juxtaposition at the other (see Fig. 53A).

4. The actual arrangement of the bases of the DNA of the nucleus acts as the template for the formation of what is called "messenger RNA"; this comes directly from the nucleus and carries the information originally "coded" in the DNA. For speculations on how this may be achieved see Watson (597). Whereas the mRNA of bacteria is regarded as unstable and very transient, that of higher organisms may need to be more stable. The relations between mRNA, the ribosome, and the adapter sRNA to produce the growing peptide are shown schematically in Fig. 53 after Zubay (636). Visual evidence for strands of mRNA attached to ribosomes has been presented by Warner *et al.* from Rich's laboratory [*Science*, **138**, 1399–1403 (1962)].

5. Granted that the sRNA-amino acids are directed to the right part of the protein-forming surface, i.e., the ribosomes or ribonucleic acid granules, they are aligned in the proper sequence for a given protein or polypeptide. Apparently the amino acids are assembled into polypeptides in a linear order, starting with a free amino group and ending with a free carboxyl. Apparently the functional units of protein synthesis are ribosomal aggregates, or "ergosomes," which consist of particles held together by mRNA.

6. The next step in peptide bond formation is not spontaneous, but requires manganese and GTP (guanosine triphosphate) [cf. Webster and Whitman (601)]. This step entails the liberation of the sRNA's and the condensation of the freed amino acids. Ergosomes are broken down during the amino acid incorporation into protein *in vitro*.

7. The final stage represents the "peeling-off" as it were of the newly formed polypeptide in the form of protein, leaving the ribosome particle free to be re-occupied by more amino acid residues. The combination of polypeptides into protein and their rolling or coiling to form their special configurations is another problem (597).

Much detail is omitted from the above summary, and much biochemical evidence which gives a measure of credence to the steps so visualized has also been omitted. Reference may be made, however, to the general summary chart (Fig. 53B), which illustrates these current concepts of protein synthesis. However, certain reservations should be made at the outset. If protein synthesis can withstand actinomycin D, which inhibits RNA and DNA synthesis [as in some systems (227a)], then apparently a stable template surface would be required instead of the usual concept of a transient mRNA-determined system. In fact the whole idea of mRNA, useful as it would be, is still hypothetical as it applies to higher plants.

Clearly the ideas outlined above derive their chief support from considerations that flow from the structure of certain special proteins, with special properties, whose amino acid sequences have actually been determined: that is, hemoglobin, insulin, and certain enzymes. Up to the present time no plant protein, especially no cytoplasmic protein, has attracted sufficient attention for this information to be obtained. Furthermore, the picture as outlined above comes into its own when it can be related to isolated systems, even though the total amount of synthesis which they produce is small, often infinitesimally small. And it also gains special credence from the relations between the nucleic acid moiety of viruses and the protein to which they give rise. All the

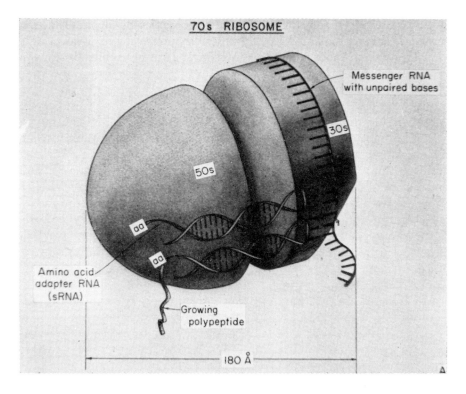

Fig. 53. (A) Visualization of adjacent adapter, or sRNA, molecules. After Zubay (636).

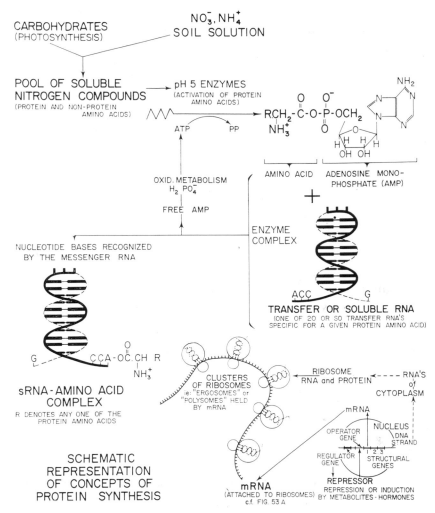

Fig. 53. (B) A schematic representation of current views on protein synthesis as visualized in molecular terms and utilizing ideas drawn from the study of bacteria and of cell-free systems. It comprises a synthesis of the views on (a) amino acid activation and recognition by specific enzymes, (b) the transfer of amino acids by sRNA and their location at specified points on a ribosome template surface, (c) the genetic (DNA) control over the nature of the template surface via mRNA, and (d) the carrying by mRNA, by a linear sequence of bases in triplets (each specific for an amino acid), of the information to arrange the protein amino acids in the linear order in which they are bound. Attention should be focused on the broad principles involved rather than upon the detail of the individual schematic diagrams, e.g., of the nature of sRNA or the precise relations of mRNA to the ribosome and to the genetic apparatus, for these may well be subject to change in this rapidly moving field.

above events, therefore, may be conceived as occurring in the *immediate vicinity of the protein synthesizing surface*, whether this is a membrane surface, a ribosome particle, or some other organelle. However, all attempts to produce continuing or net synthesis with isolated cellular preparations, which approximate the uniformity or purity of these structures, results in total protein synthesis which, though significant from the standpoint of working out reactions and mechanisms, is really negligible in comparison with the synthesizing efficiency of the whole cell. Here lies one of the major problems which the intact plant cell still seems to present.

Although pea (*Pisum sativum*) ribosomes have been claimed to synthesize proteins *in vitro* in milligram quantities (405, 600), other investigators (268) have claimed that these soluble proteins only arose by loss from the particulate fraction. A net *de novo* increase in protein was not, therefore, demonstrated. Lett and Takahashi emphasized that the conditions for optimal synthesis in Webster's system (and presumably also in Raacke's) were far removed from the optimal ones for the stability of ribosomes. Moreover, neither of the synthetic systems used by Webster or Raacke had been studied from the standpoint of RNA template destruction by the ribonuclease known to be present. Lett and Takahashi (268) refer to the fact that, up to that time, Webster had not been able to achieve even amino acid incorporation with preparations from pea varieties (including cultivar 'Progress') except in the one batch of cultivar 'Alaska' peas with which he originally worked. Raacke also worked with 'Alaska' peas.

E. The Template and Coding Problems in Protein Synthesis

The mechanism by which each amino acid recognizes, as it were, its proper place on the template seems to be as follows:

The terminal base on each sRNA chain is always a guanidyl residue (G)—and this is, therefore, not part of the *specific* amino acid code. In fact, the amino acid *itself* does not form part of the code which is entirely a property of the sRNA molecule. A minimum number of other bases (three) are so arranged that they locate the sRNA on the template surface. Such a set of three bases is termed by Crick a "codon." Each "codon" of the sRNA is similar to, but not identical with, a complementary grouping on the template surface. Thus each amino acid is directed to the template by the complementarity of two sequences of three bases (one in the sRNA, and one on the template).

Although the amino acids, still attached to sRNA which in turn is anchored to the template, are linked together by peptide α-C links (cf. Fig. 53B) there are still other problems. Where does the energy come from that forms the α-C peptide linkages? Does this condensation proceed like a wave [a sort of zipperlike motion (cf. 318)] or is it brought about one or several residues at a time? Having released its amino acid, why and how does the sRNA vacate the template to repeat the cycle? These and other questions lead one naturally to ask what is the experimental basis for this very interesting but nevertheless speculative hypothesis?

Thinking in this area was greatly influenced by the use of a cell-free enzyme system for the synthesis of polypeptides. In such systems certain polynucleotides could replace what would normally be the natural "messenger RNA." If these polynucleotides contain only uracil, the entire template surface is dominated by sites (codons) that can attract only one sRNA which, in this case, proves to be specific for phenylalanine. Thus in this system a polyphenylalanine was made and thus the "codon" for phenylalanine is believed to be the sequence UUU (where U = uracil). A similar observation for the formation of polylysine attributes the lysine code to AAA (where A = adenine).

Crick states (119), "It has yet to be shown by direct biochemical methods that the code is indeed a triplet code." On the other hand, all prevalent thinking is directed along these lines. Though ultimate proof is lacking, it is the thought that the genetic protein code is universal from bacteria to man.

On the assumptions outlined above, the laboratories of Ochoa and Nirenberg (265, 266, 298, 361–363, 468, 469) have attempted to decipher the RNA-protein code by measuring the effect of synthetic polynucleotides with known base composition, on the incorporation of amino acids into the TCA-insoluble material of well-washed ribosomal systems from *E. coli*. When polyuracil was added to the preparations of Nirenberg and Matthaei, which contained washed ribosomes, the incorporation of L-phenylalanine into protein was stimulated a 1000-fold. Of eighteen C^{14}-amino acids tried, only L-phenylalanine incorporation was stimulated. The phenylalanyl-sRNA is an intermediate in the process (363). Support for these ideas is also given by studies on the participation of viral RNA (549, 550) and enzymatically synthesized RNA in the incorporation of amino acids by a soluble protein-ribosome system from *E. coli*.

Examples of the code which designate the incorporation of specified protein amino acids into polypeptides are contained in Table XXII. The code was derived experimentally by the use of synthetic polynucleotides. The information in Table XXI has been called the "genetic code dictionary." Thus a given mRNA contains the information for a specific protein, and achieves this by means of a linear array

TABLE XXII
Genetic Code Dictionary of Code Words That
Correspond to Each of the Twenty
Common Amino Acids[a,b]

Amino acid	RNA Code Words			
Alanine	CCG	UCG■		
Arginine	CGC	AGA	UCG■	
Asparagine	ACA	AUA		
Aspartic acid	GUA			
Cysteine	UUG△			
Glutamic acid	GAA	AGU■		
Glutamine	ACA	AGA	AGU■	
Glycine	UGG	AGG		
Histidine	ACC			
Isoleucine	UAU	UAA		
Leucine	UUG	UUC	UUA	UUU□
Lysine	AAA	AAG●	AAU●	
Methionine	UGA■			
Phenylalanine	UUU			
Proline	CCC	CCU▲	CCA▲	CCG▲
Serine	UCU	UCC	UCG	
Threonine	CAC	CAA		
Tryptophan	GGU			
Tyrosine	AUU			
Valine	UGU			

[a] After Nirenberg (361).

[b] The code assumes that all the words are triplets. The sequences of the letters in the code words have not been established, hence the order shown is arbitrary. Although half of the amino acids have more than one code word, it is believed that each triplet codes uniquely for a particular amino acid. Thus various combinations of AAC presumably code for asparagine, glutamine, and threonine. Only one exception has been found to this presumed rule. The triplet UUU codes for phenylalanine and, less effectively, for leucine. A, adenine, C, cytosine, G, guanine; U, uracil. Symbols: △ Uncertain whether code is UUG or GGU; ■ need for U uncertain; □ codes preferentially for phenylalanine; ● need for G and U uncertain; ▲ need for UAG uncertain. For even further elaborations of these ideas see *Science* 14, 479–484.

of the "triplet code words" shown in Table XXII. Unfortunately, however, the code is now thought to be somewhat degenerate (at least in *in vitro* systems), i.e., there may be more than one "code word" for some of the amino acids.

If transfer sRNA were to act as an "adaptor molecule" in protein synthesis [cf. Crick (119)], it must be composed of at least 20 different molecular species, one for each protein amino acid, and even more if the specificity is not absolute. In fact, Berg *et al.* (35) have found that the transfer RNA chains which bind leucine can be separated into two fractions that have different terminal sequences:

—pGpCpApCpCpA
—pGpUpApCpCpA

where p = phosphate, G = guanine, A = adenine, C = cytosine, and U = uracil.

Again, the arrangements for incorporating leucine may depend upon two alternative sequences of sRNA nucleotide bases. Weisblum, Benzer, and Holley (602) have fractionated transfer RNA to yield two separate peaks, each containing molecules capable of combining with leucine. When added to a cell-free ribosomal system, leucine, attached to one transfer RNA, was incorporated under the stimulus of poly-UC as the synthetic messenger. No significant incorporation occurred in response to the stimulus by poly-UG or poly-U. However, leucine attached to another transfer RNA was incorporated in response to poly-UG and, to a much lesser extent, in response to poly-U, while poly-UC had no significant effect.

The results leave little doubt that *two different terminal nucleotide sequences in sRNA* both recognize leucine, and therefore leucine may be carried to different synthetic templates by two different molecules of sRNA. *Thus the incorporating mechanism may not be as absolute as was at first supposed.* Eck (142) has analyzed the possibilities inherent in 4 nucleotide bases taken 3 at a time. Sixty-four possible combinations reduce to 32 similar pairs if a given purine is replaceable by another, or a given pyrimidine by another pyrimidine. This seems to mean that there is more than enough information to code for twenty or more different protein amino acids in this way; or again, alternative, i.e., "degenerate," base combinations are possible for the same amino acid.

F. THE GENETIC CONTROL OF PROTEIN SYNTHESIS

The structure of the protein-forming surface is conceived to be genetically determined, and this is the way in which genes act insofar as they determine the intimate structure and configuration of proteins that are, in turn, to act as enzymes and thereby control metabolism. It now appears that a functional genetic unit is a length of DNA which carries the information for the assembly of one polypeptide chain. Thus a "structural gene," i.e., one that determines the structure of a polypeptide, has been defined by postulating that a point mutation in this gene will affect only one particular polypeptide chain, either by stopping its synthesis or by altering its amino acid sequence.

Most concepts of genetic control of protein synthesis, however, derive from microbiological systems in which astronomical numbers of individuals can be used. In bacteriophage, mutant subgenic units, separable by recombination, have been referred to by Benzer (33) as "cistrons." These units, though subgenic, are still considered to be large in that they may contain a hundred or more bases of the DNA. In fact ultimately, a single mutational change in a single nucleotide pair of a particular nucleic acid might cause a unit change of protein structure.

By analogy, it is now generally believed that genes, or their subunits, bring about the formation of specific proteins. Unfortunately, it has not yet proved possible to analyze the base or nucleotide sequence of DNA either by genetic or chemical methods and, at the same time, ascertain the sequence of amino acids in the corresponding protein.

The present concepts, however, provide for greater flexibility and more ability for the effect of the genes to be modified by extra or "epigenetic" factors. Waddington (589) conceives that these modifications can be cytoplasmic in nature, and they can operate at different levels in the progressive expression of the effects of genes. At one level, Waddington conceives that the cytoplasm can modify, by a "feedback" mechanism the formation of the immediate protein products of the genes. These effects are presumably quantitative. At another level, the cytoplasm acts to determine how far genes interact with each other. And, yet again, the cytoplasm may influence the way in which the proteins resulting from the action of the genes interact to produce the visible or phenotypic effects.

The regulation of genes in enzyme synthesis in terms of two classes of interacting genetic loci has been formulated by Jacob and Monod (232, 233). "Structural genes" determine the structures of individual protein molecules which are metabolically active as enzymes. Each structural gene apparently carries the information for a single polypeptide chain, and a mutation of such a gene, therefore, could result in a major change in the structure of the appropriate polypeptide or enzyme. Other genes, however, affect the intracellular concentration (i.e., rate of production or synthesis) of the metabolic enzymes but not, necessarily, their structure. Two types of such genes have been described. "Operator" genes are recognized by the fact that their mutation affects the rate at which adjacent structural genes on the same chromosome can direct the synthesis of their respective proteins. Each "operator" appears to have a specific group of structural genes which are under its immediate control, and this complex of the "operator" and the "structural" genes under its jurisdiction are collectively known as the "operon."[14] All the structural genes of an operon are affected similarly by any change in the function of the operator gene.

Each group of structural genes is believed to be affected by "operator" genes that control the activity of all the structural genes of an operon. However, each "operator gene" may also be turned on or off by a second gene and this is called a "regulator gene." A mutation in the "regulator," as indeed in the "operator," gene will affect the whole group of structural genes under its control. If the "regulator gene" produces an mRNA which is a repressor, then it interferes with the

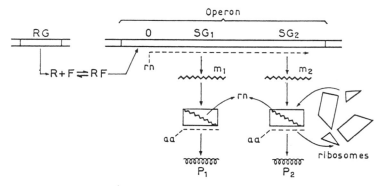

Fig. 54. General model for the regulation of enzyme synthesis in bacteria. *RG*: regulator gene; *R*: repressor which associates with effector *F* (including or repressing metabolite); *O*: operator; SG_1, SG_2: structural genes; rn: ribonucleotides; m_1, m_2: messengers made by SG_1 and SG_2; aa: amino acids; P_1, P_2: proteins made by ribosomes associated with m_1 and m_2. After Jacob and Monod *in* "Cytodifferentiation and Macromolecular Synthesis" (M. Locke, ed)., pp. 30–64. Academic Press, New York, 1963.

RNA produced by the operator; hence the effect of the operator is not transmitted to the structural genes, and this constitutes an inhibition of these structural genes. On the contrary, the structural genes are called into play if the operator gene is allowed expression, by virtue of its mRNA. This occurs when the product of the regulator gene is suppressed by some small molecule either as an end product of metabolism or, like a growth regulatory hormone, introduced into the system. These ideas are all expressed in Fig. 54. It is important to see here a means by which a variety of simple molecules may intervene in ways that would affect the proteins to be made by cells and thus provide for explanations of growth-regulating substances or even the transition to flowering [cf. Heslop-Harrison, *Brookhaven Symp.* 163, 109–125 (1963)].

[14] Further reference may be made to Stent, G. S. The operon: On its third anniversary *Science* 144, 816–820 (1964).

Mutants can arise that alter or delete the expression of the regulator, or the operator, or even of the structural genes. Perutz (389) has suggested that further development of this work may well lead to the discovery of the "organizers" for which biologists searched in vain in the 1930's and thence to an understanding of growth and differentiation.

A new and radically different concept has also emerged from genetic work. This is the concept of a single giant mRNA molecule, which may contain the code for several different proteins. For example, nine enzymes are needed for the synthesis of histidine, and some believe that all nine are determined, sequentially, by one giant mRNA molecule. It has been suggested that the whole region of DNA that codes this giant mRNA might be regarded as a gene [cf. *Cold Spring Harbor Symp.* **28** (1963). These interesting ideas are obviously still tentative and still to be evaluated. Clearly, however, the first ideas of an unstable, highly specific mRNA have had to be modified to include a possible stable mRNA system, and now there looms up the possibility of mRNA's that are very large and determine many proteins.

Some reputable workers with higher animals do not accept without reservation the ideas which follow from work on microorganisms (Harris, 202a). Work with enucleated *Acetabularia* and actinomycin D indicates that the regulatory control of protein synthesis may be mediated by means that do not originate in the nucleus and also by means that do not involve a transient form of RNA [Harris, *Nature* **202**, 249, 1301 (1964)]. Possibly DNA in organelles may be a source of mRNA; however, this would not be in accord with concepts of the nuclear origin of mRNA.[15]

G. Protein Synthesis and Morphogenesis

J. T. Bonner once said that, to understand the facts of differentiation, one needs to visualize a continuing set of "chemical conversations between nucleus and cytoplasm." Such ideas provide for the flexibility, along with the genetic concepts, that is needed to interpret differentiation and morphogenesis.

A converse idea comes out of the cytology of insect larvae and the detailed study of the chromosomes of large salivary glands during metamorphosis. Recognizable segments of these chromosomes become active during specific stages in the development of the insect. (This activation may be under chemical, i.e., hormonal control.) In brief periods of intense activity, recognizable and localized swellings or "puffs," which can be labeled by thymidine, occur on the chromo-

[15] *In* "Essays on Nucleic Acids" (Elsevier, New York, 1964) E. Chargaff has expressed reservations on the concept of mRNA.

somes. Possibly there is even a release of nucleic acids to the cytoplasm. In this way, distinctive stages in development may be characterized by a distinctive complement of RNA, which in turn could evoke special categories of protein. In fact, it has already been shown that DNA, RNA, and protein synthesis may occur in conjunction with puffing (cf. 293, 432). Although there is no known parallel for these events in plants, it is always possible that specific groups of genes may be temporarily activated even though visible "puffs" have not been seen. Perhaps a more reasonable contact may be found in the action of growth-regulating substances in plants, for they may act by locally activating genes in specific areas of chromosomes; however data are still lacking to prove these ideas.

The Cold Spring Harbor Symposia in 1961 and 1963 have dealt with "allosteric interactions" which concern alterations in the structure of enzymes. Enzymes may change their structure, and hence their specificity, in the presence of certain small molecules. The small molecules may bind to the surface of a protein and therefore induce conformational changes that affect the behavior of sites elsewhere on the protein. Thus it is possible that an enzyme may catalyze one chemical reaction on one occasion and be inhibited by, or even catalyze, a different reaction under different circumstances. In this way, the different chemical pathways of the cell may be linked together by allosteric effects and thereby be made mutually controlling. Perhaps certain hormones may eventually be shown to achieve their effects by "allosteric interaction."

H. Protein Metabolism in Higher Plants in Relation to Modern Views on Protein Synthesis

Against the general, philosophical, and biochemical background of protein synthesis outlined above there are still some major questions to be answered, particularly where flowering plants are concerned. These questions include the following:—Where are the sites of primary, or first-formed protein synthesis from nitrate in the plant body since plants can re-use their nitrogen after breakdown? How is the nitrogen required for protein synthesis brought to the vicinity where protein synthesis occurs? And how is the energy required to make the peptide bonds also donated at the site where the condensing molecules need to be activated? Furthermore, there is a question where the property of protein synthesis is localized in the cell. Is it generally in the cytoplasm or is it in inclusions or on cytoplasmic granules, or on membrane surfaces? A further question is how a cell, with a fixed complement of genes, can be caused to synthesize different proteins under different conditions and at different times. The protein comple ment of cells seems to change throughout ontogeny, and there is every reason to suppose that it is also a function of the environmental conditions (see Section VI, E).

In any cell anabolic and catabolic processes proceed simultaneously and in balance. Although the final steps of the reactions that involve the re-entry of nitrogen into protein may occur in the immediate vicinity of ribosome particles, the cell has many other compartments. These often include a large aqueous vacuole, in which a great diversity of nitrogen compounds may occur, and by no means all of these have any immediate connection with protein synthesis (cf. Section IV). Furthermore, for many cells the carbon which enters into protein readily is that which comes from carbohydrates, and this is, of course, particularly true of the photosynthesizing cells. Therefore, whatever credence is given to the general picture of protein synthesis as outlined above which stems from amino acids preformed *at the immediate site of synthesis*, and however accurately these ideas enable one to visualize the linear sequences of amino acids in the proteins ultimately formed, this picture really represents *only the last step of the long journey through nitrogen metabolism to the formation of plant protein*. From the standpoint of the plant physiologist and the cellular physiologist, there is much still to be understood in the conversion of the reduced products of nitrate or of ammonia, and the carbon compounds which represent the initial "port of entry" for that nitrogen into organic combination, into the forms which are conducive to protein synthesis and which eventually culminate in the formation of the amino acids at the actual site of synthesis. In the study of these aspects of nitrogen metabolism and protein biosynthesis the intact cell is a *sine qua non*, for the preparation of homogenates eliminates the very organization which is essential to the smooth and continuing operation of these interlocking reactions.

Much of the soluble nitrogen in the cell is certainly not present in the form that is immediately available to combine with soluble RNA in the manner indicated above. It may, however, be present in soluble nitrogen pools, which are under some circumstances accessible to the sites of synthesis though in other circumstances they may be bypassed by the metabolites that are en route to protein. It is now essential, therefore, to recognize those parts of the machinery of nitrogen metabolism that depend, not solely on the ultimate events at the template surface, but upon the smooth operation of the entire cellular organization and the interlocking of its anabolic and its catabolic phases of metabolism. In this respect, also, the metabolic machine is seen to be subject to a wide range of effects superimposed upon it during development and by the nutritional and environmental factors that determine growth and morphogenesis; and the morphogenesis is itself accompanied by very drastic alterations in the protein complement.

Thus in this chapter the problems of nitrogen metabolism are presented in their full range and complexity. This is necessary as a counterpart to much current biochemical speculation which, to be intelligible, has needed to reduce cells to their simplest terms and to avoid many of the responses of plants that are of great interest to the plant physiologist. From this standpoint the events in the immediate vicinity of protein-synthesizing ribosome particles are only a part, even though an important part, of the metabolism of nitrogen compounds in plants.

IX. The Regulation of Nitrogen Metabolism and of Protein Synthesis *in Vivo*

A. THE REGULATORY CONTROL OF PROTEIN METABOLISM

From the great range of factors, environmental, nutritional, and metabolic, that affect nitrogen metabolism and protein synthesis, it is difficult to isolate any one as the key factor or master condition that regulates the whole. Nevertheless much thought has been given by plant physiologists over the years to the regulatory control of protein metabolism.

1. Relations to Factors That Determine Growth

The most potent factors that direct the metabolism toward net synthesis of protein are the hardest to be specific about, because they represent that complex of factors which together determine growth. There is no more potent system for protein synthesis than the growing cell or organism. Beside this the potentialities of isolated particulate preparations (cf. Section VIII), often incapable of little more than brief periods of "incorporation" without net synthesis, pale into insignificance. Perhaps Chibnall (106) foresaw this when he stressed how much of the nitrogen metabolism of leaves was dictated by what he could only refer to as hormone-like activities derived from the roots. Thus, a relative lack of any of the essentials for growth may constitute a limiting factor for protein synthesis. Similarly, any factors that cause growth to be induced where it was otherwise quiescent appear as stimulants of protein synthesis, as, for example, factors such as those in coconut milk, coconut milk plus 2,4-D, or other synergists, which when added to a culture medium cause cells to grow and divide and to synthesize protein. Conversely such inhibitors as L-hydroxyproline, the naturally occurring inhibitors in maple buds, in potato tubers, etc., or such inhibitors as gramicidin-S or chloramphenicol, or a recently discovered inhibitor obtained by autoclaving urea (396), or the effect of radiation on culture media,[16] all can intervene in this complicated system to stop or restrict protein synthesis.

But what external factors promote net synthesis of protein in cells which are sufficiently well endowed with metabolites to undertake this? As shown by the work on tissue slices cut from potato tuber, a prime factor is the oxygen tension; for, in a solution in equilibrium with oxygen tensions lower than that of air, the behavior of the cells is limited by oxygen pressure. But the effects induced by oxygen can be accentuated by certain ions, e.g., by K^+, NO_3^-, $H_2PO_4^-$ and suppressed by others, i.e., Ca^{++}; the most powerful suppressor of all is an increase of CO_2/HCO_3^- in the tissue in proportions that maintain an otherwise optimum pH of the solution [(498); and cf. Chapter 4 of Volume II]. But there are other equally potent internal factors. Slices cut from tubers after prolonged storage at $+1°C$ fail to synthesize protein, even though they respire at an even higher rate than slices which are cut from normal tubers. Cells of the latter, therefore, lack the something that links metabolism constructively toward growth and even cell division, for the tubers stored at low temperature will not form dividing cells, as in a cork phellogen, when they are exposed to moist air (480). The tubers stored at $+1°C$ had, as it were, passed a low temperature-induced climacteric beyond which they became incapable of protein resynthesis and, in fact, behaved like cells of certain cotyledons which, when germination ensues, can only break down their protein reserves and lose them (e.g., to the environment or to a young seedling).

Agents, like coconut milk, which promote both growth and protein synthesis in otherwise dormant cells do so, in part, because they affect the efficiency of the RNA in those cells to act as a site of protein synthesis. Carrot cells in a basal but complete (salts, sugars, and vitamins) nutrient medium undergo a lag period of 4 days (483), and it is now known that during this period their RNA per cell increases dramatically [work of Holsten, cf. Steward et al. (489a)]. If the coconut milk growth stimulus is lacking, this high RNA content does not lead on to continued protein synthesis and growth, but if it is present, the RNA adjusts to a somewhat lower, and maintained, level characteristic of the continuously growing synthesizing cells (Fig. 55). Thus growth-regulating substances may so act that they determine the efficiency of existing RNA in protein synthesis. This shows how both exogenous and endogenous factors can be concerned in the regulation of protein synthesis in such cells.

The climacteric of fruits, such as the apple and the banana, also constitutes a point of no return. Hulme (225) did detect a small protein

[16] Holsten, R. D., and Steward, F. C. *Plant Physiol.* **40**, Suppl. (1965).

synthesis at the climacteric concurrently with the briefly enhanced respiration, but both were of short duration; there is every reason to suppose that the cells of the postclimacteric apple lack the ability for further protein synthesis. In the banana, it has been shown by Mohan Ram[17] working in this laboratory with one of us, F.C.S.) that cells of this fruit will respond in culture to coconut milk and to a suitable synergist (e.g., NAA) and grow as free cells, but they will not do so once the climacteric has been passed. There are, therefore, some very fundamental characteristics of growing cells which permit their

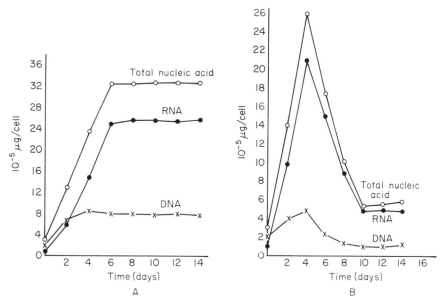

FIG. 55. The nucleic acid content of cells of carrot explants cultured on a modified White's basal medium with (B) and without (A) added coconut milk.

metabolic energy to be harnessed to protein synthesis, and this is one of the first properties to disappear when, in the course of development, age, or senescence, cells can no longer grow. It is a major task to translate these regulatory mechanisms into precise biochemical terms.

2. Nitrogen Metabolism in Relation to the Level of Carbohydrates

One of the oldest ideas is that protein synthesis in plants is regulated by the available carbohydrate. Since the schemes of Pfeffer, Schulze, and Prianishnikov, it has been recognized that certain stored nitrogen-rich compounds (amides and arginine) furnish nitrogen; therefore, to make protein they require additional carbohydrate from which to build the carbon skeletons of the protein. This idea took its more

[17] Mohan-Ram, H. Y., and Steward, F. C. *Can. J. Botany* **42**, 1559–1579 (1964).

modern form in the Paech hypothesis [which was in turn summarized by Chibnall (106)]. Although the involvement of sugar as the ultimate source of carbon in protein synthesis is beyond dispute, the regulatory role of sugar concentration cannot be sustained. In the potato tuber a substantial sugar concentration after starch hydrolysis is a feature of the system in which protein can be synthesized, but thereafter many factors that promote synthesis deplete the sugar and vice versa. Hence the changes that increase protein synthesis may often affect the sugar concentration adversely [(498), see also Volume II, Chapter 4]. Moreover, a great increase of sugar occurs in the tubers exposed to $+1°C$, but it is after such treatment that the protein synthesis from the native stored amino acids disappears entirely. No useful conclusion, therefore, comes out of attempts to place the internal sugar concentration in a role which regulates protein synthesis directly.

Lyndon (285) has studied the nitrogen metabolism of detached leaves of *Kalanchoe* in the dark with special reference to their content of glutamine. Since *Kalanchoe* is a succulent with the typical organic acid metabolism of these plants, there was an especial attempt made to see whether the formation of soluble nitrogen compounds was correlated in any distinctive way with the content of organic acids. It would have seemed reasonable if this were so, especially in the dark during proteolysis. In the outcome, however, neither glutamine nor other soluble compounds were correlated in any noteworthy manner with the content of organic acids. The only correlation that seemed meaningful was that between the content of glutamine and the oxygen uptake during the period of its synthesis, which is best explained as a need for energy (ATP) in the synthesis of the glutamine, as Speck (465, 466) and Elliott and Gale (144) showed some years ago. In fact, the leaves of the succulent behaved toward nitrogen in a very similar manner to nonsucculent leaves; i.e., protein was synthesized during the light, and broken down during the dark (especially in the excised leaves) with the formation of amides (especially glutamine).

3. Other Early Ideas on the Regulation of Protein Metabolism

An old idea was that amino acids combined by elimination of water and that at, or near, protein isoelectric points there is minimum swelling or imbibition of water, so that this pH could be thought of as favoring protein synthesis. Again there seems no modern merit in this idea, for the amino acids are surely presented for protein synthesis in some combined (sRNA-amino acid or phosphorylated) form, and it is hardly likely that their condensation is simply by elimination of water. With potato tuber slices, one can show that treatments of the slice, owing to oxygen pressure and to various salts that increase the protein content of the disks, invariably also *increased* their fresh weight, so that one should not regard free access of water and increased hydration of the cell as a whole as being in any way opposed to protein synthesis; in fact, quite commonly the reverse is true, for a high content of protein and high water content (i.e., succulence) often go hand in hand (387).

Claims that individual amino acids can regulate protein synthesis by virtue of their concentrations in the cells are also difficult to uphold. A key factor here is that the soluble nitrogen compounds are best regarded as being in a soluble phase, or pool, or compartment, which is quite remote from the actual site of synthesis (482). This being so, attempts to determine protein regulation by the total concentrations of free amino acids in the cell seem to be unprofitable. Since the effects of lack of sulfur on the size and composition of the soluble pool of nitrogen compounds are so great (see Volume III, Chapter 2, Figs. 62–63), and since the sulfur amino acids are so readily incorporated into protein that they rarely accumulate greatly in the free state, there are somewhat better grounds for attempting to interpret protein regulation in terms of the concentration of free S-amino acids (cysteine or methionine). Australian workers (26, 201, 618) have pursued this line of thought for the leaves of various grasses.

Protein regulation in cells should not be considered separately. Protein synthesis is the chief anabolic activity of cells that use energy, derived ultimately through photosynthesis and made available through respiration, to do synthetic work and to grow. Protein synthesis, like so many processes (e.g., ion accumulation) that are for all practical purposes dependent upon the full organization of living cells, will tend to follow the lead of cell division and enlargement. Hence Chibnall was in this sense correct when he attributed much of what occurred in *Phaseolus* leaves to factors that emanated from the roots to make leaves grow. Granted the ability of cells to grow, and therefore to synthesize protein, the pace of the synthesis will be affected by the over-all pace of growth: it may be increased in light, decreased in darkness, responsive to the supply of this or that essential nutrient, but essentially it is because the cell can grow that it can synthesize protein, and it is also because it can synthesize protein that it can grow. We need, therefore, to look for interrelationships between protein synthesis and the other anabolic and catabolic aspects of metabolism in cells.

B. PROTEIN METABOLISM IN RELATION TO RESPIRATION

The more physiological point of view developed above restricts what may be learned by the study of protein synthesis on isolated systems.

The latter are most useful in describing the feasible biochemical re-actions. The more physiological approach encourages one to seek unified concepts of cells and organisms in which the various functions interact and complement each other to form a smoothly working integrated metabolic machine. This approach is very necessary to elucidate the rate of continuing metabolic processes which are already in train. The view of nitrogen metabolism, now to be developed may be closer to the truth, but it is less subject to the direct proof which work on a limited or isolated part of the system can more readily yield, as, for example, in the recent impressive studies of the biochemical mechanisms of protein synthesis as made upon microorganisms and particulate preparations. However, in this section other complexities than these must be faced.

For example, in the cells of higher plants, the relations between vacuole and cytoplasm, between large soluble pools of metabolites and protein are complications that are not met with in some lower organisms. Thus protein metabolism is not only an affair of nuclei, genes, and ribosomes and their contained pentose nucleic acids. There is, in the vascular plant as a whole, the still greater degree of complexity which involves relations such as those between different organs with respect to their nitrogen metabolism.

1. Relations to Respiration: Earlier Evidence

Since respiration furnishes the stores of usable energy (as ATP or other such compounds) it is to be expected that it will often go hand in hand with protein synthesis. Since the Krebs cycle (261) can operate as a route of synthesis, producing keto acids as "ports of entry" for nitrogen, respiration is involved here also as a means of mobilizing carbon for nitrogen metabolism. In photosynthesis the nitrogen metabolism of the green cell, or leaf, has direct access to the intermediates that lie along the path of carbon in photosynthesis, to the energy-rich products of photosynthetic phosphorylation and to the reduction products from pyridine nucleotides; it is, therefore, to be expected that green cells in the light should be active in nitrogen metabolism without involving light more directly in the protein synthesis per se. (There is some recent claim that DPN, i.e., NAD, is more characteristic of leaves in the dark than is TPN, i.e., NADP, which occurs in green cells in the light, so that specific reactions mediated by these coenzyme systems could be turned on, or off, by darkness or light as the case may be.)

Growth in the most active centers of cell division (embryos, shoot and root tips, buds, cambia, developing storage organs or fruits and seeds, etc.) creates a "sink" toward which nitrogenous metabolites can converge because they are there used in synthesis. It has been pointed out also that ion accumulation may often proceed concomitantly with protein synthesis [Chapter 4 of Volume II; also (492)]. Therefore, problems of movement arise—movement across the scutellum in such grains as wheat (*Triticum*) or corn (*Zea*) from endosperm to embryo, into the cotyledons from endosperm to embryo as in buckwheat (*Fagopyrum*); or movement over long distances, predominantly in the phloem, of the "elaborated" products of nitrogen metabolism from leaves to other growing centers as investigated on the cotton (*Gossypium*) plant by Mason and Phillis (304, 305) and by Maskell and Mason (299–303). In so complex an interlocking pattern limitations may arise at many points and thus cause local accumulations of nitrogeneous products, or divert metabolism into alternative pathways.

The dilemma is that the entire interpretation cannot be made for any one plant. Therefore, the minimum number of systems that illustrate the entire range of interactions should be considered. They are presented as a series of diagrams (which though now individually somewhat out of date) have directed thought along useful channels (cf. Fig. 56).

The history of the factors that affect nitrogen metabolism in the respiration of leaves is a long one. Spoehr and McGee (470) emphasized that amino acids in leaves had more to do with their respiration than the then current ideas about the complete consumption of carbohydrate to carbon dioxide and water would lead one to suppose. Such early experiments as those of Warburg and Negelein (594) and others show that an effective way to stimulate respiration is to furnish nitrate exogenously to green plants. And in fact Yemm (629) found that as much as 40% of the carbon that emerged as carbon dioxide from starving leaves of barley (*Hordeum vulgare*) arose from protein.

In a now classical paper, Gregory and Sen (186) visualized the nutrition and metabolism of barley in the terms of Fig. 56A. Even before Vickery *et al.* (571) postulated protein turnover in leaves from their observations on the incorporation of N^{15} into tobacco leaf protein, Gregory and Sen (186) came essentially to the same conclusion. They found that at maturity some leaves which were at a somewhat deficient level of potassium or phosphorus had a relatively lower content of protein-N and a higher soluble-N than their full nutrient counterparts. If protein synthesis were the reverse of hydrolysis, Gregory and Sen expected that the high amino acid concentration would have tended to reverse this effect but, if this were not so, then the higher soluble nitrogen of these plants could be a source of carbon from

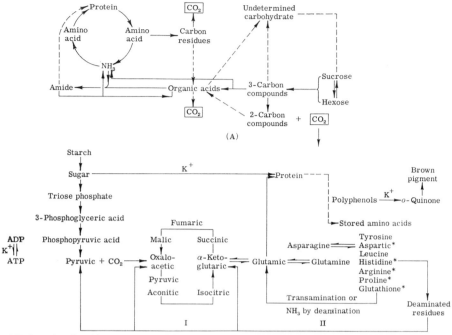

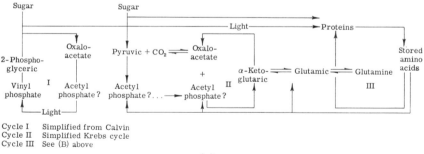

Fig. 56. Charts that show the history of schemes relating respiration to protein metabolism. Reproduced from Steward and Thompson (507). (A) After Gregory and Sen (193). (B) After Steward and Street (194) (C) After Steward and Thompson (950).

which the higher respiration of these leaves was maintained. In their view, then, a part of the leaf metabolism passed over a protein cycle, the pace of this cycle being a function of the nutrition (especially with respect to levels of K, N, P). The number of turns of the cycle (rather than the amount of protein present at any point in time) was, therefore, held to determine the pace of respiration which was mediated in this way. Thus Gregory and Sen regarded some carbon from carbohydrate as entering into the constitution of protein and, following breakdown, it was released as carbon dioxide. On this concept the large soluble pool of free nitrogen compounds which accumulate, represent, not the immediate precursors of protein, but rather the end products of its breakdown. As the products of protein breakdown became deaminated, the nitrogen was worked back into the nitrogen metabolism, while the carbon frameworks were respired away. The novelty of this scheme lay in the authors' willingness, even at that early date, to regard at least a part of the carbon dioxide which is evolved as having come from products (amino acids) which only recently were bound in the form of protein.

The second metabolic scheme (Fig. 56B) came from a different area of research using different materials, namely thin disks of storage tissue (potato tuber), which were allowed to respire under controlled conditions in either water or dilute single salt solutions. By other means, Boswell and Whiting (63) had partitioned the respiration of such disks into two moieties; one (approximately one-third of the maximum) being oxygen saturated at a low concentration level and also being quite indifferent to any conditions that suggested its involvement with the phenolic bodies and phenolases of the tissue; the other (approximately two-thirds of the total) was more responsive to treatments and conditions which did suggest that this respiration emerged from pathways mediated by the phenolic compounds. In the experiments of one of us (F.C.S.) similar tissue was exposed to various neutral salts and to other treatments, and it was noted consistently that potassium salts enhanced the slight browning which these slices undergo in aerated solution; at the same time it increased their respiration and promoted the conversion of soluble nitrogen compounds into proteins. By contrast, increasing concentrations of calcium salts with a common anion, had the converse effect. In these experiments, insofar as the respiration was affected by the salts (i.e., by the cations), the treatments that conserved the sugar concentration (high calcium) decreased the respiration, whereas the treatment that reduced the sugar concentration (high potassium) actually increased the respiration. Thus, the salts affected respiration in ways that were not intelligible if the sugar concentration played a rate-limiting role, but they did become intelligible if there were a relationship between a salt-sensitive respiration and the protein synthesized concomitantly. Furthermore, if one plots total carbon dioxide respired against total protein synthesized (as affected by many different variables), one may easily see from the regression line that approximately one-third of the maximum aerobic respiration proceeded over pathways which were oxygen saturated at a low level and in which nitrogen metabolism is not directly implicated. The remaining two-thirds of the maximum respiration, however, varied with the protein synthesized and increased as the tissue embarked upon the surface browning reaction, which is due to the activity of phenolases [cf. Steward *et al.* (500), p. 418].

At first these effects were mysterious. The best that could be said was that orthoquinones, formed from orthodiphenols during the browning reaction [in the manner worked out originally by Raper (408a)], helped in the oxidative deamination of the amino acids prior to their use in respiration. It was thought that the nitrogen would be re-used to synthesize more protein and the carbon residue could then be oxidized away. A puzzling feature, however, was that CO_2/HCO_3^- in the solution, even at the ratio required to maintain an external pH of 7.0, and even when accompanied by K^+ could, at appropriate but still somewhat low concentrations of total salt, completely inhibit all the responses in question, including the protein synthesis from stored soluble compounds. This response was attributed to a specific effect of carbon dioxide and/or HCO_3^-. Although this effect was observed long before dark fixation of carbon dioxide was known, it was quickly recognized, when the Wood and Werkman reaction was discovered, as a particular manifestation of this phenomenon. In short, an effect of the very rapid and efficient aeration (which was conducive to protein synthesis, to oxygen supply and to removal of carbon dioxide as fast as it was formed) was to maintain in these cultures a limitingly low concentration of free carbon dioxide. At these low levels of carbon dioxide, decarboxylations are favored, and those carboxylation reactions which are necessary to mediate the entry of carbon from glycolysis into the Krebs' cycle would also be limited, indeed blocked, by a virtually complete lack of carbon dioxide. However, it is under these circumstances that new protein is synthesized from carbohydrate and from nitrogen donated from the stored soluble reserves; the latter, having contributed their nitrogen, thus enter the Krebs' cycle as it were "from the back door" as keto acids and are respired away.

Thus, a part of the total metabolism and a part of the total carbon dioxide produced should be related to the metabolism of the nitrogen compounds. The fact that carbon enters protein from sugar and emerges from it again by breakdown means that the rate of this cyclical operation will determine the associated part of the respiration. Such factors as oxygen tension and potassium concentration enhance the rate of synthesis and respiration; such factors as Ca^{++} and carbon dioxide concentration tend to slow down the rate of synthesis, or to divert the path of carbon back again to the use of glycolytic end products rather than the carbon derived from metabolism of nitrogen compounds.

Later work has shown that carbon dioxide is indeed fixed by potato slices and that the principal fixation product is glutamine (493). Work has also shown that glutamine is a strategic metabolite of these cells. C^{14}-glutamine when added to growing tissue cultures, labels the carbon dioxide more quickly and, at first, more effectively even than when the tissue receives C^{14}-sugar (482). In fact, the labeling of the free sugars in the tissue with C^{14} is by no means a guarantee that the carbon dioxide which is immediately evolved will be labeled with equal intensity. The labeled carbon first accumulates in protein. Finally, C^{14} asparagine is a remarkably inert metabolite in cells of this type and it often behaves quite differently from glutamine.[18]

Although Fig. 56B is now old, it still embodies a general idea which can be adapted to the interpretation of respiration and nitrogen metabolism in cells.

As interpreted above, a key role was assigned to carbon dioxide which, by its presence or absence, could switch the path of nitrogen metabolism into one channel or the other, specifically toward protein synthesis when the CO_2 tension in the cells was low and vice versa. From this standpoint green leaves in the light and in the dark present interesting contrasts. It has long been known that the green leaf in the light is an important organ of protein synthesis, whereas in the dark, and especially so in detached or senescent leaves, protein breakdown may proceed apace (cf. Yemm, cited above).

Two other facts are necessary to understand the background of Fig. 56. First, the early experiments with $C^{14}O_2$ and *Chlorella* in the light showed how rapidly C^{14} enters, not only into certain free amino acids (alanine, aspartic and glutamic acid, serine, glycine), but also quite generally into protein from carbohydrates, or their intermediates. For example, Benson and Calvin (32) said: "The major portion of the insoluble products formed in the first few minutes by algae . . . was protein." Also, the Calvin school noted that when $C^{14}O_2$ was fixed and metabolized in the dark, the C^{14} readily entered into a variety of compounds, notably those which are intermediates of the

[18] Bidwell, R. G. S., Bon, R. A., and Steward, F. C. *Nature* 203, 367-373 (1964).

Krebs' cycle, but when the algae were fixed without having had the opportunity to metabolize $C^{14}O_2$ except in the light, the radioautographs were surprisingly free from labeled intermediates of the Krebs' cycle. At first Benson and Calvin (32) drew the physiologically unacceptable explanation that leaf respiration stops in the light, and the Krebs cycle ceases to work. To resolve this dilemma, Steward and Thompson (504) formulated a third schematic diagram, complementary to those already mentioned. Although this scheme, when formulated, made use of a then postulated carbon dioxide acceptor (vinyl phosphate), for which there is now no basis [and which may now be replaced by the system which enables ribulose diphosphate to fix carbon dioxide, to be regenerated, and to hand on 3-PGA as the first fixation product (cf. Volume IB, pages 138 et seq.)], it nevertheless serves to emphasize that there may be three reversible, interlocking, cyclical, metabolic events in green cells.

The first of these cycles is the light cycle. This generates ribulose phosphate in the light and in effect adds carbon dioxide to a 2 C unit which is not itself free at the time, but which eventually makes its stable appearance as 3-PGA. In the dark, however, sugar and glycolysis furnish two-carbon fragments which, as acetyl CoA, enter into the carboxylic acid cycle, but they require external free carbon dioxide to yield and maintain molecules, at the complexity of isocitric acid, which are built up before decarboxylations and dehydrogenations actually begin. In the light, this pathway should fail or be restricted, because the cells may be quickly deprived of the free carbon dioxide which is needed to maintain the organic acid cycle, using glycolytic products as the entering fuel. However, it is under these circumstances (i.e., the leaf in the light) that the cells are best able to make protein, using sugar and also nitrogen from their nitrogen-rich storage compounds, like glutamine. After the latter have donated their nitrogen, their carbon frameworks become available to be respired away. Therefore to label the Krebs' cycle intermediates from $C^{14}O_2$ in the light, the C^{14} would have to go first in bulk into the protein, via sugar, would need to be removed by breakdown and enter into the oxidative cycle prior to its final issue as carbon dioxide. This seems to be what the early experiments of Benson and Calvin (32) showed.

This method of approach has proved illuminating in the interpretation of other data (e.g., some recently published from Krotkov's laboratory), but by far the most interesting adaptation of this general line of thought comes from the interpretation of effects imposed on the respiration and protein metabolism of growing and nongrowing carrot explants. The growing explants received coconut milk in addition to their basal medium, but the nongrowing explants did not have this additive.

The additional and suggestive thought that is now needed to understand this system relates to its complexity as a cell. Different phases or compartments of a cell may contain the same metabolites in pools which do not mingle, but which may be carrying out very different reactions in their respective compartments.

The idea of compartments is required to interpret experiments in which C^{14}-γ-aminobutyric acid, C^{14}-glutamine and C^{14}-sugars were applied to otherwise comparable sets of carrot explants. All the arguments are too detailed to recapitulate here, though incidentally some of the key pieces of evidence have recently been confirmed by Steward and Bidwell (481, 482) using C^{14}-labeled metabolites (C^{14}-glutamine, C^{14}-sucrose, C^{14}-asparagine). The ideas and interpretations are as follows—

When the cell is in nitrogen balance, anabolism and catabolism are also in balance. Under the influence of the factors in coconut milk that cause growth and cell division several responses occur; these are (a) enhanced protein synthesis, (b) enhanced turnover, (c) increased respiration appropriate to (a) and (b). Metabolically active protein is largely labeled from C^{14}-sugar, and in so doing the C^{14} in the glutamic acid of the protein may be considerably more radioactive than the free glutamic acid of the cell. Obviously, therefore, the carbon which entered the glutamic acid of the protein had bypassed the main pool of free glutamic acid of the cell. Paradoxically, if one wishes to label the issuing carbon dioxide, one may often furnish the tissue preferentially with a radioactive nitrogen compound; if one wishes to label the protein of the cell quite generally, one may furnish, not radioactive amino acids, but C^{14}-sugar.

The data from time experiments were interpreted to mean that the growing cell turned sugar and nitrogen groups into protein without allowing any of the intermediate amino acids (which were possibly bound to RNA or present as sRNA complexes) *to mingle with the free solutes of the cell.* After protein breakdown, however, the amino acids—subject to being reworked for nitrogen-rich storage—may be, released into the vacuole, or the soluble nitrogen pool. Thereafter the nitrogenous groups may be re-used in synthesis and their carbon frameworks respired away. To maintain such a system geographical separation into cellular compartments is necessary. One must also postulate alternating breakdown and resynthesis of protein at rates which will be determined by whether the cell is in active growth, or is quiescent. Thus, net synthesis of protein is one measure of the activity of the cell system; the rate of its turnover and resynthesis is another.

2. Relations to Respiration: Some Later Evidence

Further light on the relationship between carbohydrate and nitrogen metabolism in cultured growing cells is available from experiments which have not yet been published *in extenso*. These experiments are part of the further collaboration between R. G. S. Bidwell of the University of Toronto and one of us (F.C.S.).

A symmetrically designed experiment was performed to test the ease with which C^{14} could move from sucrose and various nitrogen compounds into the protein. The medium to which the carrot explants were exposed either contained coconut milk (+CM for rapidly growing explants) or no coconut milk (—CM for slowly growing explants), but it was also supplied with one sugar (sucrose) and 7 different amino acids. In any one treatment, however, only *one* of these 8 compounds was C^{14}-labeled. In other words all the growing or nongrowing explants received the same 8 supplements, the only difference being the compound which was C^{14}-labeled.

When sucrose was the C^{14}-substrate, the specific activity of the $C^{14}O_2$ which emerged was relatively lower, in relation to the C^{14}-sucrose in the tissue, in the growing than in the nongrowing condition; hence the CO_2 was being "diluted" from nonradioactive sources when the tissue was being caused to grow. Thus the coconut milk stimulus for growth brings into play other sources of carbon than sugar for carbon dioxide production. Whereas the sucrose in the tissue was ten times more radioactive in the growing explants than in the nongrowing, the $C^{14}O_2$ which it produced was only six times more radioactive. This condition obtained despite the fact that all the soluble compounds were generally more concentrated in the nongrowing explants. Therefore, there must be some alcohol-insoluble source for carbon dioxide which is especially drawn upon in the actively growing tissue, and this could be the protein via protein turnover as previously suggested (see Table XXIII). The data of this experiment can also be calculated in such a way that the actual amount of carbon from each source which enters each protein amino acid (and its soluble counterpart) is shown (see Tables XXIV and XXV).

Some of the amino acids in protein, such as proline, arginine, and tyrosine, derived their carbon largely and directly from the same amino acid when it was supplied, whereas others were as readily, or even more readily, labeled from sucrose or some different source. In fact, glutamic acid in the protein was more effectively labeled by ex-

TABLE XXIII

Specific Activities of the Carbon of Sucrose and of Respired CO_2 in "Fast-Growing" ($+CM$) and "Slow-Growing" ($-CM$) Carrot Explants during the Supply of Sucrose-UL-C^{14} [a]

Parameter	(a) Fast-growing ($+CM$)	(b) Slow-growing ($-CM$)	$\dfrac{\text{Col. a}}{\text{Col. b}}$
Sp. act. of sucrose in the explants	5.7	0.55	10.3
Sp. act. of CO_2	1.70	0.293	5.8

[a] Sucrose-UL-C^{14} of specific activity-28 mμc per microgram of carbon.

TABLE XXIV

Amounts of Carbon That Entered the Free or Protein Amino Acids of Fast-Growing ($+CM$) Carrot Explants from the Various Substrates Present in the Medium[a]

Free (F) or protein (P) amino acid		Components of the medium and their concentration in mg/100 ml							
		Sucrose 10.4	Glutamic acid 2.13	Serine 0.73	Alanine 13.6	Valine 2.24	Proline 3.84	Arginine 0.47	Tyrosine 0.29
Aspartic	F	6.7	1.9	2.2	0	0	1.3	0	0
acid	P	2.88	0.46	0	3.09	0	0.34	0	0
Glutamic	F	17.3	11.2	0	0	0	3.7	0	0
acid	P	Trace	2.66	0	11.5	0	1.48	0	0
Serine	F	27.0	0	30.4	20	0	3.5	0.65	0.36
	P	9.3	0	8.0	0	0	0	0	0
Threonine	F	0	0	0	0	0	0	0	0
	P	1.32	0	0	1.55	0	0	0	0
Alanine	F	66	0	0	134	0	0	0	0
	P	2.56	0	0	10.8	0	0	0	0
Valine	F	0	0	3.2	0	60	9.9	3.1	0
	P	4.03	0	0	1.77	1.01	0	0	0
Leucines	F	0	0	0	0	4.2	0	0	0
	P	2.86	0	0	9.25	4.67	0.89	0.59	0
Arginine	F	0	0	1.1	0	0	0	21.6	0
	P	0	0	0	0	0	0	20.6	1.2
Tyrosine	F	0	0	0	0	0	0	0	16.9
	P	0	0	0	0	0	0	0	17.3
Proline	F	0	0	0	0	0	163	1.4	0
	P	0	0	0	0	0	4.72	0	0
Glutamine	F	7.1	8.1	0	0	0	4.0	21.8	0

[a] Eight parallel treatments were used, in each of which a different component of the medium was labeled with C^{14}. Results as micrograms carbon per 20 explants. The amount of each substrate was chosen to provide 25 μc of C^{14} in the medium that received the labeled compound.

TABLE XXV

Amounts of Carbon That Entered the Free or Protein Amino Acids of Slow-Growing (−CM) Carrot Explants from the Various Substrates Present in the Medium[a]

Free (F) or protein (P) amino acid		Components of the medium and their concentration in mg/100 ml							
		Sucrose 10.4	Glutamic acid 2.13	Serine 0.73	Alanine 13.6	Valine 2.24	Proline 3.84	Arginine 0.47	Tyrosine 0.29
Aspartic acid	F	0.150	0	0.68	0	0	0	0	0
	P	0.132	0	0	0.15	0	0	0	2
Glutamic acid	F	0.59	0.026	0	0	0	0.15	0.013	0
	P	0.215	0.042	0	0.65	0	0.34	0	0
Serine	F	0	0	1.05	0	0	0	0	0
	P	0.48	0	0.56	0	0	0	0	0
Alanine	F	0	0	0	Some	0	0.24	0	0
	P	0.075	0	0	0.75	0	0	0	0
Valine	F	0	0	0	0	0.067	0	0.10	0
	P	0	0	0	0	1.5	0	0	0
Leucines	F	0	0	0	0	0	0	0	0
	P	0.106	0	0	0	0.51	0	0	0
Arginine	F	0	0	0	0	0	0	Some	0
	P	0	0	0	0	0	0	0.83	0
Tyrosine	F	0	0	0	0	0	0.3	0	0.02
	P	0.025	0	0	0	0	0	0	1.23
Proline	F	0	0	0	0	0	2.4	0.055	0
	P	0	0	0	0	0	1.5	0	0
Glutamine	F	0	0.032	0	0	0	0.15	0.052	0

[a] Eight parallel treatments were used, in each of which a different component of the medium was labeled with C^{14}. Results as micrograms carbon per 25 explants.

ogenous alanine than even by exogenous glutamic acid. While this might be considered to be due to the relative ease of penetration of the cells by these substrates, this was *not* the explanation. An amino acid that did not readily enter the protein from the external source (like glutamic acid) nevertheless accumulated in the soluble fraction (see Tables XXIV and XXV). Alternatively, extensive labeling of an amino acid could occur in the protein without the corresponding soluble amino acid becoming labeled (e.g., threonine when C^{14}-alanine or C^{14}-sucrose was supplied; or glutamic and aspartic acids when C^{14}-alanine was supplied).

Whereas the labeled arginine predominantly entered the protein as such, it was also converted extensively to glutamine in the soluble fraction; despite this, no radioactive glutamic acid was found in either the soluble or the insoluble moieties. This is all the more significant because *exogenously* supplied glutamine is a particularly effective source of carbon for protein synthesis (482).

Whereas serine and alanine are considered to be metabolically closely related substances, their behavior in the cell nevertheless presents certain contrasts. Alanine, whether formed endogenously or supplied exogenously, distributes its carbon over many compounds. Serine on the other hand, though readily formed endogenously in the soluble fraction (from either alanine or sugar), when supplied exogenously retains its carbon and merely enters the protein as such. Also, "soluble-serine" is formed *either* from sugar *or* from alanine, but "protein-serine" is formed *only* from sugar or exogenous serine.

Therefore, any given metabolite may have a very different fate in different compartments or phases of the cell. Some amino acids (serine, arginine, tyrosine, valine) when supplied exogenously remain intact and somewhat remote from the main streams of metabolism. Others (alanine, serine, theonine) arise readily from sugar, which is already within the cell, and they do contribute carbon readily to the metabolism. Serine behaves differently if it arises endogenously from sugar or is supplied exogenously.

In another experiment, the role of amino acids supplied exogenously in the form of casein hydrolyzate, as sources of carbon or nitrogen for protein synthesis, has been investigated. Cultured carrot explants, with or without coconut milk, were supplied with radioactive sucrose, either as the sole carbon source or in the presence of a high concentration (2%) of casein hydrolyzate. If the casein hydrolyzate supplied carbon directly for protein synthesis, then the specific activity of the carbon of the protein amino acids should be decreased thereby. In point of fact, the specific activities of the carbon of most of the protein amino acids were *increased* by the presence of casein hydrolyzate as illustrated in Table XXVI. *This occurred in spite of the fact that the specific activity of the soluble amino acids, where they could be determined, actually decreased.* These results show that the casein hydrolyzate acted as a source of reduced nitrogen, and it thereby stimulated the use of carbon from sugar in the enhanced protein synthesis which follwered. The one instance in which the specific activity of carbon in a protein amino acid was decreased by the casein hydrolyzate was alanine, which has previously been shown to compete, when supplied exogenously, with sugar as a source of protein carbon. In other words, the C^{12}-alanine of the casein hydrolyzate was able to compete with the C^{14} from sugar.

These data very clearly demonstrate certain general conclusions

TABLE XXVI

Specific Activities of Soluble and Insoluble Compounds in Fast- (+CM) or Slow- (−CM) Growing Carrot Explants after 24 Hours Supplied with Sucrose-UL-C^{14} with or without 2% Casein Hydrolyzate in the Medium[a]

Compound analyzed	Fast growing (+CM)				Slow growing (−CM)			
	− Casein		+ Casein		− Casein		+ Casein	
	Sol.	Insol.	Sol.	Insol.	Sol.	Insol.	Sol.	Insol.
Sucrose	6.7	—	6.6	—	5.8	—	+	—
Glutamine	13.8	—	11.4	—	9.0	—	7.5	—
Aspartic acid	+	6.6	+	8.3	+	6.5	+	12.2
Glutamic acid	31.9	6.1	28.0	7.0	23.8	5.5	+	11.8
Serine	0	5.8	0	6.0	0	0	0	7.6
Glycine	0	6.6	0	8.9	0	0	0	10.7
Alanine	3.8	8.1	2.7	7.4	5.8	4.8	+	7.9
Valine	0	0	0	5.8	0	5.2	0	14.3
Leucines	0	2.9	0	3.1	0	0	0	2.9

[a] Results are given as millimicrocuries per micromole of carbon.

which may now be summarized: (a) The mere presence of a free amino acid in the cell is not sufficient to ensure its direct incorporation into the protein; in fact, it may well be physically debarred from this. (b) Only if the protein precursors are generated at the right site, or alternatively from the right source (i.e., one which readily reaches this site), may they be incorporated directly into the protein, and they can clearly do so without mingling with the free amino acid pool. (c) These situations can prevail only in a highly organized system with discrete compartments, the identity and separate functions of which can be physically or chemically maintained and subjected to some form of regulatory control.

The analysis of the relations between carbohydrate and nitrogen metabolism presented above ranges over effects observed with barley leaves at their point of maximum elongation (186); with slices or thin disks cut from storage organs (482); with green algae in the light or dark (32); and finally with cultured explants of carrot restored to active growth by cell division (499).

Although so broad a range of materials may justify the claim that these effects are very general, it should not be concluded that this is invariably so. There is little reason to invoke these concepts to explain the respiration and metabolism of intact, whole storage organs or of any other more or less quiescent system which is more nearly balanced internally between anabolism and catabolism. It is when the system is not so poised, when by virtue of either growth or net synthesis it is "open-ended" that these concepts apply. Among many other data that seem to invoke protein metabolism, in water uptake and growth, the following may be mentioned. Thimann and Loos (530) used potato slices to show that a small amount of heat-coagulable nitrogen which escapes precipitation with TCA disappears after 2–4 days of water uptake. Protein synthesis was rapid during the first 2 days after equilibration when water uptake was small. It fell

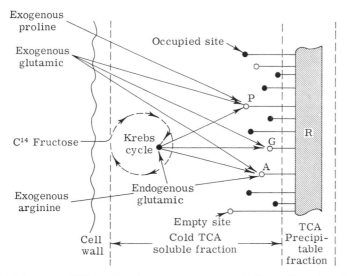

FIG. 57. The accessibility of carbon sources to metabolic pools and protein synthesis. See Cowie and Walton (117).

off in the next 2–4 days while water uptake greatly increased. Naphthaleneacetic acid, however, markedly increased protein synthesis in the first 2 days, and the increase was roughly commensurate with that of the increase of water uptake.

It is also instructive to refer to certain ideas of Wright and Anderson (623), Cowie (116), Cowie and Walton (117) which, though drawn from the study of microorganisms, nevertheless implicate different metabolic pools of the same amino acid [also see (481) for discussion of these papers]. Again, despite the minimum of morphological complexity in the experimental material, the results, as shown below, do lead to some views rather similar to those expressed above and require for their interpretation some form of compartments in the cells (see Fig. 57).

C. Effects Due to Light and the Role of Chloroplasts

The Calvin cycle, or the reductive pentose phosphate cycle (88), is implicated in the formation of carbohydrate from carbon dioxide, but this does not mean that other mechanisms of carbon dioxide fixation play a negligible role (553). This appears to be particularly true when protein synthesis is quantitatively significant, and carbon dioxide is probably assimilated via a reversal of the Krebs' cycle and to an extent of up to 10% of the total assimilatory activity (310, 326).

Arnon (13, 14) and Tagawa et al. (525) have designated an electron flow mechanism. According to Van Niel (553), the mechanism implies that the ejected electrons, by combining with hydrogen ions, yield transferable hydrogen atoms to serve for the reduction of pyridine nucleotides, which in turn function in the reductive assimilation of carbon dioxide. But, in the absence of carbon dioxide, the hydrogen can be released as gas with the mediation of a hydrogenase; or it may be used for the reduction of elementary nitrogen by nitrogen-fixing organisms, or more generally for that of imino to amino acids if the synthesis of the former is induced by the presence of an appropriate nitrogen source. The inhibition of hydrogen evolution by nitrogen or ammonia is the logical result of the availability of particular hydrogen-acceptors. It is striking that nitrogen which must undergo a primary reduction before it can serve in amino acid synthesis is a more powerful inhibitor than ammonia.

Protein, in a vigorously growing leaf is predominantly located in chloroplasts. However, a component from the expanded leaves of several species has the sedimentation constant $S = 16-19$ and is commonly referred to as "fraction 1 protein." It is generally assumed that this is a cytoplasmic component [cf. Pirie (392, 393)].

Stephenson et al. (476) have indicated that proteins can be synthesized in the chloroplast. A series of amino acids can arise directly from the intermediary products of photosynthesis (457). In addition, amino acid-activating enzymes have been demonstrated by Bové and Raacke (68) in spinach chloroplasts. A compilation of the work on chloroplast enzymes has been undertaken by Thomas (534), but as yet there is little reliable information about the total content of soluble proteins in chloroplasts (324). According to Heber (206, 207), 50% of the chloroplast protein can be soluble [for a more recent account see Smillie (454)]. In a recent communication, App and Jagendorf [Plant Physiol. 38, Suppl. vii (1963)] described their work on protein metabolism in isolated chloroplasts. They claim incorporation of C^{14} amino acids into protein of chloroplasts, which stimulated by light under nitrogen, but the presence of ninhydrin-positive soluble material in chloroplasts incubated in the dark suggests that chloroplasts break protein down in the dark and resynthesize it in the light. However the present position seems to be that any conclusion that chloroplasts synthesize proteins by a ribosomal mechanism now rests on the evidence that they contain DNA. It does not rest upon direct demonstrations of amino acid incorporation into protein by cell free preparations, for the levels of incorporation are very small, and the risks of contamination are very great.

During chloroplast development there is a sequential appearance of unique components that react antigenically with the antichloroplast serum, indicating that as chloroplast formation proceeds there is a *de novo* synthesis of materials associated with the chloroplast rather than an organization of preexisting components into the structure of the chloroplast. The source of materials for this synthesis may be an amino acid pool or there may be a degradation of existing proteins and other materials, since chloroplast development in these experiments took place in a medium containing no nitrogen or utilizable carbon sources (Schiff, personal communication).

In connection with protein synthesis in chloroplasts and the nature of the "plastome" (324) the presence of nucleic acids in plastids is noteworthy [cf. (454); see also Section VI, D]. Since chloroplasts contain only small percentages of nucleic acids, contaminations become important. Nakamura *et al.* (347) have shown chloroplasts to contain active nucleic acid depolymerases. DNA has been demonstrated in chloroplasts in addition to RNA [(67, 103); Mothes *et al.*, 1958]. In some cases (324) only RNA has been found. Lyttleton (287) has isolated ribosomes from spinach chloroplasts. Of two classes of ribosomes found in leaves, the smaller appeared to be derived from the chloroplasts.

The observations of Voskresenskaya, referred to by Cayle and Emerson (96), suggested that the balance between the amount of $C^{14}O_2$ fixed in the light which enters protein or carbohydrate is a function of wavelength. In a re-examination Cayle and Emerson noted some effects, due to wavelength, on the specific activity of carbon in certain amino acids. Hauschild *et al.* (204, 205) have shown that *Chlorella* cells produce *25–30% more aspartic acid and 20% less serine and glycine in red light plus 4% of blue light than in red light alone,* even when the comparisons were made under conditions that relate to the same amount of carbon being fixed. This important observation *shows how light may be directly implicated in nitrogen metabolism* [cf. also observations on asparagine and glutamine (see p. 516 *et seq.*)].

The plastids of etiolated bean leaves can be induced to grow in the dark by exposing the plant to brief periods of illumination with red or white light. The induction of plastid growth can be reversed by subsequent irradiation with wavelengths in the far-red, hence this phase of chloroplast development is probably controlled by the reversible pigment which controls other morphogenetic processes in the plant (210). The increase in chloroplast size as well as synthesis of proteins and lipids has thus been found by Mego and Jagendorf (319) to be activated by the red-far red pigment system.

There are two possibilities to explain how light influences the reduction of nitrate. First is Warburg and Nagelein's suggestion for *Chlorella* [(249); p. 96] that an increase of permeability to nitrate occurs upon illumination. Second, photochemical processes, instead of

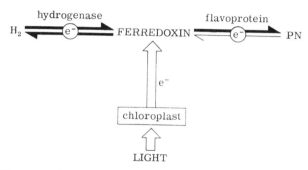

Fig. 58. Diagrammatic representation of the role of ferredoxin in photochemical pyridine nucleotide (PN) reduction and in the biological production and consumption of hydrogen gas. The flavoprotein is normally bound in the chloroplasts. After Tagawa and Arnon (524).

respiration or the hydrogenase reaction, may provide hydrogen donors in the form of reduced pyridine nucleotides as shown to occur *in vitro* by Evans and Nason (150). These authors were able to combine the photochemical reduction of pyridine nucleotides by isolated chloroplasts with the enzymatic reduction of nitrate to nitrite.

Based on the experiments of Evans and Nason (150) and of Van Niel (553), the assumption that the effect of light on nitrate reduction can be explained by a supply of photochemically reduced pyridine nucleotides has gained wide acceptance (249).

Nitrate reductase probably depends on TPNH (NADPH) as a hydrogen donor with an unidentified cofactor to mediate the transfer of hydrogen. The cofactor can be replaced *in vitro* by the reducible dye benzyl viologen. The activity of nitrate reductase can account for the rate of nitrate assimilation in intact plants (194). However, there is now a new hydrogen transport system, named "ferredoxin," which has been implicated by Carnahan and Castle (93) in the reduction of elementary nitrogen, as in nitrogen fixation, and in nitrate reduction. Tagawa and Arnon (524) and Tagawa *et al.* (525) have compared ferredoxin, believed to be present in chloroplasts, with the photosynthetic pyridine nucleotide reductase system which makes TPNH available in the light. The reducible dye methyl viologen is known to be an effective substitute for the system known as "ferredoxin," which Tagawa and Arnon now believe to mediate the transfer of hydrogen and electrons in the light to the phospho-pyridine nucleotides (Fig. 58).

D. NITROGEN COMPOUNDS IN RELATION TO TRANSLOCATION AND STORAGE

The problems of translocation and eventual storage are not unique to nitrogen compounds (cf. Swanson, Chapter 5, Volume II). The baffling rates from source to sink [estimated by Maskell and Mason (301–303) and Mason and Phillis (305) for movement in the phloem of the cotton plant to be of the order of molecular movement in gaseous systems] present a problem no less, but no greater, than the similar rates for carbohydrates. The accumulation of solutes at the "sink," or their con-version to new forms by synthesis, makes similar demands upon the activity of cells as the problem in relation to either carbohydrates or inorganic ions. In fact evidence shows that amino acids, like ions, are actively accumulated (43–45). Only in the array of molecular species that may be involved (Section IV) is nitrogen metabolism at all unique with respect to these problems. All the difficult ques-tions that surround the choice of tissue, or even its state of develop-ment when movement occurs, apply equally to carbohydrates and nitrogen compounds in transit throughout the plant body. For these more general considerations reference should, therefore, be made to Chapter 5 of Volume II for the movement of organic forms of nitrogen, and to Chapter 6 of Volume II for matters that relate to movement of inorganic salts.

In addition to the above, little more can or need be said. Asparagine, historically the first compound to attract attention as a mobile nitrogen-rich source, can with glutamine and arginine lay claim to this role. Allantoin and allantoic acid in the bleeding sap of certain trees [stressed by Mothes and Engelbrecht (344) and by Bollard (53, 55)] may also function as mobile forms of nitrogen. Although the amino acids present in phloem exudates (635) and in the bleeding sap from xylem (53, 55) have been determined chromatographically, it is doubtful whether these have any more meaning as the actual form in which the nitrogen moves than any other constituent that is present (494). The curious presence of choline phosphate and choline sulfate in the sap of barley and maize, respectively, as observed by Maizel *et al.* (290) and Nissen and Benson (364), is also of interest in this connection.

X. Epilogue

This chapter has traced the long history of knowledge of nitrogen metabolism in plants, and the very rapid additions to knowledge in recent years; it has also summarized the present status of the subject, and it now remains to look to the future.

It is safe to predict that many new, relatively simple, nitrogen compounds remain to be disclosed and unequivocally identified. Now that nitrogen metabolism can be seen to involve much more than protein metabolism, the new compounds will add to knowledge inasmuch as they are linked to metabolic pathways and to the factors that cause them to form and accumulate. Merely to describe the compounds without interpreting their origin and role will be decreasingly rewarding.

Much work still needs to be done to elucidate the enzymology that underlies the synthesis of these numerous compounds. The relations of the new compounds to genetics and to the classification of plants, as well as the factors in the environment and the conditions during development that control their formation, need to be investigated. Where there are two or more asymmetric centers in nitrogen compounds, the problem of deciding which of the various possible isomers are physiologically important, and why, is a real problem.

A large field of endeavor remains before we understand how the genetic potentialities of plants for nitrogen metabolism are controlled or modulated by such extragenetic means as in their responses to factors in the environment and in the inorganic nutrition of plants. As knowledge in this area grows, perhaps it will be necessary to concede that events in more remote parts of the cytoplasm than the ribosome granules play a more determining role than many current and somewhat schematic ideas based on protein synthesis seem to indicate. Much work with plants under controlled conditions of growth and nutrition will be required here, and the task of recording and interpreting all the interacting factors and their effects is a formidable one.

As techniques for resolving the protein moiety in plants into its components improve and as ideas upon protein synthesis, breakdown, and resynthesis become more formulated, one can look for great additions to knowledge in this area. The proteins still merit their name among the chief substances that are important to life, and it would be hard to visualize satisfying interpretations of the facts of differentiation, development, and morphogenesis which failed to take account of the special role which characteristic proteins and their metabolism must surely play.

But it is in the plant physiological implications of nitrogen metabolism, and not only in its biochemistry, that advances are to be expected. Cells and organisms are not merely an assemblage of biochemical reactions and their products; they are something more even than genes on DNA-constituted threads, or RNA templates on ribosome granules, important as all these are. Even a single parenchyma cell has, in terms of molecules, a baffling degree of complexity. In fact the problem of nitrogen metabolism in plants is typical of so many current problems of biology. As the descriptive biochemical phase yields its main results, its limitations become apparent, for it usually takes too little account of organization. Even had complete success attended the efforts initiated by Schulze and Prianishnikov to dissect and "pull apart" the system of nitrogen metabolism in plants, one would still need to assign its proper role as part of the coordinated, integrated system which growing developing cells and organisms present. Here structural heterogeneity needs to be reckoned with, and compartmentation of cells and morphological diversity in organisms provide for the simultaneous occurrence of different reaction pathways that affect the same intermediary metabolite. Anabolism and catabolism must go hand in hand. It is here that knowledge of the role in nitrogen metabolism of different cytoplasmic organelles is both needed and imminent. Work on chloroplasts, nuclei, mitochondria, membranes, etc., already discloses their complexity and begins to identify their role in nitrogen metabolism.

As the broad facts of energy storage in cells become known (see Chapter 4, Volume IB); as the ways in which energy is both released and harnessed to useful purposes are interpreted (see Chapter 3, Volume IA); as the intermediary steps of metabolism and their relations to nutrition and environment are worked out, the remaining problem is to create a synthesis of the whole in which all these several functions of cells and organism proceed concomitantly. A focal point in the metabolic machinery of cells upon which these metabolic functions impinge is the metabolism of their nitrogen compounds and of proteins.

If the nucleic acids can now be described as the chemical basis of specificity, as the way in which biological information is transmitted;

if the phosphorylated compounds can now be regarded as the molecular "currency" through which energy exchanges are mediated, then the nitrogen compounds and the proteins still seem to be the molecules that correspond to the main working parts of the machinery of life. Thus, it is in the interrelations between respiration and nitrogen metabolism, between protein metabolism and development, between protein metabolism, water relations and mineral uptake and nutrition and the ways in which these interactions are controlled and affected by environment, by nutrition, and by stimuli, exogenous and endogenous, that much understanding still remains to be achieved.

The means by which the control is exercised over the genetically determined metabolic characteristics of organisms is a gap in our knowledge which now obviously needs to be filled. This applies particularly to nitrogen and protein metabolism of plants, for one can readily discern that differentiation and morphogenesis and the environmental control over those events are associated with distinctive patterns of nitrogen metabolism and the probable formation of specific and characteristic proteins.

REFERENCES

1. Abelson, P. H. Amino acids formed in primitive atmosphere. *Science* **124,** 935 (1956).
2. Abelson, P. H. Some aspects of paleobiochemistry. *Ann. N.Y. Acad. Sci.* **69,** 255–376 (1957).
3. Albaum, H. G. Metabolism of phosphorylated compounds in plants. *Ann. Rev. Plant Physiol.* **3,** 35–58 (1952).
4. Albaum, H. G., and Cohen, P. P. Transamination and protein synthesis in germinating oat seedlings. *J. Biol. Chem.* **149,** 19–27 (1943).
4a. Alföldi, L., Stent, G. S., and Clowes, R. C. The chromosomal site of the RNA control (RC) locus in *Escherichia coli. J. Mol. Biol.* **5,** 348–355 (1962).
5. Allfrey, V. G. Amino acid transport and early stages in protein synthesis in isolated cell nuclei. *In* "Biological Structure and Function" (T. W. Goodwin and O. Lindberg, eds.), Vol. 1, pp. 264–280. Academic Press, New York, 1961.
6. Allsop, A. Chromatographical study of meristematic tissues. *Nature* **161,** 833–838 (1948).
7. Altschul, A. M. ed. "Processed Plant Protein Foodstuffs." Academic Press, New York, 1958.
8. Anderson, J. A., Ziegler, M. R., and Doeden, D. Banana feeding and the urinary excretion of 5-hydroxyindoleacetic acid. *Science* **127,** 236–238 (1958).
9. Anfinsen, C. B. "Molecular Basis of Evolution." Wiley, New York, 1961.
10. Anfinsen, C. B., and Redfield, R. R. Protein structure in relation to function and biosynthesis. *Advan. Protein Chem.* **11,** 1–100 (1956).
11. Anson, M. L. Oilseed proteins in foods. *Arch. Biochem. Biophys.* **98,** Suppl. 1, 68–77 (1962).
12. Anson, M. L., and Altschul, A. M. Introduction. *In* "Processed Plant Protein Foodstuffs" (A. M. Altschul, ed.), pp. 1–11. Academic Press, New York, 1958.

13. Arnon, D. I. Conversion of light into chemical energy in photosynthesis. *Nature* **184**, 10–21 (1959).

14. Arnon, D. I. Cell-free photosynthesis and the energy conversion process. *In* "Light and Life" (W. D. McElroy and B. Glass, eds.), pp. 489–565. Johns Hopkins Press, Baltimore, Maryland, 1961.

15. Audus, L. J., and Quastel, J. H. Toxic effects of amino acids and amines on seedling growth. *Nature* **160**, 222–223 (1947).

16. Avers, C. J. Histochemical localization of enzyme activity in the root epidermis of *Phleum pratense*. *Am. J. Botany*, **45**, 609–613 (1958).

17. Bachrach, U. Metabolism of amines. II. The degradation and synthesis of γ-aminobutyric acid by *Pseudomonas aeruginosa*. *Biochem. J.* **77**, 417–421 (1960).

18. Baglioni, C. An improved method for the fingerprinting of human hemoglobin. *Biochim. Biophys. Acta* **48**, 392–396 (1961).

19. Baglioni, C. Correlations between genetics and chemistry of human hemoglobins. *In* "Molecular Genetics" (J. H. Taylor, ed.), Vol. 4A, Part 1, pp. 405–475. Academic Press, New York, 1963.

20. Barger, G., and White, F. D. The constitution of gelagine. *Biochem. J.* **17**, 827–835 (1923).

21. Barker, W. G., and Steward, F. C. Growth and development of the banana plant. I. The growing regions of the vegetative shoot. II. The transition from the vegetative to the floral shoot in *Musa acuminata* cv. Gros Michel. *Ann. Botany (London)* [N.S.] **26**, 389–423 (1962).

22. Barnes, R. L. Formation of γ-guanidinobutyric acid in pine tissues. *Nature* **193**, 781 (1962).

23. Barnes, R. L., and Naylor, A. W. Studies on the ornithine cycle in roots and callus tissue in *Pinus serotina* and *Pinus clausa*. *Botan. Gaz.* **121**, 63–69 (1959).

24. Barnes, R. L., and Naylor, A. W. Formation of β-alanine by pine tissue supplied with intermediates in uracil and orotic acid metabolism. *Plant Physiol.* **37**, 171–175 (1962).

25. Barrales, H. Some problems of the isolation, occurrence and behavior of soluble nitrogen compounds in higher plants. Ph.D. thesis, Cornell Univ., Ithaca, New York, 1959.

26. Barrien, B. S., and Wood, J. G. Studies on the sulfur metabolism of plants. II. The effect of nitrogen supply on the amounts of protein sulfur, sulfate sulfur and on the value of the ratio of protein nitrogen to protein sulfur in leaves at different stages during the life cycle of the plant. *New Phytologist* **38**, 257–264 (1939).

27. Battersby, A. R. Alkaloid biosynthesis. *Quart. Rev. (London)* **15**, 259–289 (1961).

28. Beach, E. F. Beccari of Bologna. The discoverer of vegetable protein. *J. Hist. Med.* **16**, 354–373 (1961).

29. Beadle, G. W. Biochemical genetics. *Chem. Rev.* **37**, 15–96 (1945).

30. Bell, E. A. Canavanine and related compounds in Leguminosae. *Biochem. J.* **70**, 617–619 (1958).

31. Belozersky, A. N., and Spirin, A. S. Chemistry of the nucleic acids of microorganisms. *In* "The Nucleic Acids" (E. Chargaff and J. N. Davidson, eds.), Vol. 3, pp. 147–185. Academic Press, New York, 1960.

32. Benson, A. A., and Calvin, M. Carbon dioxide fixation by green plants. *Ann. Rev. Plant Physiol.* **1**, 25–42 (1950).

33. Benzer, S. The fine structure of the gene. *Sci. Am.* **206,** 70–84 (1962).
34. Berg, P. Specificity in protein synthesis. *Ann. Rev. Biochem.* **30,** 293–324 (1961).
35. Berg, P., Lagerkvist, U., and Dieckmann, M. The enzymic synthesis of amino acyl derivatives of ribonucleic acid. VI. Nucleotide sequences adjacent to the . . pCpCpA and groups of isoleucine- and leucine-specific chains. *J. Mol. Biol.* **5,** 159–171 (1962).
36. Bernal, J. D. "The Physical Basis of Life." Routledge & Kegan, London, 1951.
37. Bessman, M. J. The replication of DNA in cell-free systems. *In* "Molecular Genetics" (J. H. Taylor, ed.), Vol. 4A, Part 1, pp. 1–64. Academic Press, New York, 1963.
38. Beverton, R. J. H., and Holt, S. J. On the dynamics of exploited fish populations. H. M. Stationery Office, London, 1957.
39. Biemann, K., Deffner, G. G. J., and Steward, F. C. Determination by mass spectrometry of the structure of proline derivatives from apples. *Nature* **191,** 380–381 (1961).
40. Biemann, K., Seibl, J., and Gapp, F. Mass spectra of organic molecules. I. Ethyl esters of amino acids. *J. Am. Chem. Soc.* **83,** 3795–3804 (1961).
41. Biemann, K., and Vetter, W. Separation of peptide derivatives by gas chromatography combined with the mass spectrometric determination of the amino acid sequence. *Biochem. Biophys. Res. Commun.* **3,** 578–584 (1960).
42. Birdsong, B. A., Alston, R., and Turner, B. L. Distribution of canavanine in the family Leguminosae as related to phyletic groupings. *Can. J. Botany* **38,** 499–505 (1960).
43. Birt, L. M., and Hird, F. J. R. The uptake of amino acids by carrot slices. *Biochem. J.* **64,** 305–311 (1956).
44. Birt, L. M., and Hird, F. J. R. Kinetic aspects of the uptake of amino acids by carrot tissue. *Biochem. J.* **70,** 286–292 (1958).
45. Birt, L. M., and Hird, F. J. R. The uptake and metabolism of amino acids by slices of carrot. *Biochem. J.* **70,** 277–286 (1958).
46. Bisset, S. K. The non-protein nitrogen of extracts of *Pisum sativum. Biochem. J.* **58,** 225–227 (1954).
47. Blackman, F. F. "Analytical Studies in Plant Respiration." Cambridge Univ. Press, London and New York, 1954.
48. Blackman, F. F., and Parija, P. Analytic studies in plant respiration. I. The respiration of a population of senescent ripening apples. *Proc. Roy. Soc. (London)* **B103,** 412–445 (1928).
49. Blackman, G. E. The ecological and physiological action of ammonium salts on the clover content of turf. *Ann. Botany (London)* **48,** 975–1001 (1934).
50. Block, R. J. The chemical constitution of the proteins. *In* "The Chemistry of the Amino Acids and Proteins," 2nd ed., pp. 278–333. Thomas, Springfield, Illinois, 1944.
51. Block, R. J., Durrum, E. L., and Zweig, G. "Paper Chromatography: A Laboratory Manual," Part I. Academic Press, New York, 1960.
52. Blum, H. F. On the origin and evolution of living machines. *Am. Scientist* **49.** 474–501 (1961).
53. Bollard, E. G. Translocation of organic nitrogen in the xylem. *Australian J. Biol. Sci.* **10,** 292–301 (1957).
54. Bollard, E. G. Urease, urea and ureides in plants. *Symp. Soc. Exptl. Biol.* **13,** 304–329 (1959).

55. Bollard, E. G. Transport in the xylem. *Ann. Rev. Plant Physiol.* **11**, 141–166 (1960).
56. Bonner, J. Protein synthesis and the control of plant processes. *Am. J. Botany* **46**, 58–62 (1959).
57. Bonner, J., and Huang, R. C. Chromosomal control of enzyme synthesis. *Can. J. Botany* **40**, 1487–1497 (1962).
58. Bonner, J., and Zeevaart, J. A. D. Ribonucleic acid synthesis in the bud an essential component of floral induction in *Xanthium*. *Plant Physiol.* **37**, 43–49 (1962).
59. Borchers, R. Tyrosine stimulates growth on raw soybean rations. *Federation Proc.* **18**, 517 (1959).
60. Borsook, H. Protein turnover and incorporation of labeled amino acids into tissue proteins *in vivo* and *in vitro*. *Physiol. Rev.* **30**, 206–219 (1950).
61. Borsook, H. Peptide bond formation. *Advan. Protein Chem.* **8**, 127–174 (1953).
62. Borthwick, H. A., and Hendricks, S. B. Effects of radiation on growth and development. *In* "Handbuch der Pflanzenphysiologie" (W. Ruhland, ed.), Vol. XVI, pp. 299–330. Springer, Berlin, 1961.
63. Boswell, J. G., and Whiting, G. C. A study of the polyphenol oxidase system in potato tubers. *Ann. Botany* *(London)* [N.S.] **2**, 847–864 (1938).
64. Böttger, I. Stoffwechsel der Nucleinsäuren. *In* "Handbuch der Pflanzenphysiologie" (W. Ruhland, ed.), Vol. VIII, pp. 814–843. Springer, Berlin, 1958.
65. Böttger, I. Stoffwechsel der Purine und Pyrimidine. *In* "Handbuch der Pflanzenphysiologie" (W. Ruhland, ed.), Vol. VIII, pp. 763–813. Springer, Berlin, 1958.
66. Böttger, I. Änderung des Ribonucleinsäure—und Eiweisssgehaltes von Zellfraktionen isolierter Blätter von *Euphorbia pulcherrima* im Verlauf ihrer Bewurzelung. *Flora* *(Jena)* **152**, 555–569 (1962).
67. Böttger, I., and Wollgiehn, R. Untersuchungen über den Zusammenhang zwischen Nucleinsäure—und Eiweissstoffwechsel in grünen Blättern höheren Pflanzen. *Flora* *(Jena)* **146**, 302–320 (1958).
67a. Boulter, D., and Barber, J. T. Amino acid metabolism in germinating seeds of *Vicia faba* L. in relation to their biology. *New Phytologist* **62**, 301–316 (1963).
68. Bové, J., and Raacke, I. D. Amino acid-activating enzymes in isolated chloroplasts from spinach leaves. *Arch. Biochem. Biophys.* **85**, 521–531 (1959).
69. Brachet, J. Effects of ribonuclease on the metabolism of living root-tip cells. *Nature* **174**, 876–877 (1954).
70. Brachet, J. Further observations on the effects of ribonuclease on living root tip cells. *Biochim. Biophys. Acta* **16**, 611–613 (1955).
71. Brachet, J. "The Biological Role of Ribonucleic Acids." Elsevier, Amsterdam, 1960.
72. Bradstreet, R. B. A review of the Kjeldahl determination of organic nitrogen. *Chem. Rev.* **27**, 331–350 (1940).
73. Braun, A. C. A study on the mode of action of wildfire toxin. *Phytopathology* **45**, 659–664 (1955).
74. Braunstein, A. E. Transamination and the integrative functions of the dicarboxylic acids in nitrogen metabolism. *Advan. Protein Chem.* **3**, 1–52 (1947).
75. Brawerman, G., Pogo, A. O., and Chargaff, E. Induced formation of ribonucleic acids and protein in *Euglena gracilis* under the influence of light. *Biochim. Biophys. Acta* **55**, 326–334 (1962).

76. Brawerman, G., and Shapiro, H. S. Nucleic acids. *In* "Comparative Biochemistry" (M. Florkin and H. S. Mason, eds.), Vol. 4, Part B, pp. 107–183. Academic Press, New York, 1962.

77. Brohult, S., and Sandegren, E. Seed proteins. *In* "The Proteins" (H. Neurath and K. Bailey, eds.), 1st ed., Vol. 2, Part A, pp. 487–572. Academic Press, New York, 1954.

78. Brown, R., and Broadbent, D. The development of cells in the growing zones of the root. *J. Exptl. Botany* **1**, 249–263 (1951).

79. Bryant, M., and Fowden, L. Protein composition in relation to age of daffodil leaves. *Ann. Botany (London)* [N.S.] **23**, 65–74 (1959).

80. Burns, V. W. Cell division synchronization. *Progr. in Biophys. and Biophys. Chem.* **12**, 2–23 (1962).

81. Burris, R. H. Nitrogen nutrition. *Ann. Rev. Plant Physiol.* **10**, 301–328 (1959).

82. Burström, H. The nitrate nutrition of plants. *Kgl. Lantbruks-Högskol. Ann.* **13**, 1–86 (1945).

83. Büsgen, M., and Münch, E. "The Structure and Life of Forest Trees" (Translated by T. Thompson). Chapman & Hall, London, 1929.

84. Butler, J. A. V., and Davison, P. F. Deoxyribonucleoprotein, a genetic material. *Advan. Enzymol.* **18**, 161–190 (1957).

85. Calvery, H. O. The chemistry of tea leaves. II. The isolation of guanine nucleotide and cytosine nucleotide. *J. Biol. Chem.* **72**, 549–556 (1927).

86. Calvery, H. O., and Remsen, D. B. The nucleotides of triticonucleic acid. *J. Biol. Chem.* **73**, 593–597 (1927).

87. Calvin, M. "Origin of Life on Earth and Elsewhere." Univ. of California Press, Berkeley, California, 1959.

88. Calvin, M. The path of carbon in photosynthesis. *Science* **135**, 879–889 (1962).

89. Calvin, M. Communication: from molecules to Mars. *A.I.B.S. (Am. Inst. Biol. Sci.) Bull. October*, 29–44 (1962).

90. Canellakis, E. S. Metabolism of nucleic acids. *Ann. Rev. Biochem.* **31**, 271–300 (1962).

91. Cantino, E. C. The relationship between biochemical and morphological differentiation in non-filamentous aquatic fungi. *Symp. Soc. Gen. Microbiol.* **11**, 243–271 (1961).

92. Caplin, S. M., and Steward, F. C. A technique for the controlled growth of excised plant tissue in liquid media under aseptic conditions. *Nature* **163**, 920 (1949).

93. Carnahan, J. E., and Castle, J. E. Nitrogen fixation. *Ann. Rev. Plant Physiol.* **14**, 125–136 (1963).

94. Carnegie, P. R., and Synge, R. L. M. Filter-paper ionophoresis of cupric complexes of neutral amino acids and oligopeptides. *Biochem. J.* **78**, 692–696 (1961).

95. Cavalieri, L. F., and Rosenberg, B. H. Nucleic acids: molecular biology of DNA. *Ann. Rev. Biochem.* **31**, 247–270 (1962).

96. Cayle, T., and Emerson, R. Effect of wave-length on the distribution of carbon-14 in the early products of photosynthesis. *Nature* **179**, 89–90 (1957).

97. Chailakhian, M. K., Butenko, R. G., and Lyuborskaya, J. J. Effects of derivatives of nucleic acid metabolism on the growth and flowering of *Perilla nankinensis*. *Plant Physiol. (USSR) (English Summary.)* **8**, 101–113 (1961).

98. Chamberlin, M., and Berg, P. Deoxyribonucleic acid-directed synthesis of ribonucleic acid by an enzyme from *Escherichia coli*. *Proc. Natl. Acad. Sci. U.S.* **48**, 81–94 (1962).

99. Chandler, W. H. "Fruit Growing." Houghton, Boston, Massachusetts, 1925.
100. Chang, L. The amino acid and protein composition of certain varieties of wheat. Ph.D. thesis, Cornell Univ., Ithaca, New York, 1960.
101. Chang, L., Srb, A. M., and Steward, F. C. Electrophoretic separations of the soluble proteins of *Neurospora. Nature* **193**, 756–759 (1962).
102. Chargaff, E., and Davidson, J. N., eds. "The Nucleic Acids." Vols. 1–3. Academic Press, New York, 1955–1962.
103. Chiba, Y., and Sugahara, K. The nucleic acid content of chloroplasts isolated from spinach and tobacco leaves. *Arch. Biochem. Biophys.* **71**, 367–376 (1957).
104. Chibnall, A. C. Diurnal variations in the total nitrogen content of foliage leaves. *Ann. Botany (London)* **37**, 511–518 (1923).
105. Chibnall, A. C. Investigation of the nitrogenous metabolism of the higher plants. V. Diurnal variations in the protein nitrogen of runner bean leaves. *Biochem. J.* **18**, 387–394 (1924).
106. Chibnall, A. C. "Protein Metabolism in Plants." Yale Univ. Press, New Haven, Connecticut, 1939 (reprinted 1964).
107. Chibnall, A. C., Rees, M. W., and Williams, E. F. The total nitrogen content of egg albumin and other proteins. *Biochem. J.* **37**, 354–359 (1943).
108. Chibnall, A. C., and Westall, R. G. The estimation of glutamine in the presence of asparagine. *Biochem. J.* **26**, 122–132 (1932).
109. Christiansen, G. S., and Thimann, K. V. Studies on the growth and inhibition of isolated plant parts. VI. *Am. J. Botany* **36**, 821–822 (1949).
110. Clark, H. E., and Shive, J. W. The influence of the pH of a culture solution on the assimilation of ammonium and nitrate nitrogen by the tomato plant. *Soil Sci.* **37**, 459–476 (1934).
111. Clowes, F. A. L. "Apical Meristems." F. A. Davis, Philadelphia, Pennsylvania, 1961.
112. Cohen, P. P., and Brown, G. W., Jr. Ammonia metabolism and urea biosynthesis. *In* "Comparative Biochemistry" (M. Florkin and H. S. Mason, eds.). Vol. 2, pp. 161–244. Academic Press, New York, 1960.
113. Cohen, P. P., and Sallach, H. J. Nitrogen metabolism of amino acids. *In* "Metabolic Pathways" (D. M. Greenberg, ed.), Vol. 2, pp. 1–78. Academic Press, New York, 1961.
114. Consden, R., Gordon, A. H., and Martin, A. J. P. Qualitative analysis of proteins: a partition chromatographic method using paper. *Biochem. J.* **38**, 224–232 (1944).
115. Conway, E. J. "Microdiffusion Analysis and Volumetric Error," Rev. ed. Crosby, Lockwood, London, 1947.
116. Cowie, D. B. Metabolic pools and the biosynthesis of proteins. *In* "Amino Acid Pools" (J. T. Holden, ed.), pp. 633–645. Elsevier, Amsterdam, 1962.
117. Cowie, D. B., and Walton, B. P. Kinetics of formation and utilization of metabolic pools in the biosynthesis of protein and nucleic acid. *Biochim. Biophys. Acta* **21**, 211–226 (1956).
118. Crawford, R. F., and Kennedy, W. K. Nitrates in forage crops and silage. Benefits, hazards, precautions. *N.Y. State Coll. Agr. Cornell, Misc. Bull.* **37**, 15 (1960).
119. Crick, F. H. C. On the genetic code. *Science* **139**, 461–464 (1963).
120. Delgarno, L., and Hird, F. J. R. Increase in the process of accumulation of amino acids in carrot slices with prolonged aerobic washing. *Biochem. J.* **76**, 209–215 (1960).

121. Damodaran, M., Jaaback, G., and Chibnall, A. C. The isolation of glutamine from an enzymic digest of gliadin. *Biochem. J.* **26**, 1704–1713 (1932).
122. Davidson, J. N. "The Biochemistry of the Nucleic Acids." Methuen, London, 1960.
123. Davies, D. D., and Ellis, R. J. Glutamic-oxaloacetic transaminase of cauliflower. 2. Kinetics and mechanism of action. *Biochem. J.* **78**, 623–630 (1961).
124. Davies, J. W., Harris, G., and Neal, G. E. Peptide-bound polynucleotides in brewers' yeast. *Biochim. Biophys. Acta* **51**, 95–106 (1961).
125. Dekker, E. E., and Maitra, U. Conversion of γ-hydroxyglutamate to glyoxylate and alanine: purification and properties of the enzyme system. *J. Biol. Chem.* **237**, 2218–2227 (1962).
126. Delwiche, C. C., Loomis, W. D., and Stumpf, P. K. Amide metabolism in higher plants. II. The exchange of isotopic ammonia by glutamyl transphorase. *Arch. Biochem. Biophys.* **33**, 333–338 (1951).
127. DeMoss, J. A., Genuth, S. M., and Novelli, G. D. The enzymatic activation of amino acids via their acyl-adenylate derivatives. *Proc. Natl. Acad. Sci. U.S.* **42**, 325–332 (1956).
128. Dent, C. E., and Fowler, D. I. Paper chromatographic analysis of Dakin's samples of 'β-hydroxyglutamic acid.' *Biochem. J.* **56**, 54–55 (1954).
129. Dent, C. E., Stepka, W., and Steward, F. C. Detection of the free amino acids of plant cells by partition chromatography. *Nature* **160**, 682 (1947).
130. Dillemann, G. Composés cyanogénétiques. *In* "Handbuch der Pflanzenphysiologie" (W. Ruhland, ed.), Vol. VIII, pp. 1050–1075. Springer, Berlin, 1958.
131. Dixon, M., and Webb, E. "Enzymes." Academic Press, New York, 1958.
132. Dixon, R. O. D., and Fowden, L. γ-Aminobutyric acid metabolism in plants. 2. Metabolism in higher plants. *Ann. Botany (London)* [N.S.] **25**, 513–530 (1961).
133. Doi, R. H., and Spiegelman, S. Homology test between the nucleic acid of an RNA virus and the DNA in the host cell. *Science* **138**, 1270–1272 (1962).
134. Done, J., and Fowden, L. A new amino acid amide in the groundnut plant (*Arachis hypogaea*): Evidence of the occurrence of γ-methyleneglutamine and γ-methyleneglutamic acid. *Biochem. J.* **51**, 451–454 (1952).
135. Doty, P. Proteins. *Sci. Am.* **197**, 173–180 (1957).
136. Dounce, A. L. Duplicating mechanism for peptide chain and nucleic acid synthesis. *Enzymologia* **15**, 251–258 (1952).
137. Dounce, A. L. A new hypothesis for nucleic acid replication. *J. Theoret. Biol.* **2**, 152–158 (1962).
138. Duranton, H. Évolution de l'arginine dans le tubercule de topinambour au cours du cycle végétatif. *Compt. rend.* **246**, 2655–2657 (1958).
139. Duranton, H., and Morel, G. Évolution des acides aminés libres dans des fragments de tubercules de topinambours cultivés *in vitro*. *Compt. rend.* **246**, 2927–2930 (1958).
140. Eagle, H. Amino acid metabolism in human cell cultures. *Federation Proc.* **17**, 985–986 (1958); also see *Science* **148**, 42–51 (1965).
141. Eaton, S. V. Influence of sulfur deficiency on metabolism of the sunflower. *Botan. Gaz.* **102**, 536–556 (1941).
142. Eck, R. V. Genetic code: emergence of a symmetrical pattern. *Science* **140**, 477–481 (1963).
143. Eckerson, S. H. Protein synthesis by plants. I. Nitrate reduction. *Botan. Gaz.* **77**, 377–390 (1924).

144. Elliott, W. H., and Gale, E. F. Adenosine triphosphate in glutamine synthesis. *Nature* **161**, 128 (1948).

145. Engelbrecht, L. Beiträge zum Problem der Akkumulation von Aminosäuren in Blatzellen. *Flora (Jena)* **150**, 73–86 (1961).

146. Epstein, H. T., and Schiff, J. A. Studies of chloroplast development in *Euglena*. 4. Electron and fluorescence microscopy of the proplastid and its development into a mature chloroplast. *J. Protozool.* **8**, 427–432 (1961).

147. Erdtman, H. Organic chemistry and conifer taxonomy. *In* "Perspectives in Organic Chemistry" (A. Todd, ed.), pp. 453–494. Wiley (Interscience), New York, 1956.

148. Euler, H. von, and Hahn, L. Nucleic acid content of green leaves. *Arkiv Kemi, Mineral. Geol.* **25B**, 1–8 (1948).

149. Euler, H. von, Heller, L., and Hogberg, K. Nucleoproteins in birch pollen (*Betula pubescens*). *Arkiv Kemi, Mineral. Geol.* **26A**, 15 (1948).

150. Evans, H. J., and Nason, A. Pyridine nucleotide nitrate reductase from extracts of higher plants. *Plant Physiol.* **28**, 233–254 (1953).

151. Evans, W. R., Tsai, C. S., and Axelrod, B. Origin of β-aminoisobutyric acid in iris. *Nature* **190**, 809 (1961).

152. Fincham, J. R. S., and Day, P. R. "Fungal Genetics." F. A. Davis, Philadelphia, Pennsylvania, 1963.

153. Fogg, G. E., and Than-Tun. Interrelations of photosynthesis and assimilation of elementary nitrogen in a blue-green alga. *Proc. Roy. Soc. (London)* **B153**, 111–127 (1960).

154. Fowden, L. Azetidine-2-carboxylic acid: a new cyclic imino acid occurring in plants. *Biochem. J.* **64**, 323–331 (1956).

155. Fowden, L. New amino acids of plants. *Biol. Rev. Cambridge Phil. Soc.* **33**, 393–441 (1958).

156. Fowden, L. Amino acids of plants with special reference to recently discovered acids. *Symp. Soc. Exptl. Biol.* **13**, 283–303 (1959).

157. Fowden, L. Degradation of γ-methyleneglutamic acid by plants. *Nature* **186**, 897–898 (1960).

158. Fowden, L., and Gray, D. O. Some recently-characterized amino acids from plants. *In* "Amino Acid Pools" (J. T. Holden, ed.), pp. 43–53. Elsevier, Amsterdam, 1962.

159. Fowden, L., and Steward, F. C. Nitrogenous compounds and nitrogen metabolism in the *Liliaceae*. 1. The occurrence of soluble nitrogenous compounds. 2. The nitrogenous compounds of leaves of the genus *Tulipa*: environmental effects on the composition of *Tulipa gesneriana*. *Ann. Botany (London)* [N.S.] **21**, 53–67, 69–84 (1957).

160. Fowden, L., and Webb, J. A. The incorporation of C¹⁴ labelled substrates into the amino acids of the groundnut plant (*Arachis hypogaea*). *Ann. Botany (London)* [N.S.] **22**, 73–93 (1958).

161. Fox, S. W. How did life begin? *Science* **132**, 200–208 (1960).

162. Fox, S. W., Johnson, J. E., and Vegotsky, A. On biochemical origins and optical activity. *Science* **124**, 923–926 (1956).

163. Fraenkel-Conrat, H. The effect of light in the Van Slyke method for the determination of amino groups. *J. Biol. Chem.* **148**, 453–454 (1943).

164. Fraenkel-Conrat, H., Singer, B., and Williams, R. C. Infectivity of viral nucleic acid. *Biochim. Biophys. Acta* **25**, 87–96 (1957).

165. Fraenkel-Conrat, H., and Stanley, W. Implications of recent studies of a simple virus. *Chem. Eng. News* **39**, 136–144 (1961).

166. Fraenkel-Conrat, H., and Williams, R. C. Reconstitution of active tobacco mosaic virus from its inactive protein and nucleic acid components. *Proc. Natl. Acad. Sci. U.S.* **41**, 690–698 (1956).

167. Frederick, J. F. Multiple molecular forms of 4-glucosyl transferase (phosphorylase) in *Oscillatoria princeps. Phytochemistry* **1**, 153–157 (1962).

168. Freiberg, S. R., Bollard, E. G., and Hegarty, M. P. The natural occurrence of urea and ureides in the soluble nitrogen of the banana plant. *Plant Physiol.* **32**, Suppl. lii (1957).

169. Freiberg, S. R., and Steward, F. C. Physiological investigations on the banana plant. III. Factors which affect the nitrogen compounds of the leaves. *Ann. Botany (London)* [N.S.] **24**, 147–157 (1960).

170. Fruton, J. S. Proteins. *Sci. Am.* **182**, 32–41 (1950).

171. Fujimoto, D., and Tamiya, N. Incorporation of ^{18}O from air into hydroxyproline by chick embryo. *Biochem. J.* **84**, 333–335 (1962).

172. Fussel, G. E. Early 18th century crop nutrition, theory and practice. *Proc. Chem. Soc.* pp. 193–198 (1960).

173. Gaffron, H. On dating stages in photochemical evolution. *In* "Horizons in Biochemistry" (M. Kasha and B. Pullman, eds.), pp. 59–89. Academic Press, New York, 1962.

174. Gale, E. F. The synthesis of proteins and nucleic acids. *In* "The Bacteria" (I. C. Gunsalus and R. Y. Stanier, eds.), Vol. 3, pp. 471–576. Academic Press, New York, 1962.

175. Galston, A. W., and Kaur, R. An effect of auxins on the heat coagulability of the proteins of growing plant cells. *Proc. Natl. Acad. Sci. U.S.* **45**, 1587–1590 (1959).

176. Gartler, S., and Dobzhansky, T. Excretion in human urine of an unknown amino-acid derived from dates. *Nature* **174**, 553 (1954).

177. Gäumann, E., and Naef-Roth, S. Über die chelierende Wirkung einiger Welketoxine. II. Die Verschiebungen den Toxizität durch steigende Zusätze von Asche aus jungen Tomatensprossen. *Phytopathol. Z.* **23**, 147–160 (1955).

178. Ghosh, B. P., and Burris, R. H. Utilization of nitrogenous compounds by plants. *Soil Sci.* **70**, 187–203 (1950).

179. Gifford, E. M., Jr., and Tepper, H. B. Histochemical and autoradiographic studies of floral induction in *Chenopodium album. Am. J. Botany* **49**, 706–714 (1962).

180. Glasziou, K. T. The metabolism of arginine in *Serratia marcescens.* 2. Carbamyladenosine diphosphate phosphoferase. *Australian J. Biol. Sci.* **9**, 253–262 (1956).

181. Glick, D., ed. "Methods of Biochemical Analysis," Vols. 1–9. Wiley (Interscience), New York, 1955–1962.

182. Gray, D. O., and Fowden, L. N-Ethyl-L-asparagine: a new amino-acid from *Ecballium. Nature* **189**, 401–402 (1961).

183. Gray, D. O., and Fowden, L. 4-Methyleneproline: a new naturally occurring proline derivative. *Nature* **193**, 1285–1286 (1962).

184. Greenberg, D. M., ed. "Metabolic Pathways," Vols. 1–2. Academic Press, New York, 1960–1961.

185. Greenstein, J. P., and Winitz, M. "Chemistry of the Amino Acids." Vols. 1–3. Wiley, New York, 1961.

186. Gregory, F. G., and Sen, P. K. Physiological studies in plant nutrition. VI. The relation of respiration rate to the carbohydrate and nitrogen metabolism of barley leaf as determined by nitrogen and potassium deficiency. *Ann. Botany* (*London*) [N.S.] **1**, 521–561 (1937).

187. Grobbelaar, N. Growth and nitrogen metabolism of certain legumes, with special reference to the effects of controlled environments. Ph.D. thesis, Cornell Univ., Ithaca, New York, 1955.

188. Grobbelaar, N., Pollard, J. K., and Steward, F. C. New soluble nitrogen compounds (amino and imino acids and amides) in plants. *Nature* **175**, 703–708 (1955).

189. Grobbelaar, N., and Steward, F. C. Pipecolic acid in *Phaseolus vulgaris:* Evidence on its derivation from lysine. *J. Am. Chem. Soc.* **75**, 4341–4343 (1953).

190. Grobbelaar, N., and Steward, F. C. The recognition and identification of O-acetyl homoserine. *Plant Physiol.* **33**, Suppl., xiv (1958).

191. Grobbelaar, N., Zacharius, R. M., and Steward, F. C. The bulk isolation of L(-) pipecolic acid from *Phaseolus vulgaris* and its quantitative determination. *J. Am. Chem. Soc.* **76**, 2912–2915 (1954).

192. Grunberg-Manago, M. Enzymatic synthesis of nucleic acids. *Ann. Rev. Biochem.* **31**, 301–332 (1962).

193. Guggenheim, M. Die biogenen Amine in der Pflanzenwelt. *In* "Handbuch der Pflanzenphysiologie" (W. Ruhland, ed.), Vol. VIII, pp. 889–988. Springer, Berlin, 1958.

194. Hageman, R. H., Cresswell, C. F., and Hewitt, E. J. Reduction of nitrate, nitrite and hydroxylamine to ammonia by enzymes extracted from higher plants. *Nature* **193**, 247–250 (1962).

195. Hamilton, P. B., and Van Slyke, D. D. The effects of the volatile aldehydes formed on the accuracy of the manometric ninhydrin-carbon dioxide method in analysis of certain α-amino acids. *J. Biol. Chem.* **164**, 249–256 (1946).

196. Hanes, C. S. Quantitative chromatographic methods. 2. An approach to paper chromatography of improved resolving power and reproducibility. *Can. J. Biochem. Physiol.* **39**, 120–140 (1961).

197. Hanes, C. S., Connell, G. E., and Dixon, G. H. Transpeptidation and transamidation reactions. *In* "Phosphorous Metabolism" (W. D. McElroy and B. Glass, eds.), Vol. 2, pp. 95–108. Johns Hopkins Press, Baltimore, Maryland, 1952.

198. Hanes, C. S., Hird, J. F. R., and Isherwood, F. A. Synthesis of peptides in enzymic reactions involving glutathione. *Nature* **166**, 288–292 (1950).

199. Hanes, C. S., Hird, F. J. R., and Isherwood, F. A. Enzymic transpeptidation reactions involving γ-glutamyl peptides and α-aminoacyl peptides. *Biochem. J.* **51**, 25–35 (1952).

200. Hansen, R. G., and Hageman, E. The isolation of glutamic and aspartic acid derivatives of adenosine diphosphate. *Arch. Biochem. Biophys.* **62**, 511–513 (1956).

201. Hanson, E. A., Barrien, B. S., and Wood, J. G. Relations between protein-nitrogen protein sulfur and chlorophyll in leaves of Sudan grass. *Australian J. Exptl. Biol. Med. Sci.* **19**, 231–234 (1941).

202. Harris, D. N., and Davis, F. F. Electrophoresis of nucleic acids in silica gel. *Biochim. Biophys. Acta* **40**, 373–374 (1960).

202a. Harris, H. Function of the short-lived ribonucleic acid in the cell nucleus. *Nature* **201**, 863–867 (1964).

203. Hartley, H. Origin of the word "protein." *Nature* **168**, 244 (1951).
204. Hauschild, A. H. W., Nelson, C. D., and Krotkov, G. The effect of light quality on the products of photosynthesis in *Chlorella vulgaris*. *Can. J. Botany* **40**, 179–189 (1962).
205. Hauschild, A. H. W., Nelson, C. D., and Krotkov, G. The effect of light quality on the products of photosynthesis in green and blue green algae and in photosynthetic bacteria. *Can. J. Botany* **40**, 1619–1630 (1962).
206. Heber, U. Vergleichende Untersuchungen an Chloroplasten, die durch Iso-lierungs—Operationen in nicht-wäbrigem und in wäbrigem Milieu erhalten wurden. II. Kritik der Reinheit und Ferment-Lokalisation in Chloroplasten. *Z. Naturforsch.* **15b**, 100–109 (1960).
207. Heber, U. Protein synthesis in chloroplasts during photosynthesis. *Nature* **195**, 91–92 (1962).
208. Henderson, T. R., Skinner, C. G., and Eakin, R. E. Kinetin and kinetin analogues as substrates and inhibitors of xanthine oxidase. *Plant. Physiol.* **37**, 552–555 (1962).
209. Hendler, R. W. A model for protein synthesis. *Nature* **193**, 821–823 (1962).
210. Hendricks, S. B., and Borthwick, H. A. Photocontrol of plant development by the simultaneous excitations of two interconvertible pigments. *Proc. Natl. Acad. Sci. U.S.* **45**, 344–349 (1959).
211. Heslop-Harrison, J. Suppressive effects of 2-thiouracil on differentiation and flowering in *Cannabis sativa*. *Science* **132**, 1943–1944 (1960).
212. Heyes, J. K., and Brown, R. Growth and cellular differentiation. "Symposium: The Growth of Leaves" (F. C. Milthrope, ed.), pp. 31–49. Butterworths, London, 1956.
213. Heyns, K., and Grützmacher, H. F. Die Massenspektren der N-Formyl-α-aminosäuremethylester. *Z. Naturforsch.* **16b**, 293–298 (1961).
214. Hoagland, D. R. Lectures on the inorganic nutrition of plants. "Prather Lectures," Chronica Botanica, Waltham, Massachusetts, 1944.
215. Hoagland, M. The relationship of nucleic acid and protein synthesis as revealed by studies in cell-free systems. *In* "The Nucleic Acids" (E. Chargaff and J. N. Davidson, eds.), Vol. 3, pp. 349–408. Academic Press, New York, 1960.
216. Hoch, G. E., Schneider, K. C., and Burris, R. H. Hydrogen evolution and exchange, and conversion of N_2O to N_2 by soybean root nodules. *Biochim. Biophys. Acta* **37**, 273–279 (1960).
217. Hodge, A. J., McLean, J. D., and Mercer, F. V. A possible mechanism for the morphogenesis of lamellar systems in plant cells. *J. Biochem. Biophys. Cytol.* **2**, 597–608 (1956).
218. Horowitz, N. H., and Srb, A. M. Growth inhibition of *Neurospora* by canavanine, and its reversal. *J. Biol. Chem.* **174**, 371–378 (1948).
219. Horsfall, J. G., and Dimond, A. E., eds. "Plant Pathology. An Advanced Treatise," Vols. 1–2. Academic Press, New York, 1959–1960.
220. Hotchkiss, R. D. The biological role of the deoxypentose nucleic acids. *In* "The Nucleic Acids" (E. Chargaff and J. N. Davidson, eds.), Vol. 2, pp. 436–473. Academic Press, New York, 1955.
221. Hotta, Y. Protein and ribonucleic acid in the growth of fern gametophytes. *Nature* **189**, 455 (1961).
222. Huang, R. C., and Bonner, J. Histone, a suppressor of chromosomal RNA synthesis. *Proc. Natl. Acad. Sci. U S.* **48**, 1216–1222 (1962).
223. Hübsch, H., and Nehring, K. Vergleichende Untersuchungen über die Bestim-

mung des Stickstoffgehaltes nach den Methoden von Kjeldahl und Dumas. *Z. anal. Chem.* **173**, 278–284 (1960).

224. Hulme, A. C. Biochemical studies in the nitrogen metabolism of the apple fruit. The course followed by certain nitrogen fractions during the development of the fruit on the tree. *Biochem. J.* **30**, 258–268 (1936).

225. Hulme, A. C. Studies in the nitrogen metabolism of the apple during the normal and ethylene-induced climacteric rise in rate of respiration. *Biochem. J.* **43**, 343–349 (1948).

226. Hulme, A. C., and Smith, W. H. A relation between protein content and respiration in the cell of the apple. *Rept. Food Invest. Board, Dept. Sci. Ind. Res. (London)* p. 127 (1939).

227. Humphries, E. C. Mineral components and ash analysis. *In* "Moderne Methoden der Pflanzenanalyse" (K. Paech and V. Tracey, eds.), Vol. 1, pp. 468–502. Springer, Berlin, 1956.

227a. Hurwitz, J., Furth, J. J., Malmy, M., and Alexander, M. The inhibition of the enzymatic synthesis of ribonucleic acid and deoxyribonucleic acid by actinomycin D and proflavin. *Proc. Natl. Acad. Sci. U.S.* **48**, 1222–1230 (1962).

228. Ikawa, M., and Snell, E. E. Artifact production through esterification of glutamic acid during analytical procedures. *J. Biol. Chem.* **236**, 1955–1959 (1961).

229. Ingram, V. M. Abnormal human haemoglobins. I. The comparison of normal human and sickle-cell haemoglobins by "fingerprinting." *Biochim. Biophys. Acta* **28**, 539–545 (1958).

230. Ingram, V. M. Abnormal human haemoglobins. II. The chemical difference between normal and sickle-cell haemoglobins. *Biochim. Biophys. Acta* **36**, 402–411 (1959).

231. Isherwood, F. A., and Jones, R. L. Structure of the isomeric 2,4-dinitrophenylhydrazones of some alpha keto acids in relation to their infra-red spectra. *Nature* **175**, 419–421 (1955).

232. Jacob, F., and Monod, J. Genetic regulatory mechanisms in the synthesis of proteins. *J. Mol. Biol.* **3**, 318–356 (1961).

233. Jacob, F., and Monod, J. On the regulation of gene activity. *Cold Spring Harbor Symp. Quant. Biol.* **26**, 193–211 (1961).

234. Jacob, K. D. Fertilizer production and technology. *Advan. Agron.* **11**, 233–332 (1959).

235. James, W. O. "Plant Respiration." Oxford Univ. Press, London and New York, 1953.

236. Jensen, W. A. Relation of primary cell wall formation to cell development in plants. *In* "Synthesis of Molecular and Cellular Structures," 19th Growth Symp. (D. Rudnick, ed.), pp. 89–100. Ronald Press, New York, 1961.

237. Johnson, D. E., Scott, S. J., and Meister, A. Gas-liquid chromatography of amino acid derivatives. *Anal. Chem.* **33**, 669–673 (1961).

238. Johnson, S. W. "How Crops Feed." Judd, New York, 1895.

239. Jones, M. E., and Spector, L. The pathway of carbonate in the biosynthesis of carbamyl phosphate. *J. Biol. Chem.* **235**, 2897–2901 (1960).

240. Jones, M. E., Spector, L., and Lipmann, F. Carbamyl phosphate, the carbamyl donor in enzymatic citrulline synthesis. *J. Am. Chem. Soc.* **77**, 819–820 (1955).

241. Jones, M. E., Spector, L., and Lipmann, F. Carbamyl phosphate. *Proc. 3rd Intern. Congr. Biochem. Brussels, 1955*, pp. 67–70. Academic Press, New York, 1956.

242. Jones, W. 'Nucleic Acids. Their Chemical Properties and Physiological Conduct." Longmans, Green, New York, 1914.
243. Kamienski, E. S. von. Untersuchungen über die flüchtigen Amine der Pflanzen. I. Mitteilung Methodik der Trennung und des nachweises flüchtiger Amine. *Planta* **50,** 291–314 (1957).
244. Kamienski, E. S. von. Untersuchungen über die flüchtigen Amine der Pflanzen. II. Mitteilung. Die Amine von Blütenpflanzen und Moosen. *Planta* **50,** 315–330 (1957).
245. Katz, A. M., Dreyer, W. J., and Anfinsen, C. B. Peptide separation by two-dimensional chromatography and electrophoresis. *J. Biol. Chem.* **234,** 2897–2900 (1959).
246. Keil, B. The chemistry of peptides and proteins. *Ann. Rev. Biochem.* **31,** 139–172 (1962).
247. Kessler, B. Effect of methyltryptophan and thiouracil upon protein and ribonucleic acid synthesis in certain higher plants. *Nature* **178,** 1337–1338 (1956).
248. Kessler, B., *et al.* Effect of caffeine and other purines upon the RNA/DNA ratio in leaves and the suitability of these leaves for aphids. *Nature* **181,** 1595–1596 (1958).
249. Kessler, E. Reduction of nitrate by green algae. *Symp. Soc. Exptl. Biol.* **13,** 87–105 (1959).
250. Kidd, F., West, C., and Hulme, A. C. Rise in insoluble ("protein") fraction of the total nitrogen during the climacteric in apples and pears, and the effect on this phenomenon of retarding the climacteric by carbon dioxide or stimulating the climacteric by ethylene. *Rept. Food Invest. Board, Dept. Sci. Ind. Res. (London)* p. 119 (1939).
251. King, F. E., King, T. J., and Warwick, A. J. Extractives from hardwoods. III. Baikiain, an amino acid present in *Baikiaea plurijuga. J. Chem. Soc.* p. 3590 (1950).
252. Kirk, P. L. The chemical determination of proteins. *Advan. Protein Chem.* **3,** 139–167 (1947).
253. Kirk, P. L. "Quantitative Ultramicroanalysis. Wiley, New York, 1950.
253a. Kjaer, A. The distribution of sulfur compounds. *In* "Chemical Plant Taxonomy" (T. Swain, ed.), pp. 453–473. Academic Press, New York, 1963.
254. Klechkowski, K., and Kretovitch, W. L. The transamination of the amino acids of the ornithine cycle in pea and wheat seedlings. *Biochemistry (U.S.S.R.) (English Transl.)* **25,** 164–167 (1960).
255. Kobayashi, G. Metabolism of L-arginine. *J. Japan. Biochem. Soc.* **19,** 85–91 (1947).
256. Koffler, H., Horikoshi, K., and Garner, H. R. Architectural arrangement of polymers in fungal walls. *Abstr. Contrib. Papers Intern. Biophys. Congr., Stockholm, Sweden, 1961* p. 218 (1961).
257. Kornberg, A. Biological synthesis of deoxyribonucleic acid. *Science* **131,** 1503–1508 (1960).
258. Koshland, D. E., Jr. Correlation of structure and function in enzyme action. *Science* **142,** 1533–1541 (1963).
259. Kraus, E. J., and Kraybill, H. R. Vegetation and reproduction with special reference to tomato. *Oregon State Coll., Agr. Expt. Sta., Bull.* **149,** 1–90 (1918).
260. Krebs, H. A., and Henseleit, K. Untersuchungen über die Harnstoffbildungen in Tierkörper, *Z. physiol. Chem.* **210,** 33–66 (1932).

261. Krebs, H. A., and Lowenstein, J. M. The tricarboxylic acid cycle. *In* "Metabolic Pathways" (D. M. Greenberg, ed.), Vol. 1, pp. 129–203. Academic Press, New York, 1960.

262. Kun, E. The metabolism of sulfur-containing compounds. *In* "Metabolic Pathways" (D. M. Greenberg, ed.), Vol. 2, pp. 237–261. Academic Press, New York, 1961.

263. Lamport, D. T. A., and Northcote, D. H. Hydroxyproline in primary cell walls of higher plants. *Nature* **188**, 665–666 (1960).

264. Lederer, E. "Chromatographie en chimie organique et biologique," Vols. 1–2. Masson, Paris, 1959–1960.

265. Lengyel, P., Speyer, J. F., Basilio, C., and Ochoa, S. Synthetic polynucleotides and the amino acid code. *Proc. Natl. Acad. Sci. U.S.* **48**, 282–284 (1962).

266. Lengyel, P., Speyer, J. F., and Ochoa, S. Synthetic polynucleotides and the amino acid code. *Proc. Natl. Acad. Sci. U.S.* **47**, 1936–1942 (1961).

267. Leonard, M. J., and Burris, R. H. A survey of transaminases in plants. *J. Biol. Chem.* **170**, 701–709 (1947).

268. Lett, J. T., and Takahashi, W. N. Anomalies in protein synthesis: the release of soluble proteins from plant ribosomes. *Arch. Biochem. Biophys.* **96**, 569–574 (1962).

269. Lewis, E. J., and Gonzalves, E. A. The protein, peptide, and free amino-acid contents of some species of marine algae from Bombay. *Ann. Botany (London)* [N.S.] **26**, 301–316 (1962).

270. Lewis, E. J., and Gonzalves, E. A. Periodic studies of the proteins, peptides, and free amino acids in *Enteromorpha prolifera* f. *capillaris* and *Ulva lactuca* var. *rigida*. *Ann. Botany (London)* [N.S.] **26**, 317–327 (1962).

271. Liebig, J. "Chemistry in its Application to Agriculture and Physiology," p. 32. Campbell, Philadelphia, Pennsylvania, 1847.

272. Lindberg, B. Studies on the chemistry of lichens. VIII. Investigation of a *Dermatocarpon* and some *Roccella* species. *Acta Chem. Scand.* **9**, 917–919 (1955).

273. Lindberg, B. Methylated taurines and choline sulphate in red algae. *Acta Chem. Scand.* **9**, 1323–1326 (1955).

274. Linderstrøm-Lang, K. U. How is a protein made? *Sci. Am.* **189**, 100–106 (1953).

275. Linko, P., Holm-Hansen, O., Bassham, J. A., and Calvin, M. Formation of radioactive citrulline during photosynthetic $C^{14}O_2$ fixation by blue-green algae. *J. Exptl. Botany* **8**, 147–156 (1957).

276. Lipmann, F. Metabolism, generation and utilization of phosphate bond energy. *Advan. Enzymol.* **1**, 99–162 (1941).

277. Lipmann, F. Mechanism of peptide bond formation. *Federation Proc.* **8**, 597–602 (1949).

278. Loening, U. E. Messenger ribonucleic acid in pea seedlings. *Nature* **195**, 467–469 (1962).

279. Loew, O. The physiological role of mineral nutrient in plants. *U.S. Dept. Agr. Bur. Plant Ind. (Washington, D.C.), Bull.* **45**, 1–70 (1903).

280. Loomis, W. D. The synthesis of amino acids in plants. *In* "Handbuch der Pflanzenphysiologie" (W. Ruhland, ed.), Vol. VIII, pp. 224–248. Springer, Berlin, 1958.

281. Loomis, W. D., and Stumpf, P. K. Transamination and transamidation. *In*

"Handbuch der Pflanzenphysiologie" (W. Ruhland, ed.), Vol. VIII, pp. 249–261, Springer, Berlin, 1958.

282. Lovett, J. S., and Cantino, E. C. The relation between biochemical and morphological differentiation in *Blastocladiella emersonii*. II. Nitrogen metabolism in synchronous cultures. *Am. J. Botany* **47**, 550–560 (1960).

283. Lowry, O. H., Rosebrough, N. J., and Farr, A. L. Protein measurement with the phenol reagent. *J. Biol. Chem.* **193**, 265–275 (1951).

284. Lugg, J. W. H. Plant proteins. *Advan. Protein Chem.* **5**, 229–304 (1949).

285. Lyndon, R. F. Nitrogen metabolism of detached Kalanchoe leaves in the dark in relation to acidification, deacidification, and oxygen uptake. *J. Exptl. Botany* **13**, 20–35 (1962).

286. Lyndon, R. F., and Steward, F. C. The incorporation of ^{14}C-proline into the proteins of growing cells. I. Evidence of synthesis in different proteins and cellular components. *J. Exptl. Botany* **14**, 42–55 (1963).

287. Lyttleton, J. W. Isolation of ribosomes from spinach chloroplasts. *Exptl. Cell Res.* **26**, 312–317 (1962).

288. Magee, W. E., and Burris, R. H. Fixation of N_2 and utilization of combined nitrogen by *Nostoc muscorum*. *Am. J. Botany* **41**, 777–782 (1954).

289. Maitra, U., and Dekker, E. E. Enzymic reactions in mammalian metabolism of γ-hydroxyglutamic acid. *Biochim. Biophys. Acta* **51**, 416–418 (1961).

290. Maizel, J. V., Benson, A. A., and Tolbert, N. E. Identification of phosphoryl choline as an important constituent of plant saps. *Plant Physiol.* **31**, 407–410 (1956).

291. Makisumi, S. Occurrence of arginylglutamine in green alga, Cladophora species. *J. Biochem. (Tokyo)* **46**, 63–71 (1959).

292. Manske, R. H. F., and Holmes, H. L., eds. "The Alkaloids. Chemistry and Physiology," Vols. 1–4. Academic Press, New York, 1950–1954.

293. Markert, C. The origin of specific proteins. *In* "The Nature of Biological Diversity" (J. M. Allen, ed.), pp. 95–119. McGraw-Hill, New York, 1963.

294. Martin, A. J. P., and Synge, R. L. M. A new form of chromatogram employing two liquid phases. I. A theory of chromatography. II. Application to the determination of the higher monoamino acids in proteins. *Biochem. J.* **35**, 1358–1368 (1941).

295. Martin, A. J. P., and Synge, R. L. M. Analytical chemistry of the proteins. *Advan. Protein Chem.* **2**, 1–83 (1945).

296. Martin, E. M., and Morton, R. K. Enzymic and chemical properties of cytoplasmic particles from wheat roots. *Biochem. J.* **64**, 687–693 (1956).

297. Martin, N. H. Zone electrophoresis. Some basic considerations in design. *Analyst* **84**, 89–94 (1959).

298. Martin, R. G., Matthaei, J. H., Jones, O. W., and Nirenberg, M. W. Ribonucleotide composition of the genetic code. *Biochem. Biophys. Res. Commun.* **6**, 410–414 (1961).

299. Maskell, E. J., and Mason, T. G. Studies on the transport of nitrogenous substances in the cotton plant. I. Preliminary observations on the downward transport of nitrogen in the stem. *Ann. Botany (London)* **43**, 205–231 (1929).

300. Maskell, E. J., and Mason, T. G. Studies on the transport of nitrogenous substances in the cotton plant. II. Observations on concentration gradients. *Ann. Botany (London)* **43**, 615–652 (1929).

301. Maskell, E. J., and Mason, T. G. Studies on the transport of nitrogenous substances in the cotton plant. III. The relation between longitudinal move-

ments and concentration gradients in the bark. *Ann. Botany (London)* **44,** 1–29 (1930).

302. Maskell, E. J., and Mason, T. G. Studies on the transport of nitrogenous substances in the cotton plant. IV. The interpretation of the effects of ringing, with special reference to the lability of the nitrogen compounds in the bark. *Ann. Botany (London)* **44,** 233–267 (1930).

303. Maskell, E. J., and Mason, T. G. Studies on the transport of nitrogenous substances in the cotton plant. V. Movement to the boll. *Ann. Botany (London)* **44,** 657–688 (1930).

304. Mason, T. G., and Phillis, E. Studies on the transport of nitrogenous substances in the cotton plant. VI. Concerning storage in the bark. *Ann. Botany (London)* **48,** 135–333 (1934).

305. Mason, T. G., and Phillis, E. The migration of solutes. *Botan. Rev.* **3,** 47–71 (1937).

306. McCalla, A. G. Nitrogenous constituents of plants. *Ann. Rev. Biochem.* **18,** 615–638 (1949).

307. McCay, C. M. "Nutrition of the Dog." Cornell Univ. Press (Comstock) Ithaca, New York, 1949.

308. McCoy, R. H., Meyer, C. E., and Rose, W. C. Feeding experiments with mixtures of highly purified amino acids. VIII. Isolation and identification of a new essential amino acid. *J. Biol. Chem.* **112,** 283–302 (1935).

309. McElroy, W. D., and Glass, B., eds. "Inorganic Nitrogen Metabolism." Johns Hopkins Press, Baltimore, Maryland, 1956.

310. McFadden, B. A. Some products of $C^{14}O_2$ fixation by *Hydrogenomonas facilis*. *J. Bacteriol.* **77,** 339–343 (1959).

311. McKee, H. S. Review of recent work on the nitrogen metabolism of plants. *New Phytologist* **36,** 33–56, 240–266 (1937).

312. McKee, H. S. Review of recent work on nitrogen metabolism. *New Phytologist* **48,** 1–83 (1949).

313. McKee, H. S. Nitrogen metabolism in leaves. *In* "Handbuch der Pflanzenphysiologie" (W. Ruhland, ed.), Vol. VIII, pp. 516–553. Springer, Berlin, 1958.

314. McKee, H. S. "Nitrogen Metabolism in Plants." Oxford Univ. Press (Clarendon) London and New York, 1962.

315. McLeish, J. Comparative microphotometric studies of DNA and arginine in plant nuclei. *Chromosoma* **10,** 686–710 (1959).

316. McQuillen, K. Ribosomes and the synthesis of proteins. *Progr. Biophys. Biophys. Chem.* **12,** 69–106 (1962).

317. Medvedev, Z. A. Paper electrophoresis of proteins while preserving the pattern of their localization in tissues. *Nature* **191,** 1021–1022 (1961).

318. Medvedev, Z. A. A hypothesis concerning the way of coding interaction between transfer RNA and messenger RNA at the later stages of protein synthesis. *Nature* **195,** 38–39 (1962).

319. Mego, J. L., and Jagendorf, A. T. Effect of light on growth of black valentine bean plastids. *Biochim. Biophys. Acta* **53,** 237–254 (1961).

320. Meister, A. "Biochemistry of the Amino Acids." Academic Press, New York, 1957 (2nd edition, in press).

321. Meister, A., ed. "Biochemical Preparations," Vol. 8. Wiley, New York, 1961.

322. Meister, A., and Abendschein, P. A. Chromatography of α-keto acid 2,4-

dinitrophenylhydrazones and their hydrogenation products. *Anal. Chem.* **28,** 171–173 (1956).

323. Meister, A., Sober, H. A., Tice, S. V., and Fraser, P. E. Transamination and associated deamidation of asparagine and glutamine. *J. Biol. Chem.* **197,** 319–330 (1952).

324. Menke, W. Structure and chemistry of plastids. *Ann. Rev. Plant Physiol.* **13,** 27–44 (1962).

325. Mercer, F. The submicroscopic structure of the cell. *Ann. Rev. Plant Physiol.* **11,** 1–24 (1960).

326. Milhaud, G., Aubert, J. P., and Millet, J. Metabolisme du carbone dans la chimioautotrophie cycle d'assimilation de l'anhydride corbanique. *Compt. rend.* **243,** 102–105 (1956).

327. Miller, S. L., and Urey, H. C. Organic compound synthesis on the primitive earth. *Science* **130,** 245–251 (1959).

328. Mitchell, H. H., and Hamilton, T. S. "The Biochemistry of the Amino Acids." Chem. Catalog Co., New York, 1929.

329. Mohr, H. Primary effects of light on growth. *Ann. Rev. Plant Physiol.* **13,** 465–488 (1962).

330. Momotani, Y. Taxonomic study of the genus *Acer,* with special reference to the seed proteins. II. Analysis of protein. III. System of *Aceraceae. Mem. Coll. Sci., Univ. Kyoto* **B29,** 81–102, 177–189 (1962).

331. Monod, J., and Jacob, F. Teleonomic mechanism in the cellular metabolism growth and differentiation. *Cold Spring Harbor Symp. Quant. Biol.* **26,** 389–401 (1961).

332. Moore, S., Spackman, D. H., and Stein, W. H. Chromatography of amino acids on sulfonated polystyrene resins. An improved system. *Anal. Chem.* **30,** 1185–1190 (1958).

333. Moore, S., and Stein, W. Partition chromatography of amino acids on starch. *Ann. N.Y. Acad. Sci.* **49,** 265–278 (1948).

334. Morel, G. Métabolisme de l'arginine par les tissus de crown-gall de topinambour. *Bull. soc. franc. physiol. végétale* **2,** 75 (1956).

335. Morgan, C., and Reith, W. S. The composition and quantitative relations of proteins and related fractions in developing root cells. *J. Exptl. Botany* **5,** 119–135 (1954).

336. Morris, M. P., Pagan, C., and Warmke, H. E. Hiptagenic acid, a toxic component of *Indigofera endecaphylla. Science* **119,** 322–324 (1954).

336a. Morton, R. K., and Raison, J. K. A complete intracellular unit for incorporation of amino acids into storage protein utilizing adenosine triphosphate generated from phytate. *Nature* **200,** 429–433 (1963).

337. Mothes, K. Ein Beitrag zur Kenntnis des N-Stoffwechsels höherer Pflanzen. *Planta* **1,** 472–552 (1925).

338. Mothes, K. Physiologische Untersuchungen über das Asparagin und das Arginin in Coniferen. Ein Beitrag zur Theorie der Ammoniakentgiftung im Pflanzlichen Organismus. *Planta* **7,** 584–649 (1929).

339. Mothes, K. Zur Biosynthese der Säureamide Asparagin und Glutamin. *Planta* **30,** 726–756 (1940).

340. Mothes, K. Ammoniakentgiftung und Aminogruppenvorrat. *In* "Handbuch der Pflanzenphysiologie" (W. Ruhland, ed.), Vol. VIII, pp. 716–762. Springer, Berlin, 1958.

341. Mothes, K. New perspectives in the biosynthesis of alkaloids. *Symp. Soc. Exptl. Biol.* **13**, 258–282 (1959).

342. Mothes, K. The metabolism of urea and ureides. *Can. J. Botany* **39**, 1785–1807 (1961).

343. Mothes, K., and Englebrecht, L. Kinetin-induced directed transport of substances in excised leaves in the dark. *Phytochemistry* **1**, 58–62 (1961).

344. Mothes, K., and Englebrecht, L. Über Allantoinsäure und Allantoin. I. Ihre Rolle als Wanderform des Stickstoffs und ihre Beziehungen zum Eiweissstoffwechsels des Ahorns. *Flora (Jena)* **139**, 586–597 (1952).

345. Mothes, K., and Romeike, A. Die Alkaloide. *In* "Handbuch der Pflanzenphysiologie" (W. Ruhland, ed.), Vol. VIII, pp. 989–1049. Springer, Berlin, 1958.

346. Munson, R. D., and Doll, J. P. The economics of fertilizer use in crop production. *Advan. Agron.* **11**, 133–169 (1959).

347. Nakamura, H., Chow, C. T., and Vennesland, B. A note on the distribution of phosphorus in photophosphorylating chloroplast preparations. *J. Biol. Chem.* **234**, 2202–2204 (1959).

348. Nash, L. K. "Plants and the Atmosphere," Harvard Case History No. 5. Harvard Univ. Press, Cambridge, Massachusetts, 1952.

349. Nason, A. Enzymatic pathways of nitrate, nitrite and hydroxylamine metabolism. *Bacteriol. Rev.* **26**, 16–41 (1962).

350. Nason, A., and Takahashi, H. Inorganic nitrogen metabolism. *Ann. Rev. Microbiol.* **12**, 203–246 (1958).

351. Naylor, A. W. Interrelations of ornithine, citrulline and arginine in plants. *Symp. Soc. Exptl. Biol.* **13**, 193–209 (1959).

352. Needham, A. E. The origination of life. *Quart. Rev. Biol.* **34**, 189–209 (1959).

353. Neuberger, A., and Sanger, F. The nitrogen of the potato. *Biochem. J.* **36**, 662–671 (1942).

354. Neurath, H., and Bailey, K., eds. "The Proteins," 1st ed., Vols. 1–2. Academic Press, New York, 1953–1954 (2nd edition, 1963–1965).

355. Newton, W. The utilization of single organic nitrogen compounds by wheat seedlings and by *Phytophthora parasitica. Can. J. Botany* **35**, 445–448 (1957).

356. Nickerson, W. Biochemistry of morphogenesis. A report on symposium VI. *Proc. 4th Intern. Congr. Biochem., Vienna, 1958* Vol. 6, pp. 191–209, Pergamon Press, New York, 1959.

357. Nightingale, G. T. Ammonium and nitrate nutrition of dormant Delicious apple trees at 48°F. *Botan. Gaz.* **95**, 437–452 (1934).

358. Nightingale, G. T. The nitrogen nutrition of green plants. *Botan. Rev.* **3**, 85–174 (1937).

359. Nightingale, G. T. The nitrogen nutrition of green plants. *Botan. Rev.* **14**, 185–221 (1948).

360. Nightingale, G. T., and Robbins, W. R. Some phases of nitrogen metabolism in *Polyanthus narcissus. New Jersey Agr. Expt. Sta., Bull.* **472**, 1–32 (1928).

361. Nirenberg, M. W. The genetic code: II. *Sci. Am.* **208**, 80–94 (1963).

362. Nirenberg, M. W., and Matthaei, J. H. The dependence of cell-free protein synthesis in *E. coli* upon naturally occurring or synthetic polyribonucleotides. *Proc. Natl. Acad. Sci. U.S.* **47**, 1588–1602 (1961).

363. Nirenberg, M. W., Matthaei, J. H., and Jones, O. W. An intermediate in the biosynthesis of polyphenylalanine directed by synthetic template RNA. *Proc. Natl. Acad. Sci. U.S.* **48**, 104–109 (1962).

364. Nissen, P., and Benson, A. A. Choline sulfate in higher plants. *Science* **134**, 1759 (1961).

365. Noé, F. F., and Fowden, L. α-Amino-β-(pyrazolyl-N) propionic acid: a new amino-acid from *Citrullus vulgaris* (watermelon). *Nature* **184**, BA 69–70 (1959).

366. Noguchi, M., and Tamaki, E. Studies on nitrogen metabolism in tobacco plants. A. Part II. Diurnal variation in the amino acid composition of tobacco leaves. *Arch. Biochem. Biophys.* **98**, 197–205 (1962).

367. Ochoa, S. Biosynthesis of tricarboxylic acids by carbon dioxide fixation. I. The preparation and properties of oxalosuccinic acid. *J. Biol. Chem.* **174**, 115–122 (1948).

368. Ochoa, S. Enzymatic mechanisms in the transmission of genetic information. *In* "Horizons in Biochemistry" (M. Kasha and B. Pullman, eds.), pp. 153–166. Academic Press, New York, 1962.

369. Ofengand, J., and Haselkorn, R. Viral RNA-dependent incorporation of amino acids into protein by cell-free extracts of *E. coli. Biochem. Biophys. Res. Commun.* **6**, 469–474 (1962).

370. Oota, Y., and Osawa, S. Relation between microsomal pentose nucleic acid (PNA) and protein synthesis in the hypocotyl of germinating bean embryos. *Biochim. Biophys. Acta* **15**, 162–164 (1954).

371. Oota, Y., and Osawa, S. Migration of "storage PNA" from cotyledon into growing organs of bean seed embryo. *Experientia* **10**, 254–256 (1954).

372. Oparin, A. I. "The Origin of Life on Earth" (Translated by A. Synge), 3rd ed. Academic Press, New York, 1957.

373. Oparin, A. I., ed. "The Origin of Life on Earth," Reports on the International Symposium. Akad. Nauk, Moscow, 1959.

374. Oró, J. Mechanism of synthesis of adenine from hydrogen cyanide under possible primitive earth conditions. *Nature* **191**, 1193–1194 (1961).

375. Oró, J. Stages and mechanisms of prebiological organic synthesis. *In* "The Origins of Prebiological Systems" (S. W. Fox, ed.) pp. 137–162. Academic Press, New York, 1964.

376. Oró, J., and Kimball, A. P. Synthesis of purines under possible primitive earth conditions. II. Purine intermediates from hydrogen cyanide. *Arch. Biochem. Biophys.* **96**, 293–313 (1962).

377. Osborne, T. B. "The Vegetable Proteins," 2nd ed. Longmans, Green, New York, 1924.

378. Osborne, T. B., Wakeman, A. J., and Leavenworth. C. S. The proteins of the alfalfa plant. *J. Biol. Chem.* **49**, 63–91 (1921).

379. Osborne, T. B., Wakeman, A. J., and Leavenworth, C. S. The water-soluble constituents of the alfalfa plant. *J. Biol. Chem.* **53**, 411–429 (1922).

380. Pace, J. Seed proteins. *In* "Moderne Methoden der Pflanzenanalyse" (K. Paech and M. V. Tracey, eds.), Vol. 4, pp. 69–105. Springer, Berlin, 1955.

381. Paech, K., and Tracey, M. V., eds. "Moderne Methoden der Pflanzenanalyse" Vols. 1–4. Springer, Berlin, 1955–1956.

382. Pauling, L., Corey, R. B., and Hayward, R. The structure of protein molecules. *Sci. Am.* **191**, 51–59 (1954).

383. Peacocke, A. R. The structure and physical chemistry of nucleic acids and nucleoproteins. *Progr. Biophys. Biophys. Chem.* **10**, 56–113 (1960).

384. Pearsall, W. H. The distribution of the insoluble nitrogen in *Beta* leaves of different ages. *J. Exptl. Biol.* **8**, 279–285 (1931).

385. Pearsall, W. H. Nitrogen metabolism in plants. *Endeavor* **8**, 99–105 (1949).
386. Pearsall, W. H., and Billimoria, M. C. Losses of nitrogen from green plants. *Biochem. J.* **31**, 1743–1750 (1937).
387. Pearsall, W. H., and Ewing, J. The relation of nitrogen metabolism to plant succulence. *Ann. Botany (London)* **43**, 27–34 (1929).
388. Peck, R. L., Hoffhine, C. E., Peel, E. W., Graber, R. P., Holly, F. W., Mozingo, R., and Folkers, K. Streptomyces antibiotics. VII. The structure of streptidine. *J. Am. Chem. Soc.* **68**, 776–781 (1946).
389. Perutz, M. F. "Proteins and Nucleic Acids." Elsevier, Amsterdam, 1962.
390. Perutz, M. F., Rossmann, M. G., Cullis, A. F., Muirhead, H., and Will, G. Structure of hemoglobin—a three dimensional Fourier synthesis at 5.5 Å resolution obtained by X-ray analysis. *Nature* **185**, 416–422 (1960).
391. Phillips, D. M. P. The histones. *Progr. Biophys. Biophys. Chem.* **12**, 211–280 (1962).
392. Pirie, N. W., Proteins. *In* "Moderne Methoden der Pflanzenanalyse" (K. Paech and M. V. Tracey, eds.), Vol. IV, pp. 23–68, Springer, Berlin, 1955.
393. Pirie, N. W. Leaf proteins. *Ann. Rev. Plant Physiol.* **10**, 33–52 (1959).
394. Plimmer, R. H. A. "The Chemical Constitution of the Proteins," Vols. I and II. Longmans, Green, New York, 1912–1913.
395. Pollard, J. K. The nitrogenous constituents of plants: their recognition, identification and metabolic role. Ph.D. thesis, Cornell Univ., Ithaca, New York, 1955.
396. Pollard, J. K. The use of C^{14} labelled compounds in metabolic studies: effect due to autoclaving C^{14} urea. *In* "Amino Acid Pools" (J. T. Holden, ed.), pp. 69–70. Elsevier, Amsterdam, 1962.
397. Pollard, J. K., Rochow, W. F., and Steward, F. C. I. The incorporation of C^{14} labelled substrates with tobacco leaf protein. II. The incorporation of C^{14} labelled amino acids with tobacco mosaic virus. *Plant Physiol.* **33**, Suppl. xii (1958).
398. Pollard, J. K., Sondheimer, E., and Steward, F. C. New hydroxyamino acids in plants and their identification. Hydroxyvaline in *Kalanchöe daigremontiana. Nature* **182**, 1356–1358 (1958).
399. Pollard, J. K., and Steward, F. C. The use of C^{14}-proline by growing cells: its conversion to protein and to hydroxyproline. *J. Exptl. Botany* **10**, 17–32 (1959).
400. Porath, J., and Lindner, E. B. Separation methods based on molecular sieving and ion exclusion. *Nature* **191**, 69–70 (1961).
401. Porter, K. R. The endoplasmic reticulum: some current interpretations of its forms and functions. *In* "Biological Structure and Function" (T. W. Goodwin and O. Lindberg, eds.), Vol. 1, pp. 127–155. Academic Press, New York, 1961.
402. Porter, K. R. Diversity at the subcellular level and its significance. *In* "The Nature of Biological Diversity" (J. M. Allen, ed.), pp. 121–163. McGraw-Hill, New York, 1963.
403. Pratt, J. M. The fixation of nitrogen. *J. Theoret. Biol.* **2**, 251–258 (1962).
403a. Prianishnikov, D. N. "Nitrogen in the Life of Plants" (Translated by S. A. Wilde). Kramer Business Service, Madison, Wisconsin, 1951.
404. Prokofiev, M. A., Antonovitch, E. G., and Bogdanov, A. A. A study of nucleoprotein structure of RNA isolated from the pancreas. *Biochemistry (USSR) (English Transl.)* **25**, 931–936 (1961).
405. Raacke, I. D. The synthesis of proteins. *In* "Metabolic Pathways" (D. M. Greenberg, ed.), Vol. 2, pp. 263–388. Academic Press, New York, 1961.

406. Rabson, R. Some interactions of the environment and plant metabolism. Ph.D. thesis, Cornell Univ., Ithaca, New York, 1956.
407. Radhakrishnan, A. N., and Giri, K. V. The isolation of allohydroxy-L-proline from sandal (*Santalum album* L.). *Biochem. J.* **58**, 57–61 (1954).
408. Ram, H. Y. Mohan, Ram, Manasi and Steward, F. C. Growth and development of the banana plant. A. The origin of the inflorescence and the development of the flowers. B. The structure and development of the fruit. *Ann. Botany (London)* [N.S.] **26**, 657–673 (1962).
408a. Raper, H. S. The aerobic oxidases. *Physiol. Rev.* **8**, 245–282 (1928).
409. Ratner, S., and Rochovansky, O. Biosynthesis of guanidinoacetic acid. II. Mechanism of amidine group transfer. *Arch. Biochem. Biophys.* **63**, 296–315 (1956).
410. Raymond, S., and Aurell, B. Two-dimensional gel electrophoresis. *Science* **138**, 152–153 (1962).
411. Redfield, A. C. The biological control of chemical factors in the environment. *American Scientist* **46**, No. 3, 205–222 (1958).
411a. Redfield, A. C. *Sci. Progr.* (*New Haven*) **11th Ser.**, 149–170 (1957–1958).
412. Reifer, I., and Melville, J. The source of ammonia in plant tissue extracts. *J. Biol. Chem.* **178**, 715–725 (1949).
413. Reinbothe, H., and Mothes, K. Urea, ureides and guanidines in plants. *Ann. Rev. Plant Physiol.* **13**, 129–150 (1962).
414. Reinhold, L., and Powell, R. G. A stimulatory action of indole-3-acetic acid on the uptake of amino-acids by plant cells. *Nature* **177**, 658 (1956).
415. Reuter, G. *Proc. 5th Intern. Congr. Biochem., Moscow, 1961* p. 388. Pergamon Press, New York, 1961.
415a. Rhodes, M. J. C., and Yemm, E. W. Development of chloroplasts and the synthesis of proteins in leaves. *Nature* **200**, 1077–1080 (1963).
416. Rich, A. On the problem of evolution and biochemical information transfer. *In* "Horizons in Biochemistry" (M. Kasha and B. Pullman, eds.), pp. 103–126. Academic Press, New York, 1962.
417. Richards, F. J., and Coleman, R. G. Occurrence of putrescine in potassium-deficient barley. *Nature* **170**, 460 (1952).
418. Richmond, A. E., and Lang, A. Effect of kinetin on protein content and survival of detached *Xanthium* leaves. *Science* **125**, 650–651 (1957).
419. Roberts, E., Cowie, D. B., Abelson, P. H., Bolton, E. T., and Britten, R. J. Studies of biosynthesis in *Escherichia coli*. *Carnegie Inst. Wash. Publ.* **607** (1955).
420. Roberts, E., and Simonsen, D. G. Free amino acids in animal tissue. *In* "Amino Acid Pools" (J. T. Holden, ed.), pp. 284–349. Elsevier, Amsterdam, 1962.
421. Robertson, R. N., and Turner, J. S. The physiology of growth of apple fruits. II. Respiratory and other metabolic activities as functions of cell number and cell size in fruit and development. *Australian J. Sci. Res.* **4**, 92–107 (1951).
422. Robinson, E., and Brown, R. The development of the enzyme complement in growing root cells. *J. Exptl. Botany* **3**, 356–374 (1952).
423. Robinson, E., and Brown, R. Enzyme changes in relation to cell growth in excised root tissues. *J. Exptl. Botany* **5**, 71–78 (1954).
424. Rona, P., and Strauss, E. "Handbuch der biologischen Arbeitsmethoden" (E. Abderhalden, ed.), p. 715. Urban & Schwarzenberg, Berlin, 1922.
425. Ross, C. W. Nucleotide composition of ribonucleic acid from vegetative and flowering cocklebur-shoot tips. *Biochim. Biophys. Acta* **55**, 387–388 (1962).

426. Ross, J. P. Studies on the chemotherapy and physiology of the verticillium diseases of peppermint and chrysanthemum. Ph.D. thesis, Cornell Univ., Ithaca, New York, 1956.

427. Ruhland, W., and Wetzel, K. Zur Physiologie der organischen Säuren in grünen Pflanzen. I. Wechselbeziehungen im Stickstoff—und Säure—Stoffwechsel von *Begonia semperflorens*. *Planta* **1**, 558–564 (1926).

428. Ruhland, W., and Wetzel, K. Zur Physiologie der organischen Säuren in grünen Pflanzen. III. *Rheum hybridum* Hort. *Planta* **3**, 765–769 (1927).

429. Ruhland, W., and Wetzel, K. Zur Physiologie der organischen Säuren in grünen Pflanzen. V. Weitere Untersuchungen an *Rheum hybridum* Hort. *Planta* **7**, 503–507 (1929).

430. Russell, E. J. "Soil Conditions and Plant Growth" (Revised by E. W. Russell), 8th ed. Longmans, Green, New York, 1950.

431. Rydon, H. N., and Smith, P. W. G. A new method for the detection of peptides and similar compounds on paper chromatograms. *Nature* **169**, 922–923 (1952).

432. Sager, R., and Ryan, F. J. "Cell Heredity." Wiley, New York, 1961.

433. Schales, O., Mims, V., and Schales, S. S. Glutamic acid decarboxylase of higher plants. I. Distribution; preparation of clear solutions; nature of prosthetic groups. *Arch. Biochem. Biophys.* **10**, 455–465 (1946).

434. Schenk, W., and Schütte, H. R. 4-Aminopipecolinsäure, eine neue basische Iminosäure in *Strophanthus scandens. Naturwissenschaften* **48**, 223 (1961).

435. Schmalfuss, K. Die geochemische Bedeutung des Stickstoffs unter besonderer Berichsichtigung der landwirtschaftlichen Production und der Dünger. *In* "Handbuch der Pflanzenphysiologie" (W. Ruhland, ed.), Vol. VIII, pp. 1128–1146. Springer, Berlin, 1958.

436. Schmidt, C. L. A. "The Chemistry of the Amino Acids and Proteins." Thomas, Springfield, Illinois, 1938.

437. Schmucker, T., and Linnemann, G. Carnivorie. *In* "Handbuch der Pflanzenphysiologie" (W. Ruhland, ed.), Vol. XI, pp. 199–283, Springer, Berlin, 1962.

438. Schott, H. F., Rockland, L. B., and Dunn, M. S. Investigations of amino-acids, peptides and proteins. XVI. A source of error in the manometric ninhydrin method for the analysis of amino acids and its suppression by the use of hydrazine. *J. Biol. Chem.* **154**, 397–411 (1944).

439. Schramm, G., Grotsch, H., and Pollman, W. Non-enzymatic synthesis of polysaccharides, nucleosides, nucleic acids and the origin of self-reproducing systems. *Angew. Chem. Intern. Ed. Engl.* **1**, 1–7 (1962).

440. Schwab, G. Studien über Verbreitung und Bildung der Säureamide in der höheren Pflanzen. *Planta* **25**, 579–606 (1936).

441. Schweet, R., and Bishop, J. Protein synthesis in relation to gene action. *In* "Molecular Genetics" (J. H. Taylor, ed.), Vol. 4A, Part 1, pp. 353–404. Academic Press, New York, 1963.

442. Semenza, G. Chromatographie von Polyelektrolyten. VI. Protein chromatographie auf Ionenaustauscher-Cellulose derivativen. *Chimia (Aardy)* **14**, 325–343 (1960).

443. Setlow, R. B., and Pollard, E. C. "Molecular Biophysics." Addison-Wesley, Reading, Massachusetts, 1961.

444. Setterfield, G., Neelin, J. M., Neelin, E. M., and Bayley, S. T. Studies on basic proteins from ribosomes of buds of pea seedlings. *J. Mol. Biol.* **2**, 416–424 (1960).

445. Shantz, E. M., and Steward, F. C. Investigations on growth and metabolism of plant cells. VII. Sources of nitrogen for tissue cultures under optimal conditions for their growth. *Ann. Botany (London)* [N.S.] **23**, 371–390 (1959).

446. Shmuk, A. A. "The Chemistry and Technology of Tobacco" (N. I. Gavrilov, ed.). Office of Tech. Services, U.S. Dept. Commerce, Washington, D.C., 1953 (translated, 1961). Pishchepromizdat, Moscow.

447. Shooter, K. V. The physical chemistry of deoxyribosenucleic acid. *Progr. Biophys. Biophys. Chem.* **8**, 310–346 (1957).

448. Sideris, C. P., Krauss, B. H., and Young, H. Y. Assimilation of ammonium and nitrate nitrogen from solution cultures by roots of *Pandanus veitchii* Hort., and distribution of the various nitrogen fractions and sugars in the stele and cortex. *Plant Physiol.* **12**, 899–928 (1937).

449. Sideris, C. P., Krauss, B. H., and Young, H. Y. Assimilation of ammonium and nitrate by pineapple plants grown in nutrient solutions and its effects on nitrogenous and carbohydrate constituents. *Plant Physiol.* **13**, 489–527 (1938).

450. Siegel, A., and Wildman, S. G. Some aspects of the structure and behavior of tobacco mosaic virus. *Ann. Rev. Plant Physiol.* **11**, 277–298 (1960).

451. Simon, E. W. Valine decarboxylase in *Arum* spadix. *J. Exptl. Botany* **13**, 1–4 (1962).

452. Simpson, M. V. Protein biosynthesis. *Ann. Rev. Biochem.* **31**, 333–368 (1962).

453. Sissakian, N. M. Enzymology of the plastids. *Advan. Enzymol.* **20**, 201–236 (1958).

454. Smillie, R. M. Formation and function of soluble proteins in chloroplasts. *Can. J. Botany* **41**, 123–154 (1963).

455. Smillie, R. M., and Krotkov, G. Ribonucleic acid as an index of the metabolic activity of pea leaves. *Biochim. Biophys. Acta* **35**, 550–551 (1959).

456. Smirnov, A. I., Erygin, P. S., Drboglav, M. A., and Mashkovtsev, T. M. Über die biochemischen Eigentümlichkeiten des Alterns des Laubblätter. *Planta* **6**, 687–766 (1928).

457. Smith, D. C., Bassham, J. A., and Kirk, M. Dynamics of the photosynthesis of carbon compounds. II. Amino acid synthesis. *Biochim. Biophys. Acta* **48**, 299–313 (1961).

458. Smith, E. L., and Greene, R. D. The isoleucine content of seed globulins and β-lactoglobulin. *J. Biol. Chem.* **172**, 111–112 (1948).

459. Smith, I. "Chromatographic and Electrophoretic Techniques," Vol. 1, Wiley (Interscience), New York, 1960.

460. Smith, T. A., and Richards, F. J. The biosynthesis of putrescine in higher plants and its relation to potassium nutrition. *Biochem. J.* **84**, 292–294 (1962).

461. Smithies, O. Zone electrophoresis in starch gels: group variations in the serum proteins of normal human adults. *Biochem. J.* **61**, 629–641 (1955).

462. Smithies, O. Molecular size and starch gel electrophoresis. *Arch. Biochem. Biophys.* **98**, Suppl. 1, 125–131 (1962).

463. Sokolov, V. S. The influence of certain environmental factors on the formation and accumulation of alkaloids in plants. *Symp. Soc. Exptl. Biol.* **13**, 230–257 (1959).

464. Spackman, D. H., Stein, W. H., and Moore, S. Automatic recording apparatus for use in the chromatography of amino acids. *Anal. Chem.* **30**, 1190–1206 (1958).

465. Speck, J. F. The enzymic synthesis of glutamine. *J. Biol. Chem.* **168**, 403–404 (1947).

466. Speck, J. F. The enzymatic activity of glutamine: a reaction utilizing adenosine triphosphate. *J. Biol. Chem.* **179**, 1405–1426 (1949).

467. Spencer, B., and Harada, T. The role of choline sulphate in the sulphur metabolism of fungi. *Biochem. J.* **77**, 305–315 (1960).

468. Speyer, J. F., Lengyel, P., Basilio, C., and Ochoa, S. Synthetic polynucleotides and the amino acid code, II. *Proc. Natl. Acad. Sci. U.S.* **48**, 63–68 (1962).

469. Speyer, J. F., Lengyel, P., Basilio, C., and Ochoa, S. Synthetic polynucleotides and the amino acid code, IV. *Proc. Natl. Acad. Sci. U.S.* **48**, 441–448 (1962).

470. Spoehr, H. A., and McGee, F. M. Studies in plant respiration and photosynthesis. *Carnegie Inst. Wash. Publ.* **325**, 1–98 (1923).

470a. Sørensen, N. A. Chemical taxonomy of acetylenic compounds. *In* "Chemical Plant Taxonomy" (T. Swain, ed.), pp. 219–252. Academic Press, New York, 1963.

471. Srb, A. M., and Horowitz, N. H. The ornithine cycle in *Neurospora* and its genetic control. *J. Biol. Chem.* **154**, 129–139 (1944).

472. Stanley, W. The isolation and properties of tobacco mosaic and other virus proteins. *Harvey Lectures* pp. 170–204 (1937–1938).

473. Stein, W. H., and Moore, S. Chromatography of amino acids on starch columns. Separation of phenylalanine, leucine, isoleucine, methionine, tyrosine and valine. *J. Biol. Chem.* **176**, 337–388 (1948).

474. Stein, W. H., and Moore, S. The chemical structure of proteins. *Sci. Am.* **205**, 81–92 (1961).

475. Steinberg, R. A. Growth responses to organic compounds by tobacco seedlings in aseptic culture. *J. Agr. Res.* **75**, 81–92 (1947).

476. Stephenson, M. L., Thimann, K. V., and Zamecnik, P. C. Incorporation of ^{14}C amino acids into proteins of leaf disks and cell-free fractions of tobacco leaves. *Arch. Biochem. Biophys.* **65**, 194–209 (1956).

477. Stepka, W., and Takahashi, W. N. Amide constituents of tobacco mosaic virus proteins. *Science* **111**, 176–177 (1950).

478. Stetten, M. R. Some aspects of the metabolism of hydroxyproline studied with the aid of isotopic nitrogen. *J. Biol. Chem.* **181**, 31–37 (1949).

479. Steward, F. C. Effects of environment on metabolic patterns. *In* "Environmental Control of Plant Growth" (L. T. Evans, ed.), pp. 195–214. Academic Press, New York, 1963.

480. Steward, F. C., Berry, W. E., Preston, C., and Ramamurti, T. K. The absorption and accumulation of solutes by living plant cells. X. Time and temperature effects on salt uptake by potato discs and the influence of the storage conditions of the tubers on metabolism and other properties. *Ann. Botany (London)* [N.S.] **7**, 221–260 (1943).

481. Steward, F. C., and Bidwell, R. G. S. The free nitrogen compounds in plants considered in relation to metabolism growth and development. *In* "Amino Acid Pools" (J. T. Holden, ed.), pp. 25–42. Elsevier, Amsterdam, 1962.

482. Steward, F. C., Bidwell, R. G. S., and Yemm, E. W. Nitrogen metabolism, respiration, and growth of cultured plant tissue I. Experimental design, techniques, and recorded data. II. The interpretation of specific activity data in tracer experiments. III. Nitrogen metabolism and respiration of carrot tissue explants as revealed by experiments with C^{14}-labelled substrates. IV. The impact of growth on protein metabolism and respiration of carrot tissue explants. General discussion of results. *J. Exptl. Botany* **9**, 11–51, 285–305 (1958).

483. Steward, F. C., and Caplin, S. A tissue culture from potato tubers. The synergistic action of 2,4-D and coconut milk. *Science* 113, 518–520 (1951).

484. Steward, F. C., and Chang, L. O. The incorporation of C^{14}-proline into the proteins of growing cells. A note on evidence from cultured carrot explants by acrylamide gel electrophoresis. *J. Exp. Botany* 14, 379–386 (1963).

485. Steward, F. C., Crane, F., Millar, K., Zacharius, R. M., Rabson, R., and Margolis, D. Nutritional and environmental effects on the nitrogen metabolism of plants. I. Present concepts of the complement of soluble nitrogen compounds in plants. II. Effects due to day length and nutrition on the nitrogen metabolism of *Mentha piperita*. III. The effect of sulphur deficiency and recovery on the nitrogen compounds of mint. IV. Interacting effects of temperature and photoperiod on the nitrogen metabolism of *Mentha piperita*. V. Effects on the soluble nitrogen compounds in *Nicotiana* due to acute deficiency of the trace elements. VI. Effects of trace elements (Mn and Mo) upon the soluble nitrogen compounds of plants. *Symp. Soc. Exptl. Biol.* 13, 148–176 (1959).

486. Steward F. C., Freiberg, S. R., Hulme, A. C., Hegarty, M. P., Barr, R. A., and Rabson, R. Physiological investigations on the banana plant. II. Factors which affect the nitrogen compounds of the fruit. *Ann. Botany (London)* [N.S.] 24, 117–146 (1960).

487. Steward, F. C., Howe, K. J., Crane, F. A., and Rabson, R. Growth, nutrition and metabolism of *Mentha piperita* L. *Cornell Univ., Agr. Expt. Sta. Mem.* 379, 1–144 (1962).

488. Steward, F. C., Hulme, A. C., Freiberg, S. R., Hegarty, M. P., Pollard, J. K., Rabson, R., and Barr, R. A. Physiological investigations on the banana plant. I. Biochemical constituents detected in the banana plant. *Ann. Botany (London)* [N.S.] 24, 83–116 (1960).

489. Steward, F. C., and Chang, L. O. The incorporation of C^{14}-proline into the proteins of growing cells. II. A note on evidence from cultured carrot explants by acrylamide gel electrophoresis. *J. Exptl. Botany* 14, 379–386 (1963).

489a. Steward, F. C., with Mapes, M. O., Kent, A. E., and Holsten, R. D. Growth and development of cultured plant cells. *Science* 143, 20–27 (1964).

490. Steward, F. C., and Margolis, D. The effects of molybdenum upon the free amino acids and amides of the tomato plant. *Contrib. Boyce Thompson Inst.* 21, 411–421 (1962).

491. Steward, F. C., and Margolis, D. The effects of manganese upon the free amino acids and amides of the tomato plant. *Contrib. Boyce Thompson Inst.* 21, 393–410 (1962).

492. Steward, F. C., and Millar, F. K. Salt accumulation in plants: a reconsideration of the role of growth and metabolism. I. Salt accumulation as a cellular phenomenon. II. Salt accumulation in the plant body. *Symp. Soc. Exptl. Biol.* 8, 367–406 (1954).

493. Steward, F. C., and Pollard, J. K. Some further observations on glutamyl and related compounds in plants. *In* "Inorganic Nitrogen Metabolism" (W. D. McElroy and B. Glass, eds.), pp. 377–407. Johns Hopkins Press, Baltimore, Maryland, 1956.

494. Steward, F. C., and Pollard, J. K. Nitrogen metabolism in plants: ten years in retrospect. *Ann. Rev. Plant Physiol.* 8, 65–114 (1957).

495. Steward, F. C., and Pollard, J. K. C^{14}-proline and hydroxyproline in the protein metabolism of plants: an episode in the relation of metabolism to cell growth and morphogenesis. *Nature* 182, 828–832 (1958).

496. Steward, F. C., and Pollard, J. K. The soluble, nitrogenous constituents of plants. *In* "Amino Acid Pools" (J. T. Holden, ed.), pp. 25–42. Elsevier, Amsterdam, 1962.

497. Steward, F. C., Pollard, J. K., Patchett, A. A., and Witkop, B. The effects of selected nitrogen compounds on the growth of plant tissue cultures. *Biochim. Biophys. Acta* **28**, 308–314 (1958).

498. Steward, F. C., and Preston, C. The effect of salt concentration upon the metabolism of potato discs and the contrasted effect of potassium and calcium salts which have a common ion. *Plant Physiol.* **16**, 85–116 (1941).

499. Steward, F. C., Shantz, E. M., Pollard, J. K., Mapes, M. O., and Mitra, J. Growth induction in explanted cells and tissues. *In* "Synthesis of Molecular and Cellular Structures," 19th Growth Symp. (D. Rudnick, ed.), pp. 193–246. Ronald Press, New York, 1961.

500. Steward, F. C., Stout, P. R., and Preston, C. The balance sheet of metabolites for potato discs showing the effects of salts and dissolved oxygen on metabolism at 23°C. *Plant Physiol.* **15**, 409–447 (1940).

501. Steward, F. C., and Street, H. E. The soluble nitrogen fractions of potato tubers; the amides. *Plant Physiol.* **21**, 155–193 (1946).

502. Steward, F. C., and Street, H. E. The nitrogenous constituents of plants. *Ann. Rev. Biochem.* **16**, 471–502 (1947).

503. Steward, F. C., and Sutcliffe, J. F. Plants in relation to inorganic salts. *In* "Plant Physiology" (F. C. Steward, ed.), Vol. 2, pp. 253–478. Academic Press, New York, 1959.

504. Steward, F. C., and Thompson, J. F. Photosynthesis and respiration: a reinterpretation of recent work with radioactive carbon. *Nature* **166**, 593–596 (1950).

505. Steward, F. C., and Thompson, J. F. The nitrogenous constituents of plants with special reference to chromatographic methods. *Ann. Rev. Plant Physiol.* **1**, 233–264 (1950).

506. Steward, F. C., and Thompson, J. F. The analysis of the alcohol-insoluble nitrogen of plants by quantitative procedures based on paper chromatography. I. The analysis of certain pure proteins. II. The composition of the alcohol-soluble and insoluble fractions of the potato tuber. *J. Exptl. Botany* **3**, 170–187 (1952).

506a. Steward, F. C., and Thompson, J. F. Structure of asparagine. *Nature* **171**, 1062 (1953).

507. Steward, F. C., and Thompson, J. F. Proteins and protein metabolism in plants. *In* "The Proteins" (H. Neurath and K. Bailey, eds.), 1st ed., Vol. 2, Part A, pp. 513–594. Academic Press, New York, 1954.

508. Steward, F. C., Thompson, J. F., and Pollard, J. K. Contrasts in the nitrogenous composition of rapidly growing and non-growing plant tissues. *J. Exptl. Botany* **9**, 1–10 (1958).

509. Steward, F. C., Wetmore, R. H., and Pollard, J. K. The nitrogenous components of the shoot apex of *Adiantum pedatum*. *Am. J. Botany* **42**, 946–948 (1955).

510. Steward, F. C., Wetmore, R. H., Thompson, J. F., and Nitsch, J. P. A quantitative chromatographic study of nitrogenous components of shoot apices. *Am. J. Botany* **41**, 123–134 (1954).

511. Steward, F. C., Zacharius, R. M., and Pollard, J. K. Nitrogenous compounds

in plants: Recent knowledge derived from paper partition chromatography. *Ann. Acad. Sci. Fennicae: Ser. A II* **60**, 321–366 (1955).

512. Stiller, M. The path of carbon in photosynthesis. *Ann. Rev. Plant Physiol.* **13**, 151–170 (1962).

513. Stokes, J. L., Gunness, M., Dwyer, I. M., and Caswell, M. C. Microbiological methods for the determination of amino acids. II. A uniform assay for the ten essential amino acids. *J. Biol. Chem.* **160**, 35–49 (1945).

514. Street, H. E. Experimental methods available for the study of the nitrogen metabolism of plants. *New Phytologist* **48**, 84–96 (1949).

515. Street, H. E., Kenyon, A. E., and Watson, G. M. The nature and distribution of various forms of nitrogen in the potato. *Ann. Appl. Biol.* **33**, 1–12 (1946).

516. Street, H. E., Kenyon, A. E., and Watson, G. M. The assimilation of ammonium and nitrate nitrogen by detached potato sprouts. *Ann. Appl. Biol.* **33**, 369–381 (1946).

517. Street, H. E., and Trease, G. E. The discovery of asparagine. *Ann. Sci.* **7**, 70–76 (1951).

518. Stumpf, P. K., Loomis, W. D., and Michelson, C. Amide metabolism in higher plants. I. Preparation and properties of glutamyl transphorase from pumpkin seedlings. *Arch. Biochem. Biophys.* **30**, 126–137 (1951).

519. Sumner, J. B. The isolation and crystallization of the enzyme urease. Preliminary paper. *J. Biol. Chem.* **69**, 435–441 (1926).

520. Sumner, J. B., and Myrbäck, K., eds. "The Enzymes," Vols. 1, 2. Academic Press, New York, 1951 (2nd edition, 8 vols., (1959–1963).

521. Sutcliffe, J. F., Bollard, E. G., and Steward, F. C. The incorporation of ^{14}C into the protein of particles isolated from plant cells. *J. Exptl. Botany* **11**, 151–166 (1960).

522. Synge, R. L. M. Peptides (bound amino acids) and free amino acids. *In* "Moderne Methoden der Pflanzenanalyse" (K. Paech and M. V. Tracey, eds.), Vol. IV, pp. 1–22. Springer, Berlin, 1955.

523. Synge, R. L. M., and Youngson, M. A. Molecular-sieve methods for separations of polypeptides according to molecular weight. *Biochem. J.* **78**, 31P (1961).

524. Tagawa, K., and Arnon, D. I. Ferredoxins as electron carriers in photosynthesis and in the biological production and consumption of hydrogen gas. *Nature* **195**, 537–543 (1962).

525. Tagawa, K., Tsujimoto, H. Y., and Arnon, D. I. Role of chloroplast ferredoxin in the energy conversion process of photosynthesis. *Proc. Natl. Acad. Sci. U.S.* **49**, 567–572 (1963).

526. Takahashi, H., Taniguchi, S., and Egami, F. Inorganic nitrogen compounds, distribution and metabolism. *In* "Comparative Biochemistry" (M. Florkin and H. S. Mason, eds.), Vol. 5, Part C, pp. 91–202. Academic Press, New York, 1963.

527. Takebe, I. Choline sulfate as a major soluble sulfur component of conidiospores of *Aspergillus niger. J. Gen. Appl. Microbiol. (Tokyo)* **6**, 83–89 (1960).

528. Taylor, J. H., ed. "Molecular Genetics," Part 1. Academic Press, New York, 1963.

529. Taylor, J. H. The replication and organization of DNA in chromosomes. *In* "Molecular Genetics" (J. H. Taylor, ed.), Vol. 4A, Part 1, pp. 65–111. Academic Press, New York, 1963.

530. Thimann, K. V., and Loos, G. M. Protein synthesis during water uptake by tuber tissue. *Plant Physiol.* **32**, 274–279 (1957).

531. Thoai, N. van, and Olomuki, A. Arginine décarboxy-oxydase. I. Caractères et nature de l'enzyme. *Biochim. Biophys. Acta* **59**, 533–544 (1962).

532. Thoai, N. van, and Olomuki, A. Arginine décarboxy-oxydase. II. Oxydation de la canavanine et de l'homoarginine en β-guanidoxypropionamide et en γ-guanidovaleramide. *Biochim. Biophys. Acta* **59**, 545–552 (1962).

533. Thoai, N. van, and Roche, T. Dérivés guandiques biologiques. *Progr. Chem. Org. Nat. Prod.* **18**, 83–121, (1960).

534. Thomas, J. B. Chloroplast structure. *In* "Handbuch der Pflanzenphysiologie" (W. Ruhland, ed.), Vol. V, Part 1, pp. 511–565. Springer, Berlin, 1960.

535. Thompson, J. F., and Morris, C. J. Determination of amino acids from plants by paper chromatography. *Anal. Chem.* **31**, 1031–1037 (1959).

536. Thompson, J. F., Morris, C. J., Arnold, W. N., and Turner, D. H. γ-Glutamyl peptides in plants. *In* "Amino Acid Pools" (J. T. Holden, ed.), pp. 54–64. Elsevier, Amsterdam, 1962.

537. Thompson, J. F., Morris, C. J., and Gering, R. K. Purification of plant amino acids for paper chromatography. *Anal. Chem.* **31**, 1028–1031 (1959).

538. Thompson, J. F., Pollard, J. K., and Steward, F. C. Investigations of nitrogen compounds and nitrogen metabolism in plants. III. γ-Aminobutyric acid in plants, with special reference to the potato tuber and a new procedure for isolating amino acids other than γ-aminoacids. *Plant Physiol.* **28**, 401–414 (1953).

539. Thompson, J. F., and Steward, F. C. Investigations of nitrogen compounds and nitrogen metabolism in plants. *Plant Physiol.* **26**, 421–440 (1951).

540. Thompson, J. F., and Steward, F. C. The analysis of the alcohol-insoluble nitrogen of plants by quantitative procedures based on paper chromatography. II. The composition of the alcohol-soluble and insoluble fractions of the potato tuber. *J. Exptl. Botany* **3**, 170–187 (1952).

541. Tigane, E., Wade, E. H. M., Tze-Fei Wong, J., and Hanes, C. S. Quantitative chromatographic methods. 6. Improved procedure for determination of amino compounds on paper chromatograms. *Can. J. Botany* **39**, 427–438 (1961).

542. Todd, A. Synthesis in the study of nucleotides. *Science* **127**, 787–792 (1958).

543. Todd, A. Some current problems in polynucleotide chemistry. *Proc. Chem. Soc.* pp. 187–193 (1961).

544. Todd, A. Some observations on organic phosphates. *Proc. Chem. Soc.* pp. 199–204 (1962).

544a. Tolbert, N. E. Glycolate pathway. *Natl. Acad. Sci.—Natl. Res. Council, Publ.* **1145**, 648–662 (1963).

545. Towers, G. H. N., and Steward, F. C. The keto acids of the tulip (*Tulipa gesneriana*) with special reference to the keto analog of γ-methyleneglutamic acid. *J. Am. Chem. Soc.* **76**, 1959–1961 (1954).

546. Towers, G. H. N., Thompson, J. F., and Steward, F. C. The detection of the keto acids of plants. A procedure based on their conversion to amino acids. *J. Am. Chem. Soc.* **76**, 2392–2396 (1954).

547. Tschiersch, B. Über das Vorkommen von Canavanin. *Flora (Jena)* **150**, 87–94 (1961).

548. T'so, P. O. P. The ribosomes-ribonucleoprotein particles. *Ann. Rev. Plant Physiol.* **13**, 45–80 (1962).

549. Tsugita, A., and Fraenkel-Conrat, H. Contributions from TMV studies to the problem of genetic information transfer and coding. *In* "Molecular Genetics" (J. H. Taylor, ed.), Vol. 4A, Part 1, pp. 477–520. Academic Press, New York, 1963.

550. Tsugita, A., Fraenkel-Conrat, H., Nirenberg, M. W., and Matthaei, J. H. Demonstration of the messenger role of viral RNA. *Proc. Natl. Acad. Sci. U.S.* **48**, 846–853 (1962).

551. Urey, H. C. "The Planets." Yale Univ. Press, New Haven, Connecticut, 1952.

552. Vaidyanathan, C. S., and Giri, K. V. Studies on plant arginase. I. Arginase from field beans (*Dolichos lablab*). General properties and the effect of metallic ions. *Enzymologia* **16**, 167–178 (1953).

553. Van Niel, C. B. The present status of the comparative study of photosynthesis. *Ann. Rev. Plant Physiol.* **13**, 1–26 (1962).

554. Vickery, H. B. Some nitrogenous constituents of the juice of the alfalfa plant. *J. Biol. Chem.* **60**, 647–655 (1924).

555. Vickery, H. B. Some nitrogenous constituents of the juice of the alfalfa plant. *J. Biol. Chem.* **61**, 117–127 (1924).

556. Vickery, H. B. Some nitrogenous constituents of the juice of the alfalfa plant. IV. The betaine fraction. *J. Biol. Chem.* **65**, 81–89 (1925).

557. Vickery, H. B. Some nitrogenous constituents of the juice of the alfalfa plant. VI. Asparagine and amino acids in alfalfa. *J. Biol. Chem.* **65**, 657–664 (1925).

558. Vickery, H. B. End products of nitrogen metabolism in plants. *Biol. Symposia* **5**, 3–19 (1941).

559. Vickery, H. B. Liebig and the chemistry of proteins. *In* "Liebig and After Liebig," Publ. No. 16, pp. 19–29. Am. Assoc. Advance. Sci., Washington D.C., 1942.

560. Vickery, H. B. The proteins of plants. *Physiol. Rev.* **25**, 347–376 (1945).

561. Vickery, H. B. The contribution of the analytical chemist to protein chemistry. *Ann. N.Y. Acad. Sci. U.S.* **47**, 63–94 (1946).

562. Vickery, H. B. Position of Arnold in relation to the Kjeldahl methods. *J. Assoc. Offic. Agr. Chemists* **29**, 358–370 (1946).

563. Vickery, H. B. The early years of the Kjeldahl method to determine nitrogen. *Yale J. Biol. Med.* **18**, 473–516 (1946).

564. Vickery, H. B. The origin of the word protein. *Yale J. Biol. Med.* **22**, 387–393 (1950).

565. Vickery, H. B. Thomas Burr Osborne. *J. Nutr.* **59**, 3–26 (1956).

566. Vickery, H. B. Chemical investigations of the tobacco plant. XI. Composition of the green leaf in relation to position on the stalk. *Conn. Agr. Expt. Sta., New Haven, Bull.* **640**, 5–42 (1961).

567. Vickery, H. B., and Leavenworth, C. S. Some nitrogenous constituents of the juice of the alfalfa plant. III. Adenine in alfalfa. *J. Biol. Chem.* **63**, 579–583 (1925).

568. Vickery, H. B., and Osborne, T. B. A review of the hypothesis of the structure of proteins. *Physiol. Rev.* **8**, 393–446 (1928).

569. Vickery, H. B., Pucher, G. W., Clark, H. E., Chibnall, A. C., and Westall, R. G. The determination of glutamine in the presence of asparagine. *Biochem. J.* **29**, 2710–2712 (1935).

570. Vickery H. B., Pucher, G. W., and Leavenworth, C. S. Determination of ammonia and of amide nitrogen in plant tissues. *Ind. Eng. Chem., Anal. Ed.* **7**, 152–156 (1935).

571. Vickery, H. B., Pucher, G. W., Schoenheimer, R., and Rittenberg, D. The assimilation of ammonia by the tobacco plant: a preliminary study with isotopic nitrogen. *J. Biol. Chem.* **135**, 531–539 (1940).

572. Vickery, H. B., Pucher, G. W., Wakeman, A. J., and Leavenworth, C. S. Chemical investigations of the tobacco plant. VI. Chemical changes that occur in leaves during culture in the light and in darkness. *Conn. Agr. Expt. Sta., New Haven, Bull.* **399**, 757–832 (1937).

573. Vickery, H. B., Pucher, G. W., Wakeman, A. J., and Leavenworth, C. S. The nitrogen nutrition of *Narcissus poeticus*. *Conn. Agr. Expt. Sta., New Haven, Bull.* **496**, 1–93 (1946).

574. Vickery, H. B., and Schmidt, C. L. A. The history of the discovery of amino acids. *Chem. Rev.* **9**, 169–318 (1931).

575. Vickery, H. B., and Vinson, C. G. Some nitrogenous constituents of the juice of the alfalfa plant. V. The basic lead acetate precipitate. *J. Biol. Chem.* **65**, 91–95 (1925).

576. Virtanen, A. I. "Cattle Fodder and Human Nutrition." Cambridge Univ. Press, London and New York, 1938.

577. Virtanen, A. I. Neue Amino- und Ketosäuren in grünen Pflanzen und die Biosynthese der Aminosäuren. *Angew. Chem.* **67**, 381–388 (1955).

578. Virtanen, A. I. Some aspects of amino acid synthesis in plants and related subjects. *Ann. Rev. Plant Physiol.* **12**, 1–12 (1961).

579. Virtanen, A. I., and Hietala, P. K. Gamma-hydroxyglutamic acid in green plants. *Acta Chem. Scand.* **9**, 175–176 (1955).

580. Virtanen, A. I., and Kari, S. 5-Hydroxypiperidine-2-carboxylic acid in green plants. *Acta Chem. Scand.* **8**, 1290–1291 (1954).

581. Virtanen, A. I., and Kari, S. 4-Hydroxypiperidine-2-carboxylic acid in green plants. *Acta Chem. Scand.* **9**, 170–171 (1955).

582. Virtanen, A. I., and Kari, S. Free amino acids in pollen. *Acta Chem. Scand.* **9**, 1548–1551 (1955).

583. Virtanen, A. I., and Laine, T. Investigations on the root nodule bacteria of leguminous plants. XXII. The excretion products of root nodules. The mechanism of N-fixation. *Biochem. J.* **33**, 412–427 (1939).

584. Virtanen, A. I., and Linkola, H. Organic nitrogen compounds as nitrogen nutrition for higher plants. *Nature* **158**, 515 (1946).

585. Virtanen, A. L., Miettinen, J. K., and Kunttu, H. α-Ketoacids in green plants. *Acta Chem. Scand.* **7**, 38–44 (1953).

586. Virtanen, A. L., and Tarnanen, J. Die enzymatische Spaltung und Synthese der Asparaginsäure. *Biochem. Z.* **250**, 193–211 (1932).

587. Vogel, H. J. On biochemical evolution: lysine formation in higher plants. *Proc. Natl. Acad. Sci. U.S.* **45**, 1717–1721 (1959).

588. von der Decken, A. A simple paper chromatographic method for determination of amino-acid activation profiles in tissue preparations. *Nature* **189**, 145–146 (1961).

589. Waddington, C. H. "New Patterns in Genetics and Development." Columbia Univ. Press, New York, 1962.

590. Waelsch, H. Certain aspects of intermediary metabolism of glutamine, asparagine and glutathione. *Advan. Enzymol.* **13**, 236–319 (1952).

591. Wailes, P., Whiting, H. C., and Fowden, L. Synthesis of γ-methyleneglutamic acid and γ-methyleneglutamine *Nature* **174**, 130–131 (1954).

592. Wald, G. The origin of optical activity. *Ann. N.Y. Acad. Sci.* **69**, 355–376, (1957).

593. Wald, G. Life in the second and third periods; or why phosphorus and sulfur for "high-energy" bonds? *In* "Horizons in Biochemistry" (M. Kasha and B. Pullman, eds.), pp. 127–142. Academic Press, New York, 1962.

594. Warburg, O., and Negelein, W. Ueber die Reduktion des Saltpetersäure in grünen Zellen. *Biochem. Z.* **110**, 66–115, (1920).

595. Waris, H. Neomorphosis in seed plants induced by amino acids. I. *Oenanthe aquatica. Physiol. Plantarum* **12**, 753–766 (1959).

596. Warren, L. Nucleotides and nucleosides. *In* "Metabolic Pathways" (D. M. Greenberg, ed.), Vol. 2, pp. 459–524. Academic Press, New York, 1961.

597. Watson, J. D. Involvement of RNA in the synthesis of proteins. *Science* **140**, 17–26 (1963).

598. Watson, J. D., and Crick, F. H. C. Molecular structure of nucleic acid. A structure for deoxyribose nucleic acid. *Nature* **171**, 737–738 (1953).

599. Webster, G. "Nitrogen Metabolism in Plants." Row, Peterson, Evanston, Illinois, 1959.

600. Webster, G. Protein synthesis. *Ann. Rev. Plant Physiol.* **12**, 113–132 (1961).

601. Webster, G., and Whitman, S. L. Guanosine triphosphate degradation during amino acid transfer from ribonucleic acid to ribosomal protein. *Biochim. Biophys. Acta* **61**, 316–317 (1962).

602. Weisblum, B., Benzer, S., and Holley, R. W. A physical basis for degeneracy in the amino acid code. *Proc. Natl. Acad. Sci. U.S.* **48**, 1449–1453 (1962).

603. Westall, R. G. The free amino acids of body fluids and some hereditary disorders of amino acid metabolism. *In* "Amino Acid Pools" (J. T. Holden, ed.), pp. 195–219. Elsevier, Amsterdam, 1962.

604. Wettstein, D. von. The formation of plastid structures. *Brookhaven Symp. Biol.* **11**, 138–151 (1958).

605. Whaley, W. G., Mollenhauer, H. H., and Leech, J. H. The ultrastructure of the meristematic cells. *Am. J. Botany* **47**, 401–450 (1960).

606. Whitehead, E. L., and Moxon, A. I. Nitrate poisoning. *S. Dakota State Coll., Agr. Expt. Sta., Bull.* **424**, 1–24 (1952).

607. Wilkins, M. H. F. Molecular configuration of nucleic acids. *Science* **140**, 941–950 (1963).

608. Wilson, D. G., King, K. W., and Burris, R. H. Transamination reactions in plants. *J. Biol. Chem.* **208**, 863–874 (1954).

609. Wilson, L. G. Metabolism of sulfate: sulfate reduction. *Ann. Rev. Plant Physiol.* **13**, 201–224 (1962).

610. Wilson, P. "The Biochemistry of Symbiotic Nitrogen Fixation." Univ. of Wisconsin Press, Madison, 1940.

611. Wilson, P. Asymbiotic nitrogen fixation. *In* "Handbuch der Pflanzenphysiologie" (W. Ruhland, ed.), Vol. VIII, pp. 9–47, Springer, Berlin, 1958.

612. Winterstein, E. Aminosäuren. *In* "Handbuch der Pflanzenanalyse" (G. Klein, ed.), Vol. 3, pp. 1–180. Springer, Vienna, 1933.

613. Wolken, J. J. "Euglena." Institute Microbiology, Rutgers Univ., 1961.

614. Wolken, J. J. Photoreceptor structures: their molecular organization for energy transfer. *J. Theoret. Biol.* **3**, 192–208 (1962).

615. Wolken, J. J., and Gross, J. A. The development and characteristics of *Euglena* c-type cytochromes. *J. Protozool.* **10**, 189–194 (1963).

616. Wollgiehn, R. Untersuchungen über den Einfluss des Kinetins auf den Nucleinsäure—und Proteinstoffwechsel isolierter Blätter. *Flora (Jena)* **151**, 411– • 437 (1961).

617. Wood, J. G. Nitrogen metabolism of higher plants. *Ann. Rev. Plant Physiol.* **3**, 1–22 (1953).

618. Wood, J. G., and Barrien, B. S. Studies on the sulfur metabolism of plants. III. On changes in amounts of protein sulfur and sulfate sulfur during starvation. *New Phytologist* **38**, 265–272 (1939).

619. Wood, J. G., Cruickshank, D. H., and Kuchel, R. H. The metabolism of starving leaves. 1. Presentation of data; the nature of respiration rate, time curves in air and in nitrogen and the relation to carbohydrates. 2. Changes in amounts of total and chloroplast proteins, chlorophyll, ascorbic acid and soluble nitrogen compounds. 3. Changes in malic and citric acid contents and interrelations of these with soluble nitrogen compounds. *Australian J. Exptl. Biol. Med. Sci.* **21**, 37–53 (1943).

620. Wood, J. G., and Petrie, A. H. K. Studies on the nitrogen metabolism of plants. *Australian J. Exptl. Biol. Med. Sci.* **20**, 249–256 (1942).

621. Woolley, D. W., Schaffner, G., and Braun, A. C. Studies on the structure of the phytopathogenic toxin of *Pseudomonas tabaci.* *J. Biol. Chem.* **215**, 485–493 (1955).

622. Woolley, D. W., and Peterson, W. H. The chemistry of mold tissue. XIV. Isolation of cyclic choline sulfate from *Aspergillus sydowi.* *J. Biol. Chem.* **122**, 213–218 (1937).

623. Wright, B. E., and Anderson, M. L. Protein and amino acid turnover during differentiation in the slime mold. I. Utilization of endogeneous amino acids and protein. II. Incorporation of (^{35}S) methionine into the amino acid pool and into protein. *Biochim. Biophys. Acta* **43**, 62–78 (1960).

624. Wright, D. E. Amino acid uptake by plant roots. *Arch. Biochem. Biophys.* **97**, 174–180 (1962).

625. Wycoff, R. G. The electron microscopy of macromolecules. *Advan. Protein Chem.* **6**, 1–33 (1951).

626. Yanofsky, C. Genetic control of protein structure. *In* "Cytodifferentiation and Macromolecular Synthesis" (M. Locke, ed.), pp. 15–28. Academic Press, New York, 1963.

627. Yemm, E. W. The respiration of barley plants. II. Carbohydrate concentration and carbon dioxide production in starving leaves. *Proc. Roy. Soc. (London)* **B117**, 504–525 (1935).

628. Yemm, E. W. Respiration of barley plants. III. Protein catabolism in starving leaves. *Proc. Roy. Soc. (London)* **B123**, 243–273 (1937).

629. Yemm, E. W. Respiration of barley plants. IV. Protein catabolism and the formation of amides in starving leaves. *Proc. Roy. Soc. (London)* **B136**, 632–649 (1950).

630. Yocum, C. S. Nitrogen fixation. *Ann. Rev. Plant Physiol.* **11**, 25–36 (1960).

631. Zacharius, R. M. The nitrogenous constituents of plants and their relation to metabolism. Ph.D. thesis, Univ. of Rochester, Rochester, New York, 1952.

632. Zacharius, R. M., Cathey, H. M., and Steward, F. C. Nitrogenous compounds and nitrogen metabolism in the *Liliaceae.* *Ann. Botany (London)* [N.S.] **21**, 193–201 (1957).

632a. Zacharius, R. M., and Talley, E. A. Elution behavior of naturally-occurring ninhydrin-positive compounds during ion exchange chromatography. *Anal. Chem.* **34**, 1551–1556 (1962).

633. Zamecnik, P. C. Historical and current aspects of the problem of protein synthesis. *Harvey Lectures* **54**, 256–281 (1960).

634. Zechmeister, L. History, scope and methods of chromatography. *Ann. N.Y. Acad. Sci.* **49**, 145–160 (1948).

635. Zimmermann, M. H. Transport in the phloem. *Ann. Rev. Plant Physiol.* **11**, 167–190 (1960).

636. Zubay, G. Molecular model for protein synthesis. *Science* **140**, 1092–1095 (1963).

637. Zubay, G., Wilkins, M. H. F., and Blout, E. R. An X-ray diffraction study of a complex of DNA and a synthetic polypeptide. *J. Mol. Biol.* **4**, 69–72 (1962).

638. Zucker, M., and Stinson, H. T., Jr. Chloroplasts as the major protein-bearing structures in *Oenothera* leaves. *Arch. Biochem. Biophys.* **96**, 637–644 (1962).

APPENDIX I

Amino Acids and Amides Found in the Protein of Plants and Their Free Occurrence

Name and structure	Occurrence, recognition, etc.	References
Glycine (α-Aminoacetic acid) $CH_2(NH_2)COOH$	Gelatin hydrolyzate	1820 H. Braconnot, *Ann. chim. phys.* (2 Sér.) **13**, 113
	Synthesis	1858 W. H. Perkin and B. F. Duppa, *Ann. Chem.* **108**, 106; **112**, 1
	Free in sugar cane (*Saccharum officinarum*)	1897 E. C. Shorey, *J. Am. Chem. Soc.* **19**, 881
	Edestin	1903 E. Abderhalden, *Z. physiol. Chem.* **37**, 499
	Citrus grandis	1927 Y. Hiwatari, *J. Biochem. (Tokyo)* **7**, 169
L(+)-Alanine [L(+)-α-Aminopropionic acid] $CH_3CH(NH_2)COOH$	Synthesis (optically inactive)	1850 A. Strecker, *Ann. Chem.* **75**, 27
	Silk fibroin	1875 P. Schützenberger and A. Bourgeois, *Compt. rend.* **81**, 1191
	Silk fibroin	1888 T. Weyl, *Chem. Ber.* **21**, 1407, 1529
	Free in alfalfa (*Medicago sativa*)	1925 H. B. Vickery, *J. Biol. Chem.* **65**, 657
L(+)-Valine [L(+)-α-Aminoisovaleric acid] $(CH_3)_2CHCH(NH_2)COOH$	Free in pancreas	1856 E. V. Gorup-Besanez, *Ann. Chem.* **98**, 1, 15
	Silk fibroin and albumen, hydrolyzates	1879 P. Schützenberger, *Ann. chim. phys.* (5 Sér.) **16**, 289
	Synthesis (optically inactive)	1880 A. Lipp, *Ann. Chem.* **205**, 1, 18
	Free in *Lupinus luteus* seedlings	1883 E. Schulze and J. Barbieri, *J. prakt. Chem.* **27**, 337

APPENDIX I (continued)

Name and structure	Occurrence, recognition, etc.	References
	Casein (optically active) and name coined	1901 E. Fischer, *Z. physiol. Chem.* **33,** 151
l(−)-Leucine [l(−)-α-Aminoisocaproic acid] (CH₃)₂CHCH₂CH(NH₂)COOH	Decayed cheese	1819 M. Proust, *Ann. chim. phys.* (2 Sér.) **10,** 29
	Muscle fiber, wool, name leucine	1820 H. Braconnot, *Ann. chim. phys.* (2 Sér.) **13,** 113
	Free in vetch (*Vicia*) seedlings	1874 E. V. Gorup-Besanez, *Chem. Ber.* **7,** 146, 569
l(+)-Isoleucine [l(+)-α-Amino-β-methylvaleric acid]	Fibrin, free and in plant protein (corn gluten) sugar beet molasses	1904 F. Ehrlich, *Chem. Ber.* **37,** 1809
CH₃CH₂ ╲ ╱CHCH(NH₂)COOH CH₃	Synthesis	1907 F. Ehrlich, *Chem. Ber.* **40,** 2538
l(−)-Serine [l(−)α-Amino-β-hydroxypropionic acid] CH₂(OH)CH(NH₂)COOH	Silk fibroin hydrolyzate Synthesis	1865 E. Cramer, *J. prakt. Chem.* **96,** 76 1902 E. Fischer and H. Leuchs, *Chem. Ber.* **35,** 3787
	Free in alfalfa	1925 H. B. Vickery, *J. Biol. Chem.* **65,** 657
l(−)-Threonine [l(−)-α-Amino-β-hydroxybutyric acid] CH₃CH(OH)CH(NH₂)COOH	Oat protein	1925 S. Schryver and H. Buston, *Proc. Roy. Soc.* **B99,** 476
	Teozein	1925 R. Gortner and W. Hoffmann, *J. Am. Chem. Soc.* **47,** 580
	Isolation and identification (fibrin hydrolyzate)	1935 R. H. McCoy *et al., J. Biol. Chem.* **112,** 283 1936 R. H. McCoy *et al., J. Biol. Chem.* **115,** 721
	Synthesis of four stereoisomers	1937 H. D. West and H. E. Carter, *J. Biol. Chem.* **119,** 103, 109

APPENDIX I (continued)

Name and structure	Occurrence, recognition, etc.	References
L(+)-Aspartic acid [L(+)-α-Aminosuccinic acid] $CH(NH_2)COOH$ | CH_2COOH	From asparagine (heating with lead hydroxide)	1827 A. Plisson, *Ann. chim. phys.* (2 Sér.) **36**, 175 1852 L. Pasteur, *Ann. chim. phys.* (3 Sér.) **34**, 30
	Conglutin, legumin	1868 H. Ritthausen, *J. prakt. Chem.* **103**, 233 1869 H. Ritthausen, *J. prakt. Chem.* **107**, 218
	Found during N fixation with legumes	1938 A. I. Virtanen and T. Laine, *Nature* **141**, 748
L(−)-Asparagine (β-Amide of aspartic acid) $CH(NH_2)COOH$ | CH_2CONH_2	Cubic crystals in *Asparagus* *Asparagus officinalis* juice	1802 Delaville, *Ann. chim. phys.* **41**, 298 1806 L. N. Vauqelin and P. J. Robiquet, *Ann. chim.* (Paris) **57**, 88
	25% Dry weight of etiolated lupine seedlings	1876 E. Schulze *et al.*, *Chem. Ber.* **9**, 1314
	Edestin, in enzymatic digest	1932 M. Damodaran, *Biochem. J.* **26**, 235
L(+)-Glutamic acid [L(+)-α-Aminoglutaric acid] $CH(NH_2)COOH$ | CH_2 | CH_2COOH	Gluten-fibrin hydrolyzates	1866 H. Ritthausen, *J. prakt. Chem.* **99**, 454
	Sugar beet molasses	1869 C. Scheibler, *Chem. Ber.* **2**, 296
	Vetch seedlings	1877 E. V. Gorup-Besanez, *Chem. Ber.* **10**, 780
	Synthesis	1890 I. Wolff, *Ann. Chem.* **260**, 79, 118
	Seed globulin of conifers	1907 A. Kleinschmitt, *Z. physiol. Chem.* **54**, 110

APPENDIX I (continued)

Name and structure	Occurrence, recognition, etc.	References
L(+)-Glutamine (γ-Amide of glutamic acid) $CH(NH_2)COOH$ $\|$ CH_2 $\|$ CH_2CONH_2	Gourd seedlings	1877 E. Schulze and J. Barbieri, *Chem. Ber.* **10**, 199
	Beet juice	1883 E. Schulze and E. Bosshard, *Landwirtsch. Vers.-Sta.* **29**, 295
	Free in *Picea abies* (*P. excelsa*)	1878 E. Schulze, *Landwirtsch. Jahrb. Schweiz* **7**, 431
	Gliadin, enzymatic digest	1932 M. Damodaran *et al., Biochem. J.* **26**, 1704
	Synthesis	1933 M. Bergmann *et al., Chem. Ber.* **66**, 1288
L(+)-Lysine [L(+)-α,ϵ-Diaminocaproic acid] $H_2NCH_2(CH_2)_3CH(NH_2)COOH$	Casein	1889 E. Drechsel, *Z. prakt. Chem.* **39**, 425
	Free in *Lupinus, Pisum,* and *Vicia*	1899 E. Schulze, *Z. physiol. Chem.* **28**, 465 1900 E. Schulze, *Z. physiol. Chem.* **30**, 241
	Synthesis (optically inactive)	1902 E. Fischer and F. Weigert, *Chem. Ber.* **35**, 3772
	In conifer seedlings	1906 E. Schulze, *Z. physiol. Chem.* **47**, 507
	First obtained crystalline lysine	1928 H. B. Vickery and C. S. Leavenworth, *J. Biol. Chem.* **76**, 437
L(+)-Arginine [L(+)-α-Amino-δ-guanidinovaleric acid] NH_2 $\diagdown$ $\quad CNHCH_2CH_2CH_2CH(NH_2)COOH$ $\diagup\!\!\diagup$ NH	Free in etiolated *Lupinus luteus*	1886 E. Schulze and E. Steiger, *Chem. Ber.* **19**, 1177; *Z. physiol. Chem.* **11**, 43
	Horn protein hydrolyzate	1894 S. G. Hedin, *Z. physiol. Chem.* **20**, 186 1895 S. G. Hedin, *Z. physiol. Chem.* **21**, 115
	Free in *Picea, Pinus,* and *Abies* seedlings and conifer seed protein	1896 E. Schulze, *Z. physiol. Chem.* **22**, 435 1898 E. Schulze, *Z. physiol. Chem.* **24**, 276; **25**, 360

APPENDIX I (continued)

Name and structure	Occurrence, recognition, etc.	References
L(−)-Histidine [L(−)-α-Amino-β-5-imidazolyl-propionic acid] $N\text{---}C\text{---}CH_2CH(NH_2)COOH$ $HC\quad\quad CH$ $\backslash NH /$	Protamine from sturgeon sperm	1896 A. Kossel, *Z. physiol. Chem.* **22**, 176, 181
	Various proteins	1896 S. G. Hedin, *Z. physiol. Chem.* **22**, 191
	Etiolated *Lupinus luteus* seedlings	1899 E. Schulze, *Z. physiol. Chem.* **28**, 465
	Free in conifer seedlings	1906 E. Schulze, *Z. physiol. Chem.* **47**, 507
	Synthesis of (−)-histidine	1911 F. L. Pyman, *J. Chem. Soc.* **99**, 1386
	Free in *Pinus* pollen	1922 A. Kiesel, *Z. physiol. Chem.* **120**, 85
L(−)-Phenylalanine (L(−)-α-Amino-β-phenylpropionic acid) $\langle\!\!\!\bigcirc\!\!\!\rangle\text{---}CH_2CH(NH_2)COOH$	Etiolated *Lupinus luteus* seedlings (protein)	1881 E. Schulze and J. Barbieri, *Chem. Ber.* **14**, 1785
	Hydrolyzates of squash seeds	1883 E. Schulze and J. Barbieri, *Chem. Ber.* **16**, 1711
	Synthesis (optically inactive)	1882 E. Erlenmeyer and A. Lipp, *Chem. Ber.* **15**, 1006
L(−)-Tyrosine [(L(−)-α-Amino-β-(p-hydroxy-phenyl)propionic acid] $HO\text{---}\langle\!\!\!\bigcirc\!\!\!\rangle\text{---}CH_2CH(NH_2)COOH$	Casein, alkaline hydrolyzate	1846 J. Liebig, *Ann. Chem.* **57**, 127
	Flesh	1847 J. Liebig, *Ann. Chem.* **62**, 257
	Casein (HCl hydrolysis)	1849 F. Bopp, *Ann. Chem.* **69**, 16, 20
	Detected free in vetch seedlings	1877 E. V. Gorup-Besanez, *Chem. Ber.* **10**, 780
	Synthesis	1882 E. Erlenmeyer and A. Lipp, *Chem. Ber.* **15**, 1544
L(−)-Proline [L(−)-Pyrrolidine-2-carboxylic acid] $H_2C\text{-----}CH_2$ $H_2C\quad\quad CHCOOH$ $\backslash NH /$	Synthesis	1900 R. Willstätter, *Chem. Ber.* **33**, 1160
	Casein hydrolyzate (ester distillation)	1901 E. Fischer, *Z. phsiol. Chem.* **33**, 151
	Lupinus albus seedlings	1905 E. Schulze and E. Winterstein, *Z. physiol. Chem.* **45**, 38

APPENDIX I (continued)

Name and structure	Occurrence, recognition, etc.	References
L(−)-Hydroxyproline [L(−)-4-Hydroxypyrrolidine-2-carboxylic acid] HOHC——CH$_2$ H$_2$C CHCOOH NH	Gelatin hydrolyzate Synthesis (−)-compound Free (allo-) in leaves of *Santalum album* Free in pollen of certain trees	1902 E. Fischer, *Chem. Ber.* **35**, 2660 1913 H. Leuchs and J. F. Brewster, *Chem. Ber.* **46**, 986 1952 K. V. Giri *et al.*, *Nature* **170**, 579 1955 A. I. Virtanen and S. Kari, *Acta Chem. Scand.* **9**, 1543
L(−)-Tryptophan [L(−)-α-Amino-β-3-indolepropionic acid] —CH$_2$CH(NH$_2$)COOH NH	Named for an earlier breakdown product Casein (enzymatic digest) Etiolated *Lupinus albus* seedlings Synthesis *Robinia pseudoacacia* (bark albumin) D-Tryptophan in higher plants	1890 R. Neumeister, *Z. Biol.* **26**, 329 1901 F. G. Hopkins and S. W. Cole, *J. Physiol. (London)* **27**, 418 1902 F. G. Hopkins and S. W. Cole, *J. Physiol. (London)* **29**, 451 1905 F. Schulze and E. Winterstein, *Z. physiol. Chem.* **45**, 38 1907 A. Ellinger and C. Flamand, *Chem. Ber.* **40**, 3029 1925 D. B. Jones *et al.*, *J. Biol. Chem.* **64**, 655 1963 M. H. Zenk and H. Scherf, *Biochim. Biophys. Acta* **71**, 737
L(−)-Cystine Di-(α-Amino-β-thiopropionic acid) CH$_2$—S—S—CH$_2$ CHNH$_2$ CHNH$_2$ COOH COOH	Free in urinary calculi (first preparation of an amino acid) Horn Cystine and L(+)-cysteine in protein Synthesis Potato protein	1810 W. H. Wollaston, *Ann. chim. phys.* **76**, 21 1899 K. A. H. Mörner, *Z. physiol. Chem.* **28**, 595 1901 G. Embden, *Z. physiol. Chem.* **32**, 94 1903 E. Erlenmeyer, *Chem. Ber.* **36**, 2720 1912 B. Sjolleman and J. J. Rinkes, *Z. physiol. Chem.* **76**, 368

APPENDIX I (continued)

Name and structure	Occurrence, recognition, etc.	References
L(−)-Methionine [L(−)-α-Amino-γ-methylthiol-butyric acid] $CH_3SCH_2CH_2CH(NH_2)COOH$ Methionine sulfone and sulfoxide occur in extracts of soluble compounds in plants and also in hydrolyzates as derivatives of methionine	Casein hydrolyzate α-Globulin from tomato seeds; also teozein, sativin Constitution and synthesis	1922 J. H. Mueller, *Proc. Soc. Exptl. Biol. Med.* **19**, 161 1928 H. D. Baernstein, *J. Biol. Chem.* **97**, 663 1928 G. Barger and F. P. Coyne, *Biochem. J.* **22**, 1417

Addendum to Appendix I

Additional to the above well-established protein amino acids there are, in the literature, reports that a number of other amino acids occur in protein, some only in the protein of animals. These are: L(+)α-aminobutyric acid; α-aminoisobutyric acid; aminoethanolphosphoserine; O-phosphoserine; sarcosine; 3-methylhistidine; δ-hydroxylysine; ε-N-methyllysine; L(+)ornithine; L(+)-citrulline; L-α-aminoadipic acid; L(+)β-hydroxyglutamic acid; γ-hydroxyglutamic acid; α,ε-diaminopimelic acid; S-(2-amino-2-carboxyethylhomocysteine), (cystathione); (3-amino-3-carboxypropyl)-2-amino-2-carboxyethyl) selenide; 3-iodo-L-tyrosine; 3,5-diiodotyrosine; 3-5-dibromotyrosine; 3-hydroxy-L-proline (*cis* and *trans*); 3,5,3'-triiodothyronine; thyroxine.

For notes on these substances where they occur free in plants see Appendix II.

Furthermore, claims are now being made that D-amino acids often occur in nature, usually combined in peptides, especially in bacteria. Such compounds are: D-alanine, D-valine, D-leucine, D-isoleucine, D-alloisoleucine, D-aspartic acid, D-glutamic acid, D-serine, D-allothreonine, D-phenylalanine, D-proline, D-hydroxyproline, D-allohydroxyproline, D-α-aminoadipic acid, D-α,γ-diaminobutyric acid, D-ornithine, O-carbamyl-D-serine, D-cycloserine, D-α-methylserine, and also N-succinyl-D-valine. Since no valid claims have been made that these substances occur free in higher plants, no reference is made to them in Appendix II.

APPENDIX II

Amino, Imino Acids and Their Derivatives that Occur Free in Plants[a,b]

Name and Structure	Occurrence, recognition, etc.	References
Monoaminomonocarboxylic acids		
β-Alanine $NH_2CH_2CH_2COOH$	Synthesis	1870 W. Heintz, *Ann. Chem.* **156**, 26, 36
	Tomato sap	1937 W. L. Miller, *Chem. Abstr.* **31**, 1842
	Potato tuber—detected chromatographically	1947 C. E. Dent *et al.*, *Nature* **160**, 682
	Iris tingitana—isolated	1961 C. J. Morris *et al.*, *Nature* **190**, 718
L(+)-α-Aminobutyric acid $CH_3CH_2CH(NH_2)COOH$	Synthesis (optically inactive)	1862–1863 C. Friedel and V. Machuca, *Ann. Chem. Suppl.* **2**, 70
	Synthesis of L(+) compound	1900 E. Fischer and A. Mouneyrat, *Chem. Ber.* **33**, 2383, 2390
	Claimed in edestin and yeast protein	1937 E. Abderhalden and A. Bahn, *Z. physiol. Chem.* **245**, 246
	Detected free and in peptide of peas	1953 A. I. Virtanen and J. Miettinen, *Biochim. Biophys. Acta* **12**, 181

[a] This appendix excludes substances already listed in Appendix I as constituents of protein as well as alkaloids, cyanogenetic glucosides, the simpler nitrogen bases, purines and pyrimidines, and the decarboxylation products of amino acids (amines). For information on these compounds see text.

[b] A number of interesting compounds have been detected in urine. Though not yet reported in plants, these may later have a botanical significance. Among such compounds are: homocitrulline, 1-methylhistidine, 3-methylhistidine, β-L-aspartyl-histidine, S-(2-methyl-2-carboxyethyl)cysteine, S-isopropylcarboxymethylcysteine, or isovalthine, felinine. Other natural compounds primarily found in animal sources are: L threonine ethanolamine phosphate, carnitine, L-(+)-norleucine, lombricine, L-cysteinesulfinic acid, monoiodohistidine, S-(1,2-dicarboxyethyl)-L-cysteine, tyrosine-O-sulfate, β-hydroxy-γ-aminobutyric acid, D-octopine, 2-aminoethyl phosphonic acid.

APPENDIX II (continued)

Name and structure	Occurrence, recognition, etc.	References
γ-Aminobutyric acid (Piperidinic acid) $NH_2CH_2CH_2CH_2COOH$	Piperidine oxidation	1883 C. Schotten, Chem. Ber. **16**, 643
	Structure and synthesis	1890 S. Gabriel, Chem. Ber. **23**, 1767, 1770
	Bacteria	1910 D. Ackermann, Z. physiol. Chem. **69**, 273
	Corynebacterium	1949 E. Work, Bull. soc. chim. biol. **31**, 138
	Potato	1949 F. C. Steward et al., Science **110**, 439
		1953 J. F. Thompson et al., Plant Physiol. **28**, 401
	Free and bound, legume root nodules	1958 G. W. Butler and N. O. Bathurst, Australian J. Biol. Sci. **11**, 529
α-Aminoisobutyric acid $(CH_3)_2C(NH_2)COOH$	Paecilomyces antibiotic	1958 G. W. Kenner and R. C. Sheppard, Nature **181**, 48
	Free in Iris tingitana	1961 S. Asen et al., Nature **190**, 809
	Claimed in muscle protein	1961 M. Kandatsu and K. Kikuno, Agr. Biol. Chem. (Tokyo) **25**, 234
$(-)$-β-Aminoisobutyric acid $NH_2CH_2CHCOOH$ $\quad\quad\quad\mid$ $\quad\quad\quad CH_3$	Human urine	1951 H. R. Crumpler et al., Nature **167**, 307
	Iris tingitana bulbs	1959 S. Asen et al., J. Biol. Chem. **234**, 343
γ-Amino-α-methylenebutyric acid $NH_2CH_2CH_2CCOOH$ $\quad\quad\quad\quad\;\;\parallel$ $\quad\quad\quad\quad\;\;CH_2$	Isolation from Arachis hypogaea and synthesis	1953 L. Fowden and J. Done, Biochem. J. **55**, 548
γ-Amino-β-methylenebutyric acid $NH_2CH_2CCH_2COOH$ $\quad\quad\quad\;\parallel$ $\quad\quad\quad\;CH_2$	Tulip	1953 L. Fowden and J. Done, Nature **171**, 1068

Monoaminodicarboxylic acids and their amide derivatives

β-Methylaspartic acid $HOOCCHCHCOOH$ $\quad\quad\;\;\mid\quad\;\mid$ $\quad\quad\;\;H_2N\quad CH_3$	Clostridium tetanomorphum (also L-threo-β-methylaspartic acid)	1958 H. A. Barker et al., Arch. Biochem. Biophys. **78**, 468

APPENDIX II (continued)

Name and structure	Occurrence, recognition, etc.	References
D-Asparagine $NH_2COCH_2CHCOOH$ 　　　　　$\mid$ 　　　　　NH_2	Seedlings of *Lupinus albus*	1887 A. Piutti, *Gazz. chim. ital.* **17**, 519
	Questioned Piutti's isolation	1910 H. Pringsheim, *Z. physiol. Chem.* **65**, 89
	Earlier claim maintained	1923 A. Piutti, *Bull. soc. chim. France* **4**, 33, 804
Fumaryl-DL-alanine $CH_3CHCOOH$ 　　$\mid$ 　　$NHCOCH{=}CHCOOH$	*Penicillium reticulosum* and synthesis	1942 J. H. Birkinshaw *et al.*, *Biochem. J.* **36**, 829
γ-Methylglutamic acid $HOOCCH(CH_3)CH_2CH(NH_2)COOH$	*Phyllitis scolopendrium*	1955 A. I. Virtanen and A. M. Berg, *Acta Chem. Scand.* **9**, 553
γ-Methyl-γ-hydroxyglutamic acid $HOOCC(OH)CH_2CH(NH_2)COOH$ 　　　$\mid$ 　　　CH_3	*Adiantum pedatum* and synthesis	1955 J. K. Pollard and F. C. Steward, See Grobbelaar *et al.* (188)
	Phyllitis scolopendrium	1955 A. I. Virtanen and A. M. Berg, *Acta Chem. Scand.* **9**, 553
L-γ-Methyleneglutamic acid $HOOCCCH(NH_2)COOH$ 　　　$\parallel$ 　　　CH_2	*Arachis hypogaea*	1952 J. Done and L. Fowden, *Biochem. J.* **51**, 451
	Tulipa gesneriana	1954 R. M. Zacharius *et al.*, *J. Am. Chem. Soc.* **76**, 1961
	Synthesis	1956 H. Hellmann and F. Lingens, *Chem. Ber.* **89**, 77
	Synthesis and resolution	1957 Y. Nakagawa and T. Kaneko, *J. Chem. Soc. Japan, Pure Chem. Sect.* **78**, 232, 1216
γ-Methyleneglutamine $H_2NCOCCH_2CH(NH_2)COOH$ 　　　$\parallel$ 　　　CH_2	*Arachis hypogaea*	1952 J. Done and L. Fowden, *Biochem. J.* **51**, 451
	Synthesis (racemate)	1955 P. C. Wailes and M. C. Whiting, *J. Chem. Soc.* p. 3636

APPENDIX II (continued)

Name and structure	Occurrence, recognition, etc.	References
	Isolated from *Tulipa*	1954 R. M. Zacharius *et al.*, *J. Am. Chem. Soc.* **76**, 1961
γ-Amino-α-methylenebutyric acid $H_2NCH_2CH_2CCOOH$ $\parallel$ CH_2	*Arachis hypogaea*	1952 L. Fowden and J. Done, *Biochem. J.* **53**, xxxi–xxxii
N^4-Methyl-L-asparagine $CH_3NHCOCH_2CH(NH_2)COOH$		1962 L. Fowden and D. O. Gray, *in* "Amino Acid Pools (J. T. Holden, ed.), cf. Table 1, p. 46. Elsevier, Amsterdam.
N^4-Ethyl-L-asparagine $CH_3CH_2NHCOCH_2CH(NH_2)COOH$	*Ecballium elaterium* and *Bryonia dioica*	1961 D. O. Gray and L. Fowden, *Nature* **189**, 401
N^4-(2-Hydroxyethyl)-L-asparagine $HOCH_2CH_2NHOCCH_2CH(NH_2)COOH$	*Bryonia dioica*	1961 L. Fowden, *Biochem. J.* **81**, 155
L-α-Aminoadipic acid $HOOC(CH_2)_3CH(NH_2)COOH$	Free in *Vibrio comma* (*V. cholerae*)	1946 J. Blass and J. Macheboeuf, *Helv. Chim. Acta* **29**, 1315
	Free in protein of maize and *Aspergillus*	1951 E. Windsor, *J. Biol. Chem.* **192**, 595, 607
	Pisum sativum	1962 S. Hatanaka and A. I. Virtanen, *Acta Chem. Scand.* **16**, 514
α-Aminopimelic acid $HOOC(CH_2)_4CH(NH_2)COOH$	*Asplenium septentrionale* and synthesis	1954 A. I. Virtanen and A. M. Berg, *Acta Chem. Scand.* **8**, 1085, 1725

Diaminomono- and dicarboxylic acids and their derivatives

L-α,β-Diaminopropionic acid (β-Aminoalanine) $H_2NCH_2CH(NH_2)COOH$	Synthesis (optically inactive)	1893 E. Klebs, *Chem. Ber.* **26**, 2264 1894 E. Klebs, *Z. physiol. Chem.* **19**, 301, 309

APPENDIX II (continued)

Name and structure	Occurrence, recognition, etc.	References
	Viomycin from *Streptomyces*	1952 T. H. Haskell *et al., J. Am. Chem. Soc.* **74**, 599
	L(+) from seeds of *Mimosa palmeri*	1959 R. Gmelin *et al., Z. physiol. Chem.* **314**, 28
β-*N*-Oxalyl-L-α,β-Diamino-propionic acid HOOCCONH CH₂CHCOOH NH₂	Seeds of *Lathyrus sativus*	1964 S. L. N. Rao *et. al., Biochemistry* **3**, 432
L-α,β-Diaminobutyric acid (L-γ-Aminobutyrine) NH₂CH₂CH₂CH(NH₂)COOH	Synthesis	1901 E. Fischer, *Chem. Ber.* **34**, 2900
	Aerosporin (DL compound)	1948 J. R. Catch and T. S. G. Jones, *Biochem. J.* **42**, lii
D-α,γ-Diaminobutyric acid H₂NCH₂CH₂CHCOOH NH₂	Polymyxin B	1954 W. Hausmann and L. C. Craig, *J. Am. Chem. Soc.* **76**, 4892
	Conversion of asparagine to α,γ-diaminobutyric acid	1956 C. Ressler, *J. Am. Chem. Soc.* **76**, 5956
	Polygonatum multiflorum rhizome	1958 L. Fowden and M. Bryant, *Biochem. J.* **70**, 626
	Seeds of twelve species of *Lathyrus*	1962 E. A. Bell, *Nature* **196**, 1078
	Circulan A and B lipopeptides of *Bacillus circulans*	1961 H. Koffler, *Proc. 5th Intern. Congr. Biochem. Moscow* 1961, p. 36
(+)-α,α'-Diaminosuccinic acid HOOCCHCHCOOH H₂N NH₂	Synthesis of meso and racemic compound	1958 H. McKennis and A. S. Yard, *J. Org. Chem.* **23**, 980
	In Terramycin (*Streptomyces rimosus*)	1959 F. A. Hochstein, *J. Org. Chem.* **24**, 679

APPENDIX II (continued)

Name and structure	Occurrence, recognition, etc.	References
α,ϵ-Diaminopimelic acid HOOCCH(CH$_2$)$_3$CHCOOH | | NH$_2$ NH$_2$	Synthesis	1905 S. P. L. Sörensen, *Z. physiol. Chem.* **44,** 418 1908 S. P. L. Sörensen and A. C. Andersen, *Z. physiol. Chem.* **56,** 250
	Hydrolyzate from *Corynebacterium*	1950 E. Work, *Nature* **165,** 74
	Mycobacterium tuberculosis	1950 J. Asselineau *et al., Biochim. Biophys. Acta* **5,** 197
	Hydrolyzate of *Chlorella ellipsoidea*	1955 T. Fujiwara and S. Akabori, *Chem. Abstr.* **49,** 3325
	Pine pollen	1958 M. M. Cummings and P. C. Hudgins, *Am. J. Med. Sci.* **236,** 311
N-Succinyldiaminopimelic acid and N-succinyl-α-amino-ϵ-keto-pimelic acid	*Escherichia coli*	1959 C. Gilvarg, *J. Biol. Chem.* **234,** 2955
L(+)-Ornithine (α,δ-Diaminocaproic acid) H$_2$N(CH$_2$)$_3$CH(NH$_2$)COOH	As ornithuric acid in bird excreta	1877 M. Jaffe, *Chem. Ber.* **10,** 1925 1878 M. Jaffe, *Chem. Ber.* **11,** 406
	From arginine	1897 E. Schulze and E. Winterstein, *Chem. Ber.* **30,** 2879
	Free in *Alnus* root	1952 J. K. Miettinen and A. I. Virtanen, *Physiol. Plantarum* **5,** 540
	Free in *Asplenium nidus*	1955 A. I. Virtanen and P. Linko, *Acta Chem. Scand.* **9,** 531
N-Acetylornithine CH$_3$CONH(CH$_2$)$_3$CH(NH$_2$)COOH	*Corydalis ochotensis*	1937 R. H. F. Manske, *Can. J. Res.* **B15,** 84
	Asplenium species	1955 A. I. Virtanen and P. Linko, *Acta Chem. Scand.* **9,** 531

APPENDIX II (continued)

Name and structure	Occurrence, recognition, etc.	References
	Synthesis	1963 H. R. Schütte and W. Shütz, *Ann. Chem.* **665**, 203
N-Acetyldiaminobutyric acid	Latex of *Euphorbia*	1962 I. Liss, *Phytochemistry* **1**, 87
	Synthesis	1963 H. R. Schütte and W. Shütz, *Ann. Chem.* **665**, 203
L-β-Lysine (Isolysine; β,ε-diaminocaproic acid) $HOOCCH_2CH\,CH_2CH_2CH_2$ $\quad\quad\; \underset{NH_2}{\vert} \quad\quad\; \underset{NH_2}{\vert}$	Viomycin—a bacterial peptide	1952 T. Haskell *et al.*, *J. Am. Chem. Soc.* **74**, 599
	Viomycin synthesis	1952 E. E. von Tamelen and E. E. Smissman, *J. Am. Chem. Soc.* **74**, 3713
Lysopine, N^2-(D-1-carboxyethyl)-L-lysine $H_2N(CH_2)_4CH(COOH)NHCH(COOH)CH_3$	Crown galls on several higher plants	1957 C. Lioret, *Compt. rend.* **244**, 2171
	Synthesis	1960 K. Biemann *et al.*, *Bull. soc. chim. biol.* **42**, 979; *Biochim. Biophys. Acta* **40**, 369
Canaline $H_2NO(CH_2)_2CHCOOH$ $\quad\quad\quad\; \underset{NH_2}{\vert}$	Constitution	1933 M. Kitagawa and S. Monobe, *J. Biochem.* (*Tokyo*) **18**, 333
	Jack beans (*Canavalia ensiformis*) (cleavage product)	1955 W. R. Fearon and A. E. Bell, *Biochem. J.* **59**, 221
Lathyrus factor (β-*N*-(γ-L-Glutamyl)aminopropionitrile) $CN(CH_2)_2NHCO(CH_2)_2CHCOOH$ $\quad\quad\quad\quad\quad\quad\quad\; \underset{NH_2}{\vert}$	Seeds of *Lathyrus pusillus*	1954 H. P. Dupuy and J. G. Lee, *J. Am. Pharm. Assoc., Sci. Ed.* **43**, 61
	Synthesis	1954 E. D. Schilling and F. M. Strong, *J. Am. Chem. Soc.* **76**, 2848
		1955 E. D. Schilling and F. M. Strong, *J. Am. Chem. Soc.* **77**, 2843

APPENDIX II (continued)

Name and structure	Occurrence, recognition, etc.	References
β-Cyano-L-alanine CNCH₂CHCOOH │ NH₂	*Vicia sativa* var. *angustifolia*	1962 C. Ressler, *J. Biol. Chem.* **237**, 733
L-Saccharopine CH₂CH₂NHCHCOOH │ │ (CH₂)₂ CH₂ │ │ CHNH₂ CH₂COOH │ COOH	Dried bakers' yeast *Torulopsis utilis*	1951 O. Lindan and E. Work, *Biochem. J.* **48**, 337 1954 J. M. Miettinen, *Ann. Acad. Sci. Fennicae: Ser. A* **58**, 1
Lysine Glutaryl moiety moiety	Isolation from yeast Structure and synthesis	1961 S. Darling and P. Olesen-Larsen, *Acta Chem. Scand.* **15**, 743 1961 A. Kjaer and P. Olesen-Larsen, *Acta Chem. Scand.* **15**, 750

Hydroxyamino acids and their derivatives

β-Hydroxyaspartic acid HOOCCHCHCOOH │ │ H₂N OH	Extracellular material of *Azotobacter* Chromatographic evidence from alfalfa and clover roots	1957 A. I. Virtanen and N. L. Saris, *Suomen Kemistilehti* **B30**, 100 1962 M. D. Wilding and M. A. Stahlmann, *Phytochemistry* **1**, 241
α-Hydroxyalanine OH │ CH₃CCOOH │ NH₂	In peptides of ergot: ergotaminine and ergotamine	1938 L. C. Craig *et al.*, *J. Biol. Chem.* **125**, 289
α-Amino-α-hydroxybutyric acid OH │ CH₃CH₂CCOOH │ NH₂	*Escherichia coli* mutants	1955 A. I. Virtanen and P. K. Hietala *Acta Chem. Scand.* **9**, 549
α-Amino-β-hydroxybutyric acid CH₃CHCHCOOH │ │ HO NH₂	*E. coli* mutants	1953 W. W. Umbreit and P. Heneage, *J. Biol. Chem.* **201**, 15

APPENDIX II (continued)

Name and structure	Occurrence, recognition, etc.	References
α-Hydroxy-γ-aminobutyric acid $CH_2CH_2CHCOOH$ (with NH_2 and OH substituents)	*E. coli* mutants	1959 W. B. McConnell and E. Belinski, *Can. J. Biochem. Physiol.* **37**, 549
L(+)-β-Hydroxyglutamic acid OH $HOOCCH_2CHCH(NH_2)COOH$	Casein	1918 H. D. Dakin, *Biochem. J.* **12**, 290, 303
	Synthesis	1919 H. D. Dakin, *Biochem. J.* **13**, 393
[β-hydroxyglutamic acid of Dakin was shown to be a mixture; cf. L. Fowden (155)]	Globulin from *Stizolobium niveum*	1919 D. B. Jones and C. O. Johns, *J. Biol. Chem.* **40**, 435
		1941 H. D. Dakin, *J. Biol. Chem.* **140**, 874
γ-Hydroxyglutamic acid OH $HOOCCHCH_2CH(NH_2)COOH$	Free and/in protein of *Phlox* and *Linaria*	1955 A. I. Virtanen and P. K. Hietala, *Acta Chem. Scand.* **9**, 175
	Resolution of isomers	1957 L. Benoiton *et al.*, *J. Am. Chem. Soc.* **79**, 6192
	Isolated from *Phlox decussata* and identified as the allo-γ-form	1962 F. C. Steward and J. K. Pollard, *in* "Amino Acid Pools" (J. T. Holden, ed.), cf. pp. 22, 40. Elsevier, Amsterdam.
	Free in *Linaria vulgaris*	1962 S. Hatanaka, *Acta Chem. Scand.* **16**, 513
γ-Hydroxy-L-glutamine $CONH_2$ $CHOH$ CH_2 $CHNH_2$ $COOH$	Isolation from *Phlox decussata* and synthesis	1963 G. Brandner and A. I. Virtanen, *Acta Chem. Scand.* **17**, 2563

APPENDIX II (continued)

Name and structure	Occurrence, recognition, etc.	References
β,γ-Dihydroxyglutamic acid $\underset{\text{HO}}{\mid}\ \underset{\text{OH}}{\mid}$ HOOCCHCHCH(NH₂)COOH	Pepperwort (*Lepidium*) *Rheum rhaponticum*	1956 A. I. Virtanen and T. Ettala, *Suomen Kemistilehti* **B29**, 107 1957 A. I. Virtanen and T. E. Hala, *Acta Chem. Scand.* **11**, 182
α-Hydroxy-ϵ-aminocaproic acid CH₂CH₂CH₂CH₂CHCOOH $\underset{\text{NH}_2}{\mid}\qquad\underset{\text{OH}}{\mid}$	*Neurospora* mutants	1954 R. S. Schweet *et al.*, *J. Biol. Chem.* **211**, 517
α-Hydroxylysine (α,ϵ-diamino-α-hydroxycaproic acid) NH₂CH₂(CH₂)₃CH(NH₂)COOH $\qquad\qquad\underset{\text{OH}}{\mid}$	Preliminary claim in seeds of *Salvia officinalis* L.	1964 C. H. Brieskorn and J. Glasz, *Naturwiss.* **9**, 216
δ-Hydroxylysine (α,ϵ-diamino-δ-hydroxycaproic acid) $\qquad$OH $\qquad\mid$ NH₂CH₂CH(CH₂)₂CH(NH₂)COOH	Fish gelatin Fruit of date—tentative recognition Chromatographic evidence from alfalfa roots	1925 S. B. Schryver *et al.*, *Proc. Roy. Soc.* **B98**, 58 1955 N. Grobbelaar *et al.*, *Nature* **175**, 703 1962 D. W. Wilding and M. A. Stahlmann, *Phytochemistry* **1**, 241
α-Amino-γ-hydroxyadipic acid $\qquad$OH $\qquad\mid$ HOOCCH₂CHCH₂CH(NH₂)COOH	*Vibrio comma* (*V. cholerae*)	1946 J. Blass and M. Macheboeuf, *Helv. Chim. Acta* **29**, 1315 1953 A. I. Virtanen *et al.*, *Acta Chem. Scand.* **7**, 1423
γ-Hydroxyornithine NH₂CH₂CHCH₂CH(NH₂)COOH $\qquad\underset{\text{OH}}{\mid}$	Seeds of *Vicia sativa*	1964 E. A. Bell and A. S. L. Tirimanna, *Biochem. J.* **91**, 356
L-γ-Hydroxyarginine H₂NC(=NH)NHCH₂CH(OH)CH₂CH(NH₂)COOH	In the invertebrate *Polycheira rufescens*	1959 Y. Fugita, *Bull. Chem. Soc. Japan* **32**, 439; **33**, 1379

APPENDIX II (continued)

Name and structure	Occurrence, recognition, etc.	References
	Anthophopleura japonica (sea anemone)	1961 S. Makisuma, *J. Biochem.* (*Tokyo*) **49**, 284
	Seeds of genus *Vicia*	1963 E. A. Bell and A. S. L. Tirimanna, *Nature* **197**, 901
γ-Hydroxyhomoarginine and its lactone $H_2NC(=NH)NHCH_2CH_2CH(OH)CH_2CH(NH_2)COOH$	Seeds of *Lathyrus tingitanus*	1963 E. A. Bell, *Nature* **199**, 70 1964 E. A. Bell, *Biochem. J.* **91**, 358
α-Amino-γ-hydroxypimelic acid and its lactone $$\overset{\displaystyle OH}{\underset{\displaystyle |}{HOOCCH_2CH_2CHCH_2CH(NH_2)COOH}}$$	*Asplenium septentrionale*	1954 A. I. Virtanen *et al.*, *Acta Chem. Scand.* **8**, 1091
Tabtoxinine and its α-lactyl derivative (α,ϵ-diamino-β-hydroxypimelic acid) $$\overset{\displaystyle NH_2 \qquad OH}{\underset{\displaystyle NH_2}{\underset{\displaystyle |}{HOOCCH(CH_2)_2CHCHCOOH}}}$$	Toxin of *Pseudomonas tabaci* Lactone of α-lactyl derivative in the toxin	1952 D. W. Wooley *et al.*, *J. Biol. Chem.* **197**, 409 1953 D. W. Wooley *et al.*, *J. Biol. Chem.* **198**, 807 1955 D. W. Wooley *et al.*, *J. Biol. Chem.* **215**, 485
α-Hydroxyvaline $$\overset{\displaystyle OH}{\underset{\displaystyle NH_2}{\underset{\displaystyle |}{(CH_3)_2CHCCOOH}}}$$	In peptides of ergot; ergotamine and ergotaminine Crown gall of *Kalanchoe*	1938 L. C. Craig *et al.*, *J. Biol. Chem.* **125**, 289 1958 J. K. Pollard *et al.*, *Nature* **182**, 1356
$(+)$-γ-Hydroxyvaline and its lactone $$\underset{\displaystyle OH \quad CH_3NH_2}{\underset{\displaystyle | \quad | \quad |}{CH_2CH \; CHCOOH}}$$	Free in crown galls on *Kalanchoe daigremontiana*	1962 F. C. Steward and J. K. Pollard, *in* "Amino Acid Pools" (J. T. Holden ed.), cf. p. 38. Elsevier, Amsterdam.
γ-Oxo-δ-hydroxy-L-norvaline $$\underset{\displaystyle OH \quad O \quad NH_2}{\underset{\displaystyle | \quad \| \quad |}{CH_2CCH_2CHCOOH}}$$	In a tubercular substance and synthesis	1960 A. Miyake, *Chem. & Pharm. Bull.* (*Tokyo*) **8**, 1071, 1074, 1079

APPENDIX II (continued)

Name and structure	Occurrence, recognition, etc.	References
D-α-Methylserine OH CH₃ CH₂CCOOH NH₂	Synthesis In amicetin from *Streptomyces*	1945 J. Billman and E. E. Parker, *J. Am. Chem. Soc.* **67**, 1069 1953 E. H. Flynn *et al., J. Am. Chem. Soc.* **75**, 5867
L(+)-Homoserine [L(+)-α-Amino-γ-hydroxybutyric acid] CH₂CH₂CHCOOH OH NH₂	Synthesis *Pisum sativum* *Pisum* L(+) compound	1907 E. Fischer and H. Blumenthal, *Chem. Ber.* **40**, 106 1953 J. K. Miettinen *et al., Suomen Kemistilehti* **B26**, 26 1953 A. I. Virtanen *et al., Acta Chem. Scand.* **7**, 1423 1954 A. M. Berg *et al., Acta Chem. Scand.* **8**, 358
L-α-Amino-δ-hydroxyvaleric acid CH₂CH₂CH₂CHCOOH OH NH₂	Seeds of *Canavalia ensiformis*	1964 J. F. Thompson *et al., J. Biol. Chem.* **239**, 1122
O-Acetylhomoserine O CH₃COCH₂CH₂CHCOOH NH₂	*Pisum*	1958 N. Grobbelaar and F. C. Steward, *Nature* **182**, 1358
Pantonine (α-Amino-β,β-dimethyl-γ-hydroxy-butyric acid) CH₃ CH₂CCH(NH₂)COOH OH CH₃	Synthesis *Escherichia coli*	1948 W. W. Ackermann and W. Shive, *J. Biol. Chem.* **175**, 867 1948 W. W. Ackermann and H. Kirby, *J. Biol. Chem.* **175**, 483

δ-Hydroxyleucine
HOH₂C
  \
    CHCH₂CHCOOH
  /   NH₂
H₃C
	In antibiotic from *Paecilomyces*
	
	Free in *Deutzia gracilis*
		1958 G. W. Kenner and R. C. Sheppard, *Nature* **181**, 48
		1961 J. Jadot and J. Casimir, *Biochim. Biophys. Acta* **48**, 400

APPENDIX II (continued)

Name and structure	Occurrence, recognition, etc.	References
L-*threo*-β-Hydroxyleucine COOH | NH$_2$CH | HCOH | HCCH$_3$ | CH$_3$	*Deutzia gracilis*	1963 J. Jadot *et al.*, *Biochim. Biophys.* *Acta* **78**, 500
γ,δ-Dihydroxyleucine CH$_3$ | HOCH$_2$CCH$_2$CHCOOH | | OH NH$_2$	δ-Hydroxyleucenine in *Amanita phal-* *loides* (not verified) In phalloin	1955 T. Wieland and W. Schön, *Ann. Chem.* **593**, 157 1959 T. Wieland and A. Schöpf, *Ann.* *Chem.* **626**, 174
ε-Hydroxylamino-L-norleucine (L-α-Amino-ε-hydroxyaminohexanoic acid) NHCH$_2$(CH$_2$)$_3$CHCOOH | | OH NH$_2$	In mycobactin from *Mycobacterium phlei*	1954 G. A. Snow, *J.* *Chem. Soc.* pp. 2588, 4080

Aromatic amino acids and their derivatives

Name and structure	Occurrence, recognition, etc.	References
α-Amino-β-phenylbutyric acid CH$_3$ | ⬡—CHCHCOOH | NH$_2$	Bottromycin from *Streptomyces bottro-* *pensis*	1957 J. M. Wisvisz *et* *al.*, *J. Am. Chem. Soc.* **79**, 4524
m-Carboxy-α-phenylglycine COOH | ⬡—CH(NH$_2$)COOH	Iris bulbs and synthesis	1959 C. J. Morris *et* *al.*, *J. Am. Chem. Soc.* **81**, 6069
m-Carboxyphenyl-L-alanine COOH | ⬡—CH$_2$CHCOOH | NH$_2$	Iris bulbs	1961 J. F. Thompson *et al.*, *J. Biol. Chem.* **236**, 1183
L(−)3,4-Dihydroxyphenylalanine (DOPA) OH \ HO—⬡—CH$_2$CH(NH$_2$)COOH	Synthesis *Vicia faba*	1911 C. Funk, *J.* *Chem. Soc.* pp. 99, 554 1913 T. Torquati, *Arch. farmacol. sper.* **15**, 213, 308

APPENDIX II (continued)

Name and structure	Occurrence, recognition, etc.	References
	Vicia faba	1913 M. Guggenheim, *Z. physiol. Chem.* **88**, 276
	Synthesis	1921 E. Waser and M. Lewandowski, *Helv. Chim. Acta* **4**, 657

2,4-Dihydroxy-6-methylphenyl-alanine | *Agrostemma githago* seeds and synthesis | 1958   G. Schneider, *Biochem. Z.* **330**, 428

$$HO-\underset{CH_3}{\overset{OH}{\bigcirc}}-CH_2\underset{NH_2}{CHCOOH}$$

L-Tyrosine-*O*-methyl ether | Puromycin from *Streptomyces alboniger* | 1953   C. W. Waller *et al.*, *J. Am. Chem. Soc.* **75**, 2075

$$CH_3O-\bigcirc-CH_2\underset{NH_2}{CHCOOH}$$

3-(3-Carboxy-4-hydroxyphenyl)-L-alanine (*m*-Carboxy-L-tyrosine) | *Reseda odorata* seeds | 1962   P. Olesen-Larsen and A. Kjaer, *Acta Chem. Scand.* **16**, 142

| | Biosynthesis from tyrosine | 1964   P. Olesen-Larsen *Biochem. Biophys. Acta* **93**, 200 |

$$HO-\underset{}{\overset{COOH}{\bigcirc}}-CH_2\underset{NH_2}{CHCOOH}$$

Ureido and guanidino derivatives of amino acids

Canavanine [α-Amino-(*O*-guanidyl)-γ-hydroxy-butyric acid] | Seeds of *Canavalia ensiformis* | 1929   M. Kitagawa and T. Tomiyama, *J. Biochem. (Tokyo)* **11**, 265

$$\overset{NH}{\underset{\|}{NH_2CNHOCH_2CH_2}}\underset{NH_2}{CHCOOH}$$

		1933 M. Kitagawa and S. Monobe, *J. Biochem. (Tokyo)* **18**, 333
	Canavaninosuccinic acid	1954 J. B. Walker, *J. Biol. Chem.* **204**, 139
	Various legumes	1960 E. A. Bell, *Biochem. J.* **75**, 618

Deaminocanavanine | Prepared from canavanine | 1937   M. Kitagawa and J. Tsukamoto, *J. Biochem. (Tokyo)* **26**, 373

$$HN=CNHOCH_2CH_2CHCOOH$$
$$\underset{NH}{\rule{1.5cm}{0.4pt}}$$

APPENDIX II (continued)

Name and structure	Occurrence, recognition, etc.	References
	Jack bean seeds	1939 D. Ackermann and W. Apple, *Z. physiol. Chem.* **262,** 103
L(+)-Citrulline $$NH_2\overset{O}{\overset{\|}{C}}NH(CH_2)_3CH(NH_2)COOH$$	Watermelon (*Citrullus vulgaris*)	1930 M. Wada, *Biochem. Z.* **224,** 420 1933 M. Wada, *Biochem. Z.* **227,** 1
	Synthesis	1938 R. Duschinsky, *Compt. rend.* **207,** 735
	Reported in edestin, gluten, and mucin	1939 W. R. Fearon, *Biochem. J.* **33,** 902
	Free in *Alnus* root and root nodules	1952 J. K. Miettinen and A. I. Virtanen, *Physiol. Plantarum* **5,** 540
	Sap of several trees	1954 G. Reuter and H. Wolffgang, *Flora (Jena)* **142,** 146
	Sap of several trees	1957 E. G. Bollard, *Australian J. Biol. Sci.* **10,** 292
	In porcupine quills and rabbit fur	1962 G. E. Rogers, *Nature* **194,** 1149
L-Homoarginine $$H_2N\overset{NH}{\overset{\|}{C}}NH(CH_2)_4\underset{\underset{NH_2}{\|}}{C}HCOOH$$	Seeds of *Lathyrus* species	1962 A. E. Bell, *Nature* **196,** 1078 1962 A. E. Bell, *Biochem. J.* **85,** 91 1962 S. L. N. Rao *et al., Biochemistry* **1,** 298
L(−)-α-Amino-β-ureidopropionic acid (albizziine) $$\underset{\underset{NH_2}{\|}}{HOOCCH}CH_2NH\overset{\overset{O}{\|}}{C}NH_2$$	*Albizia* species and synthesis	1958 R. Gmelin *et al., Z. Naturforsch.* **13,** 252 1959 R. Gmelin, *Z. physiol. Chem.* **314,** 28 1959 A. Kjaer *et al., Experientia* **15,** 253 1959 A. Kjaer *et al., Acta Chem. Scand.* **13,** 1565

APPENDIX II (continued)

Name and structure	Occurrence, recognition, etc.	References
Diazo compounds		
Azaserine (O-Diazoacetyl-L-serine) $$N_2CH\overset{\text{O}}{\overset{\|}{C}}OCH_2\underset{\underset{NH_2}{\|}}{C}HCOOH$$	*Streptomyces* culture filtrate	1954 S. A. Fusari *et al.*, *Nature* **173**, 72; *J. Am. Chem. Soc.* **76**, 2878, 2881
	Synthesis	1954 J. A. Moore *et al.*, *J. Am. Chem. Soc.* **76**, 2884
ϵ-Diazo-δ-oxo-L-norleucine $$N_2CHC\underset{\underset{O}{\|}}{C}H_2CH_2\underset{\underset{NH_2}{\|}}{C}HCOOH$$	Isolation and synthesis	1956 D. A. Clarke *et al.*, *Angew. Chem.* **68**, 309
	Streptomyces culture fluid	1956 H. W. Dion *et al.*, *J. Am. Chem. Soc.* **78**, 3075

Acetylenic nitrogen compounds

See Sørensen (470a) for monoamides of some polyacetylenic carboxylic acids.

Name and structure	Occurrence, recognition, etc.	References
Acetylenic dicarboxylic acid diamide $$\underset{\underset{CCONH_2}{\|\|\|}}{CCONH_2}$$	Isolated from *Streptomyces chibaensis* and *S. reticuli aquamyceticus*	1958 S. Suzuki *et al.*, *J. Antibiotics (Tokyo)* **11A**

Heterocyclic compounds

Name and structure	Occurrence, recognition, etc.	References
1-Aminocyclopropane-1-carboxylic acid $$H_2C\overset{CH_2}{\underset{\underset{NH_2}{\|}}{\diagup\diagdown}}C-COOH$$	Pears and apples	1957 L. F. Burroughs, *Nature* **179**, 360
	Vaccinium vitis-idaea	1957 M. L. Vähätalo and A. I. Virtanen, *Acta Chem. Scand.* **11**, 741
β-(Methylenecyclopropyl)-L-alanine and its N-γ-L-glutamyl derivative $$CH_2{=}C\overset{\diagup\diagdown}{\underset{\underset{NH_2CHCOOH}{\overset{\|}{CH_2}}}{\overset{\|}{CH}}}CH_2$$	Hypoglycin A and B from *Blighia sapida*	1955 C. H. Hassall and K. Reyle, *Biochem. J.* **60**, 334
	Synthesis	1958 J. A. Carbon *et al.*, *J. Am. Chem. Soc.* **80**, 1002
		1958 R. S. de Ropp *et al.*, *J. Am. Chem. Soc.* **80**, 1004
	Synthesis	1959 A. Jöhl and W. G. Stoll, *Helv. Chim. Acta* **42**, 156, 716
		1959 C. H. Hassall and D. I. John, *Tetrahedron Letters* No. 3, p. 7

APPENDIX II (continued)

Name and structure	Occurrence, recognition, etc.	References
	Hypoglycin A	1960 C. H. Hassall and D. I. John, *J. Chem. Soc.* p. 4112
Azetidine-2-carboxylic acid CH₂ CH₂ CH.COOH NH	*Convallaria majalis* and synthesis In other liliaceous plants *Polygonatum officinale,* recognized but not correctly identified	1955 L. Fowden, *Nature* **176**, 347 1957 L. Fowden and F. C. Steward, *Ann. Botany (London)* [N.S.] **21**, 53 1955 A. I. Virtanen and P. Linko, *Acta Chem. Scand.* **9**, 551
α-(Methylenecyclopropyl)glycine CH₂=C——CHCH(NH₂)COOH CH₂	*Litchi chinensis*	1961 D. O. Gray and L. Fowden, *Biochem. J.* **82**, 385
4-Methylproline H CH₃---C——CH₂ COOH H₂C C N H H	Apples Configuration	1954 A. C. Hulme and W. Arthington, *Nature* **173**, 588 1961 L. F. Burroughs *et al.*, *Nature* **189**, 394
4-Methyleneproline CH₂=C——CH₂ H₂C CHCOOH N H	*Eriobotrya japonica* Synthesis	1962 D. O. Gray and L. Fowden, *Nature* **193**, 1285 1962 M. Bethell *et al.*, *Nature* **194**, 864
L-Allohydroxyproline OH H---C——CH₂ COOH H₂C C N H H	Synthesis (optically active) Leaves of *Santalum album*	1919 H. Leuchs *et al.*, *Chem. Ber.* **52**, 2086 1954 A. N. Radhakrishnan and K. V. Giri, *Biochem. J.* **58**, 57

APPENDIX II (continued)

Name and structure	Occurrence, recognition, etc.	References
4-Hydroxy-4-methylproline HOH₂CHC———CH₂ H₂C CHCOOH N H (I) OH CH₃C———CH₂ H₂C CHCOOH N H (II)	Isolated from apple, proposed structure (I) Proposed structure (II) (I) and (II) from apples identical Structure (I) finally proved	1954 A. C. Hulme, *Nature* **174,** 1055 1955 G. Urbach, *Nature* **175,** 170 1955 A. C. Hulme and F. C. Steward, *Nature* **175,** 171 1961 K. Biemann *et al.,* *Nature* **191,** 380
3-Carboxymethyl-4-isopropenyl-proline (L-α-kainic and L-α-allo-kainic acid) H₂C=C—CH——CHCH₂COOH H₃C H₂C CHCOOH N H	*Digenea simplex* Synthesis Synthesis	1953 S. Murakami *et al.,* *J. Pharm. Soc. Japan* **73,** 1026 1957 K. Tanaka *et al.,* *Proc. Japan Acad.* **33,** 47 1957 Y. Ueno *et al.,* *Proc. Japan Acad.* **33,** 53

3-Carboxymethyl-4,2-(2-carboxy-1-methyhexa-1,3-dienyl)proline (Domoic acid)
HOOC   CH₃
          H
C=C—C———CHCH₂COOH
CH₃CH₂CH=CH
          H₂C   CHCOOH
N
H
	Chondria armata	1959 K. Daigo, *J. Pharm. Soc. Japan* **79,** 353, 356

L-Pipecolic acid CH₂ H₂C CH₂ H₂C CHCOOH N H	Early attempt at synthesis Synthesis *Trifolium repens* Apple fruits	1883 H. Ost, *J. prakt. Chem.* **27,** 257 1891 A. Ladenburg, *Chem. Ber.* **24,** 640 1953 R. J. Morrison, *Biochem. J.* **53,** 474 1952 A. C. Hulme and W. Arthington, *Nature* **170,** 659

APPENDIX II (continued)

Name and structure	Occurrence, recognition, etc.	References
	In *Phaseolus vulgaris* and derivative from lysine	1954 N. Grobbelaar *et. al., J. Am. Chem. Soc.* **75**, 4341
4-Hydroxypipecolic acid	*Acacia pentadenia*	1955 A. I. Virtanen and S. Kari, *Acta Chem. Scand.* **9**, 170
		1959 A. I. Virtanen and R. Gmelin, *Acta Chem. Scand.* **13**, 1244
	Trans, 4-hydroxy configuration	1959 J. W. Clark-Lewis and P. I. Mortimer, *Nature* **184**, 1234
	Molecular conformation	1962 J. N. Shoolery and A. I. Virtanen, *Acta Chem. Scand.* **16**, 2457
5-Hydroxy-L-pipecolic acid	*Rhapis excelsa (R. flabelliformis)*	1954 A. I. Virtanen and S. Kari, *Acta Chem. Scand.* **8**, 1290
		1955 A. I. Virtanen and S. Kari, *Acta Chem. Scand.* **9**, 170
	Date pericarp	1955 N. Grobbelaar *et al., Nature* **175**, 703
	Configuration (*trans*) established	1957 B. Witkop and C. M. Foltz, *J. Am. Chem. Soc.* **79**, 192
	Molecular conformation	1962 J. N. Shoolery and A. I. Virtanen, *Acta Chem. Scand.* **16**, 2457
4-Oxypipecolic acid	In staphylomycin	1959 H. Vanderhaeghe and G. Parmetier, *17th Congr. Pure Appl. Chem.* Symp. on Peptide Antibiotics, Munich, p. 56. Butterworths, London, 1960

APPENDIX II (continued)

Name and structure	Occurrence, recognition, etc.	References
4-Aminopipecolic acid NH_2 | CH /　＼ H_2C　　CH_2 |　　　| H_2C　　CHCOOH ＼　／ N H	*Strophanthus scandens*	1961　W. Schenk and H. R. Schütte, *Naturwissenschaften* **48**, 223
Baikiain CH //　＼ HC　　CH_2 |　　　| H_2C　　CHCOOH ＼　／ N H	Wood of *Baikiaea plurijuga* Date pericarp Synthesis	1950　F. E. King *et al.*, *J. Chem. Soc.* p. 3590 1955　N. Grobbelaar *et al., Nature* **175**, 703 1958　N. A. Dobson and R. A. Raphael, *J. Chem. Soc.* p. 3642
Guvacine CH /　＼ H_2C　　C—COOH |　　　|| H_2C　　CH_2 ＼　／ N H	*Areca cathecu* Guvacine and arecaidine in betel nut	1891　E. Jahns, *Chem. Ber.* **24**, 2615 1918　K. Freudenberg, *Ber.* **51**, 976, 1669
Minaline (Pyrrole-2-carboxylic acid) HC——CH ||　　　|| HC　　C—COOH ＼　／ N H	Synthesis Purified diastase	1860　H. Schwanert, *Ann. Chem.* **116**, 257, 272 1880　G. L. Ciamician, *Monatsh. Chem.* **1**, 625 1946　T. Minagawa [cf. *Chem. Abstr.* **46**, 3102 (1952)]
Hydroxyminaline HOC——CH ||　　　|| HC　　C—COOH ＼　／ N H	*Penicillium, Aspergillus*	1945　T. Minagawa [cf. *Chem. Abstr.* **47**, 151–152 (1953)]

APPENDIX II (continued)

Name and structure	Occurrence, recognition, etc.	References
Mimosine or leucenol $HOOCCHCH_2-N$ $C=O$ $\quad$ with ring: $C=C$ (H, OH) ... $C=C$ (H H) NH_2	*Mimosa pudica* Synthesis	1936 J. Renz, *Z. physiol. Chem.* **244**, 153 1949 R. Adams and J. L. Johnson, *J. Am. Chem. Soc.* **71**, 705
Stizolobic acid β-(3-Carboxypyron-5-yl)alanine $HOOCCHCH_2-C$ $C-COOH$ $\quad NH_2 \quad HC \quad CH$ $\quad O$	Epicotyl of etiolated *Stizolobium hassjoo* seedlings	1959 S. Hattori and A. Komamine, *Nature* **183**, 1116
Willardiine [3-(1-uracyl)-L-alanine] HOC $C=O$ $HC \quad N-CH_2CHCOOH$ $\quad CH \quad NH_2$	*Mimosa pudica* Isolation *Acacia willardiana* seeds Synthesis	1936 J. Renz, *Z. physiol. Chem.* **244**, 153 1949 R. Adams and J. L. Johnson, *J. Am. Chem. Soc.* **71**, 705 1959 R. Gmelin, *Z. physiol. Chem.* **316**, 164 1961 A. Kjaer *et al.*, *Acta Chem. Scand.* **15**, 1193 1961 G. Shaw and J. H. Dewar, *Proc. Chem. Soc.* p. 216
Roseonine $HOOCHC$————$CHCHCH_2NH_2$ $\quad HN \quad N \quad OH$ $\quad C$ $\quad NH_2$	Roseothricin from streptomyces Corrected structure	1954 K. Nakanishi *et al.*, *J. Am. Chem. Soc.* **76**, 2845; *Bull. Chem. Soc. Japan* **27**, 539 1961 H. E. Carter *et al.*, *J. Am. Chem. Soc.* **83**, 4296

APPENDIX II (continued)

Name and structure	Occurrence, recognition, etc.	References

4-Imidazoleacetic acid
 	
N————CHCH₂COOH
‖   ‖
HC   CH
  \\   /
    N
    H
	Polyporus sulfureus	1958 P. H. List,
		Planta Med. **6,** 424

β-(Pyrazolyl-*N*)-L-alanine
HC————N—CH₂CHCOOH
‖ | |
HC   N   NH₂
  \\   /
   CH
	Citrullus vulgaris structure and synthesis	1959 F. F. Noé and L. Fowden, *Nature* **184,** B.A. 69
		1960 F. F. Noé and L. Fowden, *Biochem. J.* **77,** 543

Lathyrine (tingitanine)
       N
      ∕‖
H₂N—C   CH₂CH(NH₂)COOH
     | ‖
     N   CH
      \\   ∕
       CH
	Lathyrus tingitanus	1961 A. E. Bell, *Biochim. Biophys. Acta* **47,** 602
		1961 E. Nowacki and J. Przybylska, *Bull. Acad. Polon. Sci.* **9,** 279

L(+)-2-Hydroxytryptophan
(indole ring)—CH₂CH(NH₂)COOH
             —OH
    N
    H
	Isolated from phalloidin
	
	Preparation and (+)-identity
		1940 T. Wieland and B. Witkop, *Ann. Chem.* **543,** 171
		1952 M. Kotake *et al.,* *Chem. Ber.* **85,** 690

5-Hydroxytryptophan
HO—(indole ring)—CH₂CH(NH₂)COOH
    N
    H
	Bufo marinus
	
	Chromobacterium violaceum
		1953 S. Udenfriend *et al., J. Am. Chem. Soc.* **75,** 501
		1955 C. Mitoma *et al.,* *Nature* **175,** 994

N-(Indole-3-acetyl)aspartic acid
              COOH
               |
      O   CH
      ‖ |
(indole)—CH₂CNH₂CH
              COOH
    N
    H
	Roots of 'Alaska' peas
	
	Seeds of several plants
	
	Tomato
		1955 W. A. Andreae and N. E. Good, *Plant Physiol.* **30,** 380
		1960 H. D. von Klämbt, *Naturwissenschaften* **47,** 398
		1961 V. V. Row *et al.,* *Contrib. Boyce Thompson Inst.* **21,** 1
		1961 W. A. Andreae *et al., Plant Physiol.* **06, 783**

APPENDIX II (continued)

Name and structure	Occurrence, recognition, etc.	References
Ascorbigen (structure: indole with $CH_2OC_6H_7O_5$ substituent, N–H)	Bound ascorbic acid in cabbage (*Brassica oleracea* var. *capitata*)	1936 B. C. Guha and J. C. Pal, *Nature* **137**, 946
		1937 B. C. Guha and J. C. Pal, *Nature* **139**, 843
	Isolation from cabbage	1957 Z. Prochazka *et al.*, *Collection Czech. Chem. Commun.* **22**, 333, 654
	Ascorbigen formed only when cabbage is crushed	1961 R. Gmelin and A. I. Virtanen, *Ann. Acad. Sci. Fennicae: Ser. A II* **No. 107**
	Synthesis from ascorbic acid and 3-hydroxymethylindole	1962 E. Pironen and A. I. Virtanen, *Acta Chem. Scand.* **16**, 1286

N-Methylated amino and imino compounds

Name and structure	Occurrence, recognition, etc.	References
Sarcosine $CH_3 \cdot NH \cdot CH_2COOH$	Prepared from creatine	1847 J. Liebig, *Ann. Chem.* **62**, 257, 310
	Synthesis	1862 J. Volhard, *Ann. Chem.* **123**, 261
	Peanut protein (not verified)	1951 R. D. Haworth *et al.*, *Nature* **167**, 1068
	Reported free in the lichen *Cladonia silvatica*	1953 P. Linko *et al.*, *Acta. Chem. Scand.* **7**, 1310

N-Methyl-L-valine
H_3C
  $\diagdown$
    $CHCHCOOH$
    $\vert$
H_3C $NHCH_3$
	In enniatin A	1948 P. A. Plattner and U. Nager, *Helv, Chim. Acta* **31**, 2203. 2192

	Synthesis	1949 A. H. Cook *et al.*, *Proc. Chem. Soc.* p. 1022
	In actinomycin	1949 H. Brockmann and N. Grobhofer, *Naturwissenschaften* **36**, 376

N-Methyl-L-leucine
$CH_3CHCH_2CHCOOH$
  $\vert$ $\vert$
  CH_3 $NHCH_3$
	In enniatin A	1948 P. A. Plattner and U. Nager, *Helv. Chim. Acta* **31**, 2192

APPENDIX II (continued)

Name and structure	Occurrence, recognition, etc.	References
N-Methyl-L-isoleucine H_3CH_2C $\diagdown$ CHCHCOOH $\diagup$ $\mid$ H_3C NHCH$_3$	In enniatin A	1948 P. A. Plattner and U. Nager, *Helv. Chim. Acta* **31**, 665, 2192, 2203
N-Methyl-β-methyl-L-leucine H_3C CH_3 $\diagdown$ $\mid$ CHCHCHCOOH $\diagup$ $\mid$ H_3C NHCH$_3$	In etamycin	1958 J. C. Sheehan *et al.*, *J. Am. Chem. Soc.* **80**, 3349
L(+)-N-Methyltyrosine HO—⟨⟩—CH$_2$CHCOOH $\mid$ NHCH$_3$	*Andira* species; surinamine, geoffroyin, ratanhin, and angelin are all identical by synthesis	1892 O. Hiller-Bombien, *Arch. Pharm.* **230**, 513
	Synthesis	1915 E. Fischer and W. Lipschitz, *Chem. Ber.* **48**, 360
L(+)-Abrine (N-Methyltryptophan) ⟨⟩—CH$_2$CHCOOH $\mid$ N NHCH$_3$ H	Seed of *Abrus precatorius*	1932 N. Ghatak and R. Kaul, *Chem. Zentr.* II, 3730
	Abrus precatorius	1935 T. Hoshino, *Ann. Chem.* **520**, 31
	Synthesis	1935 W. G. Gordon and R. W. Jackson, *J. Biol. Chem.* **110**, 151
N-Methyl-L-phenylglycine ⟨⟩—CHCOOH $\mid$ NHCH$_3$	In etamycin	1958 J. C. Sheehan *et al.*, *J. Am. Chem. Soc.* **80**, 3349
Glycine betaine (lycin) $(CH_3)_3\overset{+}{N}CH_2COO^-$	*Lycium barbarum*	1863 A. Husemann and W. Marmé, *Ann. Chem.* Suppl. 2, 382 1864 A. Husemann and W. Marmé, *Ann. Chem.* Suppl. 3, 245
	Beta vulgaris	1869 C. Scheibler, *Chem. Ber.* **2**, 292 1870 C. Scheibler, *Chem. Ber.* **3**, 155

APPENDIX II (continued)

Name and structure	Occurrence, recognition, etc.	References

γ-Hydroxy-N-methyl-L-proline (4-Hydroxyhygric acid)
 	
HOHC——CH$_2$
  | |
H$_2$C   CHCOOH
   \\ /
    N
    |
    CH$_3$
	Bark of *Croton mega-lobotrys* (*C. gubouga*)
	
	Heartwood of *Afror-mosia elata*
		1919 J. A. Goodson and H. W. B. Clewer, *J. Chem. Soc.* pp. 115, 923
		
		1964 J. W. W. Morgan, *Chem. Ind.* p. 542

Stachydrine (L-Proline betaine)
 	
H$_2$C——CH$_2$
 | |
H$_2$C   CHCOO$^-$
  \\ /
   N$^+$
   |
  (CH$_3$)$_2$
	Stachys tuberifera
	
	Alfalfa
		1890 A. v. Planta, *Chem. Ber.* **23**, 1699
		1893 A. V. Planta and E. Schulze, *Chem. Ber.* **26**, 939
		1918 H. Steenbock, *J. Biol. Chem.* **35**, 1

3-Hydroxystachydrin
H$_2$C——CHOH
 | |
H$_2$C   CHCOO$^-$
  \\ /
   N$^+$
   |
  (CH$_3$)$_2$
	Courbonia virgata	1952 J. W. Cornforth and A. J. Henry, *Proc. Chem. Soc.* p. 597

Turcine (betaine of D-allohydroxy-proline)
Betonicine (betaine of L-hydroxy-proline)
 	
HOHC——CH$_2$
  | |
 H$_2$C   CHCOO$^-$
   \\ /
   N$^+$(CH$_3$)$_2$
	Stachys (*Betonica*) *officinalis*
	
	Stachys sylvatica
	
	Synthesis
		1912 E. Schulze and G. Trier, *Z. physiol. Chem.* **76**, 258; **79**, 235
		
		1913 A. Küng and G. Trier, *Z. physiol. Chem.* **85**, 209
		1913 A. König, *Z. physiol. Chem.* **85**, 217

L(+)-Hypaphorin (tryptophan betaine)
 	
—CH$_2$CHN$^+$(CH$_3$)$_3$
      |
    COO$^-$
	Erythrina subumbrans
	
	Synthesis
		1890 M. Greshoff, *Chem. Ber.* **23**, 3537, 3540
		1911 P. V. Romburgh and G. Barger, *J. Chem. Soc.* pp. 99, 2068

APPENDIX II (continued)

Name and structure	Occurrence, recognition, etc.	References
Hercynin (histidine betaine)	Edible mushrooms	1910 F. Kutscher, *Zentr. Physiol.* **24**, 775
	Preparation	1911 G. Barger and A. J. Ewins, *J. Chem. Soc.* pp. 99, 2336
		1913 G. Barger and A. J. Ewins, *Biochem. J.* **7**, 204
Ergothionine (betaine of thiol histidine)	Ergot	1909 M. C. Tanret, *Compt. rend.* **149**, 222
	Synthesis	1950 H. Heath *et al.*, *Nature* **166**, 106
Trigonelline (coffearin)	Seeds of *Trigonella*	1885 E. J. Jahns, *Chem. Ber.* **18**, 2518
		1887 E. J. Jahns, *Chem. Ber.* **20**, 2840
	Synthesis	1887 A. Hantzsch, *Chem. Ber.* **19**, 31

Sulfur- and selenium-containing amino acids

L(+)-Cysteine CH₂SH CHNH₂ COOH	From cystine	1884 E. Baumann, *Z. physiol. Chem.* **8**, 299
	Pisum sativum	1926 A. Kozlowski, *Biochem. J.* **20**, 1346
L-Homocysteine CH₂SH CH₂ CHNH₂ COOH	*Neurospora* mutants	1947 N. H. Horowitz, *J. Biol. Chem.* **171**, 255
(−)-S-Methyl-L-cysteine CH₃SCH₂CH(NH₂)COOH	*Phaseolus vulgaris*	1956 J. F. Thompson *et al.*, *Nature* **178**, 593

APPENDIX II (continued)

Name and structure	Occurrence, recognition, etc.	References
Methylmethionine sulfonium hydroxide $$\left[\begin{array}{c} H_3C \\ \diagdown \\ S^+CH_2CH_2CH(NH_2)COOH \\ \diagup \\ H_3C \end{array}\right] OH^-$$	Cabbage leaves	1954 R. A. McRorie *et al.*, *J. Am. Chem. Soc.* **76**, 115
	S-Adenosylmethionine serves as a methyl donor for *S*-methylmethionine	1960 R. D. Greene and S. B. Davis, *Biochim. Biophys. Acta* **43**, 362
(+)-*S*-Methyl-L-cysteine sulfoxide $$\overset{O}{\underset{\displaystyle \uparrow}{CH_2SCH_3}}$$ $$\mid$$ $$CHNH_2$$ $$\mid$$ $$COOH$$	Cabbage *Brassica rapa*	1956 R. L. M. Synge and J. C. Wood, *Biochem. J.* **64**, 252 1956 C. J. Morris and J. F. Thompson, *Chem. & Ind.* (*London*) p. 951 1956 C. J. Morris and J. F. Thompson, *J. Am. Chem. Soc.* **78**, 1605
3-Amino-3-carboxypropyldimethyl sulfonium hydroxide $(CH_3)_2S^+CH_2CH_2CH(NH_2)COOH$	Cabbage	1954 R. A. McRorie *et al.*, *J. Am. Chem. Soc.* **76**, 115
α-Amino-β-[(2-amino-2-carboxyethyl)mercapto]butyric acid $$HOOCCHCHCH_3$$ $$\underset{H_2N}{\mid}\mid$$ $$SCH_2CHCOOH$$ $$\mid$$ $$NH_2$$	In subtilin	1953 G. Alderton, *J. Am. Chem. Soc.* **75**, 2391
L-Djenkolic acid $$CH_2\left(SCH_2\overset{\displaystyle NH_2}{\underset{\displaystyle COOH}{CH}}\right)_2$$	*Pithecellobium* (*Pithecolobium*) *lobatum* (djenkol beans) Synthesis	1935 A. G. van Veen and A. J. Hijman, *Rec. trav. chim.* **54**, 493 1947 M. C. Armstrong and V. du Vigneaud, *J. Biol. Chem.* **168**, 373

APPENDIX II (continued)

Name and structure	Occurrence, recognition, etc.	References
N-Acetyl-L-djenkolic acid CH_3 | $C=O$ | NH H $HOOC\overset{:}{C}CH_2SCH_2SCH_2\overset{:}{C}COO^-$ H $^+NH_3$	*Acacia farnesiana*	1962 R. Gmelin *et al.,* *Phytochemistry* **1,** 233
Lanthionine [β,β'-Thiodi-(α-aminopropionic acid)] $S[CH_2CH(NH_2)COOH]_2$	Wool Synthesis In subtilin from bacteria (meso form)	1941 M. J. Horn and D. B. Jones, *J. Biol. Chem.* **139,** 649, 743 1942 M. J. Horn *et al., J. Biol. Chem.* **144,** 93 1941 V. du Vigneaud and G. B. Brown, *J. Biol. Chem.* **138,** 151; **140,** 767 1951 G. Alderton and H. L. Fevold, *J. Am. Chem. Soc.* **73,** 463
β-Methyllanthionine NH_2 NH_2 | | $HOOCCHCHSCH_2CHCOOH$ | CH_3	In cinnamycin Yeast Yeast	1952 R. G. Benedict *et al., Antibiot. & Chemotherapy* **2,** 591 1953 G. Alderton, *J. Am. Chem. Soc.* **75,** 2391 1957 P. F. Downey and S. Black, *J. Biol. Chem.* **228,** 171
D-Penicillamine SH | $(CH_3)_2CCH(NH_2)COOH$	In penicillin	1949 H. T. Clarke *et al., in* "The Chemistry of Penicillin." Princeton Univ. Press, Princeton, New Jersey.
N-(2-Carboxyethyl)taurine $HO_3SCH_2CH_2NHCHCOOH$ | CH_3	Red algae	1961 M. Kuriyama, *Nature* **192,** 969

APPENDIX II (continued)

Name and structure	Occurrence, recognition, etc.	References
S-(β)-Carboxyethyl-L-cysteine $CH_2SCH_2CH_2COOH$ $\mid$ $CHNH_2$ $\mid$ $COOH$	Synthesis Seeds of *Albizia julibrissin*	1956 A. Schöberl and A. Wagner, *Z. physiol. Chem.* **304**, 97 1958 R. Gmelin *et al.*, *Z. Naturforsch.* **13b**, 252
S-(2-Carboxypropyl-L-cysteine $CH_3CHCH_2SCH_2CH(NH_2)COOH$ $\mid$ $COOH$	In *tripeptide* from onion	1960 A. I. Virtanen and E. J. Matikkala, *Suomen Kemistilehti Ser. B*, **34**, 84
S-(2-Carboxyl-1-methylethyl)-L-cysteine CH_3CHCH_2COOH $\mid$ S $\mid$ $CH_2CH(NH_2)COOH$	*Acacia* seeds	1960 R. Gmelin and P. K. Hietala, *Z. physiol. Chem.* **322**, 278
Methylsulfoniummethionine $(CH_3)_2SCH_2CH_2CHNH_2$ $\mid$ $COOH$	Asparagus Cabbage	1954 F. Challenger and B. J. Hayward, *Biochem. J.* **58**, iv 1954 R. A. McRorie *et al.*, *J. Am. Chem. Soc.* **76**, 115
Ethionine $CH_3CH_2SCH_2CH_2CH(NH_2)COOH$	Isolated from bacteria	1961 J. F. Fisher and M. F. Mallette, *J. Gen. Physiol.* **45**, 1
L-Alliin (S-Allyl-L-cysteine sulfoxide) O $\uparrow$ $CH_2{=}CHCH_2SCH_2CH(NH_2)COOH$	Garlic steam distillate (isolation of an impure compound) *Allium ursinum* Synthesis	1844 I. Wertheim, *Ann. Chem.* **51**, 289 1845 I. Wertheim, *Ann. Chem.* **55**, 297 1948 A. Stoll and E. Seebeck, *Helv. Chim. Acta* **31**, 189 1949 A. Stoll and E. Seebeck, *Helv. Chim. Acta* **32**, 197 1951 A. Stoll and E. Seebeck, *Helv. Chim. Acta* **34**, 481

APPENDIX II (continued)

Name and structure	Occurrence, recognition, etc.	References
Dihydroalliin (S-n-propyl-L-cysteine sulfoxide) $$CH_3CH_2CH_2\overset{O}{\overset{\uparrow}{S}}CH_2CH(NH_2)COOH$$	Extracts of onion and garlic *Allium cepa*	1959 H. E. Renis and R. E. Henze, *Ber. wiss. Biol.* **133**, 30 1959 A. I. Virtanen and J. Matikkala, *Acta Chem. Scand.* **13**, 1898
($-$)-S-Allyl-L-cysteine $$\underset{\underset{NH_2}{\mid}}{HOOCCH}CH_2SCH_2CH{=}CH_2$$	Extracts of onion and garlic *Allium cepa*	1959 H. E. Renis and R. E. Henze, *Ber. wiss. Biol.* **133**, 30 1961 T. Suzuki *et al.*, *Chem. & Pharm. Bull. (Tokyo)* **9**, 251
Dichrostachinic acid [S-(β-Hydroxy-β-carboxyethane-sulfonylmethyl)-L-cysteine] $$\underset{\underset{OH}{\mid}}{HOOCCH}CH_2SO_2CH_2SCH_2\underset{\underset{NH_2}{\mid}}{CH}COOH$$	Seeds of *Dichrostachys glomerata* and *Neptunia oleracea*	1962 R. Gmelin, *Z. physiol. Chem.* **327**, 186
Cycloalliin (5-Methyl-1,4-thiazane-3-carboxylic acid oxide)	*Allium cepa* and synthesis	1959 A. I. Virtanen and E. J. Matikkala, *Acta Chem. Scand.* **13**, 623
Chondrine (L-1,4-Thiazane-3-carboxylic acid 1-oxide)	Red alga; *Chondria crassicaulis*	1960 M. Kuriyama *et al.*, *Bull. Fac. Fisheries, Hokkaido Univ.* **11**, 58

APPENDIX II (continued)

Name and structure	Occurrence, recognition, etc.	References
2-[S-(β-Carboxy-β-aminoethyl)-tryptophan] —CH₂CH(NH₂)COOH —SCH₂CH(NH₂)COOH	A poisonous basidio-mycete	1955 T. Wieland and W. Schön, *Ann. Chem.* **593**, 157
S-(Prop-1-enyl)-L-cysteine $CH_3CH{=}CH{-}S{-}CH_2CH(NH_2)COOH$	Free in garlic	1963 M. Sugii *et al.*, *Chem. Pharm. Bull.* (*Tokyo*) **11**, 548
S-(Prop-1-enyl)-L-cysteine sulfoxide $CH_3{-}CH{=}CH{-}S{-}CH_2CH(NH_2)COOH$ $\downarrow$ O	Free in onions	1963 C. G. Spåre and A. I. Virtanen, *Acta Chem. Scand.* **17**, 641
β-(2-Thiazole)-β-alanine HC——N HC C—CHCH₂COOH S NH₂	In bottromycin from *Streptomyces bottropensis*	1957 J. M. Waisvisz *et al.*, *J. Am. Chem. Soc.* **79**, 4524
β-Methylselenoalanine	*Astragalus bisulcatus*	1960 S. F. Trease *et al.*, *Science* **132**, 618
Glucobrassicin —CH₂C⟨ NOSO₃⁻ / SC₆H₁₁O₅ N H	*Brassica* and *Raphanus* species	1960 R. Gmelin and A. I. Virtanen, *Suomen Kemistilehti* **B33**, 172 1961 *Ann. Acad. Sci. Fennicae: Ser. A II* No. **107**
Neoglucobrassicin —CH₂C⟨ NOSO₃⁻ / SC₆H₁₁O₅ N OCH₃	*Brassica napus*	1962 R. Gmelin and A. I. Virtanen, *Suomen Kemistilehti* **B35**, 34; *Acta Chem. Scand.* **16**, 1378

Isothiocyanate-producing glucosides

For the occurrence and identity of these compounds see Kjaer (253a).

APPENDIX II (continued)

Name and structure	Occurrence, recognition, etc.	References
Halogen-containing amino acids		
Mono- and dibromotyrosines	3,5-Dibromotyrosine from *Gorgonia* species	1913 C. T. Mörner, *Z. physiol. Chem.* **88**, 138
	Monobromotyrosine	1951 E. M. Low, *J. Marine Res.* (*Sears Found. Marine Res.*) **10**, 239
Mono- and diiodotyrosines	*Ulva lactuca, Laminaria digitata, Rhodymenia palmata*	1954 R. Scott, *Nature* **173**, 1098
	Nereocystis luetkeana	1955 W. Tong and I. L. Chaikoff, *J. Biol. Chem.* **215**, 473
3,5-Diiodotyrosine (Iodogorgonic acid)	Protein of coral (*Gorgonia carolinii*)	1896 E. Drechsel, *J. prakt. Chem.* **39**, 425
	Laminaria flexicaulis	1949 J. Roche and M. Lufon, *Compt. rend.* **229**, 481
3,5,3'-Triiodothyronine	*Phaseolus vulgaris, Hordeum vulgare* (*H. sativum*), *Aster tripolium*, and *Salicornia perennis* (supplied labeled iodine—3,4-diodothyronine also formed)	1959 L. Fowden, *Physiol. Plantarum* **12**, 657
L-Thyroxine Thyronine (3,5-Diiodothyonine)	Thyroxine from thyroid protein	1919 E. C. Kendall, *J. Biol. Chem.* **39**, 125
	Rhodymenia palmata (chromatographic identification)	1955 C. B. Coulson, *J. Sci. Food Agr.* **6**, 674
γ-Glutamyl compounds		
(Moiety attached to γ-carboxyl of glutamic acid)		
L-Cysteinylglycine (Glutathione)	Isolation (yeast) and synthesis	1921 F. G. Hopkins, *J. Biol. Chem.* **15**, 286
		1929 F. G. Hopkins, *J. Biol. Chem.* **84**, 269

APPENDIX II (continued)

Name and structure	Occurrence, recognition, etc.	References
L-Phenylalanine	Onion bulb	1960 A. I. Virtanen and E. J. Matikkala, *Z. physiol. Chem.* **322,** 8
	Soybean meal	1962 C. J. Morris and J. F. Thompson, *Biochemistry* **1,** 706
	Lupinus angustifolius, Lupinus albus	1962 M. Wiewiorowski and H. Augustyniakowa, *Acta Biochim. Polon.* **9,** 399
L-Tyrosine	Soybean meal	1962 C. J. Morris and J. F. Thompson *Biochemistry* **1,** 706
	Lupinus angustifolius, Lupinus albus	1962 M. Wiewiorowski and H. Augustyniakowa, *Acta Biochim. Polon.* **9,** 399
L-Valine	Onion bulbs	1960 A. I. Virtanen and E. J. Matikkala, *Z. physiol. Chem.* **322,** 8
L-Isoleucine	Onion	1962 J. F. Thompson *et al., in* "Amino Acid Pools (J. T. Holden, ed.), p. 54. Elsevier, Amsterdam.
L-Leucine	Lima bean seeds	1958 H. Rinderknecht *et al., Helv. Chim. Acta* **41,** 1
	Onion	1961 A. I. Virtanen and E. J. Matikkala, *Suomen Kemistilehti* **B34,** 53
	Phaseolus vulgaris	1963 C. J. Morris *et al., J. Biol. Chem.* **238,** 650
Alanine	Pea seeds	1954 A. I. Virtanen and A. M. Berg, *Acta Chem. Scand.* **8,** 1089

APPENDIX II (continued)

Name and structure	Occurrence, recognition, etc.	References
L-Methionine	Onion	1961 A. I. Virtanen and E. J. Matikkala, *Suomen Kemistilehti* **B34**, 53
	Phaseolus vulgaris	1963 C. J. Morris *et al.*, *J. Biol. Chem.* **238**, 650
S-Methyl-L-sulfoxide	Lima beans	1957 H. Rinderknecht, *Chem. & Ind.* (*London*) p. 1384
S-Methyl-L-cysteine	Kidney bean seed	1959 R. M. Zacharius *et al.*, *Arch. Biochem. Biophys.* **80**, 199
	Lima bean seed	1958 H. Rinderknecht *et al.*, *Helv. Chim. Acta* **41**, 1
	Onion	1961 A. I. Virtanen and E. J. Matikkala, *Suomen Kemistilehti* **B34**, 53
S-n-Propylcysteine	*Allium sativum*	1962 A. I. Virtanen *et al.*, *Suomen Kemistilehti* **B35**, 52
S-(1-Propenyl)-L-cysteine and its sulfoxide	Onion	1961 A. I. Virtanen and E. J. Matikkala, *Suomen Kemistilehti* **B34**, 84, 114
	Seeds of *Allium schoenoprasum*	1962 E. J. Matikkala and A. I. Virtanen, *Acta Chem. Scand.* **16**, 2461
β-Alanine	Iris bulb	1962 C. J. Morris *et al.*, *J. Biol. Chem.* **237**, 2180
β-Cyanoalanine	In vetch and *Lathyrus*	1962 C. Ressler, *J. Biol. Chem.* **237**, 733 1963 C. Ressler *et al.*, *J. Am. Chem. Soc.* **85**, 2874

APPENDIX II (continued)

Name and structure	Occurrence, recognition, etc.	References
β-Aminoisobutyric acid	Iris bulbs	1961 C. J. Morris *et al., J. Biol. Chem.* **236,** 1181
Hypoglycin A [α-Amino-β-(2-methylenecyclopropyl)propionic acid] Cf. also β-(methylenecyclopropyl)-L-alanine under heading "Heterocyclic compounds," above	Akee (*Blighia sapida*) fruit	1960 C. H. Hassall and D. I. John, *J. Chem. Soc.* p. 4112
S-(β-Carboxy-n-propyl)-L-cysteinylglycine	Onion bulb	1960 A. I. Virtanen and E. J. Matikkala, *Z. physiol. Chem.* **322,** 8
S-Allyl-L-cysteine	Garlic	1962 T. Suzuki *et al., Chem. & Pharm. Bull. (Tokyo)* **10,** 345
Ethylamine (theanine)	Tea leaves	1950 Y. Sakato, *J. Agr. Chem. Soc. Japan* **23,** 262
	Mushrooms, *Xerocomus badius*	1950 Y. Sakato, *J. Agr. Chem. Soc. Japan* **23,** 269
β-Aminopropionitrile	Sweet pea seed	1954 G. F. McKay *et al., Arch. Biochem. Biophys.* **52,** 313 1955 E. D. Schilling and F. M. Strong, *J. Am. Chem. Soc.* **77,** 2843
p-Hydroxyphenylhydrazine	Reported to be in mushrooms	1961 B. Levenberg, *J. Am. Chem. Soc.* **83,** 503
	Agaricus	1961 E. G. Daniels *et al., J. Am. Chem. Soc.* **83,** 3333
p-Hydroxyaniline	Mushrooms, *Agaricus hortensis*	1960 J. Jadot *et al., Biochim. Biophys. Acta* **43,** 322

APPENDIX II (continued)

Name and structure	Occurrence, recognition, etc.	References
β-Pyrazol-1-yl-L-alanine	Cucumber seeds	1963 D. M. Dunnil and L. Fowden, *Biochem. J.* **85**, 388
	Biosynthesis	1963 D. M. Dunnill and L. Fowden, *J. Exptl. Botany* **14**, 237

Addendum to Appendix II

In addition to the above, it has recently been claimed that the following free amino acids or their derivatives occur in plants or microorganisms:

Hadacidin (*N*-formyl, *N*-hydroxyglycine), E. A. Kaczka *et al.*, *Biochemistry* **1**, 340 (1962); Lactones of L-γ-hydroxyarginine and γ-hydroxyhomoarginine, E. A. Bell and A. S. L. Tirimanna, *Biochem. J.* **91**, 356 (1964); β,γ-dihydroxyisoleucine, M. Bodanszky *et al.*, *J. Am. Chem. Soc.* **84**, 2004 (1962); β-methyl-γ, δ-dihydroxyisoleucine in the form of α-amanitive, T. Wieland and A. Haufer, *Ann.* **619**, 35 (1958); β-methyltryptophan, J. C. Sheehan *et al.*, *J. Am. Chem. Soc.* **85**, 2867 (1963); L-homolanthionine, H. T. Huang, *Biochemistry* **2**, 296 (1963); L-homomethionine, *S*-allylmercapto-L-cysteine, and γ-glutamyl-*S*-allylmercapto-L-cysteine, M. Sugii *et al.*, *Chem. Pharm. Bull.* **12**, 1114, 1115 (1964); γ-glutamyl-3,3′-(2-methylethylene-1,2-dithio)dialanine, E. J. Matikkala and A. I. Virtanen, *Acta Chem. Scand.* **18**, 2009 (1964); *N*-carbamylputrescine, T. A. Smith and J. L. Garraway, *Phytochem.* **3**, 23 (1964).

APPENDIX III

The Keto Acids that May Occur in Plants, Their Amino Acid Analogs, and Their Metabolic Roles

Keto acid	Corresponding amino acid	Possible metabolic role	Plant used in investigated cited	References: original discovery or recent observations
Glyoxylic acid	Glycine	Labeled early with C^{14} in photosynthesis	Vitis vinifera	1886 H. Brunner and E. Chuard, Chem. Ber. **19**, 595
				1951 M. Calvin et al., Symp. Soc. Exptl. Biol. **5**, 284
			Tulipa gesneriana	1954 G. N. H. Towers and F. C. Steward, J. Am. Chem. Soc. **76**, 1959
			Malus sylvestris (Pyrus malus)	1957 A. C. Hulme, Advan. Food Res. **9**, 297
			Wheat	1958 R. M. Krupka and G. N. H. Towers, Can. J. Botany **36**, 165
			Spinach	1962 P. C. Kearney and N. E. Tolbert, Arch. Biochem. Biophys. **98**, 164
Pyruvic acid	Alanine	Links glycolysis to Krebs cycle, etc.	Trifolium and Medicago	1939 A. I. Virtanen et al., Nature **144**, 597
			Allium cepa	1945 J. P. Bennett, Soil Sci. **60**, 91
			Allium cepa	1946 E. J. Morgan, Nature **157**, 512
			Solanum tuberosum	1955 J. Barber and L. W. Mapson, Proc. Roy. Soc. **B143**, 523
Mesoxalic acid	Aminomalonic acid	Possible intermediate in organic acid metabolism	Medicago sativa	1909 H. von Euler and I. Bolin, Z. physiol. Chem. **61**, 1
				1956 H. A. Stafford, Plant Physiol. **31**, 135

Keto acid	Amino acid	Source	Function	References
Hydroxypyruvic acid	Serine	*Asplenium septentrionale*	Linked to hexose-monophosphate pathway	1954 A. I. Virtanen and M. Alfthan, *Acta Chem. Scand.* **8**, 1720 1955 I. C. Gunsalus *et al.*, *Bacterial Rev.* **19**, 79
α-Keto-β-hydroxy-butyric acid	Threonine	*Vaccinium vitis-idaea*		1955 A. I. Virtanen and M. Alfthan, *Acta Chem. Scand.* **9**, 188
Oxaloacetic acid	Aspartic acid	Tobacco leaves	Krebs cycle intermediate and "port of entry" for N	1937 G. W. Pucher *et al.*, *J. Biol. Chem.* **119**, 523
		Root nodules of legumes		1943 A. I. Virtanen *et al.*, *J. prakt. chem.* **162**, 71
α-Keto-β-hydroxy-succinic acid	Hydroxyaspartic acid	*Hordeum vulgare* (*H. sativum*)	Possible intermediate in organic acid metabolism	1948 P. A. Kolesnikov, *Compt. rend. acad. sci. USSR* **60**, 1205 1956 E. Kun and M. G. Hernandez, *J. Biol. Chem.* **218**, 201
α-Keto-γ-hydroxy-butyric acid	Homoserine	*Vaccinium vitis-idaea*		1955 A. I. Virtanen and M. Alfthan, *Acta Chem. Scand.* **9**, 188
α-Keto-γ-hydroxy-glutamic acid	γ-Hydroxyglutamic acid	*Oxalis* *Phlox decussata*		1963 R. K. Morton and J. R. E. Wells, *Nature* **200**, 477 1964 G. Brandner and A. I. Virtanen, *Acta Chem. Scand.* **18**, 574
Acetoacetic acid	β-Aminobutyric acid	Flax (*Linum usitatissimum*)	Intermediate in isoprene biosynthesis and fat metabolism	1954 J. A. Johnston *et al.*, *Proc. Natl. Acad. Sci. U.S.* **40**, 1031

APPENDIX III (continued)

Keto acid	Corresponding amino acid	Possible metabolic role	Plant used in investigation cited	References: original discovery or recent observations
Succinic semialdehyde	γ-Aminobutyric acid	Glutamic acid and glutamine precursor		1956 F. C. Steward and J. K. Pollard, *in* "Inorganic nitrogen metabolism" (W. D. McElroy and B. Glass eds.), pp. 377–407. Johns Hopkins Univ. Press, Baltimore, Maryland
			Pisum sativum	1953 J. K. Miettinen and A. I. Virtanen, *Acta Chem. Scand.* **7**, 1243
			Hordeum vulgare (*H. sativum*)	1959 W. L. Kretovich and E. Galas, *Compt. Rend. Acad. Sci. URSS* **124**, 217
Aspartic-β-semialdehyde	α,γ-Diaminobutyric acid	Precursor of homoserine or azetidine-2-carboxylic acid	Yeast	1955 S. Black and N. Wright, *J. Biol. Chem.* **213**, 39
Diketosuccinic acid		Possible intermediate in organic acid metabolism	Higher plants	1954 H. Stafford *et al.*, *J. Biol. Chem.* **207**, 621
α-Acetolactic acid		Fermentation of pyruvic acid and bacterial formation of α-acetolactate	Bacteria	1950 E. Juni, *Federation Proc.* **9**, 396
				1954 Y. Kobayashi and G. Kalnitsky, *J. Biol. Chem.* **211**, 473
α-Ketoglutaric acid	Glutamic acid	Generally recognized as a Krebs cycle intermediate and "port of entry" for N	*Medicago sativa* and *Trifolium pratense*	1939 A. I. Virtanen *et al.*, *Nature* **144**, 597
				1943 A. I. Virtanen *et al.*, *J. prakt. chem.* **162**, 71

Compound	Note	Source	Year, Reference
		Tulipa gesneriana	1954 G. H. N. Towers and F. C. Steward, J. Am. Chem. Soc. **76**, 1959
α-Ketoglutaric semi-aldehyde	Enzymatic product from hydroxy-proline	Pseudomonas	1964 R. M. M. Singh and E. Adams, Science **144**, 67
		Malus sylvestris (Pyrus malus)	1957 A. C. Hulme, Advance Food Res. **9**, 297
		Wheat	1958 R. M. Krupka and G. H. N. Towers, Can. J. Botany **36**, 165
Glutamic-γ-semi-aldehyde α-Keto-δ-aminovaleric acid	Ornithine	Neurospora crassa	1954 H. J. Vogel and D. M. Bonner, Proc. Natl. Acad. Sci. U.S. **40**, 688
α-Ketoisovaleric acid	Valine	Escherichia coli	1951 H. E. Umbarger and B. Magasanik, J. Biol. Chem. **189**, 287
δ-Aminolevulinic acid	Intermediate in porphyrin bio-synthesis		1953 D. Shemin and C. S. Russell, J. Am. Chem. Soc. **75**, 4873
		Rhodopseudomonas spheroides	1958 G. Kikuchi et al., J. Biol. Chem. **233**, 1214
α-Amino-β-ketoadipic acid	Intermediate in porphyrin biosyn-thesis		1953 D. Shemin and C. S. Russell, J. Am. Chem. Soc. **75**, 4873
γ-Methylene-α-keto-glutaric acid	γ-Methyleneglu-tamic acid	Tulipa gesneriana	1954 G. H. N. Towers and F. C. Steward, J. Am. Chem. Soc. **76**, 1959
		Arachis hypogaea	1955 L. Fowden and J. A. Webb, Biochem. J. **59**, 228

Keto acid	Corresponding amino acid	Possible metabolic role	Plant used in investigation cited	References: original discovery or recent observations
γ-Methyl-α-keto-glutaric acid	γ-Methylglutamic acid		*Phyllitis scolopen-drium*	1955 A. I. Virtanen and A. M. Berg, *Acta Chem. Scand.* **8**, 1085 1956 M. E. Wickson and G. H. N. Towers, *Can. J. Biochem. Physiol.* **34**, 502
γ-Hydroxy-γ-methyl-α-ketoglutaric acid (pyruvic aldol)	γ-Hydroxy-γ-methylglutamic acid	Inhibits Krebs cycle	*Phyllitis scolopen-drium* *Adiantum pedatum*	1955 A. I. Virtanen and A. M. Berg, *Acta Chem. Scand.* **8**, 1085 1955 N. Grobbelaar et al., *Nature* **175**, 703 1955 F. C. Steward et al., *Am. J. Botany* **42**, 946
		Formed by the L-amino acid oxidase	*Proteus vulgaris*	1944 P. K. Stumpf and D. E. Green, *J. Biol. Chem.* **153**, 387
α-Keto-β-methyl valeric acid	Isoleucine		*Neurospora crassa* *Escherichia coli*	1955 R. P. Wagner and A. Bergquist, *J. Biol. Chem.* **216**, 251 1962 W. L. Kretovitch et al., *Biochemistry USSR (English Transl.)* **27**, 181
		Formed by the L-amino acid oxidase	*Proteus vulgaris*	1944 P. K. Stumpf and D. E. Green, *J. Biol. Chem.* **153**, 387
α-Ketoisocaproic acid	Leucine		*Prunus avium*	1954 P. H. Abelson, *J. Biol. Chem.* **206**, 335 1955 C. Mentzer and L. Cronenberger, *Bull. Soc. Chim. Biol.* **37**, 371
α-Ketoadipic acid	α-Aminoadipic acid	Intermediate in lysine biosynthesis	*Pisum sativum*	1954 A. I. Virtanen and M. Alfthan, *Acta Chem. Scand.* **8**, 1720

Keto acid	Amino acid	Notes	Organism	Reference
Oxalosuccinic acid	α-Aminotricarballylic acid	Krebs cycle intermediate		1948 S. Ochoa, *J. Biol. Chem.* **174**, 115
α-Keto-ε-aminocaproic acid	Lysine	Precursor of pipecolic acid		1955 R. S. Schweet *et al., in* "Amino Acid Metabolism," pp. 496–506. Johns Hopkins Press, Baltimore, Maryland
Dimethylpyruvic acid	Dimethylalanine		*Prunus avium*	1955 C. Mentzer and L. Cronenberger, *Bull. soc. chim. biol.* **37**, 371
			Tulipa gesneriana	1954 G. H. N. Towers and F. C. Steward, *J. Am. Chem. Soc.* **76**, 1959
β-Indolepyruvic acid	Tryptophan	Formed by the L-amino acid oxidase	*Proteus vulgaris*	1942 H. Waelsh and H. K. Miller, *J. Biol. Chem.* **145**, 1
				1944 P. K. Stumpf and D. E. Green, *J. Biol. Chem.* **153**, 387
		Formed by the L-amino acid oxidase	*Proteus vulgaris*	1944 P. K. Stumpf and D. E. Green, *J. Biol. Chem.* **153**, 387
Imidazolepyruvic acid	Histidine		Mussel (*Mytilus edulis*)	1954 J. Roche *et al., Compt rend. soc. biol.* **148**, 481
α-Keto-γ-methylthiolbutyric acid	Methionine		Mung bean (*Phaseolus* sp.)	1954 D. G. Wilson *et al., J. Biol. Chem.* **208**, 863
α-Keto-δ-guanidinovaleric acid	Arginine	Formed by the L-amino acid oxidase	Insects	1951 G. Ehrensvärd *et al., J. Biol. Chem.* **189**, 93
			Phlox decussata	1964 G. Brandner and A. I. Virtanen, *Acta Chem. Scand.* **18**, 574

APPENDIX III (continued)

Keto acid	Corresponding amino acid	Possible metabolic role	Plant used in investigated cited	References: original discovery or recent observations
2-Ketogluconic, 5-ketogluconic acid		Glucose oxidation intermediates		1954 D. Kulka and T. A. Walker, *Arch. Biochem. Biophys.* **50**, 169
2,5-Diketogluconic acid		Glucose oxidation intermediate		1953 H. Katznelson *et al.*, *J. Biol. Chem.* **204**, 43
2-Keto-3-deoxy-6-phosphogluconic acid		Glucose oxidation intermediate		1954 J. MacGee and M. Doudoroff, *J. Biol. Chem.* **210**, 617 1955 R. Kovachevich and W. A. Wood, *J. Biol. Chem.* **213**, 757
α-Ketopimelic acid	α-Aminopimelic acid		*Asplenium septentrionale*	1954 A. I. Virtanen and M. Alfthan, *Acta Chem. Scand.* **8**, 1720
α-Keto-γ-hydroxy-pimelic acid	γ-Hydroxyamino-pimelic acid		*Asplenium septentrionale*	1954 A. I. Virtanen and M. Alfthan, *Acta Chem. Scand.* **8**, 1720

AUTHOR INDEX

Numbers in parentheses are reference numbers and indicate that an author's work is referred to although his name is not cited in the text. Numbers in boldface show the page on which the complete reference is listed. Numbers in lightface indicate the pages in the text on which the references are cited.

Note concerning Chapter 4: Authors cited in Appendixes I, II, and III of this chapter are not listed in this index. For certain authors marked with an asterisk in this list text citations for Chapter 4 are not included.

A

Abelson, P. H., 389(1, 2), 421(419), 489(419), **604, 624**
Abendschein, P. A., 447, **619**
Ahmed, S., 129, **183**
Aisenberg, A. C., 339(1), 358, **369**
Aizenshatat, B. A., 206, **225**
Albaum, H. G., 434(3), 442, **604**
Alexander, M., 567(227a), **615**
Alföldi, L., 565(4a), **604**
Allard, H. A., 77, **189**
Allen, L. H., 205(41), 208(2), **225, 227**
Allen, M. B., 120(12), 179, **183**
Allfrey, V. G., 555, **604**
Allmendinger, D. F., 100, **183**
Allsop, A., 491, **604**
Alston, R., 429(42), **606**
Altschul, A. M., 539(12), 540(7), 541(7), **604**
Alvim, P. de T., 53, **183**
Anderson, A. J., 132(5), **183**
Anderson, J. A., 407(8), **604**
Anderson, M. L., 597, **635**
Anfinsen, C. B., 387(9, 10), 535(245), 556, **604, 616**
Anson, M. L., 535, 539(12), 542(11), **604**
Antonovitch, E. G., 434(404), **623**
Appleby, J. F., 214(5), 215(5), 223(5), **226**
Appleman, C. O., 325(3), 326(2, 4), 339(2, 3), **369**
Ap Rees, T., 362(111), 365, **375**
Archbald, H. K., 320, **373**

Arney, S. E., 271, 289, **304**
Arnold, W., 128, **188**
Arnold, W. N., 473(536), 474(536), **631**
Arnon, D. I., 21, 120, 121, 129, 131(9, 10), 132, 179, **183**, 598(14), 600, 601(524, 525), **605, 630**
Aronoff, S., 22(13), 23(13), 124, **183**
Arthur, J. M., 82, **183**
Ashton, F. M., 99, 100, 151, **183**
Aubert, J. P., 598(326), **620**
Audus, L. J., 275(3), **304**, 405(15), **605**
Aurell, B., 536(410), **624**
Avers, C. J., 545, **605**
Axelrod, B., 457(151), **611**
Aykroydt, W. R., 11, 12(17), **183**

B

Bachrach, U., 446(17), **605**
Baeyer, A. von, 20, **183**
Baglioni, C., 536(18), 560(19), **605**
Bailey, K., 535(354), **621**
Baker, D. N., 213(2a), **225**
Baker, J. E., 135, **183**
Baldwin, Ernest, 3, **5**
Ballard, L. A. T., 269, **304**
Baly, E. C. C., 20, **183**
Baptiste, E. C. D., **201**
Barber, J. T., 475, **607**
Barger, G., 429(20), **605**
Barker, H. A., 22(192), 28, 31, **183, 192**
Barker, J., 319, 323(16), 324(6, 7, 8, 9, 10, 11, 12, 13, 14, 15), 326(6), 327(6), 328, 347(59), 351(11, 12,

13), 352(8, 9, 11, 13, 14, 15), 353
(10, 11), 359(7), **369, 370, 372**
Barker, W. G., 505, **605**
Barnell, H. R., 252, 257, **304**
Barnes, R. L., 430(23), 432(24), 475, **605**
Barr, R. A.,* 4, 6, 363(29), 368(29), **371, 628**
Barrales, H., 424, 477(25), 478, 488(25), **605**
Barrien, B. S., 521(26, 201, 618), 583(26, 201), **605, 613, 635**
Barthélemy, A., 41(23), **184**
Basford, R. E., **184**
Basilio, C., 571(265, 468, 469), **617, 627**
Bassham, J. A., 414(275), 598(457), **617, 626**
Battersby, A. R., 426, **605**
Baur, E., 21(26, 27), **184**
Baur, J. R., 342(17), **370**
Bayley, S. T., 536(444), **625**
Bazyrina, K., 155(208), **192**
Beach, E. F., 386(28), **605**
Beadle, G. W., 418(29), 421(29), 558, **605**
Beevers, H., 232, 233, 235(6), 237(6), 238, 239, 242, 267, 272, 274, 278, 279, 282, **305, 306, 308,** 359, 362 (100, 111), 364(20), 365, **370, 374, 375**
Begg, J. E., 215(3), 216, 219(3), 221, 222, **226**
Bell, E. A., 429, **605**
Belozersky, A. N., 549(31), **605**
Benedict, C. R., 364(20), **371**
Benedict, F. G., 32, 33, 35, **184**
Benson, 364)
Benson, A. A., 425(364), 488(290, 364), 513(32, 290, 364), 589, 590, 596 (32), 602(290, 364), **605, 618, 622**
Ben-Yehoshua, S., 332, 333(21), 338, 339(21), 342(21), **371**
Benzer, S., 573, 574, **606, 634**
Berg, P., 550, 555, 573, **606, 608**
Bernal, J. D., 3, **5,** 388(36), **606**
Berner, E., Jr., 107(287), **196**
Bernlohr, R., 321(96), 323(96), **374**
Berry, L. J., 291(8), 292, **305,** 346, **371**

Berry, W. E., 495(480), 580(480), **627**
Berzelius, J. J., 18, **184**
Besserman, M. J., 560(37), **606**
Betz, A., 267, 293(10, 11), 295, **305**
Beverton, R. J. H., 526, **606**
Biale, J. B., 312(24, 25, 27), 313, 314(26, 141), 315(25, 26, 141), 318(27), 319(28), 321(99), 322(113, 139), 323(139), 324(123), 332(21), 333 (21), 338(21, 141), 339(21), 342 (21), 344(23, 26), **371, 374, 375, 376**
Bidwell, R. G. S.,* 4, **6,** 362(118) 363(118), 366(118), 368(29, 62, 118), **371, 373, 375, 627**
Biemann, K., 396(41), 401(40), 471(39), **606**
Bierhuizen, J. F., 215(3), 216(3) 219(3), 221(3), 222(3), **226**
Billimoria, M. C., 408(386), **623**
Birdsong, B. A., 429, **606**
Birt, L. M., 423(43, 44, 45), 601(43, 44, 45), **606**
Bishop, J., 555, **625**
Bisset, S. K., 391(46), **606**
Black, J. N., 143(34), 145, 146, 147, **184**
Blackadar, A. K., 211(23), 223(23), **227**
Blackman, F. F., 41, 42, 54, 63(32), 64, 66, 67, 88(32), 160, **184,** 237(13), 267, **305,** 311, 312, 313, 337(30), 339(30), 350(30), 354, 355, 357, 360, 364(30), **371,** 498(47, 48), **606**
Blackman, G. E., 143(35), 145, 146, 147, **184,** 408(49), **606**
Blackman, V. H., 135, **184**
Bleasdale, J. K. A., 165, **184**
Block, K., 350, **372**
Block, R. J., 393, 535, **606**
Bloom, B., 278, **305**
Blout, E. R., 552(637), **636**
Blum, H. F., 388(52), **606**
Boawn, L. C., 130(38), **184**
Boehm, J., 41, **184**
Boeri, E., 250(39), **306**
Böhning, R. H., 75(72), 78, 79, **185, 186**
Bömer, A., 73, **185**

Bogdanov, A. A., 434(404), **623**
Bolas, B. D., 73, 126, **185, 197**
Bollard, E. G.,* **606, 607, 612, 630**
Bolle-Jones, E. W., 124, **185**
Bolton, E. T., 421(419), 489(419), **624**
Bonde, E. K., 78, **185**
Bonner, J., 101, **189,** 321(99), **374,**
    431(58), 552(222), 555, **607, 614**
Bonner, W. D., Jr., 129(47), **185,** 232(38),
    233(38), 235(38), 265, 293, **306,**
    348, 349, **376**
Bonnet, 12, 160, **185**
Bonnichsen, R. K., 250(39), **306**
Bonnier, G., 28, 29, **185**
Borchers, R., 406(59), **607**
Borsook, H., 557(61), **607**
Borthwick, H. A., 385(62), 548, 600
    (210), **607, 614**
Bose, J. C., 160, **185**
Boswell, J. G., 337, 348, **371,** 587, **607**
Bottger, I., 417, 431, 434, 544(66)
    599(67), **607**
Boulter, D., 475, **607**
Boussingault, J. B., 27, 28, 41(53), **185**
Bové, J., 543, 598, **607**
Boynton, D., 118(126), **188**
Brachet, J., 4, **6,** 548(69, 70, 71), 555,
    607
Brackett, F. S., 64(180), 66(180), 67,
    150, **185, 191**
Bradbeer, J. W., 283, **305**
Bradstreet, R. B., 390, **607**
Brannaman, B. L., 167(381), **201**
Braude, E. A., 21(215), **193**
Braun, A. C., 529(73, 621), **607, 635**
Braunstein, A. E., 442, 451, **607**
Brawerman, G., 434, 544(76), 549(76),
    607, 608
Briggs, G. E., 107, **185,** 252, 270, **307**
Brittain, E. G., 102(356), **200**
Britten, R. J., 421(419), 489(419), **624**
Broadbent, D., 293, 294, 295, 422(78),
    545(78), **608**
Brohult, S., 542(77), **608**
Brown, A. H., 58, 59(57, 189), **185, 191,**
    274, **305, 306**
Brown, G. W., Jr., 438, 439, 451(112),
    609
Brown, H. T., 35(61, 62), 43, 45, 46,
    18, 64, 71, **185, 106**

Brown, J. C., 124(64, 65, 66, 67), **186**
Brown, R., 251, 257, 268, 293, 294, 295,
    297, **305,** 422, 493(78), 545(78, 212,
    422, 423), **608, 614, 624**
Brown, R. G., 326(2), 339(2), **369**
Brown, T. E., 128(125), **188**
Brun, W. A., 76, **186**
Bryant, M., 538(79), **608**
Buch, K., 33, **186**
Büsgen, M., 420(83), **608**
Burg, E. A., 318(34), 319, **371**
Burg, S. P., 318(34), 319(33, 34), **371**
Burlew, J. S., 169(69), 171, 173, 174,
    186
Burns, G. R., 83, 84(71), **186**
Burns, V. W., 434, **608**
Burnside, C. A., 75(72), **185, 186**
Burr, G. O., 56(252), 80, 90, **194, 195**
Burris, R. H., 405(178), 414(216, 288),
    439(81), 442(81, 267, 608), **608,**
    612, 614, 617, 618, 634
Burrows, F. W., 127, 129(220), 130
    (220), **193, 196**
Burström, H., 407, **608**
Burton, W. G., 324(36, 37, 38, 39, 40),
    325(39), 337(35, 36, 37), 338(36),
    340(37), 345(35, 40, 95), 347, **371,**
    374
Businger, J. A., 211(4), **226**
Butenko, R. G., 432(97), **608**
Butler, J. A. V., 537, **608**
Byers, H. R., 24, **186**

C

Calhoun, B., 134(104), **187**
Callendar, G. S., 33, **186**
Calvery, H. O., 434(86), 521(86), **608**
Calvin, M., 22, 120, **186,** 274, **310,**
    388(87, 89), 389, 414(275), 423(88),
    513(32, 88), 589, 590, 596(32), 598
    (88), **605, 608, 617**
Canellakis, E. S., 434, **608**
Cantino, E. C., 524(282), 525, **608,**
    618
Caplin, S. M., 532(92, 483), 580(483),
    608, 628
Carnahan, J. E., 601, **608**
Carnegie, P. R., 396(94), **608**
Carpenter, T. M., 32, 33, 35(76), **186**

Cartright, P. M., 297, **305**

Castle, J. E., 601, **608**

Caswell, M. C., 394(513), **630**

Cathey, H. M., 501(632), 502(632), **635**

Cavalieri, L. F., 434(95), 560(95), **608**

Cavendish, 12, **183**

Cayle, T., 599, **608**

Cerighelli, R., 72, **186**

Chailakhian, M. K., 432(97), **608**

Chamberlin, M., 549, **608**

Chance, B., 323, 349, 350, 358, 359(41), **371**

Chandler, W. H., 406(99), **609**

Chang, L. O., 494(489), 496(484), 539 (484), 540, **609, 628**

Chapman, H. W., 34, 36(78, 79), 64 (79), 66(79), 69(79), 70, 135(79), **186**

Chargaff, E., 535(102), 544(75), **607, 609**

Chesnokov, V. A., 73, **186**

Chiba, Y., 599(103), **609**

Chibnall, A. C., 260, 279, 289, 290, **305,** 391(106), 396, 397(108, 569), 398 (107), 408(106), 409(106), 419 (106), 420, 425(121), 429, 436, 439 (105), 440(105, 106), 443, 453(569), 499(104, 105), 513(105), 542(106), 579, 582, **609, 610, 632**

Childers, N. F., 70, 147, 151(172), 156, 157, **191**

Childers, W. F., 99, **197**

Chow, C. T., 599(347), **621**

Christiansen, G. S., 531, **609**

Cionco, R. M., 214(5), 215, 223(5), **226**

Clark, H. E., 397(569), 408(110), 409 (110), 453(569), **609, 632**

Clarke, F. W., 10, **186**

Clarke, L. D., 268(79), **308**

Clendenning, K. A., 97, **186,** 315(43), 334(42), **371, 372**

Clowes, F. A. L., 293, 296, **305, 609**

Clowes, R. C., 565(4a), **604**

Cocking, E. C., 301, **306**

Cohen, P. P., 415, 416, 438, 439, 442 (113), 451(112, 113), **604, 609**

Coleman, R. G., 459(417), **624**

Collin, T. R., 118(348a), 119(348a), 120(348a), **199**

Connell, G. E., 473(197), 564(197), **613**

Consden, R., 393(114), 453, 535(114), **609**

Convit, J., 170, **192**

Conway, E. J., 391(115), **609**

Corey, R. B., 400(382), **622**

Covey, W. A., 210(6), **226**

Cowie, D. B., 421(419), 489(419), 597(117), **609, 624**

Craig, F. N., 102, **187**

Crane, F. A., 459(485), 499(485), 515 (485), 516(485), 522(485), **628**

Crawford, R. F., 408(118), **609**

Cresswell, C. F., 600(194), **613**

Crick, F. H. C., 538(598), 549, 555, 560, (571), 572, **609, 634**

Crocker, W., 118(84), **187**

Crowther, F., 137(85), **187**

Crozier, W. J., 263, **306**

Cruickshank, D. H., 498(619), **635**

Cullis, A. F., 383(390), **623**

Cummings, M. B., 72, **187**

Cushing, D. H., 26, **187**

D

Daly, J. M., **306**

Damodaran, M., 425(121), **610**

Daniels, F., 23(229), 28(229), 176(229), **193**

Darwin, F., 52, **187**

Dastur, R. H., 100(90), **187**

Davidson, J. N., 434, 535(102), **609, 610**

Davies, D. D., 245(27a), 434(124), 442, **306, 610**

Davis, E. A., 170(91), 171, 172(91), 174, **187**

Davis, F. F., 536, **613**

Davison, P. F., 537, **608**

Day, A. D., 180, **187**

Day, P. R., 420(152), 421(152), 558, **611**

Decker, J. P., 57, 60(97, 98), 61(97), 62, 69, 70, 145, 149, **187, 192**

Dedrick, J., 170(91), 171(91), 172(92), 174(91), **187**

Deffner, G. G. J., 471(39), **606**

Dekker, E. E., 461, 463(125, 289), **610, 618**

DeKock, P. C., 272, 278, **307**

Delgarno, L., 423(120), **609**

Delwiche, C. C., 451(126), **610**

DeMoss, J. A., 434(127), **610**

Demoussy, E., 28(231), 29(231), 71, **187, 193**

Denmead, O. T., 206, 212(12), 214(12), 215(12), 216(12), **226**

Denny, F. E., 338(44), **372**

Dent, C. E., 393, 452, 453(129), 454, 459, **610**

Dent, K. W., 340, **375**

Desai, B. L., 100, **187**

de Saussure, N. T., 15, 27, 28, **197**

Deuber, C., 123, **187**

Dhar, N. R., 21(102), **187**

Dieckmann, M., 573(35), **606**

Dillemann, G., 435, **610**

Dimond, A. E., 529, **614**

Dixon, G. H., 473(197), 564(197), **613**

Dixon, M., 388(131), **610**

Dixon, R. O. D., 446, **610**

Dobzhansky, T., 406, **612**

Doeden, D., 407(8), **604**

Doi, R. H., 553, **610**

Doll, J. P., 404, **621**

Done, J., 472, **610**

Doty, P., 400(135), **610**

Dounce, A. L., 449, 560(136, 137), 565, **610**

Drboglav, M. A., 499(456), **626**

Dreyer, W. J., 535(245), **616**

Drosdoff, M., 112(221), 113(221), 114(221), 115(221), 129(143), **189, 193**

Ducet, G., 278, **306**, 341, **372**

Duggar, B. M., 23(229), 28(229), 82, 176(229), **193, 194**

Dugger, W. M., Jr., 54, 134(313), 135(19), **183, 187, 197**

Dunn, M. S., 397(438), **625**

Duranton, H., 495(138, 139), **610**

Durrum, E. L., 393(51), **606**

Dutrochet, R. J. H., 16(106), 40, **187, 188**

Dwyer, I. M., 394(513), **630**

Dyar, A. T., 209(8), 212(8), 216, **226**

E

Eagle, H., 386(140), **610**

Eakin, R. E., 432(208), **614**

Eaton, F. M., 76, 77, 133, **188**

Eaton, S. V., **188**, 515, **610**

Ebermayer, D., 23, 25, **188**

Eck, R. V., 573, **610**

Eckerson, S. F., 47, 48, **188**

Eckerson, S. H., 116, **188**, 412, **610**

Eckstein, O., 111, 112, **188**

Egami, F., 409(526), 410(526), 412(526), 451(526), **630**

Egle, K., 37, 38, **188**

Eisenberg, A., 170(236), 182(236), **194**

Elliott, W. H., 447(144), 526, 582, **610**

Ellis, G. H., 133(198), 134(198), **192**

Ellis, R. J., 434(123), 442, **610**

El Saifi, A. F., 324(12, 13, 14, 15), 351(12, 13), 352(13, 14, 15), **370**

Eltinge, E. T., 126, **188**

Embleton, T. W., 121, 122(284), **196**

Emerson, R., 23(121), 58, 66, 88, 89(120), 95, 128, **188**, 599, **608**

Engel, R. W., 122(275), **196**

Englebrecht, L., 289(71), **308**, 437(344), 511(145, 343), 532(343), 601(344), **611, 621**

Epstein, E., 297(30), **306**

Epstein, H. T., 542(146), **611**

Erdtman, H., 481(147), **611**

Ergle, D. R., 76, 77, **188**

Erickson, L. C., 167(381), **201**

Erygin, P. S., 499(456), **626**

Escombe, F., 35(61, 62), 43, 45, 46, 48, 54, 71, **185, 186**

Evans, G. C., 34, 39, **188**

Evans, H. G., 129, **183**

Evans, H. J., 131, 132, **188**, 600, **611**

Evans, W. R., 457, **611**

Evenari, M., 170(236), 182(236), **188, 194**

Ewing, J., 408(387), 583(387), **623**

Eyster, C., 128, **188**

F

Farr, A. L., 397(283), **618**

Fernandes, D. S., 262, **306**

Fidler, J. C., 329(46, 48), 356, 357, **372**

Fincham, J. R. S., 420(152), 421(152), 558(152), **611**
Fischer, E. G., 118(126), **188**
Fischer, Hans, 18, **188**
Fleischer, W. E., 116, 117, **189**
Florentin, D., 33, 35(130), **189**
Fischer, Hugo, 73, **188**
Fogg, G. E., 414, **611**
Folkers, K., 429(388), **623**
Folkes, B. F., 253, 254, 256(34), 257(118, 119), **306, 310**
Forward, D. F., 252, 264, 265, 266, **306**, 325, 326(51, 51a), 337(50), 338, 346, 351(50, 50a), 352(51a), **372**
Foster, J. W., 22(131), **189**
Foster, M., 2, **6**
Fowden, L., 446, 452, 457(156, 158), 469(159), 471, 472(155, 157, 159, 591), 472(160), 473(157, 158, 159, 182, 365), 477(159), 480(155, 159), 487(159), 488(159), 501((159), 538 (79), **608, 610, 611, 612, 622, 633**
Fowler, D. I., 460, **610**
Fox, S. W., 388(161), 389(161, 162), 432(162), **611**
Fraenkel-Conrat, H., 396(163), 421(165, 549, 550), 538(164, 165), 541(549), 553(164, 549), 554(550), 555, 557 (549), 571(550), **611, 612, 632**
Franck, J., 58, **189**
Fraser, P. E., 442(323), **620**
Frederick, J. F., 536(167), **612**
Freeland, R. O., 54, 97, 98, 167(134), **189**
Freiberg, S. R.,* **612, 628**
French, C. S., 84(278), 170(91), 171 (91), 172(91), 174(91), 175, **187, 190, 196**
Fritschen, L. J., 206(7), **226**
Fruton, J. S., 400(170), **612**
Fujii, R., 262, **308**
Fujimoto, D., **612**
Fujita, C., 182(193), **192**
Furth, J. J., 567(227a), **615**
Fussel, G. E., 401(172), **612**

G

Gaastra, P., 209(9), 211(9), 212(9), 225, **226**

Gabrielsen, E. L., 89, **189**
Gäumann, E., 529(177), **612**
Gaffron, H., 388(173), **612**
Gale, E. F., 447(144), 526, 555, 582, **611, 612**
Galston, A. W., 531, **612**
Gapp, F., 401(40), **606**
Garner, H. R., 469(256), 539(256), **616**
Garner, W. W., 77, **189**
Garreau, 40, **189**
Gartler, S., 406, **612**
Gaskell, W. H., 2, **6**
Gassner, G., 111, **189**
Gates, C. T., 101, **189**
Gauch, H. G., 134(313), 135(19), **183, 197**
Genuth, S. M., 434(127), **610**
Geoghegan, M. J., 171, 172(141), 174, **189**
Gering, R. K., 392(537), **631**
Gerretsen, F. C., 125, **189**
Ghosh, B. P., 405(178), **612**
Gibbs, M., 237, 278, 282, **306**
Gifford, E. M. Jr., 552, **612**
Gilbert, S. G., 112(221), 113(221), 114 (221), 115(221), 129(143, 220) 130(220), **189, 193**
Giri, K. V., 466(407), **624, 632**
Girton, R. E., 357, **375**
Gist, G. R., 92, 93, **189**
Glaser, A. H., 212(13), 221(13), **226**
Glass, B., 414(309), 442(309), **619**
Glasziou, K. T., 452(180), **612**
Gleason, L. S., 36(78), **186**
Glick, D., 535, **612**
Gloor, K., 21(25, 26), **184**
Glueckauf, E., 32(146), 33, 35(145, 146), **189**
Goddard, D. R., 232, 233, 235, 265, 269, 278, 293, **306, 307**
Godlewski, E., 42, **189**
Goeze, G., 111, **189**
Goksöyr, J., 250, **306**
Goldfine, H., 350, **372**
Gonzalves, E. A., 485(270), **617**
Gordon, A. H., 393(114), 453, 535(114), **609**
Gorham, P. R., 97, **186**
Gotaas, H. B., 180(149), **189, 190**
Graber, R. P., 429(388), **623**

Graham, A., 43, **190**
Graham, J., 43, 170(253), 182, **195**
Gray, D. O., 457(158), 471, 473(182), **611, 612**
Green, J. R., 19, **190**
Green, L., 66, 89(120), 95, **188**
Green, L. F., 129(152), **190**
Greenberg, D. M., 451(184), **612**
Greene, R. D., 399, 542(458), **626**
Greenfield, S. S., 104, 105, 106, **190**
Greenstein, J. P., 393, **612**
Gregory, F. G., 107, 108, 109, 137(169), 138, **190**, 275, **306**, 362, **372**, 441, 489, 585, 586, 596(186), **613**
Grew, N. 12, **190**
Grinfeld'd, E. G., 74, **190**
Grobbelaar, N.,* **613**
Gross, J. A., 542(615), **634**
Grotsch, H., 389(439), 432(439), **625**
Grützmacher, H. F., 401(213), **614**
Grunberg-Manago, M., 434, 549(192), **613**
Guggenheim, M., 425, 426, **613**
Gummert, F., 171, 176(158), 180(158), **190**
Gunnes, M., 394(513), **630**
Gustafson, F. G., 270, **307**

H

Hackett, D. P., 279, **306**, 347, **372**
Hackney, F. M. V., 315(58), 332(55, 56, 57, 126), 333(55, 56, 126), 334(58), 338(126), 339(57, 126), **372, 376**
Hagan, R. M., 100(357), **200**
Hageman, E., 434(200), **613**
Hageman, R. H., 600(194), **613**
Hagen, C. E., 297(29), **306**
Hahn, L., 434(148), **611**
Haldane, J. B. S., 33, 34, 35(159), **190**
Hales, Stephen, 12, **190**
Hall, E. G., 332(126), 333(126), 338(126), 339(126), **376**
Halldal, P., 175, **190**
Hamilton, P. B., 397(195), **613**
Hamilton, T. S., 392, **620**
Hanes, C. S., 347(59), **372**, 393(541), 449(199), 473(198, 199), 535, 564(198, 199), **613, 631**

Hansen, A., 19(163), **190**
Hansen, R. G., 434(200), **613**
Hanson, E. A., 521(201), 583(201), **613**
Hanson, J. B., 295(60), **306, 307**
Harada, T., 488(467), **627**
Harder, R., 63, 65, 66, 161, **190**
Harms, H., 82, **190**
Harris, A. E., 37(196), 162(196), **192**
Harris, D. N., 536, **613**
Harris, G., 434(124), **610**
Harris, H., 552, 576, **613**
Hartman, G., 250(63), **307**
Hartley, H., 388(203), **614**
Hartree, E. F., 123, **191**
Harvey-Gibson, R. J., 11, **190**
Hase, E., 171(346), 172(346), 181(346), **199**
Haselkorn, R., **622**
Hasselbring, H., 326(60), **372**
Hauschild, A. H. W., 599(205), **614**
Hawkins, L. A., 326(60), **372**
Haynes, D., 320, **373**
Hayward, R., 400(382), **622**
Hearon, J. Z., 263, **306**
Heath, O. V. S., 51(168, 170), 53, 137, 138, **190**
Heber, V., 543(206, 207), 598(207), **614**
Hegarty, M. P.,* **612, 628**
Heinicke, A. J., 70, 113, 145, 147, 149, 151(172), 156, 157, **191**
Heinze, P. H., 329, **375**
Hellebust, J. A., 368(62), **373**
Heller, L., **611**
Henderson, F. Y., 73, **185**
Henderson, T. R., 432, **614**
Hendler, R. W., 563(209), **614**
Hendricks, R. H., 70(349), 90(349), 118(348a), 119(348a), 120(348a), 121(349), 148(349), 149(349), **199**
Hendricks, S. B., 385(62), 548, 600(210), **607, 614**
Henseleit, K., 416, 427, 475(260), **616**
Heslop-Harrison, J., 432, **614**
Hess, B., 323(41, 63), 349, 350, 351, 359(41), **371, 373**
Hevesy, G., 290, **306**
Hewitt, E. J., 124, **191**, 600(194), **613**

Heyes, J. K., 545, **614**
Heyns, K., 401(213), **614**
Hiesey, W. M., 92(175, 243, 243a), 173 (243), 186(243a), **191, 194**
Hietala, P. K., 473(579), **633**
Hill, G. R., 34, 36(350), 37, 69, 70(350, 73), 89, 90(349), 91, 118(348a), 119(348a), 120(348a), 121(349), 135(350), 147, 148(349), 149(349), 150, 151(350), 152(350), 351, 352), 153, 154, 155, 156(350), 161(351), 162(331), **199**
Hill, R., 123, **191**
Hird, F. J. R., 423(43, 44, 45, 120), 449(198, 199), 473(198, 199), 564 (198, 199), 601(43, 44, 45), **606, 609, 613**
Hoagland, D. R., 121, **183**, 296, 298, **307**, 409(214), 512, **614**
Hoagland, M., 555, **614**
Hoch, G. E., 414(216), **614**
Hodge, A. J., 542(217), **614**
Hoffhine, C. E., 429(388), **623**
Hoffmann, M. B., 113(171), 145, 149, **191**
Hogberg, K., **611**
Holdheide, W., 150(177), **191**
Holley, R. W., 573, **634**
Holly, F. W., 429(388), **623**
Holmes, H. L., 426, **618**
Holmes, R. S., 124(64, 65, 66, 67), 125, **186**
Holm-Hansen, O., 414(275), **617**
Holsten, R. D., 368(119), 580(489a), **628**
Holt, A., 11, 12(178), **191**
Holt, A. S., 84(278), **196**
Holt, S. J., 526, **606**
Honda, S., 240, **307**
Hood, S. L., 128(125), **188**
Hoover, W. H., 64(180), 66(180), 67, 83, 86, **191**
Hoope-Seyler, F., 18, **191**
Horikoshi, K., 469(256), 539(256), **616**
Horowitz, N. H., 428, 430(218), **614, 627**
Horsfall, J. G., 529, **614**
Hotchkiss, R. D., 384(220), **614**
Hotta, Y., 541, **614**
Hover, J. M., 270, **307**

Howard, F. D., 322, **373**
Howe, K. J.,* **628**
Howell, R. W., 343(102), **375**
Huang, R. C., 552(57), 552, **607, 614**
Huber, B., 35, 150(177), 159(183), **191**
Hübsch, H., 390, **614**
Huelin, F. E., 329(65), **373**
Huennekins, F. M., **184**
Hulme, A. C.,* 315(72), 317(70, 71), 320(69, 70, 71), 321(67), 323(78), 324(73), 329(66), **373, 374, 615, 616, 628**
Humphreys, T. E., 134(104), **187**
Humphries, E. C., 390(227), **615**
Hurwitz, J., 567(227a), **615**
Hyde, J. C., 22(292), **196**

I

Ikawa, M., 392, **615**
Imai, K., 158(186), 211(10), 212(10), 213(10), **226**
Ingen-Houz, J., 13, 14, **191**
Ingram, V. M., 536(229, 230), 560(230), **615**
Inove, E., 158, **191**, 210(11), 211(10, 11), 212(10, 11, 12), 213, 214(11, 12), 215(12), 216, 225, **226**
Isherwood, F. A., 421(231), 445, 449(198, 199), 473(198, 199) 564(198, 199), **613, 615**
Isobe, S., 158(186), 211(10), 212(10), 213(10), **226**
Iwamura, T., 171(346), 172(346) 181(346), **199**

J

Jaaback, G., 425(121), **610**
Jacobson, L., 121, **191**, 302, **307**
Jacob, F., 385(232, 233, 331), 421(232, 233), 574(232, 233), **615, 620**
Jacob, K. D., 404, 405, **615**
Jagendorf, A. T., 600, **619**
James, A. L., 252, 259, 261, **307**
James, W. O., 232, 236, 237(50), 238, 241, 247, 251, 252(51), 253 259(51), 261, 265, 274, 277, 278, 285, 286, **307**, 347, 358, 364(76), **373**, 498(235), **615**
Jensen, W. A., 293, **307, 615**

Johnson, D. E., 396, **615**
Johnson, J. E., 389(162), 432(162), **611**
Johnson, S. W., 403, 418(238), **615**
Johnston, E. S., 64(180), 66(180), 67, 72, **191**
Johnston, J. A., 28, 31, 32, 59(189), **191**, **195**
Johnstone, G. R., 326(77), **373**
Jones, C. H., 72, **187**
Jones, J. D., 323, 324(73), **373**
Jones, L. H., 128, **191**
Jones, M. E., 416, 452(240, 241), **615**
Jones, O. W., 551(363), 571(298, 363), **618, 621**
Jones, R. L., 421(231), 445, **615**
Jones, W., 433, 534, **616**
Jones, W. W., 121, 122(284), **196**
Jorgensen, J., 170, **192**
Jost, Ludwig, 2, **6**
Jung, J., 121, **197**

K

Kamen, M. D., 22(192, 292), **192, 196**
Kanazawa, T., 182(193), **192**
Kandler, O., 267, **307**
Kari, S., 471(580), 489, **633**
Katunskii, V. M., 73(195), **192**
Katz, A. M., 535, **616**
Katz, E., 22(375), 85(375), 101(375), 167(375), 168(375), **200**
Katz, M., 37, 162, **192**
Kaur, R., 531, **612**
Keil, B., 536, **616**
Keilin, D., 129(197), **192**
Kelley, W. C., 133, 134, **192**
Kennedy, S. R., Jr., 117, **192**
Kennedy, W. K., 408(118), **609**
Kent, A. E., 368(119), **375**, 580(489a), **628**
Kenworthy, A. L., 100(3), **183**
Kenyon, A. E., 413(516), 453(515), **630**
Kephart, J. E., 295, **310**
Kessler, B., 533(248), **616**
Kessler, E., 128, **192**, 600(249), **616**
Keston, A. S., **306**
Ketchum, B. H., 169, **192**
Khan, M. A. A., 323(7), 324(7), 359(7), **370**

Kidd, F., 252, 268, 270, **307**, 313(80), 314, 315(82, 83, 84), 316, 317, 320, 329(127), 330, 333(81, 85), **373, 374, 376**, 504(250), **616**
Kimball, A. P., 389(376), **622**
King, C. G., 129(152), **190**
King, F. E., 470, **616**
King, K. W., 442(608), **634**
King, T. J., 470(251), **616**
Kirk, M., 598(457), **626**
Kirk, P. L., 390(253), 391, 397(253), 398(252), **616**
Kjaer, A., 435, **616**
Klechkowski, K., 477(254), **616**
Kluyver, A. J., 5, **6**, 22, **192**
Kobayashi, G., 395(255), **616**
Koffler, H., 469(256), 539(256), **616**
Kok, B., 57, 58, 88(372), 171(372), 172(372), 174(372), 175(205), **192, 198, 200**
Konishi, C., 170(377), 179(377), **201**
Kopp, C., **195**
Koritz, H. G., 166, **192**
Kornberg, A., 560, **616**
Koshland, D. E., Jr., **616**
Kostytschew, S. P., 28, 30, 31, 155(208), **192**
Kramer, P. J., 69, 70, **192**
Kraus, E. J., 533, **616**
Krauss, B. H., 413(448, 449), **626**
Kraybill, H. R., 533, **616**
Krebs, H. A., 416, 427, 440, 441, 475(260,261), 584(261), **616, 617**
Kretovitch, W. L., 477(254), **616**
Krotkov, G., 277, **310**, 543(455), 599(205), **614, 626**
Kuchel, R. H., 498(619), **635**
Kuijper, J., 262, 263, **307**
Kulayevo, O., 289(71), **308**
Kun, E., 488, **617**
Kunttu, H., 445(585), **633**
Kunzler, H., 21(26), **184**
Kursanov, A. L., 74, 155(210), **192, 193**
Kwong, S. S., 118(126), **188**

L

Lagerkvist, V., 573(35), **606**
Laine, E., 33(249), **194**

Laine, T., 442(583), 443, 459, **633**
Lal, K. N., 97, 112, **197**
Lamport, D. T. A., 496(263), 539(263), **617**
Lang, A., 532, **624**
Langton, R., 76(286), **196**
Laties, G. G., 312(87), 329, 342(87, 89), 343(87, 88, 89), 353, **374**
Lavoisier, 12, **183**
Leach, W., 232(98), 247, 258, 259(98), 260, 267, **307, 309**
Leavenworth, C. S., 279(106), **309**, 392(378, 379, 567), 396(570), 397(570), 408(573), 453(570), 498(572, 573), **622, 632, 633**
Lederer, E., 393(264), **617**
Ledingham, G. A., 37(196), 162(196), **192**
Leech, J. H., 537(605), 541(605), 560(605), **634**
Legendre, R., 32(211), 33, 35(211), **193**
Leggett, J. E., 297(30), **306**
Lehninger, Albert, L., 5, **6**
Lemon, E. R., 34(247), 36(247), 135(247), 139(212), 156, 159(247), **193, 194,** 204(15), 205(14, 41), 208(2), 210(14, 30), 212(12, 13, 14, 15), 213, 214(12, 14, 15), 25(3, 12), 216(3, 30), 219(3), 1(3, 13), 222(3), **225, 226, 227**
Lengyel, P., 571(265, 266, 468, 469), **617, 627**
Leonard, E. R., 330(132, 133, 134, 135), 332(135), 333(132, 135), 338, 339(133, 135), **374, 376**
Leonard, M. J., 442, **617**
Lett, J. T., 570, **617**
Lettau, H. H., 211(16), **226**
Lewis, C. M., 23(121), 58, **188**
Lewis, E. J., 485(270), **617**
Liebig, J., 17, 19, 25, **193**, 437(271), **617**
Lindberg, B., 488(272, 273), **617**
Linderstrøm-Lang, K., **306**, 400(274), **617**
Lindner, E. B., 536(400), **623**
Ling, S. C., 214(33), **227**
Linko, P., 414, **617**

Linkola, H., 405(584), **633**
Linkous, W. N., 122(275), **196**
Linnemann, G., 417, **625**
Linsbauer, K., 52, **193**
Linstead, R. P., 21, **193**
Lipmann, F., 448, 452(240, 241), 558(277), **615, 617**
Lloyd, F. E., 52, **193**
Loftfield, J. V. G., 52, **193**, 284, **307**
Loening, U. E., 549, **617**
Loew, O., 433, **617**
Long, I. F., 215, **227**
Loomis, W. D., 451(126, 281, 518), 452, **610, 617, 630**
Loomis, W. E., 34, 36(78, 79, 365), 64(79), 66(79), 69(79, 365), 70, 135(79), **186, 200**
Loos, G. M., 597, **631**
Loustalot, A. J., 100, 112, 113, 114, 115, 129, 130, **193**
Love, J. R., 206(35), **227**
Lovett, J. S., 524, **618**
Lowenstein, J. M., 440, 441, 475(261), 584(261), **617**
Lowry, O. H., 397, **618**
Ludwig, H. F., 180(149), **190**
Lugg, J. W. H., 420, 535, 542(284), **618**
Lund, H. A., 295(60), **307**
Lundegårdh, H., 72, **193**, 296(62), **307**
Lyndon, R. F.,* **618**
Lynen, F., 250, **307**
Lyons, J. M., 314(92), 319(92), 331, 332(92), **374**
Lyttleton, J. W., 543, 599, **618**
Lyuborskaya, J. J., 432(97), **608**

M

McAlister, E. D., 89, 102, 150, **193**,
McCalla, A. G., 535, **619**
McCarthy, J. F., 129(152), **190**
McCay, C. M., 387(307), **619**
McCoy, R. H., 459, **619**
MacDonald, I. R., 272, 278, **307**
McElroy, W. D., 125(225), **193**, 414(309), 442(309), **619**
McFadden, B. A., 598(310), **619**
McGee, F. M., 441, 585, **627**

McGlasson, W. B., 314(92, 98), 319(92), 331(92), 332(92), **374**

Machlis, L., 291, **307**

McIlrath, W. J., 133, **197**

McKee, H. S., 386(314), 418(314), 419(314), 426(314), 436(312, 313, 314), 452, 497(313, 314), 542(314), **619**

Mackinney, G., 124, **183**

Maclachlan, G. A., 53(306, 307), **197**

McLean, J. D., 542(217), **614**

McLean, R. C., 40, **193**

McLeish, J., 552, **619**

McQuillen, K., 538(316), 555, **619**

McVehil, 211(23), 223(23), **227**

Magee, W. E., 414(288), **618**

Magness, J. R., 330, **374**

Maitra, U., 461, 463(125, 289), **610, 618**

Maizel, J. V., 488(290), 513(290), 602, **618**

Makisumi, S., 429(291), **618**

Malmy, M., 567(227a), **615**

Malpighi, M., 11, **193**

Mangin, L., 28, 29, 41, **185, 193**

Mann, T., 129(197), **192**

Manning, W. M., 23(229), 28, 176(229), **193**

Manske, R. H. F., 426, **618**

Mapes, M. O., 368(119), **375**, 423(499), 432(499), 580(489a), 596(499), **628, 629**

Mapson, L. W., 324(8, 9, 10, 11), 345(95), 347(95), 351(11), 352(8, 9, 11), 353(10, 11), **370, 374**

Maquenne, L., 28(231), 29(231), **193**

Marchant, R. H., 274(75), **308**

Margolis, D.,* **628**

Markert, C., 421(293), 577(293), **618**

Marks, J. P., 321(96), 323, **374**

Marré, E., 348, **374**

Marsh, P. B., 269, 278, **307**

Martin, A. J. P., 392, 393(114), 453(294), 535(114), **609, 618**

Martin, E. M., 561, **618**

Martin, N. H., 536, **618**

Martin, R. G., 571(298), **618**

Mashkovtsev, T. M., 499(456), **626**

Maskell, E. J., 50, 115, **194**, 423(299, 300, 301, 302, 303), 499(299, 300), 513(299, 300, 301, 302, 303), 585(299, 300, 301, 302, 303), 610(301, 302, 303), **618, 619**

Mason, B., 10, **194**

Mason, T. G., 423(299, 300, 301, 302, 303, 304, 305), 499(299, 300), 513(299, 300, 301, 302, 303), 585(299, 300, 301, 302, 303, 304, 305), 601(301, 302, 303, 305), **618, 619**

Mathur, P. B., 326, **375**

Matthaei, G. L. C., 63(32), 88(32), **184**

Matthaei, J. H., 421(550), 551(363), 553(550), 571(298, 363, 550), **618, 621, 632**

Maximov, N. A., 155(235), **194**

Maximov, T. A. K., 155(235), **194**

Mayer, A., 182, **188**

Mayer, A. M., 170(236), 182(236), **194**

Mayer, J. R., 16, 17, **194**

Mazelis, M., 282, **308**

Medvedev, Z. A., 536, 571(318), **619**

Meffert, M. E., 171(158), 176(158), 180(158), **190**

Mego, J. L., 600, **619**

Meister, A., 396(237), 424, 442, 447, 498(237), 535, 555, 558, **615, 619, 620**

Melville, J., 427, **624**

Menke, W., 273, 598(324), 599(324), **308, 620**

Mercer, F. V., 537(325), 541(325), 542, (217), 560(325), **614, 620**

Merget, A., 41, **194**

Meyer, C. E., 459(308), **619**

Meyer, E. F. H., 19, **194**

Michelson, C., **630**

Miettinen, J. K., 445(585), **633**

Milhaud, G., 598(326), **620**

Millar, F. K.,* 297(92), **309, 628**

Miller, E. S., 90, **194**

Miller, E. V., 325(3), 339(3), **369**

Miller, L. P., 125(241), **194**

Miller, R., 159(183), **191**

Miller, S. L., **620**

Millerd, A., 321, **374**
Millet, J., 598(326), **620**
Millikin, C. R., 125(242), **194**
Milner, H. W., 92(175, 243, 243a), 170 (91, 171(91), 172(91), 173, 174 (91), 177, 179, 186(243a), **187, 191, 194, 198**
Milthorpe, F. L., 51(170), **190**
Mims, V., 458(433), **625**
Mirsky, Alfred, E., 4, **6**
Misra, D. K., 215(3), 216, 219(3), 221, 222, **226**
Mitchell, H. H., 392(328), **620**
Mitchell, J. W., 53, 54, **194**
Mitchell, P., 297, **308**
Mitra, J., 423(499), 432(499), 596(499), **629**
Mituya, A., 171(346), 172(346), 181(346), **194, 199**
Moelwyn-Hughes, E. A., 293, **308**
Mohr, H., 385(329), **620**
Molisch, H., 41, **201**
Mollenhauer, H. H., 537(605), 541(605), 560(605), **634**
Mollenhauser, H. N., 295, **310**
Momotani, Y., 491, **620**
Monod, J., 385(232, 233, 331), 421(232, 233), 574(232, 233), **615, 620**
Monteith, J. L., 205(17), 209, 212, 213, 215(17), 216(19), 221(20), 225, **225, 226**
Moore, S., 394(332, 474), 395(332, 464, 473), 400(474), 535(333, 464, 473, 474), **620, 626, 627**
Moore, W. E., 82, **194**
Morel, G., 430, 495(139), **610, 620**
Morgan, C., 545, **620**
Morris, C. J., 392(537), 393, 473(536), 474(536), **631**
Morris, M. P., 406(336), **620**
Morton, R. K., 552, 561, 562, **618, 620**
Moss, D., 34, 36(247), 135(247), 152, 156, 159(247), **194**
Moss, D. N., 213, **226**
Mothes, K., 289, **308**, 415, 417, 423(340), 426(341, 345), 427, 428(413), 429, 430(413), 437(342, 344), 477(342, 413), 498(338, 339), 511(342, 343), 513(342, 413), 532, 342, 343, 543 (338), 601(344), **620, 621, 624**

Mott, G. O., 92, 93, **189**
Moxon, A. I., 408(606), **634**
Mozingo, R., 429(388), **623**
Münch, E., 420(83), **608**
Muirhead, H., 383(390), **623**
Mukohata, Y., 274(75), **308**
Mulder, E. G., 131(248), **194**
Munson, R. D., 404, **621**
Muntz, A., 33, **194**
Murlin, J. R., 260, **308**
Musgrave, R. B., 34(247), 36(247), 135(247), 152, 156, 159(250), **194**
Myers, J., 28, 31, 32, 56(252), 80, 102, 170(91, 251, 253), 171(91), 172(91), 174(91), 182, **187, 193, 194, 195**
Myrbäck, K., 535(520), **630**

N

Naef-Roth, S., 529(177), **612**
Nakamura, H., 599, **621**
Nash, L. K., 401(348), **621**
Nason, A., 125(225), 129(220), 130(220), **193**, 408(349, 350), 412, 600, **611, 621**
Naylor, A. W., 430(23), 432(24), 475, **605, 621**
Neal, G. E., 238, **308**, 357, 362(100), **374, 375**, 434(124), **610**
Needham, A. E., 388(352), **621**
Neelin, E. M., 536(444), **625**
Neelin, J. M., 536(444), **625**
Negelein, W., 585, **634**
Nehring, K., 390, **614**
Nelson, C. D., 599(205), 614
Netter, K. F., 250(63), **307**
Neuberger, A., 391(353), 453, **621**
Neurath, H., 535(354), **621**
Newton, W., 405, **621**
Nichiporovitch, A. A., 139, 140, 143, **195**
Nicholas, D. J. D., 128(257), 129 131(257), **195**
Nickerson, W., 539(356), **621**
Niehei, T., 171(346), 172(346), 181(346), **199**
Niggli, F., 21(27), **184**

Nightingale, G. T., 408(357, 358, 359), 409(357), 422(360), 436(358, 359), 498, **621**

Nirenberg, M. W., 400(361), 421(550), 550(361, 551(361, 363), 553(550), 555, 571(298, 361, 363, 550), 572, **618, 621, 632**

Nishigaki, S., 170(377, 179(377), **201**

Nissen, P., 425(364), 488(364), 513(364), 602, **622**

Nitsch, J. P., 422(510), 460, 491(510), 492(510), **629**

Nixon, R. W., 98, **195**

Nobs, M. A., 92(243), 173(243), **194**

Noddack, W., **195**

Noé, F. F., 473(365), **622**

Nogushi, M., 499(366), **622**

Norris, W. E., 291(8), 292, **305,** 346, **371**

Northcote, D. H., 496(263), 539(263), **617**

Novelli, G. D., 434(127), **610**

Nutman, F. J., 75, **195**

Nyunoya, T., **194**

O

Ochoa, S., 475(367), 555, 571(265, 266, 468, 469), **617, 622, 627**

O'Connor, C. M., 242, **310**

Oddo, B., 123, **195**

Ofengand, J., **622**

Ohmstede, W. D., 214(5), 215(5) 223(5), **226**

Ohmura, T., 343(102), **375**

Olmstead, A. J., 319(28), **371**

Olomuki, A., 395(532), 430(531), 431 (531), **631**

Olsen, K., **306**

Oota, Y., 262, **308,** 542(370, 371), 561, **622**

Oparin, A. I., 388(373), **622**

Orchard, B., 53, **190**

Ordway, D. E., 214(22), **226**

Oró, J., 389(374, 375, 376), **622**

Osawa, S., 542(370, 371), 561, **622**

Osborne, T. B., 392(378, 379), 533, 534, **622, 632**

Oserkowsky, J., 121, **195**

Osterlind, S., 96, **195**

Oswald, W. J., 180(149), **189, 190**

Overholser, E. L., 100(3), **183**

Owen, O., 72, 168, **195**

Owen, P. C., 99, 144, 167(267), **195**

P

Pace, J, 535(380), **622**

Packer, L., 274, **308**

Paech, K., 391, 394, 417, 427, 428(381), 444, **622**

Paetz, K. W., 82, **195**

Pagan, C., 406(336), **620**

Panofsky, H. A., 211(23), 223(23), **227**

Parija, P., 498(48), **606**

Patchett, A. A., 418(497), 423(497), 430(497), 480(497), 539(497), **629**

Pauling, L., 400(382), **622**

Peacocke, A. R., 537, **622**

Pearsall, W. H., 408(386, 387), 434, 436, 498, 583(387), **622, 623**

Pearson, J. A., 315, 316, 317(104), 321, 322, **375**

Peck, R. L., 429(388), **623**

Peel, E. W., 429(388), **623**

Penman, H. L., 215, 221, **227**

Pentzer, W. T., 329, **375**

Perutz, M. F., 383(390), 550(389), 555, 575, 576, **623**

Peters, C. A., 128(190), **191**

Peterson, A. E., 206(35), **227**

Peterson, J. B., 76(286), **196**

Peterson, M. L., 100(357), **200**

Peterson, W. H., 488(622), **635**

Petrie, A. H. K., 440, **635**

Pfeffer, W., 2, 6, 19, 41, **195**

Phillips, D. M. P., 429(391), 536, **623**

Phillis, E., 423(304, 305), 601(305), **619**

Pirie, N. W., 535(393), 598 (392, 393), **623**

Pirson, A., 103, 104, 169, **195**

Platenius, H., 325(107), 326(106), **375**

Pilmmer, R. H. A., 400, **623**

Poel, L. W., 302, **308**

Pogo, A. O., 544(75), **607**

Pollacci, G., 123(272), **196**

Pollard, E. C., **625**

Pollard, J. K.,* 309, 613, 623, 628, 629, 631

Pollman, W., 389(439), 432(439), 625

Porath, J., 536(400), 623

Porter, C. W., 20(273), 196

Porter, H. K., 359, 375

Porter, K. R., 537(401, 402), 541(401, 402), 560(401, 402), 623

Powell, R. G., 531, 624

Pratt, H. K., 214(92, 98), 319(92), 331(92), 332(92), 374

Pratt, J. M., 414(403), 623

Pratt, R., 102, 196

Preston, C., 321(120), 375, 391, 409, 455(500), 493, 495(480), 580(480, 498), 582(498), 588(500), 627 629

Prevot, P., 297, 308

Prianishnikov, D. N., 403, 433a, 436, 437, 438, 623

Price, N. O., 122, 196

Priestley, J., 12(276, 13, 196

Prokofiev, M. A., 434(404), 623

Pucher, G. W., 279(106), 290(107), 309, 310, 396(570), 397(569, 570), 401(571), 408(573), 453(570), 498(572, 573), 585(571), 632, 633

Q

Quastel, J. H., 405(15), 605

Quinlan-Watson, F., 129(277), 196

R

Raacke, I. D., 543, 548, 555, 570(405), 598, 607, 623

Rabideau, G. S., 84, 196

Rabinowitch, E. I., 19, 21, 22(279), 23, 25, 63, 65, 123, 196

Rabson, R.,* 624, 628

Radhakrishnan, A. N., 466(407), 624

Raison, J. K., 552, 562, 620

Ram, A., 21(102), 187

Ram, H. Y. M., 503(408), 504, 624

Ram, M., 503(408), 504(408), 624

Ramamurti, T. K., 495(480), 580(480), 627

Ramsperger, H. C., 20(273), 196

Randall, M., 22(292), 196

Ranson, S. L., 268, 280, 283, 356(125), 364(109), 375, 376

Raper, H. S., 588, 624

Ratner, S., 430, 624

Raymond, S., 536(410), 624

Redfield, A. C., 404, 405, 624

Redfield, R. R., 387(10), 604

Reed, G. B., 277, 310

Reed, H. S., 11, 13(281), 129, 196

Rees, M. W., 398(107), 609

Reifer, I., 427, 624

Reifsnyder, W. E., 206, 227

Reinbothe, H., 415, 427, 428(413), 429, 430(413), 477(413), 513(413), 624

Reinhold, L., 531, 624

Reisenauer, H. M., 131, 196

Reith, W. S., 545, 620

Reman, G. H., 22(375), 85(375), 101(375), 167(375), 168(375), 200

Remsen, D. B., 434(86), 521(86), 608

Reuter, G., 429, 624

Reuther, W., 121, 122(319), 127, 196, 198

Reville, R., 32(285), 196

Rhoads, W. A., 167(382), 201

Rhodes, M. J. C., 258, 271, 272, 308, 543, 624

Rhykerd, C. L., 76, 196

Rich, A., 388(416), 389(416), 538(416), 624

Richards, F. J., 107(287), 108, 109, 190, 196, 255, 270, 308, 431(460), 459(417, 460), 624, 626

Richardson, J. A., 356(125), 376

Richmond, A. E., 532, 624

Rider, N. E., 209(26), 221, 227

Riedel, F., 73(289), 196

Riehl, L. A., 167(382), 201

Riley, G. A., 23, 25(291), 196

Rintelen, P., 73, 185

Rittenberg, D., 401(571), 585(571), 633

Rittenberg, O., 290(107), 310

Ritter, A., 214(22), 226

Robbins, W. R., 422(360), 498, 621

Roberts, D. W. A., 270, 308

Roberts, E., 421(419), 480(420), 489(419), 624

Robertson, R. N., 296, **308**, 315(112), 316, 317(104, 112), 321, 322, 332(21, 126), 333(21, 126), 338(21, 126), 339(21, 126), 342(21), **371, 375, 376,** 504(421), **624**

Robinson, E., 545(423), **624**

Roche, J., 395(533), **631**

Rochovansky, O., 430, **624**

Rochow, W. F., 554(397), **623**

Rockland, L. B., 397(438), **625**

Romani, R. J., 314(141), 315(141), 338(141), **376**

Romani, R. S., 322, **375**

Romeike, A., 426, **621**

Rona, P., 392, **624**

Rose, W. C., 459(308), **619**

Rosebrough, N. J., 397(283), **618**

Rosenberg, A. J., 278, **306,** 341, **372**

Rosenberg, B. H., 434(95), 560(95), **608**

Ross, C. W., 432, **624**

Ross, J. P., 486(426), 530(426), **625**

Rossmann, M. G., 383(390), **623**

Rowan, K. S., 250, **308**

Ruben, S., 22, **196**

Ruck, H. C., 126, **197**

Ruhland, W., 232, **308,** 386(427, 428, 429), 408(428, 429), **625**

Runeckles, V. C., 359, **375**

Russell, E. J., 402, 418(430), **625**

Ryan, F. J., 384(432), 385(432), 421(432), 577(432), **625**

Rydon, H. N., 395, **625**

Ryther, J. H., 26, **197**

S

Saeki, T., 206, **227**

Sager, R., 384(432), 385(432), 421(432), 577(432), **625**

Sallach, H. J., 415, 416, 442, 451(113), **609**

Sandegren, E., 542(77), **608**

Sanger, F., 391(353), 453, **621**

Sasa, T., 171(346), 172(346), 181(346), 182(193), **192, 199**

Satterwhite, L. E., 212(13), 221(13), **226**

Scarth, G. W., 52, 54, 82, 150, **197**

Schaffner, G., 529(621), **635**

Schaloo, L., 170, 608

Schales, S. S., 458(433), **625**

Scharrer, K. 121, **197**

Schenk, W., 37, 38, **188,** 471(434), **625**

Schiff, J. A., 542(146), **611**

Schleiden, M. J., 1, **6**

Schmalfuss, K., 404, **625**

Schmidt, C. L. A., 396(436), 424, 436(574), 452, **625, 633**

Schmidt, W., 158, **197**

Schmucker, T., 417, **625**

Schneider, G. W., 99, **197**

Schneider, K. C., 414(216), **614**

Schoenheimer, R., 290(107), **310,** 401(571), 585(571), **633**

Schott, H. F., 397(438), **625**

Schramm, G., 389(439), 432, **625**

Schroeder, H., 23, 25, **197**

Schrödinger, E., 246, **308**

Schütte, H. R., 471(434), **625**

Schwab, G., 409, **625**

Schwabe, W. W., 109, 110, 111, **197**

Schwann, T., 1, 2, **6**

Schweet, R., 555, **625**

Scott, S. J., 396(237), 498(237), **615**

Seaman, D. E., 250(84), **308**

Seibl, J., 401(40), **606**

Sell, H. M., 129(143), **189**

Semenza, G., 536, **625**

Sen, P. K., 275, **306,** 362, **372,** 441, 498, 585, 586, 596(186), **613**

Senebier, J., 14(304, 305), **197**

Setlow, R. B., **625**

Setterfield, G., 536(444), **625**

Shantz, E. M., 417, 423(499), 432(499), 596(499), **626, 629**

Shapiro, H. S., 434, 549(76), **608**

Shaw, M., 52, 53(306, 307), 54(297), 82, 150, **197**

Shaw, R. H., 206(7), **226**

Shepardson, W. B., 128(190), **191**

Sheppard, P. A., 211(29), **227**

Shibata, K., 171(346), 172(346), 181(346), **199**

Shimizu, T., 69, 70, **197**

Shirley, H. L., 148, **197**

Shive, J. W., 124, 127, **198,** 408(110), 409(110), **609**

Shmuk, A. A., 422(446), **626**

Shooter, K. V., 537, **626**

Shuegraf, A., 250(63), **307**

Sideris, C. P., 413(449), **626**
Sigel, A., 421(450), **626**
Sierp, H., 82, **197**
Sievers, E. G., 27, **197**
Simon, E. W., 384, **375**, 459, **626**
Simonsen, D. G., 480(420), **624**
Simpson, M. V., 555, **626**
Singer, B., 538(164), 553(164), **611**
Singh, B. H., 97, 112, **197**
Singh, B. N., 270, **308**, 326, **375**
Sisler, E. C., 134, **197**
Sissakian, N. M., 543(453), **626**
Skinner, C. G., 432(208), **614**
Skok, J., 133, **197**
Slater, W. G., 238, **307**, 364(70), **373**
Slatyer, R. O., 215(3), 216(3), 219(3), 221(3), 222(3), **226**
Small, T., 72(265), 168, **195**
Smillie, R. M., 543(455), 544(454), 598, 599(454), **626**
Smirnov, A. I., 499(456), **626**
Smith, A. M., 64, 66, 67, 160, **184**
Smith, D. C., 598(457), **626**
Smith, E. L., 66(315, 316), 161, **198**, 399, 542(458), **626**
Smith, G. L., 326(4), **369**
Smith, I., 393(459), **626**
Smith, J. H. C., 93(318), 94(318), 170(91), 171(91), 172(91), 174(91), **187, 198**
Smith, P. F., 122, **198**
Smith, P. W. G., 395, **625**
Smith, T. A., 431(460), 459(460), **626**
Smith, W. H., 314(116), 315(116), **375**, 504(226), **615**
Smithies, O., 536(461, 462), **626**
Snell, E. E., 392, **615**
Sober, H. A., 442(323), **620**
Sokolov, V. S., 426, **626**
Solomos, T., 319, 323(7, 16), 324(7), 359(7), **370**
Somers, G. F., 133(198), 134(198), **192**, 275, 276, 279, **309, 310**
Somers, I. I., 124, 127, **198**
Sondheimer, E., 463, **623**
Spackman, D. H., 394(332), 395(332, 464), 535(332, 464), **620, 626**
Specht, A. W., 122(319), **198**
Specht, A. W. II., 124(66), **186**

Speck, J. F., 447(465), 526(466), 582(466), **626, 627**
Spector, L., 416, 452(240, 241), **615**
Spence, D. A., 214(22), **226**
Spencer, B., 488(467), **627**
Speyer, J. F., 571(265, 266, 468, 469), **617, 627**
Spiegelman, S., 553, **610**
Spirin, A. S., 549(31), **605**
Spoehr, H. A., 19, 44, 170(91), 171(91), 172(91), 174(91), 177, 179, **187, 198**, 441, 585, **627**
Spragg, S. P., 247, 248, 249, 250, 258(89), 263, 264, **309**
Spruit, C. J. P., 58, **198**
Srb, A. M., 428, 430(218), 492(101), 536(101), 545(101), **609, 614, 627**
Stålfelt, M. G., **198**, 258, **309**
Stahl, E., 41, **198**
Stanley, W., 421(165), 538(165), 553, **612, 627**
Stauffer, J. F., 23(229), 28(229), 176(229), **193**
Steemann-Nielsen, E., 23(326, 334), 25(332, 333, 334), 55(326), 58, 66, 81, 96(329), 97(329), 102(331), 161(328, 329), **198**
Stefan, J., 43, **198**
Stein, W. H., 394(332, 474), 395(332, 464, 473), 400(474), 535(332, 333, 464, 473, 474), **620, 626, 627**
Steinberg, R. A., 529(475), **627**
Stent, G. S., 565(4a), **604**
Stephenson, M. L., 543(476), 598, **627**
Stepka, W.,* **610, 627**
Stern, A., 18(128), **188**
Stern, W. R., 215(3), 216(3), 219(3), 221(3), 222(3), **226**
Stetten, D., 278, **305**
Stetten, M. R., 496(478), **627**
Steward, F. C.,* 4, **6**, 291, 296(95), 297(92, 95), **308, 309**, 321(120), 362, 363(29, 118), 366(118), 368(29, 118, 119), **371, 375, 605, 606, 608, 609, 610, 611, 612, 613, 618, 623, 624, 626, 627, 628, 629, 630, 631, 635**
Stewart, W. D., 82, **183**
Stiles, W., 19, **198**, 232, 247, 248, 258(98), 259(98), 260, **309**, 340, **375**

Stiller, M., 513(512), **630**
Stiller, M. L., 283, **305**
Stinson, H. T., Jr., 542(638), 543, **636**
Stocker, O., 34, 39(338), 150(177, 339), **191, 199**
Stokes, J. L., 394(513, **630**
Stoll, A., 18(397, 398, 399), 19, 28, 30, **202**
Stoller, J., 210(30), 216(30), **227**
Stolwijk, J. A. J., 74, 85(374), 86(340, 341, 373), 87(373), **199, 200**
Stout, M., 325, **375**
Stout, P. R., 131(9), **183,** 455(500), 588(500), **629**
Stout, R. R., 321(120), **375**
Stoy, V., 85, 86, 115, **199**
Stratmann, H., 171(158), 176(158), 180(158), **190**
Strauss, E., 392, **624**
Street, H. E.,* **309, 629, 630**
Strogonova, L. E., 139, 140, 143, **195**
Stumpf, P. K., 250, **309,** 451(126, 281, 518), **610, 617, 630**
Suess, H. E., 32(285), **196**
Sugahara, K., 599(103), **609**
Sumner, J. B., 535(519, 520), 542, **630**
Sunobe, Y., 262, **308**
Sutcliffe, J. F., 291, 296(95), 297(95), **309,** 512(503), 527(521), 561(521), **629, 630**
Sutton, O. G., 204, **227**
Swinbank, W. C., 211(32), **227**
Sykes, S. M., 332(126), 333(126), 338(126), 339(126), **376**
Synge, R. L. M., 392(522), 396(94), 453(294), 536, **608, 618, 630**
Szeicz, G., 209 **226**

T

Tagawa, K., 600, 601(525), **630**
Tager, J. M., 324(123), **376**
Takahashi, H., 408(305), 409, 410, 412(526), **621, 630**
Takahashi, W. N., 451(526), 530, 570, **617, 627**
Takebe, I., 488(527), **630**
Talley, E. A., **635**
Tamaki, E., 499(366), **622**

Tamiya, H., 169(344, 345), 171, 172(346), 174(344, 345), 181(346), **194, 199**
Tamiya, N., **612**
Tan, H. S., 214(22, 33), **226**
Tani, N., 158(186), **191,** 211(10), 212(10), 213(10), **226**
Taniguchi, S., 409(526), 410(526), 412(526), 451(526), **630**
Tanner, C. B., 206, 210(36), 225, **227**
Tanner, H. A., 128(125), **188,** 220(34), **227**
Tarnanen, J., 438, **633**
Tatum, 558, **605**
Taylor, D. L., 251, **309**
Taylor, J. H., 559, **630**
Tepper, H. B., 552, **612**
Than-tun, 414, **611**
Thimann, K. V., 74, **199,** 531, 543(476), 597, 598(476), **609, 627, 631**
Thomas, J. B., 598(534), **631**
Thomas, M., 280(101), 281(101), **308, 309,** 329(124), 356, 364(109), **375, 376**
Thomas, M. D., 34, 36(350), 37, 69, 70(350), 73, 89, 90, 91, 118, 119, 120, 121, 135(347, 350), 147, 148, 149, 150, 151(350), 152(350, 351, 352), 153, 154, 155, 156(350), 161(351), 162(348, 351), 163, 164, 165(351), 166, **199**
Thomas, W. D. E., 129, **195**
Thompson, J. F.,* **629, 631**
Thorne, G. N., 141, 142, **199**
Thorpe, T. E., 33, **199**
Tice, S. V., 442(323), **620**
Tiffin, L. O., 124(67), **186**
Tigane, E., 393, 535(541), **631**
Timmons, C. J., 21(215), **193**
Tio, M. A., 60(98), **187**
Todd, A., 538(542, 543, 544), **631**
Todd, G. W., 165(354), **199**
Todd, M., 102(356), **200**
Tolbert, N. E., 488(290), 513(290), 528(544a), 602(290), **618, 631**
Tonzig, S., 34, 38, 39(355), **200**
Towers, G. H. N., 445, 447, **631**
Tracey, M. V., 391, 394, 417, 427, 428(381), 444, **622**
Trease, G. E., 424, **630**

Trelease, S. F., 102(274), **187, 196**
Trout, S. A., 329(127), 332(126) 333(126), 338(126), 339(126), **376**
Tsai, C. S., 457(151), **611**
Tschesnokow, W., 155(208), **192**
Tschiersch, B., 429, **631**
T'so, P. O. P., 537(548), 538(548), 549(548), 555, 561(548), **631**
Tsugita, A., 421(549, 550), 541(549), 553(550), 555, 571(549, 550), **632**
Tsujimoto, H. Y., 601(525), **630**
Tsuno, Y., 69, 70, **197**
Tucker, T. C., 180, **187**
Turner, B. L., 429(42), **606**
Turner, D. H., 473(536), 474(536), **630**
Turner, J. S., 102, **200**, 242, 250(84), 267(102), **308, 309**, 315(112), 317, **375**, 504(421), **624**
Tze-Fei Wong, J., 393(541), 535(541), **631**

U

Uchijima, Z., 206, 216, 221(38), **227**
Ulrich, A., 301, **309**
Ulrich, R., 312(128), **376**
Upchurch, R. P., 100, **200**
Urey, H. C., 388(551), **620, 632**

V

Vaidyanathan, C. S., **632**
Van der Honert, T. H., 64(358), 66(358), **200**
Van der Paauw, F., 28, **200**
Van der Veen, R., 57, **200**
Van Niel, C. B., 5, **6**, 10, 22, 59, **200**, 513(553), 598(553), 600, **632**
Van Oorschot, J. P. L., 88(372), 171(372), 172(372), 174(372), 178, **200**
Van Slyke, D. D., 397(195), **613**
van Thoai, N., 395(533), 429(531), 430(531, 532), 431(531, 532), **631**
Varner, J. E., 312(129), 321(96), 323(96), **374, 376**
Vatter, A E., 295(60), **307**
Vegotsky, A., 389(162), 432(162), **611**
Vennesland, B., 282, 599(347), **308, 309, 621**

Verdeil, F., 18, **200**
Verdin, J., 34, **36**(365), 45(363), 47(364), 69(365), 70, 145, 146, **200**
Vermeulen, D., 22(375), 85(375), 101(375), 167(375), 168(375), **200**
Vernadskii, V. I., 10, **200**
Vetter, W., 396(41), **606**
Vickery, H. B., 279, 281, 290, **309, 310**, 379, 386(564), 388(564), 390 (563), 392(554, 555, 556, 557, 567, 575), 396(570), 397 (560, 561, 569, 570), 398(559, 561, 568), 399(561), 400(561, 565), 401(558, 571), 408, 422, 424, 436 (559–561, 574), 452, 453(570), 459, 498(572, 573), 535, 541(559, 561), 585, **632 633**
Viets, F. G., Jr., 130(38, 367), **184, 200**
Vieweg, G. H., 150, **199**
Vines, S. H., 19, **200**
Vinson, C. G., 392(575), **633**
Virtanen, A. I., 405(576, 584), 436, 438, 439, 442(577, 578, 583), 443, 445(577, 585), 452, 459, 471, 473(579), 489, **633**
Vittorio, P. V., 277, **310**
Vogel, H. J., 487, **633**
von der Decken, A., 543(588), **633**
von Euler, H., 434, **611**
von Kamienski, E. S., 426(244), **616**
von Sachs, J., 11, 19, 40, 41, **197**
von Wettstein, D., 542(604), **634**

W

Waddington, C. H., 574, **633**
Wade, E. H. M., 393(541), 535(541), **631**
Wadsworth, R. M., 145, **201**
Waelsch, H., 474(590), **633**
Wager, H. G., 337, **376**
Waggoner, P. E., 206, **227**
Wailes P., 472, **633**
Wakeman, A. J., 279(106), **309** 392(378, 379), 408(573), 498(572, 573), **622, 633**
Wald, G., 381, 384(593), 388(592, 593), **634**
Walker, D. A., 268(79), 274, 280, 282, **308, 310**

Walker, D. R., 118(126), **188**
Waller, J. C., 107(369), **200**
Walton, B. P., **609**
Warburg, O., 22(370), 55(370), 176(370), **200**, 585, **634**
Wardlaw, C. W., 330(131, 132, 133, 134, 135, 332(135), 333(132, 135), 335, 339(132, 135), **376**
Warington, K., 133, **200**
Waris, H., 405(595), **634**
Warmke, H. E., 406(336), **620**
Warren, L., 434, **634**
Warrington, P. M., 274, **310**
Warwick, A. J., 470(251), **616**
Wassink, E. C., 22(375), 85(374, 375), 86(373), 87, 88, 101, 167, 168, 171, 172(372), 174, **200**
Watanabe, A., 170(377), 179(376, 377), **201**
Watson, D. J., 137(378), 139(380) 141(378, 379, 380), 142(379), 143, **199, 201**, 355, **376**
Watson, G. M., 413(516), 453(515), **630**
Watson, J. D., 538(597, 598), 549(597), 550(597), 555, 560, 567, **634**
Webb, E., 388(131), **610**
Webb, J. A., 472(160), **611**
Webster, G. L., 59(57), **185**, 436(600), 452, 555, 567, 570, **634**
Wedding, R. T., 98, 167(382), **195, 201**
Weeks, D. C., 296(83), **308**
Weigl, J. W., 274, **310**
Weintraub, R. L., 55(383), **201**
Weis, D. S., 274, **305**
Weisblum, B., 573, **634**
Went, F. W., 91(385), 92, 166, **192, 201**
Wessel, G., 131(10), 132, **183**
West, C., 252, 270, **307**, 313, 314, 315(82, 83, 84), 316, 317, 320, 330, 333(81, 85), **373, 374**, 504(250), **616**
Westall, R. G., 386(603), 397(108, 569), 453(569), **609, 632, 534**
Wetmore, R. H., 422(509, 510), 460, 491(509, 510), 492(509, 510), **629**
Wetzel, K., 386(427, 428, 429), 408(427, 428, 429), **625**
Whaley, W. G., 295, **310**, 537(605), 541(605), 560(605), **634**

Whatley, F. R., 120(12), 122, **183**
White, F. D., 429(20). **605**
Whitehead, E. L., 408(606), **634**
Whitehead, T., 325, **376**
Whiting, G. C., 337, 348, **371**, 587, **607**
Whiting, H. C., 472, **633**
Whitman, S. L., 567, **634**
Whittingham, C. P., 22(386), 58, 95, 185, **201**
Wiederspahn, F. E., 127(388), **201**
Wiesner, J., 11, 13(391), 41(391), **201**
Wildman, S. G., 421(450), **626**
Wilkins, M. H. F., 550(607), 552(637), **634, 636**
Wilkins, M. J., 296(83), **308**
Will, G., 383(390), **623**
Williams, E. F., 398(107), **609**
Williams, P. H., 72(265), 168, **195**
Williams, R. C., 538(164), 553(164), **611, 612**
Williams, R. F., 136, **201**
Williams, W. T., 53, **201**
Willis, A. J., 253(32), 254(32), 257(32), 298(121), 299, 300, **306, 310**
Willstätter, R., 18(397, 398, 399), 19, 28, 30, **202**
Wilson, C. C., 52, 53(394), 97, **201**
Wilson, D. G., 425(609), 442, **634**
Wilson, G. L., 143(35), **184**
Wilson, J. W., 145, **201**
Wilson, L. G., 488(609), **634**
Wilson, P., 402(610), 413(610, 611), 422(611), 436(610), **634**
Wiltshire, G. H., 290 **305**
Winitz, M., 393, **612**
Winterstein, E., 392, **634**
Wishkich, J. T., 322, 323, 348, 349, **376**
Witkop, B., 418(497), 423(497), 430(497), 480(497), 539(497), **629**
Wolken, J. J., 542(615), 543(614), **634**
Wollgiehn, R., 532, 599(67), **607, 635**
Wolstenholme, G. E. W., 242, **310**
Wood, J. G., 436, 440, 498, 521(26, 201, 618), 583(26, 201, 618), **605, 613, 635**
Woodward, R. B., 18(400), **202**
Woolley, D. W., 488(622), 529(621), **635**
Woolley, J. T., 341, 343, **376**

Wooltorton, L. S. C., 323(78), 324(73), **373**
Workman, M., 342(17), **370**
Wright, B. E., 597, **635**
Wright, D. E., 405(624), 423(624), **635**
Wright, J. L., 216, **227**
Wycoff, R. G., 535(625), **635**

Y

Yamaguchi, M., 322, **373**
Yanofsky, C., 536, **635**
Yemm, E. W.,* 247, 248, 249, 250, 253(32), 254(32), 256(34), 257(32, 118, 119), 258(89), 260(115), 263, 264, 271, 272, 275(120), 276, 277, 286(115), 289(116), 290(117), 298(121), 299, 300, 301, **306, 308, 309, 310,** 362(118), 363(118), 366(118), 368(118), **375, 624, 627, 635**

Yocum, C. S., 205(2, 41), 208(2), **225,** 414(630), **635**
Young, H. Y., 413(448, 449), **626**
Young, R. E., 312(27), 313(26), 314(141), 315(26, 141), 318(27), 319(28), 322(139), 323(139), 338(141), 344(26), **371, 376**
Youngson, M. A., 536, **630**
Yuhara, T., 182(193), **192**

Z

Zacharius, R. M.,* **613, 628, 629, 635**
Zamecnik, P. C., 543(476), 555, 598(476), **627, 636**
Zechmeister, L., 392, **636**
Zeevaart, J. A. D., 431(58), **607**
Ziegler, M. R., 407(8), **604**
Zimmerman, M. H., 601(635), **636**
Zubay, G., 552, 567, 568, **636**
Zucker, M., 542(638), 543, **636**
Zweig, G., 393(51), **606**

INDEX TO PLANT NAMES

Numbers in this index designate the pages on which reference is made, in the text, to the plant in question. No reference is made in the index to plant names included in the titles that appear in the reference lists. In general, where a plant has been referred to in the text sometimes by common name, sometimes by its scientific name, all such references are listed in the index after the scientific name; cross reference is made, under the common name, to this scientific name. However, in a few instances when a common name as used cannot be referred with certainty to a particular species, the page numbers follow the common name.

A

Abies (fir), 399, 640
Abies concolor (white fir), 98
Abrus precatorius, 667
Abutilon, 100
Acacia, 471, 672
Acacia farnesiana, 671
Acacia pentadenia, 662
Acacia willardiana, 664
Acer (maple), 436, 437
Acer negundo, 499
Aceraceae, 491
Acetabularia, 543, 576
Achillea, 30
Actinomycetes, 472
Adiantum pedatum (maidenhair fern), 422, 447, 460, 487, 492, 646, 684
Aesculus, 29, 423
Afrormosia elata, 668
Agaricus, 678
Agaricus hortensis, 678
Agave, 487
Agrostemma githago, 657
Akee (see *Blighia sapida*)
Albizia julibrissin, 658, 672
Alder (see *Alnus*)
Aleurites fordii (tung), 112–115, 121, 127, 129, 130
Alfalfa (see *Medicago sativa*)
Algae, 26, 28, 30, 56–58, 67, 78, 95–97, 103, 133, 168–170, 172–182, 274, 418, 429, 481, 485, 529, 590
Algae, blue-green, 59, 170, 176, 179, 413, 414
Algae, brown, 85, 474

Algae, green, 13, 103, 104, 487, 596
Algae, red, 488, 671, 673
Allium cepa (onion), 291–293, 432, 494, 672–674, 676–678, 680
Allium sativum (garlic), 672–674, 677, 678
Allium schoenoprasum, 677
Allium ursinum, 672
Alnus (alder), 414, 649, 658
Alnus glutinosa (alder), 419
Amanita phalloides, 656
Amaranthus, 407
Anabaena, 59, 60, 170
Anabaena cylindrica, 414
Anacharis (see *Elodea*)
Ananas comosus (pineapple), 84, 314, 418
Andira, 667
Ankistrodesmus braunii, 128
Apple (see *Malus sylvestris*)
Apricot (see *Prunus armeniaca*)
Arachis hypogaea (peanut), 446, 472, 473, 487, 645–647, 666, 683
Areca catechu (see *A. cathecu*)
Areca cathecu (*A. catechu*), 663
Artichoke, Jerusalem (see *Helianthus tuberosus*)
Arum, 459
Asparagus officinalis, 424, 639, 672
Aspergillus, 647, 663
Asplenium nidus, 649
Asplenium septentrionale, 647, 654, 681, 686
Aster tripolium, 675
Astragalus bisulcatus, 674
Atriplex hortensis (orache), 17

Avena sativa (oats), 39, 47, 48, 73, 74,
   125, 126, 407, 430, 442, 638
Avocado (see *Persea*)
Azolla, 487
Azotobacter, 133, 293, 651
Azotobacter vinelandii, 548

B

Bacillus circulans, 469, 648
Bacteria, 41, 59, 131, 170, 384, 403, 409,
   414, 434, 487, 529, 560, 563, 565,
   567, 568, 571, 575, 643, 645, 650,
   671, 672, 682
Bacteria, green, 22
Bacteria, purple, 19, 22, 167
Baikiaea plurijuga (Rhodesian teak),
   470, 477, 479, 663
Banana (see *Musa acuminata*)
Barley (see *Hordeum vulgare*)
Basidiomycetes, 481, 674
Bean, 130, 135, 440, 527, 599
Bean, castor (see *Ricinus communis*)
Bean, djenkol (see *Pithecellobium
   lobatum*)
Bean, jack (see *Canavalia ensiformis*)
Bean, lima (see *Phaseolus limensis*)
Bean, mung, 685
Bean, runner (see *Phaseolus coccineus*)
Bean, soy (see *Glycine max*)
Beet, sugar beet (see *Beta vulgaris*)
Begonia, 54
Begonia semperflorens, 409
Beta maritima (see *Beta vulgaris*)
Beta vulgaris (*B. maritima*, beet, sugar
   beet, mangold), 69–71, 74, 77, 90,
   91, 99, 137–139, 142–144, 147–149,
   156, 157, 209, 325, 424, 434, 638–640,
   667
Beta vulgaris var. *cicla* (chard), 129, 498
Betonica officinalis (see *Stachys
   officinalis*)
Betula (birch), 30
Betula pendula (*B. verrucosa*, birch), 31
Betula verrucosa (see *B. pendula*)
Birch (see *Betula*)
Blastocladiales (water molds), 524
Blighia sapida (akee), 659, 678
Bracken (see *Pteridium aquilinum*)
Brassica, 84, 406, 674

Brassica hirta (*Sinapis alba*, white
   mustard), 138, 258, 268
Brassica napus (rape), 39, 674
Brassica oleracea var. *acephala* (kale),
   141–143, 147
Brassica oleracea var. *botrytis* (broccoli,
   cauliflower), 132, 442
Brassica oleracea var. *capitata* (cabbage),
   73, 666, 670, 672
Brassica rapa, 670
Broccoli (see *Brassica oleracea* var.
   botrytis)
Broom (see *Cytisus scoparius*)
Bryonia dioica, 647
Bryophyllum (see *Kalanchoe*)
Bryophyllum calycinum (see *Kalanchoe
   pinnata*)
Bryophyllum pinnatum (see *Kalanchoe
   pinnata*)
Buckwheat (see *Fagopyrum esculentum*)

C

Cabbage (see *Brassica oleracea* var.
   capitata)
Cabomba caroliniana, 65–68
Cactus (see also *Opuntia, Phyllocactus*),
   30, 31, 280
Camellia (camellia), 52, 97
Canavalia ensiformis (jack bean), 427,
   429, 430, 535, 542, 650, 655, 657,
   658
Candida flareri (yeast), 428
Cannabis sativa (hemp), 432
Cantaloupe (see *Cucumis melo*)
Capsicum annuum (green pepper), 322
Carica papaya (papaw, papaya), 330–333,
   335, 336, 338, 339, 418
Carnivorous plants, 386, 417
Carrot (see *Daucus carota* var. *sativa*)
Carya illinoinensis (pecan), 100
Castor bean (see *Ricinus communis*)
Catalpa bignonioides (catalpa), 43, 71
Cauliflower (see *Brassica oleracea* var.
   botrytis)
Ceratophyllum demersum, 96
Cereals, cereal grains (see Gramineae)
Chard (see *Beta vulgaris* var. *cicla*)
Chenopodium, 552

Cherry, sweet (see *Prunus avium*)
Cherry laurel (see *Prunus laurocerasus*)
Chlamydomonas, 170
Chlorella, 58–60, 67, 79, 80, 82, 88, 101, 102, 104–106, 116, 117, 123, 128, 129, 146, 167, 168, 170–178, 181–183, 415, 423, 513, 589, 599, 600
Chlorella ellipsoidea, 174, 649
Chlorella pyrenoidosa, 31, 32, 95, 167, 174
Chlorella Strain A, 174, 178
Chlorella vulgaris, 174
Chondria armata, 661
Chondria crassicaulis, 673
Chromobacterium violaceum, 665
Chrysanthemum morifolium, 530
Cineraria (see *Senecio cruentus*)
Citrullus vulgaris (watermelon), 475, 658, 665
Citrus (citrus fruits), 314
Citrus grandis, 637
Citrus limon (lemon), 314, 315, 318, 338, 344
Citrus sinensis (orange), 121, 122, 166, 329
Cladonia silvatica, 666
Cladophora, 429
Cladophora insignis, 81, 96
Clostridium tetanomorphum, 645
Clover (see *Trifolium*)
Clover, alsike (see *Trifolium hybridum*)
Clover, Ladino (see *Trifolium repens*)
Clover, red (see *Trifolium pratense*)
Clover, subterranean (see *Trifolium subterraneum*)
Coconut milk, 423, 467, 484, 494–496, 531, 532, 539, 562, 579–581, 591, 592, 595
Cocklebur (see *Xanthium*)
Coffea (coffee), 60, 75
Coffee (see *Coffea*)
Coleus (coleus), 54
Colletotrichum atramentarium, 168
Colutea arborescens, 429
Compositae, 427
Conifers, 140, 399, 429, 486, 487, 518, 519, 639–641
Convallaria majalis (lily of the valley), 468–470, 660
Corn (see *Zea mays*)

Corydalis ochotensis, 649
Corynebacterium, 645, 649
Cosmos (see *Cosmos bipinnatus*)
Cosmos bipinnatus (cosmos), 77, 86, 87
Cotton (see *Gossypium*)
Courbonia virgata, 668
Cowpea (see *Vigna sinensis*)
Crassulaceae, 280
Croton gubouga (see *C. megalobotrys*)
Croton megalobotrys (*C. gubouga*), 668
Cucumber (see *Cucumis sativus*)
Cucumis melo (canteloupe), 131, 314, 319, 331–333
Cucumis sativus (cucumber), 72, 73, 138, 394, 679
Cucurbita (gourd, squash), 394, 399, 424, 473, 542, 640, 641
Cucurbita pepo (gourd, pumpkin), 251
Cucurbitaceae (cucurbits), 314, 393, 427
Cyclamen, 29
Cytisus scoparius (*Sarothamnus scoparius*, broom), 459

D

Daffodil (see *Narcissus*)
Date (see *Phoenix dactylifera*)
Daucus carota var. *sativus* (carrot), 133, 134, 269, 278, 325, 350, 362, 363, 368, 416, 423, 430, 446, 493, 494, 496, 497, 511, 512, 532, 538–540, 559, 562, 580, 581, 590–596
Delphinium, 77
Deutzia gracilis, 655, 656
Diatoms, 28, 31, 170
Dichrostachys glomerata, 673
Digenea simplex, 661
Dionaea, 417
Drosera, 417
Dryopteris, 75
Dryopteris erythrosora, 541

E

Ecballium elaterium, 647
Eggplant (see *Solanum melongena* var. *esculentum*)
Elodea (*Anacharis*), 56, 64, 66, 161
Elodea canadensis (*Anacharis canadensis*), 96

Endymion nutans (*Scilla nutans*), 138
Epilobium, 30
Ergot, 651, 654, 669
Eriobotrya japonica (loquat), 471, 660
Erythrina subumbrans, 668
Escherichia coli, 421, 438, 489, 549, 554,
  565, 571, 649, 651, 652, 655, 683, 684
Euglena, 170, 544
Euphorbia, 650
Euphorbia pulcherrima (poinsettia), 54,
  544

F

Fagopyrum esculentum (buckwheat), 48,
  77, 82, 267, 585
Fagus sylvatica (beech), 147
Ferns (see *Adiantum, Asplenium,
  Dryopteris, Nephrolepis, Phyllitis,
  Pteridium*)
Festuca elatior, 213
Ficus (fig), 84
Ficus carica (edible fig), 314
Ficus elastica (rubber plant), 54
Ficus repens, 48
Fig (see *Ficus*)
Fir (see *Abies*)
Fir, white (see *Abies concolor*)
Flax (see *Linum usitatissimum*)
Fontinalis, 161
Fontinalis antipyretica, 64–66, 68, 96
Fontinalis dalecarlica, 96
Fragaria × *ananassa* (*F. chiloensis* var.
  ananassa, strawberry), 72, 271, 289,
  314, 315
Fragaria chiloensis var. *ananassa* (see
  F. × *ananassa*)
Fungi, 409, 485, 487, 488, 529, 530, 539
Fusarium oxysporum f. *cubense* (Panama
  disease), 529, 530
Fusarium oxysporum f. *lycopersici*, 529

G

Galega officinalis, 429
Garlic (see *Allium sativum*)
"*Genista spartium*," 459
Geranium (see *Pelargonium hortorum*)
Gigartina, 89
Ginkgo biloba, 430, 487

Gladiolus, 162–166
Gleditsia, 471
Glycine hispida (see *G. max*)
Glycine max (*G. hispida*, soybean), 75,
  123–125, 128, 129, 133, 397, 406, 429,
  541, 542, 676
Gossypium (cotton), 53, 61, 75–77, 138,
  139, 147, 164–166, 423, 499, 513, 585,
  601
Gourd (see also *Cucurbita*), 424
Gramineae (grasses, cereal grains), 92,
  126, 137, 139, 213, 247, 252, 258,
  259, 285, 392, 419, 420, 427, 440,
  491, 498, 521, 541, 583
Grass, cocksfoot, 138
Grass, pasture, 122
Grasses (see Gramineae)

H

Haplopappus gracilis, 488
Helianthus (sunflower), 487, 531
Helianthus annuus (sunflower), 47–49,
  71, 74, 75, 93, 94, 100, 133, 144,
  146, 252, 260, 270, 488
Helianthus tuberosus (Jerusalem arti-
  choke), 147, 430, 495, 511, 530, 532
Hemerocallis, 473
Hemp (see *Cannabis sativa*)
Holly (see *Ilex*)
Hordeum sativum (see *H. vulgare*)
Hordeum vulgare (*H. sativum*, barley),
  74, 77, 108–111, 137, 138, 147, 166,
  252–259, 261, 264–266, 270–277, 279,
  285–291, 293, 297–302, 402, 419, 459,
  488, 512, 585, 602, 675, 681, 682
Hormidium flaccidum, 64, 66, 123
Hydrangea, 54

I

Ilex (holly), 29
Indigofera, 406
Insectivorous plants, 386, 417
Ipomoea (*Pharbitis*), 77, 552
Ipomoea batatas (sweet potato), 326, 348
"*Ipomoea coerulea*," 146
Iris, 656, 677, 678
Iris tingitana, 644, 645

J

Juglans (walnut), 423

K

Kalanchoe (*Bryophyllum*), 280–284, 582, 654
Kalanchoe daigremontiana, 463, 654
Kalanchoe pinnata (*Bryophyllum calycinum, B. pinnatum*), 90, 280, 281
Kale (see *Brassica oleracea* var. *acephala*)

L

Lactuca sativa (lettuce), 72, 84
Laminaria digitata, 675
Laminaria flexicaulis, 675
Lamium, 30
Lathyrus, 406, 648, 658, 677
Lathyrus odoratus (sweet pea), 259, 267, 678
Lathyrus pusillus, 650
Lathyrus sativus, 648
Lathyrus tingitanus, 654, 665
Legumes (see Leguminosae)
Leguminosae (legumes, leguminous plants), 76, 92, 131, 132, 259, 386, 402, 403, 413, 422, 443, 465, 472, 477, 478, 486–488, 639, 645, 657, 681
Lemna, 487
Lemon (see *Citrus limon*)
Lepidium (pepperwort), 653
Lettuce (see *Lactuca sativa*)
Leucaena glauca, 406
Lichens, 56, 146, 488, 529, 666
Ligustrum (privet), 52, 53, 97
Lilac (see *Syringa*)
Liliaceae (liliaceous plants), 461, 469, 477, 480, 487, 488, 660
Lilium (lily), 432
Lily of the valley (see *Convallaria majalis*)
Linaria vulgaris, 652
Linum usitatissimum (flax), 97, 110, 111, 260, 435, 681
Litchi chinensis, 660

Locust, black (see *Robinia pseudoacacia*)
Lolium perenne (ryegrass), 165, 443
Loquat (see *Eriobotrya japonica*)
Lotus corniculatus (trefoil), 93
Lupin, lupine (see *Lupinus*)
Lupinus (lupin, lupine), 73, 260, 397, 399, 429, 493, 542, 639, 640
Lupinus albus (white lupine), 419, 422, 491, 492, 641, 642, 645, 676
Lupinus angustifolius, 676
Lupinus luteus (yellow lupine), 419, 422, 637, 640, 641
Lycium barbarum, 667
Lycopersicon esculentum (tomato), 53, 70–73, 75, 78, 86, 87, 92, 116, 126, 134, 138, 315, 321, 323, 333, 407, 412, 522, 529, 643, 644, 665
Lycopodium cernuum, 491

M

Maize (see *Zea mays*)
Malus sylvestris (*Pyrus malus*, apple), 70, 78, 79, 99, 100, 127, 129, 130, 147, 151, 156, 157, 312–318, 320–324, 329, 332–334, 338, 339, 344, 349, 350, 353–357, 360, 361, 363, 364, 366, 415, 471, 503, 580, 659–661, 680, 683
Malus sylvestris cultivar 'Granny Smith', 503
Mangifera indica (mango), 319
Mango (see *Mangifera indica*)
Mangold (see *Beta vulgaris*)
Maple (see *Acer*)
Meadowgrass, 138
Medicago sativa (alfalfa), 70, 71, 89–91, 93, 118–120, 130, 131, 144, 147, 149, 152–157, 159, 161, 162, 166, 459, 637, 638, 651, 653, 668, 680, 682
Medicago sativa cultivar 'William Pitt', 501
Melilotus, 487
Mentha (mint), 12, 13, 443, 486, 499
Mentha piperita (mint, peppermint), 423, 499, 514, 516, 518, 520–522, 526, 529, 530, 538
Mimosa palmeri, 648
Mimosa pudica, 664
Mimulus, 61, 62

Mimulus cardinalis, 92
Mint (see *Mentha*)
Molds, 55
Molds, slime, 539
Molds, water, 524
Mosses, 146, 426
Musa (banana), 407, 485, 503
Musa acuminata (banana), 76, 314, 319, 323, 324, 330, 332, 333, 339, 342, 423, 485–488, 504–508, 510, 515, 521, 530, 580, 581
Musa acuminata cultivars: 'Cavendish', 485; 'Gros Michel', 485, 504–508; 'Lacatan', 485
Mushrooms, 55, 669, 678
Mustard, white (see *Brassica hirta*)
Mycobacterium phlei, 656
Mycobacterium tuberculosis, 649
Myriophyllum spicatum, 66, 96
Myxophyceae, 179

N

Narcissus (daffodil), 408, 422, 498, 538
Nasturtium (see *Tropaeolum majus*)
Nepenthes, 417
Nephrolepis, 75
Neptunia oleracea, 673
Nereocystis luetkeana, 675
Nerium oleander (oleander), 41, 42
Neurospora, 418, 420, 421, 428, 430, 467, 475, 486, 489, 492, 536, 537, 545, 558, 653, 669
Neurospora crassa, 537, 683, 684
Neurospora intermedia, 537
Nicotiana rustica, 511, 532
Nicotiana tabacum (tobacco), 57, 59–61, 75, 122, 123, 167, 279, 290, 359, 422, 447, 459, 498, 499, 515–517, 529, 530, 532, 553, 554, 585, 681
Nitzschia, 31, 167, 168
Nitzschia closterium, 31
Nitzschia palea, 31, 170
Nostoc muscorum, 414

O

Oats (see *Avena sativa*)
Oenothera, 542
Oleander (see *Nerium oleander*)

Onion (see *Allium cepa*)
Opuntia, 30
Orache (see *Atriplex hortensis*)
Orange (see *Citrus sinensis*)
Orchids, 531
Oryza sativa (rice), 69, 70, 158, 170, 179, 251, 265, 266
Oxalis, 75, 681
Oxalis deppei, 409

P

Paecilomyces, 645, 655
Panama disease (see *Fusarium oxysporum* f. *cubense*)
Papaw, papaya (see *Carica papaya*)
Parthenocissus quinquefolia (Virginia creeper), 530
Pea (see *Pisum sativum*)
Pea, sweet (see *Lathyrus odoratus*)
Peach (see *Prunus persica*)
Peanut (see *Arachis hypogaea*)
Pear (see *Pyrus communis*)
Pecan (see *Carya illinoinensis*)
Pelargonium (geranium), 29, 48, 52, 409
Pelargonium hortorum (geranium, pelargonium), 48, 49, 52, 54, 55, 82
Pelvetia fastigiata, 474
Penicillium, 663
Penicillium reticulosum, 646
Pennisetum typhoides (bulrush millet), 221–223
Pepper, green (see *Capsicum annuum*)
Peppermint (see *Mentha piperita*)
Pepperwort (see *Lepidium*)
Persea (avocado), 54, 315, 321–323, 332, 333, 339, 342, 344, 349
Pharbitis (see *Ipomoea*)
Phaseolus, 499, 513, 583, 685
Phaseolus coccineus (*P. multiflorus*, runner-bean), 290
Phaseolus limensis (lima bean), 676, 677
Phaseolus multiflorus (see *Phaseolus coccineus*)
Phaseolus vulgaris (bean, French bean, kidney bean), 48, 72–75, 167, 466–468, 471, 486, 662, 669, 675–677
Philodendron, 75
Phleum pratense, 213
Phlox, 473

Phlox decussata, 652, 681, 685
Phoenix dactylifera (date, date palm), 98, 406, 471, 486, 653, 662, 663
Phyllitis scolopendrium, 646, 684
Phyllocactus, 30
Phytoplankton (see Plankton)
Picea (spruce), 399, 420, 475, 487, 640
Picea abies (*P. excelsa*, Norway spruce), 640
Picea excelsa (see *P. abies*)
Picea glauca (white spruce), 487, 499, 513, 518, 519
Picea pungens (Colorado spruce), 97, 98
Pine (see *Pinus*)
Pine, Austrian (see *Pinus nigra*)
Pine, jack (see *Pinus banksiana*)
Pine, loblolly (see *Pinus taeda*)
Pine, Monterey (see *Pinus radiata*)
Pine, red (see *Pinus resinosa*)
Pine, Scotch (see *Pinus sylvestris*)
Pine, western yellow (see *Pinus ponderosa*)
Pine, white (see *Pinus strobus*)
Pineapple (see *Ananas comosus*)
Pinus (pine), 70, 97, 98, 399, 420, 432, 475, 515, 640, 641, 649
Pinus banksiana (jack pine), 518, 519
Pinus nigra (Austrian pine), 98
Pinus pinea, 429
Pinus ponderosa (western yellow pine), 98
Pinus radiata (Monterey pine), 216
Pinus resinosa (red pine), 89
Pinus serotina, 430
Pinus strobus (white pine), 84, 98, 508, 509
Pinus sylvestris (Scotch pine), 98, 147
Pinus taeda (loblolly pine), 89
Pisum sativum (pea), 15, 72–74, 247–251, 258, 259, 262–266, 268, 293–295, 297, 337, 358, 430, 443, 446, 465, 477, 486–488, 493, 536, 545, 546, 548, 549, 570, 640, 644, 647, 655, 665, 669, 676, 682, 684
Pisum sativum cultivars: 'Alaska', 'Progress', 570
Pithecellobium lobatum (*Pithecolobium lobatum*, djenkol bean), 670
Pithecolobium lobatum (see *Pithecellobium lobatum*)

Plankton, 25, 26, 40, 170, 404, 405
Poinsettia (see *Euphorbia pulcherrima*)
Polygonatum multiflorum, 648
Polyporus sulfureus, 663
Potamogeton lucens, 96
Potato (see *Solanum tuberosum*)
Potato, sweet (see *Ipomoea batatas*)
Potentilla, 30
Primula, 53
Primula sinensis, 48
Privet (see *Ligustrum*)
Proteus vulgaris, 684, 685
Prune, 530
Prunus armeniaca (apricot), 121, 127
Prunus avium (sweet cherry), 684, 685
Prunus laurocerasus (cherry laurel), 50, 275–277, 285–287
Prunus persica (peach), 121, 407
Pseudomonas, 430, 446, 683
Pseudomonas tabaci, 654
Pteridium aquilinum (bracken), 109–111
Pumpkin (see *Cucurbita pepo*)
Pyrus communis (pear), 121, 129, 314, 474, 659
Pyrus malus (see *Malus sylvestris*)

Q

Quercus falcata (*Q. triloba*), 47, 48
Quercus pedunculata (see *Q. robur*)
Quercus robur (*Q. pedunculata*), 419
Quercus triloba (see *Q. falcata*)

R

Rape (see *Brassica napus*)
Raphanus, 674
Reseda odorata, 657
Rhapis excelsa (*R. flabelliformis*), 471, 662
Rhapis flabelliformis (see *R. excelsa*)
Rheum rhaponticum (rhubarb), 653
Rhizobium, 422, 528
Rhizobium leguminosarum (*R. radicicola*), 402
Rhizobium radicicola (see *R. leguminosarum*)
Rhodopseudomonas spheroides, 683
Rhodospirillum, 423
Rhodospirillum rubrum, 59, 60

Rhodymenia palmata, 675
Rhoeo discolor (see R. spathacea)
Rhoeo spathacea (R. discolor), 54
Rhubarb (see Rheum rhaponticum)
Rice (see Oryza sativa)
Ricinus communis (castor bean), 75, 100,
    260, 430
Robinia pseudoacacia (black locust), 642
Rubber plant (see Ficus elastica)
Rye (see Secale cereale)
Ryegrass (see Lolium perenne)

S

Saccharum officinarum (sugar cane), 97,
    99, 100, 147, 151, 637
Saintpaulia, 75
Salicornia perennis, 675
Salix (willow), 11, 30
Salvia officinalis, 653
Sambucus, 29
Sambucus nigra (elderberry), 89
Santalum album, 466, 642, 660
Saraca indica, 477, 478, 487
Sarothamnus scoparius (see Cytisus
    scoparius)
Scenedesmus, 123, 170, 180, 415, 423
Scenedesmus obliquus, 133
Scenedesmus quadricauda, 96
Scilla nutans (see Endymion nutans)
Secale cereale (rye), 39, 73
Sedum, 280
Senecio cruentus (cineraria), 53, 100
Sinapis alba (see Brassica hirta)
Slime molds, 539
Solanaceae, 427
Solanum melongena var. esculentum
    (eggplant), 146
Solanum tuberosum (potato), 34, 36, 64,
    66, 69, 70, 72, 73, 124, 126, 137,
    141–143, 147, 268, 293, 324–329,
    337, 338, 340, 341, 343–345, 347,
    348, 350–353, 367–369, 391, 415,
    423, 452–455, 480–484, 488, 493–496,
    510–512, 521, 525, 532, 579, 580,
    582, 587, 589, 597, 642, 644, 645,
    680
Solanum tuberosum cultivars: 'Green
    Mountain', 'Houma', 'Katahdin',
    'King Edward', 'Sebago', 481–484

Sorghum vulgare (sorghum), 270, 435
Soybean (see Glycine max)
Spinach (see Spinacia oleracea)
Spinacia oleracea (spinach), 97, 543,
    598, 599, 680
Spiraea sorbifolia, 419
Spirodela, 123
Spirogyra communis, 30
Spruce (see Picea)
Spruce, Colorado (see Picea pungens)
Spruce, Norway (see Picea abies)
Spruce, white (see Picea glauca)
Squash (see Cucurbita)
Stachys officinalis (Betonica officinalis),
    668
Stachys sylvatica, 668
Stachys tuberifera, 668
Stichococcus, 123
Stigeoclonium, 96
Stizolobium hassjoo, 664
Stizolobium niveum, 652
Strawberry (see Fragaria × ananassa)
Streptococcus faecalis, 430
Streptomyces, 429, 431, 648, 655, 659,
    664
Streptomyces alboniger, 657
Streptomyces bottropensis, 656, 674
Streptomyces chibaensis, 659
Streptomyces reticuli aqua-myceticus.
    659
Streptomyces rimosus, 648
Strophanthus scandens, 662
Sugar beet (see Beta vulgaris)
Sugar cane (see Saccharum officinarum)
Sunflower (See Helianthus annuus)
Syringa (lilac), 30
Syringa vulgaris, 31

T

Tea, 678
Tobacco (see Nicotiana tabacum)
Tolypothrix, 170
Tomato (see Lycopersicon esculentum)
Torulopsis utilis (yeast), 651
Trefoil (see Lotus corniculatus)
Tragopogon, 530
Trifolium (clover), 39, 93, 132, 408,
    651, 680
Trifolium hybridum (alsike clover), 144

Trifolium pratense (red clover), 93, 215–220, 223–225, 682

Trifolium repens (Dutch clover, Ladino clover, white clover), 100, 144, 435, 661

Trifolium subterraneum (subterranean clover), 146, 269

Trigonella, 669

Triticum (wheat), 39, 64, 66–68, 70, 72, 83–86, 89, 97, 102, 115, 137, 139–141, 150, 159, 213, 214, 265, 270, 271, 274, 277, 285, 286, 386, 405–407, 433, 434, 477, 490, 529, 534, 536, 548, 562, 585, 680, 683

Triticum aestivum (wheat), 111, 112

Triticum sativum (wheat), 47, 48

Triticum vulgare (winter wheat), 141, 142, 251, 262

Tropaeolum majus (nasturtium), 285, 286

Tulipa (tulip), 47, 461, 472, 473, 480, 487, 488, 500–502, 545, 547, 645, 647

Tulipa gesneriana (tulip), 646, 680, 683, 685

Tung (see *Aleurites fordii*)

Turnip, 73

U

Ulva lactuca, 675

Umbelliferae, 427

V

Vaccinium vitis-idaea, 659, 681

Verticillium alboatrum, 486, 529, 530

Vetch (see *Vicia*)

Vibrio cholerae (see *V. comma*)

Vibrio comma (*V. cholerae*), 647, 653

Vicia, 47, 73, 640, 654, 677

Vicia faba (*V. vulgaris*, bean, broad bean), 48, 71, 213, 259, 274, 277, 285, 286, 293, 475, 476, 494, 656, 657

Vicia sativa (vetch), 436, 638, 639, 641, 651, 653

Vicia vulgaris (see *Vicia faba*)

Vigna sesquipedalis, 262, 542

Vigna sinensis (cowpea), 123

"*Vigna unguiculata*," 146

Vitis vinifera, 680

W

Watermelon (see *Citrullus vulgaris*)

Wheat (see *Triticum*)

Willow (see *Salix*)

X

Xanthium (cocklebur), 78, 431, 432

Xerocomus badius, 678

Y

Yeast, 55, 237, 250, 268, 304, 358, 359, 433, 434, 480, 534, 548, 644, 651, 671, 675, 682

Z

Zea mays (corn, maize), 34, 36, 39, 48, 49, 69, 73, 89, 122, 123, 130, 146, 147, 152, 156, 159, 204, 205, 207, 210, 213, 214, 216, 217, 259, 267, 270, 293, 357, 359, 405, 406, 423, 425, 427, 430, 488, 513, 536, 585, 602, 647

Zebrina pendula, 90

Zygnema stellinum, 30

SUBJECT INDEX

Owing to the large number of chemical compounds referred to in this volume, and especially in Appendixes I, II, and III of Chapter 4, they are not all separately listed in this index. Not every citation in the text is indexed even for the compounds listed. The entries relate to the more extensive treatments only. Compounds which do not appear in the index may be traced through the general categories to which they belong, e.g., amino acids, imino acids, keto acids, etc. Subjects related to specific plants, if not here listed, may be traced through the Index of Plant Names.

A

Acetaldehyde,
  in potato tubers, 329
N-Acetylglucosamine,
  in fungi, 539
O-Acetylhomoserine, 465
  occurrence of in plants, 472
Actinomycin D,
  as an inhibitor of RNA synthesis, 567
Action spectrum,
  of carbon dioxide assimilation, 85–86
  of nitrate reduction, 86
Activation energy,
  as a limiting factor, 263
Adenosine diphosphate (ADP),
  as a factor controlling respiratory rate, 242
Adenosine triphosphate (ATP),
  in amino acid activation, 566
  in nitrogen metabolism, 513–514
  in the respiratory metabolism of roots, 301
  from nicotinamide adenine dinucleotide, 235–236
  formation of, 235
  role of in protein synthesis, 565–570
  turnover of at the climacteric, 323
Agmatine (1-amino-4-guanidinobutane),
  occurrence of, 430, 431, 459
  structure of, 429
Air motion,
  factors affecting the measurement of, 206–207
Air temperature,
  profiles in red clover crop, 218
Alanine,

from hydroxyglutamate, 463
  metabolism of in cultured tissues, 595
β-Alanine,
  formation from aspartate, 459
Algae,
  as a food source, 168–169
  culture techniques for, 176–179
  effects of light on, 170
  effects of minerals on photosynthesis in, 103–106
  growth of, 169–175
  nitrogen fixation by, 179–180
Allantoic acid, 476
  as a mobile form of nitrogen, 601
  role of in plants, 428
Allantoin,
  as a mobile form of nitrogen, 601
  role of in plants, 428
  scheme for the synthesis of, 417
  synthesis of in plants, 428
Allosteric effects,
  in enzyme function, 577
Amides,
  accumulation of in detached leaves, 286
  synthesis of in leaves, 289–290
  see also Asparagine
  see also Glutamine
  see also Appendix I
Amide nitrogen,
  determination of, 397
Amino acids,
  accumulation of in leaves, 289
  as respiratory substrates, 260, 275
  chemical determination of, 398
  in seedlings, 256–257
  keto acid analogs of, 444–445, 680–686
  non-protein, 644–679

occurrence of in protein, 637–643
quantitative determination of on paper, 393
Amino nitrogen,
 determination of, 396–397
γ-Aminobutyric acid, 446
 in actively growing tissues, 532
 in cultured potato tissue, 484
 in the tulip bulb, 501
 isolation of from potatoes, 453–454
 occurrence of, 454, 457–458
 occurrence of in the potato tuber, 480
 metabolic relations of, 445–446
β-Aminoisobutyric acid,
 formation of, 457
Ammonia,
 as a measure of protein N, 390–391
 as a nitrogen source, 408–409, 413
 as a source of nitrogen, 402, 403
 as the ultimate N source, 436–437
 formation of by enzymic action, 411
 incorporation of into amino acids, 439
 in nitrogen fixation, 444
 production of by plants, 411–412
 utilization of, 410
"Ammonification," 407
Ammonium carbonate,
 as CO_2 source, 74
Anabolism,
 relation of to surface area, 526
Anaerobic metabolism,
 see Fermentation
Apparent assimilation,
 definition of, 55
Apparent photosynthesis,
 values of, 157
Apples,
 nitrogen metabolism in, 503–504
Arginase,
 in jack bean, 430
Arginine,
 accumulation of in conifers, 518
 discovery of, 429
 for TMV synthesis in tobacco leaves, 554
 in actively growing tissues, 532
 in pine cones, 509
 in the tulip bulb, 501
 relation to DNA content, 552
 structure of, 429

Arginine decarboxyoxidase,
 reactions catalyzed by, 431
Argininosuccinic acid,
 occurrence of in plants, 475
Arrhenius equation, 262
Ascorbic acid oxidase, 346, 347
 localization of, 240
 role in normal respiration, 235
Asparagine, 420
 accumulation of during germination, 260
 as nitrogen reserve in plants, 437
 biosynthesis of, 438–439
 derivatives of in plants, 473
 discovery of, 424
 effects of daylength on, 516
 from α-ketoglutaramic acid, 447–448
 in actively growing tissues, 532
 in the cotyledons of germinating seeds, 491
 in dividing cells, 494–495
 in the potato tuber, 453
 in leaves, 290, 499
 in pine cones, 509
 in vivo synthesis of, 526
 scheme for synthesis of, 435, 448
 seasonal effects on in the banana, 507
Asparagine to glutamine ratio,
 effect on environment on, 483–484
 genetic control of in potato, 482
 in cultured potato tissue, 484
Aspartase, 438
L-Aspartic acid,
 occurrence of, 639
 synthesis of, 438
Assimilation,
 and diurnal rhythms, 51
 effects of nitrogen and potassium on, 113–115
 effects of sulfur on, 118
 factors limiting, 51
Assimilation number,
 definition of, 101
Assimilation rate,
 effect of light on, 143–144
 effects of mineral deficiencies on, 107–109
Auxins,
 effects of on photosynthesis, 167
 effect on nitrogen metabolism, 531

8-Azaguanine, 541
Azetidine-2-carboxylic acid,
   as a competitive proline antagonist, 470
   detection of, 468–469
   occurrence of in plants, 477
Azide,
   effect of on leaf respiration, 278, 279

B

Baikiain (dehydropipecolic acid),
   discovery of in plants, 470–471, 477,
     486
Banana,
   development of, 504–506
   nitrogen compounds in, 485–486
   nitrogen metabolism in, 504, 507–508
Benzyl viologen,
   as a substitute for NADPH
Betaine,
   structure of, 426
Bicarbonate,
   as a trigger for morphogenesis, 525
   utilization of by plants, 96
Biological yield,
   factors affecting, 139
Blackman's schema,
   essential features of, 360–361
Boron,
   role of in plant nutrition, 133–135
Bromelin,
   occurrence of, 418

C

Caffeine,
   in flower induction, 533
Calcium,
   in the dark reactions of photosynthesis,
     118
   role in nitrogen metabolism, 521, 523
Canavanase,
   in jack bean, 430
Canavanine,
   as an inhibitor, 430
   biosynthesis of, 430
   occurrence of, 429
   structure of, 429
Carbamyl phosphate, 429

   structure of, 415
   synthesis of, 452
Carbohydrates,
   as a source of respiratory substrate,
     233, 261, 274–276
   as primary products of photosynthesis,
     31
   as regulators of protein synthesis, 581–
     582
   breakdown of in respiration, 234
   breakdown of in senescing leaves, 284
   changes in stored potatoes, 326
Carbohydrate metabolism,
   relationship to nitrogen metabolism,
     385
Carbon,
   conservation of by plants, 362
   occurrence of, 26
Carbon-14,
   occurrence in the air, 35
Carbon assimilation,
   effects of mineral deficiencies on, 103–
     104
Carbon dioxide,
   absorption of, 14
   and stomatal opening, 51–52
   as a factor controlling the rate of
     malate formation, 284
   as a regulator of nitrogen metabolism,
     524–525, 589
   as a respiratory inhibitor, 268
   assimilation of, 42
   chemical reduction of, 20–21
   content of the atmosphere, 32–40
   conversion to sugar, 19–20
   dark fixation of, 280
   diffusion of in fruits, 333
   diffusion of through multiperforate
     septa, 46
   diffusion through stomates, 49
   discovery of, 12
   effects of on fruit ripening, 314–315
   evolution of by fruits, 335–337
   exchange measurements of in the field,
     148–168
   fixation of, 25–27
   fixation of by potato slices, 589
   "insurge" and "outburst" of, 58–59
   mechanisms of absorption, 40–49
   occurrence in oceans, 32–33

production under anaerobic conditions, 337

profiles of in red clover crop, 220

rate of absorption by leaves, 43

reduction of, 16–17

role of in photosynthesis, 14, 71–73

toxic effect of on roots, 74

Carbon dioxide concentration,
    maintanence of in field studies, 206

Carbon dioxide drift,
    in detached leaves, 285–289

Carbon dioxide exchange rates,
    measurement of, 210–211

Carbon dioxide fertilization, 71–73

Carbon dioxide flux,
    equation for measurement of, 209

Carbon dioxide profiles,
    schematic representation of, 211

Carbon monoxide, 20
    effect of on leaf respiration, 278, 279

Carbon reserves, 23

Carbonic acid,
    in photosynthesis, 14

Carotenoids,
    role of in nitrate reduction, 115

Casein hydrolyzate,
    as a source of carbon or nitrogen,
        595–596

Catalase,
    in plastids and mitochondria, 123

Cellular organization,
    influence on anabolism and catabolism,
        245–246

Chlorella,
    temperature coefficient of, 88, 89

Chloroform,
    as an inhibitor of photosynthesis, 52

Chlorophyll,
    chemistry of, 18
    influence of iron on, 121–123
    role of magnesium in, 116

Chlorophyll a,
    structure of, 18–19

Chlorophyll b,
    structure of, 18–19

Chloroplast development,

Chloroplasts,
    development of, 542
    effects of manganese deficiency on, 126

protein synthesis in, 543–544, 598–599
    inhibition of, 544

Choline,
    as a carrier of phosphate, 513
    structure of, 426

Choline phosphate,
    occurrence of, 602

Choline sulfate,
    in roots. 513
    occurrence of, 602

Chromatography,
    development of, 392–393

Cistron,
    definition of, 574

Citrulline,
    as a product of N fixation, 414
    occurrence of in plants, 475

Climacteric,
    effects of on N metabolism, 510
    effects of on protein synthesis, 580
    enzyme changes during, 324
    metabolic mechanisms in, 320–324
    occurrence of in fruits, 313–315
    phosphorylation, reactions during, 323
    protein accumulation in, 321
    role of aerobic respiration in, 315
    role of ethylene in, 318–320

Cobalt,
    role of in plants, 128–131

Coconut milk,
    as a growth stimulator, 532

Codon,
    definition of, 570

Copper,
    role of in plants, 128–131

Crassulacean acid metabolism, 280

Cuticle,
    role in gas exchange, 40–41

Cutin,
    solubility of carbon dioxide in, 41

Cyanamid,
    as an artificial nitrogenous fertilizer,
        415

Cyanide,
    effect of on leaf respiration, 278, 279
    effect on oxygen uptake by freshly cut
        potato slices, 347
    use of by plants, 435

Cytochromes,
    localization of in cells, 345

Cytochrome b₇,
occurrence of in plant tissues, 346–347
Cytochrome oxidase,
oxygen saturation of, 344
in plant respiration, 235, 344, 346, 347
in the respiration of mature leaves, 279
Cytoplasm,
as the site of epigenetic factors, 574

D

Daylength,
effects on nitrogen compounds, 515–
518, 523
Deoxyribonucleic acid (DNA),
as an information carrier, 565
in chloroplasts, 543, 599
in development of root tissue, 493–494
replication mechanism for, 566
"Dephlogisticated" air,
definition of, 14
Depletion factor,
values for, 24
Detached leaves,
loss of protein in, 420
α,γ-Diaminobutyric acid,
structure of, 469
Diffusion,
effect of temperature on, 262
interference effects in, 46–47
of oxygen, 345–346
through an aperture, 45, 49
Diffusion flow,
pattern of through a stoma, 44
2,4-Dinitrophenol,
as a cell poison, 242–243
effect of on fruit ripening, 321–323
effect of on respiration, 267, 272, 302,
358–359
Diphosphopyridine nucleotide (DPN),
see Nicotinamide adenine dinucleotide
Dormant seeds,
respiration in, 247

E

Ehrlich's reagent,
for detecting compounds on paper, 395
Electrophoresis,
of carrot proteins, 540

Embryo,
metabolism of, 247
Endoplasmic reticulum, 561
Endosperm,
respiration in, 257, 258
Energy balance, 220–225
Enzyme synthesis,
model for the regulation of, 575
Epidermis,
permeability to carbon dioxide, 54–55
Ergosomes,
composition of, 567
Essential elements,
list of, 381
"Ester method,"
for isolating and separating monoamino
acids, 399–400
Ethanol,
production of by germinating pea
seeds, 250
Ethionine,
inhibitory effects of, 541
Ethylene,
in storage organs, 329
occurrence of in fruits, 318–319
role of in the climacteric, 318–320
Evaporation,
equation for, 44
Extinction point,
definition of, 266

F

Fats, 233
as sources of respiratory substrates, 261
utilization of in germination, 259, 260
Fermentation, 236–239
in higher plants, 346, 350–360
Ferredoxin,
in photochemical pyridine nucleotide
reduction, 600, 601
Fertilizers,
early use of, 419
Floral induction,
role of nucleic acids in, 431–432
Fluid velocity gradients,
equation for, 209
5-Fluorouracil,
as an inhibitor of photoperiodic induction, 431

Formaldehyde, 21
  in carbon fixation, 20
Formaldehyde theory, 22
Free diffusion,
  equation for, 44
Free energy,
  production of in respiration, 234–235
Fructose-1,6-diphosphate,
  increase of in the presence of ethylene,
    319
Fruiting,
  as affected by temperature, 92
Fruits,
  internal atmosphere of, 329–334
  as a dynamic system, 312
Fucoxanthol,
  role in photosynthesis, 85

G

Galegin (isoamylene guanidine),
  occurrence of, 429
  structure of, 429
Gas chromatography,
  in the determination of amino acids,
    396
Gene-enzyme hypothesis, 558–559
Genetic code,
  basis of, 570–573
  "dictionary of," 572
Glucose,
  as a carbon source for proteins, 362
L-Glutamic acid,
  as the first organic product of nitrogen
    fixation, 414
  occurrence of, 639
  synthesis with glutamic acid dehy-
    drogenase, 438
Glutamic acid dehydrogenase,
  occurrence of, 246
Glutamine,
  as nitrogen reserve in plants, 437
  as source of carbon for protein syn-
    thesis, 595
  discovery of, 424–425
  effects of daylength on, 516
  formation of from α-ketoglutaric acid,
    443
  in actively growing tissues, 532
  in dividing cells, 494–495

  in extracts of the potato tuber, 453
  in leaves, 499
  in pine cones, 509
  in vivo synthesis of, 526
  occurrence of, 387
  relation to organic acids in succulents,
    582
  role of in potato slices, 589
  seasonal effects on in the banana, 507
  shift to asparagine in diseased leaves,
    530
  synthesis from glutamic acid, 564
  synthesis of in aging leaves, 290
  synthesis of in roots, 299–300
Glutamine synethetase,
  action of, 447
Glutaminotransferases,
  action of, 474
γ-Glutamyl peptides,
  mechanism of synthesis, 564
  occurrence of in plants, 474
  role of in protein synthesis, 449
γ-Glutamylphenylalanine,
  occurrence of, 406
γ-Glutamyltyrosine,
  occurrence of, 406
Glutathione, 474
  in the synthesis of γ-glutamyl peptides,
    564
Glutelins, 534
Glycine,
  as a primary product of the photosyn-
    thetic cycle, 528
  first isolation of, 424
Glycolysis, 278
  factors controlling, 359–360
  localization of the enzymes of, 239–240
  regulation of, 366
  stimulation of by 2,4-dinitrophenol, 359
Glyoxalate cycle,
  conversion of fats to carbohydrates in,
    260
Glyoxylate, 476, 524
  from hydroxyglutamate, 463
  metabolic role of in nitrogen metab-
    olism, 527
Guanidino compounds,
  structure of, 429
Guanidino derivatives,
  structure of, 431

Growing points,
    nitrogen compounds in, 491–492
Growth,
    and nitrogen metabolism, 584–585
    as affected by light, 86–87
    as affected by temperature, 93
    as a limiting factor for protein
        synthesis, 579
    effect of hydrogen ion concentration
        on, 95–97
    effects of intermittent light on, 77–78
    relation to protein synthesis, 541
γ-Guanidinobutyric acid,
    occurrence of, 430
    structure of, 429
Guard cells,
    assimilation in, 53

H

Heavy water (D₂O),
    effect of on photosynthesis, 102
Hematin, 18
Hexose phosphate,
    recycling of in plants, 364–365
Histidine,
    in the banana, 485–486
    seasonal effects on in the banana, 507
Histones, 536, 552
Homoserine,
    occurrence of in plants, 465
Hormones,
    early history of, 531
"Humus" theory, 16–17
Hydantoic acid,
    synthesis from urea, 428
Hydantoin, 428
Hydrazine,
    in nitrogen fixation, 444
Hydrocyanic acid,
    incorporation into arginine, 435
Hydrogen,
    discovery of, 12
Hydrogenase, 600
    action of, 598
Hydrogen fluoride,
    effects of on photosynthesis, 162–164,
        165, 166

Hydrogen sulfide,
    as hydrogen donor, 22
Hydroxyamino acids, 459–466
γ-Hydroxyglutamic acid,
    metabolism of in mammalian systems,
        463
5-Hydroxyindoleacetic acid, 407
Hydroxylamine,
    in biosynthesis of "primary" amino
        acids, 442
Hydroxylamine reductase,
    metal ions in, 128
5-Hydroxypipecolic acid, 478, 486
    from baikiain, 471
    in human urine, 406–407
    occurrence of in plants, 477
Hydroxyproline,
    formation from proline, 496, 527
    occurrence of in cells, 539
    occurrence of in plants, 466
γ-Hydroxyvaline,
    discovery of in plants, 463
    synthesis and metabolism of, 464
Hyponitrite reductase,
    metal ions in, 128

I

Imino acids,
    occurrence of, 644–679
Indoleacetic acid (IAA),
    effects on amino acid uptake, 531
Infrared radiation,
    absorption by CO₂, 150–151
Inorganic phosphate,
    as a factor controlling respiratory rate,
        242
Inositol phosphates (phytols),
    as energy sources, 562
Insectivorous plants, 417–418
Insolation, 23, 24
Ion exchange resins, 395
Iron,
    interaction with manganese, 124
    role of in plants, 121–125
Isobutylamine, 459
Isonicotinyl hydrazide (INH),
    as an inhibitor, 528

K

Keto acids,
 as analogs of the amino acids, 444–445,
  680–686
 as ports of entry for nitrogen, 524
α-Ketoglutaramate,
 from glutamine, 447
 transamination with γ-hydroxyglutamate, 463
α-Ketoglutarate,
 content of mint leaves, 526
Kinetin (6-furfurylaminopurine),
 as a regulator of xanthine oxidase, 432
 effects of on N compounds, 511
 function of, 432
 influence on the metabolism of nucleic
  acids and protein, 532
Kjeldahl determination,
 for total nitrogen, 390–391
Kok effect, 57–58
Krebs cycle,
 see Tricarboxylic acid cycle

L

Lactic acid,
 production of in potato tubers, 351, 352
Latent heat flux,
 profiles of, 222
Leaf area index,
 definition of, 139
Leaf proteins,
 localization of, 542–543
Leaf surfaces,
 permeability of to carbon dioxide, 54
Leaves,
 nitrogen metabolism in, 497–500, 513
 protein metabolism in, 585–586, 589–
  590
 relationship between respiration and N
  metabolism in, 585–587
 respiration in, 258, 269–291
Leucenal,
 occurrence of, 406
Light,
 absorption and reflection of, 84
 as an inhibitor of photosynthesis, 79–
  82
 efficiency of utilization, 147

Light respiration,
 rates of, 62
Light saturation,
 values for field crops, 70
Limiting factors,
 principles of, 63
Lycomarasmin, 529
Lysine,
 pathway of degradation, 467
Lysopine,
 occurrence of, 530

M

Magnesium,
 in enzymes, 116
 in the dark reactions of photosynthesis,
  117–118
Malic acid,
 metabolism of in leaves, 280–283
 synthesis of *in vivo*, 283
Malic enzyme,
 role of in carbon dioxide fixation, 282
Manganese,
 effect of on net assimilation rates, 126
 roles of in plant growth, 125–128
Manganese deficiency,
 effects on nitrogen metabolism, 522
Mass and momentum,
 equations for the transfer of, 206
Mass spectrometer,
 in the study of respiration, 59
Mesophyll,
 ratio of assimilation to respiration in,
  53
Messenger RNA, 549, 567
 as a single molecule, 576
 formation from DNA, 566
 use of synthetic polynucleotides as, 571
Metabolic pools,
 and protein synthesis, 597
 existence of, 241
 function of in cells, 591
Metabolism,
 of growing cells, 494–496
Methylal, 20
Methylene groups, 18
γ-Methyleneglutamic acid, 460
 in the tulip bulb, 501
 occurrence of in plants, 477

γ-Methyleneglutamine,
  occurrence of in plants, 472
γ-Methylglutamic acid,
  occurrence of in plants, 477
γ-Methyl-γ-hydroxyglutamic acid, 492
  in the tulip bulb, 501
  isolation of, 460–461
  metabolism of, 462
  occurrence of, 447
S-Methylmethionine,
  occurrence of, 406
5-Methyltryptophan,
  inhibitory effects of, 541
Methyl viologen,
  as a substitute for ferredoxin, 601
Microclimate,
  the diurnal cycle of, 216–220
Microsomes,
Mineral deficiencies,
  effects of on nitrogen metabolism, 514–518
  role in protein synthesis, 561
Mineral elements,
  effects on nitrogen metabolism, 521–522
  roles of in photosynthesis, 103
Mitochondria,
  as highly organized multienzyme systems, 239
  in mesophyll cells, 278
  in root cells, 295
  oxidation of pyruvate in, 239
Mixing length,
  definition of, 215
Molybdenum,
  in nitrate reduction, 413, 514–515
  role of in plant nutrition, 131–133
Momentum,
  equation for the vertical transfer of, 208
Morphogenesis,
  and protein synthesis, 576–577
  relation of histones to, 552–553
  relation to protein complement, 545
Multiperforate septa,
  mutual interference by, 47

N

Naphthaleneacetic acid,
  effect of on protein synthesis, 597

Net absorption,
  definition of, 204
Net assimilation,
  definition of, 55
  values of, 146
Net assimilation rates, 102, 109–110
  determination of, 135–138
  values of, 138
Net carbon dioxide exchange rates,
  comparison of under different conditions, 213
Net radiation absorption,
  equation for, 204
Net radiation flux,
  profiles of in corn, 205
  profiles in red clover crop, 217
Nicotinamide adenine dinucleotide (NAD), 245, 246, 283
  as a hydrogen acceptor, 234
  in nitrogen metabolism, 514
  oxidation of, 235–236
Nicotinamide adenine dinucleotide phosphate (NADP), 115, 245, 246, 282, 364
  as a hydrogen acceptor, 234
  role in N metabolism in leaves, 514
Ninhydrin,
  use of as an identifying reagent, 393
Ninhydrin-reactive compounds,
  map of on paper chromatography, 455
Nitrate,
  as the inorganic source of nitrogen for plants, 407–409, 413
  occurrence of in plants, 386
Nitrate assimilation, 407
  effects of mineral deficiencies on, 103–104
Nitrate reductase,
  dependence on NADPH
  metal ions in, 128
  occurrence of in plants, 412–413
Nitrate reduction,
  action spectrum of, 115
  mechanism of, 115–116
  pathway of, 407
  role of light in, 600
Nitrite,
  occurrence of in plants, 408
Nitrogen,
  chemical determination of, 389–391

in cultured explants, 592–597
properties of, 382
sources of, 17
ultimate source of in cells, 564
Nitrogen bases,
in plants, 426
Nitrogen compounds,
in actively growing tissues, 532
translocation of, 513
Nitrogen cycle, 403–405
Nitrogen fertilizers,
use of in agriculture, 405–406
Nitrogen fixation,
history of, 402–403
occurrence of in plants, 413–415
ports of entry in, 443–444
relationship to photosynthesis, 414
Nitrogen metabolism,
effects of light on, 599–600
effects of nutrition and environment
on, 511–520
effects of plant pathogens on, 528–531
effects of potassium on, 107
in fruits, 503–509
in leaves, 513
in organs of perennation, 419–420
in seedlings, 419
in the tulip bulb, 500–503
of pine cones, 508–510
response of to environmental factors,
520–528
role of manganese in, 128
seasonal cycles of, 499–500
survey of, 384–385
Nitrogen nutrition,
early history of, 401–407
Nitrogenous substances,
secretion from pea roots, 443
β-Nitropropionic acid,
occurrence of, 406
Nonprotein nitrogen,
extraction of, 392
Nucleic acids,
electrophoresis of on silica gel, 536
history of, 433–434
in chloroplasts, 543, 599
influence of kinetin on, 532
in the root, 296
in the wheat embryo, 534

in yeast, 433
see also Deoxyribonucleic acid,
Ribonucleic acid
Nucleoproteins,
early studies on, 534
Nucleotide-peptide complexes,
occurrence of, 434
Nutrient solutions,
elements in, 122

O

Octopine,
occurrence of, 530
Ontogeny,
nitrogen metabolism during, 510–511
"Operator" genes,
function of, 574–575
"Operon,"
definition of, 575
Organic acids,
accumulation of, 30–31
metabolism of in leaves, 274–275, 281
Organization,
regulation of respiration by, 368
Origin of life, 388–389
Ornithine,
occurrence of in plants, 475
structure of, 426
Ornithine cycle,
as a metabolic pathway in plants, 475–
476
as a source of urea, 416
Orotic acid,
competitive inhibitor of 5-fluorouracil,
431
Orthoquinones, 588
Oxidative anabolism,
definition of, 363–364
function of, 363–364
Oxidative phosphorylation, 235–236
action of DNP on, 242–243
dependence of on the intact structure
of mitochondria, 239
Oximes,
formation of,
Oxygen,
as a factor limiting respiration, 339–
340, 343, 350, 366–367
diffusion of in cells, 346

diffusion of in fruits, 333
discovery of, 12
effects of on glycolysis, 361
effects of on fruit ripening, 314–315
evolution of in sunlight, 13–14
Oxygen concentration,
effect of on respiration, 266
Oxygen pressure gradient,
in cells, 341

P

Panama disease,
effects on nitrogen metabolism, 530
Papain,
occurrence of, 418
Paper chromatography,
in plant analysis, 452–454
Pasteur effect, 237–238, 267
and the conservation of carbon, 358
in seed germination, 250
Pentose phosphate pathway,
as an alternative mechanism of carbohydrate breakdown, 237
localization of the enzymes of, 239–240
occurrence of in plants, 282, 361–362
regulation of in plants, 365–366
role of in leaf respiration, 278
Peptide bonds,
theory of, 400
Peroxidase,
occurrence of in plant tissues, 346–347
pH,
effect on CO₂ supply, 95
Phenol oxidase, 346, 347
role of in potatoes, 348
Phenolic compounds,
role of in respiration, 587
Phosphoenol pyruvate (PEP) carboxylase, 282
Phosphofructokinase,
role of in the climacteric rise, 323
Phosphorylase,
role in the guard cells, 53
Phosphorylated nucleotides,
role in synthetic events, 246
Phosphorylation,
as control points in cellular respiration, 243

dependence of endergonic reactions on, 243
effect on membranes, 274
in developing pea seeds, 250
Photochemical energy flux,
profiles of in a red clover crop, 223, 224
Photooxidation, 56
bleaching of chlorophyll due to, 79
mechanism of action, 81–82
Photosynthesis,
aerodynamic measurement of, 157–160
and nitrogen metabolism, 528, 584
at various wavelengths, 83–85
effects of diseases on, 167–168
effect of environmental factors on, 63–102
effect of continuous light on, 78–79
effect of leaf age on, 97–99
effect of light intensity and carbon dioxide concentration on, 63–74
effect of light quality on, 82–88
effects of mineral elements on, 103–104, 111–112
effects of poisons on, 167
effects of sulfur deficiency on, 118–119
effect of temperature on, 88–95
effects of water supply on, 99–101
gas exchange in, 12
history of, 10–27
inhibition of, 52, 56, 102, 105
in sun and shade plants, 75
in the carboniferous period, 10–11
in vitro, 21
in water plants, 160–161
limiting factors in field experiments, 135
measurement of in the field, 135
nutritional factors in, 102–135
over-all reactions of, 16
quantitative experimentation on, 15
rates under natural conditions, 145–148
relation of to light intensity, 154
relation to N fixation, 414
role of light energy in, 16
role of magnesium in, 116–117
saturation of at different temperatures, 92
temperature coefficients of, 89, 91
theoretical yields for, 175–176
field measurements of, 69–71

Photosynthetic activity,
  in the development of young leaves,
    272–273
Photosynthetic cycle,
  in nitrogen metabolism, 528
Photosynthetic phosphorylation,
  relation to total leaf RNA, 543
Photosynthetic potential,
  definition of, 140
Photosynthetic quotients (P.Q.),
  equations for, 27
  interpretation of values, 27, 28
  values of, 28, 29, 30, 31
  determination of, 28–29
  effects of nitrate on, 32
Photosynthetic yield,
  estimates of, 25
Phytochrome, 385, 545
Phytoferritin,
  occurrence of in leaves, 543
Phytols,
  see Inositol phosphates
Picric acid,
  as protein precipitants, 392
Pipecolic acid (piperidine-2-carboxylic
    acid),
  in diseased plants, 530
  metabolism of, 467–468
  occurrence of, 467, 477, 486
  origin from lysine, 467–468
Piperidine compounds,
  in legumes, 486–487
Plankton,
  photosynthesis by, 25–26
Plant chambers,
  use in measuring photosynthesis, 151–
    152
Plant proteins,
  early studies on, 533–538
  in relation to development, 544–548
  metabolism of, 256–257, 296
  separation by electrophoresis, 536–537
Polyphenol oxidase,
  occurrence of, 129
  role in normal respiration, 235
Porometer,
  measurement of stomatal opening by,
    50–51
Potassium
  effects of on respiration, 587

influence of on iron metabolism, 124
  in nitrogen metabolism, 107, 521, 523
  role of in photosynthesis, 106–107
Potato tuber,
  nitrogen compounds of, 481–484
"Primary" absorption spectrum,
  estimation of, 85
Primitive atmosphere,
  composition of, 10, 388–389
Proline,
  antagonism by azetidine-2-carboxylic
    acid, 470
  methyl derivatives of, 471
Proteins,
  and surface phenomena, 540
  as respiratory substrates, 233, 260
  breakdown of in leaves, 274–275, 284,
    289
  categories of in leaves, 534
  classification of, 535
  complexity of structure, 383
  content of leaves in the light, 271–272
  derivation of term, 388
  dynamic state of in living cells, 385
Protein amino acids,
  occurrence of, 637–643
Protein content,
  changes in respiration with, 317
  relation of to the rate of respiration,
    320–321
Protein cycle,
  and respiration in roots, 301
  evidence for in leaves, 290–291
  scheme of, 362–363
Protein metabolism,
  early schemes of, 437–438, 582–583
  in actively growing cells, 496, 591
  in cultured explants, 592–597
  in development of root tissue, 493–494
  in leaves, 585–587
  in the seedling, 440–441
Protein nitrogen,
  analysis and determination of,
    397–401
Protein synthesis,
  by chemical condensation of amino
    acids, 557
  concepts of, 565–570
  in cell free systems, 570
  in chloroplasts, 543–544, 598

inhibitors of, 579–580
mechanism of in plants, 555–576
regulation of, 583
relationship to cell growth, 422
sites of in the cell, 560–563
stimulation of by growth substances, 532
template for, 570–573
the genetic control of, 573–576
Protein turnover,
in plant cells, 363
Proteoplast,
in cells of endosperm, 562
Protochlorophyll, 19
Putrescine,
occurrence of, 459
structure of, 426
Pyridine nucleotides,
as regulators of cellular respiration, 243
photochemical reduction of, 600, 601
Pyrrole rings, 18
Pyrrole,
role of in plant growth, 123–124
Pyruvic aldol, 461
as the precursor of γ-methyl-γ-hydroxyglutamic acid, 447

Q

Q_{10},
equation for, 262
of bulky organs, 341
values of in seedling development, 262
Quantum efficiency,
effect of wavelength on, 85
Quantum yield,
values for, 22–23

R

"R" value,
determination of, 177
Radiation exchange,
at the earth's surface, 204–206
Regulator gene,
function of, 575
Relative diffusivity,
in a red clover community, 216

Respiration,
and salt uptake, 296
and seedling development, 252–258
catabolic reactions of, 233–235
definition of, 56–57
dependence of on oxygen, 343–350
during photosynthesis, 55–62
effects of mechanical pressure on, 275–276
effects of mineral deficiencies on, 103–104, 107–109
effects of nitrate and ammonium on, 298–300
effects of nitrogen content on, 255–256
effect of oxygen on, 265, 266
effects of temperature on, 261–263
energy yield from, 238
factors limiting, 344–350
in developing seedlings, 257–258
in expanding leaves, 270
inhibitory effect of carbon dioxide on, 268
in meristematic cells, 294
in relation to oxygen supply, 355
in ripening fruits, 312–313
interrelation with photosynthesis, 274
in the potato tuber, 325
in thin disks of storage tissue, 587–588
limitation by oxygen diffusion, 341
mechanisms controlling the rate of, 249
of mature leaves, 274–284
of roots, 291–303
over-all reactions of, 16
photoinhibition of, 59
photostimulation of, 59
protein metabolism in relation to, 583–597
regulation of, 257, 368–369
relation to the internal atmosphere of fruits, 334–343
yield of carbon from, 26–27
measurement of, 241
Respiratory drift,
effects of oxygen and carbon dioxide on, 330
phases of, 252–253
Respiratory gradient,
in the root, 291–293
Respiratory index, 270

Respiratory intermediates,
  diversion of to synthetic reactions, 240–245
Respiratory rates,
  effects of minerals on, 109
  effect of temperature on, 327–328
  factors controlling, 242, 323
  for detached apples, 316–317
  in leaves, 498
  limitation by oxygen concentration, 339
Respiratory quotient (R.Q.),
  equation for, 27
  interpretation of, 27, 28
  of germinating seeds, 248, 258–261
  of mature leaves, 274
  of potato tubers, 325–326
Ribonucleic acid (RNA),
  as a single-stranded helix, 565
  in chloroplasts, 544, 599
  effects of water stress on, 101
  in cotyledons, 542
  in the development of root tissue, 493–494
  in plant viruses, 553
  metabolism of, 542, 580–581
  nitrogen bases in, 565
  occurrence of in cells, 548
  types of in cells, 549–552
  values of in the leaf, 271–272
  see also specific types of RNA's
Ribosomal RNA, 550
  composition of, 549
Ribosomes,
  in chloroplasts, 543, 599
Ribulose diphosphate, 22
Richardson number (Ri),
  equation for, 211
Roots,
  changes in the ultrastructure of, 295
  respiration and salt uptake in, 296–303
Root respiration,
  measurement of, 155–156

S

Sakaguchi's reagent,
  for detecting compounds on paper, 395
Salt uptake,
  effects of respiration on, 296–298

Sea water,
  turnover of elements in, 404–405
Seedlings,
  respiration of, 254
Seeds,
  effects of carbon dioxide on the respiration of, 268
  non-nitrogenous reserves of, 259
  proteins in, 541–542
  rate of respiration during germination in, 247–251
  soluble nitrogen in, 513
Senescent leaves,
  metabolism of, 284
Serine, 595
  relationship to photosynthesis, 528
Smog,
  effects of on plants, 164–166
Solar constant,
  definition of, 24
Soluble nitrogen,
  diurnal variation of in leaves, 499
  in growing tissues, 521
  in pine cones, 509
  occurrence of in cells, 578
  in the potato tuber, 453
  separation of into components, 393–396
Soluble nitrogen compounds,
  factors regulating, 515
  genetics effects on, 489
Soluble ribonucleic acid (S-RNA),
  structure of, 550–551
Starch,
  availability of in respiration, 240
  in freshly detached leaves, 277
Starch gel electrophoresis,
  in protein separations, 536
Stomata,
  and assimilation, 51
  density of, 47, 48
  diameter law of, 43–46
  effects of temperature on, 53
  relative rates of diffusion into, 48, 49
  roles in gas exchange, 40–44
  size of, 47, 48
Stomatal movements,
  influence of environmental factors on, 49–50
Stomatal resistance,
  of detached leaves in the dark, 277

Streptidine,
 occurrence of, 429
Structural gene,
 definition of, 573
Succinate,
 as a hydrogen donor, 235
 in the synthesis of γ-aminobutyric
  acid, 446
Succinic acid oxidase,
 inhibition by carbon dioxide, 268–269
Succinic semialdehyde, 524
Sucrose,
 breakdown of, 277
 effect on the rate of respiration, 326,
  327
 metabolism of in cultured explants,
  592–594
Sugar breakdown,
 in air and nitrogen, 355–358
Sulfosalicylic acid,
 as protein precipitant, 392
Sulfur,
 as a factor limiting protein synthesis,
  521
 deficiency effects of, 118
 in the light reactions of photosynthesis,
  120
 occurrence of in plants, 282
 role in plants, 120
 role in the accumulation of soluble
  nitrogen, 515
Sulfur dioxide,
 effects of on photosynthesis, 161–162

T

Taurine, 459
Temperature,
 as a limiting factor in photosynthesis,
  88–89
 as an ecological factor, 91–93
 effect on carbohydrate production in
  photosynthesis, 94
 effect on CO₂ content, 337–338
 effects of on nitrogen metabolism, 518–
  520
 effect on respiration rate, 327–328
 effect on stomata, 53
Temperature inversion,
 effects on carbon dioxide content, 36–37

Terminal oxidases,
 role of in plant respiration, 346–348
Thiouracil,
 as an inhibitor of floral induction, 432
Thymine,
 degradation of, 458
Tyramine,
 occurrence of, 459
Tissue cultures,
 nitrogen metabolism in, 492–493, 512
Tobacco mosaic virus,
 composition of, 553
Transamidinase,
 action of, 430
Transamination,
 discovery of, 442
 reactions concerned with, 442–443
Transfer RNA, 549, 550
Translocation,
 of nitrogen compounds, 513
Transpiration,
 effects of boron on, 135
 effects of light on, 76
Tricarboxylic acid cycle,
 function of in potato tubers, 353
 in mature leaves, 278
 in the metabolism of leaves, 279–280
 significance of in nitrogen metabolism,
  441
Trichloroacetic acid,
 as protein precipitant, 392
Triphosphopyridine nucleotide (TPN),
 see Nicotinamide adenine dinucleotide
  phosphate
True photosynthesis, 29

U

Uracil,
 degradation of, 458
 in flower induction, 533
Uracylalanine, 431
Urea,
 as an animal product, 426
 biosynthesis of, 427–428
 chemical synthesis of, 427
 conversion of to glutamine, 415
 detection of on paper chromatograms,
  476

occurrence of in plants, 427
origin of in animals, 415
role in the metabolism of the purines
    and pyrimidines, 428
Urea cycle,
    scheme of, 416
Urease, 420
    crystallization of, 535
Ureides,
    detection of on paper chromatograms,
    476

V

Vanadium,
    role of in plant nutrition, 131–133
"Vital force,"
    theory of, 16–17

W

Water vapor,
    profiles in red clover crop, 218
Willardiine, 431

Wind,
    schematic representation of, 211
Wind profile,
    within vegetation canopy, 214
Wind speed,
    equation for measurement of, 209
Wood and Werkman reaction,
    in thin slices of storage tissues, 588

X

Xanthine,
    in flower induction, 533
Xanthine oxidase,
    regulation of by kinetin, 432

Y

Yeast nucleic acid,
    isolation of, 433

Z

Zinc,
    role of in plants, 128–131